T0135332

Advances in Intelligent Systems and Computing

Volume 754

Series editor

Janusz Kacprzyk, Polish Academy of Sciences, Warsaw, Poland
e-mail: kacprzyk@ibspan.waw.pl

The series "Advances in Intelligent Systems and Computing" contains publications on theory, applications, and design methods of Intelligent Systems and Intelligent Computing. Virtually all disciplines such as engineering, natural sciences, computer and information science, ICT, economics, business, e-commerce, environment, healthcare, life science are covered. The list of topics spans all the areas of modern intelligent systems and computing such as: computational intelligence, soft computing including neural networks, fuzzy systems, evolutionary computing and the fusion of these paradigms, social intelligence, ambient intelligence, computational neuroscience, artificial life, virtual worlds and society, cognitive science and systems, Perception and Vision, DNA and immune based systems, self-organizing and adaptive systems, e-Learning and teaching, human-centered and human-centric computing, recommender systems, intelligent control, robotics and mechatronics including human-machine teaming, knowledge-based paradigms, learning paradigms, machine ethics, intelligent data analysis, knowledge management, intelligent agents, intelligent decision making and support, intelligent network security, trust management, interactive entertainment, Web intelligence and multimedia.

The publications within "Advances in Intelligent Systems and Computing" are primarily proceedings of important conferences, symposia and congresses. They cover significant recent developments in the field, both of a foundational and applicable character. An important characteristic feature of the series is the short publication time and world-wide distribution. This permits a rapid and broad dissemination of research results.

More information about this series at http://www.springer.com/series/11156

Zhengbing Hu · Sergey Petoukhov
Ivan Dychka · Matthew He
Editors

Advances in Computer Science for Engineering and Education

 Springer

Editors
Zhengbing Hu
School of Educational Information
  Technology
Central China Normal University
Wuhan, Hubei
China

Ivan Dychka
Faculty of Applied Mathematics
National Technical University of Ukraine
  "Igor Sikorsky Kiev Polytechnic Institute"
Kiev
Ukraine

Sergey Petoukhov
Mechanical Engineering Research Institute
Russian Academy of Sciences
Moscow
Russia

Matthew He
Halmos College of Natural Sciences and
  Oceanography
Nova Southeastern University
Ft. Lauderdale, FL
USA

ISSN 2194-5357   ISSN 2194-5365 (electronic)
Advances in Intelligent Systems and Computing
ISBN 978-3-319-91007-9   ISBN 978-3-319-91008-6 (eBook)
https://doi.org/10.1007/978-3-319-91008-6

Library of Congress Control Number: 2018942171

Printed on acid-free paper

This Springer imprint is published by the registered company Springer International Publishing AG
part of Springer Nature
The registered company address is: Gewerbestrasse 11, 6330 Cham, Switzerland

Preface

The rapid development of computer science provides many new effective solutions in engineering and educational methods and technologies including systems of artificial intelligence. At the present time, the level of development of different countries is largely determined by how successfully the achievements of computer science are being introduced into engineering and education applications. Therefore, the problems of computer science and its applications in engineering and education are at the center of attention of governments and scientific-technological communities. Accordingly, higher education institutions face the pressing tasks of educating new generations of specialists who can effectively use and further develop achievements of computer science and their applications. International cooperation facilitates and accelerates appropriate solves in this important field.

By these reasons, the National Technical University of Ukraine "Igor Sikorsky Kyiv Polytechnic Institute" and the International Research Association of Modern Education and Computer Science (RAMECS) jointly organized the First International Conference on Computer Science, Engineering and Education Applications (ICCSEEA2018), January 18–20, 2018, Kiev, Ukraine. The ICCSEEA2018 brought together the top researchers from different countries around the world to exchange their research results and address open issues in Computer Science, Engineering and Education Applications. The organization of such conference is one of the examples of growing Ukraine–Chinese cooperation in different fields of science and education.

The best contributions to the conference were selected by the program committee for inclusion in this book out of all submissions.

Our sincere thanks and appreciation to the board members as listed below:
Michael Zgurovsky, Ukraine
Yurii Yakymenko, Ukraine
Oleksandr Pavlov, Ukraine
Ivan Dychka, Ukraine
Valeriy Zhuykov, Ukraine
Pavlo Kasyanov, Ukraine
M. He, USA

Georgy Loutskii, Ukraine
Felix Yanovsky, Ukraine
Igor Ruban, Ukraine
Volodymyr Tarasenko, Ukraine
Prof. Janusz Kacprzyk, Poland
Prof. E. Fimmel, Germany
PhD G. Darvas, Hungary
Dr. K. Du, China
Dr. Oleksii K. Tyshchenko, Ukraine
Prof. N. A. Balonin, Russia
Prof. S. S. Ge, Singapore
Prof. A. U. Igamberdiev, Canada
Prof. A. V. Borisov, Russia
Dr. X. J. Ma, China
Prof. S. C. Qu, China
Prof. Y. Shi, USA
Dr. J. Su, China
Dr. O. K. Ban, USA
Prof. J. Q. Wu, China
Prof. A. Sachenko, Ukraine
Prof. Q. Wu, China
Prof. Z. W. Ye, China
Prof. C. C. Zhang, Taiwan
Prof. O. R. Chertov, Ukraine
Prof. N. D. Pankratova, Ukraine.

Finally, we are grateful to Springer-Verlag and Janusz Kacprzyk as the editor responsible for the series "Advances in Intelligent System and Computing" for their great support in publishing these conference proceedings.

January 2018   Zhengbing Hu
                                                                                     Sergey Petoukhov
                                                                                        Ivan Dychka
                                                                                        Matthew He

Organization

Honorary Chairs

Michael Zgurovsky   NASU Academician, Rector of the National Technical University of Ukraine "Igor Sikorsky Kyiv Polytechnic Institute," Kyiv, Ukraine

Yurii Yakymenko   NASU Academician, First Vice-Rector of Igor Sikorsky Kyiv Polytechnic Institute, Kyiv, Ukraine

Chairs

Oleksandr Pavlov   National Technical University of Ukraine "Igor Sikorsky Kiev Polytechnic Institute"

Ivan Dychka   National Technical University of Ukraine "Igor Sikorsky Kiev Polytechnic Institute"

Valeriy Zhuykov   National Technical University of Ukraine "Igor Sikorsky Kiev Polytechnic Institute"

Pavlo Kasyanov   National Technical University of Ukraine "Igor Sikorsky Kiev Polytechnic Institute"

General Co-chairs

M. He   Nova Southeastern University, USA

Georgy Loutskii   National Technical University of Ukraine "Kyiv Polytechnic Institute," Kyiv, Ukraine

Felix Yanovsky   Chair of IEEE Ukraine Section, National Aviation University, Kyiv, Ukraine

Igor Ruban   Kharkov National University of Radio Electronics, Kharkov, Ukraine

Volodymyr   National Technical University of Ukraine "Igor Sikorsky
   Tarasenko   Kyiv Polytechnic Institute," Kyiv, Ukraine

International Program Committee

Janusz Kacprzyk   Systems Research Institute, Polish Academy of Sciences
E. Fimmel   Institute of Mathematical Biology of the Mannheim
                        University of Applied Sciences, Germany
G. Darvas   Institute Symmetrion, Hungary
K. Du   National University of Defense Technology, China
Oleksii   Kharkiv National University of Radio Electronics,
   K. Tyshchenko   Kharkiv, Ukraine
N. A. Balonin   Institute of computer systems and programming,
                        St. Petersburg, Russia
S. S. Ge   National University of Singapore, Singapore
A. U. Igamberdiev   Memorial University of Newfoundland, Canada
A. V. Borisov   Udmurt State University, Russia
X. J. Ma   Huazhong University of Science and Technology, China
S. C. Qu   Central China Normal University, China
Y. Shi   Bloomsburg University of Pennsylvania, USA
J. Su   Hubei University of Technology, China
O. K. Ban   IBM, USA
J. Q. Wu   Central China Normal University, China
A. Sachenko   Ternopil National Economic University, Ternopil, Ukraine
Q. Wu   Harbin Institute of Technology, China
Z. W. Ye   Hubei University of Technology, China
C. C. Zhang   Fengchia University, Taiwan
O. R. Chertov   National Technical University of Ukraine "Igor Sikorsky
                        Kyiv Polytechnic Institute," Kyiv, Ukraine
N. D. Pankratova   National Technical University of Ukraine "Igor Sikorsky
                        Kiev Polytechnic Institute"

Steering Chairs

Vadym Mukhin   National Technical University of Ukraine "Kyiv
                        Polytechnic Institute," Ukraine
Z. B. Hu   Central China Normal University, China

Local Organizing Committee

Oleg Barabash	State University of Telecommunication, Kyiv, Ukraine
Dmitriy Zelentsov	Ukrainian State University of Chemical Technology, Ukraine
Yuriy Kravchenko	Taras Shevchenko National University of Kyiv, Ukraine
Viktor Vishnevskiy	State University of Telecommunication, Kyiv, Ukraine
Igor Parhomey	National Technical University of Ukraine "Kyiv Polytechnic Institute," Kyiv, Ukraine
Elena Shykula	Kyiv State Maritime Academy, Kyiv, Ukraine
Alexander Stenin	National Technical University of Ukraine "Kyiv Polytechnic Institute," Kyiv, Ukraine
Sergiy Gnatyuk	National Aviation University, Kyiv, Ukraine
Yaroslav Kornaga	National Technical University of Ukraine "Kyiv Polytechnic Institute," Kyiv, Ukraine
Yevgeniya Sulema	National Technical University of Ukraine "Igor Sikorsky Kyiv Polytechnic Institute," Kyiv, Ukraine
Andrii Petrashenko	National Technical University of Ukraine "Igor Sikorsky Kiev Polytechnic Institute"

Publication Chairs

Z. B. Hu	Central China Normal University, China
Sergey Petoukhov	Mechanical Engineering Research Institute of the Russian Academy of Sciences, Moscow, Russia
Ivan Dychka	National Technical University of Ukraine "Igor Sikorsky Kiev Polytechnic Institute," Ukraine
Oksana Bruy	National Technical University of Ukraine "Igor Sikorsky Kiev Polytechnic Institute," Ukraine

Contents

Computer Science for Manage of Natural and Engineering Processes

Physical and Mathematical Modeling of Permeable Breakwaters

Lidiia Tereshchenko[1]([✉]), Vitalii Khomicky[1], Ludmyla Abramova[1],
Ihor Kudybyn[1], Ivan Nikitin[1], and Igor Tereshchenko[2]

[1] Institute of Hydromechanics, National Academy of Sciences of Ukraine,
Kyiv, Ukraine
litere70@yahoo.com, {homicky,luda54,nia37}@ukr.net,
igorkud33@gmail.com
[2] National Technical University of Ukraine "Igor Sikorsky Kyiv
Polytechnic Institute", Kyiv, Ukraine
tereshchenko.igor@gmail.com

Abstract. This article presents the experimental result and mathematical analysis influences of the permeable breakwaters to dynamics of the coast on various conditions by the action of waves. Deposition of sediment behind breakwaters leads to the formation tombolo of a certain size and intensity depending on the parameters of breakwater and the presence of a weak along-shore sediment flow. The results of the experiments determined empirical dependences, which are necessary for scientific study design of permeable breakwaters. Mathematical modeling of experimental data was also carried out. Experimental data are compared with already known results of research for permeable breakwaters with frontal and oblique approaches of waves. The experiment was held in Kyiv, Ukraine at the Institute of Hydromechanics National Academy of Sciences of Ukraine.

Keywords: Permeable breakwater
Coast protection mathematical and physical modeling

1 Introduction

Recently, protecting the coast has become more common using of permeable breakwater. Many authors have conducted experimental investigation of single breakwaters. World experience show that the most effective way is the creation of artificial coastal beaches. These beaches can be use similar as on natural shore, so in artificials the cove of shore forms. There is formed a reverse circulation the sediment with using along-shore and deep bypassing. Artificial cove with capes can be of different configurations.

The range of application of permeable breakwaters are quite diverse. In practice, coastal protection structures of this type are used mainly to: (1) cover the edge of the shore or beach body from erosion by waves; (2) increase of surface planes beach and repositioning the original shoreline; (3) debarment with sedimentation of aquatorium port or creation of a reserve the soil in systems of bypassing.

Despite these advantages, the use of permeable breakwaters delayed due to the lack of probable engineering methods of calculation of such facilities. In particular, scientists have not studied the full behavior permeable breakwaters at an oblique approach of the waves; not defined the conditions of formation tombolo at blocking shore of obstacle in case of deficit sediment in alongshore stream [1–17]. Therefore, the great interest is the experimental investigatation and mathematical modeling breakwaters various of transverse profile and form.

One of the important problems is the development of mathematical methods for choosing a more effective coastal protection method. Therefore, this article helps to properly evaluate and select the optimal design of the protective structure in the form of a breakwater.

2 Experimental Set-Up

Work permeable breakwater depend on the wave regime, geomorphology, underwater coastal slope, geometry and form of the construction [10]. A series of experimental research were performed for permeable breakwater to create the engineering method of calculation. One of the experimental models is shown in Fig. 1

Fig. 1. Model of permeable breakwaters before the experiment

Figure 2 [3] shows a diagram of the interaction of waves with breakwaters. Interaction accompanied by complex processes: diffraction at a rounding moles; refraction waves in the interaction with the bottom; symptoms of multiple reflection waves breakwater and interference in the sides of construction.

Experimental researches were carried out in the large wave basin (7 × 37 × 2.5 m) in the Institute of Hydromechanics NAS of Ukraine. The depth of the basin was equal to 0.8 m.

The measurement of the parameter waves and velocity of the flow were measured models using capacitive sensors and micro turntable. A body of sandy model with underwater slope coefficient $m = 12$ made of particles $d = 0.3$ mm is developed.

The angle approach of the waves to shore was changed from 0–45°. Parameters of the waves during testing on the washing out sandy model is: $h = 0.05$–0.1 m, $\tau = 0.8$–1.2 s, $\bar{\lambda}/h = 10 - 30$. The length of the breakwaters in the experiments changing 0.7–2.1 m; distance to the original shoreline - from 0.75 to 1.75 m; the width of the cove 0.4–2.1 m.

The length of the zone of saturation flow of sediments changed from 0.4–1.85 m. The experiments were carried out at the angle approach of 45°. The conditions of the experiments at the research of wave diffraction were defined as follows:

(1) $B = 3$ m; $x = 4.5$ m; $\lambda_b = 2.06$ m; $h_b = 0.14$ m; $\tau = 1.15$ s; $L_{cr} = 3$ m; (2) $B = 3$ m; $x = 4.5$ m; $\lambda_b = 2.25$ m; $h_b = 0.1$ m; $\tau = 1.2$ s; $L_{cr} = 2.5$ m; (3) $B = 3$ m; $x = 4.5$ m; $\lambda_b = 2.85$ m; $h_b = 0.1$ m; $\tau = 1.35$ s; $L_{cr} = 2.6$ m; (4) $B = 1.5$ m; $x = 2$ m; $\lambda_b = 1.41$ m; $h_b = 0.12$ m; $\tau = 0.95$ s; $L_{cr} = 2.5$ m.

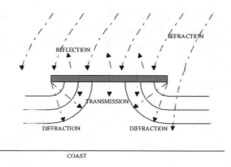

Fig. 2. Wave processes near offshore breakwaters [5]

Example of diffraction pattern for some experiments are shown in Fig. 3.

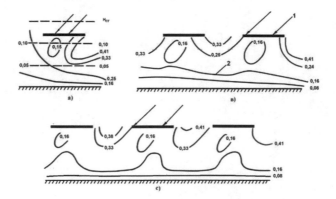

Fig. 3. Diffraction waves of single (a) two-link (b) and three-link (c) breakwaters that are in the surf zone, beam waves (1); contours diffraction coefficients (2); H_{cr} critical water depth; $h_b = 0.12$ m; $\tau = 0.05$ s; $\lambda_b = 1.4$ m; $B = 1.5$ m; $B/\lambda = 1.1$; $G = 2$ m; $G/\lambda = 1.43$

The peculiarity of the experiments is that intermittent structure was located in the surf zone at variable depths by breakwaters. Numerical solutions of diffraction of waves in such conditions has not been found. Therefore, the most reliable way to obtain parameters waves in the shade by breakwaters are experimental models on a large scale.

3   Results and Discussion

Similar to other authors [3–16] our investigation showed that the frontal action of wave, the initial shoreline eroded behind ends of breakwater. The middle part forms the accumulative ledge, which sometimes transformed into a tombolo of full profile. By using an oblique wave approach, it was observed asymmetry location head of ledge to the center of breakwater.

As shown in the experiments for a single breakwater, the most stable form of ledge appears at an angle approach of wave $40°$.

Approximation, of $S/B = f(X/B)$ by results of the experiments conducted to obtain the following empirical equations (Fig. 4):

(a) approach of frontal waves

$$S/B = -0.25 + 0.70\left(\frac{x}{B}\right) + 0.10\left(\frac{x}{B}\right)^2 \tag{1}$$

(b) approach of oblique waves

$$S/B = -0.35 + 0.65\left(\frac{x}{B}\right) + 0.10\left(\frac{x}{B}\right)^2 \tag{2}$$

$$x/B = -0.16 + 0.84\left(\frac{x}{B}\right) + 0.03\left(\frac{x}{B}\right)^2 \tag{3}$$

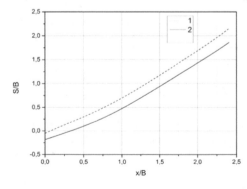

Fig. 4. Approximation $S/B = f(x/B)$: 1- frontal approach of the waves; 2- oblique approach of the wave

Dependence (1, 2) are more realistic for the boundary conditions created by tombolo at $x/B \approx 0.5$. Distance x_0 of tangent to the boundary of lines of deformations accordingly for frontal and oblique approach of waves (Fig. 5)

$$x_0/B = 0.2 + 1.0\left(\frac{x}{B}\right) \tag{4}$$

$$x_0/B = 0.1 + 1.0\left(\frac{x}{B}\right) \tag{5}$$

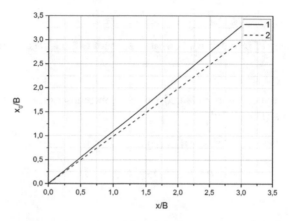

Fig. 5. Approximation $x_0/B = f(x/B)$: 1- frontal approach of the waves; 2- oblique approach of the wave

Counting the distance to the top of tombolo T from the initial of shoreline the formation conditions tombolo is evaluated accordingly (Fig. 6)

(a) approach of frontal waves

$$T/x = 1.13\left(\frac{x}{B}\right)^{-0.5} \exp\left(-\frac{x}{B}\right) \tag{6}$$

(b) approach of oblique waves

$$T/x = 1.7\left(\frac{x}{B}\right)^{-0.5} \exp\left(-\frac{x}{B}\right) \tag{7}$$

$$R_0/B = 0.17 + 1.68(x/B) \tag{8}$$

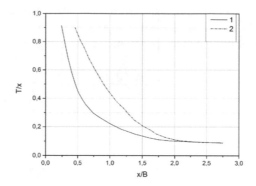

Fig. 6. Approximation $T/x = f(x/B)$: 1- frontal approach of the waves; 2- oblique approach of the wave

Results of the experiments obtained similar approximation for frontal and oblique approach of waves (Fig. 7) [9]

$$R_0/B = 0.2 + 1.2(x/B) \tag{9}$$

$$R_0/B = 0.25 + 1.25(x/B) \tag{10}$$

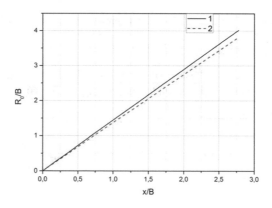

Fig. 7. Approximation $R_0/B = f(x/B)$: 1- frontal approach of the waves; 2- oblique approach of the wave

When we have R_o, value R_θ can be determined according to Figs. 8 and 9. The value R_o/R_θ for approach of wave 40° [9]. The oblique approach was used to determining the value R_o/R_θ for the several-links of breakwater to use curve 1 in Fig. 8, for internal curves according to form- 2, 3, 4.

The graphs in Figs. 8 and 9, as all the above formulas can be used for the design of single or multiple links permeable breakwaters for frontal and oblique action of wave for limit value $x/B \geq 0.5$ provided $x/B \leq 0.5$.

The results of the experiments shows that graphically it is best to define the characteristics of regulation sediment permeable of breakwaters, which must be used in the design of a single or multi links of breakwaters. The permeable of line trend in the graph shown previously explored unexplored areas of numerical values of x/B.

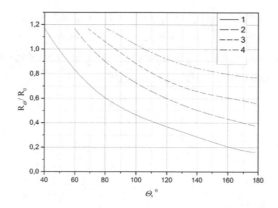

Fig. 8. Value R_θ/R_0 depending on θ waves in frontal approach and different values angle approach of wave: 1- 40°; 2- 60°; 3- 70°; 4- 80°

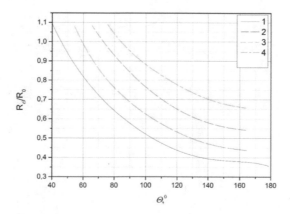

Fig. 9. Value R_θ/R_0 depending on θ waves in frontal approach and different values angle approach of wave and disclosure cove: 1- 40°; 2- $\frac{G}{B} = 0.5$; 3- $\frac{G}{B} = 1.0$; 4- $\frac{G}{B} = 2.5$

The permeable breakwater, which is located underwater of the coastal slope, have a regulating effect on the alongshore stream sediment, formed earlier in the unprotected area of the coast. Availability of breakwater reduce by 30–40% the volume of transit from protected areas compared to the transit of unprotected shore, the same length

$$\frac{Q_{ic}}{Q_{xi}} = \exp\left[\left(-1.6\frac{G}{B}\right)^{-2/3} l/L\right] \text{ provided } x/B = const = 1,0; 0.5 \le G/B \le 3.0 \quad (11)$$

$$\frac{Q_{ic}}{Q_{xt}} = \exp\left[\left(-a\frac{x}{B}\right)^{2/3} l/L\right] \text{ provided } G/B \ge 0.5 \text{ and } x/B \le 0.5 \quad (12)$$

where Q_{xt} consumption along the shore flow of sediment from unprotected areas of the coast, length L. Coefficient a in (11–12) is equal $a = \frac{5.62-0.75\frac{G}{B}}{7.5}$

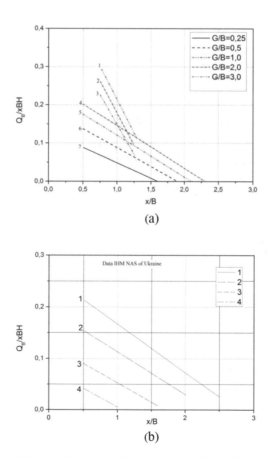

(a)

(b)

Fig. 10. Dependence of distance breakwater impact on the volume of tombolo for frontal (a) and oblique (b) approach of waves, line 1, 2, 3 data [6]; 4, 5, 6, 7 IHM NAS of Ukraine (a)

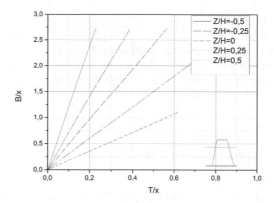

Fig. 11. The impact of shape of profile breakwater on the accumulating capacity. Data IHM NAS of Ukraine

In addition, to regulate sediment the permeable breakwater also has the wave protective ability [4].

At the frontal approach waves to structures extinction coefficient of submerged and not submerged breakwaters with stone mantles determined by the formula

$$K_t = \frac{H_M}{H_b} = \sqrt{H_z/H_b} \tag{13}$$

$$K_t = 0.75/\sqrt{h/H_b} \tag{14}$$

where as H_M and H_b wave height of sea and shore side breakwater; H_z height crest dive.

The dependence (1)–(14) and graphics Figs. 4, 5, 6, 7, 8, 9, 10 and 11 allow to predict the deformation of coastline for different exposition wind rose at the availability of permeable breakwaters.

4 Conclusion

The results of the experiment showed at variable depth behind the breakwater diffraction of wave in shade of construction accompanied by phenomena of refraction, transformation, interference and breaking of wave of underwater coastal slope. The result of the above phenomena is the formation in the shade of breakwaters tombolo. The size and intensity of formation tombolo breakwater depend on the size of breakwater, distance between the breakwaters and the axis distance of breakwater from the primary shoreline.

Approaching breakwater to the shore decreases the coefficient of diffraction for single breakwater in 1.2–1.3 times. For two-link breakwater the coefficient of diffraction decreases in 1.5 times, and for three-tier in 1.5–1.6 times.

The intermittent breakwater would be best used in a case where the retreat of edge of the coastal ledge is not acceptable, but it is necessary stock sand and frontal approach of waves. In the permeable breakwaters of all kinds of width and length backfilling of construction or shore should be sufficient to form cove erosion, preserve the jumpers and formation of sediment full or partial profile.

References

1. Abdelgwad, A.H., Said, T.M., Gody, A.M.: Microwave detection of water pollution in underground pipelines. Int. J. Wirel. Microwave Technol. (IJWMT) **4**(3), 1–15 (2014). https://doi.org/10.5815/ijwmt.2014.03.01
2. Aloui, N., Bousselmi, S., Cherif, A.: Optimized speech compression algorithm based on wavelets techniques and its real time implementation on DSP. Int. J. Inf. Technol. Comput. Sci. (IJITCS) **7**(3), 33–41 (2015). https://doi.org/10.5815/ijitcs.2015.03.05
3. CUR 97-2A: Beach Nourishment and Shore Parallel Structures. Technical report, Gouda, The Netherlands, Phase 1 (1997)
4. Herbich, I.B.: Shoreline changes due to offshore breakwaters. In: International Association for Hydraulic Research XXIII Congress IAHR, Ottava, Canada, 21–23 August, p. 245 (1989)
5. Hidding, I.: Shoreline response to offshore breakwaters, The Netherlands (2007)
6. Khuong, T.C.: Shoreline response to detached breakwaters in prototype. Dissertation Delft, Faculty of Civil Engineering and Geosciences, Department of Hydraulic Engineering (2016)
7. Lara, J.L., Losada, I.J., Jesus, M.: Modeling the interaction of water waves with porous coastal structures. J. Waterway Port Coast. Ocean Eng. **142**(6), 240 (2016). (ASCE) WW.1943-5460.0000361, 03116003
8. Lauro, E., Contestabile, P., Vicinanza, D.: Wave loadings on an innovative breakwater integrated wave energy converter. In: SCACR2015, International Short Course/Conference on Applied Coastal Research, Florence, Italy, pp. 255–266 (2015)
9. Silvester, R., Hsu, I.R.: Coastal stabilization. World Scientific Publishing, Singapore (1997)
10. Tereshchenko, L.M., Khomicky, V.V., Fomin, V.V., Abramova, L.P.: The use of numerical models SWAN spectral parameters for the calculation of wind waves in the area Kryva spit. Bulletin of Taras Shevchenko National University of Kyiv Series Physics & Mathematics, vol. 2, pp. 189-193 (2015)
11. Tsai, C., Yu, C., Chen, H., et al.: Wave height transformation and set-up between a submerged permeable breakwater and a seawall. China Ocean Eng. **26**(1), 167–176 (2012)
12. Tiwari, J., Singh, A.K., Yadav, A., Jha, R.K.: Modelling and simulation of hydro power plant using MATLAB & WatPro 3.0. Int. J. Intell. Syst. Appl. (IJISA) **7**(8), 1–8 (2015). https://doi.org/10.5815/ijisa.2015.08.01
13. Serio, F., Mossa, M.: Experimental study on the hydrodynamics of regular breaking waves. Coast. Eng. **53**, 99–113 (2006)
14. Vicinanza, D., Luppa, C., Lauro, E., Contestabile, P., Cavallaro, L., Andersen, L.: Wave loadings acting on innovative rubble mound breakwater for overtopping wave energy conversion. Coast. Eng. **122**, 60–74 (2017)
15. Zhang, S., Li, X.: Design formulas of transmission coefficients for permeable breakwaters. Water Sci. Eng. **7**(4), 457–467 (2014)
16. Zhang, Z., Liang, S., Sun, Z., Sun, J.: Study on the erosion mechanism of the Bijia Mountain Gravel Tombolo, China. J. Coast. Res. **296**(4), 851–861 (2014)
17. Guo, T., Guo, C., Li, D., Chen, A., Zhao, M.: Simulation of leaf water status. Int. J. Educ. Manage. Eng. (IJEME) **2**(9), 35–43 (2012). https://doi.org/10.5815/ijeme.2012.09.06

Risk Modeling of Accidents in the Power System of Ukraine with Using Bayesian Network

Viktor Putrenko[1(✉)] ⓘ and Nataliia Pashynska[2] ⓘ

[1] Igor Sikorsky KPI, Kyiv, Ukraine
putrenko@wdc.org.ua
[2] Taras Shevchenko National University of Kyiv, Kyiv, Ukraine

Abstract. Current studies of impact of climatic factors on overhead power lines are limited to calculations of load of climatic factors on the overhead transmission lines, so the problem of conducting a comprehensive study of accidents probability under the influence of climatic factors is important.

The paper addresses the research of approaches to spatial risk modeling of overhead power lines accidents in the power systems of Ukraine under the influence of climatic factors. The article presents the construction of a model of accidents under the influence of climatic impacts and prediction of emergencies on based geospatial data sets. Pattern recognition techniques, namely the Bayesian network, were used to simulate accidents and verification of the results. This method is based on calculation of a posteriori probabilities of model variables. As a result, a model of accidents under the influence of climatic factors was built, which constitutes a Bayesian network with given conditional probabilities and independent variables of the model.

Keywords: Risk modeling · Bayesian network · Power transmission network Spatial modeling

1 Introduction

Using geospatial data for the study of risk assessment of critical assets is one of the most important applied problems. Relevant and comprehensive spatial data on climatic conditions, engineering networks, and accident statistics should be provided by mapping services and local management companies for the purpose of selecting support solutions and predicting emergencies.

Energy security manifests itself in the ability of the country to ensure the most reliable, technically safe, environmentally acceptable and reasonably sufficient energy provision for the economy and the population under the current and forecasted conditions [1]. Safe functioning of the electric power system is one of the most important factors of energy security. Naturally, security is a complex political, economic, socio-economic, and scientific problem, which requires complex research of a wide range of issues.

The electric power system consists of a great number of objects and entities. The key ones are the power generating facilities and transmission networks. Hazardous

© Springer International Publishing AG, part of Springer Nature 2019
Z. Hu et al. (Eds.): ICCSEEA 2018, AISC 754, pp. 13–22, 2019.
https://doi.org/10.1007/978-3-319-91008-6_2

situations (accidents) in the electric power system objects are usually caused by defects in the manufacture and operation of the equipment, personnel errors, and other factors leading to the forced termination of electricity supply, which constitutes a threat to the society.

The majority of the electric power transmission networks are the overhead power lines, therefore there is a threat of adverse climatic factors impact on the power transmission network components. Extreme climatic conditions lead to power lines accidents, so analysis of climate impacts on the power transmission network and prediction of consequences of such impacts are an integral part of the system security problem.

Accident statistics shows that more than a half of the failures of overhead power lines are caused by the ice and wind overloads on wires, cables and other structures [2, 3]. In general, they are the result of an underestimation of the actual ice and wind loads. The essential difference between overhead power lines and other power system objects is their greatest length. The total length of the transmission lines with voltage of 0.4–110(150) kV is 817.9 thousand km [6].

Modeling of overhead transmission lines accidents caused by climatic factors and further forecasting of the number of accidents will be addressed in the article. The study of previous researches in the field of probabilistic accidents modeling shows that there are numerous methods of modeling, and regression analysis is used the most often. Currently, data mining methods that include neural networks, clustering algorithms, pattern recognition methods, a nearest neighbor methods, are used.

2   Problem Statement

Aging of transformer substations equipment and constituent elements of transmission lines, and deteriorating climatic conditions in Ukraine lead to higher accident rate and energy losses in the power system equipment, causing an increased number of shutdowns and failures, most of which occur in the 0,4–10 kV network [4]. The most crucial in terms of the scale of accidents caused by climatic factors are ice formations and wind load. According to 2.5.30 [4], values of ice and wind loads and climatic impacts for calculation and selection of overhead power lines design are taken from the regional climatic zoning maps of Ukraine, with further adjustment based on data gathered by meteorological observation stations and observation posts regarding wind speed, ice intensity and density, temperatures, storm activity and other climatic phenomena.

2.1   History of the Problem

In the recent years the number of power lines accidents increased significantly. A major accident in the Odesa region, accidents in the Zhytomyr and Volyn regions and others pointed out the necessity to review the impacts of climate loads on the power system in Ukraine and to create new zoning maps for climatic impacts.

Climatic zoning maps are maps of the territory divided into zones (areas) in terms of climatic impacts. In the latest version [5] it was decided to use a scale with 7 areas of loads. The pre-existing methods of climatic zoning mapping envisaged consideration of phenomena with 10 years repetition period, but in the course of further studies it was found out that periods of recurrence of many phenomena were 11–13, 34–37 years [5], therefore it was decided to take into account phenomena with 10, 25, 50-year recurrence periods. The new technique [6] contains a number of innovations: weather station selection algorithm, prior data processing approach, form of the maxima distribution function, and observation data analysis approach.

2.2 Reasons for the Study

A program of scheduled transmission line reconstruction is currently being developed in Ukraine, so the problem of accident prevention arises. Ice and wind loads often cause power lines accidents. Because of the considerable weight of ice formations on wires, accidents also involve cascading collapse of towers. The main period of ice formation and sediment is the cold season, so interruption of power supply leads to the cessation of heating, and the restoration of power lines after accidents becomes much more difficult. The main method of preventing ice-caused accidents is a method of melting ice formations on wires [7], which requires accurate knowledge of the ice deposition sizes, rate of formation and regional distribution thereof, so application of this method is considerably limited.

Thus, there is a problem to determine climatic conditions that lead to accidents. There is also a pressing issue to track ice distribution, communication and processing of real-time data to determine the need to run the ice melting process.

2.3 Input Data

To research overhead transmission lines accidents caused by climatic factors, the following input data are used:

- observations of weather stations of Ukraine regarding ice and wind events for the 1961–2012 period [8] obtained in the Central Geophysical Observatory;
- geographical database of basic layers of Ukraine from the NSDI database
- x, y, z coordinates of the meteorological stations in Ukraine from the Meteorological Service.
- statistics of accidents [9].

These observations of ice and wind events consist of data and characteristics of observations: the coordinates of the weather station, case number, type of sediment, beginning of icing (starting date, day, time of onset), growth duration of the icing event, the deposition parameters (diameter, thickness, weight), meteorological data at the beginning of the icing, and after reaching the maximum size (temperature, wind speed and direction). The period of ice observation runs from early autumn to late spring. Observations of wind speed include: characteristics of the weather station, wind speed (average and maximum), incidence rate by speed (number of cases for each value of the wind speed). The period of observation of wind speed lasts all year round. Data on accidents are presented in the following form: name, region, energy system district, name and characteristics of the line, date of the accident, absolute altitude of the accident area, characteristics of sediments, wind characteristics at the moment of failure, icing and wind observation data from the nearest weather station. In the above form, the accident rate data are available for the 1961–2015 period.

3   Bayesian Belief Network

The majority of the methods used to simulate failures require prior knowledge of the accident parameters distribution functions. In most situations, distribution data are not directly available, but instead, statistical dependency or independence between variables is known. In case of accidents, relation between the weight of ice and the number of accidents, dependency between wind speed and air temperature, and weight of ice deposits can be known. Internal connections can be represented by conditional probabilities that can be used to determine the probability of accidents under certain conditions.

3.1   The Best Option Choosing

The Bayesian network combines graphical structures (nodes representing variables and arcs expressing probabilistic dependencies between them) and corresponding conditional probabilities, which provides a comprehensive visual representation of different conditional relations between variables [10–17]. Local probability distributions are associated with each variable, and a set of independent conditions are represented in the network and can be directly combined to construct the overall probability distribution function for the entire network, which greatly simplifies the calculation of a posteriori probability variables [22].

The probabilistic structure is well illustrated [21]. For example, it is necessary to determine the probability distribution for the variables d1, d2,…… to D (Fig. 1) using the table of conditional probabilities and the network topology.

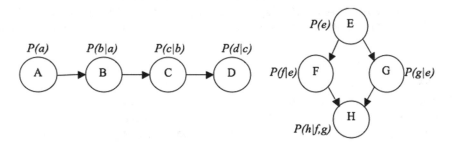

Fig. 1. Bayesian network structure

The distribution may be assessed by summing up the total general distribution, *P (a, b, c, d)* for all the variables except for d:

$$P(d) = \sum_{a,b,c} P(a, b, c, d) \sum_{a,b,c} P(a, b, c, d)$$
$$= \sum_{a,b,c} P(a)P(b|a)P(c|b)P(d|c) \tag{1}$$
$$= \sum_{c} P(d|c) \sum_{b} P(c|b) \sum_{a} P(b|a)P(a)$$

where

$$(b) = \sum_{a} P(b|a)P(a)$$
$$P(c) = \sum_{b} P(c|b) \sum_{a} P(b|a)P(a) \tag{2}$$

In this case, the network has a simple linear form; nonlinear networks (right side of Fig. 1) are calculated in a similar way [22].

3.2 Mathematical Accidents Model

To build a model of accidents caused by climatic factors, the following steps are required:

- identifying model variables and relations between them;
- building the structure of a Bayesian network and determining the possible values of variables and a priori probabilities based on equations;
- training a Bayesian network and refining its structure (variables and their probability);
- testing the model using data of accidents and meteorological observations;
- using the model for predicting the occurrence of accidents on the basis of information about accidents [23, 24].

To simulate accidents and perform forecasting, the North-West area of the country is selected, namely the area of the Zhytomyr region, because that region is represented by the majority of climatic conditions, climatic zoning areas of ice and wind values and other climatic influences that occur in Ukraine [25]. Meteorological data of Zhitomir, Kyiv and Rivne regions are used for the analysis.

Definition of the Variables for the Model. Prediction of the accidents occurrence depends on collection and processing of data on accidents, including the selection of variables used in the structure model. The variables will be selected on the basis of knowledge about the possible causes of the accident. This includes a thorough analysis of literature and intuitive engineering knowledge.

To simulate overhead transmission lines accidents under the influence of climatic factors, it is necessary to research the statistical data on accidents and meteorological data. Table 1 presents the factors affecting accidents and their possible values.

Table 1. Analysis of accidents

Factors	Description	Possible values
Year of accident	Observation period used for emergency situations 1960–2012 years	[1960–2015]
Month of accident	Ice and wind accidents often occur in cold seasons	1, 2, 3, 11, 12
Type of ice deposit	Ice and sediments type	Ice, hoarfrost, crystalline frost, etc.
Ice growth period	Defines the period of ice growth in hours from the beginning of the ice formation to the maximum size of deposit	[1, 72]
Duration of the event	Assumes values from 1 to 150 h	[1, 150]
Diameter	The diameter of ice deposit (large and small), divided into 10 mm sectors	[0, 10]… > 72
Weight	Ice weight is measured in grams. Assumes value up to >256	[1, 16]… > 256
Air temperature	Air temperature at the beginning of ice formation and at the moment it reaches the maximum size	[− 25; 2]
Wind direction	Wind direction at the beginning of ice formation. Specified in the rhumbs	[1, 8]
Wind speed	Wind speed at the beginning of ice formation and after reaching the maximum size. Given in mps	[0; 25]
Extreme events	Cases of extreme values during a year	0, 1, 2
Year of structural elements installation	Determines the effects of physical state of the structures on accident occurrence	<1960 1960–1990 1990–2015 n/a
Classification	Contains 2 classes: event considered to be an accident or not	0, 1

4   The Simulation Results

4.1   Test Examples and Model Tests

To simulate accidents, the following variables were used: ice weight, duration of the event and period of deposits growth, type of terrain, constructions lifetime, wind speed during the maximum deposition size, month of event occurrence, altitude and wind direction at the beginning and after reaching the maximum size of deposition. During the simulation, the initial relations between the variables of the model have changed. In particular, the factors that affect the accident rate were highlighted and other factors were discarded.

The remaining factors of the model and impact on accidents include: deposit diameters, ice weight, wind speed at the beginning of the event and after reaching the maximum size, altitude and lifetime of overhead lines constructions.

A Bayesian network defines different variables, dependencies between them, and conditional probabilities of these dependencies. BN can use this information to calculate the probability of various possible causes of the accident event. Conditional probabilities were calculated based on dependencies represented in the Bayesian network model that is shown in Fig. 2. The model shows that there are three nodes that require calculation of conditional probabilities. These are nodes C, F and H.

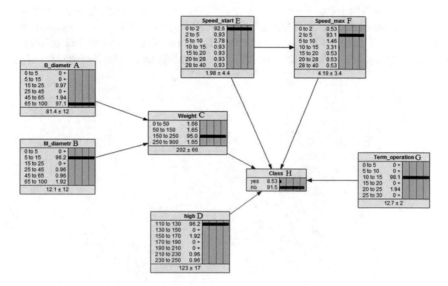

Fig. 2. Bayesian network for prediction of overhead transmission line accidents

Node C: ice weight. The ice deposition weight for each case depends on the actual deposit parameters. The conditional probability for the weight of ice (C) and for large and small diameters (A and B) was calculated by formula (3).

$$P(C|A, B) = \frac{P(A, B, C)}{P(A)P(B)} \qquad (3)$$

Node F: wind speed after reaching the maximum deposition size. Wind speed after reaching the maximum ice size depends on wind speed at the beginning of ice creation. The conditional probability for the wind speed at the ice maximum size (F), where the initial speed is (E), is calculated by formula (4).

$$P(F|E) = \frac{P(F, E)}{P(E)} \qquad (4)$$

Node H: accident occurrence. The accident occurrence depends on the parameters shown in Fig. 3. The conditional probability of an accident is calculated by formula (5)

$$P(H|A \dots G) = \frac{P(A \dots H)}{P(C|A, B)P(F|E)P(G)P(D)} \tag{5}$$

4.2   Prediction Accuracy

The forecasting accuracy was 78%. Data used for forecasting were not included into the training set. Such prediction accuracy is quite high for the network. During forecasting, errors in determination of accident classes are not equivalent, i.e. a false statement that an accident will occur is less important than the false statement that the accident will not occur. Since the follow-up steps to prevent accidents are carried out by a power

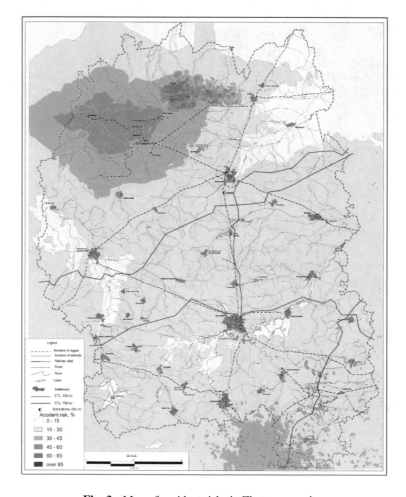

Fig. 3. Map of accident risks in Zhytomyr region

system management worker, a false alarm will not cause undesirable results. Therefore, error class definition is considered separately (class 1—accident and class 0—no accident). The above prediction accuracy corresponds to the lowest precision—error of non-recognition of an accident.

For the network training, a data set of 5,000 records was used, including 2,400 records that constituted a training sample, and 1,200 records were used for cross-testing of the model, and other records—for general model testing.

5 Conclusions

As a result of using the accident event prediction model on base Bayesian network modeling technique the following key results were obtained: wind speed and weight of ice deposits are the main causes of accidents among the climatic factors, and other influences (among the selected variables) are lifetime of line structures and altitude; two groups of characteristics that cause accidents are revealed: large ice mass in the absence of strong wind pressure, and strong wind load with medium (in terms of the model) deposition weight. The predicted places of accidents occurrence coincide with the areas where accidents actually occurred in 80% of cases.

The model allows to predict the onset of an accident. The maps created on the basis of geospatial data correspond to the events of accidents that were not used for model construction, but only for validation thereof. For the further development of the study, the following steps are proposed: inclusion into the model of other variables and relations between them to facilitate consideration of more factors and improve the prediction accuracy; extension of the results obtained in the study to all the regions of Ukraine, taking into account topography and climatic and other factors. Further studies require application of state-of-the-art techniques in the field of restoration of damaged and missing information [26] and multidimensional discretization strategies that will reduce information losses, thus increasing the effectiveness of the models using the spatial database.

References

1. Barannik, V.: Comprehensive evaluation technique and ways to ensure energy independence. Thesis, Kyiv (2008). (in Ukrainian)
2. Horokhov, E., Kazakevich, M., Shapovalov, S., Nazim, Y.: Aerodynamics of power lines constructions, Donetsk, 336 p. (2000). (in Ukrainian)
3. Horokhov, E., Kazakevich, M., Turbin, S.: Ice and wind impact on overhead transmissions lines, Donetsk, 348 p. (2005). (in Ukrainian)
4. Power grid technology policy in the construction of high voltage distribution networks. Part I, Technical Strategy in the construction and operation of electrical distribution networks, Kyiv, Ukrenergo (2011). (in Ukrainian)
5. Climatic data for determining loads on overhead transmission line. Method of processing. GRIFRE, 94 p. (2008). (in Ukrainian)

6. The Development of a US Climatology of Extreme Ice Loads: US Department of Commerce NOAA. NESDIS National Climatic Data Center Asheville, NC 28801-5696, edn. 01, 23 p. (2002)
7. Bilash, I.: Ways improvement of melting ice deposits on the overhead transmission line. In: Energy Problems and Efficiency in the Agricultural Sector of Ukraine, vol. 1, pp. 77–82 (2012). (in Ukrainian)
8. Meteorological reviews: republic hydro-meteorological management in Ukraine 1961, 326 p. (1990). (in Ukrainian)
9. Events development for Achieving reliability of power networks in influence of ice mind loads and wind pressure on the territory of Ukraine and regions (1990) Book 1, Selenergoproekt. Kiev, 110 p. (in Ukrainian)
10. Pashynska, N., Snytyuk, V., Putrenko, V., Musienko, A.: A decision tree in a classification of fire hazard factors. Eastern-Eur. J. Enterp. Technol. 5(10(83)), 32–37 (2016)
11. Rajangam, E., Annamalai, C.: Graph models for knowledge representation and reasoning for contemporary and emerging needs – a survey. Int. J. Inf. Technol. Comput. Sci. (IJITCS) 8(2), 14–22 (2016)
12. Zhang, H., Liu, M.: GIS-based emergency management system on abrupt environmental pollution accidents in counties of China. Int. J. Educ. Manage. Eng. (IJEME) 2(8), 31–38 (2012). https://doi.org/10.5815/ijeme.2012.08.06
13. Jain, E.G., Mallick, B.: A study of time series models ARIMA and ETS. Int. J. Modern Educ. Comput. Sci. (IJMECS) 9(4), 57–63 (2017). https://doi.org/10.5815/ijmecs.2017.04.07
14. Mani, K., Kalpana, P.: An efficient feature selection based on Bayes theorem, self information and sequential forward selection. Int. J. Inf. Eng. Electron. Bus. (IJIEEB) 8(6), 46–54 (2016). https://doi.org/10.5815/ijieeb.2016.06.06
15. Acharjya, D.P., Debasrita, R., Rahaman, M.A.: Prediction of missing associations using rough computing and Bayesian classification. Int. J. Intell. Syst. Appl. (IJISA) 4(11), 1–13 (2012). https://doi.org/10.5815/ijisa.2012.11.01
16. Wahid, F., Ghazali, R., Fayaz, M., Shah, A.S.: Statistical Features Based Approach (SFBA) for hourly energy consumption prediction using neural network. Int. J. Inf. Technol. Comput. Sci. (IJITCS) 9(5), 23–30 (2017). https://doi.org/10.5815/ijitcs.2017.05.04
17. Mam, M., Leena, G., Saxena, N.S.: Improved K-means clustering based distribution planning on a geographical network. Int. J. Intell. Syst. Appl. (IJISA) 9(4), 69–75 (2017). https://doi.org/10.5815/ijisa.2017.04.08
18. Heckerman, D.: Tutorial on Learning with Bayesian Networks Technical report MSR-TR-95-06, Microsoft Research (1995)
19. Cheng, J., Greiner, R.: Learning Bayesian Belief Network Classifiers: Algorithms and System, pp. 141–151. Springer, London (2001)
20. Krieg, M.L.: Tutorial on Bayesian Belief Networks. Technical Note. Electronics and Surveillance Research Laboratory PO Box 1500 Edinburgh, South Australia (2001)
21. Duda, O., Hart, P.: Pattern Classification, 2nd edn, p. 738. Wiley, New York (2000)
22. Nisbet, R.: Handbook of statistical analysis and data mining applications, p. 864. Pacific Capital Bankcorp NA, Santa Barbara (2009)
23. Gumbel, E.: Statistics of extreme values. Moscow, Mir, 452 p. (1985)
24. Farzaneh, M.: Atmospheric Icing of Power Networks, p. 397. Springer, Dordrecht (2008)
25. Climatic handbook in Ukrainian SSR. History and physical-geographical description of meteorological stations, issue 10, Kiev (1968). (in Ukrainian)
26. Neill, D., Cooper, G., Das, K., Jiang, X., Schneider, J.: Bayesian network scan statistics for multivariate pattern detection. In: Statistics for Industry and Technology, pp. 221–249 (2009)

Influence of the Deep Spherical Dimple on the Pressure Field Under the Turbulent Boundary Layer

V. A. Voskoboinick[1(✉)], V. N. Turick[2], O. A. Voskoboinyk[1],
A. V. Voskoboinick[1], and I. A. Tereshchenko[2]

[1] Institute of Hydromechanics of NAS of Ukraine, Kyiv, Ukraine
vlad.vsk@gmail.com
[2] Kyiv Polytechnic Institute, National Technical University of Ukraine, Kyiv, Ukraine

Abstract. The influence of a local dimple in the form of a deep spherical cavity on the pressure field inside the dimple and its vicinity for the turbulent flow regime is experimentally determined. Specific features of the vortex formation inside the dimple are established and the influences of vortex structures that are ejected outward from the spherical dimple on the structure of the turbulent boundary layer are shown. The antiphase oscillations of the wall pressure fluctuation field occur in the halves of the dimple separated by a longitudinal axial plane when the vortex flow "switches" from one side of the dimple to another. The spectral components of the wall pressure fluctuations on the streamlined surface of the spherical dimple have discrete components corresponding to the frequencies of the "switching" of the vortex formation inside the dimple (St $\approx$ 0.003), the frequencies of the vortex ejections from the dimple (St $\approx$ 0.05) and the frequencies of the self-oscillations of the shear layer (St $\approx$ 0.4).

Keywords: Spherical dimple · Vortex structure · Wall pressure fluctuations

1 Introduction

Modern energy-saving technologies widely use streamlined surfaces that have ensembles of cavities of various geometric shapes and locations for boundary layer control. In the majority of technical designs and structures, spherical dimples of various depressions are used, which is due not only to their manufacturability, but also to the specific formation and evolution of vortex systems and jet streams within them. Vortices and jets interact with the boundary layer, generating a vortex flow that significantly increases the thermal and hydraulic efficiency of the streamlined surface, reduces, under certain conditions, the drag and hydrodynamic noise of the dimple reliefs and also alters the conditions for the transition of the boundary layer from laminar to turbulent. The defining feature of the flow inside the dimple is the presence of a shear layer, which is formed when the incoming flow is separated from the leading edge of the cavity due to a geometric rupture of the streamlined surface. Time-averaged flows in the cavity have two- and three-dimensional features. From a large number of experimental, theoretical

© Springer International Publishing AG, part of Springer Nature 2019
Z. Hu et al. (Eds.): ICCSEEA 2018, AISC 754, pp. 23–32, 2019.
https://doi.org/10.1007/978-3-319-91008-6_3

and numerical studies carried out in recent years, it is known that the pattern of flow inside and around the cavity essentially depends on the flow parameters. Among them, the velocity of the incoming flow, the thickness of the boundary layer, the level of turbulence, and also the basic geometric characteristics of the cavity, such as, for example, the ratio of the depth to the width of the dimple or the configuration of the cavity [1–3].

The instability of the shear layer is central in the vortex flow inside the dimple and integral in the development of the oscillations. Because of the inflection point in the velocity profile, the shear layer becomes unstable to small perturbations due to the Kelvin-Helmholtz instability mechanism. As the shear layer is removed downstream of the leading edge of the cavity, the selective amplification process generates instability waves that initially grow exponentially as the distance downstream increases to such an extent that nonlinear effects become important. After this initial stage, nonlinear effects predominate, and the shear layer, as is often observed, turns off, forming coherent large-scale transversely oriented vortices (for example, for rectangular cavities). Although the formation of such structures is not essential for the development of oscillations, they usually give the necessary discrete excitation, which becomes sufficient for the generation and conservation of oscillations, and are therefore the most important in studying streams in the dimple. The time-dependent and spatially very complex nature of the shear layers in the dimple makes their numerical simulation and experimental determination difficult. In particular, since the shear layers in the cavity contain a sufficiently wide spectrum of developing scales and structures, and in addition are often characterized by significant fluctuations or cyclic variations, existing single-point measurements and time-averaged data do not fully reflect the dynamics and complexity of the shear layers inside the cavity. Because of this, a large number of data of a global and instantaneous nature is required using the multi-point research techniques.

The unstable pressure fields are generated when the boundary layer is separated from the upstream edge of the dimple. The complex feedback mechanism between the aft wall of the dimple and the separated region supports significant fluctuations, both amplitude and convective velocity of the instability waves in the shear layer, leading to an increase in hydrodynamic and acoustic phenomena. However, the mechanism by which sound and pseudo-sound pressure fluctuations are generated depends on the dynamic and spectral characteristics of the oncoming boundary layer, and also the geometry of the cavity. In addition, the instabilities of the shear layer form vortex structures, that interact with the aft wall of the dimple, and produce an acoustic response. This tendency to generate sound by the cavity due to narrowband or broadband noise emission by the shear layer is in most cases an undesirable phenomenon [4, 5].

The high pressure before the aft wall of the dimple and the low pressure behind its front wall form a return flow inside the dimple. It forms an internal circulation flow, the configuration of which depends on the flow regime and the geometry of the dimple. Vortex structures of various scales generated inside the dimple are not a chaotic group of vortices, but self-organized vortex structures that mutually coordinate their behavior. Low-frequency large-scale coherent vortex systems, that are formed from the circulation flow, and small-scale vortices, that are generated inside the dimple and also eddies of the shear layer are ejected outward from the dimple, resulting in a wide range of velocity and pressure fluctuations with intense tonal components. The generation of sound and pseudo-sound by vortex and jet flow leads to significant vibrations of the streamlined surface and the transmission of acoustic excitation to the environment. Studying and taking into account the fields of pressure fluctuations and vibro-acoustic characteristics of streamlined surfaces are of great importance at the stage of design and operation of objects moving in a liquid and also in carrying out measures to improve the environmental situation near transport highways.

The purpose of the research is to experimentally study the pressure field and the features of the formation and development of large-scale vortex structures inside a deep spherical dimple on a flat surface that is streamlined by a turbulent flow.

2   Experimental Setup

Experimental research was carried out in a narrow hydrodynamic channel of the University of Rostock (Germany). The channel had plexiglass walls about 1.2 m long, 0.2 m wide and 0.015 m deep. Water into the channel came from the soothing container 1 (Fig. 1). It was housed a height-adjustable damper 3, which ensured the necessary flow rate in the channel. The water in the soothing container fixed to the base 5 was supplied from the tank 2 by means of a pump 4. Water from the soothing container was introduced into the hydrodynamic channel through the inlet section 7 and the confuser 8, and then through an intermediate container 11 and an outlet pipeline 10 where flaps and auxiliary pump 6 were installed, entered into the tank 2. A circular hole was made at the bottom of the hydrodynamic channel, where a disk with a diameter of 0.18 m was placed. In the center of the disk was made a spherical hole 0.046 m diameter (d), and 0.012 m deep (h), respectively, h/d = 0.26. The center of the dimple was at a distance of about 0.6 m from the confusor in the axial section of the channel.

a)   b)

Fig. 1. Scheme (a) and photography (b) of the experimental setup.

Miniature sensors of absolute pressure and wall pressure fluctuations were installed flush with the streamlined surface of the dimple and in its vicinity (see, Fig. 2).

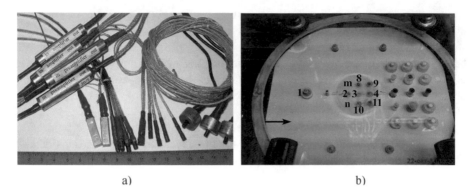

a)   b)

Fig. 2. Sensors of absolute pressure and wall pressure fluctuations (a) and their location on streamlined surface of the dimple and channel wall (b).

Sensor No. 1 was placed in front of a spherical dimple on the surface of the channel wall and recorded wall pressure fluctuations in the boundary layer, which was not disturbed by the dimple. The pressure sensors inside the dimple were mounted at a distance of 0.01 m from each other, and sensor No. 3 was located in the center of the dimple at its bottom. The sensors behind the spherical dimple were disposed at a distance of 0.015 m from each other and the first row of sensors (near field of the dimple) was located at a distance of 0.015 m from the aft edge of the dimple.

The pressure field on the streamlined surface of the dimple and the channel wall was detected by piezoresistive sensors that measured the static and dynamic pressure in the low-frequency range (up to 200 Hz). These sensors were distinguished by increased stability and accuracy and were made with the use of microelectronic technology, where the electronic components were deposited on a quartz membrane, that bented under the action of applied loads proportional to the pressure. In our studies we used sensors specially designed and manufactured for this purpose, which allowed us to measure pressures from 1 Pa (0.1 mm of water column) to 6 kPa with an error of 0.01%. Piezoresistive pressure sensors were mounted under holes of 0.8 mm in diameter on the streamlined surfaces of the dimple and the channel wall. The wall pressure fluctuation field was detected by miniature piezoceramic pressure fluctuation sensors of membrane and rod types. Sensors of 1.6 mm or 1.3 mm in diameter of a sensitive surface were flush mounted with the streamlined surface of the dimple and the flat wall of the channel. The electrical signals of the sensors through bridge and half-bridge circuits came to the monitoring and recording equipment, and then through the multichannel analog-digital converters to personal computers. The results of the measurements were processed and analyzed by specially developed programs and methods [6, 7].

The error in measuring the integral characteristics of the pressure and the wall pressure fluctuation fields was no more than 4%, the correlation dependences was no more than 6%, in the frequency range up to 1250 Hz with a confidence interval of 0.95 or 2σ.

3   Research Results

Visual studies, by supplying colorants from the holes on the streamlined surface, showed that large-scale asymmetric vortex structures (Fig. 3) were formed inside the dimple for turbulent flow, when the flow velocity in the channel was 0.86 m/s and more (Reynolds number $Re_d = Ud/\nu \geq 40,000$). Quasistable large-scale vortices were generated in the near-bottom region of the dimple closer to its frontal part. When vortices reached of certain sizes, they ejected outward from the dimple above the opposite lateral aft wall, forming a vortex system inclined with respect to the direction of flow. The angle of inclination of the vortex structures is increased with an increase in the Reynolds number of turbulent flow. The ejection of vortex systems occurs almost periodically, but at sufficiently high Reynolds numbers, substantial chaotization of this process is observed. After a number of ejections from one side of the dimple, the vortex structure "switches" to the opposite side of the dimple and new inclined vortex systems begin to form. The ejection of the vortex systems occurs at an angle $\pm(40...60)°$ relative to the median cross-section of the dimple on one side or the other side. It is found that for $Re_d = 40,000$ inclined vortices are ejected at an angle of $\pm45°$, and for $Re_d = 60,000$ the angle of formation of the vortex and its ejection increases to $\pm60°$.

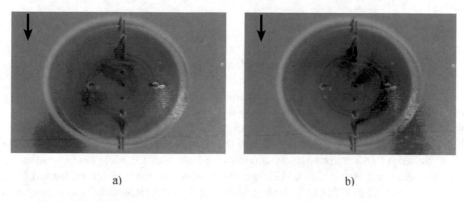

a)   b)

Fig. 3. Visualization of vortical flow inside the dimple at different time instants.

A characteristic change in the pressure inside and around the streamlined dimple is observed, both in experimental studies and in numerical calculations, as shown in Fig. 4. Here we present the results of numerical simulation of the pressure distribution along the longitudinal axis of the hydrodynamic channel, that passes through the median section of the spherical dimple (a depression of 0.26), and also for a channel whose surface is not disturbed by a local cavity (see Fig. 4a). Calculations in [3, 8] were carried out by two methods - the Large-Eddy Simulations (LES) method and the method of solving the Unsteady Reynolds-Averaged Navier-Stokes (URANS) equations. Also Fig. 4a shows the results of experimental researches, which were obtained for the Reynolds number $Re_d = 40,000$ in [9] and in our measurements [10]. It should be noted that, in general, the experimental and numerical results correlate with each other, but there are some differences due to errors in measurements and calculations, and also due to features of mathematical models and experimental facilities. The pressure field, calculated and measured inside the dimple, displays the areas of acceleration and deceleration of the flow due to the formation and development of a vortex flow inside the dimple. Along with this, it should be noted that the location of the local dimple of the investigated shape and dimensions on the wall of a narrow hydrodynamic channel does not lead to an increase in the local hydraulic drag of the channel for turbulent flow. This indicates that the reduction of frictional drag, which is formed over the opening of the dimple, is balanced by an increase in the drag of the form, which results from the deceleration of the oncoming flow on the aft wall of the dimple.

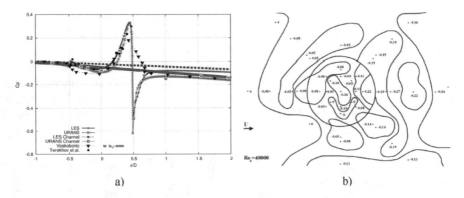

a)   b)

Fig. 4. Pressure distribution inside and around the streamlined dimple.

The pressure distribution on the streamlined surface of the dimple and in its vicinity for a turbulent flow regime corresponding to the Reynolds number of $Re_d = 40,000$ is shown in Fig. 4b. It should be noted that the pattern of pressure is clearly asymmetric. For example, the region of low pressure in the bottom area of the dimple is inclined at an angle close to 45° relative to the direction of flow (see Fig. 4b). This is consistent with the location of the quasi-stable large-scale vortex, which is observed for turbulent flow in visual studies [3, 11]. On the lateral side of the aft wall, where the vortex system is ejected into the boundary layer, the levels of the low pressure are recorded. Between the edge of the dimple above which an asymmetric vortex is ejected and the bottom region of the dimple, an area of high pressures is observed, that is associated with the flow of liquid from the aft wall (from the region of interaction of the oncoming flow with the surface of the dimple). This fluid, flowing under an asymmetric vortex, reaching the front wall, is partially carried away to the source of the vortex, and partially, forming a circulation flow, envelops a quasi-stable large-scale vortex, feeding it by the energy of the oncoming flow. As visual studies and instrumental measurements of the field of wall pressure fluctuations show [3, 11], this vortex oscillates in three mutually perpendicular planes. Part of the asymmetric vortex, and often the entire vortex, is ejected above the aft wall, moving in the form of a longitudinal vortex system into the boundary layer behind the dimple. This leads to a transformation of transversal vorticity or, more accurately, inclined at an angle of 45° to the direction of the flow vorticity into the longitudinal vorticity. The source of this asymmetric vortex (see Fig. 3) is located on the other side of the front wall of the dimple (in its bottom part), and the sink is located on the opposite lateral side of the aft wall closer to its edge, similar to the scheme proposed in work [3]. The results of visual and instrumental studies shown, the location of asymmetric vortices is not equivalent U the time. One of them is recorded most of the time of observations, which determines the asymmetric distribution of pressure inside the dimple and its vicinity for the turbulent flow regime (see Fig. 4). In addition, it should be noted that, immediately behind the aft wall of the dimple, high pressure levels are observed, and in its near wake are formed regions of low pressures, characterizing the fact that here the flow is accelerated.

It has been established that, for turbulent flow ($Re_d \geq 40,000$), wall pressure fluctuations in different parts of the front and aft walls of the spherical dimple have antiphase oscillations. At the measuring points 9 and 11, and also "m" and "n" (Fig. 5), high levels of antiphase oscillations or anticorrelations of pressure fluctuations are observed, which are caused by the features of the formation of a vortex flow inside the dimple [10, 11]. It was found that on the lateral side of the aft wall of the dimple, where the vortex structures of the shear layer interact with the latter, higher levels of wall pressure fluctuations are observed, with a predominance of positive pressures. On the surface of the opposite lateral side of the aft wall of the dimple, above which large-scale vortex structures are ejected, the levels of the wall pressure fluctuations are somewhat lower, and negative pressures prevail here.

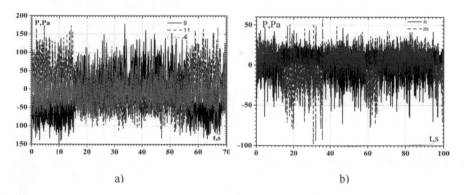

a) b)

Fig. 5. Time history of the wall pressure fluctuations at different measurement points inside the dimple on aft wall (a) and front wall (b) of the dimple (see Fig. 2b).

The spectral power densities of the wall pressure fluctuations on the streamlined surface of a deep spherical dimple and on the channel wall in front of the dimple are shown in Fig. 6 depending on the Strouhal number. The pressure spectra are nondimensionalized on the outer flow variables U and d in the form [12, 13].

$$E(St) = E(\omega)U/q^2d \quad \text{vs} \quad St = \omega d/2\pi U$$

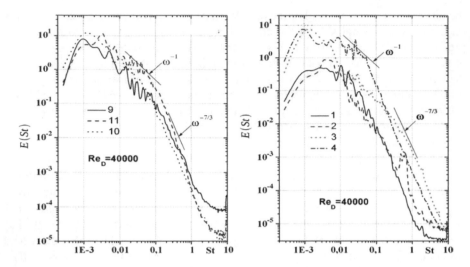

Fig. 6. Spectral power densities of the wall pressure fluctuations on the streamlined surface of the deep spherical dimple and on the channel wall at different points (see Fig. 2b).

The frequency spectrum $E(\omega)$ of the random field of the wall pressure fluctuations $p(x, t)$ is found as [14, 15]

$$E(\omega) = \frac{1}{2\pi} \int\limits_{-\infty}^{\infty} R_p(\tau)\exp(-i\omega\tau)d\tau$$

The auto-correlation of the wall pressure field is found as

$$R_{p_1p_2}(\tau) = \left\langle p_1(\vec{x}, t)p_2(\vec{x}, t + \tau) \right\rangle$$

and the dynamic pressure is found as

$$q = \rho U^2/2$$

Straight line segments corresponding to changes in the spectral components of the frequency ω^{-1} and $\omega^{-7/3}$ (where the circular frequency $\omega = 2\pi f$), which are characteristic for the behavior of the velocity and pressure fluctuation spectra in wall flows [16–18], are shown on this figure. The maximum levels of wall pressure fluctuations at frequencies $St \approx (0.002\ldots0.003)$ on the lateral sides of the aft wall are caused by the frequency of "switching" the vortex formation inside the dimple, which is consistent with visual studies and antiphase oscillations of pressure fluctuations. In the middle of the aft wall (sensor No. 4), the maximum spectral levels are observed at a frequency of $St \approx 0.05$, which corresponds to the frequency of large-scale vortex ejections from the dimple. In addition, in this section of the dimple, where the vortex structures of the shear layer interact with the wall of the dimple, the greatest levels of the spectral components of the wall-pressure fluctuations are observed practically in the entire investigated frequency range. On the front wall of the dimple where the boundary layer is separated and the shear layer is formed, a discrete rise of the pressure fluctuation spectra (curve 2 in Fig. 6) is observed in the frequency range St $\approx$ (0.3...0.5), which corresponds to the self-oscillations of the shear layer.

4   Conclusions

1. It is found that large-scale asymmetric vortex structures are formed inside the deep spherical dimple, the source of which is located on the front wall of the dimple somewhat laterally from the median section of the dimple, and the sink is located on the opposite lateral side of the aft wall of the dimple, i.e., the vortex structures are inclined and they intersect median section of the dimple. In an oscillatory motion, an asymmetric large-scale vortex is periodically ejected above the aft wall at an angle that increases with increasing flow velocity. It was found that with the Reynolds number $Re_d = 40,000$ the vortex ejection angle is approximately 45°, and at $Re_d = 60,000$ it is almost 60°.
2. It is recorded that the pressure field on the surface of the dimple, which is streamlined by the turbulent flow, is asymmetric in nature, due to the features of the generation of the vortex flow inside the dimple and the ejections of large-scale vortex structures outward from the spherical dimple. The maximum intensity of pressure and the drag are observed on the aft wall of the dimple, and the minimum - on its front wall.
3. It is established that antiphase oscillations of the wall pressure fluctuation field occur in the halves of the dimple separated by a longitudinal axial plane when the vortex flow "switches" from one side of the dimple to another.
4. It is found that the spectral components of the wall pressure fluctuations on the stream-lined surface of the spherical dimple have discrete components corresponding to the frequencies of the "switching" of the vortex formation inside the dimple (St $\approx$ 0.003), the frequencies of the vortex ejections from the dimple (St $\approx$ 0.05) and the frequencies of the self-oscillations of the shear layer (St $\approx$ 0.4).

References

1. Khalatov, A.A.: Heat transfer and fluid mechanics over surface indentations (dimples). National Academy of Sciences of Ukraine, Institute of Engineering Thermophysics, Kyiv (2005)
2. Khadivi, T., Savory, E.: Effect of yaw angle on the flow structure of low aspect ratio elliptical cavities. J. Aerosp. Eng. **232**(4), 04017002 (2017)
3. Voskoboinick, V., Kornev, N., Turnow, J.: Study of near wall coherent flow structures on dimpled surfaces using unsteady pressure measurements. Flow Turbul. Combust **90**(4), 709–722 (2013)
4. Mori, Y., Kobayashi, T., Tahara, K.: Sorting of acrylonitrile-butadiene-styrene and polystyrene plastics by microwave cavity resonance. Int. J. Wireless Microwave Technol. (IJWMT) **6**(2), 1–9 (2016). https://doi.org/10.5815/ijwmt.2016.02.01

5. Sharma, S., Kaur, D.: Measurements of dielectric parameters of aviation fuel at X-band frequencies using cavity perturbation technique. Int. J. Wireless Microwave Technol. (IJWMT) **6**(6), 48–55 (2016). https://doi.org/10.5815/ijwmt.2016.06.05

6. Voskoboinick, V.A., Makarenkov, A.P.: Spectral characteristics of the hydrodynamical noise in a longitudinal flow around a flexible cylinder. Intern. J. Fluid Mech. **31**(1), 87–100 (2004)

7. Voskoboinick, V.A., Grinchenko, V.T., Makarenkov, A.P.: Pseudo-sound behind an obstacle on a cylinder in axial flow. Intern. J. Fluid Mech. **32**(4), 488–510 (2005)

8. Turnow, J., Kornev, N., Isaev, S., Hassel, E.: Vortex mechanism of heat transfer enhancement in a channel with spherical and oval dimples. Heat Mass Transf. **47**(3), 301–313 (2011)

9. Terekhov, V.I., Kalinina, S.V., Mshvidobadze, YuM: Experimental research of flow development in channal with halfspherical cavity. Sib. Phys-Tekhn. J. **1**, 77–86 (1992). (in Russian)

10. Voskoboinick, V.A.: Pressure distribution on streamlined surface with spherical dimple. Water Transp. **18**(3), 90–96 (2014). (in Russian)

11. Voskoboinick, A.V., Voskoboinick, V.A., Isaev, S.A., Zhdanov, V.L., Kornev, N.V., Turnow, J.: Vortex flow bifurcation inside the spherical dimple in a narrow channel. Appl. Hydromech. **13**(4), 17–27 (2011). (in Russian)

12. Bull, M.K.: Wall-pressure fluctuations beneath turbulent boundary layers: some reflections on forty years of research. J. Sound Vibr. **190**(3), 299–315 (1996)

13. Goshvarpour, A., Shamsi, M., Goshvarpour, A.: Spectral and time based assessment of meditative heart rate signals. Int. J. Image Graph. Signal Process. (IJIGSP) **5**(4), 1–10 (2013). https://doi.org/10.5815/ijigsp.2013.04.01

14. Bendat, J.S., Piersol, A.G.: Random Data: Analysis and Measurement Procedures. Willey, New York (1986)

15. Elnashar, A.I., El-Zoghdy, S.F.: An Algorithm for static tracing of message passing interface programs using data flow analysis. Int. J. Comput. Netw. Inf. Secur. (IJCNIS) **7**(1), 1–8 (2015). https://doi.org/10.5815/ijcnis.2015.01.01

16. Tsuji, Y., Marusic, I., Johansson, A.V.: Amplitude modulation of pressure in turbulent boundary layer. Int. J. Heat Fluid Flow **61**, 2–11 (2016)

17. Blake, W.K.: Mechanics of Flow-Induced Sound and Vibration, vol. 2. Academic Press, New York (1986)

18. Doisy, Y.: Modelling wall pressure fluctuations under a turbulent boundary layer. J. Sound Vib. **400**, 178–200 (2017)

Structural Model of Robot-Manipulator for the Capture of Non-cooperative Client Spacecraft

D. Humennyi[(✉)] ⓘD, I. Parkhomey, and M. Tkach

National Technical University of Ukraine "Igor Kyiv Polytechnic Institute",
37, Prosp. Peremohy, Kyiv 03056, Ukraine
d.gumennuy@kpi.ua

Abstract. Reorientation, service operation of client spacecrafts etc. are very actual tasks nowadays. Most of such spacecrafts are non-cooperative. Although service operations of these satellites were held before, they are not fully automated. However, docking in manual mode needs huge amount of time and human aboard. That makes impossible to carry out the required number of service operations. In current work it is represented conceptual model of the robot - manipulator for capture and maintenance of non-cooperative client spacecraft. Mechanism of collet effector was proposed for this purpose. Payload Adapter interface PAS 1666 S, PAS 1194 C, PAS 1666 MVS, PAS 1184 VS was chosen as a docking point, because of this interface's convenience and prevalence. Proposed mechanism allows to work in conditions of client spacecraft's linear and angular dynamic position errors in a range of $\pm 5°$ per minute and ± 0.1 meters per minute respectively. This is achieved by design of the robot-manipulator. Problem of jogless docking and methods of shock prevention are examined. In addition, the berthing and girth operations need axes coincidence of non-cooperative client spacecraft and service spacecraft. Having this in mind, phases of berthing and girth are described and main functions of service spacecraft's control system for solving this problem are considered.

Keywords: Payload system · Collet effector · Telescopic console
Service spacecraft · Docking procedure

1 Introduction

Nowadays on the geostationary Earth orbit (GEO) there are about 1500 satellites. However reorientation, motion or other service operations are needed for about 750 spacecrafts [1]. Almost all spacecrafts (SC) with a weight more than 500 kg were put into GEO by medium-lift and heavy-lift launch vehicles and were equipped by Payload Adapter, which is compatible with PAS PAS1666 S, 1194 C, PAS 1666 MVS, PAS 1184 VS. In most cases such design of SC and its control system did not provide docking and orbital motion in GEO after initial adjustment. So, such vehicles are not equipped by automatic docking means (IDBS) and need special methods, instrumentality and scripts for this operations.

© Springer International Publishing AG, part of Springer Nature 2019
Z. Hu et al. (Eds.): ICCSEEA 2018, AISC 754, pp. 33–42, 2019.
https://doi.org/10.1007/978-3-319-91008-6_4

The analysis of reports Department of Mechanical Engineering Massachusetts Institute of Technology [2], International Astronautical Congress [3], University of Nebraska - Lincoln U.S. Air Force Research U.S. Department of Defense [4] and publications [1, 5, 6], showed that at this moment there are no means of docking for service spacecraft (SSC) with non-cooperative spacecraft (NCSC) in a automated or automatic mode. Reasons for this are complexity of such operations, their cost, lack of hardware-technical solutions, common standards of interfaces and functioning of equipment operation protocols in situations, which arise when preforming approach, berthing, docking, undocking, departing and projection of SC [6]. Docking with NCSC in manual mode is considered in works [1, 5]. However, the duration of such operation exceeds five hours and requires availability of the cosmonaut and complex equipment on SSC.

As follows, it is obvious that with further evolution of Astronautics execution of docking operations with NCSC will appear more often and involvement of human as an object, which controls docking process, will have no any economical, social or scientific significance.

2   Main Body

2.1   Formulation of the Problem

Process of docking NCSC with SSC is performed using the script, which is presented in Table 1. This script includes nine stages, the first five of which are essential for orbital docking of SC [5] and must be performed automatically.

Table 1.   Sequence of operations, which are performed during docking with NCSC

№	Stage name	Stage description
1	NCSC position determination	Determination of the NCSC and SSC relative position via GPS navigation and radar tools
2	NCSC and SSC approach at a distance up to 200 m	Approaching SSC at a distance of the first suspension for the docking equipment preparation
3	NCSC and SSC approach at a distance up to 20 m	Approaching SC at a distance of a second suspension for testing docking and fixing SC's continuation equipment
4	Search for a docking plane on NCSC	SSC flight-around NCSC for searching of a docking plane. Definition of the docking point position. Coordination of dynamic characteristics of devices
5	Berthing SSC to NCSC at a distance of 2 m	Approaching SSC to NCSC on the position of service with further approach. The relative position of the devices are considered
6	Definition of the position of the mechanical docking point	Optical, analytic and radio-locating determination of docking port on NCSC's docking plane. Estimation of the values of the cone of occurrence
7	Capture and maintenance of NCSC by the means of SSC	Unstressed maintenance of NCSC by mechanical means of SSC and further fixation of NCSC linear and angular movement of NCSC
8	Coordinating of electric and mechanic parameters of the devices	Coordination of the angles of the rotation in relative planes, electric level 0 V and digital interfaces
9	Departing of SSC from NCSC	Departing of SSC at a distance of 20 m with subsequent return of the predetermined orbital position

Stages 1, 2 have complied technical solution [7, 8], and stages 3, 4 were conducted within Space Orbiter programs and described in publications [7–10]. Stages 6–8, which provide the process of "soft" docking, are performed only in manual mode and the process of their automation requires the development of new control tools and control systems Table 1. Sequence of operations, which are held during docking with NCSC.

2.2 Conceptual Structure of Robot-Manipulator

Suitable Docking point (DP) in modern SC is apogee motor nozzle (Fig. 1) and Space-Craft adapter ring (S/C) Payload Adapter (Fig. 2) the technical characteristics of which are written in operating manuals [11–13]. These nodes are characterized by high rigidity and coherence of the position of center of mass of the devices what allows to move NCSC with mechanic tools, which are located in SSC [11–13].

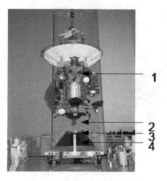

Fig. 1. Spacecraft "Cassini-Huygens" with available S/C adapter ring and apogee motor nozzle in process of connecting to PAS: 1 system block of SC; 2 nozzle of apogeum motor; 3 S/C adapter ring; 4 PAS (photo is used with permission of European Space Agency, Communication Department)

Fig. 2. Connection of SC with Payload adapter PAS 1194 C with means of S/C adapter ring: 1 - system module; SC; 2 - contact plane of SC with PAS by S/C adapter ring means; 3 - Payload adapter (photo is used with the permission of European Space Agency, Communication Department)

Capture and maintenance of NCSC through the listed DP need compensation of error of the active positioning of the spacecrafts, which is caused by the features of radar sensor system, which is characterized by low resolution at the distance of up to 20 m [15], has stationary character and is given by six degrees of freedom (DOF).

Typical structure of S/C Adapter, which remains on S/C after its departing from the NCSC's launch vehicle (Fig. 3), has mounting hardware (keys) to Payload Adapter, thank to pyrotechnic fasteners is provided "departing" of SC from launch vehicle through. After departing from the launch vehicle S/C Adapter remains on the SC and is not used repeatably for the active existence of devise on the GEO. Mechanical characteristics of pyrotechnic fasteners' mounting hardware allow the NCSC to hold the S/C Adapter 's mounting hardware in case of uncoordinated docking with the following coordination of dynamic and mechanical parameters of leaked devices.

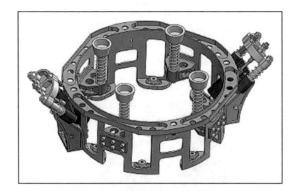

Fig. 3. Typical design of S/C Adapter

Accordingly, for providing automatic capture of SC and S/C Adapter's elements it is necessary to specify the technical capabilities of capture adapter with a diameter range of 800–1400 mm, while maintaining the coaxiality of console, finite effector, NCSC and SSC. It is also important to note the necessity of unstressed capture in conditions of dynamic positioning error of devices. In work [14], the possibility of such a capture is considered, however coaxiality of SC is not provided.

The structure of effector should provide the opportunity of capturing NCSC with consideration of such factors: the presence of errors in linear and angular desitioning; limited time of being at the distance of NCSC and SSC berthing; a large number of S/C Adapter's standards, that are different at sizes and structure; the lack of standard markers on NCSC adapter; absence of friction force between NCSC and environment.

Then, taking into account the factors above, capture of NCSC by Robot Manipulator (RM) must meet the requirements, specified in Table 2 and described in [4].

Berthing requirements for SSC, which are listed in Table 2 (points 1 and 2) are caused by specifications of NCSC, in particular, the generalized length of solar batteries of devices, arrangement of antennas, the radius of cone of occurrence of SC, errors of linear and angular positioning of the SSC to NCSC. In such conditions, delivery of RM's effector to S/C Adapter is possible with the use of telescopic console, which is equipped with two angular hinges: double-axis hinge 1 and triple-axis hinge 2, which are located in places of console mounting to NCSC (p. A) and the final effector (p. B) in accordance/ kinematic structure of console is shown on Fig. 4.

Table 2. Requirements to the RM's capability to perform the capture of NCSC

№	Description	Value
1	Distance between the base point and NCSC's S/C Adapter	1.5–3 m
2	Mutual orientation	Quasispherical
3	Relative angular velocity at capture	to 10° per 1 min
4	Relative linear velocity at capture	to 0.1 m per 1 min
5	Zone of insensitivity of relative positioning of S/C Adapter	to 0.1 m
6	Tolerances to dimensions of S/C Adapter	0.6–3 m
7	Tolerances to mass characteristics of SC	to 5000 kg

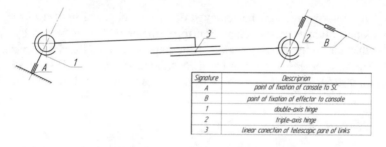

Signature	Description
A	point of fixation of console to SC
B	point of fixation of effector to console
1	double-axis hinge
2	triple-axis hinge
3	linear conection of telescopic pare of links

Fig. 4. Kinematic structure of telescopic console

Double-axis hinge 1 and telescopic link of robotic console 3 provide the work of finite effector in the polar coordinate system. Triple-axis hinge 2 provides cardan joint of finite effector, that allows to compensate for errors of its positioning. Such kinematic structure of robotic console eliminate the occurrence of singularity of the links, however does not solve problem of final effector positioning in S/C Adapter area for committing the unstressed NCSC capture. The solution of this problem is possible only by equipping the effector by sensor devices, which are suitable for recognition and relative position determination of finite SSC effector and NCSC S/C Adapter.

Taking into account the linear and angular errors of the positioning of the devices, guaranteed effector's positioning for its further girth and capture of S/C Adapter is possible in the segment of work zone (Fig. 5 p. 1). The ability to capture NCSC with deviation close to zero is also possible in zones, which are shown on Fig. 5 (p. 2 and 10). Another zones, which are shown on Fig. 5 are intended for the equipment SC or used as a part of device work in the technical operations.

Zone of insensitivity (Table 2, p. 5) and difference of the S/C Adapter's radii (Table 2, p. 6) impose restrictions on the choice of the effectore's construction and the set of sensor system tools. "Zone of insensitivity" means impossibility of the sensory system, which consisting of the SSC control system (CS), to give accurate distance coordinates and NCSC's orientation. Fault 1 m with the speed 1° per minute can cause a collision of effector with S/C Adapter and give them relative acceleration.

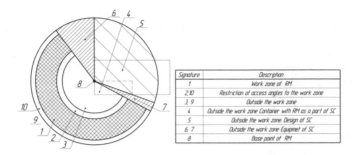

Signature	Description
1	Work zone of RM
2,10	Restriction of access angles to the work zone
3,9	Outside the work zone
4	Outside the work zone Container with RM as a part of SC
5	Outside the work zone Design of SC
6,7	Outside the work zone Equipment of SC
8	Base point of RM

Fig. 5. Operating area of RM's effector in the pale of longitudinal section

Compensation of error may be achieved due to the physical impact of the effector on S/C Adapter after its girth. Adaptation to diameter PAS is achieved by applying effector with chain compression principle. Kinematic structure of such effector is shown in Fig. 6. The contact area of effector has a constant shape due to unification of the key construction of PAS S/C Adapter. Shape of the adapter's key is shown on Fig. 7.

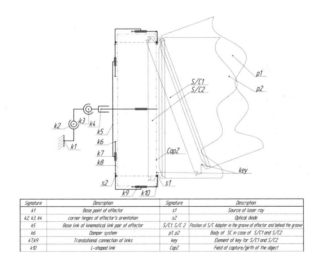

Signature	Description	Signature	Description
k1	Base point of effector	s1	Source of laser ray
k2, k3, k4	corner hinges of effector's orientation	s2	Optical diode
k5	Base link of kinematical link pair of effector	S/C1 S/C 2	Position of S/C Adapter in the groove of effector and behind the groove
k6	Damper system	pt1 p2	Body of SC in case of S/C1 and S/C2
k7,k9	Translational connection of links	key	Element of key for S/C1 and S/C2
k10	L-shaped link	CapZ	Field of capture/girth of the object

Fig. 6. Kinematic structure and main nodes of the effector of the chain type

Unstressed capture of PAS S/C Adapter is achieved by compensating force, which effector has applied, and the counteracting reactive force applied by the adapter in relation to the effector. Such phenomenon is achieved through the restrictions that are imposed by effector's links, after effector girth adapter and until the approach of capture. Monaural compression values of the S/C Adapter are determinate from the selected capture algorithm, force sensors, which are integrated in the composition of effector's actuators, and sensors of the optical leane interruption (Fig. 6, s2), which works when the ray crosses S/C Adapter elements. Prevention of incomplete capture (Fig. 6, S/C1) is ensured by a coordinated action of SSC's vision and technical means.

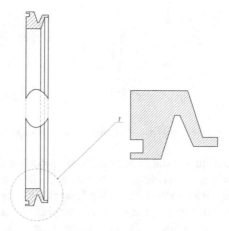

Fig. 7. Key thumbnail in the adapter PAS C/S Adapter: 1 — groove of a key

The final fixation when capture arises due to the to reduction of the free zone of the adapter movement zone within the board lines of effector's zone (Fig. 6, CapZ), which occurs as a result of translation motion in the effector units in the corresponding hinges (Fig. 6, k7, k9).

Process of S/C Adapter's fixation is accompanied by mutually rotation of the effector, which prevents the occurrence of angular momentum between effector and the adapter. Such rotation is provided by hinges, which are shown on Fig. 6 (k2, k3, k4).

2.3 Program of the Connection of the SC

In program of the connection of SCs, that involves the joint movement of devices, a required condition is preliminary alignment of their axes. Moreover, design of satellites involves the axial position of the point mass center (MCP) that provides position of a common MCP on a vector of the main SC's engine. For convenience, process of berthing is divided into two phases:

1. berthing with capture;
2. orientation of SC on one common axis.

During the first phase RM CS initiate the execution of operations of effector preparation for capture. Coordination of axes of effector and S/C Adapter is achived.

Berthing SSC to NCSC and their relative position is determinated by service spacecraft Control System (CS), however CS sets positioning faults, that can cause invariant of SC position. In this case there is a need compensation of the fault by means of the console and the effector that are part of the robot-manipulator.

Due to the pair of hinges, which are placed at the base and at the end of the RM's console and thanks to the telescopic pair of links, final effector can be positioned in the plane of S/C Adapter. Control of effector positioning is possible due vision means. The process of compensation of linear and angular faults within the plain is shown on Fig. 8.

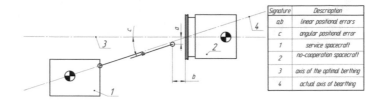

Signature	Description
a,b	linear positional errors
c	angular positional error
1	service spacecraft
2	no-cooperation spacecraft
3	axis of the optimal berthing
4	actual axis of bearthing

Fig. 8. Visualization of faults of relative position SSC and NCSC

It is important to note that at the stage of berthing robot's effector does not interact with NCSC, which leads to preservation of the Newton's first law.

Capture procedure provides alignment of SSC and NCSC mass center points (MCP) without their displacement by applying an equal number in magnitude and opposing forces in the direction of the effector. During the berthing RM can act on NCSC with essential force, which in the case of inaccurate capture, will give NCSC uncontrolled movement. For the capture of NCSC without increase acceleration to it, effector of RM has to cover structural components of S/C Adapter without contacting them. This procedure is shown on Fig. 9.

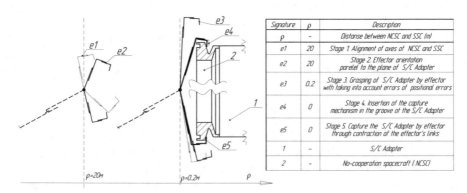

Signature	ρ	Description
ρ	-	Distanse between NCSC and SSC (m)
e1	20	Stage 1. Alignment of axes of NCSC and SSC
e2	20	Stage 2. Effector orientation parelel to the plane of S/C Adapter
e3	0.2	Stage 3. Grasping of S/C Adapter by effector with taking into account errors of positional errors
e4	0	Stage 4. Insertion of the capture mechanism in the groove of the S/C Adapter
e5	0	Stage 5. Capture the S/C Adapter by effector through contraction of the effector's links
1	-	S/C Adapter
2	-	No-cooperation spacecraft (NCSC)

Fig. 9. Capture procedure S/C Adapter of NCSC by effector of RM

First stage is held on the "decision-making" distance (20 m between NCSC and SSC). At this stage operations of initialization of sensory system, executive devices and RM CS are performed. At a distance $\rho = 20$ m, effector of RM is introduced into the work zone of manipulator. According to data from RM CS, means of technical vision and sensors and the position of links is carried out by coordinating the position of the effector in 3 coordinates. At the same time, angular orientation of effector stays random (Fig. 9, e1).

On the second stage effector places in parallel to the plane of S/C Adapter. Second stage is carried at the "berthing" distance (about two meters between NCSC and SSC).

At this stage the calibration of vision devices is performed according to sensors of orientation and position of RM's links. The classification of the S/C Adapter is carried out. Algorithm of berthing and capture is determinate (Fig. 9, e2).

At the third, forth and fifth stages (Fig. 9, e3–e5) operations of girth and capturing the S/C Adapter's construction part are performed by the effector. Taking into consideration huge number of S/C Adapter versions, effector can change zone of girth and capture in depending on the chosen algorithm, which is selected in the second stage.

During the second phase CS of RM initiate coordination of SSC and NCSC axes. For this angle c (Fig. 8) reduces to zero. After aligment of axes hinge joint of SSC, RM, NCSC design is constant, and limiting relative displacement new position of the all MCP are considered. MCP position coincides with axis of the devices (which passing through their apogee engines). This orientation of the devices allows to use apogee engine of SSC as the motor of whole construction.

3   Conclusion

Researches, which were held in this work, showed that procedure of non-cooperative capture and docking operations are economical and scientific actual engineering problems and the development of the space industry of the human management requires the technical means for carrying out satellites, their recycling, execution of mechanical operations etc.

The proposed principle and the concept of the structure of the robot-manipulator of a service spacecraft will allow the procedure of capturing spacecraft with the mass up to 5000 kg in conditions of uncertain definition of their spatial position and proposed structure of the adapter design. Working on the principle of chain will allow to carry out girth, capture and holding of SC, which is equipped by PAS 1666 S, PAS 1194 C, PAS 1666 MVS, PAS 1184 VS with positional fault $10°$ per minute and $0.1\ m^3$.

Acknowledgements. This publication would not be prepared without consultation of colleagues from SOE "Південне", PrJSC "НВК КУРС" and Department of Technical Cybernetics NTUU "Igor Sikorsky KPI".

References

1. Keshk, A.E.: Enhanced dynamic algorithm of genome sequence alignments. Int. J. Inf. Technol. Comput. Sci. (IJITCS) **6**(6), 40–46 (2014). https://doi.org/10.5815/ijitcs.2014.06.06
2. West, H., Hootsmans, N., Dubowsky, S., Stelman, N.: Experimental simulation of manipulator base compliance. Experimental Robotics I. Springer, Heidelberg (1989)
3. Jayapriya, J., Arock, M.: Aligning molecular sequences by wavelet transform using cross correlation similarity metric. Int. J. Intell. Syst. Appl. (IJISA) **9**(11), 62–70 (2017). https://doi.org/10.5815/ijisa.2017.11.08
4. Flores-Abad, A., Ma, O., Pham, K., Ulrich, S.: A review of space robotics technologies for on-orbit servicing. Prog. Aerosp. Sci. **6**, 1–26 (2014)

5. Kalaiselvi, T., Kalaichelvi, N., Sriramakrishnan, P.: Automatic brain tissues segmentation based on self initializing k-means clustering technique. Int. J. Intell. Syst. Appl. (IJISA) 9(11), 52–61 (2017). https://doi.org/10.5815/ijisa.2017.11.07

6. Janapati, R., Balaswamy, ch., Soundararajan, K.: Enhancement of indoor localization in WSN using PSO tuned EKF. Int. J. Intell. Syst. Appl. (IJISA), 9(2), 10–17 (2017). https://doi.org/10.5815/ijisa.2017.02.02

7. Langley, R.D.: Apollo experience report: The docking system. Doc. ID 19720018207, NASA Technical report (1972)

8. Bluth, B.J., Helppie, M.: Soviet space stations as analogs. Doc. ID 19870012563, NASA Technical report (1986)

9. Mukhin, V., Romanenkov, Y., Bilokin, J., Rohovyi, A., Kharazii, A., Kosenko, V., Kosenko, N., Su, J.: The method of variant synthesis of information and communication network structures on the basis of the graph and set-theoretical models. Int. J. Intell. Syst. Appl. (IJISA), 9(11), 42–51 (2017). https://doi.org/10.5815/ijisa.2017.11.06

10. Polites, M.E.: Technology of automated rendezvous and capture in space. J. Spacecraft Rockets 36(2), 280–291 (1999)

11. Eurockot. Rockot User's Guide. Eurockot Launch Service GmbH, EHB0003, Issue 5, Revision 0 (2011)

12. Krishna, S.R.M., Seeta Ramanath, M.N., Kamakshi Prasad, V.: Optimal reliable routing path selection in MANET through novel approach in GA. Int. J. Intell. Syst. Appl. (IJISA), 9(2), 35–41 (2017). https://doi.org/10.5815/ijisa.2017.02.05

13. Polar Satellite Launch Vehicle User's Manual. Antrix Corporation Limited, No: PSLV-VSSC-PM-65-87/6, Issue-6, Rev 0 (2015)

14. Pastorelli, P.P., Mauro, S., Eizaga, T.M.M., Sorli, M.: Docking mechanism concepts for the strong mission. In: 66th International Astronautical Congress, Jerusalem (2015)

15. Breger, S.L., How, J.P.: Safe trajectories for autonomous rendezvous of spacecraft. J. Guidance Control Dyn. 31(5), 1478–1489 (2008)

The Designing and Research of Generators of Poisson Pulse Sequences on Base of Fibonacci Modified Additive Generator

Volodymyr Maksymovych, Oleh Harasymchuk,
and Ivan Opirskyy[(✉)]

National University "Lviv Polytechnic", Lviv, Ukraine
volodymyr.maksymovych@gmail.com,
garasymchuk@ukr.net, iopirsky@gmail.com

Abstract. The article presents principles of optimizing the parameters of structural elements of Poisson pulses sequence generator that is based on modified additive Fibonacci generator. The results of their simulation modeling show, that statistical characteristics of output pulse sequence correspond to Poisson law of distribution. In addition, in this article have been defined the limits of control code for concrete parameters of structural scheme.

Keywords: Generators of Poisson pulse sequences
Fibonacci modified additive generator · Registers
Poisson law · Statistical characteristics · Binary code · Poisson pulse sequence

1 Introduction

The generators of Poisson pulse sequences (GPPS) are widely used in many branches of science and technology in modeling different processes, that have random time and spatial character [1, 2, 4, 9].

The Poisson law of distribution describes the characteristics of discrete random values [8]. By this distribution is described the events, that happen very seldom. In addition, to such events can be referred the amount of particles of radioactive decay, that have been registered by a counter within certain amount of time t, the quantity of incoming calls, that have been received in telephone station within time t, the flux of breakdown of energy objects in quite big system, the amount of defects in piece of fabric or in fixed length strip, the amount of accidents in industry and so on. The events that form the flux, in general case can be different, but it is necessary to examine only the flux of homogenous events, that distinguish form each other only by emergence moments. Also a simple flux of Poisson pulses can be used as output flux for receiving more complicated fluxes, for instance Erlang flux [5].

Over recent years' new works have appeared with new principles of realization of software and hardware GPPS, their structures have been suggested, that are generally based on usage of Pseudorandom Number Generators (PNG) [10], also the techniques have been suggested for evaluating the quality of their output sequences [3–7].

© Springer International Publishing AG, part of Springer Nature 2019
Z. Hu et al. (Eds.): ICCSEEA 2018, AISC 754, pp. 43–53, 2019.
https://doi.org/10.1007/978-3-319-91008-6_5

Though, we consider that in this issue there are tasks, that require further examining and solving, in particular:

- it is necessary to search for optimal parameters of structural elements of GPPS in order to receive certain parameters of output pulse sequence for concrete tasks;
- specifying the ranges and plurality of values of control code, that assign the range of values of average frequency of output pulse sequence, when it corresponds to the Poisson law of distribution
- improved GPPS, that work in binary code, because they are suitable for hardware realization and for usage in many applied tasks.

The purpose of this work is a partial solution of specified tasks using GPPS example, built on base of Fibonacci modified additive generator (FMAG), that works in binary code.

2   The Structural Scheme of GPPS and Principles of Their Work and Parameters of Output Signal, Internal Parameters of Generator and Their Interrelation

The generator [9–11] (which structural scheme is represented in Fig. 1, consists of: FMAG (that have Rg1 – Rg5 registers), combinative adders CA1 – CA3, logical scheme – LS, comparison scheme – CS and logical element I. All FMAG structural elements work in binary code.

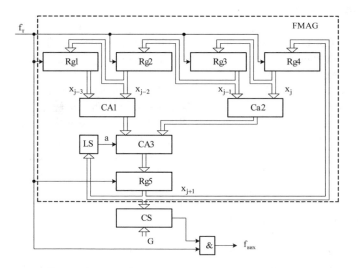

Fig. 1. Structural scheme of GPPS on base of FMAG

On FMAG output, in other words on Rg5 output, a sequence of pseudorandom numbers is formed in correspondence with an expression:

$$x_{j+1} = (x_j + x_{j-1} + x_{j-2} + x_{j-3} + a) \bmod m \tag{1}$$

where x_j, x_{j-1}, x_{j-2}, x_{j-3} are numbers in registers Rg4, Rg3, Rg2, Rg1 correspondingly, $m = 2^s$, s is the quantity of binary bits of scheme structural elements. The value of variable a is defined by logical equation

$$a = a_0 \oplus a_1 \oplus \ldots \oplus a_s \tag{2}$$

where a_i $(i = 0, 1, \ldots z; z \leq s - 1)$ is a bit value of binary number in Rg5. The quantity of Eq. (2) terms can be chosen from a range $0 \ldots z$, and s – a quantity of Rg1 - Rg5 and CA1 – CA3 binary bits.

A theoretical average value of output frequency of pulse on GPPS output

$$f_{out} = \frac{G}{2^s} f_m \tag{3}$$

where G is the control code, f_T is a frequency of clock pulses.

The main parameters of output pulse sequence of examining generator are:

- the average value of frequency f_{out};
- the range of values f_{out};
- the step of frequency change $f_{out} - \Delta f_{out}$;
- the repetition period of output pulse sequence;
- the correspondence of pulse sequence to the Poisson law.

The parameters of output pulse sequence are defined by the following internal parameters of generator (Fig. 1):

- the bits number of FMAG structural elements;
- the initial settings of registers Rg1 - Rg5;
- the quantity of terms of Eq. (2).

From these parameters depend:

- the repetition period of pseudorandom numbers on FMAG output;
- the statistical characteristics of output sequence.

The suggested structure of GPPS, is that, where the repetition period of pulse sequence is equal to the repetition period of numbers on FMAG output.

The repetition period and statistical characteristics of number sequence on FMAG output define the correspondence of output pulse sequence of GPPS with the Poisson law of distribution. Though, this correspondence depends on average frequency of output sequence f_{out} very much, which theoretical value is defined by Eq. (3), and it depends on interrelation of values of control code G and 2^S value. If G value approaches 2^s value, then f_{out} approaches f_m clock frequency, and in this condition, the output sequence loses its pseudorandom features. On the other hand, the fewer frequency of output sequence f_{out}, the longer time interval is necessary for defining its statistical characteristics. Also, this interval doesn't have to exceed the repetition period of this sequence. Theoretically, the output pulse sequence of GPPS can correspond with the Poisson law of distribution for arbitrarily small f_{out} average values, but with this its repetition period must be corresponding to a sufficiently large value. Exactly, if the average value goes to zero, the repetition period should go to infinity.

These statements have practical usage, they satisfy the majority of cases of GPPS usage, and are proved below by concrete calculations and results of modeling.

Taking into consideration everything that has been mentioned, the average frequency value f_{out}, the range of its values and a step of change can be calculated theoretically by using Eq. (3). The real values of these quantities are defined in result of imitative modeling and/or experimentally.

3   The Methodology of Evaluating of Statistical Characteristics of Output Signal and Specifying the Limits of Value Range of Control Code

The research of FMAG has been carried out with a help of generalize methodology of research of output pulse sequence parameters of GPPS in correspondence with the Poisson law of distribution with using Pearson's chi-squared test [7, 11].

In correspondence with this methodology, the flux of output GPPS pulses is divided into n similar groups, each consists of i_{max} pulses. The maximum quantity of groups is n_{max}. The groups of input pulses correspond with the groups of output impulses with number of pulses $k_1, k_2, \ldots, k_{n_{max}}$. The suggested methodology is based on classical methodology of hypothesis check considering general aggregate distribution according to the Poisson law with using Pearson's chi-squared test (χ^2 criteria). And accordingly the following additions have been proposed:

- the nominal (theoretical) average value of the numbers $k_1, k_2 \ldots k_{n_{max}} - k_c$ is fixed, independently from the value of control code G.
- the i_{max} is changeable, it depends on G value and is defined by equation

$$i_{max} = \frac{2^s}{G} k_c. \tag{4}$$

As the result of usage of this methodology is found χ_c^2 value. According to the tables of critical points of distribution χ^2, with level of importance α and a number of degree of freedom k is found the critical value χ_{cr}^2. If $\chi_c^2 < \chi_{cr}^2$, there is reason to accept the hypothesis about correspondence of pulse flux with the Poisson law.

When the statistical characteristics of output signal of GPPS are researched in the range of values of G control code, the useful operation is averaging of h values and the last (current) values of χ_c^2. The received value χ_{avr}^2 is compared with χ_{cr}^2. Taking into consideration the previous experience of carrying out imitative modeling, we can choose a value $h = 5$, which can be changed, if necessary, for more concrete (more integral) defining the range of G control code, when the output pulse sequence corresponds with the Poisson law of distribution.

The value of lower G_1 and upper G_2 limit of values of G control code, when statistical characteristics of output pulse sequence correspond with the Poisson law of distribution, can be specified from the following statement.

The time of estimating of sequence shouldn't be longer than a period of its repeat is T_n. Thus, considering the above mentioned methodology, the following inequality has to be true

$$i_{max} \cdot n_{max} \leq T_n. \tag{5}$$

From Eqs. (4) and (5) we receive

$$G \geq \frac{2^s \cdot k_c \cdot n_{max}}{T_n}. \tag{6}$$

The G_1 value is the fewest integer, that satisfies the inequality (6).

As a result of imitative modeling of GPPS has been determined that the value G_2 satisfies a condition:

$$G \leq G_2 = ss \cdot 2^s. \tag{7}$$

The value of ss coefficient defined separately for concrete positions of structural elements of FMAG, and it depends on initial settings of registers Rg1 - Rg5, the quantity of used terms of Eq. (2), and in certain conditions, it is close to 0,1.

4 GPPS Research on Base of FMAG

4.1 Defining Statistical Characteristics and the Range of Values of Control Code

In Figs. 2, 3 and 4 are represented the results of research of statistical characteristics of GPPS on base of FMAG ($s = 10$).

Here the following designations are used:

- SS_n is the χ_c^2 value;
- SS_n_pot is the averaging value of 5 last (current) values $\chi_c^2 - \chi_{avr}^2$;
- Level is the quantity of χ_{avr}^2 values bigger than χ_{cr}^2.

The results have been received by the following values of parameters of evaluating methodology of pulse sequence quality: $n_{max} = 1000$, $k_c = 10$, $\chi_{kp}^2 = 25$.

The output signal of logical scheme LS has been formed according to expression (2) for different amount of terms of equation a_i. The amount of terms i is represented in corresponding notes under the pictures.

By enumerating different options of initial states of registers Rg1 - Rg5, the answer has been found, that the most optimal is an option of values is G, 0, 0, 0, 0 correspondingly. So that the option, when the initial settings depends on control code. And for these initial settings the results are shown in Figs. 2, 3 and 4.

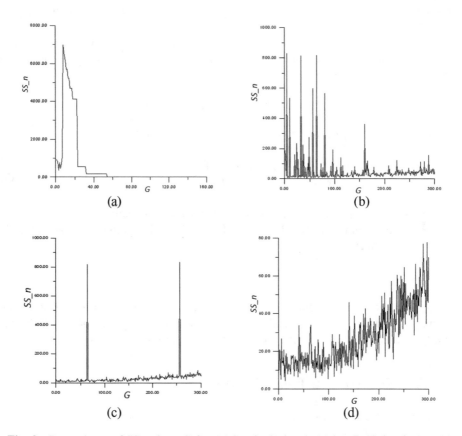

Fig. 2. Dependence of SS_n from G for: (a) i = 0, (b) i = 1, (c) i = 5, (d) i = 9, (s = 10)

As it is seen from the graphics, the growth of amount of terms of Eq. (2) influences very much on quality of pulse sequence. The optimal value i, when output pulse sequence corresponds with theoretical Poisson law of distribution, in this case (s = 10), is specified by equations:

$$G_1 = 1, \quad G_2 = 152. \tag{8}$$

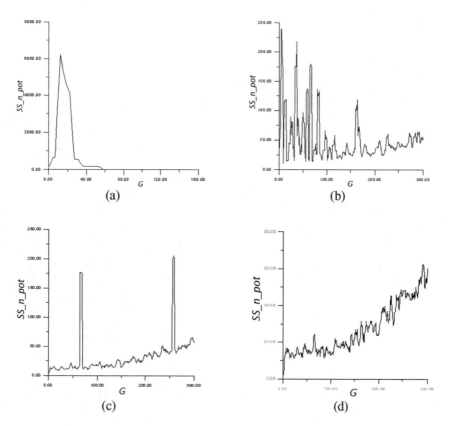

Fig. 3. Dependence of SS_n_pot from G for: (a) i = 0, (b) i = 1, (c) i = 5, (d) i = 9, (s = 10)

4.2 Defining Statistical Characteristics and the Range of Values of Control Code

The results of research of statistical characteristics of GPPS on base of FMAG $(s = 20)$ are shown in Figs. 5, 6 and 7.

The results have been received for those values of parameters of evaluating methodology of quality of pulse sequence, that have been received in previous case (for $s = 10$): $n_{max} = 1000$, $k_c = 10$, $\chi^2_{cr} = 25$.

The initial settings of registers Rg1 - Rg5 were the following: G, 0, 0, 0, 0 correspondingly. The range of values of control code $G - G_1 \div G_2$, when the output pulse sequence corresponds the Poisson law of distribution, in this case when $(s = 20)$ and i = 7, is defined by equations:

$$G_1 = 1, \; G_2 > 300. \tag{9}$$

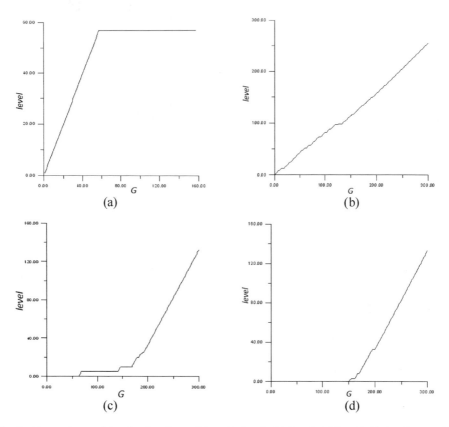

Fig. 4. Dependence of level value from G for: (a) i = 0, (b) i = 1, (c) i = 5, (d) i = 9, (s = 10)

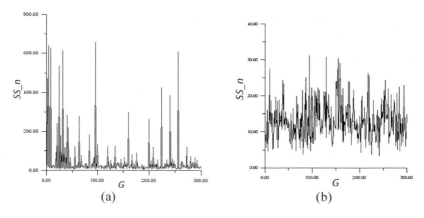

Fig. 5. Dependence of SS_n value from G for: (a) *i* = 1, (b) i = 7, (*s* = 20)

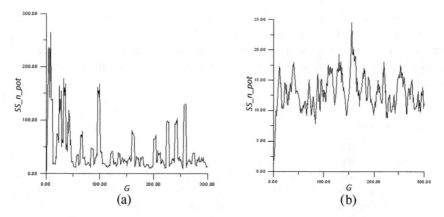

Fig. 6. Dependence of SS_n_pot value from G for: (a) $i = 1$, (b) i $= 7$, ($s = 20$)

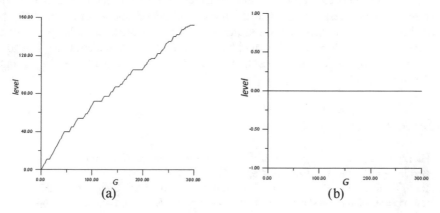

Fig. 7. Dependence of level value from G for: (a) $i = 1$, (b) i $= 7$, ($s = 20$)

4.3   Comparison of GPPS Characteristics on Base of FMAG When $s = 10$ and $s = 20$

By increasing the number of generators bits is possible to increase the repetition period of number sequence on FMAG output, and as a result the repetition period of pulse sequence on GPPS output. However, it doesn't lead to automatic increase of generator "resolution", considering setting up the average frequency value of output sequence f_{out}, which in fact provides a possibility of setting a step of change $f_{out} - \Delta f_{out}$. Certain results of research that are represented above, show that "resolution" in fixed value of position, depends on initial settings of registers Rg1 - Rg5 and also the quantity of used terms of Eq. (2), that defines the logic of forming a signal on LS output. As it is also seen from the results of research there is a certain limit, when there is no use to increase the bit's number for adding them in formula (2), because the quality of output Poisson sequence will not increase in fact, but the performance will reduce. At the same time, these parameters depend on statistical characteristics of output sequence. By taking into consideration these ideas, the ways of improving the "resolution" of generator is possible to find in further researches. In fact, by increasing the bits number of structural elements, in conditions that have been shown above, hasn't led to increasing the "resolution" of generator, so it is worth in certain cases to examine a possibility of practical usage of this generator when $s = 10$.

5   Conclusion

According to the results of carried out research is possible to make a conclusion, that by insertion in structure of Fibonacci modified additive generator the logical scheme and realization of all structural elements of FMAG in binary code, considerably improves the statistical characteristics of output Poisson pulse sequence.

As results of experimental research has been proved that main role has the amount of used terms of Eq. (2), and also initial settings in Rg1 - Rg5. The amount of terms of Eq. (2) has to be chosen with taking into consideration the criteria of quality of output pulse sequence and performance, because there is a limit of amount of terms of Eq. (2), when the quality of output Poisson sequence will not increase, and the performance will reduce. Increasing the bit's number of structural elements doesn't lead to increasing the "resolution" of GPPS, that's why due to this issue is necessary to carry out further research.

References

1. Harasymchuk, O.I., Maksymovych, V.M.: Algorithm for the formation of a Poisson pulse stream. Autom. Measur. Control **475**, 21–25 (2003)
2. Harasymchuk, O.I., Dudykevych, V.B., Maksymovych, V.M., Smuk, R.T.: Generators of test pulse sequences for dosimetric devices. Heat power Environ. Eng. Autom. **506**, 186–192 (2004)
3. Ivanov, M.A., Chugunkov, I.V.: Theory, Application and Quality Assessment of Pseudo-Random Sequence Generators. The Technical Writer's Handbook (2003). 240 p.
4. Hu, Z., Mukhin, V., Kornaga, Y., Lavrenko, Y., Barabash, O., Herasymenko, O.: Analytical assessment of security level of distributed and scalable computer systems. Int. J. Intell. Syst. Appl. (IJISA) **8**(12), 57–64 (2016). https://doi.org/10.5815/ijisa.2016.12.07
5. Harasymchuk, O.I., Maksymovych, V.M.: Generators of pseudorandom numbers, their application, classification, basic methods of construction and evaluation of quality. J. Inf. Prot. **3**, 29–36 (2003)
6. Gaeini, A., Mirghadri, A., Jandaghi, G., Keshavarzi, B.: Comparing some pseudo-random number generators and cryptography algorithms using a general evaluation pattern. Int. J. Inf. Technol. Comput. Sci. (IJITCS) **8**(9), 25–31 (2016). https://doi.org/10.5815/ijitcs.2016.09.04
7. Aluru, S.: Lagged Fibonacci random number generators for distributed memory parallel. Comput. J. Parallel Distrib. Comput. **45**, 1–12 (1997)
8. Wani, A.A., Badshah, V.H.: On the relations between lucas sequence and Fibonacci-like sequence by matrix methods. Int. J. Math. Sci. Comput. (IJMSC), **3**(4), 20–36 (2017). https://doi.org/10.5815/ijmsc.2017.04.03

9. Harasymchuk, O.I., Kostiv, Y., Maksymovych, V.M., Mandrona, M.M.: Hardware implementation and research of Fibonacci generators. Comput. Technol. Printing **29**, 167–174 (2013)
10. Mondal, A., Pujari, S.: A novel approach of image based steganography using pseudorandom sequence generator function and DCT coefficients. Int. J. Comput. Netw. Inf. Secur. (IJCNIS) **7**(3), 42–49 (2015). https://doi.org/10.5815/ijcnis.2015.03.06
11. Harasymchuk, O.I., Kostiv, Y., Maksymovych, V.M., Mandrona, M.M.: Implementation of modified additive lagged Fibonacci generator. Challenges Mod. Technol. **7**(1), 3–6 (2016)

Methods of the Statistical Simulation
of the Self-similar Traffic

Anatolii Pashko[1,2] and Violeta Tretynyk[2(✉)]

[1] Taras Shevchenko University of Kyiv, Kyiv, Ukraine
aap2011@ukr.net
[2] National Technical University of Ukraine
"Igor Sikorsky Kyiv Polytechnic Institute", Kyiv, Ukraine
viola.tret@gmail.com

Abstract. The problem of evaluating the quality of the service is one of the important tasks of analyzing the traffic of telecommunication networks. Characteristics of traffic of modern telecommunication networks vary widely and depend on a large number of parameters and network settings, characteristics of protocols and user's work. Recent studies argue that network traffic of modern networks has the properties of self-similarity. And this requires finding adequate methods for traffic simulating and loading processes in modern telecommunication networks.

The article deals with the methods of statistical simulation of fractional Brownian motion based on the spectral image. The developed methods are used for modeling of self-similar traffic and loading process of telecommunication networks. Estimates of the probability of repositioning are found.

Keywords: Fractional Brownian motion · Hurst index
Accuracy and reliability of the model · Gaussian random process
Self-similar traffic

1 Introduction

When analyzing the traffic of telecommunication networks, one of the most crucial tasks is the problem of evaluating the quality of the service. Modern telecommunication networks have a variety of features that depend on the number of parameters and network settings, characteristics of protocols and user's work.

There is a number of recent studies that argue on the topic of properties of self-similarity in modern networks. Therefore, arises a need to find adequate methods for traffic simulating and loading processes in modern telecommunication networks.

A number of Ukrainian and foreign scientists are actively engaged in the area of analysis and simulation of self-similar traffic. Work in this direction should be noted [1–3].

In the article methods of statistical modeling of fractional Brownian motion (FBM) for modifying the self-similar traffic are used. The estimation of probabilities of buffer overflow based on the properties of FBM is found.

© Springer International Publishing AG, part of Springer Nature 2019
Z. Hu et al. (Eds.): ICCSEEA 2018, AISC 754, pp. 54–63, 2019.
https://doi.org/10.1007/978-3-319-91008-6_6

The choice of the fractional Brownian motion is due to the fact that the fractional Brownian motion (often used term - fractal Brownian motion, or generalized Wiener process) has the properties of self-similarity. This property of the fractional Brownian motion is characterized by the Hurst index. In the future, we will be of the opinion of the probabilistic terminology and will use the name - fractional Brownian motion.

Two approaches to FBM modeling are considered. The first one uses the FBM representation as a random number with non-correlated terms, and the second one uses the fact that the growth of FBM is a Gaussian stationary process. The developed models are used to simulate the traffic of telecommunication networks and the loading process. The estimation of the probability of overflow of the loading process is investigated.

The list of industries using the methods of statistical simulation of FBM is expanding day by day. Thus, in addition to the problems of simulation and estimation of traffic in the theory of telecommunication networks [3–6], the models of fractional Brownian motion are used in financial mathematics [7], in solving problems of computational mathematics [8], in mass service systems [6], in medical research [9, 10], for the estimation of Hausdorff dimension [11].

The model of traffic is investigated on the basis of FBM, the time of delay and the probability of data loss are estimated in [12]. Methods of statistical simulation of fractional Brownian motion with specified accuracy and reliability in different functional spaces, namely, $L_p(T)$ spaces, Orlicz spaces and in the space of continuous functions were investigated in [13]. For modeling the FBM image was used as a random series.

We have some dimensional space (T, B, μ) with $\mu(T) = 1$.

Definition 1. A fractional Brownian motion with Hurst index $\alpha \in (0, 1)$ we call Gaussian random process $W_\alpha(t), t \in [0, T]$ with correlation function $R_\alpha(t, s) = \frac{1}{2}\left(|t|^{2\alpha} + |s|^{2\alpha} - |t - s|^{2\alpha}\right)$, such that $W_\alpha(0) = 0$ and $EW_\alpha(t) = 0$.

If $\alpha = \frac{1}{2}$ we have a standard Wiener process.

2 Method of Simulation of the Fractional Brownian Motion

Standard Wiener process $W(t)$ is the process with independent increments. The fractional Brownian motion $W_\alpha(t)$ it's the process with stationary increments. Therefore, random process $w(t) = W_\alpha(t + \Delta) - W_\alpha(t)$ is stationary Gaussian random process with correlation function [11]

$$Ew(t + \tau)w(t) = \frac{1}{2}\left(|\tau + \Delta|^{2\alpha} + |\tau - \Delta|^{2\alpha} - 2|\tau|^{2\alpha}\right)$$

and spectral density $g(\lambda) = \frac{A^2}{\pi}\left(\frac{1 - \cos(\lambda\Delta)}{|\lambda|^{2\alpha+1}}\right)$, $\lambda \in (-\infty, +\infty)$

where $A^2 = \left(\frac{2}{\pi}\int\limits_0^\infty \frac{1 - \cos(\lambda)}{\lambda^{2\alpha+1}}d\lambda\right)^{-1} = \left(-\frac{2}{\pi}\Gamma(-2\alpha)\cos(\alpha\pi)\right)^{-1}$.

Let $\xi(t)$ be a real Gaussian stationary random process with $E\xi(t) = 0$, $R(\tau)$ is a correlation function $\xi(t)$, $F(\lambda)$ is a spectral function of $\xi(t)$, and $R(\tau) = \int_0^\infty \cos(\lambda t) dF(\lambda)$. Gaussian stationary random process can be represented as

$$\xi(t) = \int_0^\infty \cos(\lambda t) d\xi_1(\lambda) + \int_0^\infty \sin(\lambda t) d\xi_2(\lambda)$$

where $\xi_1(t)$ and $\xi_2(t)$ - the centered and uncorrelated random processes with uncorrelated increments such as $0 \leq \lambda_1 < \lambda_2$ and

$$E(\xi_1(\lambda_2) - \xi_1(\lambda_1))^2 = F(\lambda_2) - F(\lambda_1),$$

$$E(\xi_2(\lambda_2) - \xi_2(\lambda_1))^2 = F(\lambda_2) - F(\lambda_1).$$

Let D_Λ be some partition of the interval $[0, \Lambda]$, $D_\Lambda : 0 = \lambda_0 < \lambda_1 < \ldots < \lambda_n = \Lambda$. The model of random process $\xi(t)$ can be obtained as $S_n(t, \Lambda) = \sum_{i=0}^{n-1} [\cos(\lambda_i t)\eta_{1i} + \sin(\lambda_i t)\eta_{2i}]$ where $\{\eta_{1i}, \eta_{2i}\}$ are centered uncorrelated strictly sub-Gaussian random variables with $E(\eta_{1i})^2 = E(\eta_{2i})^2 = F(\lambda_{i+1}) - F(\lambda_i)$.

Methods of modeling of Gaussian random processes were studied in [14–16].

Therefore, random process $w(t)$ can be represented as $w(t) = \int_0^\infty \cos(\lambda t) d\xi_1(\lambda) + \int_0^\infty \sin(\lambda t) d\xi_2(\lambda)$. And for partition $D_\Lambda : 0 = \lambda_0 < \lambda_1 < \ldots < \lambda_n = \Lambda$, the model of random process $w(t)$ can be obtained as $w_n(t, \Lambda) = \sum_{k=0}^{n-1} (\sin(\lambda_k t)X_k + \cos(\lambda_k t)Y_k)$ where $\{X_k, Y_k\}$ are uncorrelated strictly sub-Gaussian random variables with $EX_k = EY_k = 0$ and $E(X_k)^2 = E(Y_k)^2 = \int_{\lambda_k}^{\lambda_{k+1}} g(\lambda) d\lambda$.

3   The Accuracy and Reliability of Model in $L_2(T)$

Let random process $X(t)$ and all $X_n(t, \Lambda)$ belongs to certain functional Banach space $A(T)$ with norm of $\|\cdot\|$. Let the two numbers be as follow $\delta > 0$ and $0 < \varepsilon < 1$. Model $X_n(t, \Lambda)$ approximates process $X(t)$ with reliability $1 - \varepsilon$ and accuracy δ in the norm of space $A(T)$, if the following inequality holds $P\{\|X(t) - X_n(t, \Lambda)\| > \delta\} \leq \varepsilon$.

Whereas, $W_\alpha(0) = 0$, then for all Δ the model of fractional Brownian motion we constructed as

$$W_\alpha(t + \Delta) = W_\alpha(t) + w(t). \tag{1}$$

Simulation of fractional Brownian motion is reduced to simulation of stationary Gaussian random process. Modeling of FBM with the use of a spectral image was investigated in [13].

There are theorems.

Theorem 1. Model $S_n(t, \Lambda)$ approximates process $\xi(t)$ with reliability $1 - \varepsilon$ and accuracy δ in the norm of space $L_2(T)$, if for numbers Λ and n inequalities hold

$$B2_{n,\Lambda} < \delta^2 \text{ and } \exp\left\{\frac{1}{2}\right\} \frac{\delta}{\sqrt{B2_{n,\Lambda}}} \exp\left\{-\frac{\delta^2}{2B2_{n,\Lambda}}\right\} \leq \varepsilon$$

where $B2_{n,\Lambda} = \int\limits_T E(\xi(t) - S_n(t, \Lambda))^2 d\mu(t)$.

Theorem 2. Model $w_n(t, \Lambda)$ approximates process $w(t)$ with reliability $1 - \varepsilon$ and accuracy δ in the norm of space $L_2(T)$, if for numbers Λ and n inequalities hold

$$B3_{n,\Lambda} < \delta^2 \text{ and } \exp\left\{\frac{1}{2}\right\} \frac{\delta}{\sqrt{B3_{n,\Lambda}}} \exp\left\{-\frac{\delta^2}{2B3_{n,\Lambda}}\right\} \leq \varepsilon$$

where $B3_{n,\Lambda} = 2T \sum\limits_{i=0}^{n} \int\limits_{\lambda_i}^{\lambda_{i+1}} \left(1 - \frac{\sin(T(\lambda - \lambda_i))}{T(\lambda - \lambda_i)}\right) g(\lambda) d\lambda + T\left(\int\limits_{\Lambda}^{\infty} g(\lambda) d\lambda\right)$.

Let for D_Λ implemented $T(\lambda_{i+1} - \lambda_i) \leq 1$, then the corollary.

Corollary 1. Model $w_n(t, \Lambda)$ approximates process $w(t)$ with reliability $1 - \varepsilon$ and accuracy δ in the norm of space $L_2(T)$, if for numbers Λ and n inequalities hold

$$G1_{n,\Lambda} < \delta^2 \text{ and } \exp\left\{\frac{1}{2}\right\} \frac{\delta}{\sqrt{G1_{n,\Lambda}}} \exp\left\{-\frac{\delta^2}{2G1_{n,\Lambda}}\right\} \leq \varepsilon$$

where $G1_{n,\Lambda} = \frac{T}{3} \sum\limits_{i=0}^{n} \int\limits_{\lambda_i}^{\lambda_{i+1}} (\lambda - \lambda_i)^2 g(\lambda) d\lambda + T\left(\int\limits_{\Lambda}^{\infty} g(\lambda) d\lambda\right)$.

Let for D_Λ implemented $\lambda_{i+1} - \lambda_i = \frac{\Lambda}{n}$ and $\frac{T\Lambda}{n} \leq 1$, then the corollary.

Corollary 2. Model $w_n(t, \Lambda)$ approximates process $w(t)$ with reliability $1 - \varepsilon$ and accuracy δ in the norm of space $L_2(T)$, if for numbers Λ and n inequalities hold

$$G2_{n,\Lambda} < \delta^2 \text{ and } \exp\left\{\frac{1}{2}\right\} \frac{\delta}{\sqrt{G2_{n,\Lambda}}} \exp\left\{-\frac{\delta^2}{2G2_{n,\Lambda}}\right\} \leq \varepsilon$$

where $G2_{n,\Lambda} = \frac{T^3 \Lambda^2}{3n^2} \int\limits_0^{\Lambda} g(\lambda) d\lambda + T\left(\int\limits_{\Lambda}^{\infty} g(\lambda) d\lambda\right)$.

For numbers Λ and n let's put $\Lambda = \left(\frac{3n^2}{T^3}\right)^{\frac{1}{2\alpha+2}}$, then $G2_n = T^{\frac{3\alpha}{\alpha+1}}\left(1+\frac{T}{\alpha}\right)$
$\left[(3n^2)^{\frac{\alpha}{\alpha+1}}\right]^{-1}$.

Corollary 3. Model $w_n(t, \Lambda)$ approximates process $w(t)$ with reliability $1 - \varepsilon$ and accuracy δ in the norm of space $L_2(T)$, if for number n inequalities hold

$$n > \frac{1}{\sqrt{3}}\left[T^{\frac{3\alpha}{2\varepsilon+1}}\left(1+\frac{T}{\alpha}\right)\delta^{-2}\right]^{\frac{\alpha+1}{2\alpha}} \text{ and } \exp\left\{\frac{1}{2}\right\}\frac{\delta}{\sqrt{G2_n}}\exp\left\{-\frac{\delta^2}{2G2_n}\right\} \le \varepsilon$$

4 Estimation of Traffic Volume and Load Process

Let's $A(t)$ is the amount of traffic that enters the network in the time interval $[0, T]$, Let us specify that $A(s, t) = A(t) - A(s)$, $t > s > 0$. According to [1] incoming traffic has the form $A(t) = mt + \sqrt{am}W_\alpha(t)$ where m is the average traffic intensity, $W_\alpha(t)$ is the fractional Brownian motion with the Hurst index, $\alpha \in \left(\frac{1}{2}, 1\right)$, a is a constant.

If the network has one device with a service speed $C > m$, then the loading process is determined [12] by the formula $Q(t) \cong \sup_{s \le t}(A(s, t) - C(t - s))$. In the presence of n independent equivalent devices we will have $Q_n(t) \cong \sup_{s \le t}\left(\sum_{i=1}^{n} A_i(s, t) - nC(t - s)\right)$. The badge $\cong$ means the equality of distributions.

We will investigate on the segment $[0, T]$ the estimation of the probability that the value of the load $Q(t)$ exceeds a certain threshold b (the probability of overflow). Let us denote $Q \cong \sup_{t \in [0,T]}(Q(t))$, $\pi(b) = P\{Q \ge b\}$.

For the probability of overflow, we find upper bounds

$$P\{Q \ge b\} \le P\left\{\sup_{t \in [0,T]}(Q(t)) > b\right\}.$$

To find the estimates we use the results of the work [14, 15]. The load process has an estimate

$$Q(t) \cong \sup_{s \le t}(A(s, t) - C(t - s)) = \sup_{s \le t}\left(\sqrt{am}(W_\alpha(t) - W_\alpha(s)) - (C - m)(t - s)\right) \le$$
$$\le 2\sqrt{am}(|W_\alpha(t)|) - (C - m)t.$$

Then load variable is

$$Q \cong \sup_{t \in [0,T]} (Q(t)) = \sup_{t \in [0,T]} \left(2\sqrt{am}(|W_\alpha(t)|) - (C-m)t\right)$$
$$= 2\sqrt{am} \sup_{t \in [0,T]} (|W_\alpha(t)|) - T(C-m).$$

Therefore $P\{Q \geq b\} \leq P\left\{ \sup_{t \in [0,T]} (|W_\alpha(t)|) > \frac{b + T(C-m)}{2\sqrt{am}} \right\}.$

Let's $x = \frac{b + T(C-m)}{2\sqrt{am}}$, then there is a theorem.

Theorem 3. For $x \geq D$ there is an inequality

$$P\left\{ \sup_{t \in [0,T]} (|W_\alpha(t)|) > x \right\} \leq 2 \exp\left\{ -\frac{(x-D)^2}{2V} \right\}$$

where

$$D = \sqrt{2}\left(T^\alpha \ln^{\frac{1}{2}}(2^{\frac{1}{\alpha}} + 1) + 4 \int_0^{\frac{T^\alpha}{4}} \ln^{\frac{1}{2}}\left(\frac{1}{u^{\frac{1}{\alpha}}} + 1\right) du \right),$$

$$V = 4T^{2\alpha}.$$

Using the estimates obtained, we can determine what the threshold should be, so that the probability of being exceeded is less than given.

Thus, $P\left\{ \sup_{t \in [0,T]} (|W_\alpha(t)|) > x \right\} \leq \varepsilon_b$, if $x \geq D$ and $2 \exp\left\{ -\frac{(x-D)^2}{2V} \right\} \leq \varepsilon_b.$

And the threshold should be $b \geq 2D\sqrt{am} - T(C-m)$.

For $\varepsilon_b = 0.05$, $C = 100$, $m = 96$, $a = 1$, $T = 1$ i $\alpha = 0.8$ the threshold should be $b \geq 310$.

5   Simulation

By Corollary 3 results are obtained for model parameters $w_n(t, \Lambda)$ for $\varepsilon = 0.05$ (Table 1) and different values of the Hurst index.

The simulation results for different values of the Hurst index are shown in Figs. 1 and 2.

Figures 3 and 4 present the results of traffic modeling and the corresponding network loading process.

Table 1. Parameters of model

δ	n	Λ	α
0.05	$1.25\ 10^5$	2162	0.6
0.05	$3.6\ 10^4$	662	0.7
0.05	$1.5\ 10^4$	284	0.8
0.05	$7.5\ 10^3$	147	0.9

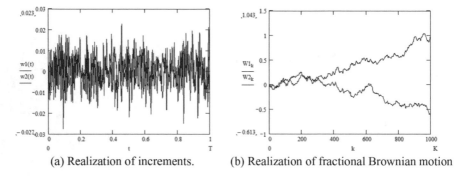

(a) Realization of increments.   (b) Realization of fractional Brownian motion

Fig. 1. Realization of fractional Brownian motion with $\alpha = 0.6$

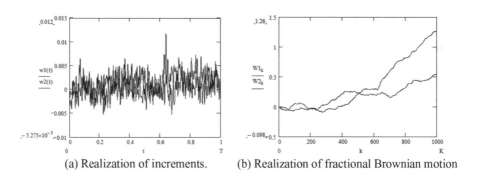

(a) Realization of increments.   (b) Realization of fractional Brownian motion

Fig. 2. Realization of fractional Brownian motion with $\alpha = 0.8$.

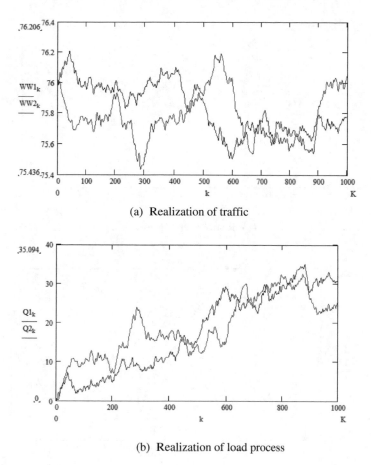

(a) Realization of traffic

(b) Realization of load process

Fig. 3. Realization of traffic and load process with $\alpha = 0.6$ and $C = 100$, $\quad m = 76$, $\quad a = 3$.

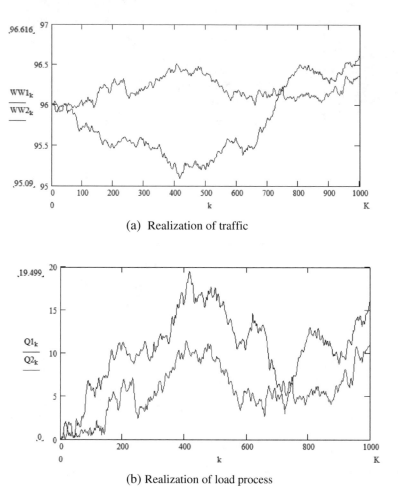

(a) Realization of traffic

(b) Realization of load process

Fig. 4. Realization of traffic and load process with $\alpha = 0.6$ and $C = 100, \quad m = 96, \quad a = 1$.

6 Conclusions

Methods of statistical simulation of fractional Brownian motion were developed. The methods are based on the spectral representation of random processes and use the property that the FBM has stationary increments. The indicated methods are used for simulation of the same traffic and loading process of telecommunication networks. Estimates of probability of overflow were found.

References

1. Norros, I.: A storage model with self-similar input. Queueing Syst. **16**, 387–396 (1994)
2. Kilpi, J., Norros, I.: Testing the Gaussian approximation of aggregate traffic. In: Proceedings of the Second ACM SIGCOMM Workshop, Marseille, France, pp. 49–61 (2002)
3. Sheluhin, O.I., Smolskiy, S.M., Osin, A.V.: Similar Processes in Telecommunication. Wiley, England (2007)
4. Chabaa, S., Zeroual, A., Antari, J.: Identification and prediction of internet traffic using artificial neural networks. Intell. Learn. Syst. Appl. **2**, 147–155 (2010)
5. Gowrishankar, S., Satyanarayana, P.S.: A time series modeling and prediction of wireless network traffic. Int. J. Inter. Mob. Technol. (iJIM) **4**(1), 53–62 (2009)
6. Kozachenko, Y., Yamnenko, R., Vasylyk, O.: φ-sub-Gaussian random process. Vydavnycho-Poligrafichnyi Tsentr "Kyivskyi Universytet", Kyiv (2008). (In Ukrainian)
7. Mishura, Y.: Stochastic Calculus for Fractional Brownian Motion and Related Processes. Springer, Berlin (2008)
8. Sabelfeld, K.K.: Monte Carlo Methods in Boundary Problems. Nauka, Novosibirsk (1989). (In Russian)
9. Goshvarpour, A., Goshvarpour, A.: Chaotic behavior of heart rate signals during Chi and Kundalini meditation. Int. J. Image Graph. Sig. Process. (IJIGSP), **4**(2), 23–29 (2012)
10. Hosseini, S.A., Akbarzadeh-T, M.-R., Naghibi-Sistani, M.-B.: Qualitative and quantitative evaluation of EEG signals in Epileptic Seizure recognition. Int. J. Intell. Syst. Appl. (IJISA) **5**(6), 41–46 (2013). https://doi.org/10.5815/ijisa.2013.06.05
11. Prigarin, S., Hahn, K., Winkler, G.: Comparative analysis of two numerical methods to measure Hausdorff dimension of the fractional Brownian motion. Siberian J. Num. Math. **11**(2), 201–218 (2008)
12. Ageev, D.V.: Parametric synthesis of multiservice telecommunication systems in the transmission of group traffic with the effect of self-similarity. Prob. Telecommun. **1**(10), 46–65 (2013). (In Russian). Electronic scientific specialized edition
13. Pashko, A.: Statistical Simulation of a Generalized Wiener Process. Bulletin of Taras Shevchenko National University of Kyiv. Physics and Mathematics, vol. 2, pp. 180–183 (2014). (In Ukrainian)
14. Kozachenko, Y., Pashko, A.: On the simulation of random fields I. Theory Probab. Math. Stat. **61**, 61–74 (2000)
15. Kozachenko, Y., Pashko, A.: On the simulation of random fields II. Theory Probab. Math. Stat. **62**, 51–63 (2001)
16. Kozachenko, Y., Pashko, A., Rozora, I.: Simulation of Random Processes and Fields. Zadruga, Kyiv (2007). (In Ukrainian)

Multidimensional Wavelet Neuron for Pattern Recognition Tasks in the Internet of Things Applications

Olena Vynokurova[1(✉)], Dmytro Peleshko[1], Semen Oskerko[1],
Vitalii Lutsan[1], and Marta Peleshko[2]

[1] IT Step University, 83a Zamarstynivs'ka str., Lviv 79019, Ukraine
vynokurova@gmail.com, dpeleshko@gmail.com, semenosker@gmail.com,
vl.brony@gmail.com
[2] Lviv State University of Life Safety, 35 Kleparivska, Lviv 79000, Ukraine
marta.peleshko@gmail.com

Abstract. Data mining and processing of Big Data is key problem in developing intelligent Internet of Things (IoT) applications. In this article, the multidimensional wavelet neuron for pattern recognition tasks is proposed. Also, the learning algorithm based on the quadratic error criterion is synthesized. This approach combines the benefits of the neural networks, the neuro-fuzzy system, and the wavelet functions approximation. The proposed multidimensional wavelet neuron can be used to solve a very large class of information processing problems for the Internet of Things applications when signals are fed in online mode from many sensors. The proposed approach is uncomplicated for computational realization and can be implemented in hardware for IoT systems. The proposed learning algorithm is characterized by a high rate of convergence and high approximation properties.

Keywords: Machine learning · Internet of Things
Multidimensional wavelet neuron · Pattern recognition · Classification
Online learning

1 Introduction

IoT produces and accumulates a lot of data of arbitrary natural, which are fed from Internet-connected sensory devices. Therefore, the development of IoT technologies requires new unique solutions for the accumulated data processing in real time, where the method of computational intelligence and machine learning have a lot of advantages as compared to conventional approaches [1–9].

The exponential growth of the stored and processed data caused a rapid transformation of the computational intelligence from a disparate set of heuristic methods into one of the most demanded applied disciplines, transforming all kinds of human activity. While using the intelligent systems becomes universal, there are growing demands for their universality, which means, in particular, stability on any type of data, adaptability to changing conditions, transparency of results interpretation [10–19].

© Springer International Publishing AG, part of Springer Nature 2019
Z. Hu et al. (Eds.): ICCSEEA 2018, AISC 754, pp. 64–73, 2019.
https://doi.org/10.1007/978-3-319-91008-6_7

As the use of intelligent systems becomes widespread, the requirements for their universality grow higher and higher, which means, in particular, a stability over any type of data, an adaptivity to changing conditions, a transparency of the results interpretation [12–23]. Strictly guarantee these properties we can only do by using rigorous mathematical methods based on the computational intelligence theory.

Nowadays, intensive researches are being carried out for the integration of IoT technologies and computational intelligence methods, among them: in [24] authors proposed 4 data mining models for processing IoT data, in [25] authors introduced a systematic manner for reviewing data mining knowledge and techniques in most common applications (classification, clustering, association analysis, time series analysis, and outline detection в IoT applications); in [26] authors ran a survey to respond to some of the challenges in preparing and processing data on the IoT through data mining techniques, in [27] authors attempted to explain the Smart City infrastructure in IoT and discussed the advanced communication to support added-value services for the administration of the city and citizens thereof.

As the analysis shows, in most cases, the existing methods are either not capable of processing the data stream in real time or cannot be implemented based on simple IoT controllers that could allow the development of the cheap IoT applications. Thus, it is important to create new high-speed methods that would have a simplicity of implementation and allow data processing in online mode.

In the paper, the architecture of multidimensional wavelet neuron and its learning algorithm are proposed. The proposed approach is characterized by simplicity of computational implementation and high speed of tuning parameters. Such systems can be used for solving tasks of the classification data, the patterns recognition, the prediction of multidimensional time series, which are generated by Internet-connected sensory devices in IoT applications.

2   The Architecture of Wavelet Neuron and Its Learning Algorithm

The architecture of the wavelet neuron [28] is shown in Fig. 1.

This architecture can be used for prediction or can be modified for binary classification of data in online mode, which are fed from the sensors of smart systems. The architecture of wavelet neuron is simple for implementation in hardware and can be used in IoT controllers and allows processing information in such systems directly.

As it can be seen, the wavelet neuron has similar architecture as neo-fuzzy neuron architecture [29, 30], but it consists of the wavelet synapses WS_i with the wavelet activation functions $(i = 1, 2, \ldots, n)$, where the synaptic weights $w_{ji}(k)$ are the adjustable parameters.

When the input of the wavelet neuron gets vector signal

$$x(k) = (x_1(k), x_2(k), \ldots, x_n(k))^T,$$

its output produces a scalar in the form

$$\hat{y}(k) = \sum_{i=1}^{n} f_i(x(k)) = \sum_{i=1}^{n} \sum_{j=1}^{h_i} w_{ji}(k-1)\varphi_{ji}(x_i(k))$$

that is defined both by adjustable weights $w_{ji}(k)$, and by the wavelet functions being used.

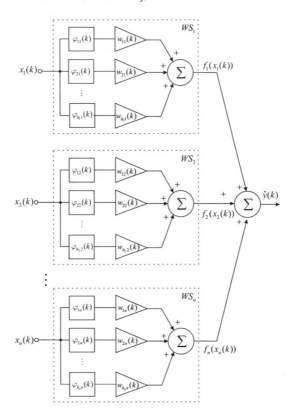

Fig. 1. Wavelet neuron

Here, we will use the one-dimensional wavelet activation function:

$$\varphi_{ji}(x_i(k)) = (1 - t_{ji}^2(k)) \exp\left(-\frac{t_{ji}^2(k)}{2}\right)$$

where $t_{ji}(k) = \left(x_i(k) - c_{ji}(k)\right)\sigma_{ji}^{-1}(k)$; $c_{ji}(k)$ is the parameter that defines a location of the function center; $\sigma_{ji}(k)$ is the parameter that defines the function width.

An additional point to emphasize here is that wavelet function oscillativity does not contradict to a membership function unipolarity since negative values can be treated as low membership levels or non-membership [31].

Learning task is to choose synaptic weights $w_{ji}(k)$ on each iteration k that will optimize the considered quality criterion.

As a learning criterion of the wavelet neuron, a quadratic error function is used that is expressed in wavelet neuron notation as follows:

$$E(k) = \frac{1}{2}(d(k) - \hat{y}(k))^2 = \frac{1}{2}e^2(k) = \frac{1}{2}(d(k) - \sum_{i=1}^{n} \sum_{j=1}^{h_i} w_{ji}(k-1)\varphi_{ji}(x_i(k)))^2 \tag{1}$$

(here $d(k)$ is the external learning signal), and its derivatives subject to adjustable parameters have the form

$$\frac{\partial E(k)}{\partial w_{ji}(k)} = -e(k)\varphi_{ji}(x_i(k)) = -e(k)(1 - t_{ji}^2(k)) \exp\left(-\frac{t_{ji}^2(k)}{2}\right) = -e(k)J_{ji}^w(k).$$

Then, by applying gradient procedure to minimize the expression (1), we can express learning method of the wavelet neuron as follows:

$$w_{ji}(k+1) = w_{ji}(k) + \eta^w e(k+1)J_{ji}^w(k+1)$$

where scalar η^w defines the learning step in adjustable parameters space.

By introducing $(h_i \times 1)$ dimension vectors of variables $\varphi_i(x_i(k)) = (\varphi_{1i}(x_i(k)), \ldots, \varphi_{h_i i}(x_i(k)))^T$, $w_i(k) = (w_{1i}(k), \ldots, w_{h_i i}(k))^T$, $J_i^w(k) = (J_{1i}^w(k), \ldots, J_{h_i i}^w(k))^T$, we can state gradient procedure of the i-th wavelet synapse learning method:

$$w_i(k+1) = w_i(k) + \eta^w e(k+1)J_i^w(k+1).$$

It is possible to improve a convergence of the learning processes by switching from gradient procedures to the Levenberg-Marquardt algorithm for adjusting of neural networks [32].

Using inverse of the sum of matrices lemma and performing obvious transformation [33], we can write the learning algorithms for real time case:

$$\begin{cases} w_i(k+1) = w_i(k) + \dfrac{e(k+1)J_i^w(k+1)}{r_i^w(k+1)}, \\ r_i^w(k+1) = \beta r_i^w(k) + PJ_i^w(k+1)P^2 \end{cases} \tag{2}$$

where β is the forgetting parameter.

It is clear that given $\beta = 1$, expression (2) acquires stochastic approximation qualities of adaptive Goodwin-Ramadge-Caines algorithm [34], and given $\beta = 0$, it becomes Widrow-Hoff algorithm that is widely used in artificial neural networks theory. Thus, modified learning methods usage does not complicate numerical implementation of the wavelet synapses adjusting procedures virtually and increases the speed of their convergence.

3   The Architecture of Multidimensional Wavelet Neuron and Its Learning Algorithm

In many cases solving the real problems in Internet of Things application is needed the prediction or the classification of multidimensional data, which are fed from some sensors at one time. For this case, we can introduce multidimensional wavelet neuron, which has n inputs, m outputs and h wavelet activation function for each input. For the task of classification and pattern recognition, the sigmoidal functions have to be added to the output layer. The architecture of the multidimensional wavelet neuron for the classification or pattern recognition tasks is shown in Fig. 2.

The input observation vector is fed to the input layer of multidimensional wavelet neuron in the form

$$x(k) = (x_1(k), x_2(k), \dots, x_n(k))^T,$$

than the output of neuron can be written in the from

$$y_j(k) = \frac{1}{1 + \exp(-\gamma u_j)},$$

$$u_j(k) = \sum_{l=1}^{h} \sum_{i=1}^{n} \varphi_{li}(x_i(k)) w_{lij}(k)$$

where k is a number of observation, y_j is j-th output of multidimensional wavelet neuron, γ is a rate of rise of sigmoidal activation function, $\varphi_{li}(x_i(k))$ - l-th wavelet activation function of i-th input of neuron, $w_{lij}(k)$ is j-th synaptic weight of l-th wavelet activation function of i-th input of neuron, $i = 1 \dots n, j = 1 \dots m, l = 1 \dots h$.

The nodes of the multidimensional wavelet neuron are wavelet neurons, which can be described above.

For the optimization of computational implementation let's rewrite the input of multidimensional wavelet neuron in the form

$$y(k) = W(k)\varphi(x(k))$$

where $\varphi(x(k)) = (\varphi_{11}(x_1), \varphi_{12}(x_2), \dots, \quad \varphi_{1n}(x_n), \varphi_{21}(x_1), \varphi_{22}(x_2), \dots, \quad \varphi_{2n}(x_n), \dots,$
$\varphi_{h1}(x_1), \dots, \varphi_{hn}(x_n))^T$ is $(hn \times 1)$ dimension vector wavelet activation functions,

$$W(k) = \begin{pmatrix} w_{111} & w_{121} & \cdots & w_{1n1} & w_{211} & w_{221} & \cdots & w_{2n1} & \cdots & \cdots & w_{hn1} \\ w_{112} & w_{112} & \cdots & w_{1n2} & w_{212} & w_{222} & \cdots & w_{2n2} & \cdots & \cdots & w_{hn2} \\ \vdots & \vdots & \ddots & \vdots & \vdots & \vdots & \ddots & \vdots & \ddots & \ddots & \vdots \\ w_{11m} & w_{12m} & \cdots & w_{1nm} & w_{21m} & w_{22m} & \cdots & w_{2nm} & \cdots & \cdots & w_{hnm} \end{pmatrix}$$

is $(m \times hn)$ matrix of synaptic weights.

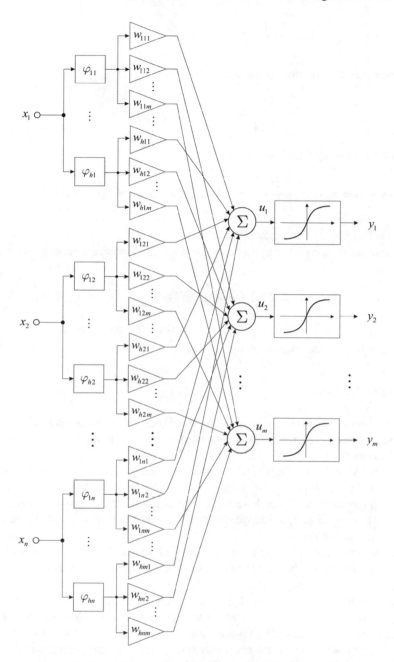

Fig. 2. The architecture of multidimensional wavelet neuron

Due to the synaptic weights of multidimensional wavelet neuron depend on the output systems linearly, we can use the stochastic approximation algorithms, which minimize criterion in the form

$$E_j(k) = \frac{1}{2}e_j^2(k) = \frac{1}{2}(d_j(k) - y_j(k))^2. \tag{3}$$

Minimizing the criterion (3) by synaptic weights $w_{lij}(k)$

$$\frac{\partial E_j}{\partial w_{lij}} = -e_j y_j (1 - y_j)\gamma\varphi_{li}(x_i),$$

we can write learning algorithm in the form

$$w_{lij}(k + 1) = w_{lij}(k) - \eta\frac{\partial E_j}{\partial w_{lij}} = w_{lij}(k) + \eta e_j(k)y_j(k)(1 - y_j(k))\gamma\varphi_{li}(x_i(k)) \tag{4}$$

where η is learning rate parameter ($0 < \eta \leq 1$).

For optimizing the learning process, we can rewrite learning algorithm (4) in matrix form

$$W(k + 1) = W(k) + \eta\gamma(e(k)ey(k)e1 - y(k))) \cdot \varphi^T(x(k))$$

where $e(k) = (e_1(k), e_2(k), \dots, e_m(k))^T$ is the errors vector, $y(k) = (y_1(k), y_2(k), \dots, y_m(k))^T$ is the outputs vector, $\odot$ is dot product.

4 Experiments

The effectiveness of proposed recognition system is examined based on both benchmark data and real data. The recognition task of handwritten digits was solved based on MNIST database [35].

The MNIST is handwritten digits database, which was taken from [35]. A training set consists of the 60.000 images, and a test set consists of the 10.000 images. The initial bi-level images from NIST database were normalized (each image was resized to a 20 × 20 pixel box. After that, the all images were centered in a 28 × 28 pixel image using the pixels' mass center.

The multidimensional wavelet neuron has 784 inputs, 10 outputs, and 10 wavelet function for each input. The initial synaptic weights values were taken zeros and learning rate parameter was taken $\eta = 0.99$·

The classifiers based on neural networks were taken for the results comparison. Table 1 shows the results of the comparison. As the quality criterion was taken the percentage of the false classified objects based on the testing data image set.

As it may be inferred from the obtained results, the multidimensional wavelet neuron has the best quality of classification among 1-layer neural network classifiers. The 2-layer perceptron has a better quality of classification but has 3.5 times more adjustable parameters than the multidimensional wavelet neuron. This fact influences on the training time of such system and as result – the impossibility of using such system in IoT controllers. The proposed multidimensional wavelet neuron has simple architecture and can be implemented based on Arduino-like controllers for many pattern recognition tasks in IoT application.

Table 1. The classification results of MNIST dataset

Neural network classifier	Preprocessing data	Test error (%)
Multidimensional wavelet neuron (1-layer NN, 10 wavelet activation function for each input)	none	6.5%
Linear 1-layer NN classifier	none	12.0%
Linear 1-layer NN classifier	deskewing	8.4%
Pairwise linear classifier	deskewing	7.6%
2-layer perceptron (300 sigmoidal activation functions in the hidden layer and 10 ones in the output layer)	none	4.7%

5   Conclusion

In this article, the architecture of multidimensional wavelet neuron is proposed. Such system can be used both a classifier and predictor of the multidimensional data sets. The main advantage of the multidimensional wavelet neuron is a simplicity of implementation in the hardware of IoT applications. Also, the learning algorithm of multidimensional wavelet neuron for solving pattern recognition task is proposed. This learning algorithm is the modification of the gradient algorithm and is characterized by high speed of information processing.

The systems based on the proposed multidimensional wavelet neuron can be implemented for solving the problems in IoT applications, Data Stream Mining, Big Data Processing.

The computational experiments are performed based on benchmark and real data sets. The obtained results have confirmed the advantages of the proposed approach in comparison with the existed methods.

References

1. Mahdavinejad, M.S., Rezvan, M., Barekatain, M., Adibi, P., Barnaghi, P., Sheth, A.P.: Machine learning for Internet of Things data analysis: a survey. Dig. Commun. Netw. (in Press, 2017)
2. Atzori, L., Iera, A., Morabito, G.: The Internet of Things: a survey. Comput. Netw. **54**(15), 2787–2805 (2010)
3. Cecchinel, C., Jimenez, M., Mosser, S., Riveill, M.: An architecture to support the collection of big data in the Internet of Things. In: Proceedings of 2014 IEEE World Congress on Services, pp. 442–449 (2014)
4. Joakar, A.: A methodology for solving problems with data science for Internet of Things. Data Science for Internet of Things. https://www.datasciencecentral.com/profiles/blogs/a-methodology-for-solving-problems-with-datascience-for-internet. Accessed 27 Nov 2017
5. Ni, P., Zhang, C., Ji, Y.: A hybrid method for short-term sensor data forecasting in Internet of Things. In: Proceedings of 2014 11th International Conference on Fuzzy Systems and Knowledge Discovery (FSKD), pp. 369–373 (2014)

6. Alam, F., Mehmood, R., Katib, I., Albeshri, A.: Analysis of eight data mining algorithms for smarter Internet of Things (IoT). Procedia Comput. Sci. **98**, 437–442 (2016)
7. Bodyanskiy, Y., Vynokurova, O., Pliss, I., Peleshko, D.: Hybrid adaptive systems of computational intelligence and their on-line learning for green it in energy management tasks In: Kharchenko, V., Kondratenko, Y., Kacprzyk, J. (eds.) Green IT Engineering: Concepts, Models, Complex Systems Architectures. Studies in Systems, Decision and Control, vol. 74, pp. 229–244. Springer, Cham (2017)
8. Peleshko, D., Peleshko, M., Kustra, N., Izonin, I.: Analysis of invariant moments in tasks image processing. In: Proceedings of 2011 11th International Conference the Experience of Designing and Application of CAD Systems in Microelectronics (CADSM), Polyana-Svalyava, pp. 263–264 (2011)
9. Ivanov, Y., Peleshko, D., Makoveychuk, O., Izonin I., Malets I., Lotoshunska, N, Batyuk, D. Adaptive moving object segmentation algorithms in cluttered environments. In.: Proceedings of 2015 15th International Conference the Experience of Designing and Application of CAD Systems in Microelectronics (CADSM), Lviv, pp. 97–99 (2015)
10. Rutkowski, L.: Computational Intelligence: Methods and Techniques. Springer, Berlin (2008)
11. Bishop, C.M.: Pattern Recognition and Machine Learning. Springer, Berlin (2006)
12. Murphy K.P.: Machine Learning: a Probabilistic Perspective. MIT Press (2012)
13. Bodyanskiy, Y.V., Vynokurova, O.A., Dolotov, A.I.: Self-learning cascade spiking neural network for fuzzy clustering based on group method of data handling. J. Autom. Inf. Sci. **45**, 23–33 (2013)
14. Hu, Z., Bodyanskiy, Y.V., Tyshchenko, O.K., Tkachov, V.M.: Fuzzy clustering data arrays with omitted observations. Int. J. Intell. Syst. Appl. (IJISA) **9**(6), 24–32 (2017). https://doi.org/10.5815/ijisa.2017.06.03
15. Bodyanskiy, Y., Vynokurova, O., Setlak, G., Peleshko, D., Mulesa, P.: Adaptive multivariate hybrid neuro-fuzzy system and its on-board fast learning. Neurocomputing **230**, 409–416 (2017)
16. Hu, Z., Bodyanskiy, Y.V., Tyshchenko, O.K., Samitova, V.O.: Possibilistic fuzzy clustering for categorical data arrays based on frequency prototypes and dissimilarity measures. Int. J. Intell. Syst. Appl. (IJISA) **9**(5), 55–61 (2017). https://doi.org/10.5815/ijisa.2017.05.07
17. Hu, Z., Bodyanskiy, Y.V., Tyshchenko, O.K., Samitova, V.O.: Fuzzy clustering data given on the ordinal scale based on membership and likelihood functions sharing. Int. J. Intell. Syst. Appl. (IJISA) **9**(2), 1–9 (2017). https://doi.org/10.5815/ijisa.2017.02.01
18. Babichev, S., Taif, M.A., Lytvynenko, V.: Inductive model of data clustering based on the agglomerative hierarchical algorithm. In: Proceedings of 2016 IEEE 1st International Conference on Data Stream Mining and Processing, DSMP 2016, pp. 19–22 (2016)
19. Babichev, S., Taif, M.A., Lytvynenko, V., Osypenko, V.: Criterial analysis of gene expression sequences to create the objective clustering inductive technology. In: Proceedings of 2017 IEEE 37th International Conference on Electronics and Nanotechnology, ELNANO 2017, pp. 244–248 (2017)
20. Bodyanskiy, Y., Vynokurova, O., Pliss, I., Setlak, G., Mulesa, P.: Fast learning algorithm for deep evolving GMDH-SVM neural network in data stream mining tasks. In.: Proceedings of 2016 IEEE 1st International Conference on Data Stream Mining and Processing, DSMP 2016, pp. 257–262 (2016)
21. Hu, Z., Bodyanskiy, Y.V., Tyshchenko, O.K., Samitova, V.O.: Fuzzy clustering data given in the ordinal scale. Int. J. Intell. Syst. Appl. (IJISA) **9**(1), 67–74 (2017). https://doi.org/10.5815/ijisa.2017.01.07

22. Bodyanskiy, Y., Setlak, G., Peleshko, D., Vynokurova, O.: Hybrid generalized additive neuro-fuzzy system and its adaptive learning algorithms. In: Proceedings of 2015 IEEE 8th International Conference on Intelligent Data Acquisition and Advanced Computing Systems: Technology and Applications, IDAACS, pp. 328–333 (2015)
23. Perova, I., Pliss, I.: Deep hybrid system of computational intelligence with architecture adaptation for medical fuzzy diagnostics. Int. J. Intell. Syst. Appl. (IJISA) 9(7), 12–21 (2017). https://doi.org/10.5815/ijisa.2017.07.02
24. Bin, S., Yuan, L., Xiaoyi, W.: Research on data mining models for the internet of things. In: Proceedings of 2010 IEEE International Conference on Image Analysis and Signal Processing, pp. 127–132 (2010)
25. Chen, F., Deng, P., Wan, J., Zhang, D., Vasilakos, A.V., Rong, X.: Data mining for the internet of things: literature review and challenges. Int. J. Distrib. Sens. Netw. 11(8), 1–14 (2015)
26. Tsai, C.-W., Lai, C.-F., Chiang, M.-C., Yang, L.T.: Data mining for internet of things: a survey. IEEE Commun. Surv. Tutor. 16(1), 77–97 (2014)
27. Zanella, A., Bui, N., Castellani, A., Vangelista, L., Zorzi, M.: Internet of 905 Things for smart cities. IEEE Internet of Things J. 1(1), 22–32 (2014)
28. Bodyanskiy, Y., Lamonova, N., Pliss, I., Vynokurova, O.: An adaptive learning algorithm for a wavelet neural network. Exp. Syst. 22(5), 235–240 (2005)
29. Yamakawa, T., Uchino, E., Miki, T., Kusanagi, H.: A neo-fuzzy neuron and its application to system identification and prediction of the system behavior. In: Proceedings of 2nd International Conference on Fuzzy Logic and Neural Networks, pp. 477–483, Iizuka, Japan (1992)
30. Miki, I., Yamakawa, I.: Analog implementation of neo-fuzzy neuron and its on-board learning. In: Computational Intelligence and Applications. pp. 144–149. WSES Press, Piraeus (1999)
31. Mitaim, S., Kosko, B.: What is the best shape for a fuzzy set in function approximation? In: Proceedings of 5th IEEE International Conference on Fuzzy Systems, Fuzz 1996, vol. 2, pp. 1237–1213 (1996)
32. Shepherd, A.J.: Second-Order Methods for Neural Networks. Springer, London (1997)
33. Bodyanskiy, Y., Kolodyazhniy, V., Stephan, A.: An adaptive learning algorithm for a neuro-fuzzy network. In: Reusch, B. (eds.) Computational Intelligence. Theory and Applications. Fuzzy Days 2001. LNCS, vol. 2206, pp. 68–75. Springer, Heidelberg (2001)
34. Goodwin, G.C., Ramadge, P.J., Caines, P.E.: A globally convergent adaptive predictor. Automatica 17(1), 135–140 (1981)
35. MNIST Homepage. http://yann.lecun.com/exdb/mnist/. Accessed 19 Nov 2017

Simulation of Shear Motion of Angular Grains Massif via the Discrete Element Method

Sergiy Mykulyak, Vasyl Kulich, and Sergii Skurativskyi[✉]

Institute of Geophysics NAS of Ukraine, Bohdan Khmelnytskyi str., 63 G, Kyiv, Ukraine
mykulyak@ukr.net, skurserg@gmail.com

Abstract. Dynamics of granular materials is a challenge for predictability of their response under dynamic load due to the complex behavior of granular systems. In this report, we consider the granular system composed of the polyhedral particles, namely cubes, and study system's dynamics via the discrete element method which allows us to describe the interactions between cubes in detail. On the base of modeling the shear motion of granular medium, the statistical laws of multiparticle system dynamics are observed. These studies are extremely important for the geophysics, material science, etc.

Keywords: Granular media · Discrete element method · Shear

1 Introduction

The dynamic behavior of granular media has many features that distinguish them from the behavior of solids, liquids and gases. Here the discreteness, heterogeneity of media, nonlinear and dissipative nature of the interaction of granules, their repacking, the presence of a network of forces that carry the load overall granular massif, and the shape of grains, rotation of them play a significant role. Characteristic features of granular media are especially brightly manifested with a shear deformation. In particular, the fluctuations of stress and velocity in granular massif sometimes are many times greater than the mean values [1, 2]. The amplitude of fluctuations depends on the size and shape of grains, degree of shape irregularity, density of granular massif, and on shear speed [3]. During shear deformation, a significant volumetric dilatation takes place. It is also very sensitive to the grains shape and degree of shape irregularity [4–6].

At present, there are no any universal mathematical model fully describing the dynamical processes in granular media. The main obstacle is concerned with the existence of long-range correlations in various complex systems [7–9] and especially in multi-particulate ones. In this case to model the dynamics of granular flows it is suitable to use the discrete elements method (DEM) [10]. The well-known advantage of this method consists in the ability to describe the dynamics of each elements of structure in detail. It is worth noting that this approach is numerically expensive, therefore, it allows to treat the granular systems with relatively small number of particles.

As a rule, DEM has been applied to granular systems composed of spherical particles or their conglomerates. Such a choice of particle's shape simplifies the description of

© Springer International Publishing AG, part of Springer Nature 2019
Z. Hu et al. (Eds.): ICCSEEA 2018, AISC 754, pp. 74–81, 2019.
https://doi.org/10.1007/978-3-319-91008-6_8

their motion and interaction but at the same time we lose the possibility to model many important properties of granular system kinematics. In particular, Zhao et al. [5] shows that the particle angularity essentially affects the deformable properties of granular massif at shear loading.

2 The Statement of the Problem and Construction of Algorithm for DEM

The present paper deals with the modeling of granular systems composed of the poly-hedral particles. The simplest example of such system is the system of cubes. Let this system of cubes be involved in the shear motion. The position of i th cube is well-defined in time via the coordinates of center of mass r_{ci} and three parameters defining its rotational motion. Thus, the granular medium composed of N cubes has $6N$ degrees of freedom and satisfies the following equations of motion:

$$m_i \frac{d^2 r_{ci}}{dt^2} = \sum_j F_{ij}, \tag{1}$$

$$\frac{dK_{ci}}{dt} = \sum_j M_c(F_{ij}). \tag{2}$$

Here m_i is the mass of i th cube; F_{ij} is the contact force appearing between i th and j th cubes; K_{ci} is the kinetic moment of i th cube with respect of its center; $M_c(F_{ij})$ is the torque with respect to the center of i th cube.

Due to the cube symmetry, the rotational motion of cube and sphere has many common features and, in particular, $K_{ci} = I_{ci}\omega_{ci}$, where ω_{ci} is the vector of angular velocity for the cube rotating about the center, I_{ci} is the corresponding moment of inertia for the cube.

The moment of contact force acting on the cube i during the contact with cube j is $M_c(F_{ij}) = r_{ij}^c \times F_{ij}$, where r_{ij}^c is the vector pointing from the cube center to the contact point of adjacent cubes.

The description of the translational motion of cube center is relatively simple task, whereas the consideration of rotational motion demands more efforts. To overcome well-known obstacles, we apply the quaternion technique [11]. In the motionless reference frame, the kinematic equation for the rotation quaternion has the following form $\frac{d\Lambda}{dt} = \frac{1}{2}\omega_{ci} \circ \Lambda$, where $\Lambda = (\lambda_0, \lambda_1, \lambda_2, \lambda_3) = \lambda_0 + \lambda_1 i_1 + \lambda_2 i_2 + \lambda_3 i_3 = \lambda_0 + \lambda$ is the rotation quaternion. During calculations, we also perform the correction of quaternion norm to 1 in accordance with the modified equation

$$\frac{d\Lambda}{dt} = \frac{1}{2}\omega_{ci} \circ \Lambda - q\Lambda(\|\Lambda\| - 1). \tag{3}$$

The parameter q belongs to the interval $(0; 1)$. We put $q = 0.5$. Thus, relations (1)–(3) form the closed system of equations for describing the interaction of pair cubes. To solve Eqs. (1) and (2), Beeman's algorithm is used [12]. Equation (3) is solved by the Runge-Kutta 4-th order method.

Let us outline the main steps of algorithm for cubes interaction. To identify a contact, we use the algorithm [13, 14] which considers the particle as a domain bounded by the half spaces $(\xi - c - a)a \leq 0$, where c is the center of mass for particle, ξ is an arbitrary point in the half space, a is the vector pointing from particle center of mass to the boundary face of half space in normal direction.

The resulting figure of intersection is a convex polyhedron. The algorithm [14] for finding his vertices consists in evaluation of all points of intersection of all possible combinations from two plates of one cube and plate of another cube.

Interaction between bodies is defined by a single force which is broken into normal force F^n and tangential friction force F^f. The coordinates of the point, where the contact force is applied, coincide with the mean values of coordinates of vertices of overlapping region $x^O = \sum_j x_j/m$, where m is the number of vertices of overlapping region.

The direction of force is defined by the contact normal n^f. To derive the contact normal, we use the vertices of contact line which is the broken line of intersection for the surfaces of two contacting particles [5]. On the base of these vertices we construct the fitting plane (FP) by means of linear square fitting. To do this, the singular value decomposition (SVD) [15, 16] is used. The normal of this plane coincides with the contact normal and define the direction for normal force of interaction between particles.

The model for normal force, which does not depend from the type of contacts, is suitable for the particle of various shapes and defines the interaction between bodies, has been proposed in [17]. Using this model, the contact interaction between bodies is described by the single force which depends on the volume V of intersection domain of interacted bodies and the depth of penetration d. Then the force $F^n = k \cdot E^* \sqrt{Vd}$, where the coefficient $k = 4 / 3\sqrt{\pi}$, $V = 1/6 \cdot d \cdot d_{max} \cdot d_p$, and d_{max} is the maximal distance between vertices of overlapping region (direction l_m), d_p is the maximal distance between vertices in the direction $n_p = n^f \times l_m$.

The effective Young's module E^* satisfies the relation

$$\frac{1}{E^*} = \frac{1 - v_1^2}{E_1} + \frac{1 - v_2^2}{E_2},$$

where $E_{1,2}$ and $v_{1,2}$ are Young's moduli and Poisson ratios for contacting bodies. We take identical $E_{1,2}$ and $v_{1,2}$. The depth of penetration d is defined as a distance of penetration of intersection domain in the direction of normal force and is derived by the following formula: $d = \max(n^f \cdot p^c) - \min(n^f \cdot p^c)$.

Besides normal forces, during contact the tangential forces arise. These forces are evaluated via the Coulomb model of friction. This model distinguishes the static friction (coupling) $F^s < \mu^s F^n$ and kinetic friction (slip) $F^k = \mu^k F^n$, where μ^s and μ^k are the empirical coefficients. At the application of this model some difficulty arises, since during the contact we don't know the type of contact and all forces applied to the particle. Another problem concerns the jumps of friction forces. To avoid this problem, authors of paper [17] offer the model for friction in which the discontinuities of friction forces are smoothed. The static friction is approximated via the slip model with very low velocities and then

$$F^f = \left((2\mu^{s*} - \mu^k)\frac{x^2}{1+x^4} + \mu^k - \frac{\mu^k}{1+x^2} \right) F^n, \quad \text{where} \quad \mu^{s*} = \mu^s \left(1 - 0.09(\mu^k/\mu^s)^4 \right),$$

$x = v^t/v^s.$

Here v^t is the tangential velocity, v^s is the velocity of transition from static to kinetic friction. The direction of friction force F^f is parallel to relative tangential velocity of cubes at contact point. By the described uniform algorithm for all contacts, we recognize the pairs of contacting cubes and derive the characteristics of their contact. On the base of outlined algorithm, the code CuBluck for the Fortran was developed.

In the modeling, we fix the following constants: the density $\rho = 1.2$ g/cm^3, Young's module $E = 3 \cdot 10^9$ N/m^2; Poisson ratio $v = 0.3$; coefficient of static $\mu^s = 0.7$ and kinetic $\mu^k = 0.4$ friction, respectively; the characteristic velocity of transition from static to kinetic friction $v^s = 0.01$ m/s.

For the numerical experiment, let us take *3000* cubes of identical size 1 cm and place them in the box composed of two parts, namely fixed lower with the high *6* cm and moving upper one with the high *50* cm. The bottom has the following size: 20 cm × 30 cm. The upper part moves with fixed velocity $V_b = 0.5$ m/s. The box with randomly placed cubes is shown in Fig. 1.

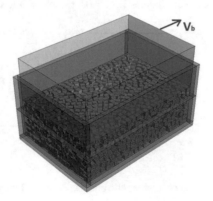

Fig. 1. Initial location of cubes inside the box.

The placing is realized by means of compacting the cubes in the region of size *20* × *30* × *50* cm. The box cover of mass *14* kg moves along vertical axis z.

Due to the volumetric dilatation taking place during shear deformation, the upper wall raises. Consecutive locations of cubes at four time moments are depicted in Fig. 2. In simulation, the total kinetic, elastic, rotational and dissipating energies are derived. The temporal profiles for these quantities and their power spectra are drawn in Figs. 3, 4, 5, and 6. It is worth noting that the spectra for all types of energy obey the power dependencies with the exponents close to 2. The type of these spectra testifies that these temporal dependencies do not contain any characteristic frequencies.

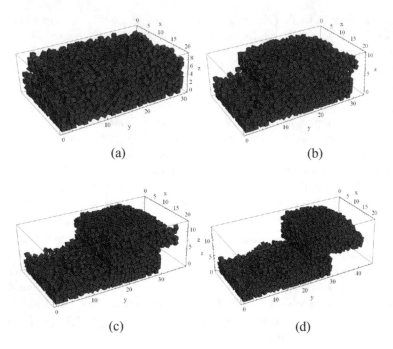

Fig. 2. The sheared granular system at times: (a) $t = 0.03$ s; (b) $t = 0.15$ s; (c) $t = 0.3$ s; (d) $t = 0.45$ s.

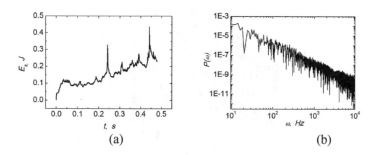

Fig. 3. Time dependence of total kinetic energy E_k (a) and its power spectrum (b).

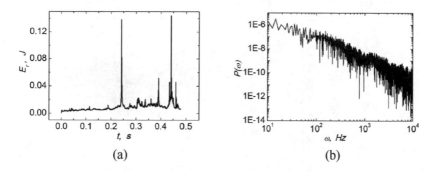

Fig. 4. Time dependence of total rotation energy E_r (a) and its power spectrum (b).

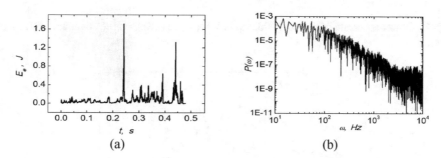

Fig. 5. Time dependence of total elastic energy E_e (a) and its power spectrum (b).

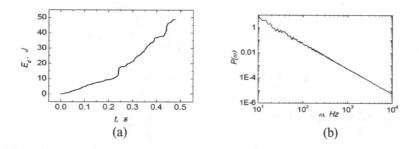

Fig. 6. Time dependence of total dissipative energy E_d (a) and its power spectrum (b).

In addition, the temporal relation for total force, with which cubes act on the upper wall of box, has the similar spectrum (Fig. 7) but exponent equals 1.4. This is an indication that there are long time correlations in this time series [18]. Besides temporal long-range correlations, the granular system possesses the spatial long-range correlations. This follows from the character of distribution functions for the forces acting between structural elements. The typical distribution for interparticle forces at the fixed time *0.3* s. having the "heavy tails" is plotted in Fig. 8. From this it follows that the attenuation obeys the power law but not Gaussian one as in uncorrelated systems.

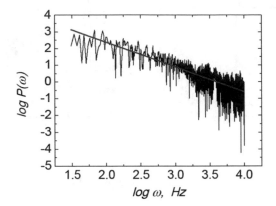

Fig. 7. Power spectrum of time dependence of total force acting on the cover and fit (straight line).

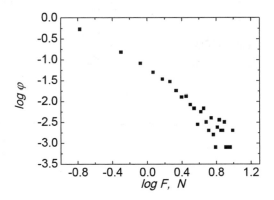

Fig. 8. Distribution of forces at time $t = 0.3$ s.

3   Concluding Remarks

Thus, we developed the computer code CuBluck for modeling the shear deformation of granular media comprised of the angular grains, namely the cubes, within the framework of DEM. The interaction of cubes incorporates the elastic interaction defining via the mutual penetration and friction including both static and kinetic one. The results obtained testify that the granular system during the shear process has the long-range correlations.

This code can be also used for the simulation of dynamical processes in natural and artificial materials, in which grains have the angular shape, in particular, sand, crushed rock, blocky geophysical media etc.

References

1. Behringer, R.P., Howell, D., Kondic, L., Tennakoon, S., Veje, C.: Predictability and granular materials. Physica D **133**, 1–17 (1999)
2. Howell, D., Behringer, R.P., Veje, C.: Stress fluctuations in a 2D granular Couette experiment: a continuous transition. Phys. Rev. Lett. **82**(26), 5241–5244 (1999)
3. Cabalar, A.F.: Stress fluctuations in granular material response during cyclic direct shear test. Granul. Matter **17**(4), 439–446 (2015)
4. Fu, Y., Wang, L., Zhou, C.: 3D clustering DEM simulation and non-invasive experimental verification of shear localisation in irregular particle assemblies. Int. J. Pavement Eng. **11**(5), 355–365 (2010)
5. Zhao, S., Zhou, X., Liu, W.: Discrete element simulations of direct shear tests with particle angularity effect. Granul. Matter **17**(6), 793–806 (2015)
6. Nguyen, D.-H., Azema, E., Sornay, P., Radjai, F.: Effects of shape and size polydispersity on strength properties of granular materials. Phys. Rev. E **91**, 032203 (2015)
7. Mahajan, A., Kaur, M.: Various approaches of community detection in complex networks: a glance. Int. J. Inf. Technol. Comput. Sci. (IJITCS) **8**(4), 35–41 (2016). https://doi.org/10.5815/ijitcs.2016.04.05
8. Gudnavar, A., Rajashekhara: Exploring an effectiveness & pitfalls of correlational-based data aggregation approaches in sensor network. Int. J. Wireless Microwave Technol. (IJWMT) **7**(2), 44–56 (2017). https://doi.org/10.5815/ijwmt.2017.02.05
9. Jain, A., Tyagi, S.: Priority based new approach for correlation clustering. I. J. Inf. Technol. Comput. Science **3**, 71–79 (2017)
10. Cundall, P.A.: A computer model for simulating progressive large-scale movements in blocky rock systems. In: Proceedings of International Symposium Rock Fracture, pp. 2–8. ISRM, Nancy (1971)
11. Gürlebeck, K., Sprössig, W.: Quaternionic and Clifford Calculus for Physicists and Engineers. Wiley, Chichester (1997)
12. Beeman, D.: Some multistep methods for use in molecular dynamics calculations. J. Comput. Phys. **20**(2), 130–139 (1976)
13. Lee, Y., Fang, C., Tsou, Y.-R., Lu, L.-S., Yang, C.-T.: A packing algorithm for three-dimensional convex particles. Granul. Matter **11**(5), 307–315 (2009)
14. Nassauer, B., Liedke, T., Kuna, M.: Polyhedral particles for the discrete element method. Granul. Matter **15**(1), 85–93 (2013)
15. Shakarji, C.V.: Least-squares fitting algorithms of the NIST algorithm testing system. J. Res. Natl. Inst. Stand. Technol. **103**(6), 633–641 (1998)
16. Forsythe, G.E., Malcolm, M.A., Moler, C.B.: Computer Methods for Mathematical Computations. Prentice Hall, Inc., Englewood Cliffs (1977)
17. Nassauer, B., Kuna, M.: Contact forces of polyhedral particles in discrete element method. Granul. Matter **15**(3), 349–355 (2013)
18. Jensen, H.J.: Self–Organized Criticality. Cambridge University Press, Cambridge (1998)

Model of Reconfigured Sensor Network
for the Determination of Moving Objects Location

Andrii Petrashenko[(✉)], Denis Zamiatin, and Oleksii Donchak

National Technical University of Ukraine "Igor Sikorsky Kyiv Polytechnic Institute",
Kyiv 03056, Ukraine
petrashenko@gmail.com

Abstract. In recent years, the implementation of distributed sensor networks has been allowing to create more and more sophisticated systems especially for environment monitoring. Among those applications are detecting, locating and tracking static or moving objects of interest. The advances of sensor technologies open up new possibilities of solving those sorts of tasks. One of the approaches is the Time Difference of Arrival algorithm, which is designed for locating an acoustic source. Moreover, sensor based systems usually produce a great deal of information streams those are subjected to further analysis: clustering, classification, regression etc. Hence, it is important to develop effective methods of determinations of moving object location based on a special model of reconfigured sensor network. The aim of the research is to propose a new model of reconfigured sensor network for the determination of the moving objects location using Time Difference of Arrival algorithm. The simulation study is presented.

Keywords: Distributed sensor networks · TDOA · Moving object location
Environment monitoring

1 Introduction

Rapid advances in information technology characterize the end of the 20th and early 21st centuries. The lowering of sensor device prices, as well as reducing their physical size lead to massive use in various sectors of the national economy and the military [1]. Given the variety of approaches to addressing the task of real-world objects tracking, the problem of effective organization for this sensor network is still relevant. Thus, increasing the efficiency of using the resources of such networks [2], as well as ensuring their fault tolerance, requires the development of a generalized algorithm for their construction and configuration [3].

Now, there are several approaches of determining the spatial coordinates of objects using sensor devices. They are image recognition in moving and static images, the construction of inertial navigation systems and directional guidance using radio waves [4].

There are several effectively applied implementations of the listed approaches but they have some important restrictions. They have a significant computational complexity (pattern recognition), require direct access to objects (inertial navigation systems) or require a complex set of equipment (radio tracking). At the same time, sound waves can be used at distances from 10 m to 10 km, so sound sensor networks might be considered as an alternative approach [5].

Therefore, the aim of the study is to develop the model based on TDOA algorithm and to evaluate the implementation of sensor networks for the object location problem. Besides, the largest possible coverage area with the least possible number of sensor devices is taken into account.

2 TDOA Algorithm Features

The idea of the TDOA algorithm, used to determine the spatial coordinates of objects, is as follows [6]:

1. Given known the positions $\left\{ (x_m, y_m, z_,) \right\}_{m=1}^{M}$, there are M ≥ 5 sound sensors.
2. Let v is the fixed speed of sound propagation in the operating environment.
3. Let $\left\{ \tau_m = t_m - t_1 \right\}_{m=2}^{M}$ be intervals between the moments of the sound wave arriving and sound sensors number m and sound sensor number 1.
4. The system of linear equations is:

$$\begin{bmatrix} A_3 & B_3 & C_3 \\ A_4 & B_4 & C_4 \\ \vdots & \vdots & \vdots \\ A_M & B_M & C_M \end{bmatrix} \begin{bmatrix} x \\ y \\ z \end{bmatrix} = \begin{bmatrix} -D_3 \\ -D_4 \\ \vdots \\ -D_M \end{bmatrix}$$

$$A_m = \frac{1}{v\tau_m}\left(-2x_1 + 2x_m\right) - \frac{1}{v\tau_2}\left(-2x_2 + 2x_1\right)$$

$$B_m = \frac{1}{v\tau_m}\left(-2y_1 + 2y_m\right) - \frac{1}{v\tau_2}\left(-2y_2 + 2y_1\right)$$

$$C_m = \frac{1}{v\tau_m}\left(-2z_1 + 2z_m\right) - \frac{1}{v\tau_2}\left(-2z_2 + 2z_1\right)D_m$$

$$= v\tau_m - v\tau_2 + \frac{1}{v\tau_m}\left(x_1^2 + y_1^2 + z_1^2 - x_m^2 - y_m^2 - z_m^2\right)$$

$$- \frac{1}{v\tau_2}\left(x_1^2 + y_1^2 + z_1^2 - x_2^2 - y_2^2 - z_2^2\right).$$

5. Having solved this system of equations, we can find the unknown coordinates (x, y, z) of the sound source.

2.1 The Solution of the Problem

Before solving the problem, we make the following assumptions [7]. The operating environment of audio sensor devices is homogeneous and characterized by a constant rate of sound waves propagation. The sound source generates spherical sound waves. A sound event is an explosion, a gunshot, a hit or other short-term sounds after a period of silence. The radius of action characterizes each sound sensor and the response to sound events of a given intensity occurring within this radius is guaranteed. Sound sensors are placed on the plane.

Thus, there are several stages to solve the problem:

1. The first stage is a placement of sound sensors on a plane. To do this, we propose to solve the following problem: "Let there be a set of circles having predefined radiuses. The point of the plane is considered to be covered if and only if it belongs to not less than five circles. The aim is to place circles on the plane, ensuring coverage of the maximum area."

2. The second stage is the partitioning of placed sound sensors into TDOA-groups. To do this, we propose to solve the following. The TDOA group considers such 5 circles or more those have at least one common point. A plane point is considered to be covered if and only if it belongs to at least one TDOA-group. The task is to find the minimum set of TDOA groups that would provide coverage of the maximum area.

3. Recognition of a sound event within one TDOA group. To do this, we need to have a special form of information reports from all controllers of sound sensors. These reports must contain the moment of recording of the sound event, as well as its intensity. After receiving reports from all sound sensors belonging to the TDOA-group, the sound source position can be calculated using the TDOA algorithm.

4. Filtering of data received by different TDOA-groups. It is possible to have situations when several TDOA-groups respond to one audio event. Following this step, we can handle such cases based on predefined tolerances in time (several groups responded almost simultaneously) and coordinates (several groups calculated close positions). Then we can assume that the data from several groups received with some predefined delay coincide with some predefined accuracy, and they relate to one sound event.

5. Reconfiguration of system in case of failure of sound sensors. The second stage is repeated, but the rejected sensors are not taken into account.

After these stages, the area covered by the TDOA-group can be calculated. It would be maximal if the sound sensors were located in the nodes of the correct pentagon.

It is easy to see that the area covered can be calculated using the following formula:

$$S = \frac{2d^2}{\left(1 + \sqrt{5}\right)\sqrt{5}},$$

where S is the area covered by the TDOA group, d is the radius of the acoustic sensor, equal to the diagonal of the pentagon.

The time the system responds to a sound event is considered as a delay between the moments of the occurrence of this event and the output of a filtered report on its recognition (Fig. 1).

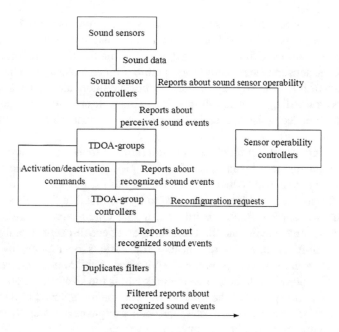

Fig. 1. The dataflow diagram

The time of the system reaction to the sound event consists of the following components:

1. Time of receipt of the sound wave to the sensors.
2. Time to report a response to a sound event from sensors controllers to TDOA-groups.
3. Additional waiting time for TDOA report groups from all related sound sensors.
4. Reducing the quota of this time may lead to the inability to recognize sound events having a poor network connection between audio sensors controllers and TDOA-groups. Excessing might prevent the recognition ability of events that have occurred with a rather short time interval between them and to increasing the response time of the system.
5. Time to execute the TDOA algorithm.
6. Time to report recognition of sound events from TDOA-groups to duplicate filters.
7. Additional waiting time for duplicate filters to recognize audio events. Reducing the quota of this time might result the erroneous recognition of one sound event whereas overestimation increases the response time of the system.

3   The Program Description

The simulation program is coded using Python 3.5. Parameters of the model are:
coordinates of sound sensors on a plane; minimum intensity of sound events fixed by sensors; coordinates of sound sources on the plane; time intervals when sound sources generate events; the intensity of the generated events; speed of sound propagation;

intervals of possible data transmission delays between controllers and the only TDOA-group server; intervals of the possible data transmission delays between TDOA-groups and the common duplicate filter; additional waiting time for TDOA report groups from all related sound sensors; the maximum acceptable sources intensity variance calculated using the data of the TDOA-group sensors; additional duplicate filter waiting time for reports on recognized sound events; the maximum possible error in coordinates, in which the duplicate filter recognizes several reports of sound events as related to the certain event.

The following model restrictions are considered. All sound sensors are placed in the fixed positions using pre-set parameters. If at a single moment of time, the sound sensor receives several sound events, then it responds only to the most intensive of them. The intensity of the sound event decreasing squared by the distance from its source. Single TDOA-group servers and duplicate filters simulate client-server architecture of the system, because the behavior and fault tolerance of a peer-to-peer system with dynamically determining the modules responsible for various operations, in the limited computing and other resources, as well as the presence of noise and interference, require a separate study. Some features of the system, such as the size of the data being transmitted, as well as the quality of the network connection are reduced to the intervals of possible data transmission delays. Each sound sensor might have status "busy" or "idle", the change of which immediately becomes known to the controller of this sensor. Situations such as "breakdown of connection between sensor and controller" are not considered.

4   Simulation

Model has been simulated using different set of parameters while the position of sound sensors remained constant. Each sound source generated 15 events and had its own constant time interval between generations of sound events, as well as their intensity, set randomly for all sources within specified limits. The behavior of the system was simulated by manipulating the number of sound sources, as well as the intervals between generations of sound events and their intensity. It is assumed that the speed of sound is constant (340.29 m/s).

4.1   Interval of Generation [100, 150)

Figures 2 and 3 depict the behavior of the system with an increase in the number of sound sources that generate events of close intensity (within [100, 200) units of intensity) with close intervals of generation (within [100, 150) units of time).

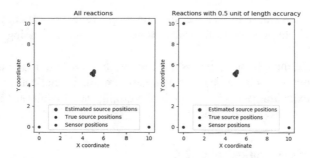

Fig. 2. One source

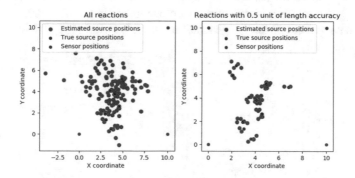

Fig. 3. Ten sources

The reaction to the event was considered correct if the calculated X and Y coordinates of the sound source differed from the true ones by no more than 0.5 units of length. The Table 1 illustrates the simulation results.

Table 1. Simulation results (generation interval is [100, 150))

Source count	Event count	Recognized events	Fraction of recognized events	Correctly recognized	Correctly recognized events fraction
1	15	15	1	15	1
2	30	30	1	29	0,97
3	45	42	0,93	39	0,87
4	60	57	0,95	36	0,6
5	75	66	0,88	30	0,4
10	150	133	0,89	63	0,42

Thus, the system is capable of operating correctly not less than 85% of the number of sources that generate sound events of close intensity not exceeding three.

A further increasing in the number of sources prevents the system to isolate their signals correctly, and consequently leads to the significant decreasing of recognition accuracy.

In any case, although the system is not capable to recognize individual sources of sound when their counts increase, the calculated coordinates indicate areas of sources accumulation, explaining a slight change in the quality of recognition with an increase in the number of sources from 5 to 10.

4.2   Interval of Generation [100, 200)

Figures 4 and 5 depict the behavior of the system when the number of sound sources is increased. Those sources generate events of close intensity (within [100, 200) of intensity units) with extended limits of generation intervals (within [50, 250) units of time).

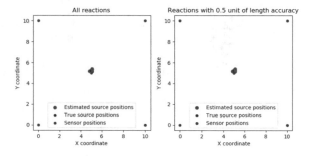

Fig. 4.   One source

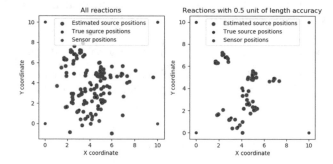

Fig. 5.   Ten sources

The reaction to the event was considered correct if the calculated X and Y coordinates of the sound source differed from the true ones by no more than 0.5 units of length. The simulation results are given in the Table 2.

Table 2. Simulation results (generation interval is [100, 200))

Source count	Event count	Recognized events	Fraction of recognized events	Correctly Recognized	Correctly recognized events fraction
1	15	15	1	15	1
2	30	30	1	29	0,97
3	45	42	0,93	39	0,87
4	60	55	0,92	49	0,82
5	75	70	0,93	54	0,72
10	150	138	0,92	71	0,47

Thus, the extension of time intervals between generations of sound events positively influenced the performance of the system with the number of sound sources greater or equal to four.

This is because signals from different sources began to come to sensors with a more noticeable delay, leading to a decrease in the number of errors such as "signals from several different sources are recognized as belonging to one wrongly."

4.3 Moving Object Tracing

Figures 6 and 7 depict the behavior of the system, when there is a single source of sound moving evenly rectilinearly, with the change in the frequency of signal generation. The results are collected in the Table 3.

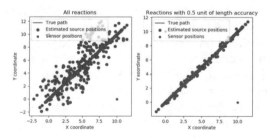

Fig. 6. Interval [10, 20] units of time

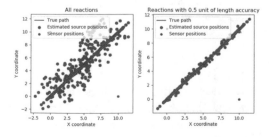

Fig. 7. Interval [1, 10] units of time

Table 3. Simulation results (generation interval is [100, 200))

Interval	Event count	Recognized events	Fraction of recognized events	Correctly recognized	Fraction of correctly recognized events
[100, 200)	30	30	1	30	1
[75, 150)	36	36	1	36	1
[50, 100)	86	83	0,97	83	0,97
[25, 50)	153	146	0,95	143	0,93
[10, 20)	285	280	0,98	248	0,87
[1, 10)	584	571	0,98	382	0,65

Thus, the presence of only one sound source completely excludes errors such as "signals from several different sources are identified as belonging to one wrongly."

On the other hand, it is still possible to have situation like "one sensor has already recorded several sound events, whereas the other is not recorded any". Therefore, having several possible combinations of reports from sensors, the system is not able to recognize the only one among them that relates to the actual event.

This explains the loss of recognition accuracy when the frequency of the sound events generation increases.

5   Discussion

The several useful results have been obtained in the study. Firstly, the TDOA algorithm is generally applicable to sensor networks in order to determine the coordinates of unknown objects. The reaction time of the system to the sound event depends on the quality of the network connection between the sensor controllers and the TDOA-groups. A satisfactory (not less than 85%) percentage of correctly recognized events possible in case of presence several near-placed sources of sound, when number of events is not more than three. A further increasing in the number of sound sources leads to the loss of the ability to isolate their signals correctly, but the ability to identify accumulation of sources accumulation retains. The applied configuration of the sensor network allows correctly recognizing sound events such as "explosion", "shot", "shock" etc. Recognition of sound sources in motion is possible, but imposes restrictions on both their speed (not higher than the sound speed in the air) and the frequency of signal generation.

6   Conclusion

In this paper, we studied the ways of effective implementation of the reconfigured sensor networks for the static and moving objects localization. The model based on the TDOA algorithm is proposed. The simulation showed that the sensor networks can be successfully applied for the different localization tasks with some restrictions.

The further work will focus on practical implementation of the proposed model and effective analysis of collected data.

References

1. Kaur, J., Kaur, K.: Internet of Things: a review on technologies, architecture, challenges, applications, future trends. Int. J. Comput. Netw. Inf. Secur. (IJCNIS) **9**(4), 57–70 (2017). https://doi.org/10.5815/ijcnis.2017.04.07
2. Lule, E., Bulega, T.E.: A Scalable Wireless Sensor Network (WSN) based architecture for fire disaster monitoring in the developing world. Int. J. Comput. Netw. Inf. Secur. (IJCNIS) **7**(2), 40–49 (2015). https://doi.org/10.5815/ijcnis.2015.02.05
3. Saxena, P., Pahuja, R., Khurana, M.S., Satija, S.: Real-time fuel quality monitoring system for smart vehicles. Int. J. Intell. Syst. Appl. (IJISA) **8**(11), 19–26 (2016). https://doi.org/10.5815/ijisa.2016.11.03
4. Accuracy Studies for TDOA and TOA Localization. http://fusion.isif.org/proceedings/fusion12CD/html/pdf/056_271.pdf. Accessed 01 Nov 2017
5. Xu, W., He, Z.-S.: A weighed least square TDOA location algorithm for TDMA multi-target. Int. J. Wirel. Microwave Technol. (IJWMT) **1**(2), 16–25 (2011). https://doi.org/10.5815/ijwmt.2011.02.03
6. TDOA Acoustic Localization. https://s3-us-west-1.amazonaws.com/stevenjl-bucket/tdoa_localization.pdf. Accessed 01 Nov 2017
7. Singh, J.: Tracking of moving object using centroid based prediction and boundary tracing scheme. Int. J. Image Graph. Signal Process. (IJIGSP) **9**(8), 59–66 (2017). https://doi.org/10.5815/ijigsp.2017.08.07

Research and Development of a Stereo Encoder of a FM-Transmitter Based on FPGA

Volokyta Artem[1], Shymkovych Volodymyr[2(✉)],
Volokyta Ivan[3], and Vasyliev Vladyslav[2]

[1] Department of Computer Engineering, National Technical University of Ukraine "Igor Sikorsky Kyiv Polytechnic Institute", 37 Prosp. Peremohy, Kyiv, Ukraine
[2] Department of Automation and Control in Technical Systems, National Technical University of Ukraine "Igor Sikorsky Kyiv Polytechnic Institute", 37 Prosp. Peremohy, Kyiv, Ukraine
shymkovych.volodymyr@gmail.com
[3] Royal Caribbean International, Florida, USA

Abstract. This paper describes the research and development of the digital stereo encoder FM transmitter including its implementation on Field Programmable Gate Array (FPGA). The development of structural schemes of the digital part of the FM transmitter exciter, the stereo encoder and the digital synthesizer of the sine wave on the FPGA is also described. Paper includes a description of development of a digital synthesizer of a sinusoidal signal on a FPGA, which uses a tabular method to generate the sinusoidal wave, fourth part of its period is written to memory and the rest of the period is calculated using the first part. To generate five sinusoids of different frequencies in a stereo encoder, only one table in memory used. The simulation of the stereocoder on the FPGA in carried out in the specialized Computer-Aided Design (CAD). The graphs of the left and right balance signals, the sum and difference of these signals, the signal of difference of the two sine waves which have their amplitude modulated and with a carrier signal of 19 kHz and the complex stereo signal are constructed. After receiving a complex stereo signal its frequency modulation is carried out with a carrier signal of 100 MHz, the FPGA chip outputs the frequency code of modulated signal.

Keywords: FPGA · Radio components · Stereo encoder
Digital signal processing

1 Introduction

At the moment a lot of forces are applied to replace all the analog components by digital parts because of their functional ability to be programed in different ways [1–8]. Same for radio systems, we see the transformation from analogue to digital.

Digital realisation has a lot of superior features compare to analogue. Major advantages are [9–14]:

– Better recurrence and stability (ageing is not a problem anymore);

© Springer International Publishing AG, part of Springer Nature 2019
Z. Hu et al. (Eds.): ICCSEEA 2018, AISC 754, pp. 92–101, 2019.
https://doi.org/10.1007/978-3-319-91008-6_10

- Possibility to effectuate functions which are not possible on an analog device (for example, FIR-filters);
- No need for fine hardware tuning due to firmware tuning;
- Development of power saving and multifunctional radio stations, which are capable to support signal modulation and different types of bandwidth
- Big potential in lowering the production cost and development time.

The goal is a development and research of a digital FM-transmitter based on FPGA, which allows increasing the speed of operation, precision and tuning possibilities.

The exciters are components of audio broadcast transmitters in the 100 MHz high frequency band using a frequency-modulated signal.

This work presents an FM-transmitter with RDS and stereo in digital execution. All the digital is executed on Field-Programmable Gate Array (FPGA).

2   Setting the Task

FM-stereo started to be common in 60's, at the time the circuit was compatible with monophonic FM-radio. The system is multiplexing two signals and then combines them into a complex signal, which is modulated by FM-frequency. Figure 1 shows us a block diagram of a typical analog stereo encoder, wich is used for FM-transmitter exciter [15–18].

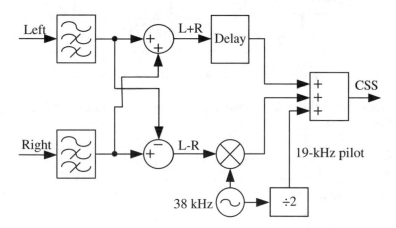

Fig. 1. Block diagram of a typical analog stereo encoder

Left and right balance channels are filtered from any disturbances by high frequency filter, and then are going into contour drive. Output signal of the first adder is a sum of two signal L + R (mono signal), but the output of the second one is their difference L − R. The output of an amplitude modulation L − R is modulated with the amplitude of 38 kHz.

Signal L + R (0–15 kHz) and signal L − R (23–53 kHz) are separated with a signal of 19 kHz. These signals are summarized to obtain the complex stereo signal (CSS).

L + R a signal that contains the frequencies of the baseband occupies a corresponding frequency range from 0 Hz up to 15 kHz. The tone signal of 19 kHz, obtained from the DSB SC signal with amplitude modulation in the tone of 38 kHz, occupies a frequency range from 23 to 53 kHz. At the output of the stereo encoder, a complex stereo signal is formed which is modulated in the FM signal with frequency modulation with a reference frequency of 100 MHz.

Digital exciter FM transmitter is proposed to consist of the following components: FPGA for digital signal processing and analogue high frequency cascade. The Fig. 2 shows the block diagram of the digital part of the FM transmitter exciter, where ADC – analog to digital converters of input sinewaves; FPGA – is a chip on which all digital signal processing is realized; DDS – is a micro-chem that generates a sinusoid at a given frequency.

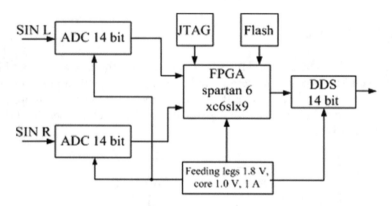

Fig. 2. Schematic structure of the digital part of the FM transmitter exciter

All digital signal processing, including frequency modulation, is performed digitally on a FPGA (Fig. 3).

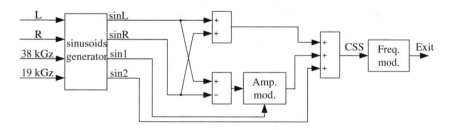

Fig. 3. The block diagram of a stereo encoder on a FPGA

3   Implementation of the Sinusoidal Generator

A direct digital synthesizer is a frequency synthesizer, in which the digital methods continuously generate the output samples (codes), which are then converted into an analog signal of a certain frequency. The functional block in which these readouts take place is called the accumulation phase (AP). Changes in AP are happening because of the clock frequency f_0 and frequency code.

The values of the signal amplitude corresponding to the current phase of the molded signal can be calculated in the synthesizer or be selected from the corresponding memorizer (ROM). The second method is better here, since the direct phase computation will reduce the speed of the circuit. The table from which the sine values are selected are most often found in ROM. The choice of ROM is carried out as follows, samples from the phase drive form the address for the ROM, which selects the cells with the desired phase, Figs. 4 and 5.

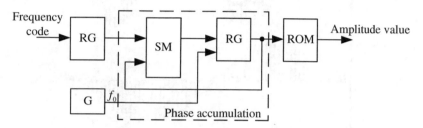

Fig. 4. The block diagram of a digital synthesizer of a sine wave on a FPGA

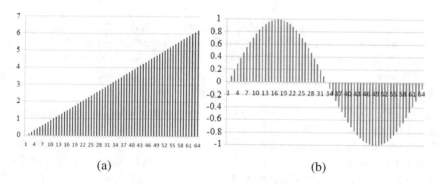

          (a)   (b)

Fig. 5. Phase accumulation (a) and converting the accumulated phase into the sinwave (b)

The main parameter of this block is the bit. This characteristic affects the parameters of the frequency synthesizer, namely: range of the input frequency is determinated by the clock frequency, same the quality of the high frequency. Phase storage works with periodic overflows which corresponds to the periodic sine wave changes. In another word the frequency of the storage overflowing is equal to the output frequency, and is determined by the formula:

$$f_{out} = \frac{K \cdot f_0}{2^n},$$

where f_{out} – output frequency, f_0 – clock frequency, K – frequency code, n – phase storage bit rate.

Here n = 5, and, consequently, whole phase circle can be divided into 32 segments, each of which corresponds to its discrete value (from 00000 to 11111). Since, depending on the frequency, we add to the content of the adder the number K, hence for the lowest frequency K will be minimal, and for the highest - at maximum. An example of two frequencies that are 2 times different is shown on Fig. 6.

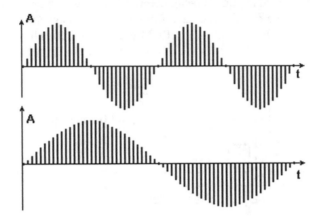

Fig. 6. Calculate the current phase for different frequencies

From this it follows that the number of points involved in the construction of the signal for a lower frequency will be greater than for a higher frequency. It follows from this that it is necessary to choose K so that at least a few points can be obtained at the highest frequency. This can be achieved by selecting a base oscillator with a clock frequency several times higher than the maximum.

K is added to the adder content $1/f_0c$, that means that with the simultaneous digit capacity augmentation of the phase storage and clock frequency, we get more samples and the same output frequency with constant K. For an example, you can calculate all the required parameters for a direct digital frequency synthesizer with such character-istics, like $f_{min} = 1$ kGz min. frequency, $f_{max} = 1$ MGz max. frequency $f_{step} = 1$ kGz step of frequency changes. As already mentioned above, take reference generator with a frequency several times higher than f_{max}. For calculations simplicity f_0 is equal to 16 MGz. From formulas:

$$n = \log_2 \frac{K \cdot f_0}{f_{min}} \quad \text{at } K = 1.$$

The number of synthesized frequencies is:

$$l = \frac{f_{max} - f_{min}}{f_{step}} + 1 = 1024.$$

So, K varies from 1 to 1024 depending on the frequency that we need to get. Every $1/f_0\,K$ is added to the adder content.

Frequency can be changed without changing the content of the adder and the register, which means that there is no phase break. And the transition to another frequency is carried out very quickly – for this it is enough to give the required number of K. Graphically this is presented on the Fig. 7.

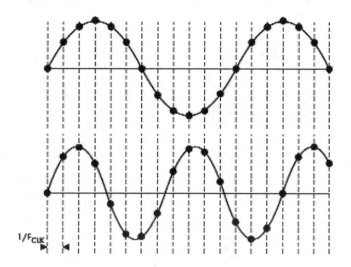

Fig. 7. Position of samples of the output signal for different frequencies

The values from the phase drive are fed to the address inputs of the ROM to select the numerical values of the sinus function.

As a block of phase calculation, as already mentioned before, we will use a permanent storage device (ROM). ROM size is determined by the number of sine function values in it. The higher the value, the clearer the signal will be received. However, in turn, this possibility is limited by the possibilities of the DAC.

You can also increase the number of values in the ROM by using the sinus function symmetry, namely half the phase or quarter of the phase. Consider the values obtained by splitting the phase sequence into 64 parts in Table 1.

As we see, the value of the first two columns (cells 1–16 and 17–32) almost coincide if you change the order of the number of entries in 1st or 2nd column. The values of the 3rd and 4th columns (cells 33–48 and 49–64) coincide with the values of the first two columns, only taken minus. Let's pretend that the ROM has a quarter of the function sin, the volume of ROM will be 256 b (2 m, m = 8). To select values, we need the ten senior disks of the phase drive (m + 2). The rest of the discharge in the synthesizer will not be taken into account, they are only needed to get more counterpart counters. As mentioned above, it is necessary to obtain a clear signal with high output frequency. Thus, it turns out that for low frequencies the selected memory cells are repeated, but in general, it does not affect the work, as the signal on the LPF is broken up. The first 2 higher digits will serve to calculate the required quarter of the sin phase. If the 2nd digit is 0, then the values for the ROM address are unchanged if it is equal to 1, then the values are inverted: thereby you can select a value from the ROM in reverse order. Similarly to the first digit: when it is equal to 0, we select the value for the first half of the phase, when it is equal to 1 - for the second. The values of the other 8 bits are the address of the ROM, which varies from 0 to 255.

Table 1. The value of the sinus function on the segment $[0;2\pi]$.

№	X	Sin(x)	№	x	Sin(x)	№	x	Sin(x)	№	x	Sin(x)
1	0,00000	0,00000	17	1,57080	1,00000	33	3,14159	0,00000	49	4,71239	−1,00000
2	0,09817	0,09802	18	1,66897	0,99518	34	3,23977	−0,09802	50	4,81056	−0,99518
3	0,19635	0,19509	19	1,76715	0,98079	35	3,33794	−0,19509	51	4,90874	−0,98079
4	0,29452	0,29028	20	1,86532	0,95694	36	3,43612	−0,29028	52	5,00691	−0,95694
5	0,39270	0,38268	21	1,96350	0,92388	37	3,53429	−0,38268	53	5,10509	−0,92388
6	0,49087	0,47140	22	2,06167	0,88192	38	3,63247	−0,47140	54	5,20326	−0,88192
7	0,58905	0,55557	23	2,15984	0,83147	39	3,73064	−0,55557	55	5,30144	−0,83147
8	0,68722	0,63439	24	2,25802	0,77301	40	3,82882	−0,63439	56	5,39961	−0,77301
9	0,78540	0,70711	25	2,35619	0,70711	41	3,92699	−0,70711	57	5,49779	−0,70711
10	0,88357	0,77301	26	2,45437	0,63439	42	4,02517	−0,77301	58	5,59596	−0,63439
11	0,98175	0,83147	27	2,55254	0,55557	43	4,12334	−0,83147	59	5,69414	−0,55557
12	1,07992	0,88192	28	2,65072	0,47140	44	4,22152	−0,88192	60	5,79231	−0,47140
13	1,17810	0,92388	29	2,74889	0,38268	45	4,31969	−0,92388	61	5,89049	−0,38268
14	1,27627	0,95694	30	2,84707	0,29028	46	4,41786	−0,95694	62	5,98866	−0,29028
15	1,37445	0,98079	31	2,94524	0,19509	47	4,51604	−0,98079	63	6,08684	−0,19509
16	1,47262	0,99518	32	3,04342	0,09802	48	4,61421	−0,99518	64	6,18501	−0,09802

However, we should take in consideration if it costs all the resources, because additional blocks can lower the speed of operation.

4 Experimental Results

To implement the stereo encoder need to create four sine waves, two sine waves are input signals, left and right balanced (Fig. 8), two more sine wave with a frequency of 38 kHz, which is used as a reference in the amplitude modulation of the difference in input signals, the second 19 kHz to obtain a complex stereo signal. For all the sinusoid, one block of ROM is created; in this unit a table with values of the fourth sinusoid with a bit of 16 bits is entered. At the input of this block, signals from phase batteries are excited, at the output the value is a sinusoid.

After the generator, the operations of adding and subtracting the left and right balance signals are performed. The sum of the two input sinus is shown in Fig. 9(a), the difference between the two signals is depicted in Fig. 9(b).

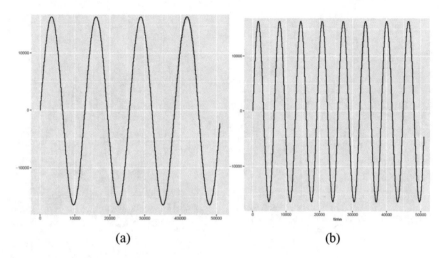

(a)   (b)

Fig. 8. Left (a) and right (b) balanced signals

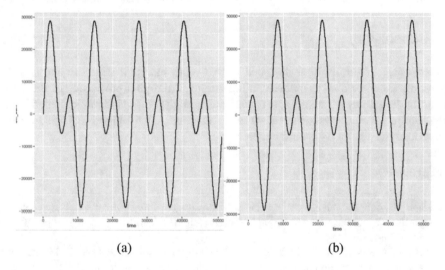

(a)   (b)

Fig. 9. Sum (a) and difference (b) of the sinewaves

The difference between two sine waves is modulated by amplitude modulation with a reference clock of 19 kHz, the received signal is shown in Fig. 10(a).

In order to obtain the complex stereo signal Fig. 10(b), we added three signals, namely sums of sine waves Fig. 9(b), modulated by amplitude of the sinusoidal difference Fig. 10(a), and the pilot signal, the sine wave with a frequency of 19 kHz.

After receiving a comprehensive stereo signal, its frequency modulation is performed. The output of the FPGA is the frequency code of the received sine wave in a sequential format for generating the sine wave using the chip Analog Devices AD9951.

Synthesis and simulation executed in software Xilinx ISE Design Suite 13.2 and ISE Simulator (ISim) using the chip family Spartan 6.

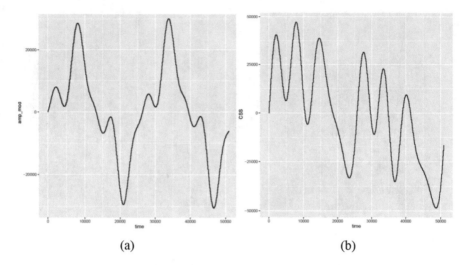

(a)   (b)

Fig. 10. Difference of the sine waves modulated by amplitude (a) and complex stereo signal (b)

5   Conclusions

Exciter of an FM-transmitter designed in this work has high accuracy and speed of operation, it can be used in modern radio systems. All digital processing of the signal, including frequency modulation, is carried out in digital form and implemented on one FPGA chip, which allows to increase its accuracy, simplify the process of its adjustment and reduce the cost of its manufacture.

References

1. Kabir, S., Ashraful Alam, A.S.M.: Hardware design and simulation of sobel edge detection algorithm. Int. J. Image Graph. Signal Process. (IJIGSP) **6**(5), 10–18 (2014). https://doi.org/10.5815/ijigsp.2014.05.02
2. Rashidi, B., Rashidi, B.: FPGA based a new low power and self-timed AES 128-bit encryption algorithm for encryption audio signal. Int. J. Comput. Netw. Inf. Secur. (IJCNIS) **5**(2), 10–20 (2013). https://doi.org/10.5815/ijcnis.2013.02.02
3. Singh, S., Saini, A.K., Saini, R.: Real-time FPGA based implementation of color image edge detection. Int. J. Image Graph. Signal Process. (IJIGSP) **4**(12), 19–25 (2012). https://doi.org/10.5815/ijigsp.2012.12.03
4. Singh, S., Saurav, S., Shekhar, C., Vohra, A.: Prototyping an automated video surveillance system using FPGAs. Int. J. Image Graph. Signal Process. (IJIGSP) **8**(8), 37–46 (2016). https://doi.org/10.5815/ijigsp.2016.08.06
5. Singh, S., Saini, R., Saurav, S., Saini, A.K.: Real-time object tracking with active PTZ camera using hardware acceleration approach. Int. J. Image Graph. Signal Process. (IJIGSP) **9**(2), 55–62 (2017). https://doi.org/10.5815/ijigsp.2017.02.07

6. Faridi, M.H., Jafari, A., Dehghani, E.: An efficient distributed power control in cognitive radio networks. Int. J. Inf. Technol. Comput. Sci. (IJITCS) **8**(1), 48–53 (2016). https://doi.org/10.5815/ijitcs.2016.01.06

7. Kravets, P.I., Shymkovych, V.M., Samotyy, V.: Method and technology of synthesis of neural network models of object control with their hardware implementation on FPGA. In: Proceedings of the 9th IEEE International Conference on Intelligent Data Acquisition and Advanced Computing Systems: Technology and Applications 21–23 September 2017, Bucharest, Romania, pp. 947–951 (2017)

8. Kravets, P.I., Volodymyr, S.: Neural network control system with direct and inverse model of control object hardware and software realization of in FPGA. In: Kulczycki, P., Kowalski, P.A., Lukasik, S. (eds.) Information Technology, Computational and Experimental Physics AGN-UST, Krakow, Poland, pp. 180–183 (2016)

9. Kravets, P.I., Shymkovych, V.M., Zubenko, G.A.: Technology of hardware and software implementation of artificial neurons and artificial neural networks by means of FPGA. Visnyk NTUU "KPI" Informatics, Operation and Computer Science, no. 55, pp. 174–180 (2012). (in Ukrainian)

10. Loutskii, H., Volokyta, A., Yakushev, O., Rehida, P., Vu, D.T.: Development of real time method of detecting attacks based on artificial intelligence. Technol. Audit Prod. Reserves **29**, 40–46 (2016). https://doi.org/10.15587/2312-8372.2016.71677

11. Volokyta, A., Rehida, P., Shyrochyn, V., Nikitiuk, A., Vu, D.T.: The effective method of distributed data storage that provides high access speed and level of safety. Bul. J. Eng. Des. **29**, 67–75 (2016)

12. Hu, Z., Mukhin, V., Kornaga, Y., Volokyta, A., Herasymenko, O.: The scheduler for distributed computer systems based on the network-centric approach to resources control. In: The 9th IEEE International Conference on Intelligent Data Acquisition and Advanced Computing Systems: Technology and Applications 21–23 September, 2017, Bucharest, Romania (2017)

13. Bonello, O.: Multiband audio processing and its influence on the coverage area of the FM stereo transmission. J. AES, (3), 145–156 (2007)

14. Li, J., Luo, Y., Tian, M.: FM stereo receiver based on software-defined radio. Int. J. Digital Content Technol. Appl. (JDCTA) **6**(1) (2012) https://doi.org/10.4156/jdcta.vol6.issue1.10

15. Minden, G.J., Evans, J.B., Searl, L., DePardo, D., Petty, V.R., Raijbanshi, R., Newman, T., Chen, Q., Weidling, F., Guffey, J., Datla, D., Barker, B., Peck, M., Cordill, B., Wyglinski, A.M., Agah, A.: KUAR: a flexible software-defined radio development platform. In: 2nd Dynamic Spectrum Access Networks Symposium, pp. 428–439. IEEE, April 2007

16. Abdullah, H.N., Valenzuela, A.: A joint Matlab/FPGA design of AM receiver for teaching purposes. In: EMNT2008 conference, Munich University of Technology, Germany (2008)

17. Braun, M., Pendlum, J.: A flexible data processing framework for heterogeneous processing environments: RF Network-on-Chip™. In: 2017 International Conference on FPGA Reconfiguration for General-Purpose Computing (FPGA4GPC), 9–10 May 2017, pp. 1–6 (2017)

18. Nivin, R., Sheeba Rani, J., Vidhya, P.: Design and hardware implementation of reconfigurable nano satellite communication system using FPGA based SDR for FM/FSK demodulation and BPSK modulation. In: 2016 International Conference on Communication Systems and Networks (ComNet), pp. 1–6 (2016)

Video Shots Retrieval with Use of Pivot Points

Dmytro Kinoshenko, Sergii Mashtalir$^{(\boxtimes)}$, Vladyslav Shlyakhov,
and Mykhailo Stolbovyi

Kharkiv National University of Radio Electronics,
Nauky Ave., 14, Kharkiv, Ukraine
{sergii.mashtalir,vladyslav.shliakhov}@nure.ua

Abstract. Intelligent analysis of video data is inextricably linked to methods aimed at reducing the amount of initial data necessary for processing in various ways. In this paper, we propose an approach that allows us to reduce the amount of processed video data by excluding it from consideration that is inappropriate for the query. This is achieved by the pivot points analysis of the original data clusters. If the pivot point to be compared is far from the query, then the entire cluster is also far from the query, respectively. Thus, it is possible to significantly reduce the number of operations of query comparison with data and, accordingly, speed up the process.

Keywords: Video clustering · Pivot points · Distance · Measure
Elimination region

1 Introduction

Intelligent video retrieval is closely associated with semantic analysis of large volumes of unstructured data specifically with queries 'ad exemplum' [1]. Therefore, one of highly developed strategies is such that unneeded information is sought out first and only then (on reduced set) required one next. Among these approaches, a metric search is now reaching some maturity since there arise possibilities to decrease online computational complexity due to taking into account distances calculated offline in image or video collections [1, 2]. Furthermore, a reasonable choice of particular temporal or spatial metrics provides growth of semantic component as linear combinations of arbitrary metrics is a metric as well. A basic unit of video retrieval is a shot and a set of shots corresponds to some event or scenario.

In general, computing of a distance between video shots (or key frames representing each shot) is a nontrivial process [3, 4], so there are approaches aimed at speed up of the whole retrieval process. These include avoiding of distance computations [5, 6], metric space transformations [7, 8] or approximate similarity search [9–11]. One of the most interesting methods are the approaches aimed at minimizing a number of distances computed under a search. In particular, this approach includes bounding strategies [5], based on the concept of pivots when under retrieval only distances from a query to pivot points are calculated. The pivot can be cluster centers, which is especially important when working with fuzzy sets [12, 13]. That is why conditions of an exclusion of shot sets from analysis (or any points in metrical spaces) are the goal of the paper.

© Springer International Publishing AG, part of Springer Nature 2019
Z. Hu et al. (Eds.): ICCSEEA 2018, AISC 754, pp. 102–111, 2019.
https://doi.org/10.1007/978-3-319-91008-6_11

Consider some finite relative (known only distances between all points) configuration $\mathbb{K} \subseteq \mathcal{F}$ in the metric space $\mathcal{F}$ of video shots. The task is to find the shot most similar to the query. Shots from the video collection will be denoted by points $x_1, \ldots, x_n \in \mathbb{K} \subseteq \mathcal{F}$, query by $y \in \mathcal{F}$, pivot points by $x_1^*, x_2^*, \ldots, x_m^* \in \mathbb{K}$. Analyzed configuration $\mathbb{K}' \subseteq \mathbb{K} \subseteq \mathcal{F}$ is represented by distance matrix $\mathcal{P} = (\rho(x_i, x_j))_{i,j=1}^n$, ($\rho(x_i, x_j)$ is a distance between points x_i, x_j). For the query $y \in \mathcal{F}$, the distance to a pivot point $x* \in \mathbb{K}$ is denoted by $\delta = \rho(y, x*)$. The threshold of objects similarity Δ characterizes the similarity measure in a metric space $\mathcal{F}$ with a current point y. If $\rho(y, x_k) > \Delta$, then the point x_k is so different from y that it is automatically excluded from further analysis.

Suppose, there is a configuration $\mathbb{K} = \{x_1, \ldots, x_n, x_1^*, x_2^*, \ldots, x_m^*\}$. In addition to the distances between n analyzed and m pivot points, the distances $\rho(y, x_i^*) = \delta_i$ $i = \overline{1, m}$ from the query y to each of the pivot points are known.

We call an elimination region such set of points that allows reducing cardinality of $\{x_1, \ldots, x_n\}$ through analysis of $x_1^*, x_2^*, \ldots, x_m^*$. Thus, it becomes possible to reduce the initial set of video under content-based information retrieval.

Attention is drawn on determining which of the two points from the configuration is closer to the query for a different number of pivot points.

2 Elimination Regions Construction by One Pivot Point

Let us start by considering the situation $n = 2$, $m = 1$, i.e. the configuration has the form $\mathbb{K} = \{x_1, x_2, x^*\}$ (Fig. 1), and a distance to the pivot point is known: $\rho(y, x^*) = \delta \geq 0$. We assume that there are reviewing points for which the threshold constraint is $\rho(y, x^*) < \Delta$. Indeed, if at least one of the configuration points $\{x_1, x_2, x^*\} \in \mathbb{K}$ does not fall into the neighborhood $U_\Delta(y)$ of the radius Δ of the query $y \in \mathcal{F}$, then two options are possible. Assume, $x^* \notin U_\Delta(y)$ means that a pivot point not satisfying the search conditions is chosen, and it is advisable to choose another. If one of the points x_1 or x_2 does not belong to a neighborhood $U_\Delta(y)$, for example x_1, then the nearest one is x_2. If both do not belong to $U_\Delta(y)$, then there is no sense in looking for the nearest one among them. However, the distances $\rho(y, x_1)$ and $\rho(y, x_2)$ are unknown, and if you do not assume that the points $x_1, x_2 \in U_\Delta(y)$, you need to take into account additional information, for example, about preliminary clustering in space $\mathcal{F}$.

The distance matrix for $\mathbb{K} = \{x_1, x_2, x^*\}$ takes the form

$$\mathcal{P} = \begin{pmatrix} 0 & \alpha_{12} & \gamma_{11} \\ & 0 & \gamma_{12} \\ & & 0 \end{pmatrix}. \tag{1}$$

The elimination region in R^3 is defined as follows. If known $\rho(y, x^*) = \delta \geq 0$ and $\Delta > 0$ the elements of the distance matrix $\mathcal{P}$ satisfy the conditions

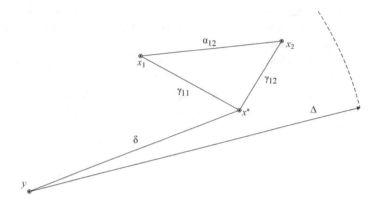

Fig. 1. Example of configuration $\mathbb{K} = \{x_1, x_2, x^*\}$

$$\begin{cases} \alpha_{12}, \gamma_{11}, \gamma_{12} > 0; \\ \left[\begin{array}{l} |\delta - \gamma_{11}| \geq \delta + \gamma_{12}, \\ |\delta - \gamma_{11}| \geq \delta + \gamma_{12}; \end{array}\right. \\ \gamma_{11} + \gamma_{12} \geq \alpha_{12}, \\ \gamma_{11} + \alpha_{12} \geq \gamma_{12}, \\ \alpha_{12} + \gamma_{12} \geq \gamma_{11}; \\ \delta \leq \Delta, \\ \delta + \gamma_{11} \leq \Delta, \\ \delta + \gamma_{12} \leq \Delta, \end{cases} \tag{2}$$

so without calculating distances $\rho(y, x_1)$ and $\rho(y, x_2)$ you can make the choice of point x_1 or x_2 closest to the query y from configuration $\mathbb{K} = \{x_1, x_2, x^*\}$.

Let's consider elements (2) in more detail. Restrictions $\alpha_{12}, \gamma_{11}, \gamma_{12} \geq 0$ are natural because the metric is nonnegative. From the triangle inequality, taking into account the query y, we have

$$\begin{cases} |\delta - \gamma_{11}| \leq \rho(y, x_1) \leq \delta + \gamma_{11}, \\ |\delta - \gamma_{12}| \leq \rho(y, x_2) \leq \delta + \gamma_{12}, \end{cases} \tag{3}$$

and taking into account $\Delta(x_1 x_2 x^*)$ (here and in what follows we shall denote triangles with corresponding vertices) $\gamma_{11} + \gamma_{12} \geq \alpha_{12}, \ \gamma_{11} + \alpha_{12} \geq \gamma_{12}, \ \alpha_{12} + \gamma_{12} \geq \gamma_{11}$.

It is easy to understand that in any case the elimination region will be that region in which at least one of the lower bounds of inequalities (3) exceeds the upper bound of the same inequalities.

On the one hand, this means $|\delta - \gamma_{11}| \geq \delta + \gamma_{12}$ or, equivalently, $\rho(y, x_1) \geq \rho(y, x_2)$. From another hand, $|\delta - \gamma_{12}| \geq \delta + \gamma_{11}$, from which it follows that the point x_1 is closer to y, than x_2. Finally, the last three inequalities of system (2) ensure that the configuration $\mathbb{K} \in U_\Delta(y)$ belongs to an allowable level of similarity.

As a result, in order for the elimination region to be non-empty, it is necessary to fulfill a fairly obvious inequality $\Delta - \delta > 2\delta$, which means that the parameters of the system (3) δ and Δ must satisfy a more strong condition $\delta < \Delta/3$ than the requirement $\delta \leq \Delta$.

3   Search Procedures Analysis at Two Pivot Points

In cases where preliminary clustering of space $\mathcal{F}$ is carried out, it is often advisable to use several pivot points, for example, cluster centers.

We first consider a configuration $\mathbb{K} = \{x_1, x_2, x_1^*, x_2^*\}$ consisting of two compared points x_1, x_2 and two pivot points x_1^*, x_2^*, as shown on Fig. 2.

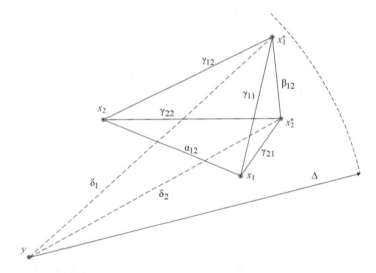

Fig. 2. Example of configuration $\mathbb{K} = \{x_1, x_2, x_1^*, x_2^*\}$

The distance matrix for $\mathbb{K} = \{x_1, x_2, x_1^*, x_2^*\}$ takes the form

$$\mathcal{P} = \begin{pmatrix} 0 & \alpha_{12} & \gamma_{11} & \gamma_{12} \\ & 0 & \gamma_{21} & \gamma_{22} \\ & & 0 & \beta_{12} \\ & & & 0 \end{pmatrix} \tag{4}$$

which means: the elimination region should be found in the six-dimensional arithmetic space.

The first type of constraint is related to the non-negativity of distances. The second type of constraint is based on the triangle inequality and, finally, the third one is an external condition concerning the similarity or difference of the query y with the compared and pivot points. The most important is the second type of constraints, from which the conditions determining the elimination region are follows.

It should be said that the constraints connected with the triangle inequality systems can be divided into two groups, namely

$$\begin{cases} |\delta_1 - \gamma_{11}| \leq \rho(y, x_1) \leq \delta_1 + \gamma_{11}, \\ |\delta_1 - \gamma_{12}| \leq \rho(y, x_1) \leq \delta_2 + \gamma_{12}, \\ |\delta_1 - \gamma_{21}| \leq \rho(y, x_2) \leq \delta_1 + \gamma_{21}, \\ |\delta_1 - \gamma_{22}| \leq \rho(y, x_2) \leq \delta_2 + \gamma_{22}, \end{cases} \tag{5}$$

that is, on a system where unknown distances $\rho(y, x_1)$ and $\rho(y, x_2)$, and a set of systems relating to the distance matrix (4) elements are directly compared. For example, for a triangle $\Delta(x_1, x_2, x_1^*)$ (Fig. 2) we have

$$\begin{cases} \gamma_{11} + \gamma_{21} \geq \alpha_{12}, \\ \gamma_{11} + \alpha_{12} \geq \gamma_{21}, \\ \gamma_{21} + \alpha_{12} \geq \gamma_{11}. \end{cases} \tag{6}$$

Figure 3 shows a number of triangles, for which the following constraints apply:

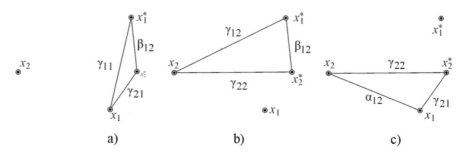

Fig. 3. Example of triangles for configuration $\mathbb{K} = \{x_1, x_2, x_1^*, x_2^*\}$

$\Delta(x_1, x_1^*, x_2^*)$ is Fig. 3(a)

$$\begin{cases} \gamma_{11} + \gamma_{12} \geq \beta_{12}, \\ \gamma_{11} + \beta_{12} \geq \gamma_{12}, \\ \gamma_{12} + \beta_{12} \geq \gamma_{11}, \end{cases} \tag{7}$$

$\Delta(x_2, x_1^*, x_2^*)$ is Fig. 3(b)

$$\begin{cases} \gamma_{21} + \gamma_{22} \geq \beta_{12}, \\ \gamma_{21} + \beta_{12} \geq \gamma_{22}, \\ \gamma_{22} + \beta_{12} \geq \gamma_{21}, \end{cases} \tag{8}$$

$\Delta(x_1, x_2, x_2^*)$ is Fig. 3(c)

$$\begin{cases} \gamma_{12} + \gamma_{22} \geq \alpha_{12}, \\ \gamma_{12} + \alpha_{12} \geq \gamma_{22}, \\ \gamma_{22} + \alpha_{12} \geq \gamma_{12}. \end{cases} \tag{9}$$

In all, we have $C_5^3 = 10$ triangles, i.e. there are still two triangles containing a point y, namely: $\Delta(y, x_1^*, x_2^*)$ and the constraints associated with it

$$\begin{cases} \delta_1 + \delta_2 \geq \beta_{12}, \\ \delta_1 + \beta_{12} \geq \delta_2, \\ \delta_2 + \beta_{12} \geq \delta_1. \end{cases}$$

but they contain one parameter β_{12}, and only the constraint on it is significant

$$|\delta_1 - \delta_2| \leq \beta_{12} \leq \delta_1 + \delta_2. \tag{10}$$

As for the triangle $\Delta(y, x_1, x_2)$, it does not put any constraints on the parameter α_{12}, since the distances $\rho(y, x_1)$ are $\rho(y, x_2)$ unknown.

Now let us return back to the inequality systems (5), which are fundamental in the construction of the elimination region. It is clear that to find the nearest point among the points x_1 and x_2 to the query y it is sufficient that the following inequality system is fulfilled.

$$\begin{bmatrix} |\delta_1 - \gamma_{11}| \geq \delta_1 + \gamma_{21}, \\ |\delta_1 - \gamma_{11}| \geq \delta_2 + \gamma_{22}, \\ |\delta_1 - \gamma_{12}| \geq \delta_1 + \gamma_{21}, \\ |\delta_1 - \gamma_{12}| \geq \delta_2 + \gamma_{22} \end{bmatrix} \tag{11}$$

In this case $\rho(y, x_1) \geq \rho(y, x_2)$ and correspondingly $\rho(y, x_1) \leq \rho(y, x_2)$, when

$$\begin{bmatrix} |\delta_1 - \gamma_{21}| \geq \delta_1 + \gamma_{11}, \\ |\delta_1 - \gamma_{21}| \geq \delta_2 + \gamma_{12}, \\ |\delta_1 - \gamma_{22}| \geq \delta_1 + \gamma_{11}, \\ |\delta_1 - \gamma_{22}| \geq \delta_2 + \gamma_{12}. \end{bmatrix} \tag{12}$$

Indeed, there are two constraints from below and two constraints from above $\rho(y, x_1)$, $\rho(y, x_2)$ and (11) mean that at least one of the constraints below the distance $\rho(y, x_1)$ exceeds the upper bound of the distance $\rho(y, x_2)$. Hence it follows $\rho(y, x_1) \geq \rho(y, x_2)$. Accordingly, the fulfillment of conditions (12) is sufficient for $\rho(y, x_1) \leq \rho(y, x_2)$.

Thus, it is established that if the elements of the distance matrix $\mathcal{P}$ of the form (4) satisfy the conditions (6)–(12), they lie in the "first quadrant" (all are positive) and in the Δ-neighborhood of the query y, i.e. satisfy the constraints

$$\begin{cases} \begin{bmatrix} \delta_1 + \gamma_{11} < \Delta, \\ \delta_2 + \gamma_{12} < \Delta, \end{bmatrix} \\ \begin{bmatrix} \delta_1 + \gamma_{21} < \Delta, \\ \delta_2 + \gamma_{22} < \Delta \end{bmatrix} \end{cases} \tag{13}$$

where if $0 < \delta_1, \delta_2 < \Delta$, then this is sufficient to select the nearest configuration $\mathbb{K} = \{x_1, x_2, x_1^*, x_2^*\}$ point to the query y. Thus, the construction of the elimination region in the case of two pivot points is completed.

4   Investigation of First-Order Elimination Regions for Configurations with M Pivot Points

After analyzing of the elimination regions for one and two pivot points, a natural extension is the analysis of the general case: the construction of elimination regions for the configuration $\mathbb{K} = \{x_1, x_2, x_1^*, x_2^*, \ldots, x_m^*\}$ ($m > 2$).

In the received notation for the configuration shown in Fig. 4, the distance matrix takes the form

$$\mathcal{P} = \begin{pmatrix} 0 & \alpha_{12} & \gamma_{11} & \cdots & \cdots & \gamma_{1m} \\ & 0 & \gamma_{21} & \cdots & \cdots & \gamma_{2m} \\ & & 0 & \beta_{12} & \cdots & \beta_{1m} \\ & & & 0 & \cdots & \cdots \\ & & & & \ddots & \beta_{m-1m} \\ & & & & & 0 \end{pmatrix}. \tag{14}$$

Just like before, the main part of the constraints that ensure the construction of the elimination region is associated with unknown distances $\rho(y, x_1)$ and $\rho(y, x_2)$. If there were 2 constraints in the case of one pivot point, in the case of two, there are 4 constraints, then at m pivot points they will be $2m$ constraints, and they will have the form

$$\begin{cases} |\delta_1 - \gamma_{11}| \leq \rho(y, x_1) \leq \delta_1 + \gamma_{11}, \\ |\delta_1 - \gamma_{21}| \leq \rho(y, x_2) \leq \delta_1 + \gamma_{21}, \\ \cdots\cdots\cdots\cdots\cdots\cdots\cdots\cdots\cdots \\ |\delta_i - \gamma_{1i}| \leq \rho(y, x_1) \leq \delta_i + \gamma_{1i}, \\ |\delta_i - \gamma_{2i}| \leq \rho(y, x_2) \leq \delta_i + \gamma_{2i}, \\ \cdots\cdots\cdots\cdots\cdots\cdots\cdots\cdots\cdots \\ |\delta_m - \gamma_{1m}| \leq \rho(y, x_1) \leq \delta_m + \gamma_{1m}, \\ |\delta_m - \gamma_{2m}| \leq \rho(y, x_2) \leq \delta_m + \gamma_{2m}. \end{cases}$$

If we group the constraints by points x_1 and x_2, namely

$$\{|\delta_i - \gamma_{1i}| \leq \rho(y, x_1) \leq \delta_i + \gamma_{1i}, \ i = \overline{1, m},$$

$$\{|\delta_i - \gamma_{2i}| \leq \rho(y, x_1) \leq \delta_i + \gamma_{2i}, \ i = \overline{1, m},$$

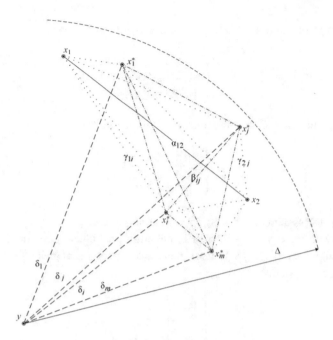

Fig. 4. Example of configuration $\mathbb{K} = \{x_1, x_2, x_1^*, x_2^*, \ldots, x_m^*\}$

then we can choose the nearest to the query y point if at least one of the lower bounds of the first inequality system exceeds at least one of the upper limits of the second inequality system and vice versa. Thus, the elimination region is a union of $2\,m^2$ inequalities having the form

$$\begin{bmatrix} \delta_i + \gamma_{1i} \leq |\delta_j - \gamma_{2j}|, \\ \delta_j + \gamma_{2j} \leq |\delta_i - \gamma_{1i}|, \end{bmatrix} \quad i,j = \overline{1,m}. \tag{15}$$

We now can take into account other restrictions related to those triangles that describe one or two pivot points and one or two of the compared points, that is, a set of pivot points $\{x_1^*, \ldots, x_m^*\}$ and a set of compared points $\{x_1, x_2\}$ are given.

From the triangle inequality, the first type of triangles $\Delta(x_i^*, x_1, x_2)$ corresponds to m constraints of the form

$$\begin{cases} \gamma_{1i} + \gamma_{2i} \geq \alpha_{12}, \\ \gamma_{1i} + \alpha_{12} \geq \gamma_{2i}, \\ \alpha_{12} + \gamma_{2i} \geq \gamma_{1i}, \end{cases} \tag{16}$$

the second type of triangles $\Delta(x_i^*, x_j^*, x_k)$ corresponds to C_m^2 constraints of the form

$$\begin{cases} \gamma_{1i} + \gamma_{1j} \geq \beta_{ij}, \\ \gamma_{1i} + \beta_{ij} \geq \gamma_{1j}, \\ \beta_{ij} + \gamma_{1j} \geq \gamma_{1i} \end{cases} \qquad (17)$$

when $k = 1$, and if $k = 2$ the constraints will look like this

$$\begin{cases} \gamma_{1i} + \gamma_{2j} \geq \beta_{ij}, \\ \gamma_{1i} + \beta_{ij} \geq \gamma_{2j}, \\ \beta_{ij} + \gamma_{2j} \geq \gamma_{1i}. \end{cases} \qquad (18)$$

Note that for systems (17), (18) $i, j = \overline{1, m}$ and $i \neq j$.

Starting with three pivot points, there are additional constraints that are associated with the triangles $\Delta(x_i^*, x_j^*, x_l^*)$. In total, there are C_m^3 such triangles and it corresponds to the inequality systems

$$\begin{cases} \beta_{ij} + \beta_{il} \geq \beta_{jl}, \\ \beta_{ij} + \beta_{jl} \geq \beta_{il}, \\ \beta_{jl} + \beta_{il} \geq \beta_{ij}. \end{cases} \qquad (19)$$

Finally, there remains the last type of constraint associated with the triangle inequality: one vertex of the triangle is a query y, and the other two are pivot points, x_i, x_j. The triangles of the form $\Delta(y, x_i^*, x_j^*)$ correspond to the system

$$\{|\delta_i - \delta_j| \leq \beta_{ij} \leq \delta_i + \delta_j. \qquad (20)$$

Summarizing, we can conclude that in the case of m pivot points, if elements of the distance matrix $\mathcal{P}$ (14) satisfy the systems of conditions (15)–(20), are strictly positive and satisfy the system of inequalities

$$\begin{cases} [\delta_i + \gamma_{1i} < \Delta, \\ [\delta_i + \gamma_{2i} < \Delta, \end{cases}$$

ensuring the hit of all configuration points $\mathbb{K} = \{x_1, x_2, x_1^*, \ldots, x_m^*\}$ in Δ-neighborhood of the query y, this is sufficient to select the nearest point without calculating distances.

5   Conclusion

It should be emphasized that the result provides simple iterative directional search procedures by pairwise comparison, which significantly improves performance in the conditions of preliminary clustering of multiple video segments.

The construction of elimination regions allows you to adhere to different search strategies, for example, to immediately find the nearest, analyzing the region, to iteratively reduce the dimension of the configuration, which is especially important for a stratified search of clustered data.

References

1. Liu, C. (ed.): Recent Advances in Intelligent Image Search and Video Retrieval. Intelligent Systems Reference Library, vol. 121, 235 p. Springer, Cham (2017)
2. Zezula, P., Amato, G., Dohnal, V., Batko, M.: Similarity Search: The Metric Space Approach. Springer, New York (2006)
3. Mashtalir, S., Mikhnova, O.: Detecting significant changes in image sequences. In: Hassanien, A., Mostafa, Fouad M., Manaf, A., Zamani, M., Ahmad, R., Kacprzyk, J. (eds.) Multimedia Forensics and Security, pp. 161–191. Springer, Cham (2017)
4. Hu, Z., Mashtalir, S.V., Tyshchenko, O.K., Stolbovyi, M.I.: Video shots' matching via various length of multidimensional time sequences. Int. J. Intell. Syst. Appl. (IJISA) 9(11), 10–16 (2017). https://doi.org/10.5815/ijisa.2017.11.02
5. Hjaltason, G.R., Samet, H.: Index-driven similarity search in metric spaces. ACM Trans. Database Syst. (TODS) 28(4), 517–580 (2003)
6. Dohnal, V., Gennaro, C., Zezula, P.: A metric index for approximate text management. In Proceedings of the IASTED International Conference Information Systems and Databases (ISDB 2002), Tokyo, Japan, 25–27 September, pp. 37–42 (2002)
7. Ciaccia, P., Patella, M.: Searching in metric spaces with user-defined and approximate distances. ACM Trans. Database Syst. (TODS) 27(4), 398–437 (2002)
8. Hafner, J.L., Sawhney, H.S., Equitz, W., Flickner, M., Niblack, W.: Efficient color histogram indexing for quadratic form distance functions. IEEE Trans. Pattern Anal. Mach. Intell. (TPAMI) 17(7), 729–736 (1995)
9. Seidl, T., Kriegel, H.-R.: Efficient user-adaptable similarity search in large multimedia databases. In: Jarke, M., Carey, M.J., Dittrich, K.R., Lochovsky, F.H., Loucopoulos, P., Jeusfeld, M.A. (eds.) Proceedings of the 23rd International Conference on Very Large Data Bases (VLDB 1997), Athens, Greece, August 25–29, pp. 506–515 (1997)
10. Wang, X., Wang, J.T.-L., Lin, K.-L., Shasha, D., Shapiro, B.A., Zhang, K.: An index structure for data mining and clustering. Knowl. Inf. Syst. 2, 161–184 (2000)
11. Ferhatosmanoglu, H., Tuncel, E., Agrawal, D., Abbadi, A.E.: Approximate nearest neighbor searching in multimedia databases. In: Proceedings of the 17th International Conference on Data Engineering (ICDE 2001), Heidelberg, Germany, 2–6 April, pp. 503–511 (2001)
12. Hu, Z., Bodyanskiy, Y.V., Tyshchenko, O.K., Samitova, V.O.: Fuzzy clustering data given in the ordinal scale. Int. J. Intell. Syst. Appl. (IJISA) 9(1), 67–74 (2017). https://doi.org/10.5815/ijisa.2017.01.07
13. Hu, Z., Bodyanskiy, Y.V., Tyshchenko, O.K., Tkachov, V.M.: Fuzzy clustering data arrays with omitted observations. Int. J. Intell. Syst. Appl. (IJISA) 9(6), 24–32 (2017). https://doi.org/10.5815/ijisa.2017.06.03

Imbalance Data Classification via Neural-Like Structures of Geometric Transformations Model: Local and Global Approaches

Roman Tkachenko[1], Anastasiya Doroshenko[1], Ivan Izonin[1(✉)],
Yurii Tsymbal[1], and Bohdana Havrysh[2]

[1] Lviv Polytechnic National University, Lviv, Ukraine
roman.tkachenko@gmail.com, anastasia.doroshenko@gmail.com,
ivanizonin@gmail.com, yurij.tsymbal@gmail.com
[2] Ukrainian Graphic Arts Academy, Lviv, Ukraine
dana.havrysh@gmail.com

Abstract. The classification task is one of the most widespread among the tasks of Data Mining - spam detection, medical diagnosis, ad targeting, risk assessment and image classification. However, all these tasks have a common feature - training dataset can be unbalanced, the number of instances of the target class can be less than one percent of all data. In this article, we compare the results of solving one of these problems using the most common classification methods (Random Forest Leaner, Logistic Regression, SVM). The article describes a new classification method based on neural-like structures of Geometric Transformations Model (local and global approaches) and compares their result with the obtained results.

Keywords: Classification · Imbalance data · Neural-like structures
Data unevenness · Geometric transformations model · Machine learning

1 Introduction

Classification is one of the most widely used techniques in machine learning, with a broad array of applications in different areas: medicine, astronomy, geology, biology, health care management, sentiment analysis, ad targeting, tax fraud detection, spam detection, risk assessment and image classification [1, 2].

However, if a linearly separable task is solved without the special difficulties with most of the known classification methods (statistical, neural networks, fuzzy logic, and heuristic algorithms), the solution of a linearly inseparable classification problem raises a number of difficulties [1–6]. It should also be noted that most data sets that describe tasks from real life are characterized by problems such as heterogeneity of data, missing data, unbalanced data sets. All these features impose additional restrictions on the process of selecting methods for solving the classification problem. The purpose of this study is to develop a new method for imbalance data classification on the basis of neural-like structure of geometric transformation model.

© Springer International Publishing AG, part of Springer Nature 2019
Z. Hu et al. (Eds.): ICCSEEA 2018, AISC 754, pp. 112–122, 2019.
https://doi.org/10.1007/978-3-319-91008-6_12

2   Problem Statement

The basis for many classification methods is the compact hypothesis [6], which assumes that objects belonging to certain classes form clusters of points-clusters in the space of signs and, therefore, can easily be separated by hypersurfaces or even hyperplanes.

However, in many cases, the functioning of the systems of data mining is carried out under conditions of uncertainty arising from the mutual overlap of classes and the blurring of the boundaries between them [7, 8]. This can be explained by the incompleteness of the information basis, the contradiction, and the unevenness of data in space, the representation of one class by several clusters [9–11].

Accordingly, classical classification methods based on the hypothesis of compactness do not provide the desired quality of the results of the problem of intellectual data analysis, which induces the creation of alternative methods.

A particularly difficult is the classifying task with linearly inseparable classes if these classes are unbalanced - the number of instances of one class (target) is many times smaller than the number of instances of other classes. The example of one of these tasks we can see on Fig. 1.

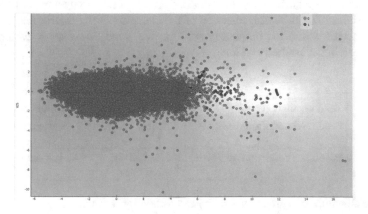

Fig. 1. The visualization of dataset [3] with significant unbalanced data in the space of two principal components.

Also, the significant problem is measuring the accuracy of classification, given the class imbalance ratio. We are recommended to combine using traditional confusion matrix with the ROC-analysis and the Area Under the Precision-Recall Curve (AUPRC) [12].

2.1   Formulation of the Classification Task

The task of classification, which is solved in the article, was formulated in [3]. The training sample consists of 284,807 lines and 31 columns and describes the transactions carried out by credit card holders within two days. For reasons of confidentiality, the dataset contains not original user data, but principal components obtained by the PCA method from the initial data (V1,..., V28). Only two features: 'Time' (the number of seconds passing through each transaction) and 'Amount' have not been transformed. Also, the dataset contains one target feature 'Class', which shows the client's affiliation to one of two classes - frauds or ordinary clients. The main feature of the dataset is that the data set is highly unbalanced - only 492 transactions out of 284807 (0.172% of all transactions) have the value of the target field 1, that is, customers are fraudulent.

The dataset has been collected and analyzed during a research collaboration of Worldline and the Machine Learning Group (http://mlg.ulb.ac.be) of ULB (Université Libre de Bruxelles) on big data mining and fraud detection.

2.2 Method of Evaluation of the Quality of Classification Results

For many real-world tasks of classification, estimating their quality using standard accuracy or error indices is not enough indicative, since, for each specific task of classification, different types of errors have different degrees of importance [13]. Therefore, this article describes the approach which combines using of confusing matrix and ROC-analysis for measuring the quality of classifiers [12, 14].

The Confusion Matrix gives the number or proportion of instances between the predicted and actual class. The selection of the elements in the matrix feeds the corresponding instances into the output signal. This way, one can observe which specific instances were misclassified and how.

		True condition	
	Total population	Condition positive	Condition negative
Predicted condition	**Predicted condition positive**	True positive	False positive, Type I error
	Predicted condition negative	False negative, Type II error	True negative

The results obtained are estimated by the following indicators:

Prevalence $= \Sigma$ Condition positive/Σ Total population
Accuracy (ACC) $= \Sigma$ True positive $+ \Sigma$ True negative/Σ Total population
Positive predictive value (PPV), Precision $= \Sigma$ True positive/Σ Predicted condition positive
False discovery rate (FDR) $= \Sigma$ False positive/Σ Predicted condition positive positive
False omission rate (FOR) $= \Sigma$ False negative/Σ Predicted condition negative
Negative predictive value (NPV) $= \Sigma$ True negative/Σ Predicted condition negative

3   Experimental Analysis of Existing Methods for Solving the Problem of Classification by Means of Computing Intelligence

The technologies of the use of neural networks to solve the problem, which are offered in commercial software for data mining, presuppose the pre-selection of the structure of the neural network in accordance with the tasks and data structure and further deter-mination of the links between the neurons-processors, which is carried out in the network learning process. Non-iterative, rapid training of neural networks is realized in auto-associative models such as Hopfield networks, the application of which, unfortunately, is rather narrow and is usually limited to tasks of constructing auto-associative memory, or optimization. The training of multilayer perceptrons, capable of solving a wide range of tasks for prediction, recognition, classification and prediction of time sequences, is carried out on the basis of multi-parameter optimization procedures, which greatly limits the possibilities of their application for tasks of large dimensions [19–21].

Experiments from Sect. 3 were performed using the Orange software, which contains widgets written in the programming language python for most popular Data Mining methods.

3.1 Logistic Regression

One of the simple but effective classifier variants is logistic regression, which is similar to perceptron based on the principle of operation - it solves linearly partitioned tasks, since the separating surface in such a system is hyperplane. The advantages of this method are better results in comparison with a linear Fisher discriminant, as well as in comparison with the delta rule and the rule of Hebb possibility to assess a posteriori probabilities and risks. Disadvantages of logistic regression are: gradient method of learning logistic regression inherits all the shortcomings, the stochastic gradient method, a practical implementation should to standardize data, screen out emissions, and regu-larize scales, selection of attributes, and other heuristics to improve convergence. It is possible to use a second-order method, but it requires the inversion of n × n-matrix at each step and can have poor convergence.

The results indicate that although the overall accuracy of the classification is very high and is equal to 99.91%, the accuracy of the recognition of the elements of the lower (target) class is average and equals 61.78% (Fig. 2a). Consequently, for data mining tasks, the goal of which is to recognize with a maximum accuracy the representatives of the lower class, applying of logistic regression requires further improvements.

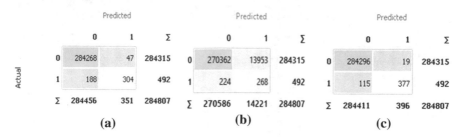

Fig. 2. Result of classification of: a) logistic regressions; b) SVM; c) Random Forest.

3.2 Support Vector Machine Classification

The support vectors machine is a linear system on which, based on the theory of statistical training, one can construct a hyperplane as a surface of solutions in order to maximally divide classes [6, 15]. Consider using a reference vector machine to solve the data classification problem, which is described in Sect. 2.1. An analysis of the results suggests that in comparison with the logistic regression SVM (Fig. 2b) more accurately classifies the representatives of both classes, but the accuracy of this classification is too low.

3.3 Random Forest Classification

Now Random Forest is one of the most popular methods of classification and nonparametric regression. The reason for this was not only high accuracy of classification provided method but also a lot of other advantages. First of all, the method guarantees overfitting protection even when the number of features significantly exceeds the number of observations. Also to create a random forest for a training sample, a task is required only two parameters that require a minimum setting Method Out-Of-Bag proposed Brayman [5] provides obtaining a natural estimate of the probability of erroneous classification of random forests based on observations not included in training bootstrap sampling used for tree construction. A training sample for constructing a random forest may contain signs, measured on different scales: numeric, order and nominal, which is unacceptable for many other classifiers.

The analysis of the classification results using Random Forest (Fig. 2c) has shown that the accuracy of the Random Forest network and the logistic regressions network for the data mining intelligence is approximately the same, and although the uniformity

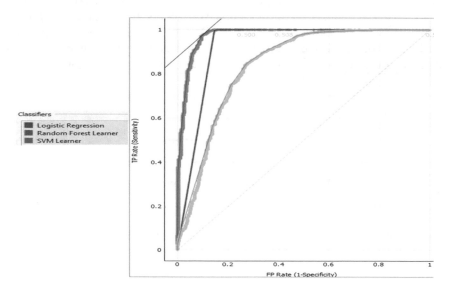

Fig. 3. ROC-analysis of the results of the applied methods in Orange.

of classifying instances of each of the classes is higher than when applied SVM, it is not enough to use such networks to solve data mining tasks (Fig. 3).

The Table 1 gives results of experiments for various methods when performing cross-validation with the following parameters: number of folds – 10; repeat train/test – 10; training set size – 66% (Table 1).

Table 1. A summary of the results of the applied methods

Evaluation Results					
Method	AUC	CA	F1	Precision	Recall
Logistic regression	0.976	0.999	0.723	0.826	0.616
Random forest learner	0.926	1.000	0.852	0.948	0.774
SVM learner	0.855	0.950	0.036	0.019	0.545

4 Data Classification Based on Neural-like Structure of Geometric Transformations Model

The geometric transformations model (GTM) is a non-iterative neural network of direct propagation [16]. The basis of the paradigm of GTM is a set of spatial-geometric representations [17, 18]. In a particular case, the architecture of the GTM coincides with the architecture of direct propagation of neural networks with braking lateral connections. Details of the advantages and disadvantages, learning algorithms and the use of neural-like structures of GTM can be found here [16].

4.1 Global Approach

The global approach to classification is realized by constructing, using the GTM, a continuous hypersurface of the response formed by the results of training of a neural-like structure on the training data. The topology of such a neural-like structures of geometric transformations model (NLS GTM) are presented in Fig. 4)

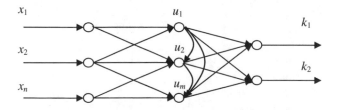

Fig. 4. Neural-like structure of GTM topology

It consists of input neurons, which are presented with the primary signs of classification objects $x_1, x_2, \ldots, x_n$, a hidden layer with lateral connections between neurons, signals on the outputs of the hidden layer $u_1, u_2, \ldots, u_n$ and output signals that specify the belonging to the specified classes - k_1, k_2. The operation of such a network can be described by the formulas:

$$k = \sum_{m=1}^{n} f(KS_j^{(m)}) \cdot x_{Ei}^{(m)},$$ (1)

where:

$$KS_j^{(m)} = \frac{\sum\limits_{m=1}^{n} x_{ji} \cdot x_{Ei}}{\sum\limits_{i=1}^{n+1} x_{Ei}^2} f,$$ (2)

x_{Ei} is vector row of the training data matrix with the greatest Euclidean's distance from the origin of coordinate, f is approximating function.

Fig. 5. Result of classification by neural-like structure of GTM topology

The Table 2 shows the settings for adjustment this network.

Table 2. Configurations of the neural-like structure of GTM debugging settings for the global approach

	Experiment Number			
	1	2	3	4
Nonlinearities of synapses	2	3	4	5
Nonlinearity of neurons	1	1	1	1
Number of hidden layers	1	1	1	1
Number of neurons	30	30	30	30

4.2 Local Approach

The local approach to classification is realized on the basis of the application of the hybrid neural structure of the geometric transformations model (Fig. 6), where the primary input-signs in the RBF inputs are pre-converted by the ratio:

$$R_j = \exp\left(-\frac{D_j^2}{2\sigma^2}\right),\tag{3}$$

where D_j is the value of the Euclidean distance between the current vector-point and j-th base; σ is parameter of the slope of the function; R_j is the magnitude of the signal corresponding to the RBF input.

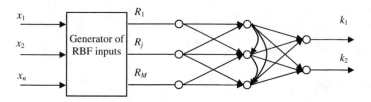

Fig. 6. Hybrid neural-like structure of GTM

The Fig. 7. gives results for various variants of the configuration (Table 3) of the hybrid neural-like structure of GTM.

#1. σ=0,1		Predicted 0	1	Σ
Ncm=30	0	283543	772	284315
	1	216	276	492
Actual	Σ	283759	1048	284807

#2. σ=0,1		Predicted 0	1	Σ
Ncm=100	0	283557	758	284315
	1	212	280	492
Actual	Σ	283769	1038	284807

#3. σ=0,5		Predicted 0	1	Σ
Ncm=30	0	284207	108	284315
	1	204	288	492
Actual	Σ	284411	396	284807

#4. σ=0,5		Predicted 0	1	Σ
Ncm=100	0	284233	82	284315
	1	201	291	492
Actual	Σ	284434	373	284807

#5 σ=1		Predicted 0	1	Σ
Ncm=30	0	284182	133	284315
	1	189	303	492
Actual	Σ	284371	436	284807

#6 σ=1		Predicted 0	1	Σ
Ncm=100	0	284199	116	284315
	1	191	301	492
Actual	Σ	284390	417	284807

Fig. 7. Result of classification by hybrid neural-like structure of GTM topology

Although in most cases, the use of RBF neural networks provides better quality of classification, in this case, the results of testing confirm the benefits of the global recognition method.

The reason for this situation should be sought in the uneven location in the space of implementation of points of observation, both for the training and for the test samples. As experiments show, the uneven distribution of realizations in space is generally inherent in phenomena that are caused by people's activities, which requires some caution when using traditional classifications.

Table 3. Configurations of the hybrid neural-like structure of GTM debugging settings for the local approach

	Experiment Number					
	1	2	3	4	5	6
	$\sigma = 0.1$		$\sigma = 0.5$		$\sigma = 1$	
Nonlinearities of synapses	3	4	3	4	3	4
Nonlinearity of neurons	1	1	1	1	1	1
Number of hidden layers	1	1	1	1	1	1
Number of centers of mass (N_{cm})	30	100	30	100	30	100
Number of neurons	60	130	60	130	60	130

5   Conclusion

The paper considers methods for solving the problem of data classification in the conditions of the incompleteness of the information base. Since such tasks are predominantly multidimensional and involve enormous amounts of data, there is a need for methods that allow for the maximum accuracy of data processing in a minimum amount of time.

A characteristic feature of modern classification tasks is the heterogeneity of data, their contradictions and incompleteness, and most importantly - the unknown law of data distribution. Because of this, the use of most known classification methods that require a priori knowledge of the law of the distribution of learning data is impossible.

One of the few methods that does not require a priori knowledge about the law of the distribution of learning data are the machines of reference vectors, but its application imposes a number of restrictions, which primarily concern the great computational complexity of the method.

The best results among the well-known classification methods are shown by the Random Forest algorithm, which not only shows a rather high accuracy of the classification, but also guarantees overfitting protection. However, for an unbalanced sample, the accuracy of the recognition of a smaller class is not high enough.

The proposed method based on the use of GTM not only showed rather high results, but also has a number of other benefits, including repeatability of results, high performance. Also, the great advantage of these networks is that they take into account the scope of tasks and degeneracy of tasks, which can be applied to data with an unknown law of data distribution.

References

1. Ting, K.M.: Encyclopedia of Machine Learning. Springer, Boston. ISBN 978-0-387-30164-8 (2011)
2. Bodyanskiy, Y.V., Vynokurova, O.A., Dolotov, A.I.: Self-learning cascade spiking neural network for fuzzy clustering based on group method of data handling. J. Autom. Inf. Sci. **45**, 23–33 (2013)

3. Pozzolo, A.D., Caelen, O., Johnson, R.A., Bontempi, G.: Calibrating probability with underdamping for unbalanced classification. In: Proceedings of the 2015 IEEE Symposium Series on Computational Intelligence, Cape Town, pp. 159–166 (2015)

4. Keilwagen, J., Grosse, I., Grau, J.: Area under precision–recall curves for weighted and unweighted data. PLoS ONE **9**(3), 92209 (2014)

5. Vapnik, V.N.: The Nature of Statistical Learning Theory. Springer, New York (1995). https://doi.org/10.1007/978-1-4757-3264-1

6. Breiman, L.: Random forests. Mach. Learn. **45**(1), 5–32 (2001)

7. Kutucu, H., Almryad, A.: Modeling of solar energy potential in Libya using an artificial neural network model. In: Proceedings of the 2016 IEEE First International Conference on Data Stream Mining & Processing (DSMP), Lviv, pp. 356–359 (2016)

8. Hu, Z., Bodyanskiy, Y.V., Tyshchenko, O.K., Samitova, V.O.: Possibilistic fuzzy clustering for categorical data arrays based on frequency prototypes and dissimilarity measures. Int. J. Intell. Syst. Appl. (IJISA) **9**(5), 55–61 (2017). https://doi.org/10.5815/ijisa.2017.05.07

9. Bodyanskiy, Y.V., Tyshchenko, A.K., Deineko, A.A.: An evolving radial basis neural network with adaptive learning of its parameters and architecture. Autom. Control Comput. Sci. **49**(5), 255–260 (2015)

10. Hu, Z., Bodyanskiy, Y.V., Tyshchenko, O.K., Boiko, O.O.: A neuro-fuzzy Kohonen network for data stream possibilistic clustering and its online self-learning procedure. Appl. Soft Comput. **14**, 252–258 (2017)

11. Hu, Z., Bodyanskiy, Y.V., Tyshchenko, O.K., Tkachov, V.M.: Fuzzy clustering data arrays with omitted observations. Int. J. Intell. Syst. Appl. (IJISA) **9**(6), 24–32 (2017). https://doi.org/10.5815/ijisa.2017.06.03

12. Bodyanskiy, Y.V., Tyshchenko, O.K., Kopaliani, D.S.: An evolving connectionist system for data stream fuzzy clustering and its online learning. Neurocomputing **262**, 41–56 (2017)

13. Bodyanskiy, Y., Vynokurova, O., Setlak, G., Peleshko, D., Mulesa, P.: Adaptive multivariate hybrid neuro-fuzzy system and its on-board fast learning. Neurocomputing **230**, 409–416 (2017)

14. Fawcett, Tom: An introduction to ROC analysis. Pattern Recognit. Lett. **27**(8), 861–874 (2006)

15. Cortes, C., Vapnik, V.: Support-vector networks. Mach. Learn. **20**(3), 273–297 (1995)

16. Tkachenko, R., Tkachenko, P., Izonin, I., Tsymbal, Y.: Learning-based image scaling using neural-like structure of geometric transformation paradigm. In: Studies in Computational Intelligence, vol. 730, pp. 537–565. Springer, Heidelberg (2018)

17. Polishchuk, U., Tkachenko, P., Tkachenko, R., Yurchak, I.: Features of the auto-associative neuro like structures of the geometrical transformation machine (GTM). In: Proceedings of the 2009 5th International Conference on Perspective Technologies and Methods in MEMS Design, Zakarpattya, pp. 66–67 (2009)

18. Medykovskyy, M., Tsmots, I., Tsymbal, Y., Doroshenko, A.: Development of a regional energy efficiency control system on the basis of intelligent components. In: Proceedings of the 2016 XIth International Scientific and Technical Conference Computer Sciences and Information Technologies (CSIT), Lviv, pp. 18–20 (2016)

19. Hu, Z., Bodyanskiy, Y.V., Tyshchenko, O.K., Samitova, V.O.: Fuzzy clustering data given on the ordinal scale based on membership and likelihood functions sharing. Int. J. Intell. Syst. Appl. (IJISA) **9**(2), 1–9 (2017). https://doi.org/10.5815/ijisa.2017.02.01

20. Hu, Z., Bodyanskiy, Y.V., Tyshchenko, O.K., Samitova, V.O.: Fuzzy clustering data given in the ordinal scale. Int. J. Intell. Syst. Appl. (IJISA) **9**(1), 67–74 (2017). https://doi.org/10.5815/ijisa.2017.01.07
21. Hu, Z., Ye, V., Bodyanskiy, O.K., Tyshchenko, A.: Deep cascade neuro-fuzzy system for high-dimensional online fuzzy clustering. In: Proceedings of the 2016 IEEE First International Conference on Data Stream Mining and Processing (DSMP 2016), Lviv, Ukraine, pp. 318–322, 23–27 August 2017

Information Technology of the System Control of Water Use Within River Basins

Pavlo Kovalchuk[1], Hanna Balykhina[1(✉)], Roman Kovalenko[1],
Olena Demchuk[2], and Viacheslav Rozhon[3]

[1] Institute of Water Problems and Land Reclamation, National Academy of Agrarian Sciences of Ukraine, 37, Vasylkivska Street, Kyiv, Ukraine
{maslova-anna,romchik89}@ukr.net
[2] National University of Water and Environmental Engineering,
11, Soborna Street, Rivne, Ukraine
[3] State Agency of Water Resources of Ukraine, 8, Velyka Vasylkivska Street,
Kyiv, Ukraine

Abstract. One of the key problems of system control of water use is the flushing of river beds with the water from reservoirs. To control the process of flushing, an information technology has been developed. Under the conditions of sustainable development, it provides the optimization of water use and contributes to the ecological rehabilitation within the river basin. The combined operational control of the distribution of water masses and pollution transformation along the river bed is implemented by the balance method. The balance difference equations describe the dynamics of water in the upper and lower layers, its movement and enable to visualize the pollution process. Optimization of the options for operational control is based on a scenario analysis. Multicriteria optimization methods are used based on economic and environmental criteria. Water quality assessment was improved by using the neural network in the monitoring system. The neural network provides a feedback coupling within the control system.

The information technology is adapted to the conditions of water supply for irrigation providing the ecological rehabilitation of the Ingulets River. A scenario analysis of some options for flushing operational control was made. The scenarios are estimated by the economic criterion for saving water resources and the ecological criterion for river rehabilitation. The decision making is based on the Pareto principle. The recommended optimal scenario provides a water supply for the period of 7 days applying small portions of water to prevent pollution sedimentation within the flood plain. The displacement of the mineralized water lens is carried out with less water consumption. It enables to provide a regulatory water supply for irrigation.

Keywords: Information technology of system control
Method of combined control · Balance method of pollution distribution
Neural networks · Structure optimization · Scenario analysis
Pareto optimal solutions

© Springer International Publishing AG, part of Springer Nature 2019
Z. Hu et al. (Eds.): ICCSEEA 2018, AISC 754, pp. 123–132, 2019.
https://doi.org/10.1007/978-3-319-91008-6_13

1 Introduction

The system control of water use refers a certain territory as a socio-ecological and economic system unit [1, 2]. The systems of territory management are being developed [3]. The integrated management within the river basin is carried out [4]. The features of integrated management in the river basin are integration by subsystems [2], by types of control (operational control, structure control, development control) [5]. Integrated management is conceptually important by its goals (economic and ecological), according to which the river basin management plans [6, 7] are developed. The development of the Plans is stipulated by the EU Directive [8], and its methodological basis is the paradigm of sustainable development [9]. The conditions of sustainable development in the river basin provide for the change from spontaneity to the control of water use along with the environmental rehabilitation of river [8].

The system control is implemented using specific information technologies [10, 11]. The most important role of such technology is providing water supply to the certain points of the basin for different needs. Water is supplied for drinking and industrial needs as well as for irrigation. An important aspect is river environmental rehabilitation. In the process of water management, system consistency is ensured by the criteria and types of management.

The structure of the information technology at the conceptual, logical-mathematical and physical levels involves combining a water distribution control with river bed flushing as well as enables to improve the models for water quality assessment within the monitoring system [12]. To meet these goals, the information technology should provide a scenario analysis, that is, the simulation modeling of the options of river flushing control along with the multicriteria optimization of these options based on ecological and economic criteria. The information technology should be adapted to the conditions of a particular object, in this case, the supply of water for irrigation, flushing and rehabilitation of the bed of the Inglets River.

The research objective is the development of a specific information technology for water management, implementation of this technology for the flushing of the Ingulets River and providing the conditions for sustainable development in the river basin.

2 Methodological Features of Information Technology for Water Use Management

2.1 Structural Optimization of Information Technology for Management Purposes

Integrated approach for reaching environmental and economic goals is one of the components of implementing system control coherence. Under sustainable development conditions, system control coherence shows through the functioning of a river basin as a system, which involves multicriteria optimization for economic goals, which are defined as a total increase on the interval T.

$$\left\{ \int_0^T F_1(X(t+1), U(t))dt, \dots, \int_0^T F_n(X(t+1), U(t))dt \right\} \rightarrow max \qquad (1)$$

where $F_1, \dots, F_n$ are economic criteria for water use evaluation; X(t) and $U(t)$ represent respectively the states of the system and control at time t.

With that the ecological goals, set as boundary conditions or as certain ecological criteria for the river restoration, are considered.

$$\begin{cases} F_{n+1}(X(t+1), U(t)) \le C_{n+1}(t) \\ F_p(X(t+1), U(t)) \le C_p(t) \end{cases} \qquad (2)$$

where $F_{n+1}, \dots, F_p$ are ecological criteria, $C_{n+1}(t), C_p(t)$ represent boundary conditions at time te [0; T] to achieve a good or excellent ecological state.

Various scenarios of the dynamics of water flow and spread of contamination are developed for the ecological and economic justifications of the options for water supply and river flushing control. The scenarios are estimated by the criteria (1, 2), the optimal solution regarding the control system is determined on the basis of multi-criteria optimization, in particular by the Pareto principle [13].

2.2 Methodological Approach to the Operational Control of River Bed Flushing

One of the important components of water use along with the ecological river restoration is flushing of the river bed using the water from reservoirs, which is expressed by formalizing the process of distribution, mixing, transformation of water masses and pollutants, as well as defining the criteria for economic efficiency of water use. The carried out analysis of the approaches to river bed flushing [14–19] determined that they are an open loop control system as the control effects are determined without considering water quality evaluation for economic needs at water intake points. The control scheme without a feedback is inefficient as the water quality at water intake points is neglected and unauthorized industrial wastes discharges as well as the filtration of contaminated wastewater from the groundwater can occur. The control process, besides the economic criteria (1), also involves defining the criteria that place limits on pollutants content (2) to achieve a good ecological state [8].

The usage of combined control systems is proposed for the operational control of river bed flushing [20]. The combined control system has both an open-circuit line and one circuit of feedback. The open-circuit line is determined by the scheduling of a given sequence of impulses in time with their water discharge from the reservoir, i.e. planning of a certain flushing regime. The circuit of feedback provides the selection or correction of water discharge from the reservoir on the basis of obtained water quality data compared to the permissible standards. The control system provides a sequence of flushing impulses depending on water quality along the river bed and the amounts of water intake for economic needs.

Making decision as to the next impulse is fulfilled by the function evaluating the control effects simultaneously using both input and output values, i.e.

$$Q_1(x, t_{n+1}) = F\left(V_i(t_n), S_i(t_n), V_j(t_n), S_j(t_n), Q_1(x, t_n)\right) \tag{3}$$

where $Q_1(x, t_{n+1})$ is water discharge (impulse) in the next $n+1$ time point; V_j, S_j are concentration of pollutants in the upper and lower layers at the observation point; V_i, S_i are concentration in the upper and lower layers along the river; $Q_1(x, t_n)$ represents water discharge at the previous time point from the reservoir; F is a function determining the decision-making algorithm for achieving economic and environmental goals.

The idea of the operational control of river bed flushing is based on the deductive models of the spread of contamination [21] as well as on the models of unsteady flow evaluation. For the modeling of the spread of contamination three-layer models are the most appropriate. In the upper flow layer the spread of contamination is at the highest rate, in the bottom area it decreases. The third equation describes the interaction with bottom sediments. In case of the spread of saline contamination it is advisable to apply a two-layer model as the process of salt settling at the bottom can be neglected.

As physical models are quite complicated and often consider irrelevant factors the system of balance models is proposed [22, 23]. The balance difference models are of independent significance. They can consider the most important physical processes and be adapted by the field survey data to the conditions of a particular river for flushing control. The proposed balance model combines the ideas of two-layer model of the spread of contamination, continuity equations and the balance equations of water flow wave propagation.

In the course of river bed flushing some polluted water is washed off by the water flow from the reservoir (removal of mineralized water plume). Some polluted water from the lower layer rises to the upper layer and is transferred faster by the flow. As a result of flushing there is the interaction of the processes of washing off polluted water and mixing upper and lower water layers.

The balance equations of water flow in the upper layer determining the continuity of flow are given as:

$$W_i^{n+1} = r_i^n\left(W_i^n + q_i^n\right) + \left(1 - r_i^n\right)\left(W_{i-1}^n + q_{i-1}^n\right)$$
$$r_i^n = c + dW_i^n; 0 \le r_i^n \le 1; i = 1, \ldots, N \tag{4}$$

where W_i^n, W_{i-1}^n are water volumes in the n-th time point in the i-th and i–1th cells; q_i^n, q_{i-1}^n are water volumes in the n-th time point in the i-th and i–1th cells coming from tributaries or filtered through soil; W_i^{n+1} presents water in the i-th cell at the n + 1th time point; r is a factor dependent on a flow rate.

The balance equations of water flow in the lower layer are given as:

$$D_i^{n+1} = D_i^n \mu + p_i^n \mu + (1 - \mu)\left(D_{i-1}^n + p_{i-1}^n\right)$$
$$0 \le \mu \le 1 \tag{5}$$

where D_i^n, D_{i-1}^n are water inflow from the i-th and i–1th cells of the lower layer; D_i^{n+1} is water in the i-th cell at the n + 1th time point; p_i^n, p_{i-1}^n present water inflow into the lower layer from tributaries or filtered through soil; μ is water velocity coefficient ($0 < \mu < 1$).

Water unsteady movement is assigned by the time variable boundary conditions in the upper layer. In the same way the impulse intensity of flushing or the sequence of impulses are assigned.

The concentration at the n + 1th time point in the i-th cell as a result of mixing in the upper layer is calculated by the equation:

$$U_i^{n+1} = \frac{r_i^n W_i^n U_i^n + r_i^n q_i^n C_{q_i} + \left(1 - r_i^n\right) W_{i-1}^n U_{i-1}^n + \left(1 - r_i^n\right) q_{i-1}^n C_{q_{i-1}} + \lambda_i^n D_i^n S_i^n - \lambda_i^n D_i^n U_i^n}{W_i^{n+1}}$$

$$\lambda_i^n = a + b W_i^n; i = 1, \dots, N, \tag{6}$$

where $r_i^n W_i^n U_i^n$ is water $r_i^n W_i^n$ with the concentration U_i^n in the i-th cell (at the n-th time point); $r_i^n q_i^n C_{q_i}$ is water inflow $r_i^n q_i^n$ from tributaries or groundwater with the concentration C_{q_i} coming into the i-th cell; $\left(1 - r_i^n\right) W_{i-1}^n U_{i-1}^n$ is water $\left(1 - r_i^n\right) W_{i-1}^n$ with the concentration U_{i-1}^n in the i-th cell; $\left(1 - r_i^n\right) q_{i-1}^n C_{q_{i-1}}$ is water inflow $\left(1 - r_i^n\right) q_{i-1}^n$ from the $i - 1$ th cell coming from tributaries or groundwater with the concentration $C_{q_{i-1}}$. The value λ_i^n describes the intensity of water exchange between upper and lower layers. Water $\lambda_i^n D_i^n$ with the concentration S_i^n comes from the lower layer into the upper one, while water $-\lambda_i^n D_i^n$ with the concentration U_i^n comes into the lower layer.

The concentration at the n + 1th time point in the i-th cell of the lower layer is calculated by the equation:

$$S_i^{n+1} = \frac{\mu D_i^n S_i^n + \mu p_i^n R_i^n + (1 - \mu) D_{i-1}^n S_{i-1}^n + (1 - \mu) p_{i-1}^n R_{i-1}^n - \lambda_i^n D_i^n S_i^n + \lambda_i^n D_i^n U_i^n}{D_i^{n+1}} \tag{7}$$

where $\mu D_i^n S_i^n$ is water μD_i^n with the concentration S_i^n that remains in the i-th cell; $\mu p_i^n R_i^n$ is water inflow μp_i^n with the concentration R_i^n from groundwater or river tributaries; $(1 - \mu) D_{i-1}^n S_{i-1}^n$ is water inflow from the $i - 1$ th cell with the concentration S_{i-1}^n; $(1 - \mu) p_{i-1}^n R_{i-1}^n$ is water inflow $(1 - \mu) p_{i-1}^n$ with the concentration R_{i-1}^n into the $l - 1$ th cell from tributaries or groundwater; $-\lambda_i^n D_i^n S_i^n$ presents water inflow with the concentration S_i^n into the upper layer (minus means the decrease in the lower layer); $\lambda_i^n D_i^n U_i^n$ is water inflow from the upper layer to the lower one with the concentration U_i^n.

To provide operational control and implement feedback coupling, the information technology involves improving the monitoring on the basis of neural networks [24]. Based on the theory of neural networks, an ecosystem model for the assessment of surface water quality in the river basin was formalized. A neural network is trained on the base of observation data. At that, the environmental evaluation methods [25] and the state standards of water suitability both for irrigation and drinking needs are also applied to build a neural network.

The developed model based on the balance difference equations and appropriate computer program enable to realize various variants of the operational control of water supply for irrigation. The scenario analysis of these variants and the selection of the optimal ones by the ecological and economic criteria provide the control of the structure [5].

3   Modeling and Analysis Results

To implement the information technology of system control the scenario analysis of different variants was realized on the example of flushing of the Inhulets River bed. The river flushing is carried out using the water from the Karachunivske reservoir aiming to the environmental rehabilitation of the river basin and providing the water supply of appropriate quality through the Inhulets irrigation system (Fig. 1).

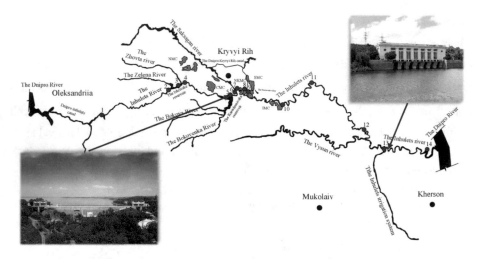

Fig. 1. The Inhulets river basin scheme with its tributaries, indicating the main tailings facilities within the Kryvyi Rih and sampling points according to the monitoring data of the State Agency of Water Resources of Ukraine: 5 - the point of water outlet from the Karachunivske reservoir; 13 - the point of water intake for the Inhulets irrigation system.

The model of scenario analysis involves the operational control of flushing by the content of toxic ions (anion-chlorine (Cl-)) to prevent soil salinization while irrigating. The balance-difference model of equations is used for these purposes that enables to estimate the movement of pollutants in time and space.

We propose several scenarios that enable to evaluate and improve the environmental and economic efficiency of flushing.

Scenario I. The Scenario simulates the existing method of flushing and confirms by the model the basic laws of the Inhulets River basin environmental rehabilitation using the existing water discharge method without feedback. Possible emergency dis charges, uncontrolled single and diffuse sources of pollution such as the flows from the tributaries of the Saksagan River and the filtration through the ground waters from the nearby tailings facilities, coming to the Inhulets River and increasing the concentration of pollutants in water are not taken into account in this Scenario. Since the Scenario doesn't involve the second impulse of water supply, the river basin runs some ecological risks of increasing the concentrations of pollution above the regulatory values.

Scenario II. It is different from the Scenario I. The first impulse of water supply lasts shorter period (only 7 days), that provides essential water savings compared to the Scenario I. When the concentration of pollutants by chlorides (280–300 mg/l) is exceeded in the monitoring points, it enables to perform the second impulse of water supply from the reservoir. The key parameters of the Scenario II are: April 5 – a gradual rise of water discharge rate up to 20 m³/s; the April 6 - April 12 period - water discharge rate is 20 m³/s; starting from April 13 - water discharge rate is 10 m³/s with regular monitoring of water quality by chlorides at monitoring points (Andriivka village in Dnipropetrovsk region and the Main Pumping Station of the Inhulets Irrigation System in Mykolaiv region.

Scenario III. This Scenario significantly improves the ecological component of the Inhulets River flushing compared to the Scenario II. Primarily to provide flushing a small impulse with a discharge rate of 8 m³/s is supplied within 7 days. It provides the river bed flushing and prevents flooding the surrounding plain. Therefore no accumulation of harmful sediments occur on the floodplain. As for the rest, the Scenario III imitates the Scenario II (Fig. 2).

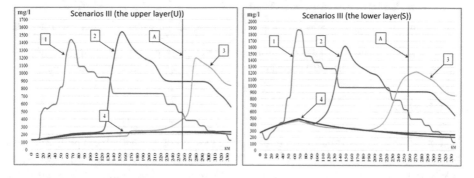

Fig. 2. The dynamics of pollution in the upper (a) and lower (b) water layers in the course of river bed flushing at different time points when applied the Scenario III: 1 - 1th day; 2 - 3th day; 3 - 15th day; 4 - 30–90th days (steady-flow operation).

Scenario IV. This Scenario is proposed to define the effect of the concentrations of pollutants of the river tributaries and the pollutant filtration from groundwater, which tend to increase the steady-state pollutant level in 30, 60 or more days reaching the values exceeding the normative ones. The flushing is performed with a rather low rate of 3–5 m³/s.

The ecological criterion sets up a limit on water quality (2), but it should be considered wider as a criterion for the selection of a control system, which minimize the pollutant sediments on floodplain. The ecological criterion F_2^* is proposed to determine in grades while conducting a comprehensive environmental assessment.

The optimization of the control system structure is carried out by two criteria: the efficiency of scenarios by the ecological criterion F_2^* and by the economic criterion

$F^* - F_1(x, t)$, which is used when comparing water saving with the maximum water discharge F^*.

In this case the selection of the variants of control system is graphically based on the Pareto two-criterion optimization model:

$$\{F^* - F_1(x, t), F_2^*\} \rightarrow max \qquad (8)$$

where F_2^* is ecological criterion, grade; $F^* - F_1(x, t)$ presents water saving, million cubic meters (Fig. 3).

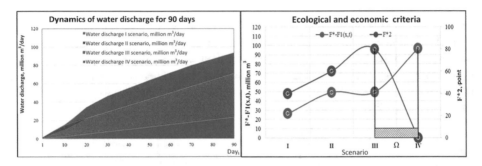

Fig. 3. Scenario analysis of the river flushing: (a) dynamics of water discharge by different flushing scenarios for 90 days; (b) ecological and economic optimization of the criteria.

4 Discussion

We represented the information technology of system control, which expands the decision making systems within a certain territory. The system feature is control of water use within the river basin. An important technological solution is the combined control system. At that, the schedule of flushing impulses is set up, which can be corrected by the circuit of feedback according to the obtained data on the concentration of pollutants at water intake point. Operating control of water dynamics and pollution transformation are implemented by the balance method. The method can be applied to new objects of water exchange such as rivers, canals, lakes and reservoirs. The information technology can be adapted to the conditions of a specific object. A specific monitoring system based on neural networks is developed. The system of operating control can be optimized by the scenario analysis and multicriteria optimization methods.

5 Conclusion

The proposed information technology provides operating control of the river bed flushing and visualization of pollution spread along the river bed as well as control capabilities of the system structure.

The control of the structure provides a scenario ecosystem analysis of flushing. The best variant is selected by the system of economic criteria of water use and environmental criteria for achieving a good ecological state of the river.

The adaptation of the information technology to the conditions of the Inhulets River enabled to conduct the scenario analysis for four variants of flushing, to select the Pareto-optimal variant by the economic criterion of water saving and ecological criterion of river restoration. The optimal variant can provide the agricultural producers of the Inhulets irrigation system with high-quality water for irrigation needs.

References

1. Kovalchuk, P., Balykhina, H., Kovalchuk, V., Matiash, T.: Water management system in the Ukrainian Danube river area for food and environmental safety. In: International Commission on Irrigation and Drainage, 2nd World Irrigation Forum (WIF2), 6–8 November 2016, Chiang Mai, Thailand, pp. 1602–1612 (2016)
2. Kovalchuk, V.: Special aspects of system management methodology of the territories water regime for protection against flooding. Inductive Model. Complex Syst. **6**, 97–106 (2014). (in Ukrainian)
3. Arezki, S.A., Djamila, H.B., Bouziane, B.C.: AQUAZONE: a spatial decision support system for aquatic zone management. Int. J. Inf. Technol. Comput. Sci. (IJITCS) **7**(4), 1–13 (2015). https://doi.org/10.5815/ijitcs.2015.04.01
4. Dukhovny, V., Sokolov, V., Manthrithilake, H.: Integrated Water Resources Management: Putting Good Theory into Real Practice. Central Asian Experience. SIC ICWC, Tashkent (2009)
5. Zgurovsky, M., Pankratova, N.: Basics of System Analysis. BHV, Kyiv (2002). (in Ukrainian)
6. Kovacs, A.: Quo vadis, Danubius? Progress and challenges of nutrient pollution control in the Danube River Basin. In: Fuchs, S., Eyckmanns-Wolter, R. (eds.) International Conference RIVER BASINS 2015, Monitoring, Modelling and Management of Pollutants, Germany, pp. 29–39. Springer, Karlsruhe (2015)
7. The Danube River Basin District Management plan. International Commission for the Protection of the Danube River ICPDR (2009)
8. UE Water Framework Directive 2000/60/EC. Definition of Main Terms
9. National paradigm of sustainable development for Ukraine. Public Institution «Institute of Environmental Economics and Sustainable Development of the National Academy of Sciences of Ukraine», Kyiv, 72p. (2016). (in Ukrainian)
10. Zhuchenko, A.I., Osipa, L.V., Cheropkin, E.S.: Design database for an automated control system of typical wastewater treatment processes. Int. J. Eng. Manuf. (IJEM) **7**(4), 36–50 (2017). https://doi.org/10.5815/ijem.2017.04.04
11. Zhang, H. Liu, M.: GIS-based emergency management system on abrupt environmental pollution accidents in counties of China. Int. J. Educ. Manag. Eng. (IJEME) **2**(8), 31–38 (2012). https://doi.org/10.5815/ijeme.2012.08.06
12. Kalburgi, P.B., Shareefa, R.N., Deshannavar, U.B.: Development and evaluation of BOD–DO model for River Ghataprabha near Mudhol (India), using QUAL2K. Int. J. Eng. Manuf. (IJEM) **5**(1), 15–25 (2015). https://doi.org/10.5815/ijem.2015.01.02
13. Podinovsky, V., Noghin, V.: Pareto Optimal Solution for Multicriterion Problems. Nauka, Moscow (1982). (in Russian)
14. Bledsoe, B., Beeby, J., Hardie, K.: Evaluation of flushing flows in the fraser river and its tributaries. Technical report, 182p. (2013)

15. Burlaka, B.: The flushing Inhulets river in 2011. Water Manag. Ukraine **5**, 17–18 (2011). (in Ukrainian)
16. Olsen, N.R.B.: Two-dimensional numerical modelling of flushing processes in water reservoirs. J. Hydraul. Res. **37**(1), 3–16 (1999)
17. Schaffranek, R.: A flow-simulation model of the Tidal Potomac River. U.S. Geological Survey Water-Supply Paper 2234: A water-quality study of the tidal Potomac river and estuary, 41p. (1987)
18. Seng Mah, D., Putuhena, F., bt Rosli, N.A.: Modelling of river flushing and water quality in a tributary constrained by barrages. Irrig. Drainage Syst. **25**(4), 427–434 (2011)
19. Varga, M., Balogh, S., Csukas, B.: GIS based generation of dynamic hydrological and land patch simulation models for rural watershed areas. Inf. Process. Agric. **3**, 1–16 (2016)
20. Ivakhnenko, A., Peka, Y., Vostrov, N.: The Combined Method of Modeling Water and Oil Fields. Naukova Dumka, Kyiv (1984). (in Russian)
21. Kovalchuk, P., Gerus, A.: Identification of the model of pollutants distribution in the surface waters based on field studies. In: Fuchs, S., Eyckmanns-Wolter, R. (eds.) International Conference RIVER BASINS 2015, Monitoring, Modelling & Management of Pollutants, Germany, pp. 101–105. Springer, Karlsruhe (2015)
22. Kovalchuk, P., Balykhina, H., Demchuk, O., Kovalchuk, V.: Modeling of water use and river basin environmental rehabilitation. In: IEEE XIIth International Scientific and Technical Conference Computer Science and Information Technologies (CSIT-2017), 5–8 September 2017, Lviv, Ukraine, pp. 468–472 (2017)
23. Kovalchuk, P., Gerus. A., Kovalchuk, V.: Perseptron model of system environmental assessment of water quality in River Basin. In: Proceedings of 4th International Conference of Inductive Modelling ICIM 2013, pp. 279–284. Springer, Kyiv (2013)
24. Saeed, N.H., Abbod, M.F.: Modelling oil pipelines grid: neuro-fuzzy supervision system. Int. J. Intell. Syst. Appl. (IJISA) **9**(10), 1–11 (2017). https://doi.org/10.5815/ijisa.2017.10.01
25. Romanenko, V., Zhukynsky, V., Oksiyuk, O.: Methods of Environmental Assessment of Surface Water Quality According to the Appropriate Categories. Symvol-T, Kyiv (1988). (in Ukrainian)

Data Mining for a Model of Irrigation Control Using Weather Web-Services

Volodymyr Kovalchuk[1(✉)], Olena Demchuk[2], Dmytro Demchuk[3],
and Oleksandr Voitovich[1]

[1] Institute of Water Problems and Land Reclamation,
Vasylkivska str., 37, Kyiv 03022, Ukraine
volokovalchuk@gmail.com, aleks@krakow.in
[2] National University of Water and Environmental Engineering,
Soborna str., 11, Rivne 33028, Ukraine
ldem1997@ukr.net
[3] National Technical University of Ukraine "Ihor Sikorsky Kyiv Polytechnic Institute",
Prospect Peremogy, 37, Kiev 03056, Ukraine
ddima199703@gmail.com

Abstract. The article deals with obtaining forecast weather data, its processing and use in mathematical models for irrigation management in the application of Decision Support System. The data obtained from the weather service databases on temperature and humidity are summarized on the basis of potential evapotranspiration calculations. Forecast data on precipitation is handled under uncertainty. On the basis of the weather forecast data, moisture transfer is modeled, soil moisture is predicted, that is, new knowledge is obtained about the state of soil moisture, on the basis of which Decision Support System generates a certain management solution. Due to the Internet and the use of the online regime, the decision maker does not directly process large arrays of weather information, but receives Decision Support System solutions as quickly and easily as possible.

Keywords: Data mining · Weather web-services · Model of irrigation control
Weather data processing under uncertainty · Calculation of evapotranspiration
DSS

1 Introduction

A great number of publications consider the questions of various aspects of irrigation management, based on both the mathematical models of moisture transfer [1–5] and other types of models [6], as well as without the use of such, only on the basis of water balances [7–10]. Weather data is the basis of any irrigation management. The weather determines the need of crops in water and, as a consequence, seasonal irrigation rates and the time and duration of irrigation periods, that is, irrigation control parameters. The decision support system (DSS) and operational control systems of watering (irrigation) are based on these principles or similar models.

© Springer International Publishing AG, part of Springer Nature 2019
Z. Hu et al. (Eds.): ICCSEEA 2018, AISC 754, pp. 133–143, 2019.
https://doi.org/10.1007/978-3-319-91008-6_14

Some mathematical models that use weather and climate data to form water regime are the part of refer to more generalized models MARS (Crop Growth Monitoring System (CGMS)) [11] or WOFOST (a simulation model of crop production) [12] for modeling and monitoring of crop development.

Development of "data mining techniques", the data obtaining and processing techniques for models was also the subject of some scientific works [13–20], the problems of irrigation control were also analyzed in some publications [15–17]. Irrigation forecast is based only on appropriate neural networks [16, 17].

Several works are also devoted to using decision support system (DSS), based on basic mathematical models of moisture transfer, which use the weather data obtained in different ways [1, 3, 4, 21, 22]. Solving the problem of obtaining and adequate processing weather data needs to have the answers to the following two questions: Which data is needed for yearly irrigation planning and operative irrigation control when using a multilayered model of moisture transfer? Which data and how can they be obtained from the weather forecast web-sites and formalized for irrigation control?

Regarding the irrigation control model proposed in our research, it is based on a balanced one-dimensional differential equation of non-stationary moisture transfer in the soil. The soil moisture balance is as follows. In the field where the agricultural crop is grown in a vertical plane there is a selected calculated layer of soil of a certain thickness. It contains the root system of agricultural plants. For a root-bearing layer, the balance of moisture is calculated: inflow due to precipitation and irrigation, increase or decrease in the moisture content of this layer of soil, flowing beyond the boundary of the calculated layer under the influence of gravity, water consumption by plants and evaporation from the soil (evapotranspiration) under the influence of weather factors. Consequently, the model needs current and forecast weather data - precipitation and other data for calculating the current and forecast potential evapotransportation of agricultural crops grown in a specific field. However, the DSS considered is not adapted to use forecast weather data.

Thus, the potential evapotranspiration (or total evaporation) is calculated using the Penman-Monteith method [23] or evaluated by the Shtoiko biophysical method [24] well-known in Ukraine. The information for both calculation methods can be found on weather websites [25–27], taking into account the probabilistic nature of weather forecasts. Consequently, formalization of such data is required for their input into the DSS. Therefore, a goal of our research is solving the task of automated online-extraction of forecast data from the sites of weather forecasting and their transfer to the DSS for operational control of irrigation is presented.

2   Research Methods

2.1   The Substance of the Irrigation Management Process Using Weather Data

Now more about irrigation management algorithms, mathematical model of moisture transfer and methods of automated obtaining (extraction) of data for it from weather websites, combined in the Decision support system (DSS). It implements seasonal irrigation norms like [1, 5, 6, 28] and operates water management similar to [7–9].

Norms of water consumption of agricultural crops to replenish the seasonal deficit of moisture are determined for seasonal planning of irrigation rules. The moisture deficit is the difference between precipitation and evapotranspiration (climatic water balance). The calculation of evapotranspiration is based on long-term retrospective weather information on a certain area. The data can be extracted from the archives of the weather service database or global climate databases [29, 30], the nearest meteorological station of the Hydrometeorological Center of Ukraine network.

In the short-term planning (up to 5 days) operative management of watering of crops is carried out. In online mode, DSS receives short-term weather data from weather service [25–27] or numerical prediction model WRF [31]. WRF is an American meso-scale model used by many weather services in Europe for its forecasts. The predicted dynamics of soil moisture is calculated on the basis of the obtained data using the mathematical model of moisture transfer. The date of reaching some critically low average moisture content of a layer of a certain thickness is predicted:

$$\theta_h^{avg} = \frac{\sum\limits_{i=1}^{m} \theta_i}{m} \tag{1}$$

where θ_i – volumetric soil moisture in the i-th layer; m is the number of horizons of the soil, forming the layer h.

These calculations determine the start date of irrigation. Considering the probabilistic nature of weather forecasts: precipitation, temperature, etc., a periodic correction of calculations is required. The feedback and the date correction are based on data from instrumental measurements or measurements of Internet meteorological stations.

2.2 Mathematical Model of One-Dimensional Moisture Transfer

The model solves the boundary problem and is based on a one-dimensional differential equation of non-stationary moisture transfer. The equation in partial derivatives of a relatively unknown humidity function $\theta(z, \tau)$ describes a saturated-unsaturated water stream in the multilayer (consisting of different physical properties of soil layers) soil profile [2, 28]:

$$\frac{\partial \theta}{\partial \tau} = \frac{\partial}{\partial z}\left[k(z, \theta)\frac{\partial\left(\psi(z, \theta)\right)}{\partial z} - k(z, \theta)\right] - I_\theta \tag{2}$$

where θ is the volume of soil moisture, $\left(m^3\,m^{-3}\right)$; I_θ is the volume of moisture that is removed from the unit volume of soil per unit time (mainly it is water consumption by plants and physical evaporation, i.e. evapotranspiration); $k(z, \theta)$ is a coefficient of moisture transfer and $\psi(z, \theta)$ is a full potential of soil moisture are functions of the volume of soil moisture θ and soil properties in each point z. Z axes is setted down.

The equation of moisture transfer is limited by the boundary conditions. The boundary conditions on the soil surface are the balance of streams: rain or watering stream (mm day^{-1}); part of evapotranspiration corresponding to the intensity of physical evaporation from the soil surface (mm day^{-1}); the intensity of infiltration into the soil (mm day^{-1}); the flow that goes to the consumption of the roots of plants that are located in the upper horizon of the soil (mm day^{-1}). On the lower boundary moisture is equal of full saturation θ_{max}. It is the constant.

The corresponding initial conditions are the initial distribution of moisture on the soil profile at the beginning of the calculation.

Evapotranspiration is mainly influenced by meteorological conditions, crop factors and soil water regime. The influence of meteorological factors is usually expressed as crop evapotranspiration. It can be calculated by the Penman–Monteith equation [23]

$$ET_o = \frac{\left(0.408 \cdot \Delta \cdot (R_N - G) + \gamma \dfrac{900}{T + 273} \cdot u_2 \cdot (e_S - e)\right)}{(\Delta + \gamma(1 + 0.34 \cdot u_2))} \tag{3}$$

where ET_0 is the crop evapotranspiration (mm day^{-1}), R_n is the net radiation at the crop surface (MJm^{-2} day^{-1}), G is the soil heat flux density (MJm^{-2} day^{-1}), T is the air temperature at 2 m height ($^{\circ}$C), u_2 is the wind speed at 2 m height (m s^{-1}), e_S is the saturation vapour pressure (kPa), e is the actual vapour pressure (kPa), Δ is the slope of vapour pressure curve (kPa $^{\circ}$C^{-1}), γ is the psychrometric constant (kPa $^{\circ}$C^{-1}).

In Ukraine, evapotranspiration is also evaluated using the Shtoiko biophysical method [24]. It is based on the hypothesis that, with optimal soil moisture, the process of evaporation is practically not regulated by the plant and soil. In these conditions, evapotranspiration is determined, mainly, by external climatic factors of evaporation (humidity of air, temperature). On this basis the total evaporation from sowing to the complete shading of the surface of plants and in the period of maturation (yellowing of leaves) is determined by the formula (m^3 ha^{-1} for a certain time n):

$$ET_1 = \sum_{i=1}^{n} t_c^i (0, 1 t_c^i - \frac{a}{100}), \tag{4}$$

where t_c^i is the average daily air temperature at 2 m height ($^{\circ}$C), a is daily average relative humidity (%). A sum of t_c^i is the sum of average daily air temperatures for n days. In other periods, with complete shading of plants and with more intensive evaporation in accordance with:

$$ET_2 = \sum_{i=1}^{n} t_c^i (0, 1 t_c^i + 1 - \frac{a}{100}). \tag{5}$$

A mathematical model with boundary conditions is solved by a numerical finite-difference method.

2.3   Obtaining and Formalizing Data Methods from Weather Forecasting Services in Conditions of Uncertainty

In seasonal irrigation planning, many years of retrospective weather data on a certain area are needed to calculate the water requirements of agricultural crops. Data on average daily or ten-day temperature and relative humidity of air and rainfall amount were obtained from the nearest located weather station of the Hydrometeorological Center of Ukraine network. The time series or the duration of the observation period must be at least the last 30 years. By Shtoiko's method (4)-(5), a daily average (average ten-day) evapotranspiration of a particular agricultural crop is calculated. There is a deficit of moisture will be climatic water balance. Results are ranked in descending order of values. The highest value corresponds to 100% security, and the smallest 1%. Norms of water demand for agricultural crops for a given territory correspond to the deficit of moisture in those years.

For short-term, up to 5 days, prediction of soil moisture based on the model of moisture transfer (2), some data are stable and are in the knowledge base: the type of soil, its water-physical properties; type and stage (phase) of the development of an agricultural plant grown there. Initial humidity is measured experimentally.

The main problem is obtaining weather forecast data for operational irrigation planning, since these data are not constant and are probabilistic. It is important to get data from the web-service in such quantity, on the basis of which it is possible to make an adequate averaging over the course of the day and to calculate the forecast value of evapotranspiration using (4)-(5). To do this, set the exact current server time and receive for each of the next 5 days the following data: daily forecast of precipitation; forecast of average daily temperature and relative humidity of air.

Weather services provide such information with a time interval of 3 h (Fig. 1). Parameters such as average daily temperature, relative humidity of air are calculated by usual averaging.

Fig. 1. Examples of weather data representation by weather services

The more complicated is the issue of forecasting precipitation, which is given in mm for every 3 h with a certain probability p_j ($j = 1, \ldots, 8$). In this case, a simple averaging will not produce the desired result. To uncover uncertainty, the following decision criteria should be used which take into account the average, most favorable and most unfavorable environmental conditions [32]. In our case, it is a complete lack of precipitation, which in case of untimely irrigation can lead to loss of crop.

Average daily precipitation values for each of the nearest k days are based on the formula for the daily average, maximum and minimum forecast values for every 3 h. If forecasting is carried out for the next five days, then the average daily precipitation values are calculated: as sum of average values (6); sum of the most favorable values (7); sum of the most adverse rainfall values every three hours (8)

$$R_K^{avg} = \sum_{j=1}^{8} p_j R_{jk}, k = 1, \ldots, 5;$$ (6)

$$R_K^{max} = \sum_{j=1}^{8} \max_j R_{jk}, k = 1, \ldots, 5;$$ (7)

$$R_K^{min} = \sum_{j=1}^{8} \min_j R_{jk}, k = 1, \ldots, 5$$ (8)

where $R_K^{avg}, R_k^{max}, R_k^{min}$ are accordingly the amount per day probable, the maximum expected and the minimum possible of precipitation; $p_j R_{jk}$ is probable for the j-th period of day (3 h) precipitation; $\max_j R_{jk}, \min_j R_{jk}$ are respectively, the maximum and minimum values in the considered 3 h.

The evaluation criterion, which is a combination (weighing) of the criteria (6)-(8) is also determined. So, use the Hodges-Lehmann criterion [32], which is a combination of average and minimum values:

$$R_{\lambda k}^{(1)} = \lambda \cdot R_K^{avg} + (1 - \lambda)R_K^{min}, k = 1, \ldots, 5$$ (9)

where $\lambda \epsilon [0;1]$ is a parameter that weighs the mean and the minimum value.

At the same time, it is important to consider the best environmental behavior - powerful, but short-term precipitation will provide the plants with the required amount of moisture. Such a criterion, which regulates the size of the level of "pessimism-optimism" of the decision maker, is the Gurvits criterion [32]:

$$R_{\lambda k}^{(2)} = \lambda \cdot R_K^{min} + (1 - \lambda)R_K^{max}, k = 1, \ldots, 5$$ (10)

where $\lambda \epsilon [0;1]$ is the Gurvits indicator is the coefficient that weighs the least and the highest value of precipitation.

The stochastic calculations of rainfall forecasting in terms of options will allow to get deadlines and norms for watering accordance with the model moisture transfer.

3   Results of Research

3.1   Program-Algorithmic Realization of the Process of Obtaining Data from Weather Services and Their Accumulation in the Database of DSS

Automatic receipt of forecast weather data from weather web services takes place using API (Application Programming Interface). The API in this case is defined by the set of HTTP request messages and response messages with the structure usually given in XML format or in the JSON object record format. This method is quite simple and less expensive in terms of load on the server.

In order to access the API response, in most cases, the API key issued by the web service developer is required regardless of which account type has been selected. For a free or non-premium version of the account, the service may provide incomplete data. There are also restrictions on the number of server requests for a certain period of time or for a period of use. An example of an incomplete amount of data may be the receipt of an API response where the air humidity data is completely absent or available only within a short period of time.

The accumulation of meteorological information in the DSS database occurs continuously after the user selects a particular object - the area where irrigation is conducted. The five-day term for obtaining weather forecasts is due to the fact that the accuracy of long-term forecasts, according to the authors, drops sharply, due to climate change. Thus, prediction of soil moisture within 5 days will enable the farmer to prepare qualitatively for irrigation. The received interval of receipt of the weather forecast is every 3 h for 5 days. This allows them to conduct an adequate averaging over the course of the day and calculate the predictive value of evapotranspiration using the Shtoiko method (4)-(5) [24].

3.2   Calculation of Evapotranspiration and Forecasting of Daily Rainfall

The calculation of evapotransportation according to weather services can be carried out at several levels of incoming information. Penman-Monteith calculation (3) requires a large number of input parameters. Among them there are those that are difficult to calculate. For example, it is difficult to numerically identify verbal descriptions of the categories of the sky (cloudy-clear) coming from weather services relative to the calculation Rn and G, relative to the degree of plant cover of the soil surface. Now the evapotranspiration is estimated by the formulas (4)-(5), which are implemented in the software complex.

For forecasting of daily precipitation using predicted rainfall data for every 3–6 h. As the baseline data, the maximum forecast value per day is used (rainfall with a probability of 100%), the average value (the sum of the probability products for the predicted value of precipitation for every 3–6 h), the minimum possible value of precipitation (the minimum value of the product in parts of the day). Daily precipitation calculations are carried out according to the formulas (6)-(10), which provide disclosure of uncertainty according to a certain criterion. The resulting value of precipitation (Table 1) is used to calculate the dynamics of soil moisture by model (2).

Table 1. The data and results of forecasting of daily precipitation according to decision criteria in conditions of uncertainty

Data	24.11	25.11	26.11				27.11				28.11			
Part of day	*1–4*	*1–4*	*1*	*2*	*3*	*4*	*1*	*2*	*3*	*4*	*1*	*2*	*3*	*4*
Probability p_j, %	2	2	8	46	77	80	73	69	64	50	45	35	39	37
Rain R_{jk}, mm	0	0	0	2	4	4	4	4	4	5	1	2	2	3
$p_j·R_{jk}$ mm	0	0	0	0.92	3.08	3.2	2.92	2.76	2.56	2.5	0.45	0.7	0.78	1.11
$\min\limits_{j}(p_j R_{jk})$, mm	0	0	0				2.5				0.45			
R_k^{max}, mm	0	0	10				17				8			
R_k^{avg}, mm	0	0	7.2				10.74				3.04			
R_k^{min}, mm	0	0	0				10				1.8			
Hodges-Lehmann, mm	0	0	3.6				10.37				2.42			
Gurvits, mm	0	0	5				13.5				4.9			

4   Analysis and Comparison of the Results

For seasonal irrigation planning, the model of moisture transfer has been tested on archival weather data of meteorological stations of the Hydrometeorological Center for the south of Ukraine [2, 28]. In simulation experiments, irrigation norms are calculated at biologically optimal and water-saving irrigation regimes (Fig. 2) of beets and winter wheat, by controlling the soil moisture by index (1).

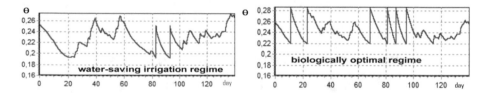

Fig. 2. An example of the seasonal irrigation norms of beets in the moderate-dry years

Similar studies on the establishment of seasonal irrigation norms of winter wheat and corn, using the modeling of moisture transfer and archival meteorological data, conducted [1]. In the work [5] in the simulation of the moisture transfer, a comparison was made of the quality of irrigation control with the balance model [10]. It is alleged that the balance models of seasonal irrigation standards are overstated. Weather data for model calculations were obtained by direct measurements, weather forecasts were not used.

Only in publications [7, 9] for operative watering presented receiving weather data from weather services or numerical weather forecasting models [31]. The balance model [7] receives data from weather services only "for today". Such models are not aimed at predicting soil moisture. In addition, they do not take into account the movement and moisture fluxes in the soil profile, since they are two-layer. Two-layered DSS models offered by [8, 33], use field weather stations and Remote Sensing data [33] to provide online advice to farmers on starting the irrigation on the basis of the fact achievement of critical soil moisture. Foreseeable calculations of soil moisture allow farmers to prepare in advance for irrigation and hold it.

DSS [3] has in its composition a model of moisture transfer, takes into account the movement of moisture in the soil profile. But it does not use predicted weather data. Such a model can not predict the date of reaching the critical humidity of the soil, but only states the "fact" of achieving a critical value of moisture.

The model of the moisture transfer, which is the basis of DSS in our study, is multi-layered. The operational forecast of soil moisture is based on the fundamental physical laws of moisture transfer and current short-term forecasts of precipitation, humidity and air temperature. In operative management of watering, a daily correction of soil moisture can be carried out by direct measurement, and the determination of the terms and norms of irrigation for forecasting weather for five days. Thus, there is a control of irrigation on a "sliding interval" using forecast data and calculations.

5 Conclusions

Consequently, the combination of a multi-layered model of moisture transfer in the DSS and the possibility of obtaining five-day forecast data from weather services brings it to a new, qualitatively higher level of operational control of irrigation.

Due to the rapid, qualitative reception and use of forecast data of weather services, the DSS will help the farmer to quickly order water of the correct volume and set the forecasted date for irrigation, to conduct watering, eliminating water and crop losses.

Mathematical processing of precipitation data in conditions of uncertainty allows qualitatively predicting the date of the irrigation, improving the quality of irrigation control, using irrigation management on the sliding interval.

Acknowledgment. Publications are based on the research provided by the grant support of the State Fund for Fundamental Research (project F76/95-2017).

References

1. Shang, S., Li, X., Mao, X., Lei, Z.: Simulation of water dynamics and irrigation scheduling for winter wheat and maize in seasonal frost areas. Agric. Water Manag. **68**, 117–133 (2004)
2. Kovalchuk, P., Balykhina, H., Kovalchuk, V., Matyash, T.: Water management system in the Ukrainian Danube river area for food and environmental safety. In: Proceedings of the 2nd World Irrigation Forum (WIF2), pp. 1–10. ICID, Chiang Mai, Thailand (2016)
3. Steduto, P., Raes, D., Hsiao, T.C., Fereres, E.: AquaCrop: concepts, rationale and operation. In: Steduto, P., Hsiao, T.C., Fereres, E., Raes, D. (Eds.) Crop Yield Response to Water. FAO irrigation and drainage paper no. 66, pp. 17–49. FAO, Rome (2012)
4. Kamble1, B., Irmak, A., Hubbard, K., Gowda, P.: Irrigation scheduling using remote sensing data assimilation approach. Adv. Remote Sens. **2**, 258–268 (2013)
5. Rallo, G., Agnese, C., Blanda, F., Minacapilli, M., Provenzano, G.: Agro-hydrological models to schedule irrigation of mediterranean tree crops. Ital. J. Agrometeorology **1**, 11–21 (2010)

6. Pulido-Calvo, I., Roldan, J., Lopez-Luque, R., Gutierrez-Estrada, J.C.: Demand forecasting for irrigation water distribution systems. J. Irrig. Drainage Eng. **129**(6), 422–431 (2003)

7. ENORASIS (Environmental Optimization of irrigation Management with the Combined use and Integration of High precision Satellite Data, Advanced Modeling, Process Control and Business Innovation). http://www.enorasis.eu/. Accessed 15 Nov 2017

8. Zhovtonog, O.I., Filipenko, L.A., Demenkova, T.F., Didenko, N.O.: Use of the information system "GIS irrigation" and the IRRIMET module for the internet weather station for operational irrigation planning in sprinkling. Taurian Sci. Bull. **92**, 159–165 (2015). (In Ukrainian)

9. Silva, D., Meza, F.J., Varas, E.: Estimating reference evapotranspiration (ETo) using numerical weather forecast data in central Chile. J. Hydrol. **382**, 64–71 (2010)

10. FAO CROPWAT 8.0. http://www.fao.org/land-water/databases-and-software/cropwat/en/, Accessed 18 Dec 2016

11. Baruth, B., Genovese, G., Leo, O.: CGMS version 9.2 - user manual and technical documentation. OPOCE, Luxembourg (2007)

12. Van Diepen, C.A., Rappoldt, C., Wolf, J., van Keulen, H.: WOFOST: a simulation model of crop production. Soil Use Manag. **5**(1), 16–24 (1989)

13. Dhore1, A., Byakude, A., Sonar, B., Waste, M.: Weather prediction using the data mining techniques. Int. Res. J. Eng. Technol. (IRJET) **4**(5), 2562–2565 (2017)

14. Liao, S.-H., Chu, P.-H., Hsiao, P.-Y.: Data mining techniques and applications – a decade review from 2000 to 2011. Rev. Expert Syst. Appl. **39**, 11303–11311 (2012)

15. Sara Kutty, T.K., Hanumanthappa, M.: Optimal water allocation using data mining techniques. a survey. Int. J. Emerg. Res. Manag. Technol. **6**(8), 226–229 (2017)

16. Khan, M., Islam, M.Z., Hafeez, M.: Irrigation water requirement prediction through various data mining techniques applied on a carefully pre-processed dataset. J. Res. Pract. Inf. Technol. **1**, 1–13 (2013)

17. Bhatt, N., Virparia, P.V.: A survey based research for data mining techniques to forecast water demand in irrigation. I.J. Comput. Sci. Mob. Appl. **3**(8), 14–18 (2015)

18. Lazri, M., Ameur, S., Brucker, J.M.: Analysis of the time trends of precipitation over mediterranean region. Int. J. Inf. Eng. Electron. Bus. (IJIEEB), **6**(4), 38–44 (2014). https://doi.org/10.5815/ijieeb.2014.04.06

19. Vamsi Krishna, G.: Prediction of rainfall using unsupervised model based approach using k-means algorithm. Int. J. Math. Sci. Comput. (IJMSC) **1**(1), 11–20 (2015). https://doi.org/10.5815/ijmsc.2015.01.02

20. Sultana, S.H., Ali, M.S., Hena, M.A., Rahman, M.M.: A simple model of mapping of land surface temperature from satellite digital images in Bangladesh. Int. J. Inf. Technol. Comput. Sci. (IJITCS) **5**(1), 51–57 (2013). https://doi.org/10.5815/ijitcs.2013.01.05

21. Rinaldi, M., He, Z.: Decision support systems to manage irrigation in agriculture. In: Sparks, D. (ed.) Advances in Agronomy, vol. 123, pp. 229–279. Academic Press, Burlington (2014)

22. Eitzinger, J., Thaler, S., Orlandini, S., Nejedlik, P., Kazandjiev, V., Vucetic, V., Sivertsen, T.H.: Agroclimatic indices and simulation models. In: Nejedlik, P., Orlandini, S. (eds.) Survey of Agrometeorological Practices and Applications in Europe Regarding Climate Change Impacts, pp. 15–115. European Science Foundation, Florence (2008)

23. Allen, R.G., Pereira, L.S., Raes, D., Smith, M.: Crop evapotranspiration–guidelines for computing crop water requirements. FAO Irrigation and Drainage Paper 56, Rome, Italy (1998)

24. Tsivinsky, G.V., Pendak, N.V., Idayatov, V.A.: Instruction on Operative Calculation of Irrigation Regimes and Forecast of Irrigation of Agricultural Crops due to Lack of Moisture Stores, 2nd edn. Ukrainian Ecological League, Kherson (2010). (In Ukrainian)

25. OpenWeatherMap. https://openweathermap.org/. Accessed 06 Nov 2017
26. AccuWeather. https://www.accuweather.com/en/. Accessed 06 Nov 2017
27. Sinoptik, U.A.: https://ua.sinoptik.ua/. Accessed 06 Nov 2017
28. Kovalchuk, V.P.: Ecological and economic optimization of irrigation regimes taking into account the quality of groundwater. In: Ways to Improve the Efficiency of Irrigated Agriculture, vol. 50, pp. 81–88 (2013). (In Russian)
29. FAO CLIMWAT 2.0. http://www.fao.org/land-water/databases-and-software/climwat-for-cropwat/en/. Accessed 11 May 2017
30. RP5. http://rp5.ua/Weather_archive_in_Kiev_airport. Accessed 01 Mar 2017
31. WRF (The Weather Research and Forecasting Model). https://www.mmm.uc-ar.edu/weather-research-and-forecasting-model. Accessed 10 Oct 2017
32. Trucharev, R.I.: Models of Decision Making in Conditions of Uncertainty. Nauka, Moscow (1981). (In Russian)
33. Car, N.J., Christen, E.W., Hornbuckle, J.W., Moore, G.A.: Using a mobile phone short messaging service (SMS) for irrigation scheduling in Australia–farmers' participation and utility evaluation. Comput. Electron. Agri. **84**, 132–143 (2012)

Hardware and Software Complex
for Automatic Level Estimation and Removal
of Gaussian Noise in Images

Serhiy V. Balovsyak[✉] and Khrystyna S. Odaiska

Yuriy Fedkovych Chernivtsi National University, Chernivtsi, Ukraine
s.balovsyak@chnu.edu.ua

Abstract. The hardware and software complex for an automatic level estimation and the removal of Gaussian noise in digital images has been developed. The complex consists of video cameras, computers and the software developed in MATLAB.

The calculation of Gaussian noise level is performed by the developed method, which is based on image filtering and iterative selection of region of interest. As the noise level, its standard deviation is considered. The developed software is designed for the video camera adjustment and is aimed at obtaining a series of images of one object, taken with video camera under the same lighting conditions, but at different values of the brightness parameter. For each image from the series, calculation of noise level and signal-to-noise ratio enable one to determine the optimal value of the brightness parameter.

The mathematical model, the method and the software for automatic removal of Gaussian noise in digital images with the use quasi-optimal Gaussian filter have been developed. A signal is described by the sum of the sinusoids, the amplitudes and periods of which are calculated on the basis of the energy spectrum of the original image. The quasi-optimal value of the standard deviation of the Gaussian filter kernel is obtained as the value at which the standard deviation between the filtered image brightness and the signal brightness is minimized. The accuracy of the developed filtration method has been verified by removing Gaussian noise in a set of 100 test images.

Keywords: Video camera · Digital image processing · Gaussian noise
Gaussian filter · Automatic image filtering

1 Introduction

Removing noise in digital images is an important stage in their pre-processing [1–4]. Currently, a fully automatic method for removing noise, which would provide optimal filtering of the resulting images, hasn't been reported yet. The development of such a method of filtration is a relevant task from both scientific and practical points of view, since manual processing a large number of images is a rather laborious process.

In this paper we'll consider the technique of removing the additive white Gaussian noise (AWGN), which is known as the widespread defect in digital images [1, 2]. The

© Springer International Publishing AG, part of Springer Nature 2019
Z. Hu et al. (Eds.): ICCSEEA 2018, AISC 754, pp. 144–154, 2019.
https://doi.org/10.1007/978-3-319-91008-6_15

noise in the AWGN model will be referred to as a Gaussian noise. The level of the Gaussian noise is described by its standard deviation σ_N. In general, the value of σ_N is unknown, therefore, the following methods are used for its determination: filter-based approaches [4], patch-based approaches or block-based methods [4], principal component analysis [4], statistical approaches [5], Wavelet transform [3], methods based on the Fourier analysis of image spectra [3]. In this paper, a highly accurate technique for noise level calculation based on filtration is used [6]. The resulting noise level enables calculating the signal-to-noise ratio (SNR) for studied images.

The process of hardware image processing is governed by such settings of the video camera, as brightness, contrast, etc., which significantly affect the noise level and the SNR for the obtained images [7, 8]. As a rule, the default settings of video camera do not provide the maximum SNR for experimental images. Therefore, to improve the visual quality of the images, the video camera customization is performed through selection of such parameters that would ensure the maximum SNR of the video signal. The next step in improving the visual quality of the obtained images includes low-pass image filtering.

There is a large number of linear and non-linear Gaussian noise filtering methods, each possessing its advantages, limitations and an application scope [1–3]. Gaussian filter is a well-known linear filter. For noise removing in images the low-pass linear filters, implemented in both spatial and frequency domains, are used. The main drawback of the low-pass linear filters is the blurring of edges in images. The common nonlinear filters include median and bilateral filters; Partial Differential Equations filters; filters based on wavelet filtration [8]. Nonlinear filters applied for noise removing reduce edge smoothing, which is achieved through rather complex procedure of image processing. Therefore, in this paper, the classical low-pass Gaussian filter is used, which is effective in removing high-pass noise in the image. The Gaussian filter is described only by one parameter, the kernel standard deviation σ_w, which simplifies the calculation of optimal filter parameters.

2 Calculation of Gaussian Noise Level in Digital Images

The experimental noise level σ_{NE} is calculated for the original digital image f_n, which is written to the rectangular matrix $f_n = (f_n(i, k))$, where $i = 1,\ldots, M$, $k = 1,\ldots, N$ [3]. All images are processed in the shades of gray, the intensity of original image is normalized within a range from 0 to 1.

The paper uses a noise level calculation method based on filtration [6], in which the selection of the noise component in the image f_n is performed using the high-pass spatial filtration f_n with the kernel w_H of the Laplacian filter:

$$w_H = \frac{1}{6}\begin{pmatrix} 1 & -2 & 1 \\ -2 & 4 & -2 \\ 1 & -2 & 1 \end{pmatrix} \tag{1}$$

The noise level calculation is performed only in the region of interest (ROI) in the image [3], where Gaussian noise is prevailing. The regions of interest are selected iteratively, taking into account the statistical characteristics of Gaussian noise, which ensures high accuracy in the noise level calculation, comparable to that of the best analogues methods.

3   Determine Optimal Camera Settings

The hardware and software system for receiving video signals consists of a computer and several video cameras connected to it. Charge-coupled device (CCD) and complementary metal-oxide semiconductor (CMOS) are used as photo sensors (Fig. 1) [8, 9]. The video signal is amplified, the image interpolation is performed, the white balancing is set and the gamma correction of the signal is carried out. Next, the video signal is digitized into analog-to-digital converter (ADC) and recorded to a computer. The obtained digital image is distorted due to atmospheric influences and imperfections of the optical system, as well as due to different types of noise (fixed pattern noise, dark current noise, shot and thermal noise, quantization noise in the ADC). The total effect of such distortions on the image is rather accurately described by the used Gaussian noise model. Connecting a video camera to a computer enables one programmatically customize the image processing in the video camera by changing image parameters such as brightness, contrast, etc.

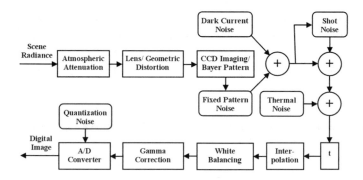

Fig. 1.   CCD camera imaging pipeline; t is exposure time [8, 9]

In general, default video camera's settings do not provide maximum SNR for the obtained images. The visual quality of experimental images can be improved by adjustment of the video camera's settings. In this paper, specification for the brightness parameter has been performed. In the adjusting process, the program receives a series of images f_n of one object, taken under the same lighting conditions, but at different values of the brightness of video camera (Fig. 2, Table 1). For each image from f_n series, the experimental noise level σ_{NE} and the SNR are calculated:

$$S_{NR} = \frac{P_{signal}}{P_{noise}} = \frac{\sigma_S^2}{\sigma_{NE}^2} = \frac{\sigma_{S0}^2 - \sigma_{NE}^2}{\sigma_{NE}^2} \qquad (2)$$

where P_{signal} is the power of the signal, P_{noise} is the noise power, σ_S is the standard deviation (SD) of the signal, σ_N is SD of noise, σ_{S0} is SD of the original image f_n.

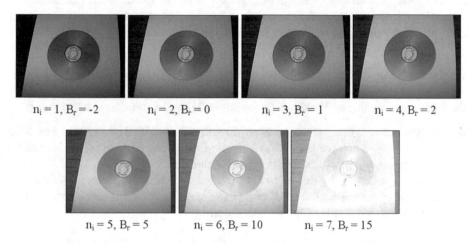

$n_i = 1, B_r = -2$ $n_i = 2, B_r = 0$ $n_i = 3, B_r = 1$ $n_i = 4, B_r = 2$

$n_i = 5, B_r = 5$ $n_i = 6, B_r = 10$ $n_i = 7, B_r = 15$

Fig. 2. A series of images obtained with the "A4Tech PK-835MJ" webcam with CMOS matrix at low illumination conditions of objects; the value of the brightness parameter (B_r) in relative units ranged from -2 to 15 (Table 1); n_i is the image number

Table 1. Parameters of series of images obtained with the webcam "A4Tech PK-835 MJ" (Fig. 2); the optimum value for the brightness parameter is 5

n_i	Brightness	σ_{NE}, %	S_{NR}, 10^9	A_{ST}	R_{SNRA}, 10^4
1	-2	0.91	0.1489	0.9999	1.2201
2	0	0.93	0.2003	0.9998	1.4151
3	1	0.91	0.2400	0.9998	1.5487
4	2	0.91	0.2838	0.9995	1.6838
5	5	0.87	0.4493	0.9974	**2.1141**
6	10	0.71	1.0301	0.6055	1.9435
7	15	0.43	3.5837	0.3022	1.8091

In experimental images obtained at high levels of brightness the saturation of the image is observed, negatively affecting the visual quality of the images (Fig. 2). Therefore, the A_{ST} parameter (Table 1) is used to estimate the saturation of the image, which is calculated as the relative number of pixels, the brightness of which does not exceed a threshold of $T_{hA} = 0.95$. Investigations of experimental images obtained at various brightness settings have shown that the visual quality of image can be described by the following parameter

$$R_{SNRA} = \sqrt{S_{NR}} \cdot A_{ST}. \tag{3}$$

On the basis of maximum of R_{NSRA} values (Table 1), the quasi-optimal value of the brightness parameter is automatically determined, which also accounts for the maximum visual quality of the image. The value of brightness is clarified periodically or at variation of the lighting conditions of the object. With the use to the described correction of camera's settings, the SNR for the obtained images can only be partially improved. Further image noise removal implies their low-pass filtration.

4   Removing Noise in Image Using Quasi-Optimal Gaussian Filter

Two simplified models of the signal in the original image f_n (Fig. 3a) [10] are used:

1. The brightness of the signal in a certain direction is described by a sinusoid with the amplitude $A_{SE}/2$ and the period T_{SE} (model of one sinusoid); the model is used for images with clear orientation of the details (quasi-uniform image).
2. A signal is described by two mutually perpendicular sinusoids with the amplitudes $A_{SE}/4$ and periods T_{SE} (model of two sinusoids); the model is used if there is no predominant orientation of the parts in the image.

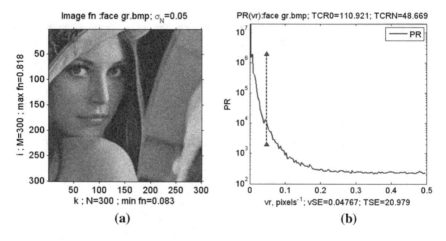

Fig. 3. Determination of signal parameters: (a) the original image f_n with noise level $\sigma_N = 5\%$; (b) radial distribution $P_R(v_r)$ for the energy spectrum P_S of the image f_n; $\sigma_{NE} = 4.81\%$

The utilized models of the useful signal can take into account its distortions that arise at image filtering in order to remove noise. The signal models allow to calculate its parameters based on the parameters of the original image, namely on the basis of the energy spectrum P_S [2, 3], which is equal to the square of the Fourier spectrum F for the image f_n. The standard deviation σ_{S0} of the image f_n is corrected, taking into account standard deviation of noise σ_{NE}, allowing for calculation of the standard deviation σ_S of the signal:

$$\sigma_S = \sqrt{\sigma_{S0}^2 - \sigma_{NE}^2} \tag{4}$$

On the basis of the energy spectrum P_S with the use of linear interpolation, its radial distribution $P_R(d)$ is calculated, where d is a distance from the spectrum element (m, n) to its canter (M_C, N_C), $d = 1, 2, \ldots N_R$, $N_R = [N/2]$ (Fig. 3b). The values of $P_R(d)$ are the arithmetic mean of $P_S(m, n)$ for the value d. Each frequency number d corresponds to the frequency value as $v_r(d) = d/N$. According to Parseval's theorem [11], the radial distribution of $P_{RN}(d)$ equals to the standard deviation of the image f_n brightness for the frequency with the number d:

$$P_{RN}(d) = \frac{1}{\sqrt{M \cdot N}} \sqrt{P_R(d)} \tag{5}$$

where $d = 1, 2, \ldots, N_R$.

On the basis of radial distribution $P_R(d)$ its average frequency v_{CR0} [12] is calculated, and using the distribution $P_{RN}(d)$ the average frequency v_{CRN} is obtained. The average frequencies v_{CR0} and v_{CRN} correspond to their average spatial periods T_{CR0} and T_{CRN}, respectively (Fig. 3b). The average frequency v_{SE} for sinusoid of signal is calculated through the empirical formula based on v_{CR0} and v_{CRN}, so that the mean square error (MSE) between the filtered signal and the one simulated by sinusoids is as close as possible to the MSE between the filtered and the initial real signals. As a result, on the basis of the original image f_n (Fig. 3a), the following parameters of the signal are calculated: the average spatial frequency v_{SE} and the range of values of the sinusoidal signal A_{SE}. The correspondence of image f_n to the model of one or two sinusoids is determined by the eccentricity E_{CE} of energy spectrum, which is calculated through its discrete central moments μ_{11}, μ_{02}, μ_{20} [13]:

$$E_{CE} = \frac{(\mu_{20} - \mu_{02})^2 + 4\mu_{11}^2}{(\mu_{20} + \mu_{02})^2} \tag{6}$$

The obtained eccentricity E_{CE} describes the orientation degree for the distribution of the image f_n brightness: $E_{CE} \approx 1$ for clear orientation, $E_{CE} \approx 0$ if the image does not have the same preferential orientation, $E_{CE} \approx 0.5$ in intermediate. If the values of a signal in the image are approximated by one sinusoid ($E_{CE} \approx 1$), the range of values of the A_{SE} sinusoidal signal is equal to the doubled amplitude of the sinusoid:

$$A_{SE} = 2\sqrt{2}\sigma_S \tag{7}$$

where σ_S is the standard deviation of the signal (4).

If the values of useful signal in the image are approximated by the sum of two mutually perpendicular sinusoids ($E_{CE} \approx 0$), the range of sinusoids A_{SE} values:

$$A_{SE} = 4\sigma_S \tag{8}$$

Considering (7) and (8), for arbitrary value of the eccentricity E_{CE} the A_{SE} sinusoidal signal range is obtained by linear interpolation of A_{SE} values for $E_{CE} \approx 0$ and $E_{CE} \approx 1$:

$$A_{SE} = 4\sigma_S \cdot (1 - E_{CE}) + 2\sqrt{2}\sigma_S \cdot E_{CE} \tag{9}$$

Thus, four experimental parameters of the original image are calculated: noise level σ_{NE}, average spatial period T_{SE}, eccentricity E_{CE}, range of A_{SE} sinusoids. On the basis of the parameters (σ_{NE}, T_{SE}, E_{CE}, A_{SE}), the quasi-optimal value of the standard deviation σ_{wRE} of the Gaussian filter kernel [10] is calculated.

Estimating σ_{wRE} will be performed on the basis of the analysis of the set of Gaussian filters kernels, the standard deviation σ_w of which takes a series of discrete values with the numbers $n_s = 1,\dots, Q_{ns}$ in the range from σ_{wMin} to σ_{wMax} (Fig. 4a). For each σ_w, the root mean square error (RMSE) for the noise and the signal is calculated:

1. R_{NwI} is RMSE of the noise component in the filtered image g.
2. R_{SwI} is RMSE of the signal in the filtered image g relative to the brightness of the simulated signal.

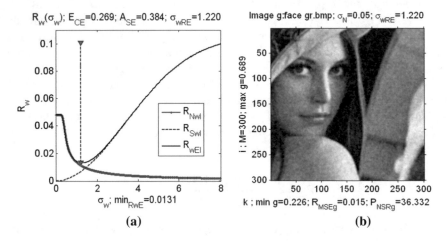

Fig. 4. Determination of quasi-optimal value of standard deviation σ_{wRE} of Gaussian filter kernel (a) to remove noise in the image f_n (Fig. 3a) and the filtered image g (b); R_{NwI} is the root mean square error (RMSE) of the noise component; R_{SwI} is the RMSE of signal; R_{wEI} is the RMSE of image g (noise component and signal); σ_w is standard deviation of Gaussian filter kernel

The values of the components R_{NwI} and R_{SwI} are calculated on the basis of the developed model filtration of Gaussian noise and sinusoidal signal [10]. The brightness of the filtered image g is the sum of the signal and noise, therefore the total RMSE of the signal and noise is given as

$$R_{wEI} = \sqrt{R_{NwI}^2 + R_{SwI}^2} \tag{10}$$

The value of σ_{wRE} is calculated as the value of standard deviation σ_w, which corresponds to the minimum of R_{wEI}. The resulting filtered image g (Fig. 4b) is obtained as a result of the convolution of the original image f_n with the Gaussian filter kernel $w(\sigma_{wRE})$. For standard deviation σ_{wRE} of filter kernel, the noise in the image will be almost completely removed, and the distortion of the signal will be negligible.

5 Testing the Developed Method of Noise Removal

The verification of the developed method of Gaussian noise removal by filtration was performed by processing the test images of the BSDS300 base [14, 15], to which the Gaussian noise with theoretical standard deviation σ_N was added via software (Fig. 5).

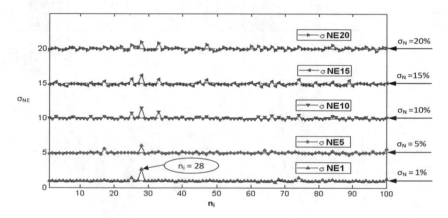

Fig. 5. Results of the determination of the experimental Gaussian noise level σ_{NE} by the proposed method for the test set (100 images) of the base BSDS300 [14, 15]; σ_N is the theoretical noise level; n_i is the number of the image, σ_{NE1}, σ_{NE2}, σ_{NE3}, σ_{NE4}, σ_{NE5} are the experimental values of standard deviation of noise for $\sigma_N = 1\%$, 5%, 10%, 15% and 20% respectively

In the test mode, the root mean square error R_{MSEg} [2] between the filtered g and original images f and the peak signal-to-noise ratio (PSNR) P_{SNRg} are calculated. The results of the determination of the experimental Gaussian noise standard deviation σ_{NE} for test images (Fig. 5, Table 2) show that the error of the proposed method ($R_{MSEN} = 0.212$) is less than the error of the best analogous methods (for example, for the PCAP method [8] $R_{MSEN} = 0.277$).

The R_{MSEg} values calculated by the proposed method exceed the optimal values R_{MSEgC} by an average value of 0.0008, indicating the effectiveness of the developed method (Fig. 6, Table 2). The optimal values of R_{MSEgC} are obtained by filtering the test images f_n at different values of standard deviation σ_w and comparing the filtered images g with the original f. Significant errors in the calculation of σ_{NE}, and consequently R_{MSEg} and P_{SNRg}, are obtained only for images with dominated textures ($n_i = 28$, Fig. 5) [16–19]. The optimal values of the peak SNR P_{SNRgC} were obtained by filtering the test images f_n at different values of standard deviation σ_w and comparing the filtered images g with the original f. The peak SNR P_{SNRg}, calculated by the proposed method, is on average less than the optimal P_{SNRgC}, by insignificant value of 0.138 dB (Table 2), so the developed filtering method is quasi-optimal.

Table 2. Results of the Gaussian noise removal in images by the proposed method for the test set (100 images) of the BSDS300 base [14, 15] (Figs. 5 and 6), to which Gaussian noise was added with theoretical standard deviation σ_N; σ_{NEA} is mean value σ_{NE}; R_{MSEN} is RMSE between σ_{NE} and σ_N

σ_N, %	σ_{NEA}	R_{MSEN}	R_{MSEg}	R_{MSEgC}	P_{SNRg}, dB	P_{SNRgC}, dB
1	1.002	0.188	0.0098	0.0096	40.233	40.373
5	5.006	0.167	0.0351	0.0338	29.385	29.651
10	10.023	0.229	0.0509	0.0496	26.291	26.450
15	14.959	0.240	0.0607	0.0598	24.822	24.905
20	20.029	0.227	0.0674	0.0669	23.942	23.982
1..20		**0.212**	**0.0493**	**0.0485**	**28.934**	**29.072**

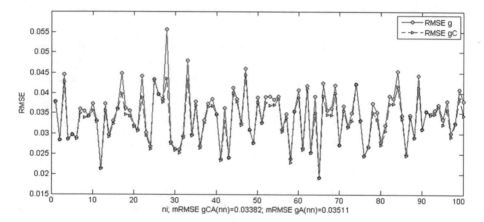

Fig. 6. The value of the RMSE between images g and f for 100 images of the BSDS300 base [14, 15], σ_N = 5%; R_{MSEg} is the obtained by the proposed method; R_{MSEgC} is the optimal RMSE; n_i is the image number

6 Discussion and Conclusion

The hardware-software complex for the automatic determination of the level and filtration of Gaussian noise in digital images obtained with video cameras has been developed. The mathematical model, the method and the software for automatic removal of Gaussian noise in digital images using a quasi-optimal Gaussian filter have been created. The method of automatic determination of the brightness parameter of a video camera, which provides the maximum visual quality of experimental images, is developed.

The novelty of this paper is to create a mathematical model that takes into account the reduction of noise and signal distortion, resulting from image filtering.

The accuracy of the developed filtration method has been verified by removing Gaussian noise in a set of 100 test images. It is shown that the developed method is quasi-optimal, since the calculated peak signal-to-noise ratio (PSNR) values are less than the optimal at 0.138 dB.

The developed method of automatic Gaussian noise filtering can be used in graphic editors and video surveillance systems. Further increase in the accuracy of the method can be achieved by the adaptive local processing of image contours.

References

1. Bovik, A.L.: The Essential Guide to Image Processing. Elsevier Inc, Burlington (2009)
2. Gonzalez, R., Woods, R.: Digital Image Processing. Prentice Hall, Upper Saddle River (2002)
3. Gonzalez, R., Woods, R., Eddins, L.: Digital Image Processing using MATLAB. Prentice Hall, Upper Saddle River (2004)
4. Liu, X., Tanaka, M., Okutomi, M.: Single-image noise level estimation for blind denoising. IEEE Trans. Image Process. **22**(12), 5226–5237 (2013)
5. Zoran, D., Weiss, Y.: Scale invariance and noise in natural images. In: Proceedings of the IEEE 12th International Conference on Computer Vision, pp. 2209–221 (2009)
6. Balovsyak, S.V., Odaiska, K.S.: Automatic highly accurate estimation of Gaussian noise level in digital images using filtration and edges detection methods. Int. J. Image, Graph. Sig. Process. (IJIGSP) **9**(12), 1–11 (2017). https://doi.org/10.5815/ijigsp.2017.12.01
7. Russ, J.C.: The Image Processing Handbook. Taylor and Francis Group, Boca Raton (2011)
8. Liu, C., Szeliski, R., Kang, S.B., Zitnick, C.L., Freeman, W.T.: Automatic estimation and removal of noise from a single image. IEEE Trans. Pattern Anal. Mach. Intell. **30**(2), 299–314 (2008)
9. Tsin, Y., Ramesh, V., Kanade, T.: Statistical calibration of CCD imaging process. In: Proceedings of the IEEE International Conference on Computer Vision, pp. 480–487 (2001)
10. Balovsyak, S.V., Odaiska, K.S.: Automatic removal of Gaussian noise in Digital images by Quasi-optimal Gauss Filter. Radioelectron. Comput. Syst. **83**(3). 26–35 (2017). (in Ukrainian). https://www.khai.edu/csp/nauchportal/Arhiv/REKS/2017/REKS317/Balovsyak.pdf
11. Solonyna, A.Y., Ulakhovych, D.A., Arbuzov, S.M., Solov'eva, E.B.: Fundamentals of Digital Signal Processing. BKhV-Peterburh, SPb (2005). (in Russian)
12. Thonhpanja, S., Phinyomark, A., Phukpattaranont, P., Limsakul, C.: Mean and median frequency of EMG signal to determine muscle force based on time-dependent power spectrum. Elektronika IR Elektrotechnika **19**(3), 51–56 (2013)
13. Jahne, B.: Digital Image Processing. Springer, Heidelberg (2005)
14. Fowlkes, C., Martin, D., Malik, J.: Local figure/ground cues are valid for natural images. J. Vis. **7**(8)(2), 1–9 (2007)
15. The Berkeley Segmentation Dataset and Benchmark: BSDS300. https://www.eecs.berkeley.edu/Research/Pro-jects/CS/vision/bsds
16. Li, J., Rich, W., Buhl-Brown, D.: Texture analysis of remote sensing imagery with clustering and Bayesian inference. Int. J. Image Graph. Sig. Process. (IJIGSP) **7**(9), 1–10 (2015). https://doi.org/10.5815/ijigsp.2015.09.01
17. Ye, Z., Yang, J., Zhang, X., Hu, Z.: Remote sensing textual image classification based on ensemble learning. Int. J. Image Graph. Sig. Process. (IJIGSP) **8**(12), 21–29 (2016). https://doi.org/10.5815/ijigsp.2016.12.03

18. Srinivasa Rao, M., Vijaya Kumar, V., Krishna Prasad, M.: Texture classification based on first order local ternary direction patterns. Int. J. Image Graph. Sig. Process. (IJIGSP) 9(2), 46–54 (2017). https://doi.org/10.5815/ijigsp.2017.02.06
19. Suresha, D., Prakash, H.N.: Data content weighing for subjective versus objective picture quality assessment of natural pictures. Int. J. Image Graph. Sig. Process. (IJIGSP) 9(2), 27–36 (2017). https://doi.org/10.5815/ijigsp.2017.02.04

Building a Generalized Peres Gate
with Multiple Control Signals

O. I. Rozhdov, I. M. Yuriychuk, and V. G. Deibuk$^{(\boxtimes)}$

Yuriy Fedkovych Chernivtsi National University, Chernivtsi, Ukraine
v.deibuk@chnu.edu.ua

Abstract. We consider a physical realization of the generalized quantum Peres and Toffoli gates with n-control signals, implemented in a one-dimensional chain of nuclear spins (one half) in a strong magnetic field coupled by an Ising interaction. Quantum algorithms in such system can be performed by transverse electromagnetic radio-frequency field using a number of resonant π-pulses on the initial states. The maximum number of π-pulses needed for the implementation of the Peres gate with n-control signals is discussed. It is found, that required number of π-pulses linearly scales with the number n of the control signals of the generalized quantum Peres gate. Comparison of our studies with the known values of the quantum cost of the generalized Peres gate allows us to suggest that proposed physical implementation of the gate is more efficient. The fidelity parameter is used to study the performance of the generalized Peres gate as a function of the relative error of the resonance frequency. The limits of an imbalance of the generator settings remaining the gate well defined are determined.

Keywords: Generalized Peres gate · Generalized Toffoli gate · Ising model
Quantum cost · Fidelity

1 Introduction

The reversible logic is increasingly used recently in low-power CMOS technology, optical computing, bioinformatics, and quantum information science [1, 2]. This is due to low energy loss at transmission and processing of information in the reversible circuits consisting of the reversible gates. As it was shown by Landauer in 1961 [3], losing every bit of information leads to the waste of $kT\ln2$ Joules of energy. The reversible gate is a gate where input information can be restored from the output states. That is a one-to-one mapping of the input signal onto the set of output signals takes place in the reversible logic gates. A characteristic feature of the reversible logic gates is that the number of inputs is equal to the number of outputs [4]. The latter requirement leads to the forbiddance of a fan-out because it violates the bijective conditions. There are the information inputs/outputs, constant (ancillary) inputs and the garbage outputs whose information should be recycled. Quantum gates correspond to the certain physical processes causing the transformation of quantum states. Such states in the quantum information science are called qubits and they have a variety of physical implementations (nuclear spins, electrons, photons, superconducting system states, etc.) [5].

© Springer International Publishing AG, part of Springer Nature 2019
Z. Hu et al. (Eds.): ICCSEEA 2018, AISC 754, pp. 155–164, 2019.
https://doi.org/10.1007/978-3-319-91008-6_16

Every logic gate of the circuit is characterized by quantum cost, which is the number of primitives from which it is built. In addition, to design a reversible circuit it is necessary to use a functionally complete (universal) basis of the gates, i.e., a set of logic gates (primitives) that allows realizing an arbitrary logic circuit. The synthesis of reversible combination devices is effective if the resulting quantum device has minimal quantum cost, delay time, a minimum number of constant inputs and garbage outputs. This is especially important nowadays when the issue of practical realization of reversible calculations in quantum computing arises [6, 7]. At the same time, possibility to build a quantum computer with a sufficient number of qubits encounters several important problems on a physical level. In particular, an interaction of a quantum system with the environment (decoherence) and technological constraints (for example, large fields and gradients in linear chains of paramagnetic ions with spins; laser control of quantum computers on ion traps; regulation of signals in NMR quantum computers, etc.). The mathematical model of a chain of paramagnetic ions with one-half spin has proved to be successful in a simulation of quantum computations [8–10].

The Peres gate is a universal, reversible quantum logic element, which has the minimal quantum cost among all three qubit gates, and therefore is widely used in combinational reversible devices [11]. In [12] it was proposed for the first time a generalization of the Peres gate to n-control signals. It is known [12, 13] that the use of multiple control signal gates may reduce the quantum cost of the reversible circuits. Quantum cost, in these works, refers to the number of primitives from which the reversible circuit is built. In particular, these are two inputs Feynman elements (CNOT) and CV/CV$^+$ (V = NOT$^{1/2}$, V$^+$ = V^{-1}), which quantum cost is equal to one [5]. By definition, a controlled U-quantum gate operates on the two input quantum bits called control and target qubit and results in U-transformation of the target qubit, in the case when the control qubit is in the state $|1>$.

In general, N-qubit generalized Peres gate has $n = N - 1$ control inputs. The state of the first output replicates the state of the first input $Y_1 = X_1$, and other control inputs determine the states of the corresponding outputs by the rule $Y_i = \bigoplus_{j=1}^{i} X_j (i = 1, ..., N - 1)$. The controlled output is determined by the expression $Y_N = X_1 X_2 ... X_{N-1} \oplus X_N$. Generalized N-qubit Peres gate with $N - 1$ control signals is presented in Fig. 1.

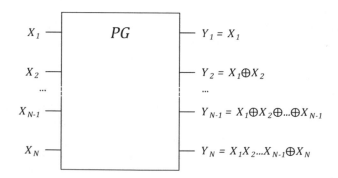

Fig. 1. Graphical symbol of generalized n-control Peres gate.

The possible implementation of the multiple control Peres gate in the base of CNOT, CV/CV+ primitives was proposed in [12]. Increase in the number of the control signals n of the Peres gate allows obtaining a device whose quantum cost is scaled as $2^{n+1} - n - 2$ [13]. In the present paper, we consider a solid-state model of the multiple control Peres gate based on a one-dimensional chain of nuclear spins in a strong magnetic field. Operation of the states in such system is carried out with the use of transverse electromagnetic radio-frequency (RF) field [8]. Although this model is physically difficult to implement, it can simulate physical processes and quantum gates in more realistic models.

2   The Model of Generalized n-Control Peres Gate

We consider a model of the quantum gate which is based on a one-dimensional Ising nuclear-spin system [14]. Such system has many of the relevant features of a physical realization of a quantum computer and is used today in liquid and solid-state nuclear magnetic resonance quantum computation with a small number of qubits.

A chain of N identical nuclear spins one half is placed in a strong constant magnetic field in z-direction and weak transverse circularly polarized in the $x - y$ plane electromagnetic radio-frequency (RF) field:

$$B = (b \cos \omega t, -b \sin \omega t, B(z)). \qquad (1)$$

Here b and ω are the amplitude and angular frequency of the RF-field. The strong magnetic field $B(z)$ has a field gradient in the z-direction in order to choose spins separately. Radio-frequency field is a time-dependent perturbation which allows to switch spins on and off and is used to implement the quantum gates.

The Hamiltonian of a chain of nuclear spins coupled by an Ising interaction in a strong magnetic field $B(z)$ (without any RF-field) has the form

$$\hat{H}_0 = -\hbar \left(\sum_{k=0}^{N-1} \omega_k I_k^z + 2J \sum_{k=0}^{N-2} I_k^z I_{k+1}^z + 2J' \sum_{k=0}^{N-3} I_k^z I_{k+2}^z \right) \qquad (2)$$

where $\omega_k = \gamma B(z_k)$ is the Larmore frequency of the k-th spin ($k = 0, 1, 2,..., N - 1$); γ is the proton gyromagnetic ratio. We assume an Ising interaction between nearest (the second term in (2)) and next-nearest (the third term in (2)) neighbors. J and J' are the coupling constants for these interactions.

The Hamiltonian $\hat{H}_0$ is diagonal on the basis of the Hilbert space of 2^N dimensionality. It consists of products of single spin (qubit) eigenstates $|0_k >$ and $|1_k >$ of the spin operator I_k^z. The eigenstates of the spin system can be represented as a combination of 2^N individual states of nuclear spins: $|00 \ldots 00 >, |00 \ldots 01 >, \ldots, |11 \ldots 11 >$, where the state $|0_k >$ corresponds to the orientation of a nuclear spin along the direction of the magnetic field $B(z_k)$, and the state $|1_k >$ corresponds to the opposite orientation. The eigenvalues $E_n = E_{i_{N-1}\ldots i_1 i_0}$ of the Hamiltonian $\hat{H}_0$ are given by

$$E_n = E_{i_{N-1} \ldots i_1 i_0} = -\frac{\hbar}{2} \left(\sum_{k=0}^{N-1} (-1)^{i_k} \omega_k + J \sum_{k=0}^{N-2} (-1)^{i_k + i_{k+1}} + J' \sum_{k=0}^{N-3} (-1)^{i_k + i_{k+2}} \right) \tag{3}$$

where $i_k = 0$ or 1. A state of the system can be represented by a linear combination $\Psi(t) = \sum_{n=1}^{2^N} C_n(t)|n\rangle$ of eigenstates $|n\rangle = |i_{N-1} \ldots i_2 i_1 i_0 \rangle$, which are associated with their binary images $(i_{N-1} \ldots i_2 i_1 i_0)$, where the content of each bit is set to 0 or 1, corresponding to the orientation of the spin along or opposite the direction of the constant magnetic field.

The implementation of any quantum gate in a chain of spins means a redistribution of the probabilities between the eigenstates of the system and is induced by monochromatic RF-field. The full Hamiltonian of a chain of spins in RF-field includes a perturbation operator $\hat{W}$ and is given as

$$\hat{H} = \hat{H}_0 + \hat{W},$$

$$\hat{W} = -\frac{\hbar \Omega}{2} \sum_{k=0}^{N-1} \left[I_k^+ \exp(i\omega t) + I_k^- \exp(-i\omega t) \right]. \tag{4}$$

Here $\Omega = \gamma b$ is a Rabi frequency; I_k^+ and I_k^- are the ascent and descent operators for a k-th spin. The dynamics of the system can be found from the solution of the Schrödinger equation

$$i\hbar \frac{\partial \Psi(t)}{\partial t} = \hat{H}\Psi(t). \tag{5}$$

Equation (5) with an initial condition $\Psi(0)$ defines the state of the system at the time t. Taking into account (4), the last equation is equivalent to a system of 2^N linear differential equations for unknown complex amplitudes C_n:

$$i\hbar \dot{C}_n = E_n C_n + \sum_{m=1}^{2^N} W_{mn} C_n, \ (n = 1, 2, \ldots, 2^N) \tag{6}$$

where W_{mn} is the matrix element of the perturbation operator (4), calculated on the eigenstates wave functions.

In order to compute the dynamics of the system, it is convenient to go over to interaction representation. The wave function $\Psi_{int}(t)$ in the interaction representation is connected to the wave function $\Psi(t)$ in the laboratory system of coordinates by the transformation $\Psi(t) = \exp\left(-i\hat{H}_0 t\right)\Psi_{int}(t)$ that gives

$$C_m = D_m \exp\left(-iE_m t\hbar^{-1}\right). \tag{7}$$

Then the system of ordinary differential equations (6) takes the form:

$$\dot{D}_m = \sum_{n=1}^{2^N} T_{mn}(t)D_n \tag{8}$$

where

$$T_{mn}(t) = -\frac{i}{\hbar}W_{mn}(t)exp(i\omega_{mn}t), \quad \text{and} \quad \omega_{mn} = \frac{E_m - E_n}{\hbar}. \tag{9}$$

Equation (8) represents a set of 2^N ordinary differential equations which can be solved numerically.

3 Results and Discussion

3.1 Quantum Cost

In order to realize quantum gates, we consider selective excitation regime, which means applying selective pulses to single spins. The spins are distinguished by imposing dependent on z-coordinate magnetic field $B(z)$ that results in different Larmore's frequencies $\omega_k = \gamma B(z_k)$ of each spin. The choice $\omega_k = 2^k\omega_0$ ($\omega_1 = 2\omega_0$, $\omega_2 = 4\omega_0$...) gives rise (ignoring Ising coupling) to a simple Zeeman spectrum with equidistant levels, where the ground level of the system is $|00\dots00>$, and the highest level is $|11\dots11>$. The spectrum of the four-qubit system is presented in Fig. 2. The value of $\omega_0 = 100$ (in units of 2π MHz) corresponds to the magnitude of the magnetic field $B = 2.3$ T, which is a reasonable one.

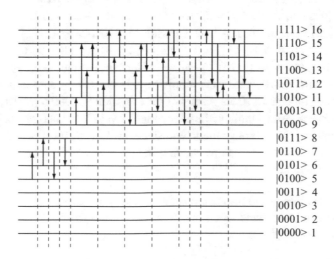

Fig. 2. Energy levels and allowed transitions for the four-qubit Peres gate.

The Ising spin-spin interaction arises, for instance, due to the exchange interaction of the neighboring qubits. Therefore, the value of the exchange interaction depends on the distance

r between qubits. In the first approximation, the exchange interaction constant for the qubits separated by the distance r is given by [9]

$$J(r) \cong 0.4 \frac{e^2}{4\pi\varepsilon_0 \varepsilon a_0} \left(\frac{r}{a_0} \right)^{5/2} \exp\left(-\frac{r}{a_0} \right). \tag{10}$$

Here ε is the dielectric constant of the medium, a_0 is the semiconductor Bohr radius, and e is the electron charge. Considering, for example, the case of donor impurity ^{31}P in silicon ($\varepsilon = 12$, $a_0 = 3$ nm, and $r = 35$ nm) we obtain $J = 5.2 \, 2\pi$ MHz. The coupling constant J' should be chosen to be at least one order of magnitude less than J since one expects the second neighbor interaction to be much weaker than the first neighbor interaction.

So, in the numerical simulations we have chosen the following parameters of the model (in units of 2π MHz):

$$\omega_0 = 100, \; \omega_1 = 200, \; \omega_2 = 400, \ldots; \quad J = 5; J' = 0.5; \quad \Omega = 0.1. \tag{11}$$

Note, that in the selective excitation regime the condition $\Omega \ll J \ll \omega$ must be satisfied. The value of Rabi frequency $\Omega = \gamma b = 0.1 \, 2\pi$ MHz corresponds the amplitude of the RF-field $b = 0.0023$ T.

In Fig. 2 we present, as an example, energy level diagram calculated for four-spin system and show transitions that simulate the operation of the four-qubit Peres gate according to the truth table (Table 1). Inputs X_1, X_2, X_3 are the control signals and X_4 is the target input. Generalization on the case of N-qubit Peres gate, as well as on N-qubit Toffoli gate (C_NNOT), is straightforward. Note, that allowed transitions are transitions with a rotation of only one spin. For four-qubit Peres gate these are transitions $5 |0100 > \to 7 |0110 >$, $6 |0101 > \to 8 |0111 >$, $7 |0110 > \to 5 |0100 >$, $8 |0111 > \to 6 |0101 >$, $13 |1100 > \to 9 |1000 >$, $14 |1101 > \to 10 |1001 >$. Direct transitions $9 |1000 > \to 15 |1110 >$, $10 |1001 > \to 16 |1111 >$, $11 |1010 > \to 13 |1100 >$, $12 |1011 > \to 14 |1101 >$, $15 |1110 > \to 12 |1011 >$, and $16 |1111 > \to 11 |1010 >$ are forbidden because they need a rotation of two spins. The operation of the Peres gate, in this case, is possible by means of two allowed transitions through an intermediate level. For the transition $9 |1000 > \to 15 |1110 >$ two ways are possible: transition through the intermediate level 13: $9 |1000 > \to 13 |1100 > \to 15 |1110 >$; and transition through the intermediate level 11: $9 |1000 > \to 11 |1010 > \to 15 |1110 >$. Such two-step transitions simulate the operation of the Peres gate also in the case of the forbidden transitions $10 \to 16$, $11 \to 13$, $12 \to 14$, $15 \to 12$, and $16 \to 11$ (see Fig. 2).

Allowed transitions from the digital state with the number l to the digital state with the number m can be realized when the frequency of the RF-field is set to resonance frequency, i.e. $\omega = |\omega_{lm}|$.

Table 1. The truth table of the four-qubit Peres gate.

X_1	0	0	0	0	0	0	0	0	1	1	1	1	1	1	1	1
X_2	0	0	0	0	1	1	1	1	0	0	0	0	1	1	1	1
X_3	0	0	1	1	0	0	1	1	0	0	1	1	0	0	1	1
X_4	0	1	0	1	0	1	0	1	0	1	0	1	0	1	0	1
Y_1	0	0	0	0	0	0	0	0	1	1	1	1	1	1	1	1
Y_2	0	0	0	0	1	1	1	1	1	1	1	1	0	0	0	0
Y_3	0	0	1	1	1	1	0	0	1	1	0	0	0	0	1	1
Y_4	0	1	0	1	0	1	0	1	0	1	0	1	0	1	1	0

Numerical calculations show that four-qubit Peres gate clearly operates in the digital states on a number of so-called π-pulses, the duration of which is $t_0 = \pi/\Omega$. As follows from the energy diagram (Fig. 2), the four-qubit Peres gate with three control signals can be implemented within two π-pulses. It is natural to determine the quantum cost as the maximum number of π-pulses required to implement the truth table of this logical gate. So defined quantum cost of the Peres gate with the multiple control signals is presented in Table 2. It is easy to see that there is a linear increase in the quantum cost of the Peres gates on the number n of the control signals. This dependence can be understood from the analysis of the logical expressions which determine the truth table of the gates. The state of the first qubit of the generalized Peres gate with $n > 2$ control signals always remains unchanged, and the rotation of the target qubit takes place only if all control inputs are in the state $|1>$. So, the quantum cost does not exceed $n-1$. In addition, for arbitrary n, there are always such combinations of input states that $n-1$-step transitions are required to obtain corresponding outputs states. Thus, the quantum cost of the generalized Peres gate scales as $n-1$.

Table 2. Number of π-pulses and quantum cost for the generalized n-control Peres gate.

Number of control signals	Number of π-pulses		Quantum cost of Peres gate	
	Toffoli gate	Peres gate	[13]	[12]
2	1	2	4	4
3	1	2	11	9
4	1	3	26	16
5	1	4	57	25
n	1	$n-1$	$2^{(n+1)} - n - 2$	n^2

For the implementation of the generalized Toffoli gate, regardless of the number of control inputs, only one-step transitions are required, which is due to the fact that the control input signals are repeated at the output, in contrast to the Peres gates.

In [12, 13] there were proposed various circuit solutions for generalized Peres gates with n-control signals on the basis of SNOT, CNOT, and CV/CV$^+$ primitives. The quantum cost of the generalized Peres gate is scaled as n^2 [12] and $2^{n+1} - n - 2$ [13]. The analysis of the presented quantum cost dependencies for the generalized Peres gate indicates the benefits of its physical implementation on a linear chain of paramagnetic atoms with the spin one half.

3.2 Fidelity

One of the serious problems in liquid and solid-state nuclear magnetic resonance (NMR) quantum computation is a destructive influence of different kinds of errors due to the external environment [15–18]. It was shown, that Peres and Toffoli gates clearly operates on digital initial states during π-pulses, i.e. probability of the realization of the final state equal to 1. One of the important points of the model under consideration is that one keeps the unchanged magnetic field in z-direction, as well as magnitude of the RF-field. An inaccuracy of the setting of the magnetic fields can cause unbalance in the Larmore frequencies ω_k, Rabi frequency Ω and resonant frequencies ω_{lm} that may be dangerous for the stability of the quantum computation protocols.

Here we consider the effect of such frequency noise, when the value of resonance frequency ω_{lm} is incorrectly set RF-field generator. That is, we study the operation of the gates when frequency of the RF-field differs from the resonance frequency: $\omega = \omega_{lm}(1 + \eta)$, where η is a relative error of the generator settings. To study the performance of the quantum gate algorithm as a function of η we will calculate the wave function $\Psi(t)$, obtained from numerical simulations with frequency ω at the end of π-pulses, and make the comparison with the ideal wave function $\Psi_0(t)$, which is expected at resonance frequency ω_{lm} ($\eta = 0$). The fidelity parameter, which is defined as follows

$$F = \langle \Psi(t) | \Psi_0(t) \rangle \tag{12}$$

is a measure of good performance of the quantum gate algorithms [16]. Figure 3 shows the fidelity parameter (12) at the end of π-pulses, as a function of the relative error of resonance frequency η for several transitions simulating the operation of the four-qubit Peres gate. This gate was taken as an example. The same results are valid for any N-qubit Peres and Toffoli gate. The fidelity is an even function of the parameter η and has, in addition to the main maximum for $\eta = 0$, a series of additional, much weaker maxima (see Fig. 3). The width of the main maximum determines the frequency interval which ensures the correct operation of the gate. Calculations show that the width of the main maximum depends on the absolute energy change during the transition. For the transition $8 \to 6$ with the energy change of ~200 2π MHz the width is $8.4 \cdot 10^{-4}$ 2π MHz and decreases to $3.6 \cdot 10^{-4}$ 2π MHz for the transition $12 \to 10 \to 14$ with the energy change of ~600 2π MHz. As follows from the calculations, the width of the main maximum for all transitions, implementing correct a operation of the four-qubit Peres gate, is within 10–100 2π kHz. Note that the fidelity in the cases of one- and two-step transition with the same energy change almost coincides. This conclusion remains valid also in the cases of many-step transitions in N-qubit Peres gate. We can conclude that the operation of the system of spins by the RF field is quite sensitive to the generator settings. The relative error of the resonant frequency should be no more than 10^{-4}. This is also true for generalized Peres and Toffoli gates. The fidelity parameter does not depend on the number of steps and remains within determined above limits.

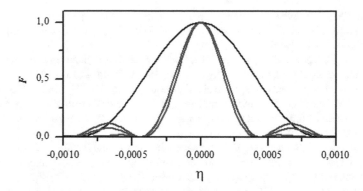

Fig. 3. Fidelity as a function of the relative error η for several transitions in four-qubit Peres gate. Transitions: $8{\rightarrow}6$ (black color); $14{\rightarrow}10$ (green color); $16{\rightarrow}15{\rightarrow}11$ (orange color); and $12{\rightarrow}10{\rightarrow}14$ (blue color).

4 Conclusion

We presented a study of the generalized Peres gate with n-control signals in the Ising model for a one-dimensional chain of nuclear spins (one half). It was shown that the maximum number of π-pulses needed for the implementation of the truth table of the Peres gate can be considered as the quantum cost of the gate and linearly scales with the number of control signals. Comparison of so-defined quantum cost with corresponding quantities studied by the other authors [12, 13], allows us to suggests that the proposed physical implementation of the generalized Peres gate is more efficient. We also examined the correctness of the operation of the generalized Peres gate taking as an example the four-qubit Peres gate. It was shown that an imbalance of the resonance frequency with the relative error $|\eta| < 10^{-4}$ remains the gate well defined with the fidelity close to unity.

References

1. Frank, M.P.: Foundations of generalized reversible computing. In: Reversible Computation, RC 2017. LNCS, vol. 10301, pp. 19–34. Springer, Cham (2017)
2. Drechsler, R., Wille, R.: Reversible circuits: recent accomplishments and future challenges for an emerging technology. In: Progress in VLSI Design and Test, pp. 383–392. Springer, Heidelberg (2012)
3. Landauer, R.: Irreversibility and heat generation in the computational process. IBM J. Res. Develop. **5**(1/2), 183–191 (1961)
4. Bennett, C.H.: Logical reversibility of computation. IBM J. Res. Develop. **17**(6), 525–532 (1973)
5. Nielsen, M., Chuang, I.: Quantum Computation and Quantum Information. Cambridge University Press, Cambridge (2000)
6. Blatt, R.: Quantum information processing with trapped ions. In: Quantum Information and Measurement, p. Th1.1 (2013)

7. Deibuk, V.G., Biloshytskyi, A.V.: Design of a ternary reversible/quantum adder using genetic algorithm. Int. J. Inf. Technol. Comput. Sci. (IJITCS) **7**(9), 38–45 (2015). https://doi.org/10.5815/ijitcs.2015.09.06

8. Berman, G.P., Doolen, G.D., Holm, D.D., Tsifrinovich, V.I.: Quantum computer on a class of one-dimensional Ising systems. Phys. Lett. A **193**(5–6), 444–450 (1994)

9. Kane, B.: A silicon-based nuclear spin quantum computer. Nature **393**(6681), 133–137 (1998)

10. Lopez, G.V., Lara, L.: Numerical simulation of controlled-controlled-not (CCN) quantum gate in a chain of three interacting nuclear spins system. J. Phys. B Atom. Opt. Mol. Phys. **39**(9), 3897–3904 (2006)

11. Bilal, B., Ahmed, S., Kakkar, V.: Optimal realization of universality of Peres gate using explicit interaction of cells in quantum dot cellular automata nanotechnology. Int. J. Intell. Syst. Appl. (IJISA) **9**(6), 75–84 (2017). https://doi.org/10.5815/ijisa.2017.06.08

12. Szyprowski, M., Kerntopf, P.: Low quantum cost realization of generalized Peres and Toffoli gates with multiple-control signals. In: Proceedings of the 13th IEEE International Conference on Nanotechnology, Beijing, China, pp. 802–807 (2013)

13. Moraga, C.: Multiple mixed control signals for reversible Peres gates. Electron. Lett. **50**(14), 987–989 (2014)

14. Berman, G.P., Doolen, G.D., Tsifrinovich, V.I.: Solid-state quantum computation – a new direction for nanotechnology. Superlattices Microstruct. **27**(2/3), 89–104 (2000)

15. Gorin, T., Lara, L., López, G.V.: Simulation of static and random errors on Grover's search algorithm implemented in an Ising nuclear spin chain quantum computer with a few qubits. J. Phys. B: Atom. Mol. Opt. Phys. **43**(8), 085508 (2010). (9 pages)

16. Mendonça, P.E.M., Napolitano, R.D.J., Marchiolli, M.A., Foster, C.J., Liang, Y.-C.: Alternative fidelity measure between quantum states. Phys. Rev. A **78**(5), 052330 (2008)

17. Moghimi, S., Reshadinezhad, M.R.: A novel 4×4 universal reversible gate as a cost efficient full adder/subtractor in terms of reversible and quantum metrics. Int. J. Mod. Educ. Comput. Sci. (IJMECS) **7**(11), 28–34 (2015). https://doi.org/10.5815/ijmecs.2015.11.04

18. Thakral, S., Bansal, D.: Novel reversible DS gate for reversible logic synthesis. Int. J. Mod. Educ. Comput. Sci. (IJMECS) **8**(6), 20–26 (2016). https://doi.org/10.5815/ijmecs.2016.06.03

Mathematical Model of Queue Management with Flows Aggregation and Bandwidth Allocation

Oleksandr Lemeshko[1] , Tetiana Lebedenko[1(✉)] ,
Oleksandra Yeremenko[1] , and Oleksandr Simonenko[2]

[1] Kharkiv National University of Radio Electronics,
Nauka Ave. 14, Kharkiv, Ukraine
oleksandr.lemeshko@nure.ua,
tetiana.lebedenko@gmail.com,
oleksandra.yeremenko.ua@ieee.org
[2] Ivan Kozhedub Kharkiv National Air Force University,
Sumska St. 77/79, Kharkiv, Ukraine
28186@ukr.net

Abstract. The flow-based mathematical model of queue management on routers of telecommunication networks on the basis of optimal aggregation of flows and bandwidth allocation in queues has been further developed. The novelty of the model is that when flows are queued, they are aggregated based on the comparison of the classes of flows and queues in the course of analyzing the set of classification characteristics. Moreover, the result of calculating the percentage of unused queues in the course of optimal aggregation of flows provided assuming the hypothesis of a uniform or normal distribution of flow service classes within the framework of the model under consideration is presented. Applying the uniform distribution law, it was possible to reduce the number of unused queues by 20%, and by 30% for the normal distribution. Research results confirmed the efficiency of the proposed model.

Keywords: Quality of Service · Mathematical model · Bandwidth allocation
Queue · Flows aggregation

1 Introduction

Based on the earlier analysis [1–6], it was found that in the currently geographically distributed telecommunications networks (TCSs), the most significant increase in packet delays and loss is usually caused by overloading the network interfaces of Layer 3 switches or routers. The main cause of overload is the excess and/or commensurability of the packet flow rate with the bandwidth (BW) of the interface through which it flows. In addition, the most important characteristics of the interface are the number of supported queues into which packets are placed when the link is overloaded, and the maximum possible number of packets in each queue. Thus, traffic management mechanisms on the interfaces of modern routers must meet the requirements for

© Springer International Publishing AG, part of Springer Nature 2019
Z. Hu et al. (Eds.): ICCSEEA 2018, AISC 754, pp. 165–176, 2019.
https://doi.org/10.1007/978-3-319-91008-6_17

ensuring the required Quality of Service (QoS) and be implemented based on the solution of the following interrelated tasks:

- classification and marking of packets;
- creation and configuration of the queue system on the interface;
- allocation of packets into the queues of the interface taking into account the parameters of the transmitted flows, the quality of service requirements, the characteristics of the queues being created, and the interface as a whole;
- determination of the order of servicing queues, i.e. determining the order of the transmission of packets from queues to the communication link;
- allocation of interface bandwidth among individual queues;
- preventive (early) limitation of the queue length.

These tasks should be solved as closely as possible to the framework of common interface mechanisms that are a part of the operating system of the router. Therefore, an actual and scientifically practical task is the development of mathematical model of queue management with flows aggregation and bandwidth allocation that can be used as a basis for promising protocols and technological solutions.

2   Related Work

In the view of the fact that the formation of queues is mainly carried out according to the settings "by default" or manually (Table 1), i.e. is assigned to the network administrator, the effectiveness of the solutions obtained depends entirely on his experience and skill level [1–3]. The same applies to the solution of problems on the allocation of the interface BW between the queues.

Table 1. Brief discription of the main mechanisms for queue servicing.

Name of mechanism	Property of mechanism		
	Number of queues	Queue forming	QoS provision
FIFO	1	Auto	No
PQ	4	Manual	No
CQ	16 + 1	Manual	No
FQ/WFQ, FB-WFQ	256 (4096)	Auto	No
CBWFQ	64 classes	Manual	Yes
LLQ	64 + 1	Manual	Yes

It is important to note, on the one hand, that the more queues are formed, the higher the differentiation that can be ensured when servicing packets. From this point of view, there is a tendency to make the number of queues correspond to the number of flows transmitted through the given interface. On the other hand, with the increase in the number of queues supported, the delays introduced by the packet scheduler also increase. Especially it is not allowed when due to non-stationary traffic some queues are unoccupied, and the scheduler unreasonably spends time analyzing their condition, introducing additional delays in servicing packets of other queues. In practice, the way out of the situation is seen in the fact that the flows are aggregated (grouped) according to a number of key parameters related to their classification characteristics and requirements to the level of quality of service. This allows to reduce the number of queues used or to limit the number of queues available by default to approximately the same level of QoS for aggregated flows, which is a key requirement for aggregation.

There are quite a lot of models describing the problem of managing the buffer and (or) link resources of a telecommunications network, taking into account certain requirements. Thus, the paper [7] describes the algorithm for efficient queue management using proportional bandwidth allocation and TCP system protection based on the multi-step increase mechanism and one-step reduction (MISD). This algorithm allows access to resource allocation for use with different priorities, and also guarantees an equitable distribution of the bandwidth for flows with equal or similar values. The disadvantage of the proposed algorithm is the ability to monitor the state in dependence on time and without taking into account the dynamics of their change in time.

In work [8] the priority technique of prevention of overloading on routers of heterogeneous networks on the basis of queues priorities for data-sensitive data is proposed for consideration. According to [8], each stage is divided into two sub-lines with high and low priorities, respectively. Based on the information about the threshold value of the packet size, the queues utilization and the probability of dropping packets, the priority of the queues is calculated and compared. The disadvantage of the proposed mechanism is the possible loss of data of low priority queues. Also, the authors note the need for further development of this approach from the position of applying it to a medium with real-time data transmission.

Paper [9] describes queue management architecture for delay tolerant networking where queue-management approach integrates connectivity status into buffering and forwarding policy, eliminating the possibility of stored data to expire and promoting applications that show potential to run smoothly. The disadvantage of [9] is seen in the absence of more detailed analysis, which incorporates total capacity and storage capacity as well, in order to highlight one major property of delay tolerant network. The model is not defined for other queuing systems. The dependence of the change in the system parameters in the real-time scale is not taken into account.

In [10] presented the dynamics analysis of multipath QoS-routing tensor model with support of different flows classes. The research results have demonstrated that it is especially important to take into account the dynamics of network state changes in the conditions of high utilization of its interfaces.

Returning to the problem of providing an agreed solution to individual interface problems, we note that the results obtained in [8–12] should be modified in such a way that during the allocation of packets in queues, the flows are aggregated according to the classification characteristics. The disadvantages of the solutions proposed in [10–12] include the fact that they did not exclude the possibility of directing flows that differed in their parameters (from the point of view of classification) in order to ensure the optimal balancing of the queue lengths, which could lead to a non-rational allocation of the interface BW and the corresponding overestimation or underestimation of the quality of service to certain flows.

3 Mathematical Model of Queue Management

The proposed work is based on the results obtained in [8–12]. Then suppose that M flows are arriving into the router interface with the following known characteristics:

a_i is the value of the average intensity of the i th flow;

$K_i = \{k_i^l, l = \overline{1, L}\}$ is a set of parameters of the i th flow, which are used to classify the network traffic within the transport technology implemented in the TCS, where L is the total number of parameters for the classification of traffic.

Assume that for each i th flow, on the basis of the analysis of the set of parameters $\{k_i^l, l = \overline{1, L}\}$, its class k_i^f is defined, which is some function of the elements belonging to the set K_i. In the general case, this function can have a nonlinear character, for example, the analog of this dependence had the form

$$k_i^f(p_i, d_i) = \frac{p_i}{v \cdot d_i}, (i = \overline{1, M}) \tag{1}$$

where p_i and d_i are the priority and average length of the packets of the i th flow, v is a certain normalization coefficient, which should smooth out the difference in the order of priority values $(0 \div 7)$ and packet length in bytes.

In the general case, the quantity k_i^f is dimensionless, and for convenience of the subsequent exposition we will assume that k_i^f is normalized from zero to one. The most important flow will have a class value of 1; i.e. $k_i^f = 1$. The smaller the class is, the closer to zero the value k_i^f will be.

Suppose that the packets arriving at the interface in accordance with the contents of the current routing table should be distributed among the queues N in the course of solving the Congestion Management tasks when calculating a set of variables of the second type $x_{i,j}$ $(i = \overline{1, M}, j = \overline{1, N})$ Each of the variables $x_{i,j}$, characterizes the share of the i th flow, aimed at servicing in the j th queue.

To solve the problems of allocating the interface BW among queues (Resource Allocation) within the proposed model, it is necessary to calculate the set of variables b_j each of which characterizes the bandwidth of the interface allocated for servicing of the j th queue.

In the framework of the following discussion, we agree that the number of flows exceeds the number of queues supported on the interface, i.e. there is an inequality,

$$M > N. \tag{2}$$

By analogy with the classification of flows (packets), we will assume that the queue management system also includes classes and queues themselves, as done, for example, in the CBQ, CBWFQ and LLQ mechanisms. Then, for each j th queue, we associate a class $k_j^q(j = \overline{1, N})$ that by analogy with the class of flows k_i^f is also a dimensionless variable, ranging from 0 (not including) to 1 (including). In practice, packets of the same flow are sent to the same queue; therefore, in accordance with the physical meaning of the problem being solved, the variable $x_{i,j}$ is Boolean:

$$x_{i,j} \in \{0,1\}. \tag{3}$$

In addition to the restrictions (3) on the control variables $x_{i,j}$, the conditions for flow conservation on the router interface are superimposed:

$$\sum_{j=1}^{N} x_{i,j} = 1, \, (i = \overline{1,M}). \tag{4}$$

Fulfillment of the condition (4) ensures that all packets of the i th flow are directed to one of the queues organized on the considered interface.

Variables b_j are positive real quantities, on which a system of constraints of the following form are imposed

$$0 < b_j, \, (j = \overline{1,N}) \tag{5}$$

where b is the total bandwidth of the interface, determined by the type of telecommunication technology used. Compliance with conditions (5) determines the correctness of the allocation of the BW interface between individual queues.

To ensure controllability of the process of preventing overloading the interface, it is necessary to fulfill the following condition:

$$\sum_{i=1}^{M} a_i x_{i,j} < b_j, \, (j = \overline{1,N}) \tag{6}$$

i.e. the total intensity of flows aimed at servicing in the j th queue should not exceed the bandwidth of the interface that is allocated to this queue.

Fulfillment of condition (6) is not sufficient to prevent the queue buffer from overloading over its length, because the intensity of the flows arriving at any queue is of an occasional and non-stationary nature. For each j th queue, we denote by $\overline{n}_j$ and $n_j^{\max}$ $(j = \overline{1,N})$ its current length (in packets) and the maximum capacity [12]. Then the conditions of controllability by the overload prevention process (6) are supplemented with the conditions for preventing queue overloading along their length:

$$\overline{n}_j \leq n_j^{\max}, \, (j = \overline{1,N}) \tag{7}$$

where the values $\overline{n}_j$ depend on the statistical characteristics of the flow, the selected packet service discipline, and the allocated BW queue of the interface.

Developing and supplementing the ideas for the concept of Traffic Engineering Queues [1–6], which regulates the issues of ensuring balanced loading of queues, we introduce a number of additional conditions into the structure of the model

$$k_j^q \overline{n}_j \leq \beta n_j^{\max}, \, (j = \overline{1,N}) \tag{8}$$

where β is the minimized dynamic upper bound for the queues utilization along their length on the interface of the router, varies from $0 \leq \beta \leq 1$. The physical meaning of conditions (8) is that the queues created on the interface are loaded in a balanced manner. In this case, the higher the queue class $\left(k_j^q\right)$ is, the shorter length it should have.

The coordinated calculation of control variables $x_{i,j}$, b_j and β will be provided in the course of solving the optimization problem associated with minimizing the objective function of the form:

$$F = \sum_{i=1}^{M} \sum_{j=1}^{N} h_{i,j}^x x_{i,j} + \beta \tag{9}$$

where $h_{i,j}^x$ is the conditional cost (metric) of serving the packets of the i th flow using the j th queue. The physical meaning of the formulated task as a whole is that the calculation of control variables should lead to minimization of the total cost of using network resources: the first term is responsible for the order of using the queue buffer, and the second is responsible for the bandwidth of the interface.

It is important to note that the criterion for the direction of a particular flow in a certain queue is the maximum coincidence (commensurability) of their classes – flow class, k_i^f and queue class, k_j^q. Then within the framework of the improvement of the model (3)–(9) it is suggested that the metric, which is responsible for the allocation of packet flows in queues, and under fulfilling the condition (2) – for aggregating flows, has been calculated according to the formula,

$$h_{i,j}^x = w_x^b (k_i^f - k_j^q)^2 + 1, (i = \overline{1,M}, j = \overline{1,N}). \tag{10}$$

Thus, the metric $h_{i,j}^x$ is a nonnegative quantity and it directly depends on the square of the distance between the classes of individual flows and queues. Using the normalization parameter w_x^b, it is possible to control the effect on the final numerical value of the objective function (9) of its first and second terms.

The proposed model with newly introduced formalisms (1), (2), (8)–(10) allows to provide a coordinated solution of tasks on aggregation and distribution of flows by queue, and also on the allocation of the interface bandwidth between the system of queues supported on it. The novelty of the model is that when flows are queued, they are aggregated based on the comparison of the flow and queue classes in the course of analyzing the set of classification characteristics.

4 Investigation of Mathematical Model of Queue Management with Flows Aggregation and Bandwidth Allocation

In this paper, we analyzed the queue management process on the router interface using the described model. The research has found that the number of queues that were not used during optimal aggregation of flows was influenced by a number of the following factors, such as: the total number of packet flows, the number of queues organized on the interface, the law of distribution of service classes between them (flows and queues). The total number of flows and queues was considered known.

The queue classes are determined and uniformly set in the range from 1 to 10. Flow classes were determined by random law within the same limits (from 1 to 10). In this paper, we used two laws of distribution of service classes between flows: a uniform and normal (truncated normal) distribution law.

Uniform distribution law is typical for the case when the probabilities of the appearance of packets of a class on the interfaces of multiservice network routers are approximately the same.

The truncated normal distribution law is characteristic in view of the restricted set of values of flow classes and is valid for the case when a certain set of classes prevails in the network, and the power of this set is given indirectly through the variance of the distribution σ^2.

Thus, for example, in Table 2, the initial data and calculation results for twenty five flows $(N = 25)$, ten queues $(M = 10)$, the interface bandwidth $(b = 100)$, normalization parameter $(w_x^b = 100)$ and hypothesis about the uniform distribution of service classes between flows are presented.

For the convenience of description, we introduce some notation:

- FN – the flow number;
- QN – the queue number;
- FC – the flow class;
- QC – the queue class;
- a_f – the average flow rate;
- a_q – the average flow rate in the queue;
- DB – the dedicated bandwidth;
- AQ/MQ – the average queue length/maximum queue length.

During the calculation, the average queue length was determined on condition that the interface was modeled by the queuing system M/M/1 [13–16]. According to the data presented in Table 2 the minimized dynamic upper bound for the queues utilization was $\beta = 0,9963$. The simulation results showed that aggregation of flows was carried out in accordance with the proximity of the flow classes and the queue class to which they were directed (10). In addition, out of ten possible, the seventh and tenth queues were not involved. In the framework of this example, it was possible to reduce the number of unused queues by 20%.

Table 2. Initial data and calculation results for $N = 25$, $M = 10$, $b = 100$

Flow characteristics			Queue parameters				
FN	*FC*	$a_f(1/s)$	*QN*	*QC*	$a_q(1/s)$	*DB*	*AQ/MQ*
10	1.1709	0.8042	1	1	2.7451	2.7724	99.6200/100
15	1.2631	1.9410					
5	1.5041	1.4065	2	2	10.8439	11.0806	44.8328/90
9	1.6042	5.5167					
13	1.8500	3.9207					
3	3.3355	4.2329	3	3	8.0895	8.3833	26.5676/80
6	3.3370	3.8566					
4	3.8974	3.0058	4	4	11.3701	11.9886	17.4351/70
11	3.9855	4.1317					
19	3.8957	4.2326					
14	5.0079	4.7179	5	5	20.8570	22.4758	11.9555/60
20	4.9014	5.7326					
23	4.9839	5.3339					
24	5.2923	5.0646					
7	5.5368	1.1793	6	6	4.8069	5.3291	8.3024/50
8	5.5029	1.2641					
18	5.6435	2.3636					
–	–	–	7	7	–	–	–
1	7.6393	3.7030	8	8	11.8491	14.4498	3.7361/30
12	8.3138	2.7988					
17	7.6665	5.3472					
2	9.4764	4.9650	9	9	17.5831	23.5202	2.2140/20
16	8.6463	0.2750					
21	8.8690	1.7542					
22	9.0505	6.3866					
25	9.0970	4.2023					
–	–	–	10	10	–	–	–

In the second example that presented in Table 3, the original data remained the same, but the law of the distribution of service classes between flows is normal with $\sigma^2 = 0.3$. The minimized dynamic upper bound for the queues utilization was $\beta = 0,8172$.

Table 3. Initial data and calculation results for $N = 25$, $M = 10$, $b = 100$

Flow characteristics			Queue parameters				
FN	FC	$a_f(1/s)$	QN	QC	$a_q(1/s)$	DB	AQ/MQ
10	0.6374	2.7336	1	1	11.7254	11.8671	81.7201/100
14	1.1192	4.1350					
23	0.2286	4.8567					
1	1.5187	5.5946	2	2	24.3278	24.9723	36.7748/90
4	1.7136	6.2721					
8	2.2219	3.8225					
18	2.1990	0.5468					
22	1.8839	4.5273					
25	2.4364	3.5644					
5	3.2998	3.7701	3	3	6.8321	7.1324	21.7923/80
20	2.6624	1.6023					
21	2.9150	1.4596					
2	3.806	3.4043	4	4	19.3050	20.57717	21.7923/70
11	4.2548	3.9107					
12	4.0434	0.4557					
13	4.0029	5.9555					
15	4.2431	0.6730					
19	3.5408	1.3812					
24	4.4940	3.5245					
–	–	–	5	5	5	–	–
16	6.1355	4.5094	6	6	4.5094	5.0955	6.8101/50
–	–	–	7	7	–	–	–
3	8.2102	5.8870	8	8	11.8539	14.9258	3.0646/30
7	8.0629	5.9669					
–	–	–	9	9	–	–	–
6	9.6647	0.7662	10	10	9.0057	15.4352	0.8172/10
9	9.9977	5.6905					
17	9.8806	2.5490					

As noted above, the number of unused queues during the flows aggregation was affected by the total number of queues organized on the interface (N), the total number of flows (M) and the law of distribution of service classes among these flows.

In the distribution of service classes between flows, the influence of the ratio of the number of flows and queues (M/N) on the percentage of unused queues during the optimal aggregation of flows within the proposed model was also estimated:

$$P_\% = \frac{N_{UQ}}{N} \cdot 100\% \qquad (11)$$

where N_{UQ} is a number of unused queues, obtained as a result of solving the queue management problem with optimal aggregation of flows.

Figure 1 shows the results of calculating the percentage of unused queues (11) in the course of optimal aggregation of flows (Table 2) provided assuming the hypothesis of a uniform distribution of flow service classes within the framework of the model under consideration.

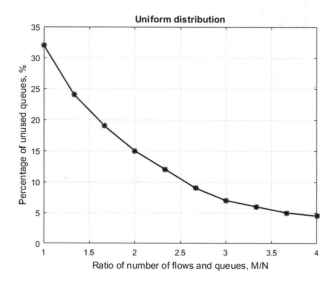

Fig. 1. Results of calculating the percentage of unused queues during the optimal aggregation of flows with the uniform distribution law of service classes among flows.

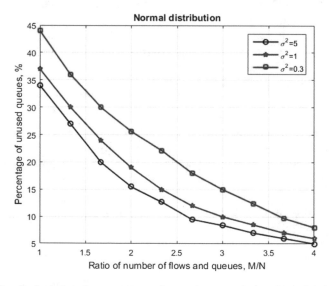

Fig. 2. Results of calculating the percentage of unused queues during the optimal aggregation of flows under the normal distribution law of service classes among flows.

Within the framework of the example considered, using the hypothesis of the uniform distribution, it was possible to reduce the number of unused queues by 20%. Minimizing the number of queues supported has resulted in a proportional decrease in the time spent on processing packets on the interface in the particular queue, on average from 5–7% to 20–25%.

Figure 2 shows a variant of calculating the percentage of unused queues (11), assuming a hypothesis about the normal distribution of service classes between flows (Table 3) for different variances $\sigma^2 = 0.3$, $\sigma^2 = 1$, $\sigma^2 = 5$.

In the second example, the original data remained the same, but the law of distribution of service classes between the flows changed to the normal distribution law. The number of unused queues was reduced by 30%. The applied model allowed to aggregate the flows in such a way that the time for their processing decreased on average

- from 12–15% to 30–35% under $\sigma^2 = 0.3$;
- from 8–12% to 28–32% under $\sigma^2 = 1$;
- from 7–10% to 25–30% under $\sigma^2 = 5$.

5 Conclusion

In this paper, the flow-based model of queue management on routers of TCS was further developed on the basis of optimal aggregation of flows and distribution of packets in queues. The novelty of the model is that when flows are queued, they are aggregated based on the comparison of flow classes and queue classes during the analysis of a set of classification features.

The optimization problem of the mixed integer nonlinear programming (MINLP) class is set in the basis of the received solutions for queue management, because control variables are both integer (3) and real (5), and among the constraints imposed on them there are both linear (4), (5), and nonlinear conditions (6)–(8).

The chosen objective function (9) to be minimized is a linear form and determines the conditional cost of an agreed solution to the problems of allocation of the flow packets among the generated queues (Congestion Management), the allocation of the bandwidth of the interface among the queues (Resource Allocation), the restriction of the intensity of the flows arriving to the interface (Congestion Avoidance).

The results of the studies confirmed the operability of the proposed model, as well as its effectiveness from the point of view of correct solution for queue management tasks. Optimization of the flow aggregation process allows to reduce the number of supported queues from 15–18% to 25–33%, depending on the ratio of the number of flows and queues, as well as the law of distribution of service classes between the flows, without reducing the level of differentiation of the quality of user flow service, which helps to reduce the processing time of packets on the interface and minimizing end-to-end packet delay in the network as a whole.

References

1. Barreiros, M., Lundqvist, P.: QOS-Enabled Networks: Tools and Foundations. Wiley Series on Communications Networking & Distributed Systems, 2nd edn. (2016)
2. Stallings, W.: Foundations of Modern Networking: SDN, NFV, QoE, IoT, and Cloud. Addison-Wesley Professional, Boston (2015)
3. Alvarez, S.: QoS for IP/MPLS Networks. Cisco Press, Indianapolis (2006)
4. Szigeti, T., Hattingh, C., Barton, R., Briley, K.: End-to-End QoS Network Design: Quality of Service for Rich-Media & Cloud Networks, 2nd edn. Cisco Press, Indianapolis (2013)
5. Varma, S.: Internet Congestion Control. Morgan Kaufmann, San Francisco (2015)
6. Haghighi, A., Mishev, D.: Delayed and Network Queues. Wiley, Hoboken (2016)
7. Lu, L., Du, H., Liu, R.P.: CHOKeR: a novel AQM algorithm with proportional bandwidth allocation and TCP protection. IEEE Trans. Ind. Inf., 637–644 (2014). https://doi.org/10.1109/TII.2013.2278618
8. John, J., Balan, R.: Priority queuing technique promoting deadline sensitive data transfers in router based heterogeneous networks. Int. J. Appl. Eng. Res. **12**(15), 4899–4903 (2017)
9. Lenas, S., Dimitriou, S., Tsapeli, F., Tsaoussidis, V.: Queue-management architecture for delay tolerant networking. Wired/Wirel. Internet Commun., 470–482 (2011). https://doi.org/10.1007/978-3-642-21560-5_39
10. Lemeshko, O., Yeremenko, O.: Dynamics analysis of multipath QoS-routing tensor model with support of different flows classes. IEEE Smart Syst. Technol. (SST), 225–230 (2016). https://doi.org/10.1109/SST.2016.7765664
11. Lemeshko, O., Semenyaka, M., Simonenko, O.: Researching and designing of the dynamic adaptive queue balancing method on telecommunication network routers. In: Proceedings of XIIth CADSM 2013, pp. 204–207 (2013)
12. Lebedenko, T., Simonenko, A., Arif, F.A.R.: A queue management model on the network routers using optimal flows aggregation. In: 13th International Conference on Modern Problems of Radio Engineering, Telecommunications and Computer Science (TCSET), pp. 605–608. IEEE (2016)
13. Wang, P.: Dynamics of Delay Differential Equations in Communications Networks: In the Framework of Active Queue Management. LAP Lambert Academic Publishing (2009)
14. Das, I., Lobiyal, D.K., Katti, C.P.: Queuing effect on multipath routing in mobile ad hoc networks. Int. J. Inf. Eng. Electr. Bus. (IJIEEB) **8**(1), 62–68 (2016). https://doi.org/10.5815/ijieeb.2016.01.07
15. Kushwaha, V., Ratneshwer: Interaction of high speed TCPs with recent AQMs through experimental evaluation. Int. J. Comput. Netw. Inf. Secur. (IJCNIS) **8**(9), 41–47 (2016). https://doi.org/10.5815/ijcnis.2016.09.06
16. Maity, D.S., Goswami, S.: Multipath data transmission with minimization of congestion using ant colony optimization for MTSP and total queue length. Int. J. Comput. Netw. Inf. Secur. (IJCNIS) **7**(3), 26–34 (2015). https://doi.org/10.5815/ijcnis.2015.03.04

Planning Automation in Discrete Systems with a Given Structure of Technological Processes

Alexander Anatolievich Pavlov$^{(\boxtimes)}$ ⓘ, Elena Andreevna Khalus ⓘ,
and Iryna Vitalievna Borysenko ⓘ

National Technical University of Ukraine
"Igor Sikorsky Kyiv Polytechnic Institute",
37, Prospekt Peremohy, Kyiv 03056, Ukraine
pavlov.fiot@gmail.com, selena.ua@gmail.com,
ireneborysenko@gmail.com

Abstract. In this paper, we consider mathematical models and algorithms for efficient planning process automation in discrete systems of a wide class (innovative software development, small-scale production). An effective solution of the proposed models is based on earlier results of M.Z. Zgurovsky, A.A. Pavlov, E.B. Misura, and E.A. Khalus in the field of intractable single stage single machine scheduling problems.

Keywords: Planning · Process automation · Scheduling
Combinatorial optimization · Just in time · Portfolio of orders
Profit maximization

1 Introduction

The results we present below relate to the field of planning process automation in systems with discrete production properties. We consider two mathematical models of systems with discrete nature of technological process and wide practical applications:

- Scheduling process automation for a two-level model of teams working on software development and maintenance. At the first level, the teams develop an innovative software. At the second level, they execute profile-specific tasks that turn the software into a finished product.
- Determining an efficient portfolio of orders, in terms of the expected profit, and automation of its execution planning process. The planning is based on an aggregated model of discrete systems of a wide class (for example, small-scale productions that account for 70% of all discrete productions). The resulting aggregated plan is a constructive basis for a subsequent operational planning. Indeed, after aggregated tasks disaggregation during an operational plan construction, regardless of the specific structure of discrete technological processes, we first execute the operations belonged to the first aggregated task of the plan, then those of the second aggregated task, and so on.

© Springer International Publishing AG, part of Springer Nature 2019
Z. Hu et al. (Eds.): ICCSEEA 2018, AISC 754, pp. 177–185, 2019.
https://doi.org/10.1007/978-3-319-91008-6_18

A constructive solution of the above formulated problems of planning process automation in discrete systems is based on earlier efficient results of M.Z. Zgurovsky, A.A. Pavlov, E.B. Misura, and E.A. Khalus in solving intractable single stage scheduling problems. Two of them are intractable combinatorial optimization problems.

Planning automation problems in discrete systems are considered in [1–8]. The results we propose are qualitatively different from those earlier works due to the originality of both the problems formulations and of their solving algorithms.

2 Formulation of Problem 1

Given a set of independent tasks $J_1 \ldots J_n$, each consisting of two successive operations. For each task J_i, we know its due date $d_i, i = \overline{1, n}$, processing time of the first operation $l_i^1, i = \overline{1, n}$ and processing time of the second operation $l_i^2, i = \overline{1, n}$. Here we denote by $\overline{p, q}$ the interval of integers from p to q, that is, $\overline{p, q} = p, p + 1, \ldots, q$.

Interrupts are not allowed during an operation execution.

There are executors of the first and second levels. Executors of the first level are the teams of developers who create an innovative software. They execute the first operations of tasks. Executors of the second level are "independent" developers who execute profile-specific tasks that turn the software into a finished product. So, they execute the second operations of tasks. Executors of the second level receive their tasks from the executors of the first level. Figure 1 shows the structure of the two-level system.

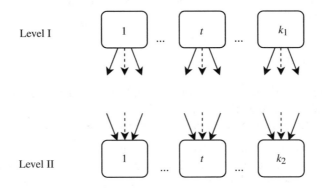

Fig. 1. Structure of a technological process in the two-level system

Level I (execution of the first operation of a task) includes k_1 objects. There are

$$J_{i1}^{1j}, \ldots, J_{il_j}^{1j}; \; J_{ip}^{1j} \in \{J_1 \ldots J_n\}, \; i = \overline{1, k_1}, \; p = \overline{1, l_j}$$

where $1j$ is jth object of the level I; ip is the number of task J_{ip}^{1j}. For each J_{ip}^{1j} we know the processing time of the first operation l_{ip}^1. The following holds:

$$\left\{ J_t^{1j}, t = \overline{1, l_j} \right\} \cap \left\{ J_t^{1s}, t = \overline{1, l_s} \right\} = \varnothing, \ \forall j \neq s, \tag{1}$$

and

$$\bigcup_{j=1}^{k_1} \left\{ J_t^{1j}, t = \overline{1, l_j} \right\} = \{ J_1, \ldots, J_n \}. \tag{2}$$

Level II (execution of the second operation of a task) includes k_2 objects. Tasks $J_{i_l}^{2j} \in \{ J_1, \ldots, J_n \}$, $l = \overline{1, p_j}$, are passed to each object of the level II to execute the second operation. Executor $2j$ executes the second operation of the task $J_{i_l}^{2j}$, here i_p is the task's number. For each such task, we know the processing time of the second operation $l_{i_l}^2$ and the due date $d_{i_l}^{2j} = d_{i_l}$. The following holds:

$$\left\{ J_{i_t}^{2j}, t = \overline{1, p_j} \right\} \cap \left\{ J_{i_t}^{2s}, t = \overline{1, p_s} \right\} = \emptyset, \ \forall j \neq s, \tag{3}$$

and

$$\bigcup_{j=1}^{k_2} \left\{ J_{l_t}^{2j}, t = \overline{1, p_j} \right\} = \{ J_1, \ldots, J_n \}. \tag{4}$$

At each object $2j$, we obtain a set of tasks $J_{i_1}^{2j}, \ldots, J_{i_{p_j}}^{2j}$, their due dates and processing times $d_{i_l}^{2j}, l_{i_l}^2, \ l = \overline{1, p_j}$.

Our goal is to execute both operations for each task and to find a schedule for each executor $1j$, $j = \overline{1, k_1}$, and $2j$, $j = \overline{1, k_2}$. Additionally, we want to execute the second operation for each task just in time, if possible. Also, we require to meet all due dates $d_j, j = \overline{1, n}$. Then, we need to find the latest start times of objects of the level I.

3 Algorithm to Solve Problem 1

To satisfy the just in time criterion, if possible, we solve a single stage scheduling problem 1 to minimize the total earliness for k_2 executors of the second level. Let us consider a general formulation of this problem.

Formulation of Single Stage Scheduling Problem 1. Given a set of n independent tasks $J = \{ 1, 2, \ldots, n \}$ to be assigned to execution without interrupts on one machine. The machine can execute no more than one task at a time. The tasks and the machine are continuously available from time zero, and the machine idle time is not allowed. For each task $j \in J$, we know processing time l_j and due date d_j. Let C_j denote the completion time of the task j and $E_j = d_j - C_j$ denote its earliness. $C_j \leq d_j$ for each task j in a feasible schedule. For any given feasible schedule σ, the total earliness of tasks in regard to their due dates is determined as

$$E(\sigma) = \sum_{j=1}^{n} d_j - C_j. \tag{5}$$

We need to find a feasible schedule with the minimum $E(\sigma)$.

An efficient approximation algorithm to solve the problem is given in [9].

We obtain k_2 schedules for objects $2j, j = \overline{1, k_2}$. Suppose that we have the following optimal schedule for object $2j$: $J_1^{2j}, J_2^{2j}, \ldots, J_{p_j}^{2j}$. We also obtained the start times $r_j, j = \overline{1, n}$, for executors $2j$ of the second level.

Let us determine due dates for objects of the level I.

r_j is the due date for the first operation of task J_1^{2j} at the corresponding object of the level I.

$r_j + l_{J_1^{2j}}^2$ is the due date for the first operation of task J_2^{2j} at the corresponding object of the level I where $l_{J_1^{2j}}^2$ is the processing time of the second operation of task J_1^{2j}.

$r_j + \sum_{t=2}^{k-1} l_{J_t^{2j}}^2$ is the due date for the first operation of task $J_k^{2j}, k = \overline{3, p_j}$, at the corresponding object of the level I.

Using these due dates, we find optimal schedules maximizing the start time for each object of the level I. Thus, we solve a single stage scheduling problem 2 with the following general formulation.

Formulation of Single Stage Scheduling Problem 2. Given a set of n independent tasks $J = \{1, 2, \ldots, n\}$. Each task consists of one operation. For a task $j \in J$ we know its processing time l_j and due date d_j. The tasks come into the system at the same time. Interruptions of tasks execution are not allowed. The execution is continuous: after the first task immediately starts the second one, etc., until all tasks are completed. We need to find a feasible schedule, i.e., such that $\forall j : C_j \leq d_j$ where the start time of tasks is maximized.

For a given schedule, by the start time of tasks we mean the start time of a task that occupies the first position in the schedule.

An efficient exact polynomial algorithm to solve the problem is given in [9].

Thus, we obtain schedules and start times for objects of the level I and schedules for objects of the level II.

Generalization to a p-Level System. In an obvious way, we can extend the solution of Problem 1 to a more general structure of objects we are planning. There are p levels of executors. At each ith level of the object, $i = \overline{1, p}$, we execute ith operation of the task $J_l, l = \overline{1, n}$. For each level, the conditions similar to (1–4) are satisfied. At the level p, we solve the corresponding single stage scheduling problem for each object with due dates $d_{i_l}^{pj} = d_{i_l}$ where $d_i, i = \overline{1, n}$, is the due date specified for the task J_j. According to the schedules obtained, we determine due dates for all objects of level $p - 1$, and so on. We solve the single stage scheduling problem 2 for all objects of levels $p - 1$ to 1.

This way, we obtain schedules for all objects of each of p levels. The structure of the p-level system is shown at Fig. 2.

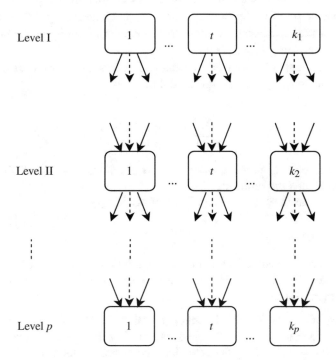

Level I

Level II

Level p

Fig. 2. Structure of a technological process in the p-level system

4 Formulation of Problem 2

Given a portfolio of orders that consists of tasks $J_1 \ldots J_n$. We suppose that we know the resources to execute each task, and that they are constant. For each task, we know its processing time l_i, the time intervals $[D_{jE}, D_{jL}]$, and costs P_{jE}, P_{jL} where $P_{jE} > P_{jL}$. P_{jE} is the cost of a task j if it is completed before the time moment D_{jE} ("early" due date), and P_{jL} is the cost of the task j if it is completed in the time moment D_{jL} ("late" due date). For each $t \in [D_{jE}, D_{jL}]$, the cost P_t is specified with a known non-increasing relation.

We define the model of a production object as a single machine. We specify for it a release time which is a completion time of all previous tasks.

We need to choose such tasks from $J_1 \ldots J_n$ that are executed within the corresponding time interval $[D_{jE}, D_{jL}]$ wherein we achieve the maximum total profit for this set of tasks.

5 Algorithm to Solve Problem 2

1. Solve the single stage scheduling problem 2 for entire portfolio of orders, using D_{jL}.
2. Determine the priority of each task: $w_j = P_{javg}/l_j$ where $P_{javg} = (P_{jE} + P_{JL})/2$. Sort all tasks in non-decreasing order of priorities, starting with the minimum priority task: $w_j \in (w_1 \dots w_n)$, $(w_{j+1} \geq w_j, j = \overline{i,n})$.
3. Eliminate sequentially from the portfolio of orders the tasks with the minimum priority from set $(w_1 \dots w_k)$, until the following is true:

$$\left| \sum_{i=1}^{k} l_{(j)} - r_f \right| \leq \varepsilon > 0$$

where ε is a specified number. After the tasks elimination, we obtain the set of tasks J_{rmd} (*rmd* means the remainder).
4. Apply Subalgorithms 1 and 2 to the set of tasks J_{rmd}.

5.1 Subalgorithm 1

We find an optimal schedule that minimizes the total tardiness for the set of tasks J_{rmd}, using due dates D_{jE}, i.e., we solve a single stage scheduling problem 3 with the following general formulation.

Formulation of Single Stage Scheduling Problem 3. Given a set of n independent tasks $J_1 \dots J_n$. Each task consists of one operation. For a task j we know its processing time l_j and due date d_j. The tasks come into the system at the same time zero. Interruptions of tasks execution are not allowed. We need to build a schedule for one machine that minimizes the total tardiness of tasks:

$$f = \sum_{j=1}^{n} \max (0, C_j - d_j).$$

A good review of solving methods for this problem is given in [10]. More approaches to solve scheduling problems, with possibility to use cloud computing, are studied in [11]. An efficient exact algorithm that covers instances of up to 1,000 tasks is given in [12, Chap. 1].

If all "late" due dates D_{jL} are met, then we further use Subalgorithm 1.1. Otherwise, we use Subalgorithms 1.2 and 1.3.

Subalgorithm 1.1. Add to the set of tasks J_{rmd} one of eliminated tasks that has the maximum priority. Solve the single stage scheduling problem 3 again. Check the obtained solution for feasibility. Execute this procedure until at least one due date D_{jL} is violated. Then the solution of Subalgorithm 1.1 is the schedule obtained at the previous iteration.

From the obtained solution, we find the total profit of the set of tasks according to the expression

$$P_{11} = \sum_{J_j \in J_{11}} P_{J_j}(C_{J_j})$$

where $P_j(C_j)$ is given non-decreasing cost function of task j with completing time C_j; J_{11} is the set of tasks that make up the solution obtained by Subalgorithm 1.1.

Subalgorithm 1.2. Determine the set J_{ovd} (*ovd* means overdue, i.e., not feasible) of the tasks that complete after their "late" due dates. Eliminate from the set J_{ovd} one task with the earliest start time on its machine. In the schedule, all tasks executed after the eliminated task are shifted to the left by its processing time. If the schedule becomes feasible, then it is the solution of Subalgorithm 1.2. Else, repeat this procedure until it yields a feasible solution.

From the obtained solution, we find the total profit of the set of tasks according to the expression

$$P_{12} = \sum_{J_j \in J_{12}} P_{J_j}(C_{J_j})$$

where J_{12} is the set of tasks that make up the solution obtained by Subalgorithm 1.2.

Subalgorithm 1.3. Eliminate from the set J_{rmd} one task that has the minimum priority. Solve the single stage scheduling problem 3 again on the new set. Repeat this procedure until it yields the first feasible schedule. It will be the solution of Subalgorithm 1.3.

From the obtained solution, we find the total profit of the set of tasks according to the expression

$$P_{13} = \sum_{J_j \in J_{13}} P_{J_j}(C_{J_j})$$

where J_{13} is the set of tasks that make up the solution obtained by Subalgorithm 1.3.

5.2   Subalgorithm 2

We solve the single stage scheduling problem 2 for the set of tasks J_{rmd}, using due dates D_{jL}. After this, we execute Subalgorithm 2.1 or 2.2.

Let r_s denote the start time of tasks and r_f denote the release time of machine.

Subalgorithm 2.1. If $r_s > r_f$, then add to the set of tasks J_{rmd} one of eliminated tasks that has the maximum priority. Solve the single stage scheduling problem 2 again, using due dates D_{jL}. Repeat this procedure until condition $r_s < r_f$ is true for the first time. Then the solution of Subalgorithm 2.1 is the solution of the single stage scheduling problem 2 obtained at the previous iteration.

From the obtained solution, we find the total profit of the set of tasks according to the expression

$$P_{21} = \sum_{J_j \in J_{21}} P_{J_j}(C_{J_j})$$

where J_{21} is the set of tasks that make up the solution obtained by Subalgorithm 2.1.

Subalgorithm 2.2. If $r_s < r_f$, then eliminate from the set of tasks J_{rmd} one task that has the minimum priority. Solve the single stage scheduling problem 2 again, using due dates D_{jL}. Repeat this procedure until condition $r_s > r_f$ is true for the first time. Then the solution of Subalgorithm 2.2 is the solution of the single stage scheduling problem 2 obtained at the previous iteration.

From the obtained solution, we find the total profit of the set of tasks according to the expression

$$P_{22} = \sum_{J_j \in J_{22}} P_{J_j}(C_{J_j})$$

where J_{22} is the set of tasks that make up the solution obtained by Subalgorithm 2.2.

The solution of Problem 2 is a schedule that has

$$\max\{(P_{11} \vee (P_{12}, P_{13})), (P_{21} \vee P_{22})\},$$

according to the logic of the algorithm (Fig. 3).

Hence, we obtained an efficient schedule aimed at profit maximization.

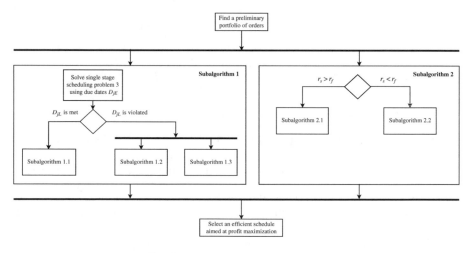

Fig. 3. Logic of the problem solving

6   Conclusion

We have proposed efficient models and algorithms for planning process automation for a fairly wide class of discrete processes (innovative software development, small-scale production). The efficiency criterion in the first case is the completion of tasks just in time. This leads to the combinatorial optimization problem of minimizing the total tardiness of tasks in a feasible solution. In the second case, the efficiency criterion is to find an efficient portfolio of orders, in terms of expected profit, and automation of its execution planning process, based on an aggregated model. Efficiency of the proposed planning automation models is determined by the efficiency of effective solutions for three single stage scheduling problems obtained earlier by M.Z. Zgurovsky, A.A. Pavlov, E.B. Misura, and E.A. Khalus. The proposed models and algorithms are now in the process of implementation within real production systems.

References

1. Alhumrani, S.A., Qureshi, R.J.: Novel approach to solve resource constrained project scheduling problem (RCPSP). Int. J. Mod. Educ. Comput. Sci. (IJMECS) **8**(9), 60–68 (2016). https://doi.org/10.5815/ijmecs.2016.09.08
2. Vollmann, T.E., Berry, W.L., Whybark, D.C.: Manufacturing Planning and Control Systems. McGraw-Hill, Boston (2005)
3. Popinako, D.: ERP-rynok glazami veterana. SK-Press, Kiev (2011). PC Week-Ukr. (in Russian)
4. Zagidullin, R.R.: Upravlenie mashinostroitelnym proizvodstvom s pomoschyu sistem MES, APS, ERP. TNT, Staryi Oskol (2011). (in Russian)
5. Zagidullin, R.R.: Planirovanie mashinostroitelnogo proizvodstva. TNT, Staryi Oskol (2017). (in Russian)
6. Swayamsiddha, S., Parija, S., Sahu, P.K., Singh, S.S.: Optimal reporting cell planning with binary differential evolution algorithm for location management problem. Int. J. Intell. Syst. Appl. (IJISA) **9**(4), 23–31 (2017). https://doi.org/10.5815/ijisa.2017.04.03
7. Soudi, S.: Distribution system planning with distributed generations considering benefits and costs. Int. J. Mod. Educ. Comput. Sci. (IJMECS) **5**(9), 45–52 (2013). https://doi.org/10.5815/ijmecs.2013.09.07
8. Garg, R., Singh, A.K.: Enhancing the discrete particle swarm optimization based workflow grid scheduling using hierarchical structure. Int. J. Comput. Netw. Inf. Secur. (IJCNIS) **5**(6), 18–26 (2013). https://doi.org/10.5815/ijcnis.2013.06.03
9. Pavlov, A.A., Misura, E.B., Khalus, E.A.: Properties' research of the scheduling problem for a single machine by minimizing the total earliness of tasks with the condition of the schedule feasibility. Visnyk NTUU "KPI" Inf. Oper. Comput. Sci. **56**, 98–102 (2012). (in Russian)
10. Koulamas, C.: The single-machine total tardiness scheduling problem: review and extensions. Eur. J. Oper. Res. **202**, 1–7 (2010). https://doi.org/10.1016/j.ejor.2009.04.007
11. Soltani, N., Soleimani, B., Barekatain, B.: Heuristic algorithms for task scheduling in cloud computing: a survey. Int. J. Comput. Netw. Inf. Secur. (IJCNIS) **9**(8), 16–22 (2017). https://doi.org/10.5815/ijcnis.2017.08.03
12. Zgurovsky, M.Z., Pavlov, A.A.: Prinyatie resheniy v setevyh sistemah s ogranichennymi resursami: Monograph. Naukova dumka, Kiev (2010). (in Russian)

Fault-Tolerant Multiprocessor Systems Reliability Estimation Using Statistical Experiments with GL-Models

Alexei Romankevich ⓘ, Andrii Feseniuk$^{(\boxtimes)}$ ⓘ, Ivan Maidaniuk ⓘ, and Vitaliy Romankevich ⓘ

National Technical University of Ukraine "Igor Sikorsky Kyiv Polytechnic Institute", Kyiv, Ukraine
romankev@scs.ntu-kpi.kiev.ua

Abstract. The article is focused on the reliability estimation of fault-tolerant multiprocessor systems with huge number of processors and complex behavior of the systems on its processors failures. A universal method for the reliability estimation of such fault-tolerant multi-processor systems for a given time period is proposed. The method is based on conducting statistical experiments (Monte-Carlo) with models that adequately reflect the behavior of the fault-tolerant multi-processor systems in the flow of failures. For that purpose, it is suggested using GL-model, which is a graph with special way formed Boolean functions assigned to its edges. The questions of synthesis, minimization and transformation of such models are considered. The article addresses the statistical estimation error (Monte-Carlo Error). The upper bound for calculating the error before conducting statistical experiments is suggested. It is shown that the error could be estimated more precisely using the results of conducted statistical experiments. Correspondent statistical estimator is proposed.

Keywords: Reliability · Fault-tolerance · Multiprocessor systems
Statistical experiments method · Monte-Carlo method · Statistical estimation error
Monte-Carlo Error

1 Introduction

The reliability theory has been developed well enough [1] however many scientific problems are still being studied. There are recent interesting results for systems where failure probability is imprecise [2], for coordinator election in distributed systems [3], reliability analysis of combat systems [4] but most of researchers nowadays are focused on software reliability (e.g. [5–8]).

Reliability becomes the most important requirement when talking about control system of critical application such as nuclear power plant, aviation, health monitoring system for critical patients etc. because the operational failure of such system could have catastrophic consequences. The critical application control systems are usually designed as multiprocessor systems with embedded testing, diagnostics, fault-tolerance and reconfiguration.

© Springer International Publishing AG, part of Springer Nature 2019
Z. Hu et al. (Eds.): ICCSEEA 2018, AISC 754, pp. 186–193, 2019.
https://doi.org/10.1007/978-3-319-91008-6_19

A lot of methods were developed for system reliability evaluation. There is parallel and series reductions method for parallel and series system reliability structures [1].

The most efficient method for k-out-of-n systems are Barlow and Heidtmann method, Rushdi method, Belfore method [1]. For general system reliability structures there are pivotal decomposition method, Delta-Star and Star-Delta transformations method, Inclusion-Exclusion method, Sum-of-disjoint-products method, Esary-Proschan method [1].

Mentioned methods in most cases have narrow sphere for application. Moreover, for the fault-tolerant multiprocessor system (FTMS) that has hundreds of processors, high level of redundancy and non-trivial behavior for its processors failures, those methods cannot be applied at all. For such systems, we introduce universal approach for system reliability evaluation based on statistical experiments (Monte-Carlo) with models that reflect system behavior in flow of its components failures. In addition, particular attention in article is paid to using method error estimation.

2   Notations

n is the number of processors.

$[0, t]$ indicates the period of time for which the reliability is evaluated.

y denotes the binary variable representing the state of the fault-tolerant multiprocessor system (0 - system failure, 1 - system works) for period $[0,t]$.

$P(y)$ indicates the reliability of the fault-tolerant multiprocessor system.

$\bar{P}(y)$ is the statistical estimator of reliability of the fault-tolerant multiprocessor system

$x_1, x_2, \ldots, x_n$ are the binary variables representing the state of processors ($x_i = 0$, if processor with index i fails during period $[0, t]$, $x_i = 1$, if processor with index i works satisfactory during period $[0, t]$, $i = 1, \ldots, n$).

p_i represents the reliability of the processor with index i.

q_i denotes the unreliability of the processor with index i.

$\mathbf{X}$ represents the system state vector, $\mathbf{X} = (x_1, x_2, \ldots, x_n)$.

$P(\mathbf{X})$ is the probability of system state vector.

$\varphi(\mathbf{X})$ corresponds to the state of the fault-tolerant multiprocessor system for vector $\mathbf{X}$, in other words $y = \varphi(\mathbf{X})$ (it can be calculated using GL-model).

$B(n)$ means the set of all binary vectors of length n.

$W(n, m)$ is the set of all binary vectors of length n which have weight (number of non-zero components) m $(0 \leq m \leq n)$.

$K(m, n)$ represents the basic FTMS which consist of n elements and keep working if not more than m elements are faulty $(0 \leq m < n)$.

3   GL-Models

When conducting statistical experiments, the state of a processor is described by a binary variable where 0 indicates a failure and 1 means that the processor works satisfactory during given period of time, the states of all processors are simulated using pseudo random binary vector (system state vector). The state of the fault-tolerant multiprocessor system (0 is system failure, 1 means system works) depends on the states of all processors and could be determined using system state vector. Using collected statistics of the system failures and reliabilities of processors we are able to estimate the reliability of the system.

In order to determine the state of the FTMS for given system state vector we need a function reflecting relationship between functioning of the system and functioning of its components. Authors consider that the most suitable model for representing this relationship is GL-model [9].

In general, GL-model (graph-logical model) is an undirected graph with a special way formed Boolean functions assigned to its edges. Boolean function corresponds to every edge of this graph. These functions depend on Boolean indicator variables x_i ($i = 1$, $2, \ldots, n$) which reflect the state of corresponding system component (0 means failure, 1 is working state), set of all such variables is called the system state vector. If an edge function takes a zero value then this edge is excluded from the graph. The connectivity of the graph corresponds to the working state of the system as a whole.

Whole set of FTMS can be divided into two types: basic and non-basic systems. Basic systems which consist of n elements keep working if not more than m elements are faulty ($0 \leq m < n$). Let's denote these FTMS as $K(m, n)$. This determination is usually associated with widely used in science literature determination k-out-of-n: G- and k-out-of-n: F-systems. All these three conceptions are similar by sense and between them the following relationship can be set: $K(m, n) \leftrightarrow (n - k)$-out-of-$n$: G- $\leftrightarrow (m + 1)$-out-of-n: F- systems. Non-basic system behavior differs from basic one on some set of system state vectors on which system can continue working in case of number of failures more than m and vice versa – fails if number of failures is less or equal to m. According to such system type separation, two types of GL-models basic and non-basic were defined.

Generally speaking, GL-model can be built on any type of graph: starting from simple edge with two vertices and finishing with full graph. However, the simpler graph not necessary gives better GL-model in a whole in terms of complexity and performance. We must take into account total amount of edge functions complexity. In [10] the next formula of GL-model complexity estimation has been suggested.

Let's consider GL-model based on cycle graph. The algorithm of forming edge functions for basic $K(m,n)$ Gl-model was proposed in [10], it can be minimized and after all gives GL-model optimized by number of edges. It can be proved that this model loses zero or one edge if corresponding system has m or less failed components, and strictly two edges in case of ($m + 1$) failures. As an example, let's demonstrate list of edge functions for basic GL-model $K(3,8)$:

$$f_1 = x_1 \vee x_2 \vee x_3 x_4$$
$$f_2 = x_1 x_2 \vee x_3 \vee x_4$$
$$f_3 = (x_1 \vee x_2)(x_1 x_2 \vee x_3 \vee x_4)(x_3 \vee x_4) \vee x_5 x_6 x_7 x_8$$
$$f_4 = x_1 x_2 x_3 x_4 \vee (x_5 \vee x_6)(x_5 x_6 \vee x_7 x_8)(x_7 \vee x_8)$$
$$f_5 = x_5 \vee x_6 \vee x_7 x_8$$
$$f_6 = x_5 x_6 \vee x_7 \vee x_8$$

The real systems are mostly non-basic. We can build non-basic GL-model by modifying basic GL-model.

According to definition when state vector with $m + 1$ zero appears it leads to excluding 2 main edges from the graph. We can prevent losing of GL-model's graph connectivity if we add internal edge (between those excluded main edges) with its own function. This is a first way to modify basic GL-model to non-basic. For instance, if it is required to save GL-model K(3,8) graph connectivity on state vector (0, 0, 0, 0, 1, 1, 1, 1) – (f_1 and f_2 take zero value) then it will be enough to add internal edge with a function: $f' = x_1x_2x_3x_4$ between first and second base edges, as its shown on Fig. 1. Of course, the more non-basic state vectors the more difficult to form functions to additional edges and graph connectivity solving task complexity increases.

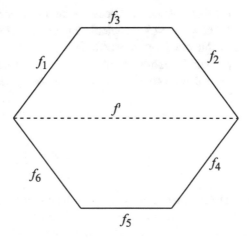

Fig. 1. K(3,8) cycle GL-model example

Second way to modify of basic GL-model is changing of main edge functions [11]. It is worth noticing that to find such functions is nontrivial task.

For basic system reliability calculation is not really complicated [1], but mentioned methods not always can be applied to non-basic systems. It's suggested universal reliability calculation method which is statistical experiments with above mentioned GL-model.

4   Statistical Estimation of the FTMS Reliability

There are different ways to conduct statistical experiments and they result in different accuracy. Direct approach is not efficient as modern processors are high reliable and it requires too many experiments to simulate a failure of a processor. The "statistical experiments acceleration" method [12] is more efficient.

Best results are achieved with recently proposed approach [13]. It suggests dividing statistical experiments into several stages by system state vector weight (number of non-zero components which corresponds to the number of working processors).

The fault-tolerant multiprocessor system reliability could be written as [13]:

$$P(y) = \sum_{\mathbf{X} \in B(n)} \varphi(\mathbf{X}) P(\mathbf{X})$$

where the notation $\mathbf{X} \in B(n)$ means that the sum is taken for all binary vectors of weight n (which corresponds to all possible states of all processors); $P(\mathbf{X})$ is the probability that processors are in states described by system state vector $\mathbf{X}$:

$$P(\mathbf{X}) = \prod_{i=1}^{n} \left(\left(1 - x_i\right) q_i + x_i p_i \right)$$

$\varphi(\mathbf{X}) = 1$, if the system works when its processors are in states described by system state vector $\mathbf{X}$, $\varphi(\mathbf{X}) = 0$ otherwise. $\varphi(\mathbf{X})$ is calculate using GL-model.

The number of all possible binary vectors of length n equals $|B(n)| = 2^n$. Therefore, direct calculation of $P(y)$ could be extremely time-consuming if number of processors is big enough (note that real fault-tolerant multiprocessor system could consist of hundreds of processors).

Let's consider unbiased consistent statistical estimator proposed in [13]:

$$\bar{P}(y) = \sum_{m=0}^{n} \left(\frac{C_n^m}{L_m} \cdot \sum_{\mathbf{X} \in \Omega(n,m)} \varphi(\mathbf{X}) P(\mathbf{X}) \right)$$

where the notation $\mathbf{X} \in \Omega(n, m)$ means that the sum is taken for binary vectors of length n and weight m which were produced by pseudo random generator that do not repeat vectors and the probability of each vector generating equals [14]

$$P_{gen}(\mathbf{X}) = \frac{1}{C_n^m}$$

$\Omega(n, m)$ denotes the set of the generated binary vectors of length n and weight m ($\Omega(n, m) \subset W(n, m)$), $L_m = |\Omega(n, m)|$ indicates the statistical experiments number (for the current stage), $C_n^m = |W(n, m)|$ stands for the number of different binary vectors of length n and weight m (the combinations of m ones out of n binary units). As could be observed from the expression for $\bar{P}(y)$ the statistical experiments are divided into $(n + 1)$ stages where the parameter of each stage is the number m of non-zero components in generated binary vectors.

The idea of FTMS reliability estimation with statistical experiments has been developed well enough. For now, we have results on reliability estimation of hierarchical FTMS [15], hierarchical FTMS which subsystems have processor in common [16].

5   Statistical Estimation Error

One of the most important aspects of statistical experiments (Monte-Carlo) method is statistical estimation error (Monte-Carlo Error). According to the central limit theorem, statistical estimator $\bar{P}(y)$ is approximately normally distributed and consequently its error could be estimated using so-called three-sigma rule:

$$P\big(|P(y) - \bar{P}(y)| < 3\sigma\big) \geq 0.997$$

where $\sigma = \sqrt{D\big(\bar{P}(y)\big)}$ and $D\big(\bar{P}(y)\big)$ denotes the dispersion of $\bar{P}(y)$.

In accordance with article [13] the dispersion of the statistical estimator $\bar{P}(y)$ is viewed as:

$$D\big(\bar{P}(y)\big) \leq$$
$$\sum_{m=0}^{n} \left(\frac{C_n^m - L_m}{L_m} \cdot \left(\frac{C_n^m - 2}{C_n^m - 1} \cdot \sum_{\mathbf{X} \in W(n,m)} (\varphi(\mathbf{X})P(\mathbf{X}))^2 - \frac{1}{C_n^m - 1} \cdot \left(\sum_{\mathbf{X} \in W(n,m)} \varphi(\mathbf{X})P(\mathbf{X}) \right)^2 \right) \right)$$

As we can see the dispersion approaches infinity as the variables L_m are simultaneously reduced to zero. In other words, the less statistical experiments are conducted the bigger is the statistical estimator error. On the other hand, the dispersion shrinks to zero if variables L_m simultaneously approach C_n^m for each $0 \leq m \leq n$, which corresponds to the situation when statistical estimator deteriorates to exhaustive calculation algorithm.

The dispersion above can be estimated before conducting statistical experiments. Let Φ_m denotes the number of system state vectors of length n and weight m on which system works. Let P_m^{min} and P_m^{max} denotes correspondingly the minimal and maximal probability of system state vectors of length n and weight m. If in the dispersion formula every positive appearance of $P(\mathbf{X})$ we replace with P_m^{max} and every negative appearance of $P(\mathbf{X})$ we replace with P_m^{min} then we can receive the upper bound of the dispersion:

$$D\big(\bar{P}(y)\big) \leq \sum_{m=0}^{n} \left(\frac{C_n^m - L_m}{L_m \cdot (C_n^m - 1)} \cdot \left((C_n^m - 2) \cdot \Phi_m \cdot \big(P_m^{max}\big)^2 - \Phi_m^2 \cdot \big(P_m^{min}\big)^2 \right) \right)$$

Values of Φ_m may be unknown but we can exclude them if we find the maximum value of the upper bond expression:

$$D\big(\bar{P}(y)\big) \leq \sum_{m=0}^{n} \left(\frac{C_n^m - L_m}{L_m \cdot (C_n^m - 1)} \cdot \left(\frac{(C_n^m - 2) \cdot \big(P_m^{max}\big)^2}{2 \cdot P_m^{min}} \right)^2 \right)$$

The proposed upper bound of the dispersion can be used to define the required number of statistical experiments for given statistical estimation error.

Statistical estimation error can be estimated more precisely after conducting statistical experiments. We suggest using the following unbiased consistent statistical estimator:

$$\bar{D}(\bar{P}(y)) = \sum_{m=0}^{n} \left(\frac{C_n^m - L_m}{L_m} \cdot \left(\frac{C_n^m - 2}{C_n^m - 1} \cdot \frac{C_n^m}{L_m} \cdot \sum_{X \in \Omega(n,m)} (\varphi(X)P(X))^2 - \frac{1}{C_n^m - 1} \cdot \left(\frac{C_n^m}{L_m} \cdot \sum_{X \in \Omega(n,m)} \varphi(X)P(X) \right)^2 \right) \right)$$

The Fig. 2 represents computer-modeling results of statistical estimation error. "Err.Stat.Est. of ABM(ap)" means the error of "statistical experiments acceleration" method, "Err.Stat.Est. of EVGW(ap)" is the error of proposed approach. As we can see proposed method has better accuracy.

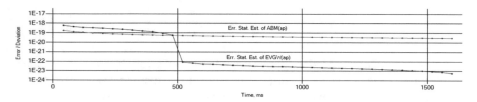

Fig. 2. Statistical estimation error for FTMS with 20 processors which unreliabilities are 10^{-5}–10^{-4}

The statistical experiments error has been studied in details. There are results on reliability estimation error of hierarchical FTMS [15] and hierarchical FTMS if subsystems have processors in common [16], there are results for optimization of statistical experiments number for given error [17].

6   Conclusion

In the article a review of known types of fault-tolerant multiprocessor systems and corresponding methods of its reliability's evaluation is given. Conclusion made that for the fault-tolerant multiprocessors systems with big enough the number of processors and complex behavior of the system under its processors failures the most suitable method is statistical experiments (Monte-Carlo).

The upper limit of the statistical evaluation error (Monte-Carlo) is proposed. It allows to calculate the error for given number of statistical experiments or define required the number of statistical experiments for given error before conducting these experiments.

Reliable statistical evaluator is proposed for statistical estimation error of the system reliability. It allows to calculate the error more precisely using the results of executed statistical experiments and three-sigma rule.

References

1. Kuo, W., Zuo, M.J.: Optimal reliability modeling: principles and applications. Wiley, New Jersey (2003)
2. Suo, B., Cheng, Y.S., Zeng, C., Li, J.: Calculation of failure probability of series and parallel systems for imprecise probability. Int. J. Eng. Manuf. (IJEM) **2**(2), 79–85 (2012)
3. Rahdari, D., Rahmani, A.M., Aboutaleby, N., Karambasti, A.S.: A distributed fault tolerance global coordinator election algorithm in unreliable high traffic distributed systems. Int. J. Inf. Technol. Comput. Sci. (IJITCS) **7**(3), 1–11 (2015). https://doi.org/10.5815/ijitcs.2015.03.01
4. Wang, X., Li, S., Liu, F., Fan, X.: Reliability analysis of combat architecture model based on complex network. Int. J. Eng. Manuf. (IJEM), **2**(2), 15–22 (2012)
5. Keshtgar, S.A., Arasteh, B.B.: Enhancing software reliability against soft-error using minimum redundancy on critical data. Int. J. Comput. Netw. Inf. Secur. (IJCNIS) **9**(5), 21–30 (2017). https://doi.org/10.5815/ijcnis.2017.05.03
6. Kaswan, K.S., Choudhary, S., Sharma, K.: Software reliability modeling using soft computing techniques: critical review. Int. J. Inf. Technol. Comput. Sci. (IJITCS) **7**(7), 90–101 (2015). https://doi.org/10.5815/ijitcs.2015.07.10
7. Wason, R., Soni, A.K., Rafiq, M.Q.: Estimating software reliability by monitoring software execution through opcode. Int. J. Inf. Technol. Comput. Sci. (IJITCS) **7**(9), 23–30 (2015). https://doi.org/10.5815/ijitcs.2015.09.04
8. Thomas, M.O., Rad, B.B.: Reliability evaluation metrics for internet of things, car tracking system: a review. Int. J. Inf. Technol. Comput. Sci. (IJITCS) **9**(2), 1–10 (2017). https://doi.org/10.5815/ijitcs.2017.02.01
9. Romankevich, A.M., Karachun, L.F., Romankevich, V.A.: Graph-logic models for analysis of complex fault-tolerant computing systems. Elektron. Model **23**(1), 102–111 (2001). (in Russian)
10. Romankevich, A.M., Ivanov, V.V., Romankevich, V.A.: Analysis of fault-tolerant multi-processor systems with complex fault distribution based on cyclic GL-models. Elektronnoe modelirovanie **26**(5), 67–81 (2004). (in Russian)
11. Romankevich, V.A., Morozov, K.V., Feseniuk, A.P.: On a method for modifying the edge functions of GL-models. Radioelektronni i kompiuterni systemy, **6**, 95–99 (2014). (in Russian)
12. Ushakov, I. (ed.): Reliability of Technical Systems: Handbook. Radio i Sviaz, Moskov (1985). (in Russian)
13. Romankevich, A.M., Romankevich, V.A., Feseniuk. A.P.: About one method of calculation of reliability indicators of fault-tolerant multiprocessor systems. USiM no. 6, pp. 14–18, 37 (2011) (in Russian)
14. Romankevich, V.A., Maidaniuk, I.V.: Structural method for the formation of binary pseudo-random vectors of a given weight USiM no. 5, pp. 28–33, 58 (2011). (in Russian)
15. Romankevich, A.M., Grol, V.V., Romankevich, V.A., Feseniuk, A.P.: Estimation of the error of the reliability statistical calculation of the FTMS, which correspond to the hierarchical GL-models. Radioelektronni i kompiuterni systemy 7, 142–146 (2010). (In Russian)
16. Romankevich, V.A., Feseniuk, A.P.: About the reliability calculation of fault-tolerant multiprocessor systems, the subsystems of which have common processors. Radioelektronni i kompiuterni systemy **3**, 62–67 (2010). (In Russian)
17. Tarassenko, V.P., Romankevich, V.A., Feseniuk, A.P.: Statistical experiments error minimization for fault-tolerant multiprocessor system reliability estimation. J. Qafqaz Univ.Math. Comput. Sci. **4**(2), 140–146 (2016)

Optimal Control of Point Sources in Richards-Klute Equation

A. Tymoshenko$^{(\boxtimes)}$, D. Klyushin, and S. Lyashko

Faculty of Computer Science and Cybernetics,
Taras Shevchenko National University of Kyiv, Kyiv, Ukraine
inna-andry@ukr.net

Abstract. This article represents an approach to humidity control in porous media, which combines linearization with numerical methods. The main problems and several ideas to solve them are mentioned. The main interest for our research is humidity regulation, described by Richards-Klute equation. The optimization problem is to minimize the difference between a reached state and a desired state.

Modelling moisture transport in porous media requires taking into account the processes of heat transfer, chemical and physical processes, as they have a considerable influence on characteristics of the medium.

First of all, a mathematical model is developed with a number of simplifications: moisture incompressibility, constant external pressure, limitations on transfer or isothermal requirements. Then, it is often preferable to make a transition to linear problem, as this case is explored and it allows us to use more theoretical background for the research. After that, numerical models and time and space discretization are constructed according to the related problem. When the process is represented in a suitable way, control and optimization problems arise and should be solved.

Keywords: Control · Optimization · Richards-Klute equation

1 Introduction

Richards equation and Darcy's law form the basis for mathematical simulation of water flow through isotropic saturated porous media [1]. Usually, to apply the linear theory some simplifications and assumptions are made. Van Genuchten [2] proposed an approach to mathematical simulation of humidity, which allows formulating a quasi-linear or nonlinear problem, depending on the choice of approximation operator. Another approach to linearization was proposed in [3] by Donald Kirk. It uses the Hamilton-Jacobi-Bellman equation for the optimal control problem, leading to a system, where numerical methods, mentioned in [4], can be applied.

Empirical calculations of humidity for several types of ground with buried point sources of water can be found in [5], where Shatkovskiy explains and demonstrates suitable humidification to increase the crop of cultivated plants.

Optimization and control problems for linear functions were discussed in details by Lyashko [6, 7], where results for convergence, stability and necessary conditions can

© Springer International Publishing AG, part of Springer Nature 2019
Z. Hu et al. (Eds.): ICCSEEA 2018, AISC 754, pp. 194–203, 2019.
https://doi.org/10.1007/978-3-319-91008-6_20

be found. Vabishchevich [8] offers a conjugate operator approach, which is used for solving the optimization problem.

Analytic results for the process of ground saturation using several point sources refer to [9]. These results include several types of sources: linear, spherical, point as well as several models of the ground and their parametrization. Transition to linear problem, described in this work, is used in our research.

As an alternative to the approach described here, several recent works can be mentioned. A method for solving a class of nonlinear optimal control problems, based on replacing the dynamic nonlinear optimal control problem by a sequence of quadratic programming problems, using a finite length Chebyshev series with unknown parameter, is described in [10]. A new generalized derivative for non-smooth functions to convert a class of non-smooth optimal control problem to the corresponding smooth form is used in [11]. The method of structural identification of nonlinear dynamic systems is offered in [12] in the conditions of uncertainty. The method of construction the set containing the data about a nonlinear part of system is developed. A single component, two-phase flow simulator is developed in [13] to investigate the behavior of isothermal two-phase fluid flow in porous media. The simulator is based on the Lattice-Boltzmann method and the Shan–Chen multiphase model of non-ideal fluids that allow coexistence of two phases of a single substance.

2 Process Description and Its Model

Consider a mathematical model of the process [1]:

$$\frac{\partial v}{\partial t} = div(K \, grad\Psi) - \frac{\partial K}{\partial z}. \tag{1}$$

Here Ψ is hydraulic moisture potential, K is a moisture conductivity function, v is humidity, t is time, z is vertical space coordinate taken positive downwards. This equation can be applied to homogeneous and heterogeneous medium. If K and Ψ are functions depending only on v, then diffusivity can be applied resulting into a new equation:

$$\frac{\partial v}{\partial t} = div(K \, grad\Psi) - \frac{\partial K}{\partial z}, \quad D(v) = K(v)\frac{d\Psi}{dv}. \tag{2}$$

Functions $K(v), \Psi(v), D(v)$ are often nonlinear, so it is convenient to add the Kirchhoff's potential:

$$\Theta = \int_{v_1}^{v} D(v)dv \tag{3}$$

As a result, the equation can be written the following way:

$$\frac{1}{D(v)}\frac{\partial\Theta}{\partial t} = div(grad\Theta) - \frac{1}{D(v)}\frac{\partial K}{\partial z}\frac{\partial v}{\partial z}. \tag{4}$$

3   Linearization

Our approach is based on Novoselskiy's explorations [9] on humidity distribution under irrigation from point sources. Consider the problem in the following form:

$$
\frac{\partial \omega}{\partial t} = \frac{\partial}{\partial x}\left[K_x(\omega)\frac{\partial H}{\partial x}\right] + \frac{\partial}{\partial y}\left[K_y(\omega)\frac{\partial H}{\partial y}\right] + \frac{\partial}{\partial z}\left[K_z(\omega)\frac{\partial H}{\partial z}\right]
$$
$$
+ \sum_{j=1}^{N} Q_j(t)\delta\left(x - x_j\right)\delta\left(y - y_j\right)\delta\left(z - z_j\right) - I(\omega, x, y, z, t), \tag{5}
$$
$$
(x, y, z, t) \in \Omega_0 \times (0, \infty);
$$

$$
z = z_0, K_z(\omega) - D_z(\omega)\frac{\partial \omega}{\partial z} = \varepsilon(x, y, t); \tag{6}
$$

$$
z \rightarrow \pm\infty, \omega(x, y, z, t) = \omega_0 = const, (x, y, z, t)\epsilon\Gamma^0 \times [0, \infty); \tag{7}
$$

$$
x, y \rightarrow \pm\infty, \frac{\partial \omega}{\partial x} = \frac{\partial \omega}{\partial y} = 0; \tag{8}
$$

$$
\omega(x, y, z, 0) = \varphi(x, y, z), (x, y, z)\epsilon\bar{\Omega}. \tag{9}
$$

Here I is the intensity of liquid absorption by plants, $P = \Psi(\omega) - z$ is velocity head, ε is evaporation intensity, $D_z(\omega) = K_z(\omega)\frac{d\psi}{d\omega}$ is diffusion along z axis, $\Omega_0 = [(x, y, z) : -\infty < x, y < \infty]$, $z = z_0$ is the plane at the ground surface level, Γ^0 is the border of Ω_0 domain.

Following [9], $K_x(\omega) = k_1 k(\omega), K_y(\omega) = k_2 k(\omega), K_z(\omega) = k_3 k(\omega)$, where k_1, k_2, k_3 are filtration coefficients along axes Ox, Oy, Oz, $k(\omega)$ stands for humidity function for the ground. To make a transition to an equivalent dimensionless equation, some more scaling variables are used:

$$
\xi = \beta_1 x, \eta = \beta_2 y, \zeta = \beta_3 z, \tau = \alpha t, \beta_3 = 0.5l, \beta_1 = \sqrt{\tfrac{k_3}{k_1}}\beta_3, \beta_2 = \sqrt{\tfrac{k_3}{k_2}}\beta_3,
$$
$$
\alpha = \langle D_z \rangle \beta_2^2. \tag{10}
$$

The Kirchhoff's potential will be:

$$
\Theta = \frac{4\pi\sqrt{k_1 k_2}}{Q^* k_3 \beta_3} \int_{\omega_0}^{\omega} D_z(\omega)d\omega. \tag{11}
$$

As a result, we can make a transition to linearized problem

$$
L\Theta = \frac{\partial \Theta}{\partial \tau} - \frac{\partial^2 \Theta}{\partial \xi^2} - \frac{\partial^2 \Theta}{\partial \eta^2} - \frac{\partial^2 \Theta}{\partial \varsigma^2} + 2\frac{\partial \Theta}{\partial \varsigma} = -f(\xi, \eta, \varsigma, \tau)
$$
$$
+ 4\pi \sum_{j=1}^{N} q_j(\tau)\delta(\xi - \xi_j)\delta(\eta - \eta_j)\delta(\varsigma - \varsigma_j), (\xi, \eta, \varsigma, \tau)\epsilon\Omega \times (0, \infty).
$$
(12)

The boundary conditions over new axes can be written as:

$$
\varsigma = \varsigma_0, \Theta - 0.5\frac{\partial \Theta}{\partial \varsigma} = \varepsilon_0(\xi, \eta, \tau) + \tilde{k}_0;
$$
(13)

$$
\varsigma \to \infty, \Theta = 0; \xi, n \to \pm\infty, \frac{\partial \Theta}{\partial \xi} = \frac{\partial \Theta}{\partial \eta} = 0, (\xi, \eta, \varsigma, \tau)\epsilon\Gamma \times [0, \infty).
$$
(14)

At the start of simulation, the initial condition will be:

$$
\Theta(\xi, \eta, \varsigma, 0) = \varphi_1(\xi, \eta, \varsigma), (\xi, \eta, \varsigma)\epsilon\bar{\Omega}.
$$
(15)

Here $\varepsilon_0 = \frac{2\pi\sqrt{k_1 k_2}\varepsilon}{k_3\beta_3^2 Q^*}$, $\tilde{k} = \frac{2\pi\sqrt{k_1 k_2}K_z(\omega_0)}{k_3\beta_3^2 Q^*}$, $q_j = \frac{Q_j}{Q^*}$, Q^* is a vector parametric scale multiplier, Ω, Γ are dimensionless analogues to Ω_0, Γ^0 and D_z is an average value of $D_z(\omega)$, $\varphi_1 = \Theta(\varphi)$.

To make such transition, several conditions are required:

- The relation between $\Theta(\omega)$ and $K_z(\omega)$ is linear: $\frac{1}{D_z(\omega)}\frac{dK_x(\omega)}{d\omega} = l = const$;
- $f(\xi, \eta, \varsigma, \tau) = \frac{4\pi\sqrt{k_1 k_2}}{Q^* k_3\beta_3^3}I\left(\frac{\xi}{\beta_1}, \frac{\eta}{\beta_2}, \frac{\varsigma}{\beta_3}, \frac{\tau}{\alpha}\right)$,
- Linearization is achieved by assuming $\frac{\partial \omega}{\partial t} = \frac{Q^* k_3\beta_3}{4\pi\sqrt{k_1 k_2}}\frac{1}{D_z(\omega)}\frac{\partial \Theta}{\partial t} \simeq \frac{Q^* k_3\beta_3^3}{4\pi\sqrt{k_1 k_2}}\frac{\partial \Theta}{\partial \tau}$.

4 Control and Optimization

For the control problem we will use an approach offered in [7] so that we can apply results on existence and uniqueness of the solution. Consider the optimal control belongs to Hilbert space with the following inner product:

$$
\langle X, Y \rangle = \sum_{\beta=1}^{P} \int_0^T x_\beta(t)y_\beta(t)dt.
$$
(16)

The points r_β define locations of sources with undefined (they should be found) capacities $q_\beta(t), \beta = \overline{1,P}$. Measurements of concentration $\varphi(z,t)$ are given as a known function. The task is to minimize the deviation $u(z,t)$ of $\varphi(z,t)$.

By determining $\alpha > 0$ as a regularization parameter according to possible measurement errors, the objective functional will be:

$$J_\alpha(Q) = \sum_{m=1}^{M} \int_o^T \int_\Omega (\varphi(t) - u(t,x))^2 dx\, dt + \alpha \|Q\|^2. \tag{17}$$

As a result, the optimal control can be found as $J_\alpha(Q^*) = min_{Q \in H} J_\alpha(Q)$. The iterative algorithm proposed in [7] consists of three stages. As for the conjugated operator, it is tested and explained by Vabishevich [8].

1. Solve the state problem

$$\frac{\partial u^{(k)}}{\partial t} + Lu^{(k)} = f^{(k)}, 0 < t \le T, u^{(k)}(0) = g(x);$$

2. Solve the conjugate problem

$$-\frac{\partial \psi^{(k)}}{\partial t} - L^* \psi^{(k)} = 2(u^k - \varphi(t)), 0 < t \le T, \psi^{(k)}(T) = 0;$$

3. Define the new approximation for the optimal control

$$\frac{Q^{(k+1)} - Q^{(k)}}{\tau_{k+1}} + \Psi^{(k)} + \alpha Q^{(k)} = 0, k = 0, 1, ..$$

5 Solving the Example Task

Assume having a dry area with a point source of water, inserted into the ground with a fixed intensity. Some cultivated plants grow there and their roots need some water, which is described by the humidity function. Our optimization task is to find the source intensity, which leads to humidity indexes as near to humidity function as possible at the last time step.

In the research, a one-dimensional linearized according to [9] problem

$$L\Theta = \frac{\partial \Theta}{\partial \tau} - \frac{\partial^2 \Theta}{\partial \varsigma^2} + 2\frac{\partial \Theta}{\partial \varsigma} = -f(\varsigma, \tau) + q_j(\tau)\delta(\varsigma - \varsigma_j) \tag{18}$$

is solved. The permeability coefficients are taken equal $k_1 = k_2 = k_3 = 0.1$, while $\Theta(\omega) = \frac{4\pi\sqrt{k_1 k_2}}{Q^* k_3 \beta_3} \cdot (-2 + 2e^{0.5\omega})$ and the scale multiplier for source intensity Q^* is equal to 1.

Assume having a line segment with a length of 100, with the axis taken downwards. The boundary conditions will be:

$$\Theta(100) = 0, \Theta(0) - 0.5\frac{\partial\Theta}{\partial\varsigma}\Big|_0 = \frac{2\pi\sqrt{k_1 k_2}K_z(\omega_0)}{k_3\beta_3^2 Q^*}. \tag{19}$$

At the start of simulation, the initial condition is given by the equation:

$$\Theta(\varsigma, 0) = \varphi_1(\varsigma), \varsigma\epsilon\bar{\Omega}. \tag{20}$$

Currently, 10 positions in time are used and space discretization is made with a constant step $h = \frac{1}{100}$. In case we have one source with unknown intensity q_1, using an implicit difference scheme to get stability and convergence, the following equation can be written:

$$\frac{\Theta_i^n - \Theta_i^{n-1}}{\tilde{\tau}} - \frac{\Theta_{i+1}^n - 2\Theta_i^n + \Theta_{i-1}^n}{h^2} + 2\frac{\Theta_{i+1}^n - \Theta_i^n}{h} \tag{21}$$
$$= -f^n(\zeta_i, \tau) + 4\pi q_1(\tau)\delta(\zeta_i - \zeta_1).$$

It can be transformed by combining coefficients into:

$$(2h\tilde{\tau} - \tilde{\tau})\Theta_{i+1}^n + (h^2 + 2\tilde{\tau} - 2h\tilde{\tau})\Theta_i^n + (-\tilde{\tau})\Theta_{i-1}^n \tag{22}$$
$$= h^2\Theta_i^{n-1} - f^n(\zeta_i, \tau) + 4\pi q_1(\tau)\delta(\zeta_i - \zeta_1).$$

Using tridiagonal matrix algorithm with $\Theta_{i-1}^n = \alpha_i^n\Theta_i^n + \beta_i^n$, the equation becomes:

$$(2h\tilde{\tau} - \tilde{\tau})\Theta_{i+1}^n + (h^2 + 2\tilde{\tau} - 2h\tilde{\tau})\Theta_i^n + (-\tilde{\tau})(\alpha_i^n\Theta_i^n + \beta_i^n)$$
$$= h^2\Theta_i^{n-1} - f^n(\zeta_i, \tau) + 4\pi q_1(\tau)\delta(\zeta_i - \zeta_1),$$
$$(2h\tilde{\tau} - \tilde{\tau})\Theta_{i+1}^n + (h^2 + 2\tilde{\tau} - 2h\tilde{\tau} - \tilde{\tau}\alpha_i^n)\Theta_i^n \tag{23}$$
$$= h^2\Theta_i^{n-1} + \tilde{\tau}\beta_i^n - f^n(\zeta_i, \tau) + 4\pi q_1(\tau)\delta(\zeta_i - \zeta_1).$$

Finally, it is possible to transform the equation to get $\alpha_{i+1}^n, \beta_{i+1}^n$:

$$\Theta_i^n = \underbrace{\frac{\tilde{\tau} - 2h\tilde{\tau}}{h^2 + 2\tilde{\tau} - 2h\tilde{\tau} - \tilde{\tau}\alpha_i^n}}_{\alpha_{i+1}^n}\Theta_{i+1}^n$$
$$+ \underbrace{\frac{h^2\Theta_i^{n-1} + \tilde{\tau}\beta_i^n - f^n(\zeta_i, \tau) + 4\pi q_1(\tau)\delta(\zeta_i - \zeta_1)}{h^2 + 2\tilde{\tau} - 2h\tilde{\tau} - \tilde{\tau}\alpha_i^n}}_{\beta_{i+1}^n}. \tag{24}$$

As a result, all values can be found from initial and boundary conditions, but they depend on q_1 . This variable is unknown, but it will be found from optimization functional after inserting $\Theta(i, \tau_{end}, q_1)$ as a function with one unknown variable.

The approach for the conjugated problem [7] is almost identical, it uses the condition describing the state at the last time step. But instead of point approximation, our goal will be to minimize the optimization functional

$$min_{q_1} \left(\sum_{i=1}^{N} (\Theta(i, \tau_{end}; q_1) - \Theta_{desired}(i))^2 + \alpha q_1^2 \right), \tag{25}$$

where i denotes indexes according to the location of points, in order to get humidity values as near to desired ones as possible. In our case, $\Theta_{desired}(i)$ is given by a known function values: $\Theta_{desired}(i) = \varphi(\varsigma_i)$. Because of $\Theta \geq 0$, by finding derivative for this functional by q_1, it is possible to find the optimal intensity. When we found the value of q_1, it can be checked by the third stage, described in [7].

The desired humidity level is given by the functional $\varphi(x) = \begin{cases} 0.2, 1 \leq x \leq 10 \\ 0, x > 10 \end{cases}$.

Here 0 stands for "not important", as the roots are located in the area $1 \leq x \leq 10$. The point source is located at $x = 1$ and works downwards. Choosing $\varphi_1(\varsigma) = 0$, the calculated source intensity with these conditions led to the following result:

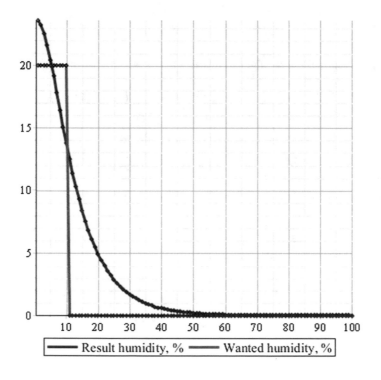

As we can see, the solution has only 7.65% deviation although the humidity function was really hard to achieve. The regularization parameter was chosen equal to 0.01.

Taking the humidity function $\varphi(x) = \begin{cases} 0.2 - 0.015x, 1 \leq x \leq 10 \\ 0, x > 10 \end{cases}$ changes the

optimal control leading to the following result:

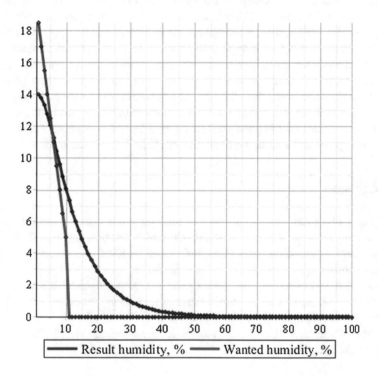

Now the deviation equals 4.66%. The graph of optimisation functional for this case is shown below, where x stands for unknown source intensity.

These results prove the efficiency of our approach for different humidity functions. In or research, permeability coefficients have a great impact on the model. For the ground with a higher vertical coefficient, much less intensity is required. Apart from that, the desired humidity function should be realistic so that the deviation could be minimized all over the area.

6 Discussion

Although the stated problem is well-known, some nonlinear cases still attract attention. In some of them, the proposed transition to a linear problem is possible. The transition can modify the initial problem into a scaled over space and time one, more suitable for numerical methods and further discretization.

Although there are many ideas on solving nonlinear systems, most of them require lots of additional calculations and are also limited to some cases. Our approach is relatively simple and has many possibilities to be improved. Moreover, it is possible to convert the results into resulting humidity, so comparison to expected values is not so difficult.

As for the optimization theory, works by Lyashko allow us to check the stability of iterative methods for linear problems and to prove existence and uniqueness of the solution. For some cases, alternative optimization criteria can be applied, according to the particular problem but their application should base on theoretical results.

Numerical methods and difference schemes can also be changed to reduce the time or increase accuracy. The current implicit scheme is numerically stable and convergent, but its errors are linear over the time step and quadratic over the space step.

Our work states that the optimal control of source intensity can be found for some problems, where transition is possible. The desired humidity function should take into account the type of soil and its permeability coefficients for best results.

7 Conclusion

The proposed algorithm for identifying the optimal point source power is an opportunity to solve a number of nonlinear problems, which can be transformed into linear in case they fulfil the transition restrictions. It is based on Kirchhoff transformation and is easy to implement.

Increasing the permeability coefficient results in lower optimal intensity.

The desired humidity function limits the accuracy of the proposed method, but it is caused by physical limitations of soils and water.

It is possible to check the intensity by putting it as a known constant and getting humidity at the last time step.

References

1. Pullan, A.J.: The quasilinear approximation for unsaturated porous media flow. Water Resour. Res. **26**(6), 1219–1234 (1990)
2. Van Genuchten, M.T.: A closed-form equation for predicting the hydraulic conductivity of unsaturated soils. Soil Sci. Soc. Am. J. **44**, 892–898 (1980)
3. Kirk, D.E.: Optimal Control Theory. An Introduction Optimal Control Theory: An Introduction. Dover Books on Electrical Engineering (2004)
4. List, F., Radu, F.: A Study on Iterative Methods for Richards' Equation. Math.NA. http://arxiv.org/abs/1507.07837
5. Shatkovskiy, A.P.: Parameters of the mods of drip irrigation and productivity of sugar beets in the area of Ukrainian Steppes (in Ukrainian). Tsukrovi Buryaky **3**, 15–17 (2016)
6. Lyashko, S.: Generalized Control of Linear Systems. Naukova Dumka, Kyiv (1998). (in Russian)
7. Lyashko, S., Klyushin, D., Semenov, V., Shevchenko, K.: Identification of point contamination source in ground water. Int. J. Ecol. Develop. **5**(F06), 36–43 (2006)

8. Vabishchevich, P.N.: Numerical solution of the problem of the identification of the right-hand side of a parabolic equation (in Russian). Russ. Math. (Iz. VUZ) **47**(1) (2003)

9. Novoselskiy, S.N.: Solution of some boundary value problems of moisture transport with irrigation sources (in Russian). Dissertation for the academic degree of a Candidate in Mathematics and Physics, 01 February 2005, Kalinin Polytechnic Institute, Kalinin (1981)

10. Jaddu, H., Majdalawi, A.: An iterative technique for solving a class of nonlinear quadratic optimal control problems using Chebyshev polynomials. Int. J. Intell. Syst. Appl. (IJISA) **6**(6), 53–57 (2014). https://doi.org/10.5815/ijisa.2014.06.06

11. Noori Skandari, M.H., Erfanian, H.R., Kamyad, A.V., Farahi, M.H.: Solving a class of non-smooth optimal control problems. Int. J. Intell. Syst. Appl. (IJISA) **5**(7), 16–22 (2013). https://doi.org/10.5815/ijisa.2013.07.03

12. Karabutov, N.: Structural identification of nonlinear dynamic systems. Int. J. Intell. Syst. Appl. (IJISA) **7**(9), 1–11 (2015). https://doi.org/10.5815/ijisa.2015.09.01

13. Zhang, X.: Single component, multiphase fluids flow simulation in porous media with Lattice Boltzmann method. Int. J. Eng. Manuf. (IJEM) **2**(2), 44–49 (2012). https://doi.org/10.5815/ijem.2012.02.07

Method of Calculation of Averaged Digital Image Profiles by Envelopes as the Conic Sections

Serhiy V. Balovsyak[✉], Oleksandr V. Derevyanchuk, and Igor M. Fodchuk

Yuriy Fedkovych Chernivtsi National University, Chernivtsi, Ukraine
s.balovsyak@chnu.edu.ua

Abstract. The method of calculation of averaged digital image profiles has been developed. The image profile is dependence of the value of the pixel brightness on the image coordinate along the specified line segment. The corresponding software was developed in the MATLAB system.

Profile analysis is widely used in the processing of experimental and simulated digital images, especially if the images contain band-shaped objects. The presence of bands is characteristic of electron diffraction images, X-ray moire images, images of scanning probe microscopy, optical medical images, and others. Cross-section profiles contain important information about the explored object, since they describe the one-dimensional distribution of object brightness.

A single band profile may contain an appreciable noise component. Therefore, in order to increase the signal-to-noise ratio, a series of band profiles were obtained, on the basis of which the averaged profile was calculated. The calculation of the average profile is relatively easy to implement in cases when all the band profiles have the same scale, and line consisting of their starting points is parallel to line consisting of their ending points. However, the most of the experimental images undergo the geometric distortions, and the lines consisting of starting or ending points of the profiles correspond to conic-shaped curves. Therefore, in this paper we proposed firstly to approximate the curves consisting of starting/ending points by two envelopes, and then to calculate a series of profiles on the basis of these envelopes. Circles, ellipses, parabolas and hyperbolas were used as envelope functions.

The mathematical model, algorithm and software for calculating enveloping profiles in images are developed. The envelopes are calculated on the basis of the coordinates of the base points, which are determined by the user or calculated through the contours of the band. The high accuracy of the developed method for calculating averaged profiles has been confirmed in the processing of images of electron and X-ray diffraction, atomic force microscope, optical and medical images etc.

Keywords: Digital image processing · Signal-to-noise ratio · Profile · Envelope
Ellipse · Parabola · Hyperbola

© Springer International Publishing AG, part of Springer Nature 2019
Z. Hu et al. (Eds.): ICCSEEA 2018, AISC 754, pp. 204–212, 2019.
https://doi.org/10.1007/978-3-319-91008-6_21

1   Introduction

Digital image profiles that describe the brightness of pixels along line segment are widely used in the processing of experimental and simulated images [1–5]. Particularly these profiles are useful when processing images contain band-shaped fragments. The presence of the bands is proper to electron [6] and X-ray (in particular, X-ray moiré) diffraction images [7], images of scanning probe microscopy, optical medical images etc. [5]. Profiles of band cross-section contain important information about the explored object, since the one-dimensional distribution of the brightness determined by properties of an object. Further analysis of such one-dimensional brightness distributions is much simpler than two-dimensional images. For example, profiles of Kikuchi bands on electron diffraction images contain information about structure perfection of the explored samples [6]. The profiles of bands on X-ray moiré images allow calculate the spatial distribution of deformations for the investigated crystals [7]. Vascular profiles on human retinal photographs can be used to diagnose diseases.

A single profile of a band may contain a significant noise component, therefore, in order to increase the signal-to-noise ratio, a series of band profiles were obtained, on the basis of which the averaged profile was calculated. The calculation of the averaged profile is relatively easy to implement in cases where all the band profiles have same scale, and line consisting of their starting points is parallel to line consisting of their ending points. However, for most experimental images undergo geometric distortions and edge lines of bands described by conic-shaped curves (curves of the second order). For example, electron diffraction images with Kikuchi bands are obtained in a gnomonic projection, in which points of the sphere from its centre projected to the tangent plane, so the edges of the Kikuchi bands can be described by hyperbolas [6]. Circles or ellipses can describe fragments of bands on X-ray moiré images. Some trajectories of body motion, for example, the body movement in the gravitation field, approach to parabolas. If the band edges are described by conic-shaped curves, than all profiles along band are presented on different scales. In addition, coordinates of the starting/ending points of the profiles are unknown as well as function that describe corresponding curve.

Therefore, in this paper we proposed firstly to approximate the curves consisting of starting/ending points of band profiles by two envelopes, and then to calculate a series of profiles on the basis of these envelopes. Circles, ellipses, parabolas and hyperbolas were used as envelope functions. In the simplest case, the envelope is a straight line (the curvature radius goes to infinity). The function of envelope is chosen depending on the geometric distortion of the bands in the image (for example, the envelopes of Kikuchi bands are hyperbolas). Use the envelope allows brings all band profiles to same scale and get the maximal signal-to-noise ratio for the average profile.

Thus, the calculation of the averaged profile consists of two tasks: decomposition and synthesis of signals:

1. Decomposition of signals is to partition the image of a band into a series (set) of profiles.
2. Synthesis of signals is to calculate the averaged profile based on a series of profiles.

A task of decomposition is rather complicated it can be simplified by approximating the band edges with two envelopes.

2   Mathematical Model of Calculating Envelopes and Profiles in Images

The initial digital image f is written to a rectangular matrix f = (f (i, k)), where i = 1, ..., M, k = 1, ..., N (Fig. 1) [1, 2]. Image processing is performed in shades of gray, and the intensity of the initial image is normalized in the range from 0 to 1.

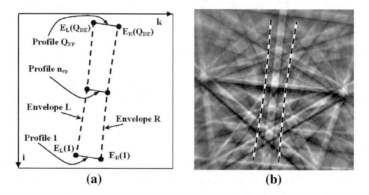

(a)   **(b)**

Fig. 1.   The mathematical model of the profile envelopes in the coordinate system of the image ki (a) and the example of the electron diffraction image of the Kikuchi bands [6] with envelopes (b)

The profiles of the band are contoured by two envelopes (Fig. 1). The envelope, which describes the starting points of the profiles, is conventionally referred as the left (Envelope L) (Fig. 1a). Correspondingly, the envelope, which describes the ends of the profiles, is conventionally referred as the right (Envelope R). The left and right envelopes are described by the base points $E_L(n_{be})$ and $E_R(n_{be})$, respectively, where $n_{be} = 1, ...,$ Q_{BE}, Q_{BE} is the number of base points. Base points are determining by the user or calculating through the contours of the band.

In the simplest case, the number of base points of the envelope $Q_{BE} = 2$, and the envelopes are approximated by straight lines.

If the number of envelope base points $Q_{BE} = 3$, then each envelope is approximated by the circle arc

$$(k - a_C)^2 + (i - b_C)^2 = r_C^2 \tag{1}$$

where (a_C, b_C) are the coordinates of the circle centre in the coordinate system of the image ki (Fig. 1a), r_C is the radius of the circle.

Using the coordinates of the three basic points (k_1, i_1), (k_2, i_2) and (k_3, i_3), the parameters of the circle are calculated by the formulas:

$$b_C = \left(\frac{k_3^2 - k_1^2 + i_3^2 - i_1^2}{2(k_3 - k_1)} - \frac{k_2^2 - k_1^2 + i_2^2 - i_1^2}{2(k_2 - k_1)}\right) \bigg/ \left(\frac{i_3 - i_1}{k_3 - k_1} - \frac{i_2 - i_1}{k_2 - k_1}\right) \qquad (2)$$

$$a_C = \frac{k_2^2 - k_1^2 + i_2^2 - i_1^2}{2(k_2 - k_1)} - \frac{i_2 - i_1}{k_2 - k_1} \cdot b_C \qquad (3)$$

$$r_C = \sqrt{(k_1 - a_C)^2 + (i_1 - b_C)^2} \qquad (4)$$

Q_{EP} points evenly placed on each envelope, $n_{ep} = 1, ..., Q_{EP}$ is number of profile. $k_{EL}(n_{ep})$, $i_{EL}(n_{ep})$ are the coordinates of starting points of profiles, and $k_{ER}(n_{ep})$, $i_{ER}(n_{ep})$ are the coordinates of ending points. For example, the first basic points of the envelope $E_L(1)$ and $E_R(1)$ are the starting and ending points of the first profile (Fig. 1a). The length Q_p of the profile with the number n_{ep} is calculated as the distance between the corresponding points of the envelopes. The profile n_{ep} is calculated as the dependence of brightness of the initial image on coordinate along Q_p line. The profile points are evenly distributed between starting and ending profile points. The calculation of the brightness of each profile point is performed by the method of bicubic interpolation. The values of the brightness for a series of profiles are written to an array $z_f(p, n_{ep})$, where $p = 1, ..., Q_p(n_{ep})$, p is the point number on the profile, n_{ep} is the profile number.

Minimal Q_{pMin} and maximal Q_{pMax} profile lengths are determined for the resulting series of profiles. Based on, the averaged profile $z_{fS}(p)$ is calculated using the array of profile series z_f, where $p = 1, ..., Q_{pMin}$. Before calculating averaged profile z_{fS}, all profiles in series transform to same Q_{pMin} size. Averaging a series of profiles is an effective way to increase the signal-to-noise ratio. Each point of the averaged profile can be considered as the arithmetic mean of Q_{EP} equally distributed mutually independent random variables [8]. If σ_N is the standard deviation of noise in the initial image then the standard deviation σ_{NS} of noise in the averaged profile z_{fS} is equal to

$$\sigma_{NS} = \frac{\sigma_N}{\sqrt{Q_{EP}}} \qquad (5)$$

If the number of base points of the envelope $Q_{BE} > 3$, then each envelope is approximated by an arc of ellipse, parabola or hyperbola [8]. The choice of a suitable curve of the second order is performed on the basis of apriori data about a digital image or on the basis of band analysis. The conic-shaped curves (Fig. 2) is described by the equation in the polar coordinate system $\rho\varphi$:

$$\rho = \frac{\mu}{1 - \varepsilon \cdot \cos \varphi} \qquad (6)$$

where ρ is the radial coordinate, φ is the polar angle, μ is the conic section parameter, ε is the conic section eccentricity ($0 \leq \varepsilon < 1$ for ellipse, $\varepsilon = 1$ for parabola, $\varepsilon > 1$ for hyperbola).

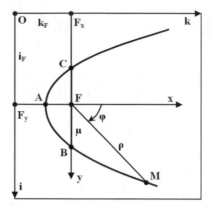

Fig. 2. The mathematical model of the conic section in the coordinate system of the image ki; A is the vertex of a conic section, $\mu = BF = FC$ is the conic section parameter

The pole of the polar coordinate system $\rho\varphi$ coincides with the focus F of the conic section (Fig. 2). Axis x of the conic section and perpendicular to it axis y pass through the focus F. In Fig. 2, the x axis is parallel to the k axis, and in the general case, the axis x forms the angle θ with the axis k. The distance between the arbitrary point M of the conic section and the focus F is ρ. The pole of the polar coordinate system $\rho\varphi$ in the coordinate system of the image ki has the coordinates $k_F = OF_x$ and $i_F = OF_y$ respectively.

According to the mathematical model of a conic section described above, the coordinates of its arbitrary point M in the coordinate system of the image ki are described by the formulas:

$$k_M = k_F + \rho \cdot \cos\varphi = k_F + \frac{\mu \cdot \cos\varphi}{1 - \varepsilon \cdot \cos\varphi} \tag{7}$$

$$i_M = i_F + \rho \cdot \sin\varphi = i_F + \frac{\mu \cdot \sin\varphi}{1 - \varepsilon \cdot \cos\varphi} \tag{8}$$

Thus, to determine the envelope, which is described by the conic section, it is necessary to calculate the focus coordinates (k_F, i_F), the parameter μ and the eccentricity ε of the conic section. As the initial approximation of the k_F and i_F coordinates, the coordinates of the circle centre are used. The parameters of this circle are calculated according to (2–4) using the coordinates of the three basic points. The radius of the circle is the initial approximation of μ. The initial values of the eccentricity ε are: $\varepsilon = 0.5$ for the ellipse, $\varepsilon = 1.0$ for the parabola, $\varepsilon = 1.5$ for the hyperbola. The envelope parameters $(k_F, i_F, \mu, \varepsilon)$ are calculated by the method of coordinate descent by minimizing the root of the mean square error between the envelope and the base points.

3   Algorithm and Software for Calculation of Averaged Image Profile

The software of the method is developed in MATLAB [2]. The algorithm for calculating the average profile is as follows. At first, Q_{BE} base points are established for each band envelope in the image f (Fig. 3a). Basic points are choosing by the user interactively with the mouse manipulator or calculating through coordinates.

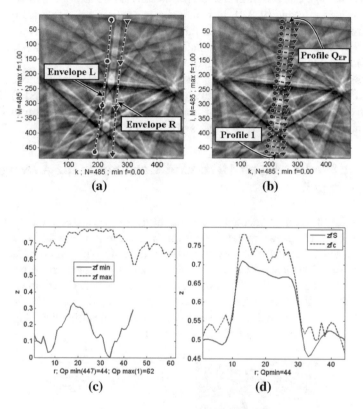

Fig. 3. Calculation of the averaged profile of the experimental electron diffraction image of Kikuchi bands [6]: (a) selection of $QBE = 4$ base points for the left and right envelopes; (b) calculation of series of $QEP = 900$ profiles based on envelopes (envelopes are approximated by hyperbolas, each 20th profile is shown); (c) zfmin profile with a minimum length of QpMin and a zfmax profile with a maximum length of QpMax; (d) reference profile zfc and averaged profile zfS; r is the length of the profile

Base points can also be calculated through the edges of the band (methods of Prewitt, Sobel or Canny [1] are used for the selection band edges). The Canny method provides the highest accuracy, but has relatively low performance. The accuracy of edges calculation mainly depends on the correctness of selected threshold in each method. A low-pass image filtration with Gaussian filter kernel is performing before the edge calculation in the case of noisy image.

Parameters of two conic sections, which approximate the band envelopes in the image, were calculated using base points. Based on these envelopes, a series of Q_{EP} profiles (Fig. 3b) was calculated. The z_{fmin} and z_{fmax} profiles with a minimal Q_{pMin} and maximal Q_{pMax} lengths, respectively, were selected from series of profiles (Fig. 3c). Based on a set of Q_{EP} profiles, the averaged z_{pS} band profile with Q_{pMin} length was calculated (Fig. 3d). The program also performs the calculation of the averaged profile in the classification mode [9, 10], in which the least distorted reference profile z_{fc} is selected from the series of profiles. In the classification mode, the averaged profile determined by those series profiles for which the root of the mean square error relatively the reference profile z_{fc} does not exceed the threshold of R_{qT}.

The classification mode allows read the averaged band profiles without significant distortion, even if the band is being intersected by other bands (Fig. 3b and d). To reduce the negative effects of superposition of the bands, the oriented filtration of the image along the direction of the explored band is also used [11–14].

4   Testing the Developed Method of Calculating the Averaged Image Profile

The test of the developed method for calculating the averaged image profile was performed by processing a simulated electron diffraction image with Gaussian noise, the edges of which are described by hyperbolas (Fig. 4). The averaged z_{fS} profile is computed in the classification mode (comparing series profiles with reference profile z_{fc}) to reduce the negative effects of superposition of the bands. Due to the averaging of the profiles, the standard deviation of the Gaussian noise σ_N decreased by 23.9 times, at the same time, the root signal-to-noise ratio increased. Thus, the averaging allowed to accurately restoring the profile of the band (Fig. 4b).

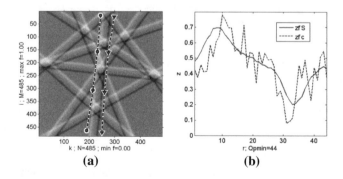

(a)   (b)

Fig. 4. The averaged profile determination for the electron diffraction image (Kikuchi bands) with standard deviation $\sigma_N = 10\%$ of Gaussian noise [10]: (a) selection of QBE = 4 basic points for the left and right envelopes (the envelopes are approximated by hyperbolas), the series consists of QEP = 900 profiles; (b) reference profile zfc and averaged profile zfS; r is the length of the profile

Experimental X-ray moiré images were processed by the developed method of averaged profiles calculating with a high signal-to-noise ratio (Fig. 5). Due to the averaging of profiles the root of signal-to-noise ratio increased by 23.3 times. The developed method provides for the exact restoration of the profile of not only one band, but an entire set of bands, which is required for further analysis of moiré images (Fig. 5b).

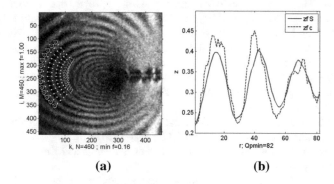

(a)   (b)

Fig. 5. Calculation of the averaged profile for an experimental X-ray moiré image [7]: (a) calculation of a series of QEP = 543 profiles based on envelopes (envelopes are approximated by ellipses, each 20th profile shown), envelopes constructed on the basis of QBE = 4 base points; (b) reference profile zfc and averaged profile zfS; r is the length of the profile

5   Discussion and Conclusion

A mathematical model, algorithm and software for calculating averaged digital image profiles that are calculated through the envelopes of bands in the image are developed. The envelopes are calculated on the basis of the coordinates of the base points, which are determined by the user or calculated through the contours of the band. Circles, ellipses, parabolas and hyperbolas were used as envelope functions. The envelope parameters are calculated by the method of coordinate descent by minimizing the root of the mean square error between the envelope and the base points. The accuracy of the developed method is confirmed in the processing of images of electron and X-ray diffraction, atomic force microscope, optical and medical images etc. The root of signal-to-noise ratio for the investigated images on the average profiles increased by $\approx$23 times, relative to the series profiles.

The novelty of the work is to create a mathematical model of envelopes of profiles like to conic sections, as well as to develop an algorithm for calculating the envelope parameters. The envelope model allows calculate averaged profiles with high accuracy for band fragments of arbitrary shape. Using the classification mode, in which the calculation of the averaged profile is performed only on the basis of the least distorted series profiles, also is novel.

References

1. Gonzalez, R., Woods, R.: Digital Image Processing. Prentice Hall, Upper Saddle River (2002)
2. Gonzalez, R., Woods, R., Eddins, L.: Digital Image Processing Using MATLAB. Prentice Hall, Upper Saddle River (2004)
3. Bovik, A.L.: The Essential Guide to Image Processing. Elsevier Inc., Burlington (2009)
4. Louban, R.: Image Processing of Edge and Surface Defects. Theoretical Basis of Adaptive Algorithms with Numerous Practical Applications. Springer, Heidelberg (2009)
5. Russ, J.C.: The Image Processing Handbook. Taylor and Francis Group, Boca Raton (2011)
6. Borcha, M.D., Balovsyak, S.V., Fodchuk, I.M., Khomenko, V.Y., Tkach, V.N.: Distribution of local deformations in diamond crystals according to the analysis of Kikuchi lines profile intensities. J. Superhard Mater. **35**(4), 220–226 (2013). http://link.springer.com/article/10.3103/S1063457613040035
7. Fodchuk, I.M., Novikov, S.M., Yaremchuk, I.V.: Direct and inverse problems in X-ray three-crystal LLL-interferometry. Appl. Opt. **55**(12), 120–125 (2016)
8. Korn, G., Korn, T.: Mathematical Handbook. For Scientists and Engineers. McGraw-Hill Book Company, New York (1968)
9. Ye, Z., Yang, J., Zhang, X., Hu, Z.: Remote sensing textual image classification based on ensemble learning. Int. J. Image Graph. Sig. Process. (IJIGSP) **8**(12), 21–29 (2016). https://doi.org/10.5815/ijigsp.2016.12.03
10. Balovsyak, S.V., Harabazhiv, Y.D., Fodchuk, I.M.: Oriented filtration of digital electron diffraction images. Radioelectron. Comput. Syst. **77**(3), 26–35 (1992). (in Russian)
11. Bandyopadhyay, A., Banerjee, S., Das, A., Bag, R.: A relook and renovation over state-of-art salt and pepper noise removal techniques. Int. J. Image Graph. Sig. Process. (IJIGSP) **7**(9), 61–69 (2015). https://doi.org/10.5815/ijigsp.2015.09.08
12. Balovsyak, S.V., Odaiska, K.S.: Automatic highly accurate estimation of Gaussian noise level in digital images using filtration and edges detection methods. Int. J. Image Graph. Sig. Process. (IJIGSP) **9**(12), 1–11 (2017). https://doi.org/10.5815/ijigsp.2017.12.01
13. Srinivasa Rao, M., Vijaya Kumar, V., Krishna Prasad, M.: Texture classification based on first order local ternary direction patterns. Int. J. Image Graph. Sig. Process. (IJIGSP) **9**(2), 46–54 (2017). https://doi.org/10.5815/ijigsp.2017.02.06
14. Gourav, T.S.: Various types of image noise and de-noising algorithm. Int. J. Mod. Educ. Comput. Sci. (IJMECS) **9**(5), 50–58 (2017). https://doi.org/10.5815/ijmecs.2017.05.07

On Formalization of Semantics of Real-Time and Cyber-Physical Systems

Ievgen Ivanov, Taras Panchenko$^{(\boxtimes)}$, Mykola Nikitchenko,
and Fabunmi Sunmade

Taras Shevchenko National University of Kyiv, Kyiv, Ukraine
ivanov.eugen@gmail.com, taras.panchenko@gmail.com

Abstract. In the article we describe theoretical foundations of a framework for formalizing semantics of real-time and cyber-physical systems in interactive theorem proving environments. The framework is based on viewing a system as a predicate on system's executions which are modeled as functions from the continuous time domain to a set of states. We consider how it can be applied to the safety verification problems. The proposed framework may be useful in verification of software for real-time and cyber-physical systems and of the corresponding development tools.

Keywords: Formal methods · Real-time systems · Cyber-physical systems
Formal semantics · Proof assistant

1 Introduction

A real-time system [1, 2] is an information processing system whose specification includes both logical and temporal correctness requirements. A cyber-physical system is a system which integrates computational and physical processes and heavily depends on their interaction. Examples of such systems include flight control systems, automotive control systems, robotics, etc. Many of them are controlled by software, which may be implemented in real-time extensions of universal programming languages or in specialized real-time programming languages [1–3]. Because of complexity of these systems and high reliability and safety requirements to them, methods of verification of real-time programs and methods of automated program generation from a real-time system specification play important role in the fields of real-time system design and implementation [4]. Rigorous development of such methods relies on precise definition of semantics of real-time programming languages and real-time system specification languages.

Usually, semantics of a language is defined as a transformation from the language syntax to a class of semantic models of real-time systems [5] like timed automata, timed Petri nets, timed process algebras, timed temporal logics, etc. More general models like hybrid automata [5] may also be used, if real-time programs interact with a continuously changing environment (e.g. physical environment). In most cases, semantics of a real-time system model can be defined as a set of its possible traces (executions, behaviors) [5]. A trace is considered as a mapping from a time domain to a

© Springer International Publishing AG, part of Springer Nature 2019
Z. Hu et al. (Eds.): ICCSEEA 2018, AISC 754, pp. 213–223, 2019.
https://doi.org/10.1007/978-3-319-91008-6_22

value domain, which represents evolution of system's state. A formal definition of semantics can be used in verification of properties of individual programs or common properties of classes of programs, verification of program translators, generators, etc.

In the previous work [6] we described formalization of the runtime behavior of a distributed, almost synchronous multiuser presentation system Infosoft e-Detailing 1.0 [7–9], a formalization of its correctness property, and a computer-checked proof of its correctness in Isabelle proof assistant.

In this paper we generalize these results and consider the problems of formalization of semantics of real-time and cyber-physical systems and proving common properties of classes of programs which control systems (mostly, safety properties of program executions). A common approach to such problems relies on interactive theorem proving [10]: a language syntax and semantics, (if necessary) a translator algorithm and target properties are specified in an interactive theorem proving environment like Isabelle, PVS, Coq, etc. [10], and the target properties are proved in this environment. However, these environments rarely support real-time system models directly or in the form of extensions, and therefore, to represent language semantics, one has to encode these and corresponding proof methods in the environment's underlying logic.

Here we propose theoretical foundations of a framework for representing semantics of real-time systems and programming languages. It can be implemented in the interactive theorem proving environment like Isabelle [11].

A high-level objective of our framework is to provide a sufficiently general semantic model of real-time systems and programs and corresponding methods of reasoning about properties of system's/program's behaviors. We consider many traditional real-time models (e.g. timed automata) not enough general for our purposes, because of their limited ability to represents systems' interaction with continuously changing environment. Hybrid system models [5] are more appealing, but we consider them too detailed because of their focus on specific methods of modeling of continuous dynamics. Ordinary differential equations are often used to describe continuous dynamics, but such a description may not always be adequate. Therefore, we decided to develop a sufficiently abstract real-time system model, which can represent discrete-state and mixed (discrete-continuous) state systems, which may have non-deterministic behavior. We describe this model in the next section.

2   Traces and Their Properties

2.1   Informal Considerations

The notion of a trace is one of the main notions in our approach. Informally, we consider a trace as an evolution of the state of a real-time system. We assume that a trace can be represented as a function on a time scale and that all system's traces are defined on the same (global) time scale. We assume that this global time scale is the set of non-negative real numbers, however, it is possible to generalize the constructs described below to the case, when the time scale is a dense bounded-complete totally ordered set. Also, we assume that a trace is a total function (a real-time system's/program's execution never terminates). Traces take values in a set of states.

A state may include a real-time system's/program's execution state (e.g. program's memory content), its components are not necessarily discrete-valued or controllable. For example, a state may include a continuously increasing real clock variable.

We consider semantics of a real-time system/program as a set of traces, or alternatively, as a property of traces.

2.2 Properties of Traces

Let T be the set of non-negative real numbers and Y be a set of values. We assume that T is equipped with a topology induced by a standard topology on real numbers.

Let us introduce the following notions:

- A trace is a total function $T \to Y$. The set of all traces is denoted by Tr.
- A trace property is a predicate $p: Tr \to Bool$

  where $Bool = \{True, False\}$. The class of all trace properties is denoted by TP.

- A trace s satisfies a trace property p, if $p(s) = \text{True}$.
- A trace s satisfies a trace property p on a set $A \in T$, if there exists a trace s' which satisfies p such that $s(t) = s'(t)$ for each $t \in A$.

Consider the class of trace properties p such that:

(1) If a trace s satisfies p on $\{t \in T | t \leq t_o\}$ for each $t_0 \in T$, then s satisfies p.
(2) If a trace s satisfies p on $\{t \in T | t \leq t_0\}$ and $\{t \in T | t \geq t_0\}$ for a $t_o \in T$, then s satisfies p.

We call properties which satisfy (1) *finitary*, and properties which satisfy (2) – *Markov-like*. A finitary Markov-like property (FM property) is a property, which satisfies (1) and (2). Denote by *FM* the class of all FM properties.

Informally, (1) means that a property of a trace is determined by properties of the finite-length prefixes of a trace and (2) means that the set of possible continuations (forward and backward) of a trace depends only on the current value of a trace and the current time moment. The class of FM properties can be considered as a generalization of the class of sets of infinite execution sequences of (discrete-time) transition systems to the real-time domain. We chose this class as a class of high-level semantic models of real-time systems and programs.

3 Finitary Markov-Like Properties

We call a predicate $ps: Tr \times T \times T \to Bool$

- left-segment-monotone, if $ps(s, t_1, t_2) \Rightarrow ps(s, t_1', t_2)$ for all $t_1, t_2 \in T$ such that $t_1 \leq t_2$ and $t_1' \in [t_1, t_2]$ (here $\Rightarrow$ denotes implication);
- right-segment-monotone, if $ps(s, t_1, t_2) \Rightarrow ps(s, t_1, t_2')$ for all $t_1, t_2 \in T$ such that $t_1 \leq t_2$, and $t_2' \in [t_1, t_2]$;
- coherent, if $ps(s, t_1, t_2) \Rightarrow ps(s, t_1, t_2)$ for each $s_1, s_2 \in Tr$ and $t_1, t_2 \in T$ such that $t_1 \leq t_2$ and $s_1(t) = s_2(t)$ for all $t \in [t_1, t_2]$.

Coherence means that for each $t_1 \leq t_2$, the value $ps(s, t_1, t_2)$ depends only on the values of s on the segment $[t_1, t_2]$.

Denote by Cm the set of all pairs (lc, rc) where $lc: Tr \times T \times T \to Bool$ is a left-segment-monotone and coherent predicate and $rc: Tr \times T \times T \to Bool$ is a right-segment-monotone and coherent predicate.

We write $(l, r) = loc(lc, rc)$ where $(lc, rc) \in Cm$, if $l: Tr \times T \to Bool$ and $r: Tr \times T \to Bool$ are predicates such that

1. $l(s, t_0)$ iff $t_0 = 0$, or there exists $t_1 \in [0, t_0)$ such that $lc(s, t_1, t_0)$;
2. $r(s, t_0)$ iff $t_o = 0$, or there exists $t_2 > t_0$ such that $rc(s, t_1, t_0)$.

Theorem 1 *(About representation of an FM property)*

(1) *Let $(lc, rc) \in Cm$ and $(l, r) = loc(lc, rc)$. Let $p: Tr \to Bool$ be a predicate such that $p(s) \Leftrightarrow \forall t \in T(l(s, t) \wedge r(s, t))$ for each $s \in Tr$. Then $p \in FM$.*

(2) *If $p \in FM$, then there exists $(lc, rc) \in Cm$ such that $p(s) \Leftrightarrow \forall t \in T(l(s, t) \wedge r(s, t))$ for each $s \in Tr$ where $(l, r) = loc(lc, rc)$.*

Here $\Leftrightarrow$ denotes (logical) equivalence. If $p \in FM$, $(l, r) = loc(lc, rc)$ for some $(lc, rc) \in CM$ and $p(s) \Leftrightarrow \forall t \in T(l(s, t) \wedge r(s, t))$ for each $s \in Tr$, then we call a pair (l, r) a *representation* of an *FM* property p.

Theorem 1 implies that each FM property has at least one representation. In general case, an FM property can have more than one representation.

Let $Rep(p)$ be the set of all representation of an FM property p.

Theorem 2. *For each FM property p there exists a (unique) representation $(l_*, r_*) \in Rep(p)$ such that for each $(l, r) \in Rep(p)$, $s \in Tr$ and $t \in T$, $l_*(s, t) \Rightarrow l(s, t)$ and $r_*(s, t) \Rightarrow r(s, t)$.*

We call the representation $(l_*, r_*) \in Rep(p)$ described in Theorem 2 *the least representation* of p and denote it as $rep_*(p)$.

It is possible to represent an FM property using a single predicate (instead of a pair of predicates) in the following way. Let $sRep(p)$ be the set of all predicates $g: Tr \times T \to Bool$ such that $g(s, t) = l(s, t) \wedge r(s, t)$ for each $s \in Tr, t \in T$. We call such a predicate g a symmetric representation of p. We call a predicate g_* such that $g_*(s, t) = l_*(s, t) \wedge r_*(s, t)$ for each $s \in Tr, t \in T$ where $(l_*, r_*) = rep_*(p)$ the least symmetric representation of p and denote it as $srep_*(p)$.

Theorem 3. *Let p be an FM property. Then for each $g \in sRep(p)$ there exists a unique pair $(l, r) \in Rep(p)$ such that $g(s, t) = l(s, t) \wedge r(s, t)$ for all s, t.*

3.1   Examples of FM and not-FM Properties

In this subsection we give examples of FM and not-FM properties of traces. Let $Y = R$. For each predicate $p: Tr \to Bool$ denote by $\neg p$ a predicate such that $(\neg p)(s) = \neg(p(s))$ for each $s \in Tr$.

Consider the following predicates on traces $T \to R$:

1. $p_{all}(s) = True$ for all s;
2. $p_{cont}(s)$ iff s is continuous (on T);
3. $p_{bnd}(s)$ iff s is bounded, i.e. $\{s(t) | t \in T\}$ is a bounded subset of R;
4. $p_{cvg}(s)$ iff $s(t)$ converges to 0, when $t \to +\infty$;
5. $p_{aff}(s)$ iff there exist $a, b \in R$ such that $s(t) = at + b$ for each $t \in T$.

Lemma 1

(1) $p_{all}, \neg p_{all}$, and p_{cont} are finitary Markov-like;
(2) p_{bnd}, p_{cvg}, and $\neg p_{cvg}$ are Markov-like, but not finitary;
(3) p_{aff} is finitary, but not Markov-like;
(4) $\neg p_{cont}, \neg p_{bnd}, \neg p_{aff}$ are neither finitary, nor Markov-like.

Let $J: T \to 2^{Y \times Y}$ be a given function and $lc^J: (T \to Y) \times T \times T \to Bool$, $rc^J: (T \to Y) \times T \times T \to Bool$ be predicates defined as follows:

- $lc^J(s, t_1, t_2)$ iff $t_1 \le t_2$ and s is constant on the segment $[t_1, t_2]$;
- $rc^J(s, t_1, t_2)$ iff $t_1 \le t_2$ and there exists $y \in Y$ such that $(s(t_1), y) \in J(t_1)$; and $s(\tau) = y$ for all $\tau \in (t_1, t_2]$.

Let $p^J: (T \to Y) \to Bool$ be a predicate such that $p^J(s) \Leftrightarrow \forall t \in T(l^J(s, t) \wedge r^J(s, t))$ for each trace $s: T \to Y$ where $(l^J, r^J) = loc(lc^J, rc^J)$.

Lemma 2. $(lc^J, rc^J) \in Cm$ and p^J is an FM property.

We call p^J a jump FM property generated by J. Note, that if $p^J(s) = True$, then s is a piecewise-constant function. Jump FM properties can be useful for representing semantics of real-time programs.

As an example, let us consider a simple programming language with a syntax defined by the following grammar (in which *italic* elements denote non-terminals, **bold** elements denote terminals, $\{x\}$ denotes 0 or more repetitions of x and $[x]$ means that x is optional):

program	::=	*{task} schedule*
task	::=	*header { stmt_list }*
header	::=	**task** *tname* [*local_defs*] **wcet** *timedur*
local_defs	::=	**local** (*lvardecl*{;*lvardecl*})
lvardecl	::=	*loc_id* :*type*
stmt_list	::=	*statement* {; *statement*}
statement	::=	| *loc_id* := *expr* | **if** *bexpr* **then** *stmt_list* **else** *stmt_list* **fi**
bexpr	::=	*expr* = *expr* | *expr*! = *expr*
expr	::=	*loc_id* | *boolean* | *natural*
type	::=	**bool** | **nat**
boolean	::=	**true** | **false**
natural	::=	**0** | **1** | **2** | ...
schedule	::=	**schedule** *frame* {; *frame*}
frame	::=	**duration** *timedur* [**task** *tname*]
timedur	::=	**1** | **2** | **3** | ...

We do not define non-terminals *tname* and *loc_id*. Informally, they range over sets of admissible task names and local identifiers correspondingly.

A program in this language consists of a list of tasks and a schedule. A task has a header and a body (a sequence of statements). A header includes a task name (*tname*), a list of local variable definitions and a worst-case execution time – *wcet* (in some time units). A schedule consists of a (non-empty) list of frames, each of which has a duration and optionally, an associated task name. The language includes usual constructs of imperative programming: sequential execution (;), assignment (:=), branching (**if**), however it does not have a loop statement. There is an empty statement, which does no operation. We will denote this statement as ε. Branching depends on a boolean expression (*bexpr*), which has a form of equality or inequality of expressions and may include local variables and constants.

A well-formed program in our language should satisfy several additional conditions: tasks should have different names; local variables defined in a header of a single task should have different names; each local variable which is used in a task's body, should be declared in a header of this task; types of values on the left- and right-hand sides of assignments, equalities and inequalities should be the same; each variable should be initialized before use (it is possible to introduce a sufficient syntactic criterion to check this condition), if a frame specifies a task, the worst-case execution time of this task should not exceed the frame's duration.

Informally, semantics of a program can be described as follows. An execution of a program is cyclic. Each cycle consists of the same sequence of execution phases (frames), which is specified by the program's schedule. The duration of each frame is fixed and specified in the schedule. A frame starts immediately after (end of) the previous frame. If a frame starts at time t (during a given cycle), has a duration d and an associated task *tsk*, then a program invokes a task *task* at time t. The values of the task's local variables are not preserved between different invocations (note, that the language does not support global variables). A task finishes execution at some time $t' \in [t, t + wcet)$. Then the program remains idle till the end of the given frame. If no task is associated with a frame, the program remains idle during this frame.

The execution model described above is rather similar to the execution model of Safety Critical Java (SCJ) Level 0 – a proposed standard for Java applications for mission-critical real-time systems [12] (however, in SCJ Level 0 several handlers may execute sequentially in one frame and a memory model is more complex).

We give a small-step operational semantics to an individual task. Let S be the set of all (non-empty, finite) lists of statements (*stmt_list*) of our language. We will make no distinction between 1-element statement lists and statements. Let P be a task with a body b_0 and a set of local variables L. For each variable $x \in L$ denote by $dom(x)$ the domain of x (boolean values or natural numbers, depending on the type of x). Let D_L be the set of all variable assignments of the form $[x \mapsto a_x | x \in L]$ where $a_x \in dom(x)$ for each $x \in L$. Such variable assignments are total functions on L, so we will use the functional notation $d(y) = a_y$, if $y \in L$ and $d = [x \mapsto a_x | x \in L]$.

For any expression e in our language (defined by the non-terminal *expr*) and $d \in D_L$ we will write $\langle e, d \rangle = d(e)$, if e is a name of a local variable in L; $\langle e, d \rangle = e$ if e is a boolean or natural constant and $\langle e, d \rangle = \bot$ (undefined), if e is a variable name, but

$e \notin L$. In other words, $\langle e, d \rangle$ is a value of an expression e which corresponds to a variable assignment d.

Then semantics of a task is based on a binary relation $\rightarrow$ (a small step, or a reduction) on pairs $(b, d), b \in S, d \in D_L$. This relation can be defined similarly to a small-step semantics of other imperative programming languages. We define a (complete) execution of a task as a finite sequence of pairs $(b_1, d_1), \ldots, (b_n, d_n), b_i \in S, d_i \in D_L$ such that b_1 is the task's body, $(b_i, d_i) \rightarrow (b_{i+1}, d_{i+1})$ for each $i = 1, 2, \ldots, n-1$ and $b_n = \varepsilon$, i.e. an empty statement (or a list of one empty statement). We also assume that a pair (b, d) cannot be reduced, if $b = \varepsilon$.

Now we define semantics of a program. Suppose that the schedule of a program prg is a finite sequence of frames $F_1, F_2, \ldots F_n, n \geq 1$. Let us introduce the following notation for $i = 1, 2, \ldots, n$: $Dur(F_i)$ is the duration of the frame F_i, $Task(F_i)$ is a name of the task, associated with F_i, $TBdy(F_i)$ is the body of the task $Task(F_i)$ and $Twcet(F_i)$ is the worst-case execution time of the task $Task(F_i)$. We assume that $Task(F_i), TBdy(F_i), TWcet(F_i)$ are undefined, if no task is associated with the frame F_i. Let TN be the set of names of all tasks of prg.

Let $Y = TN \times S \times D_L \cup \{\bot\}$ where $\bot \notin TN \times S \times D_L$ is an element, which denotes an idle state. Denote $o_i = Dur(F_i) + \ldots + Dur(F_{i-1})$ for each $i = 1, 2, \ldots, n$ (we assume $o_1 = 0$) and $P = Dur(F_i) + \ldots + Dur(F_n)$.

Let $J : T \rightarrow 2^{Y \times Y}$ be a function such that if $t \in [o_i + kP, o_i + kP + Dur(F_i))$ for some $i = 1, 2, \ldots, n$ and $k = 0, 1, 2, \ldots$, then

1. if $Task(F_i)$ is defined and $t = o_i + kP$, then

$$J(t) = \{(\bot, (Task(F_i), TBdy(F_i), d)) | d \in D_L\} \cup$$
$$\{((tn, \epsilon, d'), (Task(F_i), TBdy(F_i), d)) | tn \in TN, d, d' \in D_L\};$$

2. if $Task(F_i)$ is defined and $o_i + kP < t < o_i + kP + TWcet(F_i)$, then $J(t) = \{((Task(F_i), b, d), (Task(F_i), b', d'))|.$

$$(b, d) = (b', d') \text{ or } (b, d) \rightarrow (b', d')\} \cup \{((Task(F_i), \epsilon, d), \bot) | d \in D_L\} \cup \{(\bot, \bot)\};$$

3. if $Task(F_i)$ is defined and $t \geq o_i + kP + TWcet(F_i)$, then $J(t) = \{((Task(F_i), \epsilon, d), \bot) | d \in D_L\} \cup \{(\bot, \bot)\};$

4. if $Task(F_i)$ is undefined, then $J(t) = \{(\bot, \bot)\}$.

Then we can formally define semantics of prg as an FM property p^J, which was defined above (in Lemma 2).

3.2 Operations and Relations on FM Properties

Let $p_i, i \in I$ be a (possibly empty) indexed set of trace properties. Then $\bigwedge_{i \in I} p_i$ (conjunction) denotes a trace property p such that $p(s)$ iff $p_i(s)$ for all $i \in I$.

Theorem 4. *Let $p_i, i \in I$ be an indexed family of FM properties. Then*

(1) $\bigwedge_{i \in I} p_i \in FM$

(2) if $(l_i, r_i) \in Rep(p_i)$ for each $i \in I$ then $(l, r) \in Rep(\wedge_{i \in I} p_i)$ where $l(s, t)$ iff $l_i(s, t)$ for all $s \in Tr, i \in I$, and $r(s, t)$ iff $r_i(s, t)$ for all $s \in Tr, i \in I$;

(3) if $(l_i, r_i) = rep_*(p_i)$ for each $i \in I$, then $(l, r) = rep_*(\wedge_{i \in I} p_i)$ where $l(s, t)$; iff $l_i(s, t)$ for all $s \in Tr, i \in I$, and $r(s, t)$ iff $r_i(s, t)$ for all $s \in Tr, i \in I$.

Let $p : (T \to Y) \to Bool$ be an FM property and $f : Z \to Y$ be a (total) function. Denote by $f \cdot p$ a trace property $p' : (T \to Z) \to Bool$ such that for each $s, p'(s) = p(s')$ where $s'(t) = f(s(t))$ for all $t \in T$.

Theorem 5

(1) $f \cdot p \in FM$;

(2) if $(l, r) \in Rep(p)$, then $(l', r') \in Rep(f \cdot p)$ where $l'(s, t) = l(f(s), t)$ and $r'(s, t) = r(f(s), t)$ for all $s : (T \to Z) \to Bool$ and $t \in T$;

(3) if $(l, r) = rep_*(p)$, then $(l', r') = rep_*(f \cdot p)$ where $l'(s, t) = l(f(s), t)$ and $r'(s, t) = r(f(s), t)$ for all $s : (T \to Z) \to Bool$ and $t \in T$.

Let p_1, p_2 be trace properties. We write $p_1 \Rightarrow p_2$ (p_1 implies p_2) if $p_1(s) \Rightarrow p_2(s)$ for each $s \in Tr$.

Theorem 6. Let p_1, p_2 be FM properties. Then the following statements are equivalent:

(1) $p_1 \Rightarrow p_2$;

(2) there exists $(l_1, r_1) \in Rep(p_1)$ and $(l_2, r_2) \in Rep(p_2)$ such that for each $s \in Tr$ and $t \in T$, $l_1(s, t) \Rightarrow l_2(s, t)$ and $r_1(s, t) \Rightarrow r_2(s, t)$;

(3) for each $s \in Tr$ and $t \in T$, $l^1_*(s, t) \Rightarrow l^2_*(s, t)$ and $r^1_*(s, t) \Rightarrow r^2_*(s, t)$, where $(l^1_*, r^1_*) = rep_*(p_1)$ and $(l^2_*, r^2_*) = rep_*(p_2)$;

(4) there exists $g_1 \in sRep(p_1)$ and $g_2 \in sRep(p_2)$ such that for each $s \in Tr$ and $t \in T$, $g_1(s, t) \Rightarrow g_2(s, t)$;

(5) for each $s \in Tr$ and $t \in T$,

$$g^1_*(s, t) \Rightarrow g^2_*(s, t), \text{ where } g^1_* = srep_*(p_1) \text{ and } g^2_* = srep_*(p_2).$$

Theorem 6 shows that to check the implication between FM properties, it is important to find the least (symmetric) representations of FM properties. However, in some cases, other representations may be sufficient.

We propose a characterization of the least symmetric representation which is based on fixed points of special operators. Let $Rf : (Tr \times T \to Bool) \to (Tr \times T \to Bool)$ be a function (an operator) such that for each $h : Tr \times T \to Bool, s \in Tr$ and $t \in T$, $(Rf(h))(s, t)$ iff the following conditions are satisfied:

1. $\forall t' \in [0, t) h(s, t') \vee \exists s' \in Tr \exists \tau \in (t', t) s'(\tau) = s(\tau) \wedge \forall t'' \in [t', \tau] h(s', t'')$;

2. $t = 0 \vee \exists t' \in [0, t) \forall \tau \in [t', t] h(s, \tau)$;

3. $\exists t' > t \forall \tau \in [t, t'] h(s, \tau)$;

4. $\forall t' > t \, h(s, t') \vee \exists s' \in Tr \exists \tau \in (t, t') s'(\tau) = s(\tau) \wedge \forall t'' \in [\tau, t'] h(s', t'')$.

Theorem 7 *(About symmetric representation). Let p be an FM property, $g \in sRep(p)$. Then if a predicate $g_* : Tr \times T \to Bool$ satisfies $g_* = srep_*(p)$ then the following conditions are satisfied:*

(1) *for each $s_1, s_2 \in Tr$ and $t_0 \in T$,*

$$if\ s_1(t) = s_2(t_0)\ and\ g_*(s_1, t_0)\ and\ g_*(s_2, t_0),\ then\ g_*(s_1, t) \Rightarrow g_*(s, t)$$
$$for\ each\ t \leq t_0\ and\ g_*(s_2, t) \Rightarrow g_*(s, t)\ for\ each\ t \geq t_0$$

where $s(t) = s_1(t)$ if $t \leq t_0$ and $s(t) = s_2(t)$ if $t \geq t_o$;

(2) *$srep_*(p)(s, t) \Rightarrow g_*(s, t) \Rightarrow g(s, t)$ for each $s \in Tr, t \in T$;*

(3) *$Rf(g_*) = g_*$.*

If we introduce a partial order $\leq$ on the predicates $Tr \times T \to Bool$ such that $g_1 \leq g_2$ iff $g_1(s, t) \Rightarrow g_2(s, t)$ for all s, t, and denote by G the set of all predicates g_* which satisfy condition (1) of the Theorem 7, then we get the following result:

Theorem 8. *Let p be an FM property, $g \in sRep(p)$. Then*

(1) *Rf is a monotonic and decreasing operator, i.e. $Rf(g_1) \leq Rf(g_2)$ if $g_1 \leq g_2$ and $Rf(g) \leq g$;*

(2) *Rf maps G to G;*

(3) *$g \in G$ and $srep_*(p) \in G'$*

(4) *$srep_*(p)$ is (with respect to $\leq$) a fixed point of Rf such that $srep_*(p) \in G$ and $srep_*(p) \leq g$.*

4 Applications

The implication on FM properties can be used in the methods of real-time system/program verification. For example, suppose that we have a real-time programming language L. Let L_{syn} be the set of all well-formed programs (as syntactic entities) in this language. Suppose that semantics of the language defines an FM property $p_L : (T \to Y) \to Bool$ where Y is a set of states, T is a time scale (the set of non-negative real numbers) and a mapping $init : L_{syn} \to Y$. This mapping assigns an initial state to each program prg and semantics $Sem(prg)$ of the program prg is defined as a property of traces $p = Sem(prg)$ such that $p(s)$ iff $s(0) = init(prg)$ and $p_L(s)$. Note, that p' is an FM property.

Suppose that we want to check, whether some property holds for each trace of each well-formed program. If this property is an FM property q, we can check the implication $p_L \Rightarrow q$ to solve this problem (we can assume that p_L and q are specified using representations described in the previous sections). Note, that if q does not belong to FM, in some cases it may be possible to over approximate it using a suitable FM property. If we want to check q for an individual program, we can check if $Sem(prg) \Rightarrow q$. The following example shows an example of a useful property that may be checked for all programs using this approach.

Let us call an FM property p "forward-deterministic", if for all traces s_1, s_2 such that $p(s_1)$ and $p(s_2)$, if $s_1(t_0) = s_2(t_0)$ for $t_o \geq 0$, then $s_1(t) = s_2(t)$ for all $t > t_0$. If semantics of a program prg is a forward-deterministic FM property, then prg has not more than one trace (execution).

The class of all forward-deterministic FM properties can be expressed as

$$\{p \in FM | p^2 \Rightarrow Det\}$$

where

- p^2 is a trace property such that $p^2(s)$ iff there exist traces s_1, s_2 such that $p(s_1), p(s_2)$ and $s(t) = (s_1(t), s_2(t))$ for all $t \in T$;
- Det is a trace property such that $Det(s)$ iff there exist traces s_1, s_2 such that $s(t) = (s_1(t), s_2(t))$ for all $t \in T$, and if $s_1(t_0) = s_2(t_0)$ for $t_o \in T$, then $s_1(t) = s_2(t)$ for all $t > t_0$.

Note, that Det is an FM property, and if p is an FM property, then p^2 is an FM property. Hence, to check that some FM property p is forward-deterministic, it is sufficient to check validity of the implication $p^2 \Rightarrow Det$. Such checking can be done within composition-nominative logics [13, 14].

5   Conclusions

We have described theoretical foundations of a framework for representing semantics of real-time and cyber-physical systems. This framework may be useful in verification of real-time software and language tools like real-time language translators or compilers. We have described theoretical foundations of a framework for representing semantics of real-time and cyber-physical systems. More specifically, in Sect. 2 we have introduced the notion of a trace as a state-valued function on the real time scale, and considered a predicate on traces which satisfies a special condition called finitary Markov-likeness as a model of a dynamical behavior of a real-time or cyber-physical system. Informally, the truth domains of Markov-like predicates generalize the sets of runs of discrete-time state transition systems to continuous time. In Sect. 3 we have shown a representation theorem for Markov-like predicates. We have also introduced a simple real-time programming language and formalized its semantics in terms of Markov-like predicates. In Sect. 4 we have formulated a safety verification problem for the programs in this language and proposed an approach to solving it.

We plan to describe an implementation of this framework in the interactive theorem proving environment Isabelle in forthcoming papers.

Also applications in other domains e.g. [15] will be investigated further.

References

1. Sifakis, J.: Modeling real-time systems – challenges and work directions. In: Proceedings First International Workshop on Embedded Software, LNCS 2211, pp. 373–389. Springer, Heidelberg (2001)
2. Burns, A., Wellings, A.: Real Time Systems and Programming Languages, 3rd edn. Addison Wesley Longman, Reading (2001). 611 p.
3. Kirsch, C.M.: Principles of real-time programming. In: proceedings Second International Workshop on Embedded Software, LNCS 2491, pp. 61–75. Springer, Heidelberg (2002)
4. Merz, S., Navet, N. (eds.): Modeling and Verification of Real-Time Systems: Formalisms and Software Tools. INRIA Lorraine, Nancy (2008). 400 p.
5. Furia, C.A., Mandrioli, D., Morzenti, A., Rossi, M.: Modeling time in computing: a taxonomy and a comparative survey. ACM Comput. Surv. **42**, 1–59 (2010)
6. Panchenko, T., Ivanov, I.: A formal proof of properties of a presentation system using Isabelle. In: 2017 IEEE First Ukraine Conference on Electrical and Computer Engineering UKRCON, pp. 1155–1160 (2017)
7. Polishchuk, N., Kartavov, M., Panchenko, T.: Safety property proof using correctness proof methodology in IPCL. In: Proceedings of the 5th International Scientific Conference "Theoretical and Applied Aspects of Cybernetics", Bukrek, Kyiv, pp. 37–44 (2015)
8. Kartavov, M., Panchenko, T., Polishchuk, N.: Infosoft e-detailing system total correctness proof in IPCL. Bull. Taras Shevchenko National Univ. Kyiv Ser. Phys. Math. Sci. **3**, 80–83 (2015). (in Ukrainian)
9. Kartavov, M., Panchenko, T., Polishchuk, N.: Properties proof method in IPCL application to real-world system correctness proof. Int. J. Inf. Models Anal. **4**(2), 142–155 (2015). ITHEA, Sofia, Bulgaria
10. Geuvers, H.: Proof assistants: history, ideas and future. Sadhana **34**, 3–25 (2009)
11. Paulson, L.C.: Isabelle: A Generic Theorem Prover; with Contributions by Tobias Nipkow, Lecture Notes in Computer Science, vol. 828. Springer, Heidelberg (1994). 321 p.
12. Henties, T., Hunt, J., Locke, D., Nilsen, K., Schoeberl, M., Vitek, J.: Java for Safety-Critical Applications. Electronic Notes in Theoretical Computer Science (2009). 11 p. http://www.jopdesign.com/doc/safecert2009.pdf
13. Nikitchenko, M., Shkilniak, S.: Applied Logic. Publishing House of Taras Shevchenko National University of Kyiv, Kyiv (2013) 278 p. (in Ukrainian)
14. Nikitchenko, M., Shkilniak, S.: Algebras and logics of partial quasiary predicates. Algebra Discrete Math. **23**(2), 263–278 (2017)
15. Gladun, A., Rogushina, J.: Use of semantic web technologies and multilinguistic thesauri for knowledge-based access to biomedical resources. Int. J. Intell. Syst. Appl. (IJISA) **4**(1), 11–20 (2012). https://doi.org/10.5815/ijisa.2012.01.02

Optimization of Operation Regimes of Irrigation Canals Using Genetic Algorithms

V. O. Bohaienko[1(✉)] and V. M. Popov[2]

[1] VM Glushkov Institute of Cybernetics of NAS of Ukraine, Kyiv, Ukraine
sevab@ukr.net
[2] Institute of Water Problems and Land Reclamation of NAAS of Ukraine, Kyiv, Ukraine

Abstract. The paper focuses on the problem of irrigation canals operation regimes optimization which is important for minimizing operating expenses and ensuring stable water supply for agriculture. Regarding the complexity of the optimization problem we propose to solve it using genetic algorithm that searches for per-hour pumping station units' operation regimes, their pumping rates and the heights of controllable weirs and gate structures that should guarantee water levels and flow velocities needed by farmers. As an underlying direct problem we use an initial-boundary value problem for one-dimensional Saint-Venant equations system discretized by finite difference scheme. The algorithms of direct and optimization problems solution was applied to model water flow and determine optimal water supply rates for North Crimean canal.

Keywords: Saint-Venant equation · Inverse problems · Genetic algorithms
Irrigation systems

1 Introduction

Efficient management of irrigation systems requires accurate determination of water supply regimes depending on the needs of consumers and other economic factors. Irrigation canals can be modelled by different means including transfer functions [1, 2] or hydraulic models [3]. When dynamic water flow models are used, operation regimes can be optimized by solving inverse problems for them. Given the complexity of such problems and different types of parameters to be optimized, usage of heuristic algorithms is urgent similarly to the case of field level irrigation planning [4, 5].

The goal of optimization is to find such an hourly operation regime of pumping stations units, pumping rates of units that allow their change, and heights of controllable weirs or gate structures that assure given water level and flow velocity on given hydraulic structures. Additionally, the total cost of electricity needed to pump water should be minimal.

We model the flow using the system of one-dimensional Saint-Venant equations and solve the optimization problem by genetic algorithm [6–10] with a coding of optimization parameters that depends on their type. Hourly operation regimes are coded as binary string while pumping rates and gate heights are coded as floating point numbers.

© Springer International Publishing AG, part of Springer Nature 2019
Z. Hu et al. (Eds.): ICCSEEA 2018, AISC 754, pp. 224–233, 2019.
https://doi.org/10.1007/978-3-319-91008-6_23

2   Direct Problem Statement

Simulation of water flow in a single canal of hierarchical irrigation system is carried out according to the following system of equations [11]:

$$\frac{\partial A}{\partial t} + \frac{\partial Q}{\partial x} = q - E, \tag{1}$$

$$\frac{\partial Q}{\partial t} + \frac{\partial(\beta Q^2/A)}{\partial x} = gA(sin\theta - cos\theta\frac{\partial h}{\partial x} - S_f) \tag{2}$$

where

- $A = \frac{1}{2}\left(2w_0 + \frac{(h-h_0)(w_1-w_0)}{h_1-h_0}\right)(h-h_0) = a + bh + ch^2$ is a wetted cross section area at absolute water level h and trapezoidal canal cross section parameters h_0 (canal bed absolute level), h_1 (absolute water level at full capacity), w_0 (canal bed width), w_1 (canal width at h_1);
- $Q = VA$ is a discharge, V is a flow velocity;
- q is a source function (discharge per unit length of the canal) that simulates the operation of the stations pumping water out of the canal and free flow of water between canals;
- E are losses considered constant over the canal;
- $S_f = \frac{n_m^2 Q^2}{A^2 R^{4/3}}$ is a friction [7] where n_m is Manning coefficient and $R = \frac{(h_1-h_0)(w_1+w_0)}{2(h_1-h_0)+0.5(w_1+w_0)}$ is a hydraulic radius;
- θ is a slope angle;
- g is the acceleration of gravity.

The Eqs. (1) and (2) will have the following form in terms of discharge and absolute water level:

$$(b + 2ch)\frac{\partial h}{\partial t} + \frac{\partial Q}{\partial x} = q - E, \tag{3}$$

$$\frac{\partial Q}{\partial t} + \frac{\partial(\beta QV)}{\partial x} = gA(sin\theta - cos\theta\frac{\partial h}{\partial x} - S_f). \tag{4}$$

Boundary conditions for the system (3) and (4) are as follows:

- $\left.\frac{\partial h}{\partial x}\right|_{x=0} = 0$, $Q|_{x=0} = 0$ in the absence of canal headwork;
- when the canal has a headwork, $Q|_{x=0} = Q_0(t)$ condition is set where $Q_0(t)$ is the sum of pumping rates for all units that pumps water at the moment of time t taking their linear change during start-up and switch-off into account;

- $\dfrac{\partial h}{\partial x}\Big|_{x=L} = 0$, $Q|_{x=L} = 0$ where L is the length of the canal, if there are no pumping station at the tail of the canal;

- $Q|_{x=L} = Q_L(t)$ when pumping station with $Q_L(t)$ total pumping rate is located at the tail of the canal.

When pumping station that pumps out water from the canal is located at the internal point x, $Q = -Q_i(t)$ condition is set where $Q_i(t)$ is total pumping rate of the station at the moment of time t per unit length of the canal.

In the presence of free flow of water into another canal at the internal point x, an internal condition $Q = -k_1(A_1 - A_2)$ is set where k_1 is a given parameter of flow rate, A_1 and A_2 are wetted cross section areas of the canals. At the point $x = 0$ of the canal in which the water flow, $Q|_{x=0} = k_1(A_1 - A_2)$ condition is set.

When weir or gate structure is located at the internal point x, an internal condition $Q = k_2 A_0 \sqrt{2g\Delta h}$ is set where A_0 is a cross section area of water flowing through the structure [12], Δh is a water level drop across the structure, k_2 is a given constant.

Initial conditions are set according to given values of relative or absolute water levels and flow velocities at hydraulic structures.

3 Numerical Method for the Direct Problem

Equations (3) and (4) are discretized using finite difference method with linear systems solved by simple iteration method. Discretization in time variable is based on an explicit scheme of the first order of accuracy. Discretization in space variable is done by a fully implicit scheme described in [13].

The nonlinearity in flow velocity was iteratively taken into account by

- solving the system of equations on one time step with a fixed V;
- further calculation of a new V value using the calculated Q and A;
- system (3) and (4) re-solution until the value of water levels and discharges will not stop changing with a given accuracy ε.

Dry zones of the canal are modelled as follows. After each iteration, $h = h_0$, $Q = 0$, $V = 0$ is set in the nodes where $h < h_0$. Similarly, the limitations imposed by closed sections of the canal are taken into account: $h = h_1$ is set if $h > h_1$. Since such procedure can violate the mass conservation law, we use the following water level correction mechanism. The volume of water $A_0 = -A(h)$ for the dry zone or $A_0 = A(h) - A(h_1)$ for the closed section of the canal is redistributed to the adjacent cells (for the dry zone to the cell $i + 1$ if $Q_i \geq 0$ or to the cell $i - 1$ if $Q_i < 0$ where i is the index of current cell, for the closed zone the same is done in reverse way). The value of water level in the adjacent cell is set to follow the mass conservation law: $A(h_n^{(1)}) = A(h_n^{(0)}) + A_0$ where $h_n^{(0)}$ is a current and $h_n^{(1)}$ is a modified value of water level.

To reduce an error of flow velocity calculation in near dry zones, it was calculated as [14] $V = \sqrt{2}AQ/\sqrt[4]{A^4 + \max(A^4, k_3)}$ where k_3 as a given constant. In addition, a mechanism for reducing the value of the convective term coefficient β in the Eq. (2) [15]

can be used for the same purpose. For transcritical flow, when the Froude number $F_n = |V|/\sqrt{g(h - h_0)} > 1$, the coefficient β takes value $1/F_n^{k_4}$, $k_4 \geq 2$. For subcritical flow β is set to 1. The use of these mechanisms can result in mass conservation law violation.

When linear system solution algorithm or iterative procedure that deals with nonlinearity diverges or significant errors arise when applying mechanisms of total water volume preservation, the time step decreases by half and calculations for the current time step are repeated.

4   Genetic Algorithm for Solving the Optimization Problem

We propose to solve two-criterion canal infrastructure parameters optimization problem using a genetic algorithm [6]. The optimization parameters are hourly operation regimes of pumping stations' units, encoded binary, the pumping rates of pumps that allow their change, and the heights of controllable weirs and gate structures, which are encoded with floating point numbers.

The first optimization criterion is the sum of squares of differences between minimal required and current water levels and flow velocities for

- all hydraulic structures where constraints on them are imposed;
- all moments of time in the case when hydraulic structure is not a pumping station;
- the moments of time when at least one unit works on pumping station.

The second optimization criterion was the total cost of electricity needed to pump the required amount of water.

The multicriteria optimization problem is reduced to unicriterion problem by scalarization. Weights are separately found on the base of total deviations of water levels and flow velocities. The weight factor for electricity costs is taken as 1 and the coefficients for water level and flow velocity are determined the way to make their maximal possible contribution to goal function equal to the contribution of maximal possible expenses on electricity. The maximal deviation values are estimated from scenarios when all pumps are turned on or off for the complete simulation time interval.

The proposed genetic algorithm for solving problem of irrigation canal operation regime optimization can be stated as follows:

- initialization stage: goal function weights are calculated and initial population is filled with randomly generated solutions;
- algorithm's iterations that consist of
  - Selection operation: two potential solutions are selected randomly from the population with inverse values of goal function as weight coefficients;
  - Crossover operation defined as a combination of binary strings representing pumping units working time for two potential solutions by randomly generated mask. For parameters described by floating point numbers, the resulting crossover value is a linear combination with random coefficients. Crossover operands are chosen by selection operation;

- Mutation operation defined as a combination of binary string with randomly generated string upon randomly generated mask. Floating-point described parameters are mutated changing their values by a given per cent of their maximal value. Mutation is applied with a given probability to the result of crossover operation.
- Replacement operation: a new potential solution, generated by crossover and potentially modified by mutation operation, replaces the worst solution within the population if the worst goal function value is greater than the one of a generated solution. Lexicographic approach to multicriteria optimization with a priori known priorities is used for performing the comparison.
- Checking of the following search process completion criteria:
  - a solution that fully satisfies the constraints is found;
  - given maximal number of iterations is exceeded;
  - difference between maximal and minimal values of goal function within the population became less than a given value.

5 Software Implementation of the Algorithm

The above-described algorithm was implemented within the web-based decision support system in agriculture. User interface and visualization tools were implemented using HTML and Javascript languages on the client side of client-server application. Numerical algorithms were implemented in C++ as an extension of php-interpreter on the server side.

Network of canals and hydraulic structures descriptions are stored as multiline and point shapefiles and are displayed using an online map software (Fig. 1), using which they can be selected by user. Characteristics of canals and hydraulic structures (canal cross sections, types of hydraulic structures, pumping rates, weir and gate heights, etc.) are stored in server-located database and are loaded for objects selected on an online map.

User-specified parameters of direct problem solution algorithm are

- discretization steps for time and space variables;
- ending simulation time;
- maximal number of iterations and maximal permissible error of iterative procedures;
- minimal volume of water redistributed by mass conservation ensuring mechanisms;
- k_3 coefficient used in high velocities reduction mechanism;
- losses E due to filtration, evaporation, and other physical processes;

  Parameters of the genetic algorithm are

- maximal number of iterations;
- permissible error (difference between the maximal and minimal values of goal function within the population);
- population size;
- probability of mutation and maximal changes in floating point described parameters values made by mutation operation.

6   Result of North Crimean Canal Modelling

Algorithms were used to model actual operational situations on continental part of North Crimean canal (Fig. 1) using data collected under the supervision of Dr. V.M. Popov (Institute of Water Problems and Land Reclamation of NAAS of Ukraine) in the spring and summer of 2017.

Fig. 1. North Crimean canal on an online map

Accuracy testing and parameters identification for the direct problem was carried out by modelling wave propagation in the canal on 15.06.2017.

Input data for simulation were as follows:

- Starting time - June 15, 2017, 12:00;
- Ending time - June 16, 2017, 6:00;
- Fitted value of Manning coefficient: 0.05;
- Fitted losses due to evaporation, filtration and other factors per unit length of the canal: 0.0245;
- Initial values of water levels taken from logs of measurements:
  - on headwork - 13.68 m;
  - at M-5 mark - 13.46 m;
  - at 61 km mark - 13.46 m;
  - in ПС-1 gate structure upstream - 13.46 m;
- Water discharge through ПС-1 gate structure was set according to the opening level equal to 12 cm;
- Water discharge on headwork was set according to logged values as follows:
  - From 12.00 to 1.00 - 50 m³/s;
  - From 1.00 to 2.00 - linear increase up to 55 m³/s (source of the modelled wave);
  - From 2.00 to 3.00 - linear decrease to 40 m³/s;
  - From 3.00 to 6.00 - 40 m³/s.

Modelled and actual values of water levels at 78 km mark are shown on Fig. 2 (red chart depicting actual data, blue chart – simulated data). Actual levels were taken from the automatic meter log. The data were compared starting from the time of 15/06/2017, 20:00, because the first 8 h of the simulation were used to compensate inaccuracies in the initial conditions. The modelled wave height and propagation velocity agrees with the measured data with a sufficient accuracy.

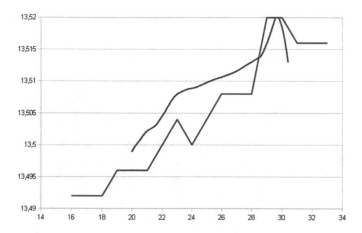

Fig. 2. Water level changes in time at 78 km mark (Color figure online)

In the second computational experiment we simulated the charging of the canal on 01.04.2017.

Input data were as follows:

- Starting time - April 1, 2017, 0:00;
- Ending time - April 1, 2017, 20:00;
- Fitted value of Manning coefficient was equal to 0.05 that is the same value as determined when modelling wave propagation;
- Fitted losses per unit length of the canal was equal to 0.005 that is less than the value determined when modelling processes in the condition of higher temperatures in June;
- Initial values of water levels taken from logs of measurements:
  - on headwork - 11.40 m;
  - at M-5 mark - 10.50 m;
  - at 61 km mark - 10.60 m;
  - in ПС-1 gate structure upstream - 10.57 m;
  - at PM-2 mark near the tail of the canal - 10.06 m;
- Opening level of ПС-1 gate structure was equal to 10 cm;
- Water discharge on headwork was set as follows:
  - From 0.00 to 4.00 - 25 m³/s;
  - From 4.00 to 8.00 - 30 m³/s;
  - From 8.00 to 12.00 - 20 m³/s;
  - From 12.00 to 16.00 - 15 m³/s;
  - From 16.00 to 20.00 - 20 m³/s.

The values of water levels near canal's headwork are shown on Fig. 3 (red chart depicting actual data, blue chart – simulated data). The same values at PM-2 mark are presented on Fig. 4.

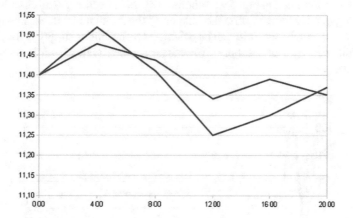

Fig. 3. Water level changes in time at canal's headwork (Color figure online)

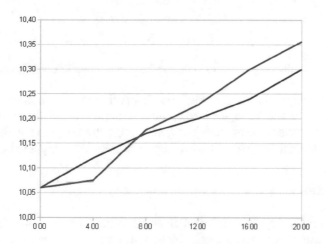

Fig. 4. Water level changes in time at PM-2 mark (Color figure online)

The errors here were within 25 cm and could be caused by the same order errors in the model of canal's geometry. Charging rates within the experiment were nearly equal to the measured ones.

Genetic algorithm efficiency was tested by selecting optimal headwork operation regime that maintains the required water level in the canal. The testing was conducted upon the data that was used for wave propagation modelling.

The headwork water discharge function was determined to maintain the water level at 61 km mark not lower than the initial value. At the same time, the total headwork water discharge had to be minimized.

The population size of the genetic algorithm was equal to 15. The difference between the maximal and minimal values of goal function in the population required to stop the search process was equal to 2.

The solution depicted in Fig. 5 was obtained by 294 iterations. Red chart here depicts the actual logged discharges while the blue one - the simulated values. The logic of the simulated water supply scenario can be described as an initial supply of large volumes of water followed by water supply, if necessary, when water level is reduced to the critical value.

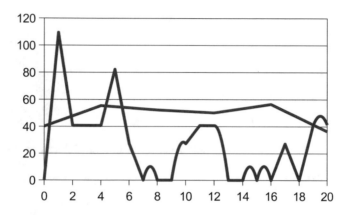

Fig. 5. Water discharge changes in time on canal's headwork (Color figure online)

Total water supply for the simulated time interval according to the actual data was 4188 thousand m^3 while in the obtained optimal scenario it was equal to 2080 thousand m^3, which demonstrates the efficiency of genetic algorithm application for solving the considered problem along with potential inconsistencies in the input data.

7   Conclusions

Genetic algorithm for optimizing irrigation canal operation regime was developed and tested in the case of continental part of North Crimean canal. One-dimensional Saint-Venant equation discretized by finite difference method was used to model water flow in the canal. We assessed Manning coefficient of the considered part of the canal and rate of losses in it by modelling using available measurements of water levels and discharges. The modelling showed changes in current state of the canal comparing with used design canal cross sections parameters. Determination of optimal water supply in the canal to maintain measured water levels showed that statistically accounted rate of water intake can be different from an actual one.

References

1. Barbosa de Oliveira, J., Pinho, T.M., Coelho, J.P., Boaventura-Cunha, J., Oliveira P.M.: Optimized fractional order sliding mode controller for water level in irrigation canal pool. IFAC-PapersOnLine **50**(1), 7663–7668 (2017)
2. Ding, Z., Wang, C.: Research on canal system operation based on controlled volume method. Int. J. Intell. Syst. Appl. (IJISA) **1**(1), 19–29 (2009). https://doi.org/10.5815/ijisa.2009.01.03
3. Lozano, D., Arranja, C., Rijo, M., Mateos, L.: Simulation of automatic control of an irrigation canal. Agric. Water Manag. **97**(1), 91–100 (2010)
4. Srinivasa Raju, K., Nagesh Kumar, D.: Irrigation planning using genetic algorithms. Water Resour. Manag. **18**(2), 163–176 (2004)
5. Mathur, Y.P., Sharma, G., Pawde, A.W.: Optimal operation scheduling of irrigation canals using genetic algorithm. Int. J. Recent Trends Eng. **1**(6), 11–15 (2009)
6. Goldberg, D.E.: Genetic algorithms in search, optimization and machine learning. Addison-Wesley, Reading (1989)
7. Umbarkar, A.J., Joshi, M.S., Sheth, P.D.: Dual population genetic algorithm for solving constrained optimization problems. Int. J. Intell. Syst. Appl. (IJISA) **7**(2), 34–40 (2015). https://doi.org/10.5815/ijisa.2015.02.05
8. Seshadri Sastry, K., Prasad Babu, M.S.: Adaptive population sizing genetic algorithm assisted maximum likelihood detection of OFDM symbols in the presence of nonlinear distortions. Int. J. Comput. Netw. Inf. Secur. (IJCNIS) **5**(7), 58–65 (2013). https://doi.org/10.5815/ijcnis.2013.07.07
9. Boeira, J.N.R.: The effects of "Preferentialism" on a genetic algorithm population over elitism and regular development in a binary F6 fitness function. Int. J. Intell. Syst. Appl. (IJISA) **8**(9), 38–46 (2016). https://doi.org/10.5815/ijisa.2016.09.05
10. Zhu, Y., Qin, D., Zhu, Y., Cao, X.: Genetic algorithm combination of boolean constraint programming for solving course of action optimization in influence nets. Int. J. Intell. Syst. Appl. (IJISA), **3**(4), 1–7 (2011)
11. Malaterre, P.O., Baume, J.P.: Modeling and regulation of irrigation canals: existing applications and ongoing researches. In: 1998 IEEE International Conference on Systems Man and Cybernetics, vol. 4, pp. 3850–3855 (1998)
12. Meselhe, E.A., Sotiropoulos, F., Holly, F.H.: Numerical simulation of transcritical flow in open channels. ASCE J. Hydr. Eng. **123**(9), 774–783 (1997)
13. Agu, C.E., Elseth, G., Lie, B.: Simulation of transcritical flow in hydraulic structures. In: Proceedings of the 56th Conference on Simulation and Modelling (SIMS 56), 7–9 October 2015, pp. 369–375. Linköping University, Sweden (2015)
14. Brodtkorb, A.R., Hagen, T.R., Lie, K.A., et al.: Simulation and visualization of the Saint-Venant system using GPUs. Comput. Vis. Sci. **13**, 341–353 (2010)
15. Abedini, M.J., Hashemi, M.R.: Effect of convective term suppression in numerical simulation of trans-critical open channel flow. Iran. J. Sci. Technol. Trans. B Eng. **30**(1), 85–96 (2006)

Monitoring of Laser Welding Process Using Its Acoustic Emission Signal

Volodymyr Shelyagin, Ievgen Zaitsev, Artemii Bernatskyi[✉],
Vladyslav Khaskin, Ivan Shuba, Volodymyr Sydorets,
and Oleksandr Siora

E.O. Paton Electric Welding Institute of the NAS of Ukraine,
11, K. Malevicha Street, 150, Kyiv 03680, Ukraine
avb77@ukr.net

Abstract. Processing and identification of signals informatively characterizing any technological process, especially the signals of acoustic and electromagnetic emission, is a complex multi-level task. Many studies are devoted to solving it. However, no comprehensive system for analyzing these signals has been represented yet. This study found out that there is a relationship between the parameters of laser welding technological process and registered acoustic emission signals, and that relationship is independent on investigation methods. The signals of acoustic and electromagnetic emission have been chosen as feedback signals of laser welding technological process, because their transmission and analysis require minimal time and computational resources. The paper describes the peculiarities of channels shielding for transmitting the technological data, in particular, the circuits of applying the shields for electromagnetic protection are given and their efficiency is shown.

Keywords: Laser welding · Acoustic emission · Signal registration
Signal identification · Quality · Defect detection

1   Introduction

A service technological communication, namely communication between discrete modules, programs or subsystems in automatic mode without human involvement, is a separate direction in the area of electrical communication systems. Propagation of the signals providing such a communication must comply with the criteria of reliability, interference protection and transfer rate in accordance with the requirements of the data communication protocol or specifications of service transmission channel [1]. Service information of technological processes includes the control commands (direct channel) as well as feedback data (feedback channel) that make it possible to control the state of the process or of the entire remote system [2].

The investigation of the properties of feedback channel data types showed that the data of visual control of the signals of electromagnetic emission EE in the range of visible, infrared or gamma radiation as well as the signals of acoustic emission AE in the range of audio and ultrasonic radiation, taking place during the process, provide the

© Springer International Publishing AG, part of Springer Nature 2019
Z. Hu et al. (Eds.): ICCSEEA 2018, AISC 754, pp. 234–243, 2019.
https://doi.org/10.1007/978-3-319-91008-6_24

highest information value. The data of these types require preprocessing, particularization and classification of the signals in order to introduce a correction in the control channel [3]. In this study, the main attention is paid to feedback channel of information transmission, since the complexity and multivaluedness of AE and EE signals often require human involvement for their identification and introduction of appropriate corrections in the system operation. The task of the study is to develop the algorithms of processing and transmitting the technological data, which will provide the increase of the rate and reliability of a technological process control.

The analysis of known studies showed that quite many different modern approaches are applied for AE and EE signals identification depending on identification tasks and accuracy requirements. AE analysis method makes it possible to identify the state of signal source. It is applied in many areas of science, i.e. medicine, for separation of sounds of normal breathing and deviations from the norm [4, 5], in zoology, for classification of animal species [6, 7], in defectoscopy, for detection of the cracks and breakdowns in mechanisms or leakages [8–10], in agriculture, for detecting the defects in wheat kernels or failures of transportation technology [11, 12], etc. The identification of AE signals caused by material processing, for example, laser welding, is highly relevant today [2, 13–15]. The registered AE signals depend on the sensor type, the examined object kind, and the changes of the object state during processing.

The aim of this work is to extract the reliable information from the signals of acoustic emission during monitoring of the technological process of laser welding. The main part of the study is devoted to the acoustic component of the signals entering the feedback channel.

2   Research Methods and Equipment

2.1   System for Analysis of AE Signal Characteristics

It may be stated that the means applied for processing AE signals of audio range use the sensors, which have the characteristics that do not exceed the average human hearing abilities. Moreover, in some cases [16–18], signal processing was carried out in such a way as to approximate the audio characteristics of personal computer (PC) to human hearing characteristics, thereby reducing frequency resolution capacity of the system. Following the results of the studies [16–18], it was concluded that the system for analysis of AE signal characteristics can be developed using only an electret microphone, with a circular directional pattern, for example, FM-4F with the parameters shown in Table 1.

However, the disadvantage of application of such a microphone can be the interferences that appear in the production environment, since it is connected by a circuit that has low protection from electromagnetic field interferences. Therefore, in order to compare the characteristics, the collection of the information from the feedback channel was also carried out with the help of dynamic microphone Sony WM-800 (Table 1). Such a microphone requires a pre-amplifier for PC connection, since the PC sound cards are not designed for operation with such sources, however, it allows signal cable shielding using one of the most efficient circuits. A balance connection of the dynamic microphone to the operational amplifier removes in-phase electrical interference. Rational shielding provides elimination of an interference magnetic component.

An important element of data collection with interference protection was application of the shielded cables. Shielded cable BENIAMIN Microphone cable OFC OD 6.2 with XLR connectors was used in the study.

Table 1. Characteristics of electret microphone FM-4F and dynamic microphone Sony WM-800

Characteristics	Electret microphone FM-4F	Dynamic microphone Sony WM-800
Sensitivity	-62 ± 2 dBV/0.1 Pa	-72 ± 2 dBV/0.1 Pa
Impedance	Low (less than 600 Ω)	600 Ω
Directivity	Omnidirectional	Cardioid
Frequency band	20 Hz ... 16 kHz (by 3 dB level)	10 Hz ... 10 kHz (by 3 dB level)
Operating voltage	1.5 ... 10 V	
Current consumption	0.3 mA	
S/N ratio	More than 60 dB	More than 60 dB

2.2 Choice of Investigation Method

Table 2 represents the comparison of accuracy of application of time-frequency and neural network methods of signal identification [19].

The data are obtained after 5000 control measurements. A comparative analysis was carried out since a defectless welded joint is a necessary condition for its quality. The first column of the results shows the percentage of false identification in the case of defect-free welding. The second column shows the percentage of unfound defects when they were present in the process.

Therefore, the results [19] show that develop of the model for processing and identification of signals should be based on time-frequency methods.

Table 2. Comparative analysis of identification methods

Method of analysis	Test for analysis of defect-free welding (%)	Test for defect detection (%)
Neural network	2.6	8.7
Time-frequency	3.2	0.9

LabVIEW (Laboratory Virtual Instrument Engineering Workbench) [20, 21] is the most convenient environment for development of the model for processing and analyzing the investigated signals.

3   Experimental Investigations and Results

The experimental investigations on processing and identification of the signals of acoustic and electromagnetic channels for technological processes control have been carried out for emission signals of laser welding at the E.O. Paton Electric Welding Institute of the National Academy of Sciences of Ukraine.

The structural model of carrying out of the experiment on registration of the signals of feedback channel is given in Fig. 1.

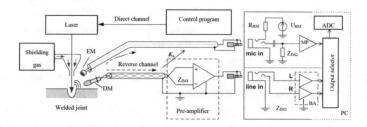

Fig. 1. Structural model of experiment

The following symbols are used in Fig. 1: EM is the electret microphone; DM is the dynamic microphone; PC is the personal computer; BP is the buffer amplifier; DP is the differential amplifier; MP is the microphone amplifier; U_{BM} is the bias voltage of electret microphone ($R_{BM} = 2\ \Omega$); Z_{IM} is the input impedance ($Z_{IM1} = 1\ M\Omega$; $Z_{IM2} = Z_{IM3} = 10\ k\Omega$); ADC is the analog-digital converter (PC sound card); line in is the linear input of PC sound card; mic in is the mic input of PC sound card.

The signals are registered in ".wav" format, since it ensures the minimum distortion of the output information at acceptable audio compression. Signal registration is carried out by sound card AC-97 at the following modes:

- Dynamic microphone signal (sensitivity range 10–10000 Hz) was digitized with sampling frequency 11.025 kHz, sampling rate 16 bits, pre-amplification was carried out to −10 dB level);
- Signal from the electret microphone (23–16000 Hz range) is digitized with a frequency of 44.1 kHz, no additional amplification was applied.

Recording software is Spectrum Laboratory v.2.7. It allows realizing a wide range of signal recording parameters, select a sampling rate, level limit, input source, coding depth, degree of signal decimation. The program interface allows observing a signal instantaneous spectrum thanks to Short Time Fourier Transform and its displaying with a "waterfall" type diagram or 3D model, which is quite convenient for tracking the dynamics of signal parameters. The following was selected during the analysis process, i.e. the window duration is 20 ms, the window type is the Hann window, and the maximum number of readings for the Fourier transform is 16384.

Laser welding was taken as a technological process for investigation. During it a metal surface layer is molten and left to cool at ambient temperatures, which changes its structure and mechanical properties. The investigation was focused on detection of a relationship between the spectral characteristics of the signal and result of the technological process (first of all, defect-less of welded joint). The monitoring was used to determine a deviation from a standard signal. The presence of this deviation correlated with the position of given deviation relative to workpiece coordinates. Any deviation from the standard signal indicated a need in additional quality control (visual, radiographic, etc.) at this particular area of the weld. Thus, this method does not allow determining defect type (pore, crack, lack of penetration or other), but permits to indicate the possible location of defect.

Two S235J0 steel samples of 1 and 3 mm thickness were processed. The parameters of technological process were the following: CO_2 laser radiation power made 200 W; length of radiation wave was 10.6 µm; beam velocity made 5 cm/sec and beam diameter in the point of contact with the metal was 0.3 mm.

Air of 5 atm pressure and 1.5 mm nozzle diameter were used for 1 mm thick specimen instead of shielding gas. For 3 mm thickness specimen the works were carried out without gas for noise reduction during recording.

Figure 2(a) shows a diagram (waveform) of registered AE and the result of metal treatment compared with the scale. It can be noted that there are some dependencies between the corresponding surface fragments and AE signal content. Given AE signal fragments were taken during long-term registration of AE signals and extracted by means of the Sound Forge 10 software package.

This approach allowed detecting the fragments of the registered signal, which are the most appropriate for further analysis, avoid analysis of the signals that were limited by amplitude and contain excessive information.

Appearance of "non-informative" and noise fragments in the registered signal is caused by long-term experimental studies as for optimum positioning of the microphone based on a criterion of minimum acoustic noises and maximization of registered signal level.

Figure 2(b) presents the 3D diagrams on the results of Short Time Fourier Transform of AE signal extracted fragments obtained in the environment of Spectrum Laboratory v.2.7 program. Since the fragments are of insignificant duration, it was decided to use multiple repetition to form a visible periodic time-frequency characteristic of the registered signals. The results of analysis indicate a clear correlation of the conditions and result of technological process effecting a behavior of the registered signal. In particular, it is observed in the areas covered with rust and ones with uneven treatment.

The obtained results are of visual nature and only confirm the data given in analysis of the reference [1–19]. However, a purpose of the work is to determine informative characteristics of the signals for their application in control of TP by the feedback channel. Therefore, LabView 2009 software package was used for development of a system model that registers the AE signals, extracts their characteristics, stores, and provides a decision on termination or continuation of TP based on comparison of the accumulated results with the current signal characteristics.

The developed block diagrams are joined in a model front panel and provide a mathematical basis for its operation.

The front panel of the development is designed in a way that allows a user having the maximum effect on examined signal and identify its informative characteristics.

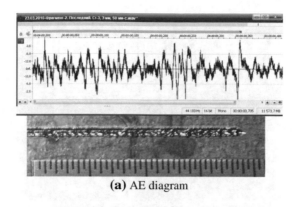

(a) AE diagram

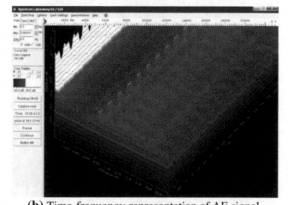

(b) Time-frequency representation of AE signal

Fig. 2. AE diagram (a) at the end of processing area registered using electret microphone. Scale division is 0.5 mm. The fragment full length makes 0.7 s, and time-frequency (b) representation of AE signal at the end of processing area. Duration (frequency) of the examined fragment is 0.7 s.

Figure 3(a) shows a fragment of front panel of the developed model with the results of analysis of the registered signal.

The following, namely normalization (diagram 1, Fig. 3), shift of the signal into positive values region (diagram 2), division into segments of 70 ms duration of set duration and applying the Gaussian window function to each segment (diagram 3) were carried out with the signal saved in "23.03.-Fragment-2.wav" file, the diagram of which is also shown in Fig. 3. Thanks to the window function, all time fragments of the signal are smoothed, and all transient components, which appeared during division in the beginning at the fragment end, were removed. The range of investigated frequencies is set within 20 Hz–20 kHz limit, filter bandpass makes 200 Hz. The recorded signal, saved in "23.03.-Fragment-1.wav" file, was taken as a reference signal to demonstrate model operation. Figure 3(b) represents a lower front panel of the developed virtual model, which shows the results of comparison of given file with the reference one. The maxima of the reference signal are represented by white rectangles and the minima are red. The current signal divided into frequency fragments includes the components that exceed

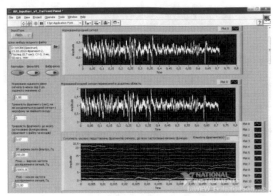

(**a**) Top part and front panel of the model

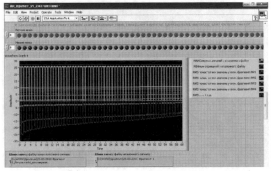

(**b**) Lower part of the model

Fig. 3. Model front panel. Top part (a) and front panel of the model. Lower part (b)

the reference signal level within the whole frequency range. The experiments determined that appearance of the signal fragments exceeding the reference level is more probable in the cases of application of a narrow time window. This phenomenon was provoked by the fact that averaging of very narrow time-frequency regions of the signal can be completed close to the signal maximum and does not take into account the features of neighboring signal elements.

The excessive narrowing of time-frequency elements results in a surplus of received information and analysis complication.

The developed system during performance analysis showed that the calculation time on average is 2.5 s under following initial conditions, i.e. signal band frequency is 20 Hz–20 kHz, bandwidth of signal frequency division makes 200 Hz; duration of signal fragment: 1 s; signal window duration is 0.07 s; algorithm for level calculation: amplitude averaging; window type: Hamming; PC parameters: 1.6 GHz Celeron M380, 1024 MB RAM, 333 MHz.

Such results indicate the necessity of improvement of developed model operation in order to achieve a performance of no more than 0.5 s.

In general, implementation of the software model shows the possibilities of application of such systems and algorithms for extraction and identification of the informative characteristics of complex acoustic or electromagnetic signals appearing as technological process feedback signals.

4   Discussion

The acoustic emission signals were selected in the laser welding technological process feedback channel, since they can be transmitted and analyzed at minimal time, computing resources and other factors. The reliability of processing and identification of the signals depends, first of all, on the method, which takes into account distribution and mutual communication of the informative characteristics in the signal.

The time parameters of the signal (duration, frequency, rise time, decay time etc.), frequency ones (signal composition, distribution of amplitudes in the spectrum), time-frequency (amplitude value of the signal on a certain frequency at given time point), or cepstral parameters (rate of occurrence of some frequency components in the signal) can be used as the informative characteristics. Calculation of the corresponding parameters requires appropriate conversion methods, duration and calculation complexity of which increase in proportion to informative utility of form of presentation.

Following the results of previous analysis of the signals by means of Spectrum Laboratory v.2.7, Sound Forge 10 and analysis of references, the concept of the system for analyzing accumulation and identification of signal informative characteristics was proposed and realized. The model, developed by software package LabVIEW 2009 graphical tools, allowed investigating the effect on the recorded signal of such time-frequency signal transform parameters as bandwidth, analysis frequency range, window time duration and window function.

However, model performance under conditions of increased detail of the analysis indicates the necessity in optimizing the algorithms of processing by time characteristics and conducting additional experiments with model registration and identification of the signals of TP of performed at different modes of laser welding.

5   Conclusion

It was found out that there is a relationship between the parameters of technological process and the registered acoustic emission signals following the process, and that relationship is valid regardless the research method.

The relationship was illustrated with the dependency of the result of laser welding technological process (first of all, the quality of welded joints) on the registered acoustic emission signals.

References

1. Borkar, P., Malik, L.G.: Acoustic signal based traffic density state estimation using SVM. Int. J. Image Graph. Sig. Process. (IJIGSP) **5**(8), 37–44 (2013). https://doi.org/10.5815/ijigsp. 2013.08.05

2. Jager, M., Humbert, S., Hamprecht, F.A.: Sputter tracking for the automatic monitoring of industrial laser-welding processes. IEEE Trans. Industr. Electron. **55**(5), 2177–2184 (2008)

3. Nagashree, R.N., Aswini, N.: Approaches of buried object detection technology. Int. J. Wirel. Microwave Technol. (IJWMT) **4**(2), 31–37 (2014). https://doi.org/10.5815/ijwmt.2014.02.04

4. Sheela Selvakumari, N.A., Radha, V.: Voice pathology identification: a survey on voice disorder. Int. J. Eng. Manuf. (IJEM) **7**(2), 39–49 (2017). https://doi.org/10.5815/ijem. 2017.02.04

5. Bahoura, M.: Pattern recognition methods applied to respiratory sounds classification into normal and wheeze classes. Comput. Biol. Med. **39**(9), 824–843 (2009)

6. Huang, C.J., Yang, Y.J., Yang, D.X., Chen, Y.J.: Frog classification using machine learning techniques. Expert Syst. Appl. **36**(2), 3737–3743 (2009)

7. Mporas, I., Ganchev, T., Kocsis, O., Fakotakis, N., Jahn, O., Riede, K.: Integration of temporal contextual information for robust acoustic recognition of bird species from real-field data. Int. J. Intell. Syst. Appl. (IJISA) **5**(7), 9–15 (2013). https://doi.org/10.5815/ijisa.2013.07.02

8. Loutas, T.H., Sotiriades, G., Kalaitzoglou, I., Kostopoulos, V.: Condition monitoring of a single-stage gearbox with artificially induced gear cracks utilizing on-line vibration and acoustic emission measurements. Appl. Acoust. **70**(9), 1148–1159 (2009)

9. Anami, B.S., Pagi, V.B.: Acoustic signal based fault detection in motorcycles – a comparative study of classifiers. Int. J. Image Graph. Sig. Process. (IJIGSP) **5**(1), 8–15 (2013). https:// doi.org/10.5815/ijigsp.2013.01.02

10. Ahadi, M., Bakhtiar, M.S.: Leak detection in water-filled plastic pipes through the application of tuned wavelet transforms to acoustic emission signals. Appl. Acoust. **71**(7), 634–639 (2010)

11. Pearson, T.C., Cetin, A.E., Tewfik, A.H., Haff, R.P.: Feasibility of impact-acoustic emissions for detection of damaged wheat kernels. Digit. Sig. Proc. **17**(3), 617–633 (2007)

12. Zhao, Y., Wang, J., Lu, Q., Jiang, R.: Pattern recognition of eggshell crack using PCA and LDA. Innovative Food Sci. Emerg. Technol. **11**(3), 520–525 (2010)

13. Park, Y.W., Park, H., Rhee, S., Kang, M.: Real time estimation of CO_2 laser weld quality for automotive industry. Opt. Laser Technol. **34**(2), 135–142 (2002)

14. Conesa, S., Palanco, S., Laserna, J.J.: Acoustic and optical emission during laser-induced plasma formation. Spectrochim. Acta Part B **59**(9), 1395–1401 (2004)

15. Duley, W.W., Mao, Y.L.: The effect of surface condition on acoustic emission during welding of aluminium with CO_2 laser radiation. J. Phys. D Appl. Phys. **27**(7), 1379–1383 (1994)

16. de la Rosa, J.G., Puntonet, C.G., Lloret, I.: An application of the independent component analysis to monitor acoustic emission signals generated by termite activity in wood. Measurement **37**(1), 63–76 (2005)

17. Yella, S., Gupta, N.K., Dougherty, M.S.: Comparison of pattern recognition techniques for the classification of impact acoustic emissions. Transp. Res. Part C Emerg. Technol. **15**(6), 345–360 (2007)

18. Saini, D., Floyd, S.: An investigation of gas metal arc welding sound signature for on-line quality control. Weld. J. **77**(4), 172–179 (1998)

19. Molino, A., Martina, M., Vacca, F., Masera, G., Terreno, A., Pasquettaz, G., D'Angelo, G.: FPGA implementation of time-frequency analysis algorithms for laser welding monitoring. Microprocess. Microsyst. **33**(3), 179–190 (2009)

20. Rao, Y., Sarwade, N., Makkar, R.: Denoising and enhancement of medical images using wavelets in LabVIEW. Int. J. Image Graph. Sig. Process. (IJIGSP) **7**(11), 42–47 (2015). https://doi.org/10.5815/ijigsp.2015.11.06
21. Vinoth Kumar, K., Suresh Kumar, S.: LabVIEW based condition monitoring of induction machines. Int. J. Intell. Syst. Appl. (IJISA) **4**(3), 56–62 (2012). https://doi.org/10.5815/ijisa.2012.03.08

QoS Ensuring over Probability of Timely Delivery in Multipath Routing

Oleksandra Yeremenko$^{(\boxtimes)}$ ⓘ and Oleksandr Lemeshko ⓘ

Kharkiv National University of Radio Electronics, 14 Nauka Avenue, Kharkiv, Ukraine
oleksandra.yeremenko.ua@ieee.org, oleksandr.lemeshko@nure.ua

Abstract. This article proposes a solution aimed at improving such an important Quality of Service indicator as the probability of timely delivery of packets based on optimization of multipath routing processes in infocommunication networks. The novelty of the proposed solution consists in using a dynamic model of the operation of the network routers interfaces, which allowed to take into account the nonstationary nature of the change in their state and to track the nonlinear dynamics of the average delay both on individual network interfaces and along the used set of routes in the network. This allowed formulating in an analytical form an expression for calculating the probability of timely delivery of packets, the maximum of which was the criterion of optimality in solving the problem of multipath routing in the infocommunication network. The results of the research have shown that the application of the proposed routing solution significantly improves the probability of timely delivery of packets in comparison with the results obtained during the implementation of the Traffic Engineering technology.

Keywords: Quality of Service · Multipath routing · Flow-based model
Router interface · Dynamics · Probability of timely delivery · Packet rate
End-to-end delay

1 Introduction

The main task in the operation of infocommunication systems is providing of the end-to-end Quality of Service (QoS) to the customers. In this case, the process of routing is one of the solutions for supporting the QoS over the multiple miscellaneous parameters simultaneously [1, 2]. In turn, QoS parameters can be conditionally divided into three main groups. The first group includes particular indicators (network performance parameters), namely, rate, time (average packet delay and jitter), and reliability indicators (probability of packet loss). The second group consists of generalized indicators that connect a number of particular indicators. The probability of timely delivery of packets can be an example of such an indicator. This indicator connects the time and reliability parameters of packets delivery. The general QoS indicators can be represented, for example, by the total value of the transferred data, system performance when satisfying the requirements for a set of particular QoS indicators, the perceived quality of service at the user level (Quality of Experience, QoE), which are also complex functions from a variety of QoS indicators.

© Springer International Publishing AG, part of Springer Nature 2019
Z. Hu et al. (Eds.): ICCSEEA 2018, AISC 754, pp. 244–254, 2019.
https://doi.org/10.1007/978-3-319-91008-6_25

Providing the given values of the listed QoS indicators is quite a complex scientific and applied task, and its solution involves the integrated use of technological capabilities of all Open Systems Interconnection (OSI) layers. A special place among these tools belongs to QoS routing protocols, which are directly responsible for the numerical values of the end-to-end values of such indicators. In turn, solutions regarding QoS-based routing [3] can be conditionally combined in the following directions. The first direction includes solutions that are based on load balancing in the implementation of multipath routing [4–7], which, as it is known, contributes to an implicit improvement in network performance. The second direction of solutions is based on the use of routing metrics, which are directly connected either to individual QoS indicators, or to their convolutions. Similar metrics, as a rule, are used during the formation of weight coefficients in the optimality criterion of routing solutions. Solutions of the first and second directions are currently implemented in most existing routing protocols [1, 2]. As shown by the analysis, the most effective direction of solutions in the field of QoS routing is Constraint Based Routing, when the conditions for performing QoS requirements are formalized explicitly, for example, by introducing QoS constraints into the structure of the optimization model of routing [8].

In this regard, the scientific and applied problem, which is related to the improvement in the field of multipath routing solutions in order to increase such an important indicator of the QoS like probability of timely delivery, based on the account of nonstationary nature of the changes in the state of network interfaces, is considered to be relevant.

2   Related Work

Let us analyze the existing approaches and solutions related to ensuring the timely delivery of packets in infocommunication networks. Thus, in [9], a study of the processes of timely delivery of HTTP video streaming packets in wireless networks such as LTE was conducted. In this work, an expression is introduced in the analytical form to calculate the probability of timely delivery of streaming video packets depending on the bandwidth allocated to the user. In addition, this expression takes into account the Application, Transport, and Physical Layers parameters related to the rate and quality of the communication link used by the video stream. The proposed solution allows calculating the level of bandwidth that must be allocated to maintain acceptable system performance in accordance with QoE requirements, as well as video rate and probability of timely delivery of video packets.

In [10], a heuristic approach was proposed to maximize the timely delivery of content in Delay Tolerant Networks and a corresponding framework was suggested. In doing so, the content properties are used initially to obtain the optimal number of routing hosts for each type of flow (content) in order to maximize the number of nodes required. The next step is to collect and process data on the capacity and location of mobile devices. Then the distributed forwarding scheme uses the optimal number of hosts for routing and the data from the nodes for the timely delivery of content to the necessary nodes. The advantages of the solution include achieving a high delivery ratio and reducing the amount of overhead, and the disadvantage is the excessive use of network resources (nodes, communication links).

In [11] a heuristic solution of increasing the probability of timely and reliable delivery of messages in Vehicular Communication also presented. It is proposed to introduce a real-time layer containing a deterministic medium access control protocol and retransmissions at the transport layer over IEEE 802.11p in order to ensure guaranteed delivery of data in real time and increase overall reliability.

In [12], a hybrid scheme of staircase and conservative staircase schemes is proposed that provide timely broadcast delivery of video data to users and has better performance than a conservative staircase scheme for such an indicator as waiting time of the user. In turn, the framework for providing timely delivery of critical data in Mobile Ad Hoc Networks (MANET) was proposed in [13].

Thus, as the results of the analysis have shown, it can be concluded that solutions on the calculation and use of the probability of timely delivery indicator are as much as possible tied to the content of the solving network task and to the parameters of the transmitted content. The disadvantages of the solutions presented are, as a rule, their heuristic nature, as well as the orientation to a particular type of transmitted content in the network. In this regard, in the presented paper, the approach for calculating and analyzing the probability of timely delivery of packets when solving multipath routing problems in infocommunication networks is proposed. The solution is a further development of the approaches outlined in [14, 15], and is based on the use of the dynamic model of the network routers interfaces [16].

3   Model of Multiflow Routing

Consider the routing model for transmitting the multiple flows $k \in K$, where K is the set of flows in the network, which is represented by the graph $G = (R, E)$. Let the set of nodes $R = \left\{ R_i; i = \overline{1, m} \right\}$ correspond to network routers, and the set of edges $E = \left\{ E_{i,j}; i,j = \overline{1, m}; i \neq j \right\}$ denotes the links. It is assumed that link $E_{i,j} \in E$ capacity $\varphi_{i,j}$ is known and measured in the number of packets per second (1/s).

In order to determine the fraction of intensity of each kth flow from the specified ith node to the jth node through the assigned jth interface, the routing variables $x_{i,j}^k$ should be calculated within the multiflow model. The conditions of flow conservation from the viewpoint of network nodes overload prevention must be necessarily met. It can be written for the source, transit, and destination nodes for the multiflow case as [4, 8]:

$$\begin{cases} \sum_{j:E_{i,j} \in E} x_{i,j}^k - \sum_{j:E_{j,i} \in E} x_{j,i}^k = 0; \; k \in K, \; R_i \neq s_k, d_k; \\ \sum_{j:E_{i,j} \in E} x_{i,j}^k - \sum_{j:E_{j,i} \in E} x_{j,i}^k = 1; \; k \in K, \; R_i = s_k; \\ \sum_{j:E_{i,j} \in E} x_{i,j}^k - \sum_{j:E_{j,i} \in E} x_{j,i}^k = -1; \; k \in K, \; R_i = d_k \end{cases} \quad (1)$$

where s_k is the source node for the kth flow, and d_k is the corresponding destination node.

The multipath routing strategy in this model is implemented depending on the fulfillment of the condition $0 \leq x_{i,j}^{k} \leq 1$, related to the control variables. In addition, the capacity constraints on the network links utilization take place:

$$\sum_{k \in K} \lambda^{k} x_{i,j}^{k} \leq \varphi_{i,j}, E_{i,j} \in E,$$ (2)

where λ^{k} is the average intensity of the kth flow.

4   Conditions of QoS Ensuring over Probability of Timely Delivery

For obtaining the conditions of QoS ensuring over probability of timely delivery, the proposed model uses Pointwise Stationary Fluid Flow Approximation PSFFA M/G/1 model [16], and it is a special case of M/M/1, as a model of the dynamics of changes in the network interface of the router. According to this approximation, the dynamics of average packet delay on the router interface is described by the following differential equations of the network state [8]:

$$\frac{d\tau_{i,j}(t)}{dt} = 1 - \varphi_{i,j} \left(\frac{\tau_{i,j}(t)}{\lambda_{i,j}\tau_{i,j}(t) + 1} \right)$$ (3)

where $\lambda_{i,j} = \sum\limits_{k \in K} \lambda^{k} \cdot x_{i,j}^{k}$ is the total intensity of flows in the $E_{i,j}$ link. Therefore, using the expression (3), it is possible to estimate the dynamics of changes in time for the average packet delay on the interfaces of network nodes. In addition, given a steady state operation of the interface ($t \rightarrow \infty$), the well-known formula for the M/M/1 queuing system $\tau_{i,j} = 1/(\varphi_{i,j} - \lambda_{i,j})$ has been obtained.

Suppose that the probability of timely delivery of the kth flow of packets P_{TD}^{k} can be defined in the following way [15]:

$$P_{TD}^{k} = \frac{T(\tau_{MP}^{k} \leq \tau_{req}^{k})}{T_{U}}$$ (4)

where T_{U} is the update time of routing table, and τ_{MP}^{k} is the average end-to-end delay over the multiple paths calculated for the kth flow. Then, τ_{MP}^{k} can be determined using the expression

$$\tau_{MP}^{k} = \sum_{i \in I^{k}} x_{p_{i}}^{k} \tau_{p_{i}}^{k},$$ (5)

where $x_{p_{i}}^{k}$ is the fraction of intensity of the kth flow in the individual path p_{i} from the set I^{k} of all paths, calculated for this flow transmission; $\tau_{p_{i}}^{k}$ is the average end-to-end delay for the kth packet flow transmitted along the path p_{i}. Moreover, the following condition must be met in the case of different demands of transmitted flows in relation to probability of timely delivery:

$$P_{TD}^k \geq \delta_{TD}^k \cdot P_{TD}^{k\,req} \tag{6}$$

where

$$\delta_{TD}^k = \begin{cases} 1, & \text{if timely delivery is guaranteed;} \\ 0, & \text{otherwise;} \end{cases}$$

and $P_{TD}^{k\,req}$ is the required value of the probability of timely delivery for the kth packet flow.

5   Set of Optimality Criteria

Depending on the network type, the set of optimality criteria for QoS ensuring under the probability of timely delivery constraint in multipath routing can be formulated. For example, in the communication operator's network the following objective function should be minimized in order to find the routing variables:

$$J_1 = \sum_{E_{i,j} \in E} \sum_{k \in K} h_{i,j} \cdot \lambda^k \cdot x_{i,j}^k \tag{7}$$

where $h_{i,j}$ is the routing metric for the link between the ith and the jth network nodes. Whereas in enterprise networks another type of the optimality criterion is suggested to use, which is associated with maximization of the next objective function within the calculation of the routing variables:

$$J_2 = \sum_{k \in K} w_k P_{TD}^k \tag{8}$$

where $w_k = IP_{pr}^k + 1$ is the weighting coefficient determined by the priority value IP_{pr}^k of packets of the kth flow. In the IP network, in the case of using the 3 bits of IP precedence in the IP packet header for prioritization, the value of priority IP_{pr}^k is of the range from 0 to 7, while for the DSCP (Differentiated Services Code Point) priorities IP_{pr}^k will vary from 0 to 63.

Then the task of QoS routing by the parameter of the probability of timely delivery was reduced to the solution of the optimization problem with the objective function (7) or (8) under constraints (1)–(6). The optimization problem belongs to the class of nonlinear programming problems, because the constraint conditions (6) due to the nonlinearity of the forms (3)–(5) are also nonlinear.

6   Numerical Study of the Proposed Model

Let us demonstrate the features of the proposed QoS solution of multipath routing using the model (1)–(8) for the example of the network structure shown in Fig. 1. The network consists of 8 nodes (routers) and 12 links. During the research, the influence on the probability of timely delivery of packets (4) of the type of mathematical model used, as well as the intensity of the transmitted flows (network utilization) was analyzed. The routing solution obtained using the proposed model (1)–(8) was compared with the solution based on the Traffic Engineering (TE) technology [4].

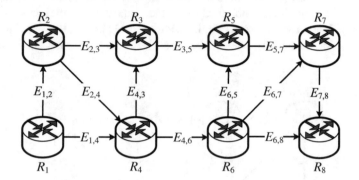

Fig. 1. The modeled network structure.

In the modeling a routing solution corresponding to the TE technology, the conditions of the flow conservation (1) were supplemented with the modified conditions for preventing overload of the network links

$$\sum_{k \in K} \lambda^k x_{i,j}^k \le \alpha \cdot \varphi_{i,j}, E_{i,j} \in E, \tag{9}$$

and the optimality criterion of routing solutions took the form

$$\alpha \to \min \tag{10}$$

where parameter α characterized the upper bound of the links utilization of the network as a whole. The solution proposed in this work, based on the use of the model (1)–(6), (8), is denoted as "TD" (Timely Delivery).

Features of solving routing problems using two compared models (TD and TE) are demonstrated in the following example. Let us assume that between the nodes 1 and 8 it is necessary to transmit a single flow with the intensity 420 1/s. Links capacities and initial delays on the routers interfaces are shown in Table 1.

Table 1. Investigation input data.

Link	$E_{1,2}$	$E_{1,4}$	$E_{2,3}$	$E_{2,4}$	$E_{4,3}$	$E_{3,5}$
Link capacity (1/s)	500	300	200	300	800	150
Initial delay, τ_0 (ms)	10	20	20	10	30	20
Link	$E_{4,6}$	$E_{6,5}$	$E_{5,7}$	$E_{6,7}$	$E_{6,8}$	$E_{7,8}$
Link capacity (1/s)	300	900	500	400	200	500
Initial delay, τ_0 (ms)	10	20	30	10	40	30

The solution for the problem of multipath routing of the initial flow with the intensity of 420 1/s and $\tau_{req} = 100$ ms, obtained using the proposed Timely Delivery model, is shown in Fig. 2, and for the Traffic Engineering model (1), (9), (10) – in Fig. 3. For the solution according to the TE model, the boundary value α (10) corresponded to the value 0.933.

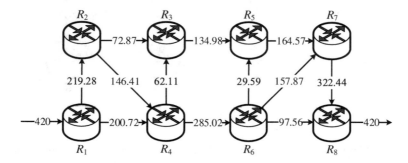

Fig. 2. The order of routing the flow with the intensity of 420 1/s using the Timely Delivery model.

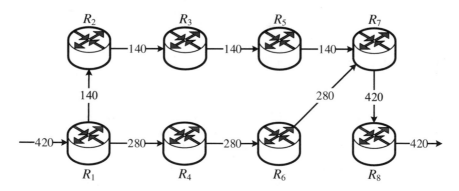

Fig. 3. The order of routing the flow with the intensity of 420 1/s using the Traffic Engineering model.

Based on the solutions obtained, using expressions (3) and (5), for each of the routing models in Fig. 4 the dynamics of changing in the average end-to-end packet delay in time with the initial delays at the routers interfaces (τ_0), given in Table 1, is demonstrated.

As shown in Fig. 4, the solution based on the use of TD model provides average end-to-end packet delay less than $\tau_{req} = 100$ ms throughout the observation time period, which, for example, was selected for 30 s. In turn, the solution based on the TE model ensured that the requirements for the average end-to-end packet delay (τ_{req}) were fulfilled only for one second within the observation period of 30 s (Fig. 4). In this case, the probability of timely delivery (4), which corresponded to this solution, is approximately equal to 0.033.

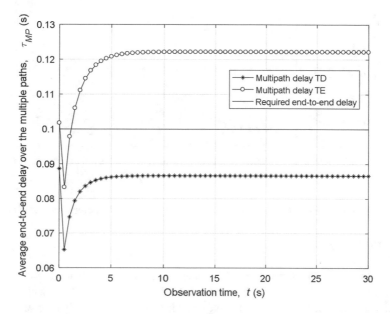

Fig. 4. Dynamics of changing in the average end-to-end packet delay (flow intensity 420 1/s).

In general, the tightening of the requirements for the average end-to-end packet delay (τ_{req}) adversely affects the probability of timely delivery of packets. However, the use of the TD multipath routing model proposed in this work provided the best probability of timely delivery of packets in comparison with solutions based on the TE model (Fig. 5).

As shown in Fig. 5, for example, when the flow intensity is 420 1/s, the use of the TD model provided the value of the probability of timely delivery (4) equal to one at $\tau_{req} = 90$ ms and higher, whereas the routing solution based on the TE model only at $\tau_{req} \geq 125$ ms.

Thus, the application of the model (1)–(8) proposed in the work, as the above example showed, allowed to ensure the maximum probability of timely delivery of packets with an increase in requirements to the end-to-end delay on average by 20–28% (Fig. 5). The advantages of the proposed solution were manifested as much as the utilization of the communication links and the network as a whole increased, which determines the area of the preferred use of the model (1)–(8).

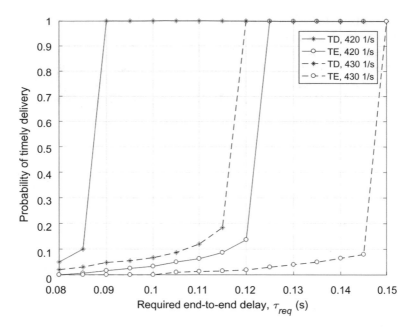

Fig. 5. Dependence of the probability of timely delivery on the required end-to-end delay.

7 Conclusion

When providing infocommunication services, especially multimedia services, it is important to ensure QoS requirements that are relatively high with respect to time and reliability indicators, the numerical values of which are very closely connected. Therefore, the tasks related to the improvement of the generalized QoS indicator – the probability of timely delivery of packets – come to the fore. The solution of such problems is quite laborious and requires the use of practical functions at all OSI levels.

In this paper, in order to improve the probability of timely delivery of packets, it is proposed to improve the process of multipath routing in infocommunication networks. The proposed solution is a further development of the approach described in [14, 15]. The advantage of the proposed solution is that the calculation of the set of desired routes is based on the nonstationary nature of the operation of the router interfaces. This allows evaluating and maximizing the probability of timely delivery of packets, taking into account the dynamics of network state changes over a certain time interval, for example, the update timer of routing tables.

As the results of the study have shown, due to the nonlinear nature of the change in the average delay, both on individual network interfaces and along the multiple routes used in the network, the application of the proposed model (1)–(8) allows ensuring the maximum probability of timely delivery of packets with increasing requirements for the end-to-end delay on average by 20–28% compared to the solution derived from the Traffic Engineering model. The application of the proposed model of multipath routing is most effective in the range of high loads in the network, as well as the level of requirements for the average end-to-end packet delay.

References

1. Barreiros, M., Lundqvist, P.: QOS-Enabled Networks: Tools and Foundations, 2nd edn. Wiley, Chichester (2016)
2. Szigeti, T., Hattingh, C., Barton, R., Briley Jr., K.: End-to-End QoS Network Design: Quality of Service for Rich-Media & Cloud Networks. Cisco Press (2013)
3. Crawley, E., Sandick, H., Nair, R., Rajagopalan, B.: A Framework for QoS-based Routing in the Internet. RFC 2386 (1998)
4. Lee, Y., Seok, Y., Choi, Y., Kim, C.: A constrained multipath traffic engineering scheme for MPLS networks. In: 2002 IEEE International Conference on Communications, Conference Proceedings. ICC 2002 (Cat. No.02CH37333), vol. 4, pp. 2431–2436. IEEE (2002)
5. Sharma, G., Singh, M., Sharma, P.: Modifying AODV to reduce load in MANETs. Int. J. Mod. Educ. Comput. Sci. (IJMECS) 8(10), 25–32 (2016). https://doi.org/10.5815/ijmecs.2016.10.04
6. Zaman, R.U., Shehnaz Begum, S., Ur Rahman Khan, K., Venugopal Reddy, A.: Efficient adaptive path load balanced gateway management strategies for integrating MANET and the Internet. Int. J. Wirel. Microwave Technol. (IJWMT) 7(2), 57–75 (2017). https://doi.org/10.5815/ijwmt.2017.02.06
7. Das, I., Lobiyal, D.K., Katti, C.P.: An analysis of link disjoint and node disjoint multipath routing for mobile ad hoc network. Int. J. Comput. Netw. Inf. Secur. (IJCNIS) 8(3), 52–57 (2016). https://doi.org/10.5815/ijcnis.2016.03.07
8. Lemeshko, O.V., Yeremenko, O.S.: Dynamics analysis of multipath QoS-routing tensor model with support of different flows classes. In: 2016 International Conference on Smart Systems and Technologies (SST) Proceedings, pp. 225–230. IEEE (2016)
9. Colonnese, S., Russo, S., Cuomo, F., Melodia, T., Rubin, I.: Timely delivery versus bandwidth allocation for dash-based video streaming over LTE. IEEE Commun. Lett. 20(3), 586–589 (2016)
10. Rao, W., Zhao, K., Zhang, Y., Hui, P., Tarkoma, S.: Towards maximizing timely content delivery in delay tolerant networks. IEEE Trans. Mob. Comput. 14(4), 755–769 (2015)
11. Jonsson, M., Kunert, K., Böhm, A.: Increasing the probability of timely and correct message delivery in road side unit based vehicular communication. In: 15th International IEEE Conference on Intelligent Transportation Systems (ITSC) Proceedings, pp. 672–679. IEEE (2012)
12. Chand, S., Om, H.: Timely delivery of video data in staircase scheme. In: 2012 National Conference on Communications (NCC) Proceedings, pp. 1–5. IEEE (2012)
13. Luqman, F.: TRIAGE: applying context to improve timely delivery of critical data in mobile ad hoc networks for disaster response. In: 2011 IEEE International Conference on Pervasive Computing and Communications Workshops (PERCOM Workshops) Proceedings, pp. 407–408. IEEE (2011)
14. Lemeshko, A.V., Evseeva, O.Y., Garkusha, S.V.: Research on tensor model of multipath routing in telecommunication network with support of service quality by great number of indices. Telecommun. Radio Eng. 73(15), 1339–1360 (2014)

15. Lemeshko, O., Yeremenko, O., Hailan, A.M.: Design of QoS-routing scheme under the timely delivery constraint. In: 2017 14th International Conference the Experience of Designing and Application of CAD Systems in Microelectronics (CADSM) Proceedings, pp. 97–99. IEEE (2017)
16. Xu, K., Tipper, D., Qian, Y., Krishnamurthy, P., Tipmongkonsilp, S.: Time-varying performance analysis of multihop wireless networks with CBR traffic. IEEE Trans. Veh. Technol. 63(7), 3397–3409 (2014)

Icing Thickness Prediction of Overhead Power Transmission Lines Using Parallel Coordinates and Convolutional Neural Networks

Baiming Xie[1(✉)], Chi Zhang[2], Qing-wu Gong[3], Koyamada Koji[4], Hua-rong Zeng[1], Li-jin Zhao[1], Hu Qiao[3], and Liang Huang[1]

[1] Guizhou Electric Power Research Institute, Guizhou Power Grid Limited Liability Company, Guiyang, China
102131694@qq.com
[2] Graduate School of Engineering, Kyoto University, Kyoto, Japan
[3] School of Electrical Engineering, Wuhan University, Wuhan, China
[4] Academic Center for Computing and Media Studies, Kyoto University, Kyoto, Japan

Abstract. In this paper, a model used for predicting icing thickness of power transmission lines is proposed. An algorithm derived from parallel coordinates is applied to convert the high-dimensional source data, which includes relevant factors about the icing thickness of power transmission lines, to two dimensional images. Then the images are used for training convolutional neural networks (CNNs). Finally, the icing thickness is predicted by the trained CNNs. In this way, our system combines the advantages of information visualization and CNNs. It provides an universal method to process multi-dimensional numerical data with CNNs algorithmically and in a real sense.

Keywords: Icing thickness · Parallel coordinates
Convolutional neural networks · Visualization · Deep learning

1 Introduction

CAUSED by high humidity and low temperature, icing of overhead power transmission lines increases the risk of power system accidents such as line trip, break line, pole collapse and insulator flashover [1, 2]. It threatens the normal operation of power systems and significantly decreases the reliability of power system networks. To decide which transmission line should be de-iced, the icing thickness of the specified line needs to be measured or calculated.

Due to the climate, power transmission lines in the southwest of China usually suffer from icing accidents in winter [2]. Based on previous research [3], power grid corporations in southwestern China set up a monitoring system to record the related micro-meteorology data and tower status data such as temperature, humidity, maximum pulling strength, angle of wind deflection at the maximum pull and angle of inclination at the maximum pull. The purpose of our research is to analyze the relationship between these

© Springer International Publishing AG, part of Springer Nature 2019
Z. Hu et al. (Eds.): ICCSEEA 2018, AISC 754, pp. 255–267, 2019.
https://doi.org/10.1007/978-3-319-91008-6_26

parameters and the measured values of icing thickness and find a way to predict the icing thickness by using these parameters.

To achieve the goal, a convolutional neural network (CNN) is applied to analyze the monitoring data. As a typical deep learning algorithm, CNN has been successfully used in the fields of speech recognition and image identification [4–6]. It automatically learns relationships and representations of input data without the dependence of prior knowledge. Comparing with other types of deep neural networks, the features of CNN such as local connectivity and shared weights reduce the computational complexity. Thus the computing efficiency is improved and the problem of overfitting is alleviated to some extent.

However, there are some problems when processing multidimensional numerical data with a conventional CNN model. Specifically, multi-dimensional data is usually stored as a two dimensional array or matrix when processed by computer. In the array or matrix, each column indicates a dimension of data and each row indicates a data point of multi-dimensional data. In the supervised learning used by CNNs, if a data point is considered as a training sample, there need to be a label matched for it. In the case of the monitoring data, a training sample includes a group of values about temperature, humidity, maximum pulling strength, angle of wind deflection at the maximum pull and angle of inclination at the maximum pull, and the label data is the icing thickness or icing thickness level in the corresponding meteorological environment. Unlike processing images, the matrix consisting of multi dimensional data cannot be input to a CNN directly as an image because there are multiple data points and multiple labels in the matrix, whereas an image represents only one data point and it has only one label as well when processed by a conventional CNN model. Moreover, images such as photos or paintings usually have their actual meanings. Hence, colors of the images also distribute with regularity. For example, an object with a single color shows continuous and gradually variational color distribution in an image. If the color changes dramatically, it is possible that the outline of the object is reached. But if a numerical matrix is treated as an image and its values are treated as color values, it is apparent that the color distribution will not follow a certain pattern, which increases the difficulty of training and prediction. In our model, an algorithm derived from parallel coordinates is used for transforming each data point in multi-dimensional data to a simplified parallel coordinates plot, which is available for further processing by CNN as an image. Meanwhile, a data point is represented as a polyline in a parallel coordinates plot rather than noise-like color dots, which makes it easier for a CNN model to discover the features and complete the analysis work. The details of the relevant algorithms will be described in the following sections.

2   Related Work

Parallel coordinates plot is a common way of visualizing high-dimensional geometry and analyzing multivariate data [7–9]. Data points in a typical parallel coordinates plot are represented as polylines connecting corresponding positions on a set of vertical axes. Some interactive features can be added to parallel coordinates so as to enable users to customize some details such as axes or color maps in real time [10, 11]. Scatter points and clustering algorithms can also be applied to parallel coordinates to optimize the visualization results of large scale data [12–14]. Furthermore, linked view between parallel coordinates and other visualization methods makes it possible for users to switch among different visualization results smoothly, which helps to deepen their understanding about the data from different angles [15].

As a type of deep learning [16] algorithm, CNNs are usually used for recognizing speeches and identifying images [4–6]. Based on the neocognitron [17], which is the embryonic form of the CNN, the LeNet model [18, 19] is considered as the first typical architecture of CNN. The AlexNet model proposed in 2012 optimizes the LeNet model by applying the layer of rectified linear units (ReLU) and the dropout method [20–22]. The usage of ReLU substantially accelerates the training speed of CNNs without making a significant difference to generalisation accuracy, while the dropout method prevents overfitting and improves the performance of CNNs.

Since the ice disaster happened in China on the beginning of 2008, research on the icing thickness prediction of power transmission lines has become more and more significant. Some laboratories and research institutions have presented their models for predicting icing thickness. For example, a model based on data-driven has good accuracy of prediction according to the simulation on the condition that the training data and prediction data are in the same icing process [23]. The fuzzy logic theory is also applied to the prediction of icing thickness. To build a prediction model based on fuzzy logic theory, the most crucial part is to set fuzzy rules, which needs vast professional experience and analysis of relevant data. It is possible for a highly optimized fuzzy prediction model to limit the error in an acceptable range [24]. Another method proposed by Zeng et al. calculates the average icing thickness based on a re-closure transient travelling wave. It uses travelling wave location equipments to collect travelling wave signals and calculate the length of iced transmission lines. Then the average icing thickness can be calculated by using the formula of the relationship between the compressive conductor load ratio and the thickness of ice covering on transmission lines [25]. Moreover, a combination model based on a wavelet support vector machine (w-SVM) and a quantum fireworks algorithm (QFA) is proposed for predicting icing thickness. In this model, the wavelet kernel function is applied to a support vector machine instead of a Gaussian kernel function. The QFA model is used for optimizing the parameters of w-SVM. Temperature, humidity, wind speed, wind direction and sunlight are selected as main impact factors of the icing thickness prediction [26].

Compared with the prediction models mentioned above, our model has advantages in the following aspects. There is no limitation that the training data and prediction data must be in the same icing process. Our model is trained by the monitoring data from a certain area and it is capable of predicting the icing thickness within the area in the specified meteorological environment. Meanwhile, as a distinguished feature of deep learning models, our model optimizes its parameters automatically during the training procedure. There is no need of prior knowledge such as relevant professional experience or specific formulas. Wide universality and high extensibility make our model not only can be applied to icing thickness prediction, but also available for other types of numerical data classification. Because the interface of our model is available for various types of multi-dimensional data, it is possible to test and discover new impact factors.

3   Proposed Method

Our system can be considered as a combination of a simplified parallel coordinates generator and a deep convolutional neural network. The former is a converter used for reflecting multi-dimensional data points to simplified parallel coordinates plots one by one. The latter is a conventional CNN model used for processing the images converted from parallel coordinates plots so as to learn to classify them to different levels based on the predictive values of icing thickness. The two parts of the system are explained respectively as follows.

3.1   Simplified Parallel Coordinates Plot

In a parallel coordinates plot, each dimension of the input data is plotted as a vertical axis. Different with the Cartesian coordinate system, all of the axes are parallel with each other and a data point is plotted as a polyline connecting the relevant positions on these axes. Hence, it is possible to show more than three dimensions in a two-dimensional plane.

Figure 1a shows the traditional two-dimensional Cartesian coordinate system. In this system, a data point with two dimensions x and y is plotted as a dot with the corresponding coordinate values x_1 and y_1 on the two mutually perpendicular axes x and y respectively. If this data point is converted to the parallel coordinate system, which is illustrated in Fig. 1b, the data point becomes a line connecting the relevant coordinate values x_1 and y_1 on the two mutually parallel axes x and y respectively. If the number of dimensions increases, the data point changes from a line to a polyline connecting all of the corresponding coordinate values on each axis, as shown in Fig. 2a. There are usually many data points in a multidimensional data. When all of these data points are plotted in a parallel coordinate system, it becomes a parallel coordinates plot as Fig. 2b.

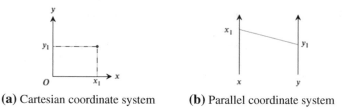

(a) Cartesian coordinate system   (b) Parallel coordinate system

Fig. 1.   From Cartesian coordinate system to parallel coordinate system

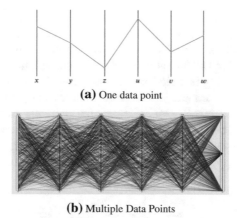

(a) One data point

(b) Multiple Data Points

Fig. 2. Visualization of multi-dimensional data by parallel coordinates

In our system, parallel coordinates plot is used for converting each data point to an image. However, it is a simplified version because there is only a polyline in an image and other elements such as axes and values are not plotted. Some samples of converted images are illustrated in Fig. 3.

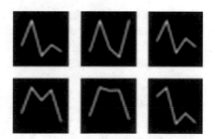

Fig. 3. Samples of converted images

The most prominent advantage of this algorithm is that not only is the conversion result very simple, but also it contains all necessary information about the original data point. Although comparing with the typical parallel coordinates, the value of the data point on each axis is omitted, it is detailed enough for the CNN model to discern these data points via the relative positions of them in the images.

After being converted to images, all data points are imported to a CNN for the further processing, which is explained in the next sub section.

3.2 Convolutional Neural Network

Essentially, a CNN is a mapping from its input to output for the purpose of classification. It automatically learns the relationships between the input and the output. The features of local connectivity and parameter sharing reduce the computational burden substantially. Multiple convolution kernels are used for detecting features from the input data. These feature maps are finally assembled to represent the global features.

A typical LeNet-5 CNN model [18, 19] is illustrated in Fig. 4. In this model, first, the input data is convolved by trainable filters called convolution kernels. The output of the convolution processing is called feature maps, which are considered as features extracted by convolution kernels. Second, these feature maps are sub-sampled in order to decrease the size of them. Meanwhile, it can also be treated as a procedure of feature extraction. Thus the output of this step is some feature maps as well with smaller size. Then, the previous two steps are repeated so as to extract features and decrease the size of feature maps ulteriorly. The last three layers include two hidden layers and an output layer, which are fully connected layers.

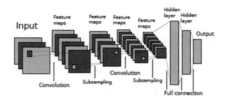

Fig. 4. Architecture of convolutional neural network

Specifically, in a convolution layer, an input image or feature map is convolved by a number of convolution kernels. Hence, the output consists of some feature maps with the same number of convolution kernels. If a node j in convolution layer c connects to feature maps, $x_1\ x_2,\ \cdots,\ x_i$, the output can be represented as

$$o_j = \phi\left(\sum_i X_i * \Theta_j + b_j\right) \tag{1}$$

where Θ_j is the weight matrix of the convolution kernel, b_j is the bias and ϕ is the activation function.

In a sub-sampling layer, a max pooling algorithm is used for decreasing the size of the input feature map, meaning that adjacent $n \times n$ pixels in a feature map are mapped to a new pixel with its color value equaling to the highest color value of these pixels. For the output o_j of the node j, the result of sub-sampling is

$$S_j = downsample(O_j) \tag{2}$$

Before input to a hidden layer, feature maps are rasterized, meaning being reshaped into a one-dimensional vector. For feature maps $x_1,\ x_2,\ \cdots,\ x_i$, the output of rasterization is

$$O_k = [x_{111}, x_{112}, \ldots, x_{11n}, x_{121}, x_{122}, \ldots, x_{12n},$$
$$\ldots, x_{1mn}, \ldots, x_{2mn}, \ldots, x_{jmn}]^T \tag{3}$$

In a hidden layer, each pixel or node in the previous layer is connected to all of the nodes in the current layer by different weight values, called full connection. The dropout method [21, 22] is applied to the two hidden layers. Besides, the output layer is also fully connected with the previous hidden layer. The nodes number in the output layer equals to the number of classified types. The output of a node j in a hidden layer can be represented as

$$f_j(x) = \phi(\theta_j^T x + b_j)$$

(4)

where x is the input vector, θ_j is the weight vector, b_j is the bias and ϕ is the activation function.

The training of a CNN can be divided into two procedures: feed forward pass and back propagation pass. The feed forward pass passes the training data from the input layer to the output layer. The input data is calculated by corresponding algorithms in each layer and finally we get the actual output. The back propagation pass compares the actual output with the ideal output and propagates the errors backwards so as to adjust weights and biases by gradient descent method. In our model, the cross-entropy is used as the cost function, which is given by

$$C = -\frac{1}{n} \sum_x [y \ln a + (1 - y) \ln(1 - a)]$$

(5)

where n is the total number of items of training data, the sum is over all training inputs x, y is the corresponding desired output and $a = \sigma(z) = \sigma(\sum_j w_j x_j + b)$ is the actual output. Then, the gradient can be calculated as

$$\frac{\partial C}{\partial w_j} = \frac{1}{n} \sum_x x_j(\sigma(z) - y)$$

(6a)

$$\frac{\partial C}{\partial b} = \frac{1}{n} \sum_x (\sigma(z) - y)$$

(6b)

where x_j is the input of the node, w_j is the corresponding weight, σ is the activation function and b is the bias [27].

The loop of the above training procedures continues until the specified stopping criterion is satisfied. After the training is completed, the testing data will be input to the model and the corresponding predicted values will be calculated.

4   Experiment

The data used in our system is from the monitoring system of power transmission lines. There are 5 dimensions in the data including temperature, humidity, maximum pulling strength, angle of wind deflection at the maximum pull and angle of inclination at the maximum pull. As the labels being necessary for supervised learning, each sample in the dataset corresponds to a measured value of icing thickness on the power transmission line. In this dataset, 70000 data points with the icing thickness over 0 mm from the year of 2013 to 2016 are selected as the input data. Among these data points, 60000 points are used for training and 10000 points are used for testing.

First, data points are converted images by simplified parallel coordinates algorithm. As shown in Fig. 3, the background is black and the polyline is white in each image. Therefore, the black pixel and the white pixel can be represented as 0 and 1 respectively. Since the size of each image is 28 pixels × 28 pixels, an image is treated as a matrix with 28 rows and 28 columns.

Second, the labels of all data points are calculated. The range of measured icing thickness is from 0 mm to 35.95 mm and it is split evenly into 10 levels. The corresponding label of each data point is the icing thickness level, from 0 to 9. After the processing of the previous two steps, the input data for the CNN and its labels are prepared.

Third, the input data and its labels are imported to the CNN model. The input data includes training data and testing data. The training data is used for training the CNN model, while the testing data is used for verifying it. In the first convolution layer, 32 convolution kernels with the size of 5 × 5 are applied to each input image, generating 32 feature maps, which can be considered as different channels of the input image. Then, these feature maps are down-sampled in the first sub-sampling layer and their size decreases to 14 × 14 pixels. In the second convolution layer, 64 convolution kernels with the size of 5 × 5 are applied to the feature maps output by the previous subsampling layer, generating 64 feature maps. Finally, feature maps are ulteriorly down-sampled in the second sub-sampling layer and their size decreases to 7 × 7 pixels.

Fourth, the feature maps generated in the second subsampling layer are reshaped into a one-dimensional vector. Then, it is input to the first fully connected layer. There are 1024 nodes in both the first and the second fully connected layers, and 10 nodes in the output layer. In the output layer, an output value of a node represents the probability of the corresponding icing thickness level. Therefore, the icing thickness level with the highest probability is considered as the predicted icing thickness level.

Finally, the difference between the output of the system and the ideal output are propagated backwards to adjust weights and biases of all layers. Then, the new output is calculated and the weights and biases are updated again. The training procedure continues until the specified training times are reached. Then, the testing data is input to the model and the corresponding predicted icing thickness levels are calculated. By comparing the predicted levels with the measured values, we can get the testing accuracy. The system flow chart is illustrated as Fig. 5, which shows the processes of our model.

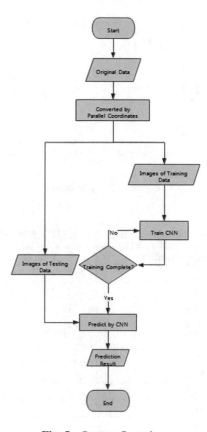

Fig. 5. System flow chart

The setups of the experiment are listed as Table 1. In our experiment, the batch size is set to 50, meaning that there are 50 data points being loaded every time during a training procedure. The number of training steps is set to 400 and the change of training accuracy is illustrated in Fig. 6, where the horizontal axis indicates training steps and the vertical axis indicates training accuracy. The results show that the training accuracy reaches 100% and oscillates slightly near 100% after about 140 training steps. The whole computation of the system loops 10 times. The testing accuracy and the corresponding training time are listed in Table 2.

Table 1. Setups of experiment

CPU	Intel Xeon E3-1230 v5, 3.4 GHz, 4 Cores
Memory	16 GB DDR4, 2133 MHz
Operating System	Ubuntu 16.10

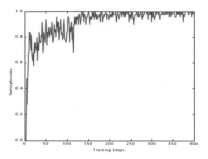

Fig. 6. Change of training accuracy (10 levels)

Table 2. Testing accuracy and training time (10 levels, 400 training steps)

Testing accuracy	Training time (s)
90.85%	46.230
91.05%	46.192
89.92%	46.157
92.20%	46.314
91.17%	46.227
92.00%	46.225
91.52%	46.248
91.35%	46.271
91.03%	46.297
91.54%	46.179

According to the results above, the average testing accuracy is 91.26% and the average training time is 46.234 s. In consideration of the exactly divided icing thickness levels, the training procedure completed in a relatively short time and the prediction accuracy is acceptable. Comparing with the MNIST (Mixed National Institute of Standards and Technology) handwritten digit database, the difference between the images under different icing thickness levels is relatively not obvious in our dataset, especially for the images in adjacent levels. Therefore, the testing accuracy cannot reach to a very high level even if the training steps are increased, which is also confirmed by our experiments.

On the other hand, in engineering practice, there are standards for designed line icing thickness in different icing areas. When monitoring icing thickness of transmission lines, we only focus on the icing thickness values which exceed design constraints. Hence, it is practicable to decrease the number of predicted icing thickness levels in our model. If the original 10 icing thickness levels are combined to 5 levels (Level 1: 0 mm to 7.19 mm. Level 2: 7.19 mm to 14.38 mm. Level 3: 14.38 mm to 21.57 mm. Level 4: 21.57 mm to 28.76 mm. Level 5: 28.76 mm to 35.95 mm), the training accuracy reaches to 100% in no more than 50 training steps, as shown in Fig. 7. With the new settings of 5 output levels and 300 training steps, the system also runs 10 times continuously and the results of testing accuracy and training time is listed in Table 3. As the results, we get an average testing accuracy of 96.875% and the average training time is 35.308 s. Generally speaking, the prediction accuracy is high enough and the training time is short enough to satisfy the needs of engineering.

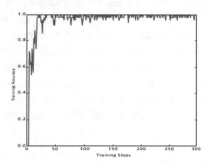

Fig. 7. Change of training accuracy (5 levels)

Table 3. Testing accuracy and training time (5 levels, 300 training steps)

Testing accuracy	Training time (s)
96.69%	35.304
98.70%	35.281
97.92%	35.412
97.28%	35.316
96.35%	35.279
95.42%	35.306
97.29%	35.301
95.57%	35.292
97.06%	35.286
96.47%	35.302

5   Conclusion

The prediction model for icing thickness of power transmission lines proposed in this paper can be considered as an integration of a data conversion system based on parallel coordinates and a data analysis system based on CNN. The data conversion system maps multi-dimensional input data to simplified parallel coordinates plots, while the data analysis system loads the converted images and trains itself so as to predict icing thickness levels. According to the results, our model completes the training procedure in about 35 s and the prediction accuracy is over 96.8%, which satisfies the needs of engineering.

The interface of our model is available for various types of multi-dimensional numerical data. Therefore, it provides an universal way to process multi-dimensional numerical data with CNNs. Moreover, because the input data of the CNN is a series of images, the CNN can make full use of its advantage in processing two-dimensional pictures.

As the future work, in order to increase the accuracy, our model can be improved in the following aspects. First, besides to the simplified parallel coordinates, other algorithms can be applied to the data conversion system. The combination of multiple conversion algorithms makes converted images include more features, which helps to improve the discernment of the analysis system. Second, more variates from the micro-meteorology data and the tower status data can be selected and tested so as to find more impact factors about icing thickness, which helps to increase the prediction accuracy. Third, more deep learning algorithms [28, 29] can be tried being used for data analysis. The improvement for analyzing algorithms helps to increase the efficiency and accuracy of the data analysis system.

References

1. Yi, H.: Analysis and countermeasures for large area accident cause by icing on transmission line. High Volt. Eng. **4**, 005 (2005)
2. Li, Q.-F., Fan, Z., Wu, Q., Gao, J., Su, Z.-Y.: Zhou, W.J.: Investigation of ice-covered transmission lines and analysis on transmission line failures caused by ice-coating in China. Power Syst. Technol. **9**, 009 (2008)
3. Yang, L., Hao, Y., Li, W., Li, Z., Dai, D., Li, L., Luo, B., Zhu, G.: Relationships among transmission line icing, conductor temperature and local meteorology using grey relational analysis. Gaodianya Jishu/High Volt. Eng. **36**(3), 775–781 (2010)
4. Le Cun, Y., Jackel, L., Boser, B., Denker, J., Graf, H., Guyon, I., Henderson, D., Howard, R., Hubbard, W.: Handwritten digit recognition: applications of neural network chips and automatic learning. IEEE Commun. Mag. **27**(11), 41–46 (1989)
5. Lawrence, S., Giles, C.L., Tsoi, A.C., Back, A.D.: Face recognition: a convolutional neural-network approach. IEEE Trans. Neural Netw. **8**(1), 98–113 (1997)
6. Sainath, T.N., Kingsbury, B., Saon, G., Soltau, H., Mohamed, A.R., Dahl, G., Ramabhadran, B.: Deep convolutional neural networks for large-scale speech tasks. Neural Netw. **64**, 39–48 (2015)
7. Inselberg, A.: The plane with parallel coordinates. Vis. Comput. **1**(2), 69–91 (1985)
8. Inselberg, A.: Parallel Coordinates: Visual Multidimensional Geometry and Its Applications. Springer-Verlag, New York (2009). https://doi.org/10.1007/978-0-387-68628-8
9. Wegman, E.J.: Hyper dimensional data analysis using parallel coordinates. J. Am. Stat. Assoc. **85**(411), 664–675 (2012)
10. Zhang, C., Uenaka, T., Sakamoto, N., Koyamada, K.: Extraction of vortices and exploration of the Ocean Data by Visualization System. In: Xiao, T., Zhang, L., Ma, S. (eds.) ICSC 2012. CCIS, pp. 114–123. Springer, Heidelberg (2012). https://doi.org/10.1007/978-3-642-34396-4_14
11. Wang, S., Yang, Y., Chang, J., Lin, F.: Using penalized regression with parallel coordinates for visualization of significance in high dimensional data. Int. J. Adv. Comput. Sci. Appl. **4**(10) (2013)
12. Yuan, X., Guo, P., Xiao, H., Zhou, H., Qu, H.: Scattering points in parallel coordinates. IEEE Trans. Vis. Comput. Graph. **15**(6), 1001–1008 (2009)

13. Guo, P., Xiao, H., Wang, Z., Yuan, X.: Interactive local clustering operations for high dimensional data in parallel coordinates. In: Visualization Symposium, pp. 97–104 (2010)
14. Zhang, C., Uenaka, T., Sakamoto, N., Koyamada, K.: Extraction Of vortices and exploration of the ocean data by visualization system. In: Xiao, T., Zhang, L., Ma, S. (eds.) ICSC 2012. CCIS, pp. 114–123. Springer, Heidelberg (2012). https://doi.org/10.1007/978-3-642-34396-4_14
15. Blaas, J., Botha, C.P., Post, F.H.: Extensions of parallel coordinates for interactive exploration of large multi-time point data sets. IEEE Trans. Vis. Comput. Graph. 14(6), 1436–1451 (2008)
16. Lecun, Y., Bengio, Y., Hinton, G.: Deep learning. Nature 521(7553), 436–444 (2015)
17. Fukushima, K.: Neocognitron: a self-organizing neural network model for a mechanism of pattern recognition unaffected by shift in position. Biol. Cybern. 36(4), 193–202 (1980)
18. Lecun, Y., Boser, B., Denker, J.S., Henderson, D., Howard, R.E., Hubbard, W., Jackel, L.D.: Backpropagation applied to handwritten zip code recognition. Neural Comput. 1(4), 541–551 (1989)
19. Lecun, Y., Bottou, L., Bengio, Y., Haffner, P.: Gradient-based learning applied to document recognition. Proc. IEEE 86(11), 2278–2324 (1998)
20. Krizhevsky, A., Sutskever, I., Hinton, G.E.: Image net classification with deep convolutional neural networks. In: Advances in Neural Information Processing Systems, pp. 1097–1105 (2012)
21. Hinton, G.E., Srivastava, N., Krizhevsky, A., Sutskever, I., Salakhutdinov, R.R.: Improving neural networks by preventing coadaptation of feature detectors. Comput. Sci. 3(4), 212–223 (2012)
22. Srivastava, N., Hinton, G., Krizhevsky, A., Sutskever, I., Salakhutdinov, R.: Dropout: a simple way to prevent neural networks from overfitting. J. Mach. Learn. Res. 15(1), 1929–1958 (2014)
23. Li, P., Li, N., Li, Q.M., Cao, M., Chen, H.X.: Prediction model for power transmission line icing load based on data-driven. Adv. Mater. Res. 143–144, 1295–1299 (2010)
24. Huang, X.B., Jia-Jie, L.I., Ouyang, L.S., Li-Cheng, L.I., Bing, L.: Icing thickness prediction model using fuzzy logic theory. Gaodianya Jishu/High Volt. Eng. 37(5), 1245–1252 (2011)
25. Zeng, X.J., Luo, X.L., Lu, J.Z., Xiong, T.T., Pan, H.: A novel thickness detection method of ice covering on overhead transmission line. Energy Procedia 14, 1349–1354 (2012)
26. Ma, T., Niu, D., Fu, M.: Icing forecasting for power transmission lines based on a wavelet support vector machine optimized by a quantum fireworks algorithm. Appl. Sci. 6(2), 54 (2016)
27. Nielsen, M.A.: Neural networks and deep learning. Determination Press (2015)
28. Rather, N.N., Patel, C.O., Khan, S.A.: Using deep learning towards biomedical knowledge discovery. Int. J. Math. Sci. Comput. (IJMSC) 3(2), 1–10 (2017). https://doi.org/10.5815/ijmsc.2017.02.01
29. Sharma, D., Kumar, B., Chand, S.A.: A survey on journey of topic modeling techniques from SVD to deep learning. Int. J. Mod. Educ. Comput. Sci. (IJMECS), 9(7), 50–62 (2017). https://doi.org/10.5815/ijmecs.2017.07.06

Prediction of Dissolved Gas Concentration in Oil Based on Fuzzy Time Series

Jun Liu[1(✉)], Lijin Zhao[1], Liang Huang[1], Huarong Zeng[1],
Xun Zhang[1], and Hui Peng[2]

[1] Electric Power Research Institution of Guizhou Power Grid,
Guiyang 550002, China
whwdwb@126.com
[2] School of Electrical Engineering, Wuhan University,
Wuhan 430072, China

Abstract. The prediction of dissolved gas content in transformer oil is helpful for early detection of latent faults in transformer, and it has important guiding significance for better condition based maintenance. In view of the abundant data of transformer DGA, and that the trend of the change of dissolved gas content in oil under normal running condition is not obvious, a prediction method based on fuzzy time series model is proposed. Consider that the change in dissolved gas content in oil is interaction and influenced, in this paper, the classical fuzzy time series model is improved from the view of domain division, and propose a multi factor fuzzy time series model based on spatial FCM domain partition. The example analysis shows that the method can well fit the changing trend of DGA data, and compared with the classic fuzzy time series model and the one-dimensional FCM fuzzy time series model, the superiority of the improved model in prediction is verified.

Keyword: Power transformer · Dissolved gas analysis · Fuzzy time series Data prediction

1 Introduction

Power transformer is the key equipment in power system that assume the role of switching voltage, distribution and transmission of electrical energy, it's operation condition will directly influence the operation state of power system. Once the power transformer fails, the users and power systems will be affected and endangered, and serious fire and even personal injury accidents will occur, which will cause huge economic losses. The dissolved gas content in transformer oil can effectively reflect the type of transformer fault because it is not disturbed by the external electromagnetic field, it's an effective mean to identify transformer faults and latent faults. In document [1], a transformer fault prediction model based on particle swarm optimization radial basis function neural network is presented. In document [2], a transformer fault prediction model based on particle swarm optimization support vector regression (SVM) algorithm is presented. Neural network method has the disadvantage of slow training and easy to fall into local minimum, the modeling accuracy of SVM depends too much on the choice of kernel function, and the prediction accuracy is not significantly improved with the increase of training samples.

With the development of smart grid technology, more and more DGA data can be collected [3]. The concentration of dissolved gas in oil [5] is affected by many factors, but its change law is not obvious, and its randomness and fluctuation are strong. Fuzzy time series model has great advantages in dealing with multi sample data, the more the sample data, the higher the prediction accuracy of the model. The core of the fuzzy time series method is fuzzy matching [6], so that its prediction of the change trend is not restricted by the function model, and it's suitable for the DGA data with randomness and volatility. Taking the objective conditions of mutual influence and function of dissolved gas concentration in oil into account, in this paper, a fuzzy time series model based on the spatial FCM domain partition is adopted to predict the DGA data of transformers.

2 Classical Fuzzy Time Series Model

2.1 The Concepts of Fuzzy Time Series

In 1965, Professor Zadeh [7], an American expert on automatic control, published a paper entitled "Fuzzy Sets", which marked the birth of fuzzy mathematics. The introduction of fuzzy time series model is based on fuzzy mathematics and fuzzy set theory. Fuzziness refers to the extension of the concept of subjective uncertainty of understanding, in the classical set theory [4], an element does not belong to a set is not belong to this set, fuzzy set theory with classical concepts cannot be described because it cannot use "belong" or "does not belong" to describe a set of description, the set of elements with the concept of absolute degree is a real number between 0 and 1 between.

Fuzzy sets: given a domain U, μ_A for any mapping of elements from the domain U to the interval [0,1], $\mu_A:U \rightarrow [0,1]$. A is called a fuzzy set on the domain, μ_A is the membership function of the fuzzy set. Suppose x is an element in the domain, $\mu_A(x)$ is called membership degree of fuzzy set A. A represents the extent to which an element belongs to a fuzzy set, and is valued on the interval [0, 1]. The membership function is usually expressed as a triangle membership function, as shown in the diagram (Fig. 1).

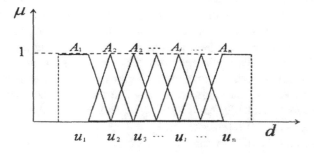

Fig. 1. Variable-span triangular membership function

Suppose that U is the domain of time series, U is divided into n ordered subsets, $U = \{u_1, u_2, \ldots, u_n\}$, each subset u_i is a semantic set on a domain, also called a fuzzy set of time series.

Fuzzy time series: suppose that $Y(t), t = 1, 2, \ldots, n$ is a set of real number sets, the domain of the set of real numbers is divided into k fuzzy sets and represented by $f_1, f_2, \ldots f_n$. The fuzzy set B corresponding to the real number A is determined according to the maximum principle of membership degree, the sequence of fuzzy sets $\{f_{i_1}(1), f_{i_2}(2), \ldots f_{i_t}(t), \ldots\}$ obtained by time ordering is fuzzy time series.

Fuzzy logic relation FLR: suppose that $F(t)$ is a set of time dependent fuzzy set sequences, if the change relationship between $F(t)$ and $F(t-1)$ can be expressed by $F(t) = F(t-1) \cdot R(t-1, t)$, that $F(t)$ can be deduced by $F(t-1)$. The fuzzy logic relation can be expressed by $F(t-1) \to F(t)$, which is abbreviated as FLR. $F(t-1)$ is called the left part of FLR, $F(t)$ is called the right part of FLR.

Fuzzy logic relation set FLRS: The fuzzy logical relation FLR, which has the same left part, is divided together, for example, we can merge three fuzzy logical relations $A_j \to A_{k_1}, A_j \to A_{k_2}, A_j \to A_{k_3}$ with the same left pieces into one FLRS, express as $A_j \to A_{k_1}, A_{k_2}, A_{k_3}$.

2.2   The Establishment of the Classical Fuzzy Time Series Model

The core idea of classical fuzzy time series model [8] is the first domain partition of original data sequence into several fuzzy intervals, each fuzzy interval consists of a fuzzy set, the original data according to the principle of maximum membership degree of fuzzy processing fuzzy time series corresponding. Then explore its development regularity and describe the fuzzy relations by fuzzy relation matrix R, and the future fuzzy set is predicted according to the fuzzy relation matrix R. Finally, the real reduction is carried out according to the corresponding de-fuzzification rule, and the prediction result is obtained.

The classical fuzzy time series model generally adopts the interval division method, this is a domain division method represented by Song, Chen, Hwang, Lee [9–12] in the early stage. Let U be the domain of sample data, where U_{min}, U_{max} are the minimum and maximum of the domain respectively, appropriately expand the field of discourse into interval $[U_{min} - D_1, U_{max} - D_2]$. Among them, the value of A makes the broadened domain interval be integer interval so that it can be calculated later. Then the average interval of the universe is divided into k fuzzy intervals $[U_{min} - D_1, U_{max} - D_2] = [u_1, u_2, \ldots, u_k]$, the intervals are $u_1 = [m_1, m_2]$, $u_2 = [m_2, m_3], \ldots$, $u_k = [m_k, m_{k+1}]$, where $m_1, m_2, \ldots, m_k, m_{k+1}$ are interval boundary values, and satisfy $|m_2 - m_1| = |m_3 - m_2| = \ldots = |m_{k+1} - m_k| = \ldots = |m_{k+1} - m_k|$.

In fuzzy time series, two adjacent fuzzy sets form a set of fuzzy logic relations $F(t) = F(t-1) \cdot R(t-1, t)$, all fuzzy logic relations are expressed in the same matrix, according to the following principles: assuming that there is a fuzzy set on the domain U, a k order matrix R is established, if $F(t-1) = A_i, F(t) = A_j$, there is a kind of fuzzy logic relation $A_i \to A_j$, it's represented as $R_{ij} = 1$ on the matrix R. If there is no fuzzy logical relation, then $R_{ij} = 0$. In summary, the fuzzy relation matrix R is obtained.

After the fuzzy relation matrix R is forecasted by the fuzzy set, the predicted fuzzy set should also be restored. The restore rule is as follows:

(1) If fuzzy set A_t corresponds to time t, each fuzzy logic relation matrix only has a corresponding fuzzy logic relation, the corresponding relation result is fuzzy set A_m, fuzzy set A_m corresponds to fuzzy interval U_m, the predicted result obtained after defuzzification is the midpoint k_m corresponding to interval U_m.

(2) If fuzzy set A_t corresponds to time t, each fuzzy logic relation matrix has multiple fuzzy logic relations $A_t \rightarrow A_{m_1}, A_{m_2}, \ldots, A_{mp}$. Where the fuzzy interval of the fuzzy set $A_{m_1}, A_{m_2}, \ldots, A_{mp}$ and the midpoint of the interval are $U_{m_1}, U_{m_2}, \ldots, U_{mp}$ and $k_{m_1}, k_{m_2}, \ldots, k_{mp}$ respectively, the result obtained after defuzzification is $(k_{m_1} + k_{m_2} + \ldots + k_{mp})/p$.

(3) If fuzzy set A_t corresponds to time t, each fuzzy logic relation matrix does not have any fuzzy logic relation, the prediction result is the midpoint k_t of the fuzzy set corresponding to the fuzzy set A_t.

The flow chart of the classical fuzzy time series model is shown in Fig. 2.

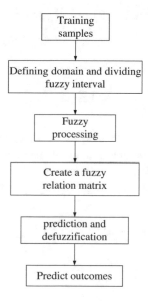

Fig. 2. The flow chart of the classical fuzzy time series model

3 Improvement of Classical Fuzzy Time Series Model from the Perspective of Domain Division

The classical fuzzy time series of medium interval domain method is simple to calculate, but the method has some subjectivity, it is difficult to use a specific "fuzzy" semantic set to explain it. Clustering algorithm can cluster data with strong similarity or the same characteristics, the clustering algorithm can improve the scientificity, rationality and interpretability of the domain. In this paper, fuzzy C means clustering algorithm is used to divide the domain.

3.1   Fuzzy C - Means Clustering Algorithm (FCM)

Fuzzy C - Means Clustering Algorithm [13] is a fuzzy clustering algorithm, is an improvement on the method of hard clustering, its specific algorithm process is as follows.

The core idea of FCM is to find the appropriate cluster center and membership value [14], the clustering cost function and the iterative error of the non-similarity index are minimized. Suppose there are n data elements $X = \{x_1, x_2, \ldots, x_n\}$, divide it into c clusters, the cluster center is expressed as $V = \{v_1, v_2, \ldots, v_c\}$, the degree of membership of element x_j to the extent to which the cluster center belongs is expressed as u_{ij}, the membership matrix can be expressed as $u = \{u_{ij}\}, 1 \le i \le c, 1 \le j \le n$.

The FCM algorithm can be transformed into the following quadratic programming problem:

$$\begin{cases} MinJ(U, V) = \sum_{i=1}^{c} \sum_{j=1}^{n} u_{ij}^m d_{ij}^2 \\ s.t. 0 \le u_{ij} \le 1, 1 \le i \le c, 1 \le j \le n \\ \sum_{i=1}^{c} u_{ij} = 1, 1 \le j \le n \\ \sum_{j=1}^{n} u_{ij} > 0, 1 \le i \le c \end{cases} \tag{1}$$

Where d_{ij} is the distance from element x_j to cluster center v_i, represented by Euclidean distance. m is the weighted index of membership degree, it determines the degree of blur of the classification results and is generally equal to 2. Then construct Lagrange function to solve the two programming problem:

$$J(U, V, \lambda_1, \lambda_2, \ldots, \lambda_n) = J(U, V) + \sum_{j=1}^{n} \lambda_j \left(\sum_{i=1}^{c} u_{ij} - 1 \right)$$

$$= \sum_{i=1}^{c} \sum_{j=1}^{n} u_{ij}^m d_{ij}^2 + \sum_{j=1}^{n} \lambda_j \left(\sum_{i=1}^{c} u_{ij} - 1 \right) \tag{2}$$

The input parameters are derived, and obtain the minimum value of formula (1):

$$v_i = \frac{\sum_{j=1}^{n} u_{ij}^m x_j}{\sum_{j=1}^{n} u_{ij}^m}, i = 1, 2, \ldots c \tag{3}$$

$$u_{ij} = [\sum_{r=1}^{c} \frac{d_{ij}}{d_{rj}}]^{-1}, i = 1, 2, \ldots c, j = 1, 2, \ldots, n \tag{4}$$

Formula (3) and Formula (4) represent the iterative expression of clustering center and membership value respectively.

Calculation steps of FCM algorithm are as follows:

(1) Set the number of clusters as c, $m > 1$, usually takes 2 or 3, initialization cluster center is $V(0) = \{v_{01}, v_{02}, \ldots, v_{0c}\}$, set the termination condition $\varepsilon = \|V(k+1) - V(k)\|$ as a particular threshold, initialization iterations k = 0;

(2) Calculate membership degree matrix;

(3) Update the iteration of the cluster center;

(4) Updating iterations of membership functions

(5) Number of iterations k = k+1;

(6) $\|V(k+1) - V(k)\|$, determine whether $\|V(k+1) - V(k)\| < \varepsilon'$ is true, ε' is the convergence degree threshold. If $\|V(k+1) - V(k)\| < \varepsilon'$ is true, the iteration ends and outputs the clustering center and membership matrix of the last iteration. If $\|V(k+1) - V(k)\| < \varepsilon'$ is not true, return to step (2).

3.2 Domain Division Method Based on Spatial FCM

Spatial FCM method [15] is used to divide the multidimensional data into multidimensional domain, its divide steps are as follows:

(1) Divide K dimensional time series $S(t)$ into domain U1, U2,... Uk, express as follows:

$$U_1 = [D_{1.\min} - \delta_{1,1}, D_{1.\max} + \delta_{1,2}], U_2 = [D_{2.\min} - \delta_{2,1}, D_{2.\max} + \delta_{2,2}],$$

$$U_K = [D_{K.\min} - \delta_{K,1}, D_{K.\max} + \delta_{K,2}], \ D_{i.\min} \text{ and } D_{i.\max}$$

are the maximum and minimum values of the i dimension data respectively.

(2) Use fuzzy C mean algorithm to divide sample data into C cluster centers, so as to form a cluster center matrix:

$$V_{cK} = \begin{vmatrix} v_{11} \ldots v_{K1} \\ \cdots \cdots \cdots \\ v_{1c} v_{Kc} \end{vmatrix} \tag{5}$$

(3) Arrange each column of the central clustering matrix according to the order of small to large. Each row of cluster centers is arranged as $sortcenter(:j) = [v_{ji_1}, v_{ji_2}, \ldots, v_{ji_c}]$.

(4) Domain division. Regard the midpoint of the adjacent cluster center as the boundary of the domain, divide the domain into c intervals $U_j = \{u_{j1}, u_{j2}, \ldots, u_{jc}\}$, $u_{j1} = [D_{j.\min} -\delta_{j,1}, (u_{j1} + u_{j2})/2]u_{j2} = [(u_{j1} + u_{j2})/2, (u_{j2} + u_{j3})/2]u_{jc} = [(u_{jc-1} + u_{jc})/2, \ D_{j.\max} + \delta_{j,2}]$. $u_{j1}, u_{j2}, \ldots, u_{jc}$ are the coordinates of the clustering centers of n-dimensional data respectively.

3.3   Fuzzy Time Series Model Based on Spatial FCM Domain Division

Improve the classical fuzzy time series model from the perspective of domain division and obtain fuzzy time series model based on spatial FCM domain partition as shown in the following Fig. 3 [16].

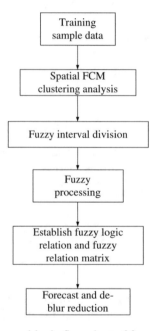

Fig. 3. Improved basic flow chart of fuzzy time series

4   DGA Content Forecasting Process Based on Spatial FCM Domain and Fuzzy Time Series Model

4.1   Selection of DGA Feature

A lot of theoretical research and practice show that the dissolved gas content in oil can reflect the running state of the transformer to a certain extent. Select A, B, C, D, E, these six gas content CH_4, C_2H_2, C_2H_4, C_2H_6, H_2, CO as characteristic quantities.

4.2 Data Acquisition and Preprocessing

To reduce the impact of measurement error and randomness error, the data need to be de-noising and smoothing.

(1) Eliminate noise data and reduce the impact of measurement errors.

The basic idea of rejecting noise data is to determine a confidence limit, any error that exceeds this limit is considered as abnormal value, which should be removed. Generally take three times of the standard deviation as a confidence level.

The average of the samples is:

$$\bar{x} = \frac{1}{n} \sum_{i=1}^{n} x_i \tag{6}$$

The standard deviation of the sample is:

$$S_x = \left(\frac{1}{n} \sum_{i=1}^{n} (x_i - \bar{x})^2\right)^{\frac{1}{2}} \tag{7}$$

If $\left| x_i - \bar{x} \right| > 3 S_x$, it is considered noise data, need to be removed.

(2) Smoothing process of data

Smoothing the sample data, on the one hand, it reduces the randomness of data and the influence of noise, on the other hand, the distortion of the signal is reduced. In this paper, we use five point two smoothing method as smoothing method. Its basic idea is to compute the weighted average of sample data in the sample center interval, the closer the data is to the sample center, the greater its weight is [17, 18].

The formula is:

$$y_0' = \frac{1}{35} (-3y_{-2} + 12y_{-1} + 17y_0 + 12y_1 - 3y_2) \tag{8}$$

4.3 Criteria Standard for Prediction Accuracy

The average absolute percentage error (MAPE) [19, 20] is used as the criterion of prediction accuracy, the formula is as follows:

$$\text{MAPE}_{\text{Tr}} = \frac{1}{n_{Tr}} \sum_{i=1}^{n_{Tr}} \frac{\left| y_{Tri} - x_{Tri} \right|}{y_{Tri}} \times 100\% \tag{9}$$

$$\text{MAPE}_{\text{Te}} = \frac{1}{n_{Te}} \sum_{i=1}^{n_{Te}} \frac{\left| y_{Tei} - x_{Tei} \right|}{y_{Tei}} \times 100\% \tag{10}$$

$$\text{MAPE} = \text{MAPE}_{\text{Tr}} + \text{MAPE}_{\text{Te}} \tag{11}$$

y_{Tri}, x_{Tri}, n_{Tr} represent the fitting values, actual values and sample number values of training data respectively, y_{Tei}, x_{Tei}, n_{Te} represent the fitting values, actual values and sample number values of test data respectively. $MAPE_{Tr}$ reflects the fitting error of training samples, $MAPE_{Te}$ reflects the extrapolation error of test samples.

4.4   Prediction Process of Model

The prediction procedure of transformer fault based on fuzzy time series model is as follows:

Step 1   Extracting the content of dissolved gas in the related oil from the on-line monitoring system of transformer, eliminate the error data and smoothing the sample data periodically;

Step 2   Extracting information about six characteristic gases, divide the sample data into training set and test set in chronological order;

Step 3   Use spatial FCM clustering method to calculate the clustering center matrix of characteristic gas data

Step 4   Divide intervals of data domain according to the clustering center matrix, establish fuzzy logic relation and fuzzy relation matrix after fuzzy treatment of the characteristic gas content;

Step 5   According to the fuzzy relation matrix, obtain the fitting result of the training set and extrapolate the test set

5   Case Study

Selected a part of the oil chromatographic monitoring data of a 500 kV transformer in Guizhou Power Grid from 2012 to 2013. The data were smoothed in ten days as a cycle, resulting in 32 sets of data, as shown in Table 1.

Table 1. Content of dissolved gas in the oil

Numbers	H_2	CO	C_2H_4	C_2H_6	C_2H_2	CH_4
3	5.898	241.606	7.056	1.416	0.656	7.753
6	4.042	132.832	6.581	1.314	0.599	6.854
9	7.310	208.891	7.848	1.717	0.841	8.638
12	7.064	204.819	7.675	1.647	0.863	8.338
15	8.920	213.995	8.308	1.792	0.884	9.406
18	9.933	234.195	8.963	2.059	0.974	10.014
21	8.852	236.045	8.757	1.966	0.875	9.366
24	8.644	237.504	8.680	1.922	0.822	8.796
27	7.034	221.996	8.305	1.822	0.799	7.954
30	6.275	206.411	7.782	1.622	0.766	7.241
31	6.500	211.427	8.101	1.768	0.733	7.412
32	6.980	181.381	7.802	2.078	0.729	7.621

Take ethylene concentration as an example, the first 30 sets of data were selected as training set and the last two sets of data were used as test sets. Use Equal interval division method, One - dimensional FCM domain division method and Fuzzy Time Series Model for Spatial Division of FCM Domain to predict and analyze ethylene concentration respectively. The model parameters of the three domain partitioning methods are shown in Table 2.

Table 2. Model parameters for different domain partitioning methods

Name of model	Interval boundary point	Fuzzy relation matrix
Equal interval division method	{6,6.4,6.8,7.2,7.6,8.0 8.4,8.8,9.2,9.6,10.0}	$\begin{bmatrix} 0 & 0 & 0 & 0 & 0 & 0 & 0 & 0 & 0 & 0 \\ 0 & 1 & 1 & 0 & 0 & 0 & 0 & 0 & 0 & 0 \\ 0 & 1 & 1 & 1 & 1 & 0 & 0 & 0 & 0 & 0 \\ 0 & 0 & 1 & 0 & 1 & 0 & 0 & 0 & 0 & 0 \\ 0 & 0 & 1 & 0 & 0 & 1 & 0 & 0 & 0 & 0 \\ 0 & 0 & 0 & 0 & 1 & 1 & 1 & 0 & 0 & 0 \\ 0 & 0 & 0 & 0 & 0 & 1 & 1 & 1 & 0 & 0 \\ 0 & 0 & 0 & 0 & 0 & 0 & 1 & 1 & 0 & 0 \\ 0 & 0 & 0 & 0 & 0 & 0 & 0 & 0 & 0 & 0 \\ 0 & 0 & 0 & 0 & 0 & 0 & 0 & 0 & 0 & 0 \end{bmatrix}$
One - dimensional FCM domain division method	{6.0,6.61,6.86,7.11,7.18,7.51, 7.98,8.17,8.30,8.60,10}	$\begin{bmatrix} 0 & 1 & 0 & 0 & 0 & 0 & 0 & 0 & 0 & 0 \\ 0 & 0 & 1 & 0 & 0 & 0 & 0 & 0 & 0 & 0 \\ 1 & 0 & 1 & 0 & 1 & 1 & 0 & 0 & 0 & 0 \\ 0 & 0 & 0 & 0 & 0 & 0 & 0 & 0 & 0 & 0 \\ 0 & 0 & 1 & 0 & 0 & 1 & 0 & 0 & 0 & 0 \\ 0 & 0 & 1 & 0 & 0 & 0 & 0 & 1 & 0 & 0 \\ 0 & 0 & 0 & 0 & 0 & 1 & 1 & 0 & 0 & 0 \\ 0 & 0 & 0 & 0 & 0 & 0 & 0 & 0 & 1 & 0 \\ 0 & 0 & 0 & 0 & 0 & 0 & 1 & 0 & 1 & 1 \\ 0 & 0 & 0 & 0 & 0 & 0 & 0 & 0 & 1 & 1 \end{bmatrix}$
Fuzzy Time Series Model for Spatial Division of FCM Domain	{6.0,6.82,7.13,7.46,7.75,7.94, 8.18,8.39,8.61,8.87,10.0}	$\begin{bmatrix} 1 & 1 & 0 & 0 & 0 & 0 & 0 & 0 & 0 & 0 \\ 1 & 1 & 1 & 0 & 1 & 0 & 0 & 0 & 0 & 0 \\ 0 & 1 & 0 & 1 & 0 & 0 & 0 & 0 & 0 & 0 \\ 0 & 0 & 0 & 0 & 0 & 0 & 1 & 0 & 0 & 0 \\ 0 & 1 & 0 & 0 & 0 & 0 & 0 & 0 & 0 & 0 \\ 0 & 0 & 0 & 0 & 1 & 1 & 0 & 0 & 0 & 0 \\ 0 & 0 & 0 & 0 & 0 & 1 & 0 & 1 & 0 & 0 \\ 0 & 0 & 0 & 0 & 0 & 0 & 1 & 1 & 0 & 1 \\ 0 & 0 & 0 & 0 & 0 & 0 & 0 & 1 & 1 & 0 \\ 0 & 0 & 0 & 0 & 0 & 0 & 0 & 0 & 1 & 1 \end{bmatrix}$

The prediction results of the three methods are shown in Table 3.

From the results of Table 3, we can see that the fitting error of the training data of space FCM is the smallest, and the error is 2.81%. The extrapolation error of the test data of space FCM is also the smallest, and the error is 2.81%. Thus, the feasibility and scientificity of improving the fuzzy time series model from the view of domain division are verified.

Table 3. Comparison of predicted results (unit: μL/L)

Numbers	Actual measurement values	Predictive values of spatial FCM	Predictive values of one-dimensional FCM	Predictive values of equal interval method
3	7.06	7.13	7.10	7.20
6	6.58	7.13	7.10	7.20
9	7.84	7.13	7.10	7.20
12	7.68	7.29	7.37	7.40
15	8.31	8.74	8.61	8.60
18	8.96	9.09	8.88	8.80
21	8.76	9.09	8.88	8.80
24	8.68	8.62	8.88	8.80
27	8.31	8.74	8.61	8.60
30	7.78	7.95	7.91	8.20
31	8.10	7.82	7.61	7.60
32	7.80	7.62	7.61	7.40
$MAPE_{Tr}$ (%)		2.81	2.98	3.50
$MAPE_{Te}$ (%)		2.88	4.24	5.65
MAPE (%)		5.79	7.22	9.15

6 Conclusion

In this paper, an improved fuzzy time series model based on spatial FCM domain partition is constructed from the view of domain division. It is proved that the model can fit the changing trend of DGA data well, and its prediction accuracy is improved greatly compared with the classical fuzzy time series model. The model can better meet the needs of the project and can be used as an effective tool for DGA data prediction.

References

1. Fu, B.: Research on Transformer Fault Diagnosis and Prediction Based on Particle Swarm Optimization. Hua Qiao University (2012)
2. Zhang, Y.: Transformer Fault Diagnosis and Prediction Based on Particle Swarm Optimization Support Vector Machine. Xihua University (2011)
3. Azizzadeh, L., Zadeh, L., et al.: Information and Control. Fuzzy Sets 8(3), 338–353 (1965)
4. Li, G.: Intelligent Predictive Control and Its Realization of MATLAB. Publishing House of Electronics Industry Beijing (2010)
5. Song, Q., Chissom, B.S.: Forecasting enrollments with fuzzy time series—part I. Fuzzy Sets Syst. 62(1), 1–8 (1994)
6. Chen, S.M.: Forecasting enrollments based on fuzzy time series. Fuzzy Sets Syst. 81(3), 311–319 (1996)
7. Qiu W.: Fuzzy time series model and its application in stock index trend analysis. Dalian University of Technology (2012)

8. Song, Q., Chissom, B.S.: Fuzzy time series and its model. Fuzzy Sets Syst. **54**(3), 269–277 (1993)
9. Chen, S.M.: Forecasting enrollments based on fuzzy time series. Fuzzy Sets Syst. **81**(3), 311–319 (1996)
10. Hwang, J.R., Chen, S.M., Lee, C.H.: Handling forecasting problems using fuzzy time series. Fuzzy Sets Syst. **100**(1–3), 217–228 (1998)
11. Lee, M.H., Efendi, R., Ismail, Z.: Modified weighted for enrollment forecasting based on fuzzy time series. J. Artif. Intell. **25**(1) (2009)
12. Wang, J., Wang, S., Bao, F.: A fast fuzzy C mean clustering algorithm based on spatial distance. Comput. Eng. Appl. **51**(1), 177–183 (2015)
13. Ma, Y.: The study of multi factor fuzzy time series forecasting model. Dalian Maritime University (2016)
14. Zhenyong, Yang: Discussion on judgement of transformer fault by using guidelines for the analysis and judgement of dissolved gases in transformer oil. Transformer **45**(10), 24–27 (2008)
15. Donghua, Zhou: Data processing of gas well production based on fusion of smoothing algorithm and wavelet transform. Oil Gas Field Surf. Eng. **30**(4), 33–35 (2011)
16. Chen, X.: Research on Transformer Fault Prediction Method Based on Limit Learning Machine. North China Electric Power University (2015)
17. Wang, T., Xia, T., Cao, X.: Feature dimension reduction algorithm based prediction method for protein quaternary structure. Int. J. Wirel. Microwave Technol. (IJWMT) **2**(5), 28–33 (2012)
18. Man, D.-P., Li, X.-Z., Yang, W., Wang, W., Xuan, S.-C.: A Multi-step attack recognition and prediction method via mining attacks conversion frequencies. Int. J. Wirel. Microw. Technol. (IJWMT) **2**(2), 20–25 (2012). https://doi.org/10.5815/ijwmt.2012.02.04
19. Sarailoo, M., Rahmani, Z., Rezaie, B.: Fuzzy predictive control of step-down DC-DC converter based on hybrid system approach. Int. J. Intell. Syst. Appl. (IJISA) **6**(2), 1–13 (2014). https://doi.org/10.5815/ijisa.2014.02.01
20. Zhao, Y., Jin, H., Wang, L., Wang, S.: Rough neuron network for fault diagnosis. Int. J. Image Graph. Signal Process. (IJIGSP) **3**(2), 51–58 (2011). https://doi.org/10.5815/ijigsp.2011.02.08

Perfection of Computer Algorithms and Methods

A Comprehensive Analysis
of the Bat Algorithm

Yury Zorin[(⊠)] [iD]

National Technical University of Ukraine "Igor Sikorsky Kyiv
Polytechnic Institute", Kyiv, Ukraine
yzorin@gmail.com

Abstract. Optimization is one of the most challenging problems that has
received considerable attention over the last decade. The bio-inspired evolutionary optimization algorithms due to their robustness, simplicity and efficiency
are widely used to solve complex optimization problems. The Bat algorithm is
one of the most recent one from this category. Given that the original Bat
algorithm is vulnerable to local optimum and unsatisfactory calculation accuracy, the paper presents detailed analysis of its main stages and a measure of
their influence on the algorithm performance. In particular, the global best
solution acceptance condition, the way a new solution is generated by random
flight and the local search procedure implementation have been studied. The
ways to overcome the original algorithm's flaws have been suggested. Their
effectiveness has been proved by numerous computational experiments.

Keywords: Optimization problem · Metaheuristic algorithm · Bat algorithm
Lévy flight

1 Introduction

Most of the heuristic and metaheuristic algorithms have been derived from the behavior
of biological systems and/or physical systems in nature. A group of metaheuristic are
referred to as swarm intelligence based algorithms [11].

For example, particle swarm optimization was developed based on the swarm
behavior of birds and fish, while the firefly algorithm was based on the flashing pattern
of fireflies [12] and cuckoo search algorithm was inspired by the brood parasitism of
some cuckoo species [13].

The original Bat Algorithm developed by Yang [1] is inspired by echolocation
characteristic of bats. Echolocation is typical sonar which bats use to detect prey and to
avoid obstacles. A bat emits a very loud sound and listens for the echo that bounces
back from the surrounding objects. Thus, a bat can compute how far it is from an
object. Furthermore, a bat can distinguish the difference between an obstacle and a prey
even in complete darkness [2].

Subsequently some modifications aimed at improving the algorithm performance in
sense of results accuracy and robustness have been proposed. In particular, the use of
chaotic maps instead of uniform distribution to generate solutions by random fly was
proposed [7]. In order to improve Bat Algorithm behavior for higher-dimensional

© Springer International Publishing AG, part of Springer Nature 2019
Z. Hu et al. (Eds.): ICCSEEA 2018, AISC 754, pp. 283–291, 2019.
https://doi.org/10.1007/978-3-319-91008-6_28

problems, the original Bat Algorithm was hybridized with differential-evolution strategies [14]. In [15] it was proposed to improve exploration mechanism of the algorithm by modifying the equation of pulse emission rate and loudness of bats.

In this paper we present a detailed analysis of original Bat Algorithm and propose ways to improve its performance.

2 Bat Algorithm

The original Bat Algorithm uses the following idealized rules.

1. All bats use echolocation to sense distance, and they also 'know' the difference between food/prey and background barriers in some magical way.
2. Bats fly randomly with velocity v_i at position x_i with a fixed frequency f_i, varying wavelength and loudness A_0 to search for prey. They can automatically adjust the wavelength (or frequency) of their emitted pulses and adjust the rate of pulse emission depending on the proximity of their target.

Although the loudness can vary in many ways, it is assumed that the loudness varies from a large (positive) A_0 to a minimum value A_{min}.

Initial population of n bats is randomly generated from real-valued vectors with dimension d, by taking into account lower and upper boundaries

$$x_{ij} = x_{\min j} + rand(0, 1)(x_{\max j} - x_{\min j}).$$

Where $i = 1, 2, \ldots, n$, $j = 1, 2, \ldots, d$, x_{minj} and x_{minj} are lower and upper boundaries for dimension j respectively.

Velocity v_i of solution is proportional to frequency f_i and a new solution depends on its new velocity

$$f_i = f_{\min} + (f_{\max} - f_{\min})\beta. \tag{1}$$

$$v_i^t = v_i^{t-1} + (x_i^t - x_*)f_i. \tag{2}$$

$$x_i^t = x_i^{t-1} + v_i^t. \tag{3}$$

Where $\beta \in [0, 1]$ is randomly generated number.

For the exploitation (local search) part of the algorithm, one solution is selected among the best solutions and random walk is applied

$$x_{new} = x_{old} + \varepsilon A^t. \tag{4}$$

Where A^t is average loudness of all bats, $\varepsilon \in [-1, 1]$ is random number and represents direction and intensity of random walk.

The loudness and pulse emission rate must be updated as iterations proceed. As a bat gets closer to its prey, loudness A usually decreases while pulse emission rate increases. Loudness A and pulse emission rate r are updated by the following equations

$$A_i^{t+1} = \alpha A_i^t.$$

$$r_i^{t+1} = r_i^0(1 - e^{-\gamma t}).$$

Where α and γ are constants, r_i^0 and A_i are random value factors, and A_i^0 is typically [1, 2], while r_i^0 is typically [0,1].

Thus, here is the pseudo code of the original Bat Algorithm.

```
Initialize bat population x_i (i=1,2,...,n) and v_i

Define pulse frequency f_i at x_i

Initialize pulse rates r_i and the loudness A_i

while (iter < Max number of iterations){
        Generate new solutions by adjusting frequency,
        and updating velocities and locations/solutions
        [equations (1) to (3)]
        Select a solution among the best solutions
        Generate a local solution around the selected best
        solution
        Generate a new solution by flying randomly
        if (rand < A_i & f(x_i)<f(x_*) ){
                Accept the new solutions
                Increase r_i and reduce A_i
        }
        Rank the bats and find the current best x_*
}
```

3   Test Suit

To perform the algorithm analysis well known benchmark functions have been chosen [3].

The Ackley function is widely used for testing optimization algorithms. In its two-dimensional form it is characterized by a nearly flat outer region, and a large hole at the center. The function poses a risk for optimization algorithms to be trapped in one of its many local minima.

$$f(x) = -ae^{-b\sqrt{(\frac{1}{d}\sum_{i=1}^{d} x_i^2)}} - e^{\frac{1}{d}\sum_{i=1}^{d} \cos(cx_i)} + a + e.$$

Variable values are $a = 20, b = 0.2, c = 2\pi$ and $d = 20$

The function has been evaluated on the hypercube $x_i \in [-32.768, 32.768]$ for $i = 1, 2, \ldots, d$. Its global minimum is $f(x^*) = 0$ at $x^* = (1, \ldots, 1)$.

The Sphere function has d ($d = 50$ in our tests) local minima and the global one. It is continuous, convex and unimodal. The function has been evaluated on the hypercube $x_i \in [-100, 100]$ for $i = 1, 2, \ldots d$.

$$f(x) = \sum_{i=1}^{d} x_i^2.$$

Its global minimum is $f(x^*) = 0$ at $x^* = (1, \ldots, 1)$.

The Griewank function has many widespread local minima which are regularly distributed.

$$f(x) = \sum_{i=1}^{d} \frac{x_i^2}{4000} - \prod_{i=1}^{d} \cos(\frac{x_i}{\sqrt{i}}) + 1.$$

The function has been evaluated on the hypercube $x_i \in [-600, 600]$ for $i = 1, 2, \ldots d$
($d = 50$ in our tests). Its global minimum is $f(x^*) = 0$ at $x^* = (1, \ldots, 1)$.

The Rastrigin function has several local minima. It is highly multimodal, but locations of the minima are regularly distributed.

$$f(x) = 10d + \sum_{i=1}^{d} (x_i^2 - 10\cos(2\pi x_i)).$$

The function has been evaluated on the hypercube $x_i \in [-5.12, 5.12]$ for $i = 1, 2, \ldots d$
($d = 30$ in our tests). Its global minimum is $f(x^*) = 0$ at $x^* = (1, \ldots, 1)$.

The Rosenbrock function is unimodal and the global minimum lies in a narrow parabolic valley. However, even though this valley is easy to find, convergence to the minimum is difficult.

$$f(x) = \sum_{i=1}^{d-1} [100(x_{i+1} - x_i^2)^2 + (x_i - 1)^2].$$

The function has been evaluated on the hypercube $x_i \in [-5, 10]$ for $i = 1, 2, \ldots d$
($d = 30$ in our tests). Its global minimum is $f(x^*) = 0$ at $x^* = (1, \ldots, 1)$.

4   The Global Best Solution Acceptance

We will start with the global best solution acceptance condition in the pseudo code above. It is stated as

$$if(rand < A_i \,\&\, f(x_i) < f(x_*)).$$

Where x_* is the global best solution found by all bats so far. It means that a new solution x_i, which might be of better quality than x_*, is accepted only if the i-th bat loudness A_i meets certain conditions. It seems reasonable to accept x_i regardless of the value of A_i since it is current global best solution and there is risk to lose it in the next iterations. So we will accept new global best solution unconditionally. Table 1 presents the test results that have been obtained on the benchmark functions. The results obtained by the algorithm where a new global best x_i is accepted unconditionally are compared against those obtained by the original algorithm. The result standard deviations are shown in brackets. Taking into account the stochastic nature of the algorithm, their values might be considered as an indication of results' robustness. The tests have been run 50 times upon all of the benchmark functions.

Table 1. The influence of the global best solution acceptance condition.

Test function	Best result accuracy gained	Mean result accuracy gained
Ackley	5.91% (6.23e−04)	1.8% (1.21e−4)
Sphere	6.8% (4.82e−05)	7,5% (1.55e−06)
Griewank	0.2% (1.19e−21)	0.2% (1.19e−21)
Rastrigin	19% (1.29e−06)	24% (6.41e−06)
Rosenbrock	3.3% (4.59e+00)	8.1% (1.61e+00)

It's worth to mention, that there was almost no accuracy gaining when optimizing the Griwank function. The cause of it is that both algorithms have been quickly trapped in one of the multiple local minima.

5 On Solutions Generated by Random Flight

An issue of vital importance for every heuristic algorithm is to keep balance between exploration (also called diversification) and exploitation (also called intensification). Several random distributions can be helpful in this line. For example, uniform distribution generates each point of the search space with equal probability. The Gaussian distribution is biased towards obtaining new solutions where smaller modifications occur more often than larger ones [4]. The chaotic maps [5–7] and Lévy Flight [8] are widely used for algorithm randomization as well. The original Bat Algorithm uses a uniform distribution.

Lévy Flight is a random walk in which the step lengths have a stable probability distribution

$$Levy \sim u = t^{-\lambda}, (1 < \lambda \le 3).$$

When defined as a walk in a space of dimension greater than one, the steps made are in isotropic random directions. Due to the remarkable properties of stable distribution it is now believed that Lévy statistics provides a framework for the description of many natural phenomena in biological, physical, and chemical systems from a common point of view. The step length of Lévy Flight depends on the value of λ. The less the value of λ, the bigger is the step length. This fact might be used to get fair balance between exploration and exploitation. To achieve it, we will use controlled (to some extent) step length [9, 10]. In order to increase the quality of the Bat Algorithm in searching the problem space efficiently and find the optimal solution we assume that 'worse' bats should perform Lévy Flight with greater swing (smaller value of λ) to explore remote parts of the search space. On the other hand, 'better' bats are intended to search more thoroughly the perspective areas found in the previous iterations. This means that their step length should be smaller, or the value of λ is greater. It may be achieved if λ belongs to the interval $[\lambda_{min}; \lambda_{max}]$ and the dependence of λ value on the bat's rank is as follows

$$\lambda = \lambda_{max} - i(\lambda_{max} - \lambda_{min})/(m - 1).$$

Where m is the colony size, i is the rank of i-th bat.

Table 2 presents the test results that have been obtained with the use of two kinds of random flight implementations compared against those obtained by the original algorithm.

Table 2. The influence of the random flight implementation.

Test function	Gaussian distribution		Lévy Flight	
	Best result accuracy gained	Mean result accuracy gained	Best result accuracy gained	Mean result accuracy gained
Ackley	5.1% (4e−02)	9.3% (1e−03)	8.9% (1e−02)	6.1% (2e−03)
Sphere	11% (1e−04)	6.2% (4e−04)	8.2% (2e−05)	5.2% (5e−05)
Griewank	3.2% (1e−05)	5.9% (3e−05)	6.9% (1.e−05)	7.2% (1e−06)
Rastrigin	8.1% (1e−05)	10.2% (4e−05)	16.2% (4e−05)	9.4% (1e−06)
Rosenbrock	2.3% (1e−02)	5.3% (8e−03)	6.1% (4e−03)	8.3% (5e−04)

The sphere function is convex and unimodal, so that might be the reason why Gaussian distribution outperforms Lévy distribution when applied to this function optimization.

6   Local Search Issues

Finally, let's study the local search part of the algorithm defined by (4). It uses the uniform distribution which is, generally speaking, more preferable for the stochastic search process when the fitness landscape is flat, whilst in rougher fitness landscapes Lévy Flight might be more appropriate.

$$x_{new} = x_{old} + \alpha_t \oplus Levy(\lambda)A^j.$$

Let us divide the population of bats into q equal groups. Here A^j is average loudness of bats in the j–th group. Assume that $best\_bat_j$ is a bat with the best solution found so far in the j–th group. Again, like in previous section, we suppose that another bat s in this group whose current position is far from $best\_bat_j$ should perform Lévy Flight with greater step length for more extensive exploration of the search space. Thereby, we will calculate the value of λ for the s–th bat as follows

$$\lambda = \Phi e^{(\Psi(-dist_{s,best\_bat_j}/L))}.$$

Where $dist_{s,best\_bat_j}$ is a distance from s-th bat to $best\_bat_j$, L is the maximum distance in all dimensions, Φ and Ψ are the algorithm parameters [10]. Thus, the bats move to several better current solutions (the best in their group) instead of one which prevents premature conversion to local minimum.

Additional way to increase the thoroughness of the perspective solution neighbourhood exploration is reduction of α_t value form α_{max} to α_{min} during the iterations so, that it varies with the iteration counter t as follows

$$\alpha_t = \alpha_0 \delta^t.$$

Where $0 < \delta < 1$ and is calculated as

$$\delta = \sqrt[k]{\frac{\alpha_{min}}{\alpha_{max}}}.$$

Where k is maximum number of iterations and $\alpha_0 = \alpha_{max}$,

The influence of the local search implementation on algorithm performance is presented in Table 3. Again, the smallest accuracy improvement was gained when optimizing the Sphere function due to its smooth convex landscape.

Table 3. The influence of the local search via Lévy Flight.

Test function	Best result accuracy gained	Mean result accuracy gained
Ackley	2.9% (1.8e−05)	4.1% (2.6e−05)
Sphere	1.4% (2.1e−06)	2.8% (5.1e−06)
Griewank	3.1% (1.1e−05)	3.2% (2.4e−06)
Rastrigin	4.8% (4.3e−05)	7.1% (1.5e−06)
Rosenbrock	7.1% (3.2e−04)	9.1% (6.1e−04)

Now let us summarize our findings. Table 4 presents comparison of the results obtained by the original algorithm and the algorithm that combines all three modifications suggested above.

Table 4. The overall comparison of two algorithms.

Test function	Original Bat Algorithm		Modified algorithm	
	Best result	Mean result	Best result	Mean result
Ackley	6.4e−07 (3.e−08)	9.6e−07 (5e−08)	2.3e−12 (1e−13)	7.1e−12 (1e−13)
Sphere	9.4e−05 (1e−04)	1.7e−04 (7−04)	4.2e−08 (2e−07)	2.6e−05 (1e−04)
Griewank	7.5e−04 (3e−05)	9.7e−04 (1e−05)	4.2e−10 (5e−11)	9.1e−10 (3e−11)
Rastrigin	1.1e−01 (2e+00)	4.4e−01 (4e+00)	2.3e−06 (1e−07)	8.5e−06 (6e−07)
Rosenbrock	3.7e−02 (6e−02)	9.9e−02 (1e−01)	1.1e−05 (5e−04)	8.8e−05 (1e−04)

7   Conclusion

The statistical results obtained from the computational experiments demonstrate the superiority of the modified algorithm to the original Bat Algorithm in regard of results accuracy. The values of the result standard deviation demonstrate more stable behavior of the modified algorithm. The greatest performance improvement has been achieved when optimizing functions with rough landscape and multiple minima. Thus, the introduction of Lévy distribution both in random flight and local search procedures helps proposed algorithm to find globally and locally improved solutions.

In future research more detailed studies should be conducted in regard of the comprehensive choice of f_{max} and f_{min} parameter values depending on the problem length scales. As it follows from Eq. (1), these values have a vital impact on velocity calculation and, consequently, on the bat movement pattern.

References

1. Yang, X.S.: A new metaheuristic bat-inspired algorithm. Nat. Inspired Coop. Strat. Optim. **284**, 65–74 (2010)
2. Altringham, J.D.: Bats: Biology and Behaviour. Oxford University Press, New York (1996). p. 379
3. Virtual Library of Simulation Experiments: Test Functions and Datasets. http://www.sfu.ca/~ssurjano/index.html
4. Farahani, S.M., Abshouri, A.A., Nasiri, B., Meybodi, M.R.: A Gaussian firefly algorith. Int. J. Mach. Learn. Comput. **1**(5), 448–453 (2011)
5. dos Santos Coelho, L., Mariani, V.C.: Use of chaotic sequences in a biologically inspired algorithm for engineering design optimization. Expert Syst. Appl. **34**, 1905–1913 (2008)
6. Dhal, K.G., Quraishi, I., Das, S.: A chaotic Lévy flight approach in bat and firefly algorithm for gray level image enhancement. Int. J. Image Graph. Signal Process. (IJIGSP) **7**(7), 69–76 (2015). https://doi.org/10.5815/ijigsp.2015.07.08
7. Abdel-Raouf, O., Abdel-Baset, M., El-Henawy, I.: An improved chaotic bat algorithm for solving integer programming problems. Int. J. Mod. Educ. Comput. Sci. (IJMECS) **6**(8), 18–24 (2014). https://doi.org/10.5815/ijmecs.2014.08.03
8. Reynolds, A.M., Rhodes, C.J.: The Levy flight paradigm: random search patterns and mechanisms. Ecology **90**, 877–887 (2009)

9. Zorin, Y.: A metaheuristic algorithm for multimodal functions optimization. In: Proceedings of the International Scientific Conference Intellectual information analysis IIA 2015, Kyiv, Ukraine on 20–22 May, pp. 88–92 (2015)
10. Zorin, Y.: An improved cuckoo search algorithm. In: System Analysis and Information Technology SAIT 2016, Kyiv, Ukraine on 30 May–2 June, pp. 48–49 (2016)
11. Roy, S., Biswas, S., Chaudhuri, S.S.: Nature-inspired swarm intelligence and its applications. Int. J. Mod. Educ. Comput. Sci. (IJMECS) 6(12), 55–65 (2014). https://doi.org/10.5815/ijmecs.2014.12.08
12. Abdel-Raouf, O., Abdel-Baset, M., El-henawy, I.: Chaotic firefly algorithm for solving definite integral. Int. J. Inf. Techn. Comput. Sci. (IJITCS) 6(6), 19–24 (2014). https://doi.org/10.5815/ijitcs.2014.06.03
13. Roy, S., Chaudhuri, S.S.: Cuckoo search algorithm using Lèvy flight: a review. Int. J. Mod. Educ. Comput. Sci. (IJMECS) 5(12), 10–15 (2013). https://doi.org/10.5815/ijmecs.2013.12.02
14. Fister Jr., I., Fister, D., Yang, X.-S.: A hybrid bat algorithm. Elektrotehnitski Vestnik 80(1–2), 1–7 (2013)
15. Yılmaz1, S., Kucuksille, E.U., Cengiz, Y.: Modified bat algorithm. Elektronika ir Electrotechnika 20(2), 36–43 (2014)

Complex Steganalytic Method for Digital Videos

A. V. Akhmametieva[(✉)]

Odessa National Polytechnic University, Odessa, Ukraine
a.v.akhmametieva@opu.ua

Abstract. In work the analysis of steganalytic algorithms realizing methods which are earlier developed by the author aimed to detect the additional information attachments in digital videos is carried out. Comparison of efficiency of steganalytic algorithms allowed to develop practical recommendations for use of the proposed methods which formed the basis of a complex steganalytic method of detecting attachments of additional information in digital videos. The complex method proposed in work is capable to detect hidden information embedded into digital videos under different conditions, including small values of hidden capacity, irrespective of a format and color representation of an original container.

Keywords: Steganalysis · Spatial domain · Digital video
Frame of digital video

1 Introduction

Continuous development and penetration of information technologies into various spheres of public and political life of the state leads to accumulation of large volumes of the electronic information containing both the public, and confidential data which are subject to protection. Leakage of the protected information including disclosure of the fact of leak, can do irreparable harm for the state and the companies, in particular to significant material losses, bankruptcy and loss of reputation.

Confidential data transmission to extraneous persons can be carried out using the steganography providing the organization of a secret communication channel through the exchange of imperceptible at first glance containers, as which digital images, audio or video can act. The result of embedding a secret message (additional information) into a container will be called stego.

If necessary to transfer significant amount of confidential data as a container, as a rule, apply digital videos thanks to the following advantages:

- due to the large number of frames it is possible to provide immersion of significant amount of additional information at small values of hidden capacity that allows to save visual integrity of digital content and will complicate process of steganalysis aimed at the detecting the fact of the presence/absence of hidden information in any digital content [1];
- thanks to features of compression of the video sequences by MPEG codecs it is possible to ensure safety of a secret message embedded not only in the DCT [2] or

DWT [3] domain, but also in the spatial domain by the LSB method [4] that is unstable to small perturbations by embedding additional information only into I-frames of digital videos that will allow to save embedded information when coding.

The considered scheme of embedding of additional information allows to provide simplicity of implementation and to reduce computing complexity of a steganographic algorithm.

However, despite advantages of using of digital videos as a container, steganalysis of digital videos it is less developed in comparison with steganalysis of digital images therefore increase of efficiency of steganalysis of digital videos is an urgent task.

Modern methods of a steganalysis of digital images and digital videos [5–7], as a rule, perform analysis of transformation domain of digital contents that leads to additional accumulation of a computational error, and this, in turn, significantly reduces the efficiency of detection stego formed at low hidden capacity, in particular, in the [7] the accuracy of correctly identified stego and original videos at a hidden capacity of 0.0625 bpp is only 77.74%.

In [6] an adaptation of steganalytic developments aimed to detect the presence of additional information in digital images for digital videos was proposed, which allowed obtaining a detection accuracy of 93% in case of hidden capacity of 0.1 bpp.

It should be noted that in the vast majority of cases modern authors make computational experiments on the basis of the small number of test videos (up to 26 digital videos) that in principle does not allow to evaluate efficiency of the offered methods objectively.

Among the steganalytic methods that analyzes the spatial domain of digital content we can note Close Color Pair Analysis. In [8] good results of detection stego at hidden capacity not lower than 0.2 bpp (type I errors and type II errors are 0.5%) are received, however this method is directed to the analysis of digital images and not adapted for video.

Not less important problem is that the existing developments are not able to perform the frame analysis of digital videos, if confidential information was embedded only into small number of frames. Such experiment was carried out in [9] according to which the detection accuracy in the case of 40% of the filled frames is 79.74%, in the case of 20% of the filled frames is 66.88% at hidden capacity 0.1 bpp that does not allow to speak about the efficiency of the proposed solution.

Thus, the problem of detection stego formed by embedding of additional information into digital videos with small values of hidden capacity remains unresolved. In order to increase the efficiency of detection such attachments of additional information in digital images and digital video in [10–12] the steganalytic methods were proposed by the author. Proposed methods analyze the spatial domain of digital contents due to which high results of detecting stego were obtained. Nevertheless, each of the methods has advantages and disadvantages therefore to ensure high identification accuracy of stego and unfilled containers it is advisable their complex using that will allow to detect stego formed in various conditions: at different values of hidden capacity, at different number of color components of digital videos used for embedding of additional information (in 1, 2 or 3 color components), when using as containers of color or grayscale digital videos irrespective of a format (losses or lossless) and the size of frames of digital videos.

The aim of the work is development of a complex steganalytic method for detecting the attachment of confidential information in digital videos on the basis of the steganalytic methods which are earlier proposed by the author.

2   Proposed Steganalytic Methods in Spatial Domain

In the basics of steganalytic methods for digital videos [10, 11] developed earlier by the author the analysis of number of blocks with same color brightness values in matrixes of color components of digital contents lies. It was established that $n \times n$-blocks look

$$A = \left(a_{ij}\right)_{i,j=1}^{n}, \; a_{ij} = a, \; i,j = \overline{1,n}, \; a \neq 0 \tag{1}$$

have one only nonzero singular number $\sigma_1 > 0$, $\sigma_2 = \ldots = \sigma_n = 0$, that belongs to a set of natural numbers $\left(\sigma_1 \in N\right)$ and multiple the block size $n \left(\sigma_1 : n\right)$. Such feature is sensitive to small perturbations because in general case singular numbers for any matrix are non-negative real numbers [13]. Changing only one element of matrix (1) which can arise as a result of embedding of additional information into digital content will lead to the fact that $\sigma_1 \notin N$. On the basis of the established feature of perturbation of the unique nonzero singular number σ_1 $n \times n$-block of a look (1) accordance between domain of singular decomposition of the corresponding matrixes and spatial domain of digital contents is established that allowed not only to avoid additional accumulation of a computational error, but also the additional computational spending connected with transition to transformation domain and back when developing steganalytic algorithms.

Steganalytic method [10] is based on the analysis of relative (in relation to total quantity $n \times n$-blocks in a matrix of the color component of a digital video frame) number of blocks with the same values of brightness in matrixes of digital contents as a result of primary and repeated embedding of additional information that is caused by a hypothesis that the quantity of $n \times n$-blocks look (1) in stego will be insignificant (perhaps such blocks will be absent) that will result in small values of relative number of such blocks after repeated embedding of additional information in comparison with primary embedding. The main steps of the algorithm SAA1 realizing the proposed method match with brought in [10] where threshold values are $T_R = 27.5$, $T_G = 19.5$, $T_B = 11.5$ for digital videos in losses format, $T_R = 1.6, T_G = 1.7, T_B = 1.6$ for digital videos in lossless format.

The method developed in [11] is based on the accounting of character of change in the number of blocks with equal values of brightness as a result of primary and repeated embedding of additional information. It has been established that as a result of primary embedding of additional information in the majority of matrixes of the color components of digital videos frames (irrespective of a format of an original container) a reduction of number of blocks with the same values of brightness is observed (up to 0, especially if hidden capacity was 0.25–0.5 bpp while primary embedding) and as a result of repeated embedding of additional information into stego - either insignificant increase in such blocks or their invariance. The main steps of an algorithm SAA2 realizing the developed method match with brought in [11] where the threshold is $T = 3$. This algorithm, unlike SAA1, allowed to effectively detect stego formed by embedding of additional information into video with a small frame size at values of hidden capacity not less than 0.167 bpp.

In [12] the steganalytic algorithm (hereinafter SAA3) for detecting stego in digital images formed by embedding of additional information into one only arbitrary color component of digital images in losses format, based on the accounting of quantity of sequential triads of color triplets in a matrix of unique colors UCT by the size $U \times 3$ which contains unique color triplets (r_i, g_i, b_i), $i = \overline{1, U}$ contained in the image is proposed. The developed algorithm was adapted for digital videos (SAA3v), for analysis of separate frames of video sequence it is necessary to use an algorithm SAA3 which main steps are given in [12].

3 Analysis of Efficiency of Developed Steganalytic Algorithms

For definition of efficiency of the algorithms developed by the author computational experiments on the basis of the following digital contents were carried out:

- 200 videos of frame size 320×240 received by cameras of mobile devices with expansion *.3gp or *.mp4 (compression uses H.263 or H.264 codecs) - Set 1;
- 167 video sequences of frame size 320×240 received by the video camera with the *.avi expansion (compression uses Xvid codec) - Set 2;
- 49 videos of frame size 176×144 received by the camera of outdated model of the mobile phone (Sony Ericsson W700i - 2006 release) with expansion *.3gp - Set 3.

Each video contains 250 frames, after embedding of additional information videos was saved as *.avi files without loss. In the computational experiment aimed at the approbation of the developed algorithms, embedding of additional information was carried out by LSB method into a randomly selected color component of digital videos with different values of hidden capacity.

By results of experiment type I errors (False Negative FN) - the pass stego in case of its presence and type II errors (False Positive FP) - false detection stego in case of its absence (Table 1) were received.

As can be seen from Table 1 the best results of detection are achieved by the SAA3v method, however, as it was noted above, this algorithm can be used only in case when additional information embeds into one only color component of color containers. Unlike SAA3v algorithms SAA1 and SAA2 can be used both for color and grayscale digital videos, besides they are capable to detect the presence of additional information which was embedded into two or three color components, as the analysis of color matrixes happens separately for each color component. It should be noted that the developed methods allow not only to answer a question whether confidential information in digital content was hidden, but also to define into what color component embedding of additional information was carried out.

Thus, on the basis of the conducted researches it can be concluded that each of the developed algorithms (SAA1, SAA2, SAA3v) has advantages:

- SAA1 allows to detect stego with high efficiency, that formed by embedding of additional information with hidden capacity not less than 0.125 bpp into one (two,

Table 1. The effectiveness of detecting the presence/absence of additional information in digital videos by algorithms SAA1, SAA2, SAA3v, %

Algorithm	Errors	Hidden capacity, bpp					
		0.5	0.25	0.167	0.125	0.1	0.05
Videos from Set 1							
SAA1	FN	0	0	2.19	11.68	51.09	90.88
	FP	0.36	0.36	0.36	0.36	0.73	6.14
SAA2	FN	0	0	0	64.23	97.08	100
	FP	0	0	0	0	0	0
SAA3v	FN	0	0	0	0.73	2.19	32.12
	FP	0	0	0	0	0	0
Videos from Set 2							
SAA1	FN	0	0	19.56	98.26	100	100
	FP	0	0	0	0	0	0
SAA2	FN	0	0	0	80	100	100
	FP	0	0	0	0	0	0
SAA3v	FN	0	0	0	0	0	10
	FP	0	0	0	0	0	0
Videos from Set 3							
SAA1	FN	0	0	0	0	0	0
	FP	94.90	94.90	94.90	91.84	91.84	91.84
SAA2	FN	0	0	0	6.12	40.82	90.82
	FP	5.10	5.10	5.10	5.10	5.10	5.10
SAA3v	FN	79.59	83.67	85.71	95.92	93.88	85.71
	FP	1.02	7.14	5.10	13.27	12.24	6.12

three) color components of digital videos received by cameras of mobile devices and with hidden capacity not less than 0.167 bpp in case of other videos;

- SAA2 effectively detects stego, formed by embedding of additional information with hidden capacity not less than 0.167 bpp under other similar conditions that is worse than SAA1, however SAA2 is single from the developed algorithms that allows detect video with a small frame size effectively;
- SAA3 and SAA3v can be used only for color digital contents and are effective when additional information embeds only into one color component including with small values of hidden capacity.

Comparison of the algorithms SAA1, SAA2, SAA3v among themselves according to various criteria are given in Table 2.

Comparison of the developed steganalytic algorithms SAA1, SAA2 and SAA3v with analogues V1, V2, V3 [6], V4 [7] and V5 [9] is shown in Fig. 1 where detection accuracy characterizing the proportion of correctly identified videos is determined by the formula

$$AD = \frac{TP + TN}{TP + FN + TN + FP}, \qquad (2)$$

where *TP* is the quantity of correctly identified stego, *TN* is the quantity of correctly identified unfilled containers.

Table 2. Comparative analysis of algorithms SAA1, SAA2, SAA3v

Comparison criterion		SAA1	SAA2	SAA3v
Color representation of digital videos		Color, grayscale		Color
Container format of digital videos		Losses, lossless		Losses
Frame size limits		Ineffective at a small frame size	no	Ineffective at a small frame size
Degree of filling frames of digital video		Full		Partial/full
The number of color components of color digital videos used for embedding of additional information		One, two or three color components		One color component only
Category of digital contents with better detection efficiency		Digital videos from set 1 (received by cameras of mobile devices)	Digital videos from set 3 (with a small frame size)	Digital videos from set 2
High detection efficiency of additional information, embedded with hidden capacity (bpp):	0.5	yes	yes	yes
	0.25	yes	yes	yes
	0.167	yes	yes	yes
	0.125	no/yes[1]	no/yes[2]	yes
	0.1	no	no	yes
	0.05	no	no	yes

[1]For digital videos, received by cameras of mobile devices
[2]For digital videos with a small frame size

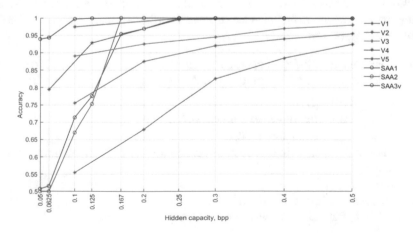

Fig. 1. Comparison of detection accuracy of the developed steganalytic algorithms SAA1, SAA2 and SAA3v with modern analogues

4   Practical Recommendations for Application of the Proposed Steganalytic Algorithms

Based on the identified advantages and disadvantage we formulate practical recommendations for the using of the developed algorithms:

1. If frame size of digital video (color or grayscale) is small (less than 320×240), it is necessary to use SAA2 algorithm.
2. If there are grayscale digital videos, it is recommended to use algorithms SAA1 and SAA2.
3. For primary analysis of color video of average and big frame size it is necessary to use the algorithms SAA1 and SAA2 allowing to detect the presence of additional information in one, two and three color components.
4. If losses format was determined as a format of an original container for color digital video and as a result of primary analysis the presence of additional information was defined only in one color component, or was not defined at all, it is necessary to apply the algorithm SAA3v to check of the result received at the previous stage.
5. If it is supposed that additional information was embedded into not all frames but only a part of color video frames, it is necessary to carry out the frame analysis of digital videos by SAA3 algorithm.
6. If it is supposed that the color digital image for which the original format was determined as losses format contains attachments only in one color component, it is necessary to use SAA3 algorithm.

5   Complex Steganalytic Method for Detection of Additional Information Attachments in Digital Videos

Taking into account the comparison of developed earlier by the author algorithms and restrictions for their using, the complex steganalytic method for detection of additional information attachments in digital videos that realizes the two-stage analysis of the video sequences was developed:

Step 1. Determination the quantity of frames, frame size of digital videos (M, N), calculation of total quantity of 4×4-blocks in a frame by the formula $cb = \lceil M/4 \rceil \cdot \lceil N/4 \rceil$.

Step 2. (*Detecting of digital videos with a small frame size*)
If $cb < T_{cb}$, where $T_{cb} = 4800$,
then analysis of digital video performs by algorithm SAA2 and transition to a step 6,
else step 3.

Step 3. (*Detecting of grayscale digital videos*).
If the digital video has only one color component, *then*:

    3.1. Definition of a format of an original container by MIFOCB [14] algorithm.

3.2. The choice of threshold values T_R, T_G, T_B, depending on a certain format of a container.

3.3. Detecting the presence of additional information in digital video by SAA1 and SAA2 algorithms.

3.4. Making decision about presence/absence of additional information in a digital video (on condition of opposite results of detecting, the most effective algorithm for a probable source of video has advantage - for the digital videos received by cameras of mobile devices SAA1 is preferable), transition to a step 6.

Else step 4.

Step 4. (*Detecting of color digital videos*).

Definition of a format of an original container by MIFOCA [15] algorithm.

The choice of threshold values T_R, T_G, T_B, depending on a certain format of a container.

Detecting the presence of additional information in digital video by SAA1 algorithm.

If additional information was revealed in two or three color components of a digital video or the lossless format of a container was defined in a step 4.1, *then* additional check by SAA2 algorithm carries out,

else additional check by an algorithm SAA3v carries out.

4.5. Making decision about presence/absence of additional information in a digital video (on condition of comparable results received by SAA1 and SAA3v algorithms, the last has advantage; when results received by SAA1 and SAA2 algorithms compares, that algorithm has advantage, which is most effective for a probable source of a digital video), transition to a step 6.

Step 5. (optional). (*Frame analysis of digital videos*).

If in a step 4.1 the losses format was defined and in a step 4.3 the presence of additional information was revealed in one color component or was not revealed,

then for each frame of a digital video the analysis by SAA3 algorithm carries out.

Step 6. Output of the final result.

Algorithmic implementations of all algorithms developed by the author have polynomial computational complexity: for algorithms SAA1 and SAA2 computational complexity is defined as $\underline{O}(M^2)$, where $M \times M$ is the size of a matrix of the digital image/frame of a digital video, for algorithms SAA3, SAA3v - $\underline{O}(M^4)$, that gives the chance for their practical use for a steganalysis of digital images/digital videos of any, including, the considerable size.

6   Conclusions

In work the complex steganalytic method for detection of additional information attachments in digital videos embedded under different conditions, namely in case of different values of hidden capacity, when using as containers of color or grayscale digital videos irrespective of a format (with losses or without loss) and a size of frames of digital videos is developed that allowed to increase efficiency of detection stego, formed in case of small values of throughput (less than 0.2 bits/pixel).

References

1. Bohme, R.: Advanced Statistical Steganalysis. Springer, Heidelberg (2010)
2. Das, P., Ray, S., Das, A.: An efficient embedding technique in image steganography using lucas sequence. Int. J. Image Graph. Sig. Proces. (IJIGSP) 9(9), 51–58 (2017). https://doi.org/10.5815/ijigsp.2017.09.06
3. Joshi, K., Yadav, R.: A LL subband based digital watermarking in DWT. Int. J. Eng. Manuf. (IJEM) 7(2), 50–63 (2017). https://doi.org/10.5815/ijem.2017.02.05
4. Kaur, R., Kaur, S.: XOR-EDGE based video steganography and testing against chi-square steganalysis. Int. J. Image Graph. Sig. Proces. (IJIGSP) 8(9), 31–39 (2016). https://doi.org/10.5815/ijigsp.2016.09.05
5. Punithalatha, M., Amsaveni, A.: Reversion-based features video steganalysis for images and images fusion technique using fuzzy logic. In: International Conference on Innovations In Intelligent Instrumentation, Optimization and Signal Processing "ICIIIOSP-2013", pp. 15–22 (2013)
6. Pankajaksan, V., Ho, A.T.S.: Improving video steganalysis using temporal correlation. In: Third International Conference on Intelligent Information Hiding and Multimedia Signal Processing, vol. 1, pp. 287–290 (2007)
7. Li, S., Liu, P., Dai, Q., Ma, X., Deng, H.: Detection of information hiding by modulating intra prediction modes in H.264/AVC. In: Proceedings of the 2nd International Conference on Computer Science and Electronics Engineering (ICCSEE 2013), pp. 590–593 (2013)
8. Geetha, S., Sindhu, S., Kamaraj, N.: Close color pair signature ensemble adaptive threshold based steganalysis for LSB embedding in digital images. Trans. Data Priv. 1(3), 140–161 (2008)
9. Wu, K.: Research of video steganalysis algorithm based on H265 protocol. In: MATEC Web of Conferences, vol. 25, pp. 03003-1–03003-7 (2015)
10. Kobozeva, A.A., Akhmamet'ieva, A.V., Efimenko, A.A.: Steganalytic method for digital containers stored in a lossless format, that works in the spatial domain of the container. Inf. Secur. 1(13), 31–42 (2014). (in Russian)
11. Maevsky, D.A., Akhmametieva, A.V.: Steganalytic algorithm, based on the analysis of the spatial domain of digital containers. Inf. Mathe. Meth. Simul. 6(1), 52–60 (2016). (in Ukrainian)
12. Akhmametieva, A.V.: Steganalysis of digital images stored in losses formats, Protection of information, Special edition 23, pp. 135–145 (2016). (in Russian)
13. Demmel, J.: Computational Linear Algebra. Theory and Applications. Moscow (2001)
14. Bobok, I.I.: Identification of digital images that have been resaved into lossless format from a lossy format as part of a steganalysis. Mod. Spec. Tech. 3(34), 64–70 (2013). (in Russian)
15. Akhmametieva, A.: Method of detection the fact of compression in digital images as an integral part of steganalysis. Inf. Mathe. Meth. Simul. 6(4), 357–364 (2016)

Video Data Compression Methods in the Decision Support Systems

V. Barannik[1], O. Yudin[1,2], Y. Boiko[1,2(✉)], R. Ziubina[1,2],
and N. Vyshnevska[2]

[1] Kharkiv National Air Force University, Kharkiv, Ukraine
{yak333, julia_boiko2010}@ukr.net
[2] National Aviation University, Kyiv, Ukraine

Abstract. The paper presents the developed methods of video data compression and decompression providing the maximum degree of intersecting information flows of critical video information at the given quality levels of digital video images. Due to the developed methods, mathematical models and techniques, the technology of video data compression has been improved on the basis of reducing the structural redundancy under limited loss of visualization quality. The proposed technology provides an increased level of effective functioning of communication channels and critical video information processing, as well as presents an opportunity for information support and improved quality of decision-making in crisis situations.

Keywords: DSS-systems · Video data compression · Image compression
Structural redundancy · Psycho-visual redundancy · Adaptive coding
Discrete cosine transformation

1 Introduction

Decision support system is an integral part of information and telecommunication systems that focus on improving the efficiency of managerial decision-making in terms of operational management in crisis situations. In such conditions, the nature of data and information flow for the processing and analysis by the decision support systems is considerably changing. [1]. Video data support for decision support systems provides more complete and clear information to the decision maker in a form convenient for analysis and optimal decision making, which is highly important in crisis situations [2]. Organization of video information servicing for decision support systems is significantly complicated by the fact that in modern conditions the growth rate of video data volumes is several orders of magnitude higher than the rate of data transmission speed in telecommunication systems [3].

For efficient transfer, processing and display of critical video data in DSS systems, the following class of problems should be solved: reducing the processing time and increasing the overall speed of video transmission with modern communication channels [4].

The solution to this problem should be implemented on the basis of reducing the total amounts of data based on new compression methods. However, the theoretical and

© Springer International Publishing AG, part of Springer Nature 2019
Z. Hu et al. (Eds.): ICCSEEA 2018, AISC 754, pp. 301–308, 2019.
https://doi.org/10.1007/978-3-319-91008-6_30

practical characteristics of existing methods and technologies do not ensure the processing and timely delivery of large volumes of video information with an adjustable quality level at required or specified time intervals. Depending on the available data transfer speed, the compression ratio should be increased on average by 10–50%. This means that increasing the efficiency of the delivery process of high resolution video information by improving the compression techniques in the framework of compression methods with adjustable loss of image visualization quality is considered as an actual scientific and applied task [6].

In this regard, the objective of research is to develop the video data compression methods at the required level of visualization.

2 Theoretical Research

In modern conditions, the character of data and information flows in crisis situations varies considerably. The requirements for rational decision making on the basis of high-quality video data are growing [7].

Both wired and wireless communication channels are used in order to provide interconnection at different levels. In the case of transmission of video data streams consisting of high-resolution images, the delivery time increases by several orders, since the existing data rates are at least ten times lower than the required speeds for video traffic. Therefore, it is recommended to decrease the volumes of images, which are transferred to DSS-systems, using the advanced compression technology [8].

While developing video data compression methods to reduce their volume in conditions of limited quality imaging impairment, it is necessary to exclude uncontrollable decrease in the quality of restored images [9]. This decrease is caused by both processes of reducing psycho-visual redundancy and insufficient selection of service data to characterize structural components at the stages of data transformation. It should be noted that the time of data analysis ought to be reduced during the processes of direct and inverse transformation at the phases of preliminary data analysis and coding which are performed to reduce spatial redundancy.

It is appropriate to reduce psycho-visual redundancy at the initial stage of video data compression [10]. This stage provides a converted form of images taking into account the following features of human visual perception: high sensitivity to the characteristics of the image brightness, low sensitivity to colour components, extremely high sensitivity to objects of low frequency nature, and extremely low sensitivity to distortion of objects operating in the high frequency spectral range [11].

Psycho-visual redundancy has two features, in particular transformation of triple-coloured images into the signal used in video systems to convey the colour information of the picture, and two-dimensional discrete cosine transformation (DCT) [12]. As a result, three components are formed. One of them is brightness which is considered to have the greatest significance for visualization, while the other two are additional monochrome components. The initial RGB image is transformed into a YUV colour space. This transformation results in the appearance of three components with identical sizes in vertical and horizontal areas.

Owing to the divisibility property of the basic function core, the discrete cosine transformation (DCT) is carried out in two phases. At phase 1, one-dimensional DCT is conducted for the matrix columns of obtained images. As a result, an array of one-dimensional DCT transforms is formed.

DCT allows to achieve the re-transformation of output signal energy and obtain areas which are filled with compressed data on transformed images depending on the quality of visualization. This fact makes it possible to eliminate psycho-visual redundancy [10].

The proposed data compression method consists of six phases:

Phase 1. Transforming RGB image into YUV colour space.

Phase 2. Discrete cosine transformation (DCT) is performed in two stages. At first, one-dimensional DCT column array of source image is executed resulting in one-dimensional array $Z'(\xi, \chi)$ of DCT transform $Z'(\xi, \chi) = F(\xi) X(i,j)_{\xi,\chi}$, where $X(i,j)_{\xi,\chi}$ - an array of video output image; ξ, χ an index row and column array element $X(i,j)_{\xi,\chi}$, $\xi = \overline{1, m}$; $\chi = \overline{1, n}$.

$$Z'(\xi, \chi) = \sqrt{\frac{2}{m}} \sum_{\xi-1}^{m} X_{\xi,\chi}^{i,j} \cos \frac{(2m+1)\xi\pi}{2m} \quad \xi = \overline{2, m} \tag{1}$$

The second stage is performed for DCT-dimensional array of rows $Z'(\xi, \chi)$. resulting in a transform $Z''(\xi, \chi)$ of two-dimensional DCT, that is $Z''(\xi, \chi) = Z'(\xi, \chi) F(\chi)^{(-1)}$, where $F(\chi)^{(-1)}$ - a transposed vector of discrete values of DCT basic functions.

$$A_v = \begin{cases} A_{v,k} = \{v_{1,k}, \cdots, v_{S,k}\}, \to k \leq K - 1 \\ A_{v,k} = \{v_{1,K}, \cdots, v_{S_K,K}\} \to k \leq K \end{cases} \tag{2}$$

Phase 5. Adaptive coding of positional numbers formed for the columns of segments of binary elements. These columns are described as follows: $A_v^{(k)} = \{v_{s + (k-1)S,k}\}$

$$C(p)_k = \sum_{s-1}^{s'} v_{s + (k-1)S,k} p^{S'-s}, \text{ where } p = \max_{1 \leq s \leq S}\{p_s\} \tag{3}$$

Adaptive coding ensures the simultaneous reduction of additional data and exclusion of uncontrollable information loss at the transformation stages of the binary form, as well as the formation of a compressed transform expressed with the help of code words. Besides, the size of a code word can be determined only once for the first column of length arrays of binary element segments after calculating the positional number code, using only one multiplying operation.

Phase 6. Constructing a compressed transform pattern based on the sequence of code words containing data on the code of positional numbers.

To ensure the specified quality of image visualization obtained as the outcome of decompression, it is necessary to exclude the following:

Uncontrollable information loss as a result of both inaccurate determination of the endings of code words containing the values of positional number codes, and inaccurate positioning of patterns of adjacent transforms;

Errors in calculating of binary element series, which contain the main data on transform components, including low frequency components of the spectral form.

The developed decompression method consists of the following phases:

Phase 1. Defining the characteristics of compressed form patterns of data series. This phase includes the process of coding renewable transforms in compressed images and constructing the series of code words to calculate codes of adaptive one-based numbers.

Phase 2. Restoring array elements $A_v = \{A_v^{(k)}\}$ and the series of binary elements by performing the adaptive coding.

Phase 3. Forming absolute values of DCT transform components by using the polynomial value.

$$Z_{\xi,\chi} = w(d-1)_{\xi,\chi}2^{d-1} + \cdots + w(0)_{\xi,\chi} \tag{4}$$

Phase 4. Two-dimensional inverse DCT. Taking into consideration that $Z'''(\xi,\chi) \neq Z''(\xi,\chi)$, the restoration of values $X'_{i,j}$ can differ from the obtained value $X_{i,j}$ of video data series $X(i,j)$. This leads us to conclusion that $X'_{i,j} \neq X_{i,j}$.

Phase 5. Obtaining a fragment of RGB image. Based on the developed methods of adaptive coding and decoding, it is possible to improve the compression of ITS (information telecommunication systems) digital image, which helps to reduce redundancy under the limited decrease in the quality of visualization, as well as reduce delays in data processing (Fig. 1).

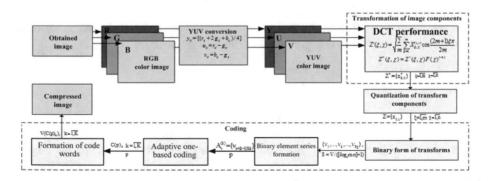

Fig. 1. The improved technique of image compression

3 Experimental Research

The method of video data compression can be assessed when reducing data volumes taking into account that the exclusion of uncontrolled loss of quality is ensured without using data on the vector of binary element series in the columns.

Average volume $\overline{V}_{ci}$ of video data having the compressed form depends on loss of quality. The level of image visualization under limited quality loss corresponds to the ratio 'signal/noise' and equals $\sigma = 50$ dB. The normal level of image visualization equals $\sigma = 30$ dB. The largest reduction in the volume of video data is achieved with increasing loss of quality [15]. The average volume $\overline{V}_{ca}$ of video data for one series is equal to $\overline{V}_{ca} = \overline{V}_{mz} + \overline{V}_{mo}$, consequently the average volume $\overline{V}_{\Sigma}$ of compressed video data for one series is equal to

$$\overline{V}_{\Sigma} = \frac{mn(([S_{\Sigma}(\log_2 mn)] + 1) + 1)V_c/V_{mc} + \overline{V}_{mz} + \overline{V}_{mo}}{MN} \tag{5}$$

where $\overline{V}_{mz}$ - the average volume of marks matrix; $\overline{V}_{mo}$ - the average volume of adaptive bases; V_c - the codeword length; V_{mc} - the array length of the column; S_{Σ} - the number of long series of binary elements in bit description transforms; $M \times N$ - the image size; $m \times n$ - the size of array.

Figure 3 shows the comparative assessment regarding average quality losses in compressed video data for imbalance positional coding (IPC) and adaptive coding (AC) [12].

The level of quality losses corresponds to the maximum signal/noise ratio $\sigma = 27$ dB, and processed images are selected in three types, depending on their degree of correlation. When selecting the images, only the errors made at the stage of quantization of transform components are taken into account.

Upon analyzing the diagrams in Figs. 3, the following conclusions can be made:

- If limited quality loss takes place under $\sigma = 27$ dB, the average loss of bits for one series varies from 20 to 50 bits, depending on the degree of image coherence;
- Adaptive coding allows to decrease average volumes of compressed video data by 12% for low-correlated images and by 7% for highly correlated images.

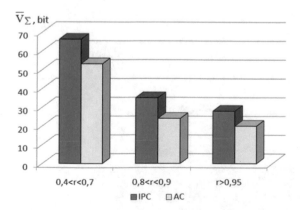

Fig. 2. The value of volume $\overline{V}_{\Sigma}$ for imbalance positional coding (IPC) and adaptive coding (AC) with the maximum signal/noise ratio $\sigma = 27$ dB, depending on the degree of image correlation

According to the developed method, for specified volume $\overline{V}_\Sigma$, compression degree η is calculated based on the following ratio:

$$\eta = \frac{dMN}{([S_\Sigma([\log_2 mn] + 1)] + 1)V_c/V_{mc} + \overline{V}_{mz} + \overline{V}_{mo}} \qquad (6)$$

where d – the depth of sampling picture element.

Taking into consideration the data in Fig. 2, obtained for the average volume of compressed images, a compressed fragment can be estimated using compression coefficient η, while the parameters of block coding [13] are as follows: $m \times n = 64$ and $d = 8$ bits. Then, depending on the degree of correlation, the average compression coefficient for "signal/noise ratio" equals as follows: for low-correlated images - 12 times; for medium-correlated images - 28 times; for highly correlated images - 35 times.

Figures 3 and 4 show the comparative analysis regarding the level of image compression carried out in accordance with the developed method which allows to apply different techniques of coding positional numbers obtained for the columns of binary elements and JPEG images [14].

Upon analyzing the diagrams in Figs. 3 and 4, the following conclusions can be made:

(1) Using the proposed method, under reduced loss of quality in a renewable image that corresponds to 'signal/noise ratio' 50 dB, it is possible to achieve a decrease in the compressed volume of video data as follows: 15% for low and medium correlated images; 20% for highly correlated images;

(2) In comparison with the methods applied for JPEG images, the proposed method helps to achieve a decrease in the volume of compressed images under loss of quality that corresponds to 'signal/noise ratio' 27 db, when low-correlated images are processed for 20%, and medium-correlated images for 18% respectively. At the same time, the compression of highly correlated images can be carried out according to both the proposed method and methods applied for JPEG images.

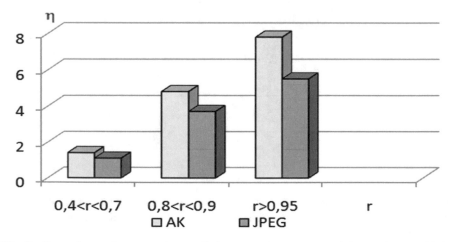

Fig. 3. Dependence of compression coefficient η on the image correlation under $\sigma = 27$ dB

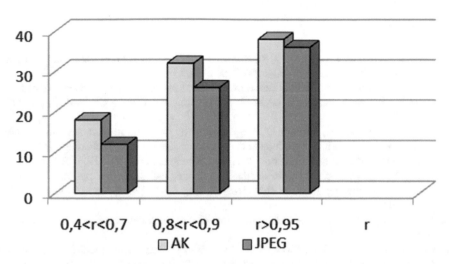

Fig. 4. Dependence of compression coefficient η on the image correlation under $\sigma = 50$ dB

4 Conclusions

The method of video data compression includes the previous quantization of DCT transform components and implementation of adaptive coding. This method allows researchers to increase the level of compression; to exclude an uncontrolled decrease in the quality of renewable images caused by both the elimination of psycho-visual redundancy and insufficient selection of data to describe the structural components at the stages of data conversion; to reduce the processing time at the pre-selection and coding stages to decrease redundancy.

In order to provide a specified quality of image visualization derived as a result of image restoration, the method of image decompression has been proposed. This method is intended to restore images with the limited decrease in the quality of visualization and to exclude uncontrolled loss of data.

The developed methods provide improvement of adaptive coding and decoding, compression of digital video images, which, in turn, allows to reduce compressed volumes of data as follows: 15% for low-correlated and medium-correlated images; 20% for highly correlated images at a signal/noise ratio equal to 50 dB, as well as minimize delays in data processing.

The proposed method enables to increase the efficiency of communication channels and video data processing performed by DSS-systems, and provides information support and high quality of decision making in crisis situations.

References

1. Rees, L.P., Deane, J.K., Rakes, T.R., Baker, W.H.: Decision support for cybersecurity risk planning. Decis. Support Syst. **51**(3), 493–505 (2011)
2. Lai, C.L., Chen, Y.S.: The application of intelligent system to digital image forensics. In: 2009 International Conference on Machine Learning and Cybernetics, vol. 5, pp. 2991–2998. IEEE (2009)
3. Yudin, O., Boiko, Y., Frolov, O.: Organization of decision support systems for crisis management. In: Problems of Infocommunications Science and Technology (PIC S&T), 2015 Second International Scientific-Practical Conference, pp. 115–117. IEEE (2015)
4. Tank, D.M.: Enable better and timelier decision-making using real-time business intelligence system. Int. J. Inf. Eng. Electron. Bus. (IJIEEB) **7**(1), 43–48 (2015). https://doi.org/10.5815/ijieeb.2015.01.06
5. Kumari, N., Agarwal, A.: A review on wireless data center management. Int. J. Comput. Appl. **138**(13) (2016)
6. Cuzzocrea, A., Song, I.Y., Davis, K.C.: Analytics over large-scale multidimensional data: the big data revolution! In: Proceedings of the ACM 14th International Workshop on Data Warehousing and OLAP, pp. 101–104. ACM (2011)
7. Shang, J., Ding, W., Shi, Y., Sun, Y.: Fast intra mode decision algorithm based on texture direction detection for H.264/AVC. Int. J. Educ. Manage. Eng. (IJEME) **1**(5), 70–77 (2011). https://doi.org/10.5815/ijeme.2011.05.12
8. Lakhno, V., Boiko, Y., Akhmetov, B., Mishchenko, A.: Designing a decision support system for the weakly formalized problems in the provision of cybersecurity. Eastern Eur. J. Enterpr. Technol. **1**(2(85)), 4–15 (2017). https://doi.org/10.15587/1729-4061.2017.90506
9. Garae, J., Ko, R.K.L.: Visualization and data provenance trends in decision support for cybersecurity. In: Palomares Carrascosa, I., Kalutarage, H., Huang, Y. (eds.) Data Analytics and Decision Support for Cybersecurity, Data Analytics, pp. 243–270. Springer, Cham (2017). https://doi.org/10.1007/978-3-319-59439-2_9
10. Ziubina, R., Boiko, Yu.: Video data compression methods in the aviation crisis management. Inżynier XXI wieku: VI Międzynarodowa Konferencja Studentów oraz Doktorantów. 73 (2016)
11. Jaiswal, S., Dhavale, S.: Video forensics in temporal domain using machine learning techniques. Int. J. Comput. Netw. Inf. Secur. (IJCNIS) **5**(9), 58–67 (2013). https://doi.org/10.5815/ijcnis.2013.09.08
12. Kaushik, M.: Comparative analysis of exhaustive search algorithm with ARPS algorithm for motion estimation. Int. J. Appl. Inf. Syst. (IJAIS) **1**(6), 16–19 (2012)
13. Nayak, A., Biswal, B., Sabut, S.K.: Evaluation and comparison of motion estimation algorithms for video compression. Int. J. Image Graph. Signal Process. (IJIGSP) **5**(10), 9–18 (2013). https://doi.org/10.5815/ijigsp.2013.10.02
14. Jeengar, V., Omkar, S.N., Singh, A., Yadav, M.K., Keshri, S.: A review comparison of wavelet and cosine image transforms. IJIGSP **4**(11), 16–25 (2012). https://doi.org/10.5815/ijigsp.2012.11.03
15. Wang, W., Farid, H.: Exposing digital forgeries in video by detecting double MPEG compression. In: Proceedings of the 8th Workshop on Multimedia and Security, pp. 3–47. ACM (2006)

Method of Searching Birationally Equivalent Edwards Curves Over Binary Fields

Zhengbing Hu[1], Sergiy Gnatyuk[2], Maria Kovtun[2(✉)],
and Nurgul Seilova[3]

[1] Central China Normal University, Wuhan, China
hzb@mail.ccnu.edu.cn
[2] National Aviation University, Kyiv, Ukraine
s.gnatyuk@nau.edu.ua, mg.kovtun@gmail.com
[3] Satbayev University, Almaty, Republic of Kazakhstan
seilova_na@mail.ru

Abstract. This paper is devoted to increasing of performance of digital signature algorithms based on elliptic curves over binary fields. Group operations complexity on Edwards curves are less than on Weierstrass curves and have immunity to some side channel attacks. Hence, it is interesting to search birationally equivalent curves in Edwards form for curves in Weierstrass form presented in NIST recommended curves list. It allows using operations over points on Edwards curve in intermediate computations in scalar multiplications over curves in Weierstrass form. This approach improves the performance and security of digital signature.

Keywords: Cryptography · Random binary elliptic curves · Weierstrass curves
Edwards curves · Digital signature · Cube root

1 Introduction

Today elliptic curves (EC) successfully used in modern cryptosystems, for example digital signature (DS), shared secret generation protocols. There are international standard of DS, which describes algorithms ECDSA, ECKCDSA, ECGDSA according to [1, 2]. Performance is one of main requirement for modern implementation of DS algorithm. Especially, performance is important in information systems with intensive exchange of large amount of e-documents, for example in critical information infrastructure [16, 17]. The problem of most widely used DS algorithm ECDSA is not enough performance and weakness to some attacks [3]. The most complex operation in DS is scalar multiplication of EC points on a large integer [1]. All EC cryptosystems, such ECDSA, have recommended common parameters, which includes EC parameters.

Edwards curves are new normal form for elliptic curve which has some cryptographically desirable properties and advantages over the Weierstrass form. The group law on Edwards curve is complete and unified; implementation has resistance to some attacks. Hence, Edwards's curves provide a better platform for cryptographic primitives. Moreover, in many cases the group law involves less operations, meaning that the more secure computations can also be faster [5, 7]. For increasing performance of existing cryptosystems based on EC, it is proposed to use Edwards curves in intermediate computations of scalar multiplication. This approach is applied not only for ECDSA, but also for other DS algorithms [5–11]. Authors of [10] presented an algorithm of birationally equivalent mapping Weierstrass curves to Edwards curve and vice versa. Hence, the actual scientific and technical problem is a search of birationally equivalent Edwards curves for Weierstrass curves over binary fields from NIST recommended list [14]. Main goal of this paper is present of algorithm for search of birationally equivalent Edwards's curves to Weierstrass curves from NIST recommended list [14].

2   Mathematical Background

An existent and standardized cryptosystems uses random binary EC: B-163, B-233, B-283, B-409, B-571 in Weierstrass form recommended by NIST [14] in polynomial basis. Canonical elliptic curve over $\mathbf{GF}(2^m)$ in Weierstrass form [8]:

$$v^2 + uv = u^3 + au^2 + b, \tag{1}$$

where a and b from $\mathbf{GF}(2^m)$. Corresponding birationally equivalent Edwards curve in form [8]:

$$d_1(x+y) + d_2(x^2 + y^2) = xy + xy(x+y) + x^2 y^2, \tag{2}$$

where d_1 and d_2 satisfy conditions $d_1 \neq 0$ and $d_2 \neq d_1^2 + d_1$.

Curve (2) is birationally equivalent to curve (1) as

$$v^2 + uv = u^3 + \left(d_1^2 + d_2\right) u^2 + d_1^4 \left(d_1^4 + d_1^2 + d_2^2\right), \tag{3}$$

where condition $d_1^4 \left(d_1^4 + d_1^2 + d_2^2\right) \neq 0$ shows a non-singularity of curve.

Birationally equivalent mapping $(u, v) \mapsto (x, y)$ point on Weierstrass curve to point on Edwards curve defined as [7, 8]:

$$x = \frac{d_1 \left(u + d_1^2 + d_1 + d_2\right)}{u + v + \left(d_1^2 + d_1\right)\left(d_1^2 + d_1 + d_2\right)}, y = \frac{d_1 \left(u + d_1^2 + d_1 + d_2\right)}{v + \left(d_1^2 + d_1\right)\left(d_1^2 + d_1 + d_2\right)}.$$

An inverse map $(x, y) \mapsto (u, v)$ as [9]:

$$u = d_1 \left(d_1^2 + d_1 + d_2\right) \frac{x+y}{xy + d_1(x+y)}, v = d_1 \left(d_1^2 + d_1 + d_2\right) \left(\frac{x}{xy + d_1(x+y)} + d_1 + 1\right).$$

According to (3), if ordinary EC over $\mathbf{GF}(2^m)$ has an equivalent complete Edwards curve, then exists correspondence:

$$\begin{cases} a = d_1^2 + d_2 \\ b = d_1^4 \left(d_1^4 + d_1^2 + d_2^2\right) \end{cases}$$

Definition. Let k be a field with $\mathrm{char}(k) = 2$. Let d_1, d_2 be elements of $\mathbf{GF}(2^m)$ with $d_1 \neq 0$. Assume that there is no element $t \in \mathbf{GF}(2^m)$ satisfies $t^2 + t + d_2 = 0$. The complete binary Edwards curve with coefficients d_1 and d_2 is the affine curve (2).

The complete case has the extra requirement that $t^2 + t + d_2 = 0$ for all $t \in \mathbf{GF}(2^m)$, not just for $t = d_1$. An equivalent requirement for $t \in \mathbf{GF}(2^m)$ is that $\mathrm{Tr}(d_2) = 1$ [8].

Theorem 1. Let n be an integer with $n \geq 3$. Each ordinary elliptic curve over $\mathbf{GF}(2^m)$ is birationally equivalent over $\mathbf{GF}(2^m)$ to a complete binary Edwards curve [8].

The authors [10] proposed and demonstrated the following statement by analogy with Theorem 1.

Theorem 2. Over $\mathbf{GF}(2^m)$, an elliptic curve $W_{a,b}$ in Weierstrass form has the equivalent complete binary Edwards curve E_{d_1,d_2} with a and b in (3) if and only if $\mathrm{Tr}\left(a^6/b\right) = 0$ [10].

If (x_1, y_1) and (x_2, y_2) are two points on the binary Edwards curve E_{d_1,d_2}, then the mapping $(x_1, y_1) + (x_2, y_2) \rightarrow (x_3, y_3)$ turns points on this curve into Abelian group, where

$$x_3 = \frac{\left(d_1 \cdot (x_1 + x_2) + d_2 \cdot (x_1 + y_1) \cdot (x_2 + y_2) + \left(x_1 + x_1^2\right) \cdot \left(x_2 \cdot (y_1 + y_2 + 1) + y_1 \cdot y_2\right)\right)}{\left(d_1 + \left(x_1 + x_1^2\right)(x_2 + y_2)\right)},$$

$$y_3 = \frac{\left(d_1 \cdot (y_1 + y_2) + d_2 \cdot (x_1 + y_1)(x_2 + y_2) + \left(y_1 + y_1^2\right)\left(y_2 \cdot (x_1 + x_2 + 1) + x_1 \cdot x_2\right)\right)}{\left(d_1 + \left(y_1 + y_1^2\right)(x_2 + y_2)\right)}.$$

After simplification [10]:

$$w_1 = x_1 + y_1, w_2 = x_2 + y_2, A = x_1^2 + x_1, B = y_1^2 + y_1, C = d_2 w_1 \cdot w_2, D = x_2 \cdot y_2,$$

$$x_3 = y_1 + (C + d_1 \cdot (w_1 + x_2) + A \cdot (D + x_2))/(d_1 + A \cdot w_2),$$

$$y_3 = x_1 + (C + d_1 \cdot (w_1 + y_2) + B \cdot (D + y_2))/(d_1 + B \cdot w_2).$$

Computational complexity of the affine addition: $2\mathbf{I} + 8\,\mathbf{M} + 2\mathbf{S} + 3\mathbf{D}$, where $\mathbf{I}$ is the cost of field inversion, $\mathbf{M}$ is the cost of field multiplication, $\mathbf{S}$ is the cost of a field squaring, and $\mathbf{D}$ is the cost of multiplication by a curve parameter. Affine point representation is not efficient and on practice uses projective point addition and mixed point addition.

Montgomery ladder is widely used technique for scalar multiplication with resistance to attacks on implementation. For efficient implementation of group low on elliptic curves uses Lopez-Dahab coordinates with addition formulas in projective representation [19]. In [7, 19] use formulas with $6\mathbf{M} + 5\mathbf{S}$ operations for a mixed addition and doubling. After several speedups it allows to decrease count of operations to $6\mathbf{M} + 1\mathbf{S}$ and mixed addition takes $4\mathbf{M} + 1\mathbf{S}$, doubling takes $1\mathbf{M} + 3\mathbf{S} + 1\mathbf{D}$. The mixed addition and doubling on each step of Montgomery scalar multiplication take $5\mathbf{M} + 5\mathbf{S} + 1\mathbf{D}$ [4, 7, 8].

Montgomery ladder is widely used technique for scalar multiplication with resistance to attacks on implementation. For efficient implementation of group low on Edward curve uses ladders for differential projective addition and doubling via explicit formulas with $8\mathbf{M} + 4\mathbf{S} + 4\mathbf{D}$. It is evident that the implementation of the group low on Weierstrass curve has slightly lower complexity than Edwards curve. In common, Edwards curve have simpler, unified and complete group low. It allows having large perspective in further curve-based cryptosystem implementation.

In existent cryptosystems with scalar multiplication algorithm of points on Weierstrass EC could represent with intermediate calculations on Edwards curve:

1. Elliptic curve and point mapping.
   1.1. Search the birationally equivalent Edwards curve.
   1.2. Compute mapping of base point P on Weierstrass curve to the point P' on Edwards curve.
2. Scalar multiplication of point on Edwards curve.
   2.1. Compute $Q' = k \cdot P'$ via Montgomery ladder with differential w-projective addition and doubling.
   2.2. Mapping point Q' on Edwards curve to point Q on Weierstrass curve.
3. Return Q.

This approach increases performance and security of scalar multiplication for binary EC by using w-coordinates in Montgomery ladder [7, 8].

3   Proposed Method

Papers [5–9] analysis shown the theoretical existence of several algorithms for searching equivalent Edwards curve to Weierstrass curve. These algorithms shown in [8, 10]. based on complex field operations: square root [8] and cube root [10].

The algorithm for searching birationally equivalent Edwards curves based on the cube root operations [10] shown below:

Algorithm 1. Searching birationally equivalent Edwards curves by using cube root.

Input: $a, b \in \mathbf{GF}(2^m)$, $X_W = (u, v) \in \mathbf{GF}(2^m)$.

Output: $d_1, d_2 \in \mathbf{GF}(2^m)$, $X_E = (x, y) \in \mathbf{GF}(2^m)$.

1. $a_2 \leftarrow a$.

2. While $\left(\mathrm{Tr}\left(a_2^6 / b\right) \neq 0 \right)$.

2.1. Random (λ), $\lambda \in \mathbf{GF}(2^m)$.

2.2. $a_2 \leftarrow a_2 + \lambda^2 + \lambda$.

3. Solving quadratic equation, over r: $r^2 + r + a_2^6 / b = 0$.

4. $d_1 \leftarrow \sqrt[3]{\sqrt{b} \cdot r} + \sqrt[3]{\sqrt{b} \cdot (r+1)} + a_2$, $\quad d_2 \leftarrow d_1^2 + a_2$.

Calculating the birationally equivalent point on binary Edwards curves.

5. $x' \leftarrow x$, $y' \leftarrow y + \lambda x'$.

6. $u \leftarrow d_1 \left(d_1^2 + d_1 + d_2\right) \dfrac{x' + y'}{x'y' + d_1(x' + y')}$,

$v \leftarrow d_1 \left(d_1^2 + d_1 + d_2\right) \left(\dfrac{x'}{x'y' + d_1(x' + y')} + d_1 + 1 \right)$.

Calculating the birationally equivalent point on ordinary curves.

7. $x' \leftarrow \dfrac{d_1 \left(u + d_1^2 + d_1 + d_2\right)}{u + v + \left(d_1^2 + d_1\right)\left(d_1^2 + d_1 + d_2\right)}$, $\quad y' \leftarrow \dfrac{d_1 \left(u + d_1^2 + d_1 + d_2\right)}{v + \left(d_1^2 + d_1\right)\left(d_1^2 + d_1 + d_2\right)}$.

8. $x \leftarrow x'$, $y \leftarrow y' + \lambda x$.

Algorithm 1 [9] has several drawbacks, which should resolved in proposed modified algorithm:

- Decreased operations count by pre-computations.
- Increased performance exponentiation in $\mathbf{GF}(2^m)$ in cube root extraction by using chain for representation of special kind of exponent.
- Using one inverse for Edwards point [6].

Algorithm 2. (Modified algorithm) Searching birationally equivalent Edwards curves by using cube root.

Input: $a, b \in \mathbf{GF}(2^m)$, $X_W = (u, v) \in \mathbf{GF}(2^m)$.

Output: $d_1, d_2 \in \mathbf{GF}(2^m)$, $X_E = (x, y) \in \mathbf{GF}(2^m)$.

Precomputations.

1. $SR_b \leftarrow \sqrt{b}$, $\gamma \leftarrow 1/b$.

Main algorithm.

2. $a_2 \leftarrow a$, $\lambda \leftarrow 0$.

3. $\left(\mathrm{Tr}(a_2^6 \cdot \gamma) = 0\right)$ then

3.1. Random (λ), $\lambda \in \mathbf{GF}(2^m)$.

3.2. $a_2 \leftarrow a_2 + \lambda^2 + \lambda$.

4.   While $\left(\mathrm{Tr}(a_2^6 \cdot \gamma) \neq 0\right)$
4.1. Random (λ), $\lambda \in \mathbf{GF}(2^m)$.

4.2. $a_2 \leftarrow a_2 + \lambda^2 + \lambda$.

4.3. Solving quadratic equation, over r: $r^2 + r + a_2^6 \cdot \gamma = 0$.

$g_1 \leftarrow SR_b \cdot r$, $g_2 \leftarrow SR_b \cdot (r+1)$, $SR_1 \leftarrow \sqrt[3]{g_1}$, $SR_2 \leftarrow \sqrt[3]{g_2}$, $d_1 \leftarrow SR_1 + SR_2 + a_2$, $\delta \leftarrow d_1^2$, $d_2 \leftarrow \delta + a_2$.

5.   Solving quadratic equation, over g: $g^2 + g + \left(d_1^2 + d_2 + a\right) = 0$.

Calculating birationally equivalent Edwards curve point if defined point on Weierstrass curve.

$C_0 \leftarrow a_2 + d_1$, $C_1 \leftarrow C_0 + d_2$, $C_2 \leftarrow C_1 \cdot C_0$, $C_3 \leftarrow \delta \cdot C_0$, $C_4 \leftarrow g \cdot u$, $C_5 \leftarrow d_1 \cdot u$, $C_6 \leftarrow u^2$, $z \leftarrow \frac{1}{(C_6 + C_5 + C_3)}$, $C_7 \leftarrow C_4 + v + C_2$, $x \leftarrow z \cdot d_1 \cdot C_7$, $y \leftarrow z \cdot d_1 \cdot (C_7 + u)$.

Calculating birationally equivalent point on Weierstrass curve, if point on Edwards curve have found already to check the condition of birational equivalence.

6.
$$u \leftarrow d_1 \left(d_1^2 + d_1 + d_2\right) \frac{x+y}{xy + d_1 (x+y)}, \quad v \leftarrow d_1 \left(d_1^2 + d_1 + d_2\right) \left(\frac{(g+1) \cdot x + g \cdot y}{xy + d_1 (x+y)} + d_1 + 1\right).$$

7.   Return $(d_1, d_2, (x, y))$.

In algorithm 1 cube root exponent operation may be represented as rational exponent $1/3 = \sum\limits_{j=0}^{(m-1)/2} 2^{2 \cdot j} \bmod (2^m - 1)$ [15]. In this formula, authors propose to use decomposition of exponent in chain with fewer multiplications [13, 16]. Example of steps of exponent decomposition in chain with reduced number of multiplication for $\mathbf{GF}(2^{163})$ is shown in Table 1.

Table 1. Decomposition in $\mathbf{GF}(2^{163})$

#	Simplified ordered set	Parameters	Restriction
1	$1 + 2^2 + 2^4 + \ldots + 2^{162}$	$k = 83, \ n = 2$	$k - 1 \equiv 0 \bmod 2$
2	$(1 + 2^2)(1 + 2^4 + \ldots + 2^{(83-3) \cdot 2})$	$k = 42, \ n = 4$	$k - 1 \equiv 1 \bmod 2$
3	$\ldots (1 + 2^4(1 + 2^4)(1 + 2^8 + \ldots + 2^{(42-4) \cdot 4}))$	$k = 21, \ n = 8$	$k - 1 \equiv 0 \bmod 2$
4	$\ldots (1 + 2^8)(1 + 2^{16} + \ldots + 2^{(21-3) \cdot 8})$	$k = 11, \ n = 16$	$k - 1 \equiv 0 \bmod 2$
5	$\ldots (1 + 2^{16})(1 + 2^{32} + \ldots 2^{(11-3) \cdot 16})$	$k = 6, \ n = 32$	$k - 1 \equiv 1 \bmod 2$
6	$\ldots (1 + 2^{32}(1 + 2^{32})(1 + 2^{64}))$		

Thus, rational exponent $\frac{1}{3}$ should represent as

$$(1 + 2^2)(1 + 2^4(1 + 2^4)(1 + 2^8)(1 + 2^{16})(1 + 2^{32}(1 + 2^{32})(1 + 2^{64}))).$$

The process of extracting cube root via Algorithm 1 by using binary exponentiation algorithm [10, 15] for $\mathbf{GF}(2^{163})$ require $((m-1)/2) \mathbf{M} = 81 \mathbf{M}$ and $(m-1) \mathbf{S} = 162 \mathbf{S}$. Author's proposed Algorithm 2 takes only $8 \mathbf{M}$ and $162 \mathbf{S}$. Sample representation of exponentiation in $\mathbf{GF}(2^{163})$ via chain approach available below.

Algorithm 3. Extracting cube root in $\mathbf{GF}(2^{163})$ using decomposition.

Input: $c \in \mathbf{GF}(2^{163})$.

Output: $\sqrt[3]{c} \in \mathbf{GF}(2^{163})$.

1. $t_0 \leftarrow c$, $t_0 \leftarrow (t_0^2)^2 \cdot t_0$, $t_1 \leftarrow t_0$, $t_0 \leftarrow (t_0^2)^4$, $t_0 \leftarrow (t_0^2)^4 \cdot t_0$, $t_0 \leftarrow (t_0^2)^8 \cdot t_0$.

2. $t_0 \leftarrow (t_0^2)^{16} \cdot t_0$, $t_2 \leftarrow t_0$, $t_0 \leftarrow (t_0^2)^{32}$, $t_0 \leftarrow (t_0^2)^{32} \cdot t_0$, $t_0 \leftarrow (t_0^2)^{64} \cdot t_0 \cdot t_2 \cdot t_1$.

3. Return t_0.

Chain's decompositions according method described in [13] of exponents for cube root extraction for NIST recommended binary fields is shown in Table 2.

Table 2. Chain's decomposition of exponents for cube root extraction

Field	Decomposition of chain
$GF(2^{163})$	$(1+2^2)(1+2^4(1+2^4)(1+2^8)(1+2^{16})(1+2^{32}(1+2^{32})(1+2^{64})))$
$GF(2^{233})$	$1+2^2(1+2^2)(1+2^4)(1+2^8(1+2^8)(1+2^{16})(1+2^{32}(1+2^{32})(1+2^{64})(1+2^{128})))$
$GF(2^{283})$	$(1+2^2)(1+2^4(1+2^4)(1+2^8(1+2^8)(1+2^{16}(1+2^{16})(1+2^{32})(1+2^{64})(1+2^{128}))))$
$GF(2^{409})$	$1+2^2(1+2^2)(1+2^4)(1+2^8(1+2^8)(1+2^{16}(1+2^{16})(1+2^{32})(1+2^{64})(1+2^{128}(1+2^{128}))))$
$GF(2^{571})$	$(1+2^2)(1+2^4(1+2^4)(1+2^8(1+2^8)(1+2^{16}(1+2^{16})(1+2^{32}(1+2^{32})(1+2^{64})(1+2^{128})(1+2^{256})))))$

4 Experiments

In experiments used workstation with Intel Core i7 2600 CPU, 8 Gb RAM, Microsoft Windows 7 x86-64 operation system and Microsoft Visual C ++ 2015 compiler.

Computational complexity and timings of cube root extraction via both methods available in Table 3. In software implementation of field arithmetic is used modular reduction of special moduli [18].

Table 3. Computational complexity and performance of cube root extraction

Field	Complexity		Time	
	Known method	Proposed method	Known method, ms	Proposed method, ms
$GF(2^{163})$	162S + 81M	162S + 8M	0.0233	0.0093
$GF(2^{233})$	232S + 116M	232S + 10M	0.0481	0.0173
$GF(2^{283})$	282S + 141M	282S + 10M	0.0745	0.02425
$GF(2^{409})$	408S + 204M	408S + 11M	0.1634	0.0695
$GF(2^{571})$	570S + 285M	570S + 12M	0.2655	0.0823

Edwards curve parameters birationally equivalent to EC in Weierstrass curves from NIST recommended list [14] shown in Table 4.

Timings of mapping point on Weierstrass curve to the point on birationally equivalent Edwards curve is shown in Table 5.

Table 4. Parameters of corresponding birationally equivalent Weierstrass and Edwards curves

Parameters of Edwards curve	
B-163	
d1=7b4a71eb0b1d71928895953f95d cbbdc24e4aa671; d2=30255967341afd682ab9b0bc10f d4898d81eba969;	X=5bbfa782cd266f3f31ac80c6c8d 99ceeda3b58f51; Y=6abdb295945ac13407079ada410 bc2482bef892c4;
B-233	
d1=182c53703916641c1fcbede90d9 be6ce7b0f7b74ede6ea9e10a124ea2 93; d2=100dedc0cd0f399e29ee55d788f 7be9d8f557470c7a54f6c1a564948c ba;	X=e1824b0552ec931073163db498b 593308d9fe4a42953799454e0ff7aa9; Y=dc7821990f8cc98cecc96f40a8e 40271c0f412b63d6cbb951daa49e457;
B-283	
d1=431a3cfe686eec8ef61c5c33d4b eafd0eec0dadb724c969d5c18c9bca 5f430c7bffbde9; d2=5352c278302eacc156b9f2bf877 8ca562c71bac9ef03533fe37a57e27 5823edd8b69ed3;	X=ee6aafe7457d61062947fa5ef76 f370548690a5f77620bbd1da45063 3b412de60baa; Y=3f066d1a9ebb8da4fb8070a23d4 a4ac2eee5110eec01c4ac377440b8 b03d7d7c41175c8;
B-409	
d1=125be71600f8f3e6d6cd5e91dd2 c0700007cae2c5f4563585e4f2a28c eeb13df776b5fa585f8f0538801394 ae631e26cb0fcff8; d2=9e5275c828b501ccb0bd585f561 915d94f89473d012f6557508e2c80e bcff05716a176cd7312a8997e101c1 89e599f61a3a859;	X=dee1bfc43f5163b2082603a5ebd 7ae2409866598780bb98cad540887 ada03d5f3d6571be7e907ffc28f41 7a0a33f4d128; Y=b5157db7000394d8541b2f85dc0 b1ab4fe473579a6a28cdc88e3b8af 9a784baf5fea15794485703418d49 545ea013a5b816c5e;
B-571	
d1=26e8eaf072f1073d873b1946105 c9dc8be77da0f2ca1e5b3bd35767aa 59fcef150e0c6edb5c2386d34090ae ddbb79a630962011e57990ed39481a 4ec1731d4756a1358081f3efa9; d2=50b9858e33c089e7f0ef98d0ffa 994f3de25c4970351fcbda1db5de1a 4ff0838aebc7aa6e7b2ac61bea6d47 161cc5b24d75ce7332b3d5bcdac809 5fb19f3538cfa98812f9483fc5;	X=538ebc0899f19093426e0732d12 48e8202159599b39104655fd89850 557922ec36582763574106a5fe83d e6d4dbba51561100a2fc086a5bd7c 52fdf9cd66e8dbdfe13272343d788; Y=41494c33470a85508b434eea3c9 a3bfb8e26c04da1c1eff04722de14 768d7e8946e9b49cb349b315864e5 d3ed4438b41d18bc6c93cb72f4867 390cbec18415e510e0758f331d585;

Table 5. Performance comparison of search birationally equivalence curve and point

Field	Known method, ms	Proposed method, ms
$\mathbf{GF}(2^{163})$	0.1202	0.0805
$\mathbf{GF}(2^{233})$	0.2741	0.2054
$\mathbf{GF}(2^{283})$	0.3254	0.2351
$\mathbf{GF}(2^{409})$	0.7319	0.4285
$\mathbf{GF}(2^{571})$	1.1254	0.8125

5   Conclusions

It is proposed a modified algorithm for searching the birationally equivalent curve Edwards from Weierstrass curve over $\mathbf{GF}(2^m)$ fields. For fast exponentiation in cube root extracting used method with exponent decomposition in chain $1 + 2^2 + 2^4 + \ldots + 2^{m-1}$ in field $\mathbf{GF}(2^m)$ where m is odd. This modification may significantly reduce the amount of field multiplication, thus allow increasing speed of cube root extraction. For EC from [14] NIST recommended list, it is found birationally equivalent Edward curves and mapped base points.

Proposed algorithm of search birationally equivalent curve and point mapping showed performance in 1.1–1.7 times faster when known algorithm.

Acknowledgment. This scientific work was financially supported by self-determined research funds of CCNU from the colleges' basic research and operation of MOE (CCNU16A02015).

References

1. IEEE working group: IEEE 1363-2000: Standard Specifications For Public Key Cryptography. IEEE standard. IEEE, New York, NY 10017 (2000). http://grouper.ieee.org/groups/1363/P1363/
2. ISO/IEC. ISO/IEC 14888-3:2006, Information technology – Security techniques – Digital signatures with appendix – Part 3: Discrete logarithm based mechanisms (2006)
3. Bernstein D.J., Lange T.: Failures in NIST's ECC standards (2016). https://cr.yp.to/newelliptic/nistecc-20160106.pdf
4. Kovtun, V., Tevyashev, A., Zbitnev, S.: Algorithms of scalar multiplication in group of elliptic curve points and some of their modifications. Radiotekhnika **141**, 82–96 (2005). (in Russian)
5. Bernstein, D.J., Lange, T.: Analysis and optimization of elliptic-curve single-scalar multiplication. In: Gary, L., Mullen, D. (eds.) Finite Fields and Applications, Contemporary Mathematics, vol. 461, pp. 1–19. American Mathematical Society (2008)
6. Moloney, R., O'Mahony, A., Laurent, P.: Efficient implementation of elliptic curve point operations using binary Edwards curves. IACR Cryptology ePrint Archive, Report 2010/208 (2010). http://eprint.iacr.org/2010/208.pdf
7. Kwang, H., Chol, O., Christophe, N.: Binary Edwards curves revisited. In: INDOCRYPT 2014. LNCS, vol. 8885, pp. 393–408 (2014)

8. Bernstein, D.J., Lange, T., Rezaeian Farashahi, R.: Binary Edwards curves. In: Oswald, E., Rohatgi, P. (eds.) Cryptographic Hardware and Embedded Systems – CHES 2008. LNCS, vol. 5154, pp. 244–265. Springer, Heidelberg (2008)
9. Bernstein, D.J.: Batch binary Edwards. In: Halevi, S. (eds.) Advances in Cryptology - CRYPTO 2009. LNCS, vol. 5677, pp. 317–336. Springer, Heidelberg (2009)
10. Ming, L., Ali, M., Daming, Z.: Fast algorithm for converting ordinary elliptic curves into binary Edward Form. Int. J. Dig. Content Technol. Appl. **6**(1), 405–412 (2012)
11. Bernstein, D.J., Birkner, P., Joye, M., Lange, T., Peters, C.: Twisted Edwards curves. In: Vaudenay, S. (eds.) Progress in Cryptology – AFRICACRYPT 2008. LNCS, vol. 5023, pp. 389–405. Springer, Heidelberg (2008)
12. Bernstein, D.J., Lange, T.: Inverted Edwards coordinates. In: Boztaş, S., Lu, H.F. (eds.) Applied Algebra, Algebraic Algorithms and Error-Correcting Codes, AAECC 2007. LNCS, vol. 4851, pp. 20–27. Springer, Heidelberg (2007)
13. Kovtun, M., Gnatyuk, S., Trofimenko, V.: Accelerated r-th root extraction in binary field. In: 2nd International Scientific Conference: Information and Telecommunication Technologies: Education, Science and Practice, pp. 547–551. Almaty (2015). (in Russian)
14. Digital signature standard (DSS). Federal Information Processing Standard 186-4. National Institute of Standards and Technology (2015)
15. Barreto, P.S.L.M., Voloch, J.F.: Efficient computation of roots in finite fields. Des. Codes Crypt. **39**, 275–280. https://doi.org/10.1007/s10623-005-4017-5
16. Bluhm, M.: Software optimization of binary elliptic curves arithmetic using modern processor architectures, Ph.D. RUHR-Universitat Bochum (2013)
17. Hu, Z., Gnatyuk, S., Koval, O., Gnatyuk, V., Bondarovets, S.: Anomaly detection system in secure cloud computing environment. Int. J. Comput. Netw. Inf. Secur. (IJCNIS) **9**(4), 10–21 (2017). https://doi.org/10.5815/ijcnis.2017.04.02
18. Gnatyuk, S., Okhrimenko, A., Kovtun, M., Gancarczyk, T., Karpinskyi, V.: Method of algorithm building for modular reducing by irreducible polynomial. In: 16th International Conference on Control, Automation and Systems. Gyeongju, Korea, pp. 1476–1479 (2016)
19. Explicit-Formulas Database. http://www.hyperelliptic.org/EFD

Criteria for Evaluating the Effectiveness of the Decision Support System

V. Tolubko[1,2], S. Kozelkov[1,2], S. Zybin[1], V. Kozlovskyi[1,2(✉)], and Y. Boiko[2]

[1] State University of Telecommunications, Kyiv, Ukraine
vvkzeos@gmail.com
[2] National Aviation University, Kyiv, Ukraine

Abstract. The task of increasing effectiveness for decision-making support in the condition of information protection is considered. The main criteria for assessing the processes effectiveness of forming an information security system in conditions of limitations and uncertainties are described. The integral criteria of effectiveness consists of sub-criteria: efficiency, quality, continuity, reliability, uniqueness, risk. In this article, the author suggests using risk criteria for implementing threats to assess effectiveness.

Keywords: Decision making · Evaluating criteria · Information security
Security program · Decision making · Protection system · DSS
Decision support system · Evaluation · Simulating and judgement

1 Introduction

To assess the efficiency of the decision support systems (DSS), it is necessary to determine the indicator. It is necessary to make a numerical analysis of the indicator and provide answers to the following questions. Is the result good? Is the decision acceptable? This indicator is called the performance criterion [1].

Based on the analysis of numerous papers [2], it is possible to determine the basic requirements for the efficiency criterion. The criterion should be consistent and complex. The criterion should depend on the system structure and the values of its parameters, the kind of the outside ambient influence and internal factors. The criterion should be complete. It is mean, that the criterion should display the main types of expenses. The criterion should be representative. It is mean that the criterion should reflect the main goal of the management. The criterion should allow comparison of the efficiency and efforts. The criterion should provide a clear understanding of the physical sense for each variant in comparison with degree of the goal achievement.

2 Analysis of Recent Achievements and Publications

The aspects of the design and implementation of the DSS are discussed in detail in following papers [3, 4]. The analysis of their development and field of application was carried out. The most common DSS are described in [5]. The necessary conditions for the effectiveness of decisions are timeliness, complexity and optimality. The first of these conditions is a limitation and others are the defining fundamental conditions [6]. The requirement of complexity requires a complete and comprehensive account of the impact of external and internal factors on decisions, as well as their interconnections [7].

3   Statement of the Problem and Its Solution

The criteria should calculate simple and fast. The criteria contents should be clear to understanding. It must have a saturation property [8]. The duplication of the same indicator in a generalized criterion is unacceptable. It is the reason that this leads to an overestimation of its role in comparison with others. The criteria should have a minimum dimension [9].

At present time, the criteria are widely used to assess the effectiveness of the funds use, as well as the degree of accomplishment of the task.

The most often used are two groups of criteria: operational and economic. By their significance, depending on the nature of the system, they can be main and auxiliary.

As a rule, the operational criterion is chosen as the main criterion. It allows you to assess the degree of the goal achievement or the degree of the task accomplishment.

$$K_{OP} = \frac{F}{F_H} \tag{1}$$

where: K_{OP} is operational criteria; F i F_H are the actual and normative value of the efficiency indicator.

This kind of criterion is widely used in evaluating the effectiveness of different control systems.

Economic criteria allow you to estimate the amount of resources to achieve the goal. The criteria can be expressed in temporal indicators: in costs, energy costs. This occurs in the process of planning and finding a solution. Comparison of generally accepted evaluation criteria indicates the need for their transformation taking into account the specificity of DSS [10].

The analysis of methods for evaluating the functioning effectiveness of the DSS allows us to make the following conclusions: existing criteria are diverse and do not represent a harmonious system, choosing the best criteria causes some difficulties; there are no clear recommendations for choosing integral criterion; the criteria are simple and, as a result, there are no complicated criteria [11].

All this determines the need to develop criteria for evaluating the functioning effectiveness of DSS. The determination of evaluation objectives is important in assessing the functioning effectiveness of DSS [12]. You can raise the question of evaluating the functioning effectiveness of system. The construction of criteria system for assessing the functioning effectiveness of management was carried out based on defining the goals and constructing the "tree" of DSS objectives (Fig. 1), i.e. the criteria are created in the same way as the goal, based on any formal approach.

By the degree of formalization, the goals must be clearly formulated and subjected to a formalized description. They should express only the general direction of the system's actions. In presence at one level of a few descriptions which have a special

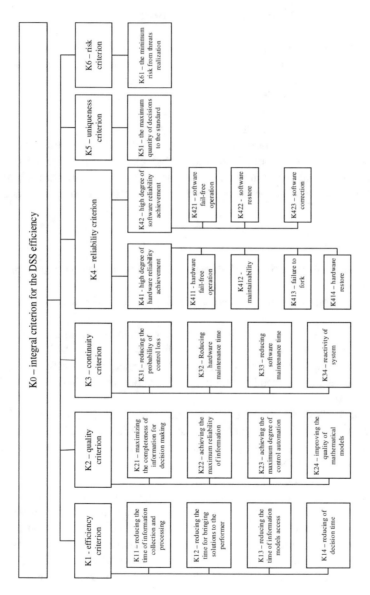

Fig. 1. Criterion for evaluating the effectiveness the functioning effectiveness

purpose state of the system two approaches are possible: the first one is following: as one of the most important characteristics of the system's goal is taken; the second is following: when it is impossible to distinguish the main characteristic as a goal, several of them are taken. The optimal state of DSS will match their optimal ratio [13].

There are goals at each level of DSS, which are inherent only to this level [14]. Higher-level goals contain more uncertainty; they are less structured and cannot be used to select a specific mode of action. Therefore, the main goal is to be decomposed into sub-goals until they are so concise that they can be implemented in the management process.

The analysis of the criteria (Fig. 1) made it possible to reveal a number of regularities. The main goal of K_0 is to be decomposed, that is to say, it splits into sub-goals, which, in turn, are broken down into subgoals from the top to down until they become concrete and can be implemented in the management process. Vertical connections of decomposition are necessary and most important, since their violation leads to the impossibility of achieving the ultimate goal. They are the strongest in the upper part of the "tree" and weaken as the goals are decomposed down. The analysis of goals allowed to identify the sub-goals and distribute them in order of preference $(K_1, K_2, \ldots, K_n)$. Each goal splits again on the subgoals $(K_{11}, \ldots, K_{1n}, K_{21}, \ldots, K_{2k}, K_{l1}, \ldots, K_{ln})$ from the top to down by using vertical links.

The timeliness of decisions and their quality are generally accepted indicators of the ergatic system effectiveness [15].

The goal tree can include the following goals and their indicators. K_0 is an integral efficiency indicator, which consists of $(K_1, K_2, \ldots, K_n)$ criteria.

K_1 is the efficiency criterion. The purpose of the efficiency criterion K_1 is to increase the timeliness of decisions. The criterion of timeliness consists of criteria (sub-goals) $K_{11}, K_{12}, K_{13}, K_{14}$. Where K_{11} is the sub-goal of reducing the time of information collection and processing; K_{12} is the sub-goal of reducing the time to bring solutions to the performer; K_{13} is the sub-goal of reducing the time of access to information models; K_{14} is the sub-goal of decision time reduction.

K_2 is the criterion for the decisions quality or a criterion for the decisions validity. The purpose of the criterion is to increase the validity of the decisions. The criterion of the decisions validity consists of the criteria (sub-goals) $K_{21}, K_{22}, K_{23}, K_{24}$. Where K_{21} is the sub-goal of achieving the maximum completeness of information for decision-making; K_{22} is the sub-goal of achieving the maximum reliability of information; K_{23} is the sub-goal of achieving the maximum degree of automation for control elements; K_{24} is the sub-goal for improving the quality of mathematical models.

K_3 is the continuity criterion. The purpose of the criterion is to ensure the continuity of the information security system. The criterion K_3 consists of the criteria (sub-goals) $K_{31}, K_{32}, K_{33}, K_{34}$.

Where K_{31} is the sub-goal of reducing the loss probability of management in the event of equipment or automation devices failure; K_{32} is the sub-goal of reduction in the hardware maintenance time; K_{33} is the sub-goal of reducing software maintenance time; K_{34} is the sub-goal for the reactivity of the system, that is, it is necessary to ensure a high level of probability of solving management problems for given time limits.

K_4 is the reliability criterion. The purpose of the criterion is to increase the reliability of the system's operation. The criterion K_4 consists of the criteria (sub-goals). K_{41} is sub-goal of achieving a high degree of hardware reliability (consists of sub-goals $K_{411}, K_{412}, K_{413}, K_{414}$) and K_{42} is the sub-goal of achieving a high degree of software reliability (consist of sub-goals $K_{421}, K_{422}, K_{423}$). Where K_{411} is the sub-goal of increasing in the time of hardware fail-free operation; K_{412} is the sub-goal for increasing the degree of maintenance for equipment; K_{413} is the sub-goal of increasing the error-free running time; K_{414} is the sub-goal to reduce the time for technical component recovery; K_{421} is the sub-goal of increasing the software's fail-free time; K_{422} is the sub-goal to reduce the software recovery time; K_{423} is the sub-goal to increase the degree of the software correctness. That is, compliance with the specifications.

K_5 is the uniqueness criteria. The purpose of the criterion is to increase the uniqueness of system management, consists of the one sub-goal. K_{51} is the sub-goal to reach the maximum quantity of executable decisions to the quantity of decisions. It is can be take into account at a certain time.

K_6 is the risk criterion. The purpose of the criterion is to reduce the risks of implementing threats for information security [16].

The risk criterion of information security should use to assess the DSS functioning effectiveness.

The most widespread assessment of information risks is:

$$R = pC$$

where R is the information risk; p is the probability of information security violation; C is the cost of information resources.

When we are building a security system, it's difficult to calculate the overall probability of a security threats, but it is possible to find the value of the probability for execution of individual threats. Properly, the overall risk value can be found as the sum of the risk values for the execution of all threats:

$$R = \sum_{i=1}^{n} R_i = \sum_{i=1}^{n} (p_i L_i) \qquad (2)$$

where R_i is the information risk for the execution of i-th threat;

p_i is the probability of i-th threat execution;
L_i is the loss value for the i-th threat.

Each type of threats in the DSS, that works under conditions of limitations and uncertainties can be represented as a set of such objects:

$$T = \langle N, S, V, W, O, A \rangle \qquad (3)$$

where N is the type of threats;

S is the set of possible sources for this type of threats;

V is the set of possible vulnerabilities which are associated with this type of threat;

W is the set of possible executing ways for threats of this kinds;

O is the objects set of influence for threats;

A is the destructive actions set for threats of this type that are executed.

The content of the listed sets for each type of threats is determined by the experts and is contained in the knowledge base.

The sets S, V, O, A are composed by selecting from a set of acceptable values for the given type of threats that are determined by the experts in accordance with the properties and conditions of the DSS functioning. The set W is formed depending on the composition of other sets. In this case, the set A consists of five subsets, according to the security information criteria [19]:

$$A = \langle A_K, A_C, A_D, A_S, A_G \rangle$$

where A_K is a set of destructive actions aimed at violation of confidentiality;

A_C is a set of destructive actions aimed at breaking the integrity;

A_D is a set of destructive actions aimed at violation of accessibility;

A_S is a set of destructive actions aimed at observation;

A_G is a set of destructive actions aimed at guaranteeing.

Most known methods of risk analysis suggest to input the value of the threat probability p_i by expert [12]. The use of statistical data does not allow determining how the value changes when implementing a particular mechanism of protection. For this reason, the method for calculating probabilities based on the information that are contained in the threat description is proposed (3).

Identification of a possibility ways set for a threat implementation is the key to calculating of the probability. The more possible ways to implement a threat, the higher the probability that an intruder will try to implement the threat. The W-set is formed according to the potential sources of a threat and vulnerabilities and objects are presented in the DSS. Thus, the implementation probability of the i-th threat can be found according to the formula (4)

$$p_i = \left(\sum_{k=1}^{m} a_k p_{r_k} \right) / m \tag{4}$$

where p_{r_k} is the probability of an intruder's successful use of the k-th variant of the i-th threat;

a_k is the coefficient of reliability, which determines the degree of assurance that the intruder can use the k-th variant of the i-th threat;

m is the quantity of possible ways for implementation of the i-th threat.

In the absence of statistical or other information, the coefficients a_k are considered to be identical in terms of probability. The coefficient a_k takes on two possible values:

$a_k = 0$, if there are no objective preconditions for using a threat variant and $a_k = 1$, if such preconditions exist.

Each variant of threat implementation can be completely blocked by a certain set of countermeasures. At the same time, the probability of a successful use of the variant of threat implementation depends on the quantity of possible implemented countermeasures, and can be calculated according to the following formula:

$$p_{r_k} = 1 - \frac{K_{r_k}}{K_{u_k}} \tag{5}$$

where K_{r_k} is the quantity of countermeasures that have been taken that block the k-th variant of threat implementation;

K_{u_k} is the total quantity of countermeasures that block the k-th variant of threat implementation.

The calculation of the losses volume for the implementation of i-th threat is carried out according to the following formula:

$$L_i = \left(\omega_k + \omega_c + \omega_d + \omega_s + \omega_g \right) C \tag{6}$$

where $\omega_k, \omega_c, \omega_d, \omega_s, \omega_g$ are coefficients of significance that determine the importance of ensuring the respective criteria: confidentiality, integrity, availability, observation and assurance;

C is the cost of information resources.

The calculation of the losses volume can be made according to formula (6) separately for each information resource.

4 Experimental Research

Currently, information and analytical support is impossible without use of various networks. Therefore, we will calculate the risk value for the threat of "Network Traffic Analysis" [18].

The purpose of such attacks is the tapping of communication channels, as well as the analysis of transmitted data and service information in order to study of the topology and architecture of system construction, and also the obtainment of critical user information [17, 20]. Protocols, such as FTP and Telnet, are subject to such attacks.

Traffic analysis attacks can be preceded by network intelligence, which is the gathering of information about the network using public data and applications. Network intelligence is conducted in the form of DNS queries, ping sweeps and port scanning. The result is extraction of information that can be used for hacking. Note that network intelligence is impossible to get rid of completely. In general, the threat of "Network Traffic Analysis" has six possible defined implementation variants P1, …, P6 (m = 6):

P1 is the usage of a malicious program;
P2 is the implementation of software and hardware bugs;

P3 is the interception of transmitted information over internal communication channels, using the vulnerabilities of channel level protocols;

P4 is the interception of transmitted information over internal communication channels, using the vulnerabilities of network level protocols;

P5 is the interception of transmitted information over external communication channels, using the vulnerabilities of network level protocols;

P6 is the usage of vulnerabilities in application and special software.

Wireless access tools are the most vulnerable spots of any computing or distributed system. Let's assume that a computing system is constructed using network switches and data transmission over external communication channels, but without the use of wireless access networks. No statistical information is available. In this case, there are no objective prerequisites for the use of the P3 variant, and the coefficients have the following values: $a_3 = 0, a_1 = a_2 = a_4 = a_5 = a_6 = 1$.

In order to block the remaining variants of threat implementation (P1, P2, P4, P5, P6), the following countermeasures are applied:

for P1, it is the antivirus protection programs;

for P2, it is the tools for ensuring integrity and physical security of internal communication channels;

for P4, it is the tools for intruder detection within internal communication channels;

for P5, it is the tools for intruder detection within external communication channels;

for P6, it is the tools for ensuring integrity and security analysis tools.

If the given countermeasures are not implemented the probability of a successful use of all variants threat implementation is equal to one. In this case, the probability of a threat is determined according to the formula (4):

$$p_i = \frac{1+1+1+1+1}{6} = \frac{5}{6} = 0,8333.$$

When using the tools for ensuring integrity, the probability of a successful use of the variants of threat implementation take on the following values: $p_{r_1} = p_{r_4} = p_{r_5} = 1$, $p_{r_2} = p_{r_6} = 0, 5$.

Therefore, the probability of threat implementation takes on the following value: $p_i = 0, 6667$.

Thus, it is possible to calculate the value of the probability of threat implementation when using different protection tools and their variations.

Afterwards, experts need to determine the value of significant coefficients that determine the importance of ensuring the respective criteria.

By its nature of influence, network traffic analysis is a passive influence. Implementation of the given attack without feedback leads to a breach of confidentiality of information within one network segment within the channel level of an OSI model. Accordingly, we will, only take into account the significant coefficient of confidentiality.

So long as the sum of significant coefficients cannot exceed the value of one, and the value of information resources can be considered constant (or unchangeable over a certain period of time) and, as a consequence, not taken into account during relative comparison, then the value of risk cannot exceed the value of probability of a threat implementation.

Table 1 contains examples for calculating maximum and relative values of the risk of threat implementation when applying various countermeasures for information protection.

Table 1. Probability of implementation of a threat to information security

Countermeasures	Probability of threat implementation
Unavailable	0,8333
Antivirus protection tools	0,5
Tools for ensuring integrity from internal perpetrators	0,54
Tools for ensuring integrity from external perpetrators	0,25
Physical security of internal communication channels	0,58
Security breach detection tools	0,5
Security analysis tools	0,58
Antivirus protection tools and tools for ensuring integrity	0,33
Antivirus protection tools, tools for ensuring integrity from internal perpetrators, physical security of internal communication channels, security analysis tools	0,23

The proposed method of risk analysis allows to determine the efficiency of the already taken or planned decisions. In addition, the value of probability of threat implementation depends on the level of separation of security and protection.

The method does not take into account the effectiveness of blocking the means of countering threats. In addition, there are no means of determining the value of information resources. However, this method allows to evaluate the efficiency and correctness of generated solutions based on the results of prototype DSS testing.

The risk analysis and performance assessment module is a main component of the DSS that is being developed.

The result of a DSS application is an increase in efficiency of the decisions taken and a decrease in costs during the creation and exploitation of information security systems.

5   Conclusion

The author proposes to use the method for evaluating the effectiveness of the decisions and risk analysis of information security. The proposed method of risk analysis allows to determine the efficiency of the already taken or planned decisions. In addition, the value of probability of threat implementation depends on the level of echelonment of security and protection.

The method does not take into account the effectiveness of blocking the means of countering threats. In addition, there are no means of determining the value of information resources. However, this method allows to evaluate the efficiency and correctness of generated solutions based on the results of prototype DSS testing.

References

1. Otero, A.R., Otero, C.E., Qureshi, A.: A multi-criteria evaluation of information security controls using boolean features. Int. J. Netw. Secur. Appl. (IJNSA), **2**(4), 1–11 (2010)
2. Mellado, D., Fernández-Medina, E., Piattini, M.: A common criteria based security requirements engineering process for the development of secure information systems. Comput. Stand. Interfaces **29** (2), 244–253 (2007)
3. Larichev, O.I., Kortneva, A.V., Kochin, D.Y.: Decision support system for classification of a finite set of multicriteria alternatives. Decis. Support Syst. **33**, 13–21 (2002)
4. Zybin, S., Khoroshko, V.: Support for decision making in the formation of state information security programs: evaluation of program effectiveness. Inform. Math. Methods Model. **5**(2), 122–128 (2015)
5. Koshal, J., Bag, M.: Cascading of C4. 5 decision tree and support vector machine for rule based intrusion detection system. Int. J. Comput. Netw. Inf. Secur. **4**(8), 8 (2012)
6. Rannenberg, K.: Recent development in information technology security evaluation-the need for evaluation criteria for multilateral security. In: Security and Control of Information Technology in Society, pp. 113–128, August 1993
7. Kruger, R., Eloff, J.H.: A common criteria framework for the evaluation of information technology systems security. In: Information Security in Research and Business, pp. 197–209. Springer, Boston (1997)
8. Ling, A.P.A., Masao, M.: Selection of model in developing information security criteria on smart grid security system. In: 2011 Ninth IEEE International Symposium on Parallel and Distributed Processing with Applications Workshops (ISPAW), pp. 91–98. IEEE, May 2011
9. Zopounidis, C., Doumpos, M.: Multi-criteria decision aid in financial decision making: methodologies and literature review. J. Multi-Criteria Decis. Anal. **11**(4–5), 167–186 (2002)
10. Lakhno, V., Kozlovskii, V., Mishchenko, A., Boiko, Y., Pupchenko, O.: Development of the intelligent decision-making support system to manage cyber protection at the object of informatization. East. Eur. J. Enterp. Technol. **2**(9), 53–61 (2017). https://doi.org/10.15587/1729-4061.2017.96662
11. Rees, L.P., Deane, J.K., Rakes, T.R., Baker, W.H.: Decision support for cybersecurity risk planning. Decis. Support Syst. **51**(3), 493–505 (2011)
12. Larichev, O., Asanov, A., Naryzhny, Y.: Effectiveness evaluation of expert classification methods. Eur. J. Oper. Res. **138**(2), 260–273 (2002)
13. Hashemi, A., Pilevar, A.H., Rafeh, R.: Mass detection in lung ct images using region growing segmentation and decision making based on fuzzy inference system and artificial neural network. Int. J. Image, Graph. Sig. Process. (IJIGSP) **5**(6), 16–24 (2013). https://doi.org/10.5815/ijigsp.2013.06.03
14. Mir, I.A., Quadri, S.M.K.: Analysis and evaluating security of component-based software development: a security metrics framework. Int. J. Comput. Netw. Inf. Secur. (IJCNIS) **4**(11), 21–31 (2012). https://doi.org/10.5815/ijcnis.2012.11.03
15. Shameli-Sendi, A., Shajari, M., Hassanabadi, M., Jabbarifar, M., Dagenais, M.: Fuzzy multi-criteria decision-making for information security risk assessment. Open Cybern. Syst. J. **6**(1), 26–37 (2012)
16. Alharbi, E.T., Qureshi, M.R.J.: Implementation of risk management with SCRUM to achieve CMMI requirements. Int. J. Comput. Netw. Inf. Secur. (IJCNIS) **6**(11), 20–25 (2014). https://doi.org/10.5815/ijcnis.2014.11.03
17. Filali, F.Z., Yagoubi, B.: Global trust: a trust model for cloud service selection. Int. J. Comput. Netw. Inf. Secur. (IJCNIS) **7**(5), 41–50 (2015). https://doi.org/10.5815/ijcnis.2015.05.06

18. Rostami, M., Koushanfar, F., Karri, R.: A primer on hardware security: models, methods, and metrics. Proc. IEEE **102**(8), 1283–1295 (2014). https://doi.org/10.1109/JPROC.2014.2335155
19. Liang, G., Weller, S.R., Zhao, J., Luo, F., Dong, Z.Y.: The 2015 Ukraine blackout: implications for false data injection attacks. IEEE Trans. Power Syst. **32**(4), 3317–3318 (2017). https://doi.org/10.1109/TPWRS.2016.2631891
20. Stoneburner, G., Goguen, A., Feringa, A.: Sp 800-30 Risk Management Guide for Information Technology Systems. NIST Special Publication (2002)

Prospects for the Application of Many-Valued Logic Functions in Cryptography

Artem Sokolov[1](✉) and Oleg Zhdanov[2]

[1] Odessa National Polytechnic University, Shevchenko avenue, 1,
Odessa 65044, Ukraine
radiosquid@gmail.com
[2] Siberian State Aerospace University named after Academician
Mikhail F. Reshetnev, Krasnoyarsky Rabochy Av., 31,
Krasnoyarsk 660014, Russia

Abstract. The paper considers development of cryptographic methods based on the principles of many-valued logic. The results concerning the construction of block and stream cryptographic algorithms based on functions of many-valued logic are presented. The synergy of the principles of many-valued logic and the variable fragmentation of the block made it possible to construct an effective block symmetric cryptographic algorithm. The results of computational experiments confirm its high cryptographic quality and easily scalable number of protection levels. As shown by experiments, the principles of many-valued logic are an excellent basis for the construction of gamma generators (the basis of stream ciphers), which are based on the use of triple sets of ternary bent-sequences. The paper outlines the scope of the tasks, the solution of which is necessary for the further development in this direction of cryptography.

Keywords: Cryptography · Many-valued logic · Variable block fragmentation

1 Introduction

The importance of cryptographic methods in the protection of information is well known. The rapid growth of computer processing power and the new results obtained by researchers in the field of cryptanalysis necessitates the need of further increase of the cryptographic strength of existing cryptographic algorithms, as well as the development of new algorithms.

The purpose of increasing of cryptographic strength is usually achieved by the significant complication of non-linear dependencies between the key and the plaintext and between plaintext and ciphertext (confusion), as well as effective destruction of open-text statistics by cryptographic algorithm (diffusion) [1], which requires, corresponding, often significant computational cost.

A block of plain text has its own structure and more or less inherits the features of texts in natural language. Therefore, working with blocks of the same size in all rounds cannot guarantee the dispersion of such patterns in the cipher text. To improve the quality of the cryptographic algorithm, we have to either complicate round block operations or increase the number of rounds [2].

© Springer International Publishing AG, part of Springer Nature 2019
Z. Hu et al. (Eds.): ICCSEEA 2018, AISC 754, pp. 331–339, 2019.
https://doi.org/10.1007/978-3-319-91008-6_33

Earlier [3], we proposed an encryption algorithm with dynamic resizing of cryptographic primitives in various rounds.

In other words, it was suggested that the plain text is encrypted using substitution tables of different sizes in different rounds.

Proposed cryptographic algorithm [3] is the combination of substitution and permutation blocks, addition with a round key. The change in the size of the block, according to the authors' intention, should effectively destroy the links between the elements of the plaintext and increase the diffusion.

The results of research of the statistical parameters of this algorithm are presented in [4]. After the fifth round, the statistics of the text differ little from the statistics of the random sequence. In addition, the nonlinearity of the conversion and the correlation characteristics are comparable with those for full-round algorithms, such as GOST 28147-89 and AES after 7 or 8 rounds. The algorithm is programmed and adapted in [5].

The use of a binary system of representation of data is traditional in modern computing. However, annual conferences held by IEEE confirm the attention of specialists to developments based on non-binary numbering systems. And the analysis of publications allows to draw a conclusion: the main efforts today are concentrated in the field of construction of data processing algorithms based on non-binary logic.

At the moment, the methods of many-valued logic have found their application in many practical information systems for the synthesis of signal constructions, error-correcting and effective codes [6], to ensure data confidentiality.

The application of many-valued logic is obviously best suited to the paradigm of multi-core data processing, and they are of special interest for promising quantum cryptographic algorithms. Thus, in [7, 8] effective algorithms for generating pseudorandom key sequences based on functions of many-valued logic are proposed.

We believe that approach of the concept of variable fragmentation of a block and many-valued logic is one of the best ways in future developing of cryptography. In this paper we present our block symmetric cryptographic algorithm as well as some results in many-valued logic pseudorandom sequence generators based on many-valued perfect algebraic constructions.

2 Block Symmetric Cryptographic Algorithm Based on the Principles of Many-Valued Logic and Variable Block Length

As it is known, many modern encryption algorithms use S-boxes and operation of addition modulo with the round key. Wherein the size of the S-box and its quality, that affects the performance of confusion and diffusion, are of importance.

For the most effective organization of the encryption algorithm with dynamic resizing of cryptographic primitives the length of input block of the algorithm must be a composite number. For example, in [3], the plaintext block has the length $L = 120 = 2^3 \cdot 3 \cdot 5$ bits. Partitioning of the plaintext block for the encryption algorithm may be performed in various ways, such as segments convenient from a computational point of view of the length of $\sigma = 6, 8, 10, 12, 15, 20, 30$ bits, wherein within one procedure the segment size is not changed.

The proposed encryption algorithm involves three basic procedures: Substitution, Permutation and Gamma (adapted to non-binary case).

Successively described, each of these procedures, which are reversible, can therefore be used both in the encryption algorithm and in the decryption algorithm.

Encryption is performed iteratively, while it is possible to vary the number of rounds. Round transform consists of implementation of Gamma, Substitution or Permutation procedures. The basic version of the encryption algorithm includes five rounds, the first round consists of the Gamma and Permutations procedures, while other rounds include Substitution and Gamma Procedures. The scheme of the proposed encryption algorithm is shown in Fig. 1. We are focusing on the differences of this version of the algorithm:

1. Ternary logic is used.
2. The block size is increased to 240 elementary units, in this case they are trits.

Alphabet: $A = \{0, 1, \ldots, q - 1\}$, $q > 2$. In this paper we consider in details the case $q = 3$.

Input text: $\{x_i\}$, $i = 0, 1, \ldots N - 1$. In this paper we consider $N = 240$.

The key: $K = \{g_i, Q_l, E, a_i\}$, where a_i — variables of splitting which are chosen as different values for each procedure, that are the parts of the encryption round; Q_l— are the substitution sequences; E— is the permutation sequence; $\{g_i\}$— is the gamma sequence.

Output text: $\{y_i\}$, $i = 0, 1, \ldots N - 1$.

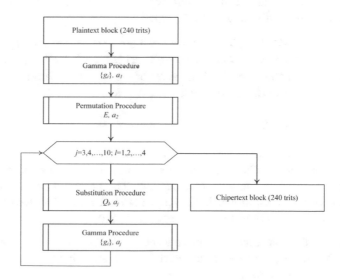

Fig. 1. Proposed encryption algorithm

For the organization of the encryption rounds, the values $\{a_i\}$ must be selected from the divisors of the number N, as well as the substitution Q_1, Q_2, Q_3, Q_4 sequences and permutation E sequence must be determined.

Decryption algorithm is completely similar to the encryption algorithm. More information about this algorithm can be found in [9].

Experiments have shown that the proposed encryption algorithm can be successfully implemented in the modern high-level programming languages, such as MatLab, which allows the encryption/decryption of data on a binary computer. We note in particular that during experiments the encryption was done in a mode of "Electronic Codebook", nonetheless the structure of the original image was completely destroyed, which is a very strong result for a symmetric block cryptographic algorithm.

3 Generation of Pseudorandom Sequences and Stream Ciphers Based on the Principles of Many-Valued Logic

The principles of many-valued logic are an excellent basis for the construction of stream ciphers. The basic components of modern stream ciphers are the generators of pseudo-random key sequences (gamma generator), which in many respects determines their performance and cryptographic strength. In fact, the development of stream cipher today means the development of a high-quality generator of gamma, which, subsequently, with the help of the operation of summation modulo p is added element-wise to the plain text.

The construction of the gamma generator based on the principles of many-valued logic was carried out by using the classical scheme for binary gamma generators, using many-valued LFSR (Linear Feedback Shift Register), as well as such highly nonlinear functions of many-valued logic as a bent-sequences.

Definition 1. A ternary sequence $H = [h_0, h_1, \cdots, h_i, \cdots, h_{n-1}]$ of length $N = 3^m, m \in \mathbb{N}$, where the coefficients $h_i \in \pm 1 \{e^{j0^\circ}, e^{j2\pi/3}, e^{j4\pi/3}\}$, is called a bent-sequence, if it has uniform absolute values of the Vilenkin-Chrestensen spectrum, which is representable in the matrix form

$$|\Omega_B(\omega)| = |H \cdot V_N| = const, \omega = \overline{0, N-1}, \tag{1}$$

where V_N is the Vilenkin-Chrestensen matrix of order N.

We have proposed an efficient method for synthesizing bent-sequences, which are necessary for constructing the gamma generators. The proposed method is based on three constructions for each of which the rules of reproduction have been found. More information about proposed method can be found in [10].

Construction 1. To construct a bent-sequence of length $N = 9$ in the Vilenkin-Chrestensen basis, we carry out a sequence of concatenations of the rows of the Vilenkin-Chrestensen matrix

$$V_3 = \begin{bmatrix} 0 & 0 & 0 \\ 0 & 1 & 2 \\ 0 & 2 & 1 \end{bmatrix} \rightarrow \begin{bmatrix} e^{j0} & e^{j0} & e^{j0} \\ e^{j0} & e^{j2\pi/3} & e^{j4\pi/3} \\ e^{j0} & e^{j4\pi/3} & e^{j2\pi/3} \end{bmatrix}, \tag{2}$$

resulting

$$B = [000012021] = \begin{bmatrix} e^{j0} & e^{j0} & e^{j0} & e^{j0} & e^{j2\pi/3} & e^{j4\pi/3} & e^{j0} & e^{j4\pi/3} & e^{j2\pi/3} \end{bmatrix}. \tag{3}$$

In order to verify that the constructed sequence (3) is indeed a bent-sequence, we multiply it by the Vilenkin-Chrestensen matrix of order 9, with the help of the following construction

$$V_9 = \begin{bmatrix} V_3 & V_3 & V_3 \\ V_3 & (V_3 + 1) \bmod 3 & (V_3 + 2) \bmod 3 \\ V_3 & (V_3 + 2) \bmod 3 & (V_3 + 1) \bmod 3 \end{bmatrix}. \tag{4}$$

As a result, we obtain the spectrum of the sequence (3)

$$S = \begin{bmatrix} 3 \cdot e^{j0} & 3 \cdot e^{j0} & 3 \cdot e^{j0} & 3 \cdot e^{j0} & 3 \cdot e^{j4\pi/3} & 3 \cdot e^{j2\pi/3} & 3 \cdot e^{j0} & 3 \cdot e^{j2\pi/3} & 3 \cdot e^{j4\pi/3} \end{bmatrix}. \tag{5}$$

So, the sequence (3) really satisfies the Definition 1 and is a bent-sequence in the Vilenkin-Chrestensen basis.

Here are the rules of reproduction:

Rule 1. When the rows of the Vilenkin-Chrestensen matrix are concatenated, their permutation can be performed in $\mu!$ ways, where $\mu!$ is the number of rows in Vilenkin-Chrestensen matrix.

Rule 2. With respect to each row of the Vilenkin-Chrestensen matrix, a re-coding operation can be applied using all possible rules by adding the number 0, 1 or 2.

By combining *Rule 1* and *Rule 2*, we obtain the first class of bent-sequences in the Vilenkin-Chrestensen basis of the cardinality $J_1 = 6 \cdot 27 = 162$. This class is an analog of the Majorana-McFarland class for binary bent-sequences.

We note that the use of this construction and the reproduction rules for any order of the Vilenkin-Chrestensen matrix determines the possibility of constructing bent-sequences in the Vilenkin-Chrestensen basis of any length $N = 3^{2k}$.

For example, on the basis of the Vilenkin-Chrestensen matrix of order $\mu = 3^2 = 9$ (4), bent-sequences in the Vilenkin-Chrestensen basis of length $N = 81$ can be constructed. Taking into account the rules of reproduction, their number will reach $J_{1,81} = \mu! 3^\mu = 9! \cdot 3^9 = 7142567040$.

Construction 2. It is found that the synthesis of bent-sequences of length $N = 9$ in the Vilenkin-Chrestensen basis can be performed on the basis of the following two regular constructions

$$C_1 = \begin{bmatrix} r \\ r \\ (r+1) \bmod 3 \end{bmatrix}, \quad C_2 = \begin{bmatrix} r \\ r \\ (r+2) \bmod 3 \end{bmatrix}, \tag{6}$$

where r is any ternary sequence of length 3, except the rows of the Vilenkin-Chrestensen matrix and their linear combinations. Thus, as the r $27 - 9 = 18$ such sequences can be used.

Obviously, we can rearrange the last line of the construction in three different ways, for example: for the first construction

$$C_1 = \begin{bmatrix} r \\ r \\ (r+1) \bmod 3 \end{bmatrix}, \quad C_1' = \begin{bmatrix} r \\ (r+1) \bmod 3 \\ r \end{bmatrix}, \quad C_1'' = \begin{bmatrix} (r+1) \bmod 3 \\ r \\ r \end{bmatrix}. \tag{7}$$

Thus, the cardinality of the class is $J_2 = 3 \cdot 2(3^3 - 3^2) = 6(27 - 9) = 108$.

Construction 3. Another construction that allows the synthesis of bent-sequences of length $N = 9$ in the Vilenkin-Chrestensen basis can be written as

$$C_1 = \begin{bmatrix} r \\ r \leftarrow 1 \\ (r+1) \bmod 3 \leftarrow 2 \end{bmatrix}, \quad C_2 = \begin{bmatrix} r \\ r \leftarrow 1 \\ (r+2) \bmod 3 \leftarrow 2 \end{bmatrix}, \tag{8}$$

where $\leftarrow$ is the operator of cyclic left shift by the corresponding number of positions; r is any ternary sequence of length 3, except the rows of the Vilenkin-Chrestensen matrix and their linear combinations.

The Construction 3 can be subjected to all $3! = 6$ possible permutations. Thus, the total cardinality of the third class of bent-sequences in the Vilenkin-Chrestensen basis is $J_3 = 2 \cdot 6 \cdot (3^3 - 3^2) = 216$.

Combining all three classes of bent-sequences in Vilenkin-Chrestensen basis, we obtain the full class of bent-sequences of cardinality $J = 162 + 108 + 216 = 486$. Previously, such results could only be obtained by a brute force method.

Note that, unlike binary analogs, bent-sequences of many-valued logic exist for lengths $N = 2^k$, $k \in \mathbb{N}$. Of particular interest is the construction of bent-sequences for odd values of k.

The research carried out in [10] made it possible to classify the complete set of bent-sequences of length $N = 9$ in the Vilenkin-Chrestensen basis, depending on their weight structure into 6 classes

$$\begin{bmatrix} \{1,4,4\}(54); \\ \{4,1,4\}(54); \\ \{4,4,1\}(54); \end{bmatrix} \quad \begin{bmatrix} \{5,2,2\}(108); \\ \{2,5,2\}(108); \\ \{2,2,5\}(108), \end{bmatrix} \tag{9}$$

where the numbers in curly brackets indicates, respectively, the number of characters "0", "1" and "2" in the bent-sequence, and the numbers in parentheses indicates the number of bent-sequences having this structure.

Definition 2. A set of three bent-sequences in Vilenkin-Chrestensen basis is called a triple set if the concatenation of their truth tables is symbolically balanced, i.e. number of characters "0" is equal to the number of characters "1" and is equal to the number of characters "2".

We propose a gamma generator scheme based on the properties of a complete class of bent-sequences divided into 2 types of triple sets (Fig. 2).

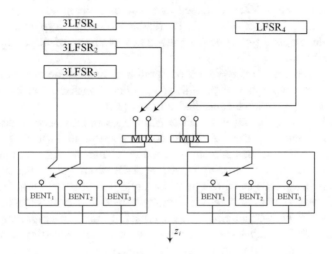

Fig. 2. Gamma generator based on triple sets of bent-sequences

The gamma generator scheme based on triple sets of bent-sequences, shown in Fig. 2 consists of two ternary LFSR that generates input values for a bent-sequence, as well as a single ternary LFSR that selects a 3-bent sequence within the triple set. The circuit contains one binary LFSR, which selects one of the two possible triple sets at each clock cycle.

Researches of the proposed scheme confirmed it's high stochastic and cryptographic quality [10].

4   Conclusion

The performed research made it possible to establish that the use of the principles of many-valued logic is a promising basis for further development of both block and stream cryptographic algorithms. In this case, the functions of many-valued logic are best suited to modern concepts of multi-core data processing and quantum computing.

The task of future research in this direction seems to be important for us.

5 Open Problems and Directions for Further Research

Despite the positive results obtained in the application of the principles of many-valued logic in cryptography, this question undoubtedly still requires significant attention of the scientific community.

Currently, issues of building of high-quality S-boxes of length $N = p^k$ are poorly covered in the literature. Research of the dynamics of change in parameters of S-boxes with the growth of their length is also an actual problem.

Methods of construction of P-boxes and methods of diagnosis of their cryptographic quality are not available in the literature, not only for many-valued logic, but even for the binary case. Vital for the current state of development of cryptography is the task of creating of a set of criteria for diagnostic of the cryptographic quality of P-boxes and development of effective methods of synthesis of these important cryptographic primitives.

The major problem of many-valued encryption is the development of many-valued schemes of such cryptographic primitives as affine transform blocks, cyclic shifts blocks, as well as perspective controlled F-boxes [11].

On the other hand, the development of gamma generators necessitates research of the already determined perfect algebraic constructions and further search for new perfect algebraic constructions of many-valued logic. Thus, an important problem is the description of the classes of many-valued bent-functions (in particular, of an odd number of variables), the generalization of the definition of perfect binary arrays to the many-valued case, as well as research of many-valued de Bruijn sequences.

In recent years cellular automata was successfully used in cryptographic applications [12, 13]. Their many-valued analogues seem to be interesting and perspective constructions that are yet not researched well enough.

Note also the task of developing more efficient methods of representation of binary data for processing in many-valued logic systems, and representation of many-valued logic data in binary form. The development of these methods may be connected to the fundamental researches performed in [14].

The solution to these problems, of course, will not only significantly improve the proposed concept of many-valued logic encryption algorithm with a variable block length, but will lead to a better understanding and the generalization of the theory of binary cryptography and further development of systems of binary data encryption.

References

1. Shannon, C.E.: A Mathematical Theory of Cryptography. Bell System Technical Memo. MM 45-110-02 (1945)
2. Goyal, R., Khurana, M.: Cryptographic security using various encryption and decryption method. Int. J. Math. Sci. Comput. (IJMSC) **3**(3), 1–11 (2017). https://doi.org/10.5815/ijmsc.2017.03.01
3. Zhdanov, O.N., Sokolov, A.V.: The encryption algorithm with variable block fragmentation, collection of scientific papers on the results of international scientific-practical conference. Probl. Achiev. Sci. Technol. **2**, 153–159 (2015). [Original text in Russian]

4. Zakharova, K.O.: Research of statistical parameters of the encryption algorithm with different block fragmentation. In: Materials of the XXI International Scientific and Practical Conference, Dedicated to the Memory of the General Designer Of Rocket-Space Systems Academician M.F. Reshetnev, Krasnoyarsk, Part 2, pp. 400–401 (2017)
5. Mitrashchuk, V.V.: Protocol of secure data exchange based on encryption algorithm with alternating block fragmentation. Mod. Sci. Technol. Innov. **16**, 299–301 (2017)
6. Fang, Z., Liu, Y.: Ternary Error Correcting Codes, Chinese Science Abstracts Series A, p. 54 (1995)
7. Gnatyuk, S.O., Zhmurko, T.O., Kinzeryavy, V.M., Siyilova, N.A.: Method for quality evaluation of trit pseudorandom sequence to cryptographic applications. Inf. Technol. Secur. **3**(2), 108–116 (2015)
8. Arshad Ali, M., Ali, E., Ahsan Habib, M., Nadim, M., Kusaka, T., Nogami, Y.: Pseudo random ternary sequence and its autocorrelation property over finite field. Int. J. Comput. Netw. Inf. Secur. (IJCNIS) **9**(9), 54–63 (2017). https://doi.org/10.5815/ijcnis.2017.09.07
9. Zhdanov, O.N., Sokolov, A.V.: Block symmetric cryptographic algorithm based on principles of variable block length and many-valued logic. Far East J. Electron. Commun. **16**(3), 573–589 (2016)
10. Mazurkov, M.I., Sokolov, A.V., Barabanov, N.A.: Synthesis method for bent sequences in the Vilenkin-Chrestenson basis. Radioelectr. Commun. Syst. **59**(11), 510–517 (2016)
11. Zui, H.N., Moldovyan, N.A., Fakhrutdinov, R.S.: New class of controlled elements f 2/3 for the synthesis of high-speed block ciphers. Probl. Inf. Secur. **1**, 10–18 (2011). [Original text in Russian]
12. Kumaresan, G., Gopalan, N.P.: An Analytical study of cellular automata and its applications in cryptography. Int. J. Comput. Netw. Inf. Secur. (IJCNIS) **9**(12), 45–54 (2017). https://doi.org/10.5815/ijcnis.2017.12.06
13. Nandi, S., Roy, S., Dansana, J., Karaa, W.B.A., Ray, R., Chowdhury, S.R., Chakraborty, S., Dey, N.: Cellular automata based encrypted ECG-hash code generation: an application in inter human biometric authentication system. Int. J. Comput. Netw. Inf. Secur. (IJCNIS) **6**(11), 1–12 (2014). https://doi.org/10.5815/ijcnis.2014.11.01
14. Stakhov, A.: Brousentsov's ternary principle, Bergman's number system and ternary mirror-symmetrical arithmetic. Comput. J. **45**(2), 221–236 (2002)

Suspicious Object Search in Airborne Camera Video Stream

Artem Chyrkov[(✉)] [ID] and Pylyp Prystavka [ID]

National Aviation University, Kosmonavt Komarov ave., 1, Kyiv, Ukraine
a.chyrkov@nau.edu.ua, chindakor37@gmail.com

Abstract. In some areas of drone application an object search task arises. Also there are cases where usage of standard approaches, e.g. object detection methods or fully manual video view, could be complicated or problematic. However it is possible to find local image (video frame) areas where suspicious object potentially can be present in such cases. We propose (i) an algorithm for suspicious object search in real time and (ii) an automated system (drone and ground control station) based on it, show brief results of its testing, make conclusions about further research direction.

Keywords: Aerial reconnaissance · Unmanned aircraft
Suspicious object search

1 Introduction

Technical level of drones (unmanned aerial vehicles, UAVs) at the moment of publication of this paper makes them able to be used for a wide range of tasks. The suspicious objects search in a video data recorded at UAV during the flight over specified territory is one of the typical tasks needed to be solved. The standard approach requires operator to view all video data in offline mode (after UAV landing) or online at the ground control station (GCS) during UAV flight.

And one of the significant issues of this approach is a presence of human factor. When the operator performs big amount of routine work his capability to process video data efficiently decreases due to attention dissipation. This in turn increases probability of non-finding an object of interest. And due to tunnel vision effect of human eye it is unable to process the whole video frame, but only some part of it. Therefore automation of suspicious object search in video from UAV is relevant and interesting research topic. The most obvious way is to use detector (algorithm which finds some particular objects). However, this approach is not appropriate to be used directly: at first, objects of interest can belong to different classes; at second, detectors need a training dataset (for each object-of-interest class) with sufficient amount of the corresponding objects recorded in different conditions. In practice it is quite challengeable to prepare such datasets for each class.

The specific feature of video data from UAV is an absence of static background, which complicates significantly use of majority image processing methods. Below we list some widely used approaches and corresponding issues of their application.

Background subtraction approach is based on the analysis of last frames for presence of areas which are changed insignificantly [1]. However, this approach requires video data with the static background (a lot of papers, e.g. [2–5], describe usage of this approach for processing a video from CCTV cameras), thus it is not appropriate for video data from the moving camera.

Template matching is a task of objects search which matches a template [6]. Therefore it is necessary to solve here additionally task template preparation. Hence, this approach could be used only as a part of some suspicious object search method.

Another feature of UAV video data is the presence of textured regions (grass, sand, forest, water etc.) which introduce additional challenges in video processing algorithms. E.g. edge detection methods (Canny edge detector [7] and others [8]) produce a large amount of wrong edges inside the textures: they focus on the high-frequency image component which presents largely inside textures. Also due to vibrations UAV video data has some distortions which cannot be eliminated at video recording stage. These features complicate usage of mentioned methods class.

Paper [9] describes clustering-based approach for moving object detection in video surveillance systems. It is developed for usage with a static camera, thus it is not appropriate for the UAV data. Papers [10, 11] describe a complex approach for man-made object detection which includes image preprocessing, clustering, line and corner detection. The speed of the proposed approach is not specified here, but as practice shows, each of these steps (implemented in a described way) aren't fast; also there is no remark about an ability of this approach to be used in real time systems. Paper [12] proposes segmentation approach for detection of suspicious objects in aerial and long-range surveillance applications. The fastest processing speed given in the testing section is more than 6 min for 2560×1920 pixels image, thus it is not appropriate for real time usage too.

Considering the above facts, it makes sense to replace a task of particular object detection by identification task of video frame local regions potentially containing suspicious objects. Also some practical tasks (e.g. aerial reconnaissance or rescue operations) require real time object search. Thus corresponding algorithm must be fast enough to be implemented in real time automated systems. In [13] mathematical model of the above task has been formulated and the strategy of its solution in terms of mathematical statistics has been presented. Paper [14] describes a simplified algorithm for suspicious object search and formulates the further research direction.

In this paper we propose (i) an improved version of histogram-based algorithm (from [14]) for suspicious object search in real time and (ii) an automated system for suspicious object search which consists of UAV and GCS. Section 2 describes the algorithm. Section 3 presents the system. Section 4 gives conclusions and direction for further research.

2 Histogram-Based Algorithm for Suspicious Object Search in Video Stream from UAV Airborne Camera

This section is aimed to briefly describe a histogram-based algorithm for suspicious object search. The proposed algorithm works with grayscale images, therefore the colored ones (e.g. RGB video frames) must be converted preliminary.

2.1 Local Histograms

Local histogram h is an array of N elements. The value of each element is calculated as a number of pixels inside local region L of image which have brightness value from the corresponding interval I_i, i.e.

$$h = \left\{ h_i; i = \overline{0, (N-1)} \right\}; h_i = \sum_{(x,y) \in L} \psi\left(p(x,y), I_i\right); \psi(a, I) = \left\{ \begin{array}{l} 1,\ a \in I \\ 0,\ a \notin I \end{array} \right. ;$$

$$I_i = [i\,\Delta i; (i+1)\,\Delta i); N\Delta i = B_{\max},$$

where $p(x, y)$ is a brightness value of a pixel with coordinates (x, y); B_{max} is a maximal possible brightness value, in most practical cases $B_{max} = 255$.

Set of pixels brightness inside the local region $\{p(x, y)|\ (x, y) \in L\}$ can be considered as a statistical sample. Mean value m and standard deviation σ of this sample can be calculated by using the local histogram:

$$m = \frac{1}{H} \sum_{i=0}^{N-1} i h_i; \sigma = \sqrt{\frac{1}{H} \sum_{i=0}^{N-1} (i - m)^2 h_i}; H = \sum_{i=0}^{N-1} h_i.$$

Thus we will name σ of a corresponding sample as "σ of histogram" in the below text.

2.2 Algorithm Description

We propose to divide histograms into 3 types (see Fig. 1: (a) unimodal with $\sigma \le \sigma_1$ (where σ_1 is some threshold value), (b) unimodal with $\sigma > \sigma_1$, (c) multimodal. Each type is considered as a class which potentially relates to some object class: texture (grass, sand,

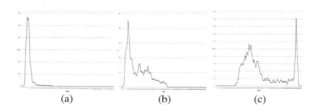

(a)   (b)   (c)

Fig. 1. Histogram types: (a) unimodal with $\sigma \le \sigma 1$, (b) unimodal with $\sigma > \sigma 1$, (c) multimodal.

water etc.), combination of textures, suspicious region; and is processed in a different way (see flowchart of the proposed algorithm in Fig. 2). Image (video frame) is divided into local regions and algorithm is applied to each of them.

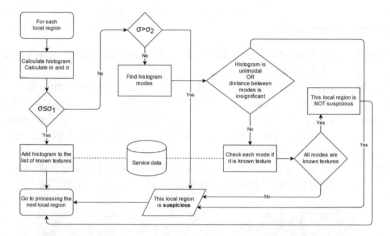

Fig. 2. Flowchart of the proposed algorithm.

The histogram modes search is one of the separate tasks in the proposed algorithm. The corresponding approach must be fast (because the proposed algorithm for suspicious object search has to work at a real time) and appropriately accurate (because the proposed algorithm is sensitive to the subpart of histogram mode finding). We propose to solve this task at the following way (see the below pseudocode).

```
input: histogram h[i], i=0..(N-1)
separate_index ← 1
optimal_separate_index ← 1
optimal_F ← +∞
while separate_index < N-1:
  σ1 = std( h[0..separate_index) )
  σ2 = std( h[separate_index..N) )
  F = σ1 + σ2
  if F < optimal_F:
    optimal_F = F;
    optimal_separate_index = separate_index
  end if
  separate_index += 1
end while
// m(x) is either mean(x) or max(x)
m1 = m( h[0.. optimal_separate_index) )
m2 = m( h[optimal_separate_index..N) )
return  m1, m2
```

If the distance between the returned "candidate-to-be-mode" values is insignificant, corresponding histogram is considered as unimodal. If this distance is significant, the returned values are considered as the histogram modes.

2.3   Quality of Algorithm

Taking into account the specifics of the proposed algorithm usage, it makes sense to consider its qualitative and quantitative characteristics.

The main goal of the proposed algorithm is to find suspicious objects on a video stream and to form a list of them for operator. It is not required to find a particular object an all video frames, finding it in at least one frame will be a sufficient result. Thus presence of all objects (real and/or test) from video stream in the result list can be used as the qualitative characteristic.

Example of a part of video frame is shown in Fig. 3 (video has been taken from [15]). Example of object list found in the corresponding video is shown in Fig. 4.

Fig. 3. Example of a part of video frame.

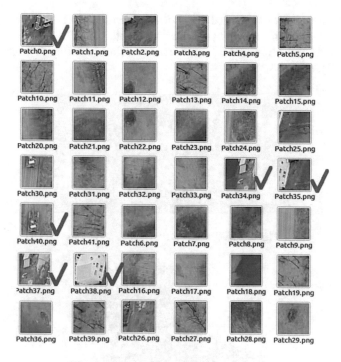

Fig. 4. List of objects found in test video.

As Fig. 4 shows, algorithm has found suspicious objects (tank and parts of building), but also has returned false regions (e.g. trees, spots at field). Also it is fast enough to be used in real time (see description of automated system in Sect. 3).

In order to compare different algorithm versions between each other or with its analogs on a research phase a quantitative characteristics should be given. Type I and type II errors can be used for this purpose. For its calculation a labeled dataset must be available. Labeled dataset is a set of video files for each of them suspicious objects are marked at each frame. Creating such dataset from scratch or from another available datasets is a separate task with a significant time cost.

3   Automated System for Suspicious Object Search

Based on the algorithm described in Sect. 2 we have developed an automated system for suspicious object search. Its detailed technical description is given in [16]. Below we give the brief one.

The automated system consists of (i) UAV with target payload (camera and single-board computer) and (ii) laptop as a ground control station (GCS). The automatic part is implemented on a single-board computer and works on UAV in real-time mode. It forms a list of suspicious objects with their GPS coordinates (possible calculation algorithms are described in [17–19]) and sends it to the GCS. The operator selects object(s) from the list manually (on GSC) and can choose actions for the UAV (not implemented yet): detect particular object, track particular object etc.

Photo of UAV which is used for experiments is shown in Fig. 5. Example of GCS is shown in Fig. 6.

Fig. 5. UAV which is used for experiments.

Fig. 6. Laptop; it can be used as GCS.

Automated system components and data flow are shown in Fig. 7. Telemetry in this case means GPS coordinates and UAV's orientation (yaw, pitch, roll).

The developed automated system has two main advantages: first, it can work efficiently in real-time mode, and second, it reduces data traffic between UAV and GCS (total size of patch list is much less than the total size of all video stream frames).

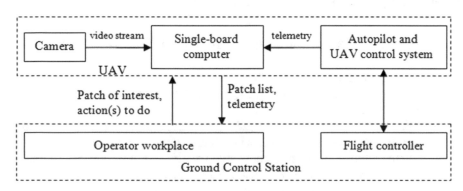

Fig. 7. Automated system components and data flow.

The corresponding field tests of the automated object search system have been performed. The obtained results confirm that the system is able to work in real time and successfully identify all test objects (see Figs. 3 and 4; results of the automated system on test objects are the same).

4   Conclusions

The proposed histogram-based algorithm for suspicious object search can be used as a part of automated real time systems. It finds all suspicious objects in camera field of view. Number of false suspicious objects (type II error) is appropriate.

The automated system for suspicious object search consisting of UAV and laptop as GCS has been developed. Its main advantages are (i) transfer suspicious object list instead of whole video frame for reducing data traffic between UAV and GCS; (ii) ability for real time usage if our algorithm is used for suspicious object search.

Further research can be related to decreasing type II error of the proposed algorithm, improvement of the automated system and adding new functionality to it (e.g. detector, tracker, integration to new or existing combat control systems).

References

1. Piccardi, M.: Background subtraction techniques: a review. In: IEEE International Conference on Systems, Man and Cybernetics, pp. 3099–3104. IEEE, The Hague (2004)
2. Krishna, M.T.G., Ravishankar, M., Babu, D.R.R.: Automatic detection and tracking of moving objects in complex environments for video surveillance applications. In: 2011 3rd International Conference on Electronics Computer Technology, pp. 234–238. IEEE, Kanyakumari (2011)
3. Mohan, A.S., Resmi, R.: Video image processing for moving object detection and segmentation using background subtraction. In: 2014 First International Conference on Computational Systems and Communications (ICCSC), pp. 288–292. IEEE, Trivandrum (2014)
4. Risha, K.P., Kumar, A.C.: Novel method of detecting moving object in video. Procedia Technol. **24**, 1055–1060 (2016)
5. Appiah, O., Ben Hayfron-Acquah, J.: A robust median-based background updating algorithm. Int. J. Image Graph. Signal Process. (IJIGSP) **9**(2), 1–8 (2017). https://doi.org/10.5815/ijigsp.2017.02.01
6. Brunelli, R.: Template Matching Techniques in Computer Vision: Theory and Practice. Wiley, Chichester (2009)
7. Canny, J.: A computational approach to edge detection. IEEE Trans. Pattern Anal. Mach. Intell. **PAMI-8**(6), 679–698 (1986)
8. Poobathy, D., Chezian, R.M.: Edge detection operators: peak signal to noise ratio based comparison. Int. J. Image Graph. Signal Process. (IJIGSP) **6**(10), 55–61 (2014). https://doi.org/10.5815/ijigsp.2014.10.07
9. Singh, S., Saurav, S., Shekhar, C., Vohra, A.: Moving object detection scheme for automated video surveillance systems. Int. J. Image Graph. Signal Process. (IJIGSP) **8**(7), 49–58 (2016). https://doi.org/10.5815/ijigsp.2016.07.06

10. Cai, F., Chen, H., Ma, J.: Man-made object detection based on texture clustering and geometric structure feature extracting. Int. J. Inf. Technol. Comput. Sci. (IJITCS) **3**(2), 9–16 (2011)
11. Cai, F., Chen, H., Ma, J.: Man-made object detection based on texture visual perception. Int. J. Eng. Manuf. **2**(3), 1–8 (2012)
12. Papic, V., Turic, H., Dujmic, H.: Two-stage segmentation for detection of suspicious objects in aerial and long-range surveillance applications. In: 10th WSEAS International Conference on Automation & Information, WSEAS, Athens, Greece, pp. 1–5 (2009)
13. Prystavka, P., Rogatyuk, A.: The mathematical foundations of foreign object recognition in the video from unmanned aircraft. Proc. Natl. Aviat. Univ. **3**(64), 133–139 (2015)
14. Chyrkov, A.: Suspicious object search on video stream from unmanned aircraft camera by using histogram analysis. Probl. Creat. Test. Appl. Oper. Complex Inf. Syst. **13**, 126–135 (2016). (in Ukrainian)
15. Kasyanov, Y.: – YouTube. https://www.youtube.com/channel/UCTG6kc99C4U6GN Ex7h_apwA
16. Prystavka, P., Sorokopud, V., Chyrkov, A.: Experimental version of an automated system for suspicious object search on video stream from unmanned aircraft. Weapon Syst. Military Equip. **2**(50), 26–32 (2017). (in Ukrainian)
17. Nichikov, E., Chyrkov, A.: Information technology of unmanned aircraft camera field of view definition. Probl. Inf. Control **4**(52), 106–112 (2015). (in Ukrainian)
18. Buryi, P., Pristavka, P., Sushko, V.: Automatic definition the field of view of camera of unmanned aerial vehicle. Sci.-Based Technol. **2**(30), 151–155 (2016)
19. Piskunov, O., Yurchuk, I., Bilyanska, L.: Camera field of view definition for aerial photography. Sci.-Based Technol. **3**(35), 204–208 (2017). (in Ukrainian)

Stochastic Optimization Method in Computer Decision Support System

Andrey Kupin⬥, Ivan Muzyka$^{(\boxtimes)}$⬥, Dennis Kuznetsov⬥,
and Yurii Kumchenko⬥

Kryvyi Rih National University, 11, Vitaliya Matusevycha str.,
Kryvyi Rih 50027, Ukraine
kupin.andrew@gmail.com, musicvano@gmail.com,
kuznetsov.dennis.1706@gmail.com,
y.kumchenko@gmail.com

Abstract. The questions of parameter optimization are considered in the article. These parameters describe technological processes at ore-dressing plants. There are many problems in creating of automated control systems for such technological processes. Among them are a large number of industrial parameters, the nondeterministic nature of physical and mechanical properties of the raw materials, and many disturbing influences. Thus, to solve these problems authors propose to use an intelligent computer decision support system. It allows calculating a regression equation of some optimization criterion. The complete Kolmogorov-Gabor polynomial is used as a regression model. In addition, the possibility of applying the differential evolution method is analysed. The stochastic optimization method has been compared with the full enumeration method, gradient descent method and Monte Carlo. A simple function with an extremum has been used as an example for demonstrating of search efficiency.

Keywords: Computer decision support system · Optimization algorithm
Method of differential evolution

1 Introduction

Various information systems today have become an integral part of almost all industrial enterprises. They have found wide application in mining, metallurgical, chemical, machine-building and other industries. Such information systems are created in order to automate the control of complex technological processes. In modern market conditions, automation of production becomes the main means of increasing labour productivity, improving the quality of products and reducing its cost. A characteristic feature of automatic control systems for individual industrial facilities is their integration into whole complexes, complicated organized SCADA-systems (Supervisory Control and Data Acquisition). In addition, alternative ways of describing mathematical models with the help of neural networks and models using fuzzy logic are gradually added to the classical theory of automatic control, which is now developed at a sufficiently high level. However, in recent years, the special attention of scientists has been attracted by the potential possibilities of using intelligent computer decision support systems

© Springer International Publishing AG, part of Springer Nature 2019
Z. Hu et al. (Eds.): ICCSEEA 2018, AISC 754, pp. 349–358, 2019.
https://doi.org/10.1007/978-3-319-91008-6_35

(CDSS) and control systems that are based on artificial intelligence models, evolutionary computations and genetic algorithms, neural networks, situational analysis, cognitive modelling, etc.

Therefore, the main goal is developing a structure of an automated system that will be able to calculate a mathematical model of the technological process in real time. It should be stable to stochastic disturbances and have efficient optimization mechanism as well.

2 Problem Statement

Technological processes at mining and processing plants of Ukraine are largely automated with modern programmable logic controllers by Schneider Electric and Siemens. First and foremost, powerful industrial controllers are used in those cases when the technological process proceeds at high speed, is dangerous for human life and health and requires the control of a large number of input and output parameters.

However, such automated control systems provide only support and control of the parameters of individual industrial units, that is, local control, irrespective of the global strategy of the enterprise. This aspect is very important, since most of the production redistribution is now considered from the point of view of at least three indicators, namely: the productivity of work, the cost of production and its quality. Therefore, the main attention should be paid to technological optimization, which is now the main reserve for increasing the competitiveness and profitability of the industrial sector of the economy.

The second problem of control at the shop or factory level is an extremely complex mathematical model that describes the entire technological process. Analyzing, for example, operation of beneficiation section of mining and beneficiation complex, one can see that for effective management of such a facility it is necessary to control dozens of technological parameters. This is further complicated by the fact that the characteristics of industrial equipment are not permanent, but change with time. For example, during the operation of crushers, the lining steel is erased, the size of the discharge opening changes, and the process of grinding the ore changes. Any machines and mechanisms in the process of wear change not only their technical and economic characteristics, but also the probability of failure-free operation. In other words, it is hardly possible to talk about creating a single, unchanging mathematical model of the whole production. Even if it is possible to accurately describe all industrial objects of the technological process with the help of analytical functions, the model will still require correction after a certain period of time. Therefore, synchronization of the mathematical model with the actual status of the production equipment is an integral part of the optimal management of the whole enterprise.

The third problem that significantly affects the course of production processes is the random nature of the quality indicators of incoming raw materials. This task is now solved by mixing different grades of ore, which differ not only in the content of iron, but also in abrasiveness, strength and ability to enrich. For example, the strength of Kryvyi Rih iron ore basin ranges in a wide range of 6–20 on M. M. Protodyakonov scale, and the iron content in them is 32–35%. Such a procedure is carried out with the aim of averaging the physics-mechanical characteristics of the raw material, which is transported from different parts of the open pit. That is, the concentrator is set up to process ore with certain characteristics, only in this case the desired quality of concentrate can be achieved. It is quite difficult to create such a technological process, which would not depend on the quality of input raw materials and would produce concentrate with a constant level of iron content at the output.

The fourth, not less significant, problem of modern enterprises is energy resources saving. For example, a large crushing mill KPD-1500/120 in full load mode can consume power up to 400 kW. It is clear that this indicator largely depends on the physical and mechanical characteristics of the ore, but the question arises about the efficiency of any industrial equipment. If you analyze the operation of the crusher, it becomes clear that the efficiency of its operation depends on how fully the energy of the electric drive is converted to work on rock material destruction. For example, depending on the lumpiness of the incoming ore, the productivity of the first stage KPD-1500/180 crushers varies by 25–40%. Thus, at the enterprise, it would be quite useful to monitor the efficiency of all energy-intensive units. This would give an opportunity to monitor the rational use of energy resources in real time and identify the weakest areas of the technological process.

3 Literature Review

Many scientific papers have been devoted to the problems of constructing adaptive and optimal control systems [1–7]. In some of them, situational control or neural networks approaches are applied. To a large extent, scientists are trying to create such systems that could work under conditions of partial uncertainty and under the influence of stochastic disturbances. However, the progressive development of methods and algorithms of artificial intelligence causes the emergence of new approaches in the construction of intelligent CDSS. Over the past decade, many works have been devoted to the topic of the CDSS. However, each of these systems is highly specialized, focusing on a specific subject area, it satisfies only the needs of a particular branch of the economy. In addition, most of the considered CDSS were created primarily for organizational management of operational resources, optimal work planning, analysis of financial and economic indicators of production, management of fixed assets of the enterprise, and the like.

With regard to the issue of the application of intelligent CDSS in the mining industry, it remains relevant today. Now most systems with elements of artificial intelligence are built on the basis of neural networks, so the main attention is paid to finding solutions to the problem of decision-making with the help of alternative approaches.

Proceeding from the foregoing, it was decided to develop architecture of intelligent CDSS that would monitor local automatic control systems and subordinate their algorithms to a certain global criterion, for example, a criterion for minimizing the cost of production. We should propose an approach to solving the optimization problem with a large number of incoming and outgoing technological parameters describing the work of a separate redistribution of the enterprise. It is also necessary to take into account the absence or complexity of constructing of a priori mathematical models of industrial facilities. In addition, various stochastic disturbances that adversely affect the quality of the source product must necessarily be monitored and tracked by CDSS in order to minimize their influence.

4   Discussion of Research Results

Figure 1 represents a block diagram of an automated control system based on an intelligent CDSS. The decision-making system performs higher-order controls. It monitors the value of the global control criterion $f(X)$ and specifies the input vector of signals X for the local automatic control system, which operates according to the classical stabilization scheme and maintains the output parameter vector Y at a constant level. This architecture is due primarily to the fact that the CDSS only works with the static mode of the facility operation, when all the transient processes have ended. The preservation of stability, maintaining the quality of regulation in dynamic modes is performed by a local control system, which can be built on an ordinary proportional-integral-differential regulator.

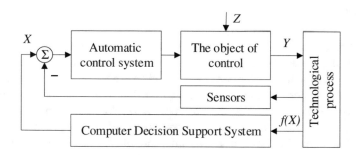

Fig. 1. Structural scheme of the automatic control system based on CDSS.

Scientific works [8, 9] substantiate the effectiveness of applying regression analysis for technological process mathematical model construction in real time on the basis of statistical data of industrial parameters. As the regression equation, the complete polynomial by Kolmogorov-Gabor of the second order was chosen, which for most technical problems according to [10] gives a rather accurate approximation

$$y(x_1, x_2, \ldots, x_m) = \beta_0 + \sum_{i=1}^{m} \beta_i x_i + \sum_{i=1}^{m} \sum_{j=i}^{m} \beta_{ij} x_i x_j, \tag{1}$$

where y is the objective function; $x_1, x_2, \ldots, x_m$ are independent predictor variables; m is the number of variables in the model; $\beta i, \beta ij$ are regression coefficients received by method of the least squares.

After constructing the regression mathematical model, the CDSS uses it to find the optimal parameters of the technological process. Since the m-dimensional surface of function (1) is continuous and differentiated, it is possible to find its minimum in analytical form. For this, according to the theorem on the necessary condition for the existence of an extremum, after computing the partial derivatives with respect to all variables, we should solve the following system of equations

$$
\begin{cases}
\frac{\partial f}{\partial x_1} = \beta_1 + 2\beta_{11}x_1 + \beta_{12}x_2 + \ldots + \beta_{1m}x_m = 0 \\
\frac{\partial f}{\partial x_2} = \beta_2 + \beta_{13}x_1 + 2\beta_{22}x_2 + \ldots + \beta_{2m}x_m = 0 \\
\ldots \\
\frac{\partial f}{\partial x_m} = \beta_m + \beta_{1m}x_1 + \beta_{2m}x_2 + \ldots + 2\beta_{mm}x_m = 0
\end{cases}
\tag{2}
$$

Therefore, we get a stationary point

$$
X_0 = M_\beta^{-1} \cdot V_\beta,
\tag{3}
$$

$$
X_0 = \begin{pmatrix} x_1 \\ x_2 \\ \ldots \\ x_m \end{pmatrix}, \quad
M_\beta = \begin{pmatrix} 2\beta_{11} & \beta_{12} & \cdots & \beta_{1m} \\ \beta_{12} & 2\beta_{22} & \cdots & \beta_{2m} \\ \ldots & \ldots & \ldots & \ldots \\ \beta_{1m} & \beta_{2m} & \cdots & 2\beta_{mm} \end{pmatrix}, \quad
V_\beta = \begin{pmatrix} -\beta_1 \\ -\beta_2 \\ \ldots \\ -\beta_m \end{pmatrix}.
$$

According to the theorem on a sufficient condition for an extremum: if the quadratic form $A(X_0)$ of the second differential of the function at the point X_0 is positive, then the point X_0 is a point of strict minimum, that is

$$
A(X_0) = \left\| \frac{\partial^2 f(X_0)}{\partial x_i \partial x_j} \right\|_{i,j=1}^m .
\tag{4}
$$

Having found the second partial derivatives with respect to all variables, we obtain a sufficient condition for the stationary point X_0 to be the minimum point of the function $f(x_1, x_2, \ldots, x_m)$. In this case, according to the Sylvester criterion, the quadratic form is positive-definite if all its principal (corner) minors are positive, so

$$
\Delta_i = \begin{vmatrix} 2\beta_{11} & \beta_{12} & \cdots & \beta_{1i} \\ \beta_{12} & 2\beta_{22} & \cdots & \beta_{2i} \\ \ldots & \ldots & \ldots & \ldots \\ \beta_{1i} & \beta_{2i} & \cdots & 2\beta_{ii} \end{vmatrix} > 0, i = 1, 2, \ldots, m.
\tag{5}
$$

This method of determining the minimum point has the greatest efficiency on m-dimensional surfaces of parabolic form (Fig. 2a). In this case, the extremum of a function is most often its minimum value from the domain of definition.

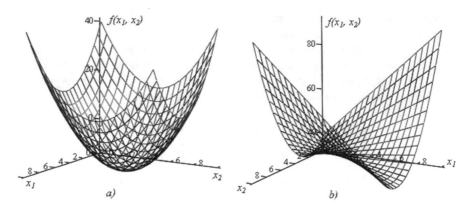

Fig. 2. Graphs of the function, (a) of paraboloid type (b) of saddle type.

However, studies of regression models (1) obtained by analyzing technological statistics show that condition (5) is often not satisfied [10]. Usually, there are functions that have a surface of the "saddle" type. The graph of the surface of such a function is shown in Fig. 2(b). In this case, we need to look for a minimal value in a given domain rather than a strict extremum.

In connection with the above, research was conducted to find the most effective optimization algorithm. Table 1 lists the advantages and disadvantages of some multiparameter optimization methods.

Table 1. Comparison of multi-factor numerical optimization methods

Optimization method	Advantages	Disadvantages
1. Full enumeration method	Simplicity of realization, achievement of predetermined accuracy of the result, guaranteed solution	A large number of computations (over 10^{14}), convergence only if all iterations are performed
2. Gradient descent method	For quadratic functions having one local extremum, the solution can be found for a finite number of steps	Has a slow convergence in the case of elastic functions (possible looping), requires the presence of derivative function, does not work with poor conditionality of statistical data
3. Monte-Carlo method	Much less number of computations than with complete overview (up to 10^{12})	Stochasticity of the result, accuracy and convergence are random variables
4. Method of differential evolution	The target function can be discontinuous and undifferentiated, the presence of a suboptimal solution even with the multi-extremity function	The need to configure at least two parameters of the algorithm, the difficulty of assessing the accuracy of the result, which has a stochastic nature

Finding the extremum of multifactor dependence is a rather complicated calculation procedure that requires considerable computing power. For example, applying a full search for 10 variables, whose values are in the range from 5 to 15 in steps of 0.1, you need an algorithm with 10 nested cycles. At the same time saving mode is calculated 10^{20} times. Even if 0.1 µs is consumed by a processor with a clock frequency of 3 GHz, the entire search process will last more than 316,000 years.

Let's consider the application of the algorithm of differential evolution for multi-dimensional mathematical optimization, which was developed by scientists Storn and Price in 1995 [11, 12]. This algorithm belongs to the class of stochastic optimization algorithms and uses some ideas of genetic algorithms. An important advantage of this method is that it requires only the opportunity of calculating the objective function without taking into account its derivatives. The differential evolution method is designed to find the global extremum of undifferentiated, nonlinear functions of many variables that can have a large number of local extrema.

This method is quite easy to implement on a multi-core computer system, since it is easy to imagine it in the form of parallel computations. A small number of control parameters of this algorithm allows you to configure it for the fastest possible search for an extremum. Figure 3 shows the basic form of the differential evolution algorithm for finding the minimum.

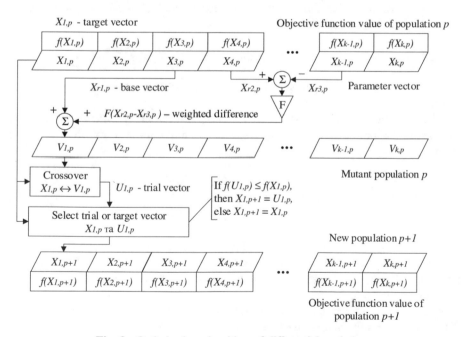

Fig. 3. Optimization algorithm of differential evolution.

First, for the system a certain set of input signal vectors $\{X_{1,p}, X_{2,p}, ..., X_{k,p}\}$ is generated, which is called a generation. A vector is to be understood as the point of the n-dimensional space in which the objective function $f(X)$ is defined, the control criterion. At each iteration the algorithm generates a new generation of vectors, randomly combining the vectors of the previous generation. The number of vectors in each generation is constant and is one of the parameters of the algorithm. The new generation is generated as follows. For each vector $X_{i,p}$ from the old generation, three different random vectors $X_{r1,p}$, $X_{r2,p}$, $X_{r3,p}$ are selected, except for the vector $X_{i,p}$ itself, and the so-called mutant vector is generated by the formula

$$V_{i,p+1} = X_{r1,p} + F(X_{r2,p} - X_{r3,p}), \tag{6}$$

where F is one of the parameters of the method, some positive real number in the interval [0, 2].

The "crossover" operation is performed over the mutant vector $V_{i,p}$, which consists in replacing some of its coordinates with the corresponding coordinates from the initial vector $X_{i,p}$. Each coordinate is replaced with some certain probability, which is also another parameter of the algorithm. The vector V, obtained after crossing, and p, is called a trial vector. If it is better than the vector $X_{i,p}$, that is, the value of the objective function is less, then in the new generation the vector $X_{i,p}$ is replaced by the test vector $V_{i,p}$, and otherwise - $X_{i,p}$ remain.

We simulate the work of the CDSS on the basis of the optimization method of differential evolution. Let the objective function of multidimensional control, which should be minimized, has the form

$$f(X, Z) = 3 + 0.25(x_1 - z_1 - 20)^2 + 0.85(x_2 - 15)^2 + 0.7(x_3 - 10)^2 - 0.3z_2 \rightarrow \min \tag{7}$$

where X is a vector of input signals, Z is a perturbation vector having a random character.

The extremum of the function (7) can easily be found by differentiation. Equation (7) is a paraboloid in n-dimensional space, the minimum value of which is reached when

$$f_{\min} = 3 - 0,3z_2, \begin{cases} x_1 \rightarrow z_1 + 20 \\ x_2 \rightarrow 15 \\ x_3 \rightarrow 10 \end{cases}. \tag{8}$$

Figure 4 represents the process of finding the optimum for the function $f(X, Z)$. For this purpose, the following algorithm settings were chosen: the number of vectors in the population $K = 8$, the weighted difference coefficient $F = 1$, and the probability of permutations $I = 0.5$.

The graphs show the values of the input parameters x_1, x_2, x_3, changing which the CDSS tries to find the minimum of the performance criterion. In this case, for each unit of time, a new population of vectors was generated. The values of the objective function are also presented during the simulation process. It can be seen that the algorithm coincides in approximately 10–15 steps and the error of achieving an

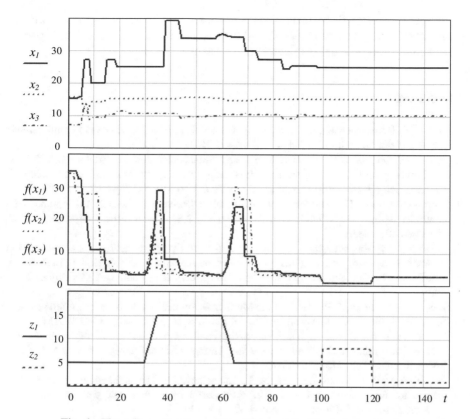

Fig. 4. Time diagrams of searching the target function minimum value.

optimum does not exceed 5–10%. In addition, it should be noted that the action of various perturbations z_1, z_2 is quickly compensated by this algorithm.

5   Conclusion

Having considered modern problems of technological processes automation at mining enterprises, we listed the possible ways of their solution, and also proposed an approach to creation of intelligent CDSS. As shown in the article it is possible to use the complete Kolmogorov-Gabor polynomial for the creation of the mathematical model. It provides sufficient level of accuracy for a major part of technological processes at ore dressing plants. Utilising of CDSS allows calculating regression equation in real time by using technological statistics. This feature helps to minimize the influence of nondeterministic disturbances.

The stochastic optimization method has been compared with the full enumeration method, gradient descent method, and Monte Carlo. The results of the simulation justify the effectiveness of optimization methods based on genetic algorithms. This allows finding an optimum in real-time with an error not exceeding 5–10%.

Further research will be focused on implementing of parallel calculations in the mathematical subsystem of CDSS. It will help to accelerate the procedure of parameters identification on multi-core computing systems.

References

1. Golik, V., Komashchenko, V., Morkun, V.: Geomechanical terms of use of the mill tailings for preparation. Metall. Min. Ind. **4**, 321–324 (2015)
2. Wahid, F., Ghazali, R., Fayaz, M., Shah, A.S.: Statistical features based approach (SFBA) for hourly energy consumption prediction using neural network. Int. J. Inf. Technol. Comput. Sci. (IJITCS) **9**(5), 23–30 (2017). https://doi.org/10.5815/ijitcs.2017.05.04
3. Puangdownreong, D.: Multiobjective multipath adaptive tabu search for optimal PID controller design. Int. J. Intell. Syst. Appl. (IJISA) **7**(8), 51–58 (2015). https://doi.org/10.5815/ijisa.2015.08.07
4. Faradian, A., Manjafarashvili, T., Ivanauri, N.: Designing a decision making support information system for the operational control of industrial technological processes. Int. J. Inf. Technol. Comput. Sci. (IJITCS) **7**(9), 1–7 (2015). https://doi.org/10.5815/ijitcs.2015.09.01
5. Das, T.K., Acharjya, D.P., Patra, M.R.: Multi criterion decision making using intuitionistic fuzzy rough set on two universal sets. Int. J. Intell. Syst. Appl. **7**(4), 26–33 (2015)
6. Roszkowska, E., Burns, T.: Decision-making under conditions of multiple values and variation in conditions of risk and uncertainty. Human-Centric Decision-Making Models for Social Sciences **502**, 315–338 (2014)
7. Kupin, A., Vdovichenko, I., Muzyka, I., Kuznetsov, D.: Development of an intelligent system for the prognostication of energy produced by photovoltaic cells in smart grid systems. East. Eur. J. Enterp. Technol. **8**(89), 4–9 (2017)
8. Kupin, A., Muzyka, I.: Mathematical model of drilling and blasting operations in optimization decision support system. East. Eur. J. Enterp. Technol. **6**(8), 56–59 (2010). (in Ukrainian)
9. Wahba, Y., ElSalamouny, E., ElTaweel, G.: Estimating the sample size for training intrusion detection systems. Int. J. Comput. Netw. Inf. Secur. (IJCNIS) **9**(12), 1–10 (2017). https://doi.org/10.5815/ijcnis.2017.12.01
10. Norman, R., Smith, H.: Applied Regression Analysis, 3rd edn. Wiley-Interscience, New York (2014)
11. Storn, R., Price, K.: Differential evolution – a simple and efficient heuristic for global optimization over continuous spaces. J. Global Optim. **11**(4), 341–359 (1997)
12. Altinoz, O.T.: A comparison of crowding differential evolution algorithms for multimodal optimization problems. Int. J. Intell. Syst. Appl. (IJISA) **7**(4), 1–10 (2015). https://doi.org/10.5815/ijisa.2015.04.01

Optimal Control of Retrial Queues with Finite Population and State-Dependent Service Rate

Vadym Ponomarov[(✉)] and Eugene Lebedev

Taras Shevchenko National University of Kyiv, Kyiv, Ukraine
vponomarov@gmail.com

Abstract. The research of wide class of retrial queuing systems faces the problem of calculating characteristics of the system in stationary regime. Markov chain that describes service process in such system is multidimensional and its transaction matrix usually does not have special properties that would streamline the explicit solution of Kolmogorov set of equations. In addition, the probabilities of transition between states of the controlled system depend on its current state that complicates their obtaining even more. Therefore, only the simplest models are explicitly researched on this moment.

In this paper we consider a finitesource retrial queue with c servers and controlled parameters. The primary calls arrive from n customers. Each customer after some random period of time which is exponential distributed random variable tries to get service and is served immediately if there is any free server. Service times are also exponentially distributed. The customer who finds all servers busy leaves the system and returns after an exponential time. Two- and three-dimensional Markov models that describe threshold and hysteresis control policies are taken into account. Explicit vector-matrix representations of stationary distributions are main results in both cases. Also we state and give an algorithm for solving a multi-criteria problem of maximization of total income from the system.

Keywords: Queue · Repeated calls · Stationary regime · Optimal control

1 Introduction

Classic queuing systems and networks are often used on practice for modeling, analysis and optimization of communication networks [1], mobile cellular networks [2], cloud computing systems [3], traffic management problems [4, 5], etc. In classic queuing theory it is usually assumed that the incoming flow of requests is Poisson. This means that the number of independent sources (potential requests) is infinite and each of them appeals to the system very rarely. It is also assumed that if the request cannot be served immediately, it goes to the queue or it is lost forever. In fact, the assumption of losing request is just preliminary approximation of the real-world situation. Usually the client who cannot get service returns to the system after some random period of time and tries to get service again. The classic queuing systems do not take into account such retrial phenomenon and thus cannot be applied to the variety of practically important problems. To remove that defect the retrial queues were developed.

© Springer International Publishing AG, part of Springer Nature 2019
Z. Hu et al. (Eds.): ICCSEEA 2018, AISC 754, pp. 359–369, 2019.
https://doi.org/10.1007/978-3-319-91008-6_36

Consider a group of c servers in which a flow of calls arrives from n sources. Each source after some random period of time, that is exponentially distributed with rate λ tries to get service. The service time is exponentially distributed variable with parameter v_j, that depends on number of retrials. If all servers are engaged, the call becomes a source of repeated calls and returns to the system after exponentially distributed with rate μ time period.

Detailed comprehensive analysis of standard and retrial queues, main results and literature review can be found in [6, 7]. The phenomenon of repeated calls changes the service process in such a way that classical approaches for steady state analysis become inapplicable. Most of the researches are dedicated to the case of single server [8, 9]. In general case of N servers, the computer modeling [10] or computation algorithms [11] are usually used. Such approaches allow to obtain numerical characteristics for the given system, but do not allow to perform high-level analysis of system's behavior in general case.

So the problem of finding the explicit representation of retrial queues probability characteristics is actual but in general case is an opened problem. The goal of this paper is to give formulas of the steady state probabilities of the system, that can be used for further analysis and to show how these results can be applied for solving the problem of optimization of total income from the system.

First, let us research in more details the functioning of the system in the steady state when it is controlled by different types of policies.

2 Threshold Control Policy

Consider continuous time Markov chain $X(t) = (C(t);N(t))$, where $C(t) \in \{0, 1, ..., c\}$, $N(t) \in \{0, 1, ..., n - c\}$, that is specified by infinitesimal transition rates $a_{(i,j)(i',j')}$, $(i,j), (i',j') \in S(X) = \{0,1,...,c\} \times \{0,1,...,n-c\}$:

1. if $i = \{0, 1, ..., c - 1\}, j = \{0, 1, ..., n - c\}$, then

$$
a_{(i,j)(i',j')} = \begin{cases} (n-i-j)\lambda, & (i',j') = (i+1,j); \\ j\mu, & (i',j') = (i+1,j-1); \\ iv_j, & (i',j') = (i-1,j); \\ -[(n-i-j)\lambda + j\mu + iv_j], & (i',j') = (i,j); \\ 0, & \text{otherwise.} \end{cases} \tag{1}
$$

2. if $i = c, j = \{0, 1, ..., n - c\}$, then

$$
a_{(i,j)(i',j')} = \begin{cases} (n-c-j)\lambda, & (i',j') = (c,j+1); \\ cv_j, & (i',j') = (c-1,j); \\ -[(n-c-j)\lambda + cv_j], & (i',j') = (c,j); \\ 0, & \text{otherwise.} \end{cases} \tag{2}
$$

Markov chain $X(t)$ defines the behavior of the system described above. The number of busy servers at any time is defined by its first component and the number of retrials – by the second.

Let us research the system's characteristics in a stationary regime.

Define π_{ij}, $i = 0, ..., c$, $j = 0, ..., n - c$ stationary probabilities of the system:

$$\pi_{ij} = \lim_{t \to \infty} P\{C(t) = i, N(t) = j\}.$$

Consider the following definitions.

$$e_i(c) = (\delta_{i0} \quad \delta_{i1} \quad \cdots \quad \delta_{ic-1})^T, \quad \delta_{ij} = \begin{cases} 1, & i = j, \\ 0, & i \neq j; \end{cases}$$

$1(c)$ is a vector of length c that consists from 1

$$A_j = \left\| a_{ik}^j \right\|_{i,k=0}^{c-1}, \quad \text{where } a_{ik}^j = \begin{cases} -(n+1-i-j)\lambda, & k = i-1, \\ (n-i-j)\lambda + j\mu + iv_j, & k = i, \\ -(i+1)v_j, & k = i+1, \\ 0, & \text{otherwise}; \end{cases}$$

if $i \neq 0, c - 1$. When $i = 0$

$$a_{0k}^j = \begin{cases} (n-j)\lambda + j\mu, & k = 0, \\ -v_j, & k = 1, \\ 0, & \text{otherwise}, \end{cases}$$

and when $i = c - 1$

$$a_{c-1k}^j = \begin{cases} -(n-c+2-j)\lambda, & k = c-2, \\ (n-c+1-j)\lambda + j\mu + (c-1)v_j, & k = c-1, \\ 0, & \text{otherwise}, \end{cases}$$

$$B_j = \left\| b_{ik}^j \right\|_{i,k=0}^{c-1}, \quad \text{where } b_{ik}^j = \begin{cases} (j+1)\mu, & k = i-1, \\ 0, & \text{otherwise}; \end{cases}$$

if $i \neq 0, c - 1$. When $i = 0$, $b_{ik}^j = 0$, $k = 0, 1, ..., c - 1$, and when $i = c - 1$

$$b_{c-1k}^j = \begin{cases} \dfrac{c(j+1)\mu v_j}{(n-c-j)\lambda}, & k \neq c-2, \\ \dfrac{(j+1)\mu[(n-c-j)\lambda + cv_j]}{(n-c-j)\lambda}, & k = c-2, \end{cases}$$

$$C = \left\| c_{ik} \right\|_{i,k=0}^{c-1}, \quad \text{where } c_{ik} = \begin{cases} 1, & k = 0, i = 0, \\ a_{i-1k}^{n-c}, & \text{otherwise}; \end{cases}$$

$$\Phi_j = \left(\prod_{i=j}^{n-c-1} A_i^{-1} B_i \right) C^{-1} e_0(c).$$

Vector Φ_j is defined correctly by the last equation. As $|C| = (-1)^{c-1}$ $(c-1)! v_{n-c}^{c-1} \neq 0$, so C^{-1} always exists. Matrices A_j, $j = 0, 1, \ldots, n-c$ are not singular because they satisfy the Adamar column condition.

Probabilities π_{ij}, $(i,j) \in S(X)$ can be explicitly expressed in terms of the system's parameters.

Theorem 1. *Stationary probabilities of the system are defined by the following set of equations:*

$$\pi_j = \Phi_j \pi_{0n-c}, j = 0, \ldots, n-c,$$

$$\pi_{cj} = \frac{(j+1)\mu}{(n-c-j)\lambda} 1(c)^T \Phi_{j+1} \pi_{0n-c}, j = 0, \ldots, n-c-1,$$

$$\pi_{cn-c} = \frac{[\lambda + (n-c)\mu + (c-1)v_{n-c}]e_{c-1}^T(c) - 2\lambda e_{c-2}^T(c)}{cv_{n-c}} C^{-1} e_0(c) \pi_{0n-c}$$

where

$$\pi_j = \left(\pi_{0j} \quad \pi_{1j} \quad \ldots \quad \pi_{c-1j} \right)^T,$$

$$\pi_{0n-c} = \left\{ \sum_{j=0}^{n-c} \frac{(n-c+1-j)\lambda + j\mu}{(n-c+1-j)\lambda} e(c)^T \Phi_j \right.$$

$$\left. + \frac{[\lambda + (n-c)\mu + (c-1)v_{n-c}]e_{c-1}^T(c) - 2\lambda e_{c-2}^T(c)}{cv_{n-c}} \Phi_{n-c} \right\}^{-1}.$$

The proof of Theorem 1 is based on the equations of probability flows balance for special subset of the phase space [12]. It should be noted that in case of one or two servers the above vector-matrix equations turn into explicit formulas of the scalar type.

Let us apply the received representation of the stationary probabilities to the problem of maximization of the total income from the finitesource retrial queue that is controlled by threshold policy.

Consider the essence of threshold control policy. With such type of control if the number of retrials in the system does not exceed H, then the system operates in the first mode with rate $v^{(1)}$. If the number of retrials becomes greater than H, the system switches into the second mode with the service rate $v^{(2)}$. Threshold H can have values $-1, 0, \ldots, n-c$. If $H = -1$ the system operates in the second mode all the time, $v = v^{(2)}$. If $H = n-c$, then for any value of $N(t)$ the system operates in the first mode, $v = v^{(1)}$.

Formally threshold control policy means that the service rate is defined as

$$v_j = \begin{cases} v^{(1)}, & when\ j = 0, 1, \ldots, H; \\ v^{(2)}, & when\ j = H+1, \ldots, n-c. \end{cases}$$

Let us define an optimization problem.

Let

$S_i(t,H)$, $i = 1,2$ be the number of requests that were served in the i-th operating mode;

$S_3(t,H)$ be the number of requests that were rejected and became the sources of retrials;

$S_4(t,H)$ be the number of the service intensity switches.

We will define limits $\lim\limits_{t\to\infty} t^{-1}S_i(t, H)$ as $S_i(H)$, $i = 1, \ldots, 4$.

Consider the following optimization problem:

$$C_1 S_1(H) + C_2 S_2(H) - C_3 S_3(H) - C_4 S_4(H) \to \max,$$
$$H = \{-1, 0, 1, \ldots, n-c\} \tag{3}$$

where C_i, $i = 1, 2$ is income from serving one request while operating in the i-th mode;

C_3 is a penalty for request rejection;

C_4 is a penalty for intensity switching.

Solution of the problem (3) gives such threshold H, that maximizes the total income from the given system. Similar optimization problems for the classic single server queues where discussed in [13].

Boundary functionals $S_i(H)$, $i = 1, \ldots, 4$ can be expressed in terms of stationary probabilities:

$$S_1(H) = v^{(1)} \sum_{j=0}^{H} \sum_{i=1}^{c} i\pi_{ij}(H),$$

$$S_2(H) = v^{(2)} \sum_{j=H+1}^{n-c} \sum_{i=1}^{c} i\pi_{ij}(H), S_3(H) = \lambda \sum_{j=0}^{n-c-1} (n-c-j)\pi_{cj}(H),$$

$$S_4(H) = \lambda(n-c-H)\pi_{cH}(H) + (H+1)\mu \sum_{i=0}^{c-1} \pi_{iH+1}(H).$$

Taking into account the result of Theorem 1 one can find the solution of (3).

Now let us take a deep look at another type of control policies - hysteresis policy.

3 Hysteresis Policy

This class of control policies is a generalization of threshold policies. As in the previous case the current operating mode of the system depends on the number of retrials. But intensity switching goes with some delay (hysteresis). The existence of this delay is caused by the fact that the switching is expensive procedure and the best operating mode could not be fully determined only by the number of sources of repeated calls at current moment of time.

The main behavioral model of the system remains the same as in case of threshold policies. The main difference lies in the control process.

The control process based on hysteresis policies can be described as follows. Two integer numbers H_1 and H_2 (that are called thresholds) are fixed. $H_1 \leq H_2$. If the current number of retrials in the system does not exceed H_1 then the system operates in the first mode with service rate $v^{(1)}$. If the number of retrials is greater than H_2 the system is in the second mode and service rate is equal to $v^{(2)}$. And if the number of retrials lies in $(H_1, H_2]$ then the system keeps its previous operating mode. Thresholds H_1 and H_2 take values $-1, 0, \ldots, n - c$. In case $H_1 = H_2$ the hysteresis policy turns into the threshold policy.

For the given hysteresis policy $H = (H_1, H_2]$ the state of the system at any time t can be described in terms of threevariate process $Y(H,t) = \{(C(H,t); N(H,t); R(H,t));$ $t \geq 0\}$, where $C(H,t)$ is the number of occupied servers, $N(H,t)$ is the number of retrials and $R(H,t)$ is current operating mode. If $R(H,t) = 1$ the system operates in the first mode with service rate $v^{(1)}$. If $R(H,t) = 2$ the system operates in the second mode with rate $v^{(2)}$. $Y(H,t)$ is a continuous time Markov chain with $S(Y) = \{0, 1, \ldots, c\} \times \{0, \ldots, n - c\} \times \{1, 2\}$ as a state space. Its infinitesimal characteristics b_{ij}, $i = (i_1, i_2, i_3)$, $j = (j_1, j_2, j_3)$, $(i, j) \in S(Y)$ are defined with the following conditions:

1. if $[(i_1 = 0, \ldots, c - 1) \wedge (i_2 = 0, \ldots, H_2) \wedge (i_3 = 1)] \vee [(i_1 = 0, \ldots, c - 1) \wedge (i_2 = H_1 + 2, \ldots, n - c) \wedge (i_3 = 2)]$, then

$$
b_{ij} = \begin{cases}
(n - i_1 - i_2)\lambda, & \text{when } j = (i_1 + 1, i_2, i_3); \\
i_2\mu, & \text{when } j = (i_1 + 1, i_2 - 1, i_3); \\
i_1 v^{(i_3)}, & \text{when } j = (i_1 - 1, i_2, i_3); \\
-[(n - i_1 - i_2)\lambda + i_2\mu + i_1 v^{(i_3)}], & \text{when } j = i; \\
0, & \text{otherwise;}
\end{cases}
$$

2. if $[(i_1 = c) \wedge (i_2 = 0, \ldots, H_2 - 1) \wedge (i_3 = 1)] \vee [(i_1 = c) \wedge (i_2 = H_1 + 1, \ldots, n - c) \wedge (i_3 = 2)]$, then

$$
b_{ij} = \begin{cases}
(n - c - i_2)\lambda, & \text{when } j = (c, i_2 + 1, i_3); \\
c v^{(i_3)}, & \text{when } j = (c - 1, i_2, i_3); \\
-[(n - c - i_2)\lambda + c v^{(i_3)}], & \text{when } j = i; \\
0, & \text{otherwise;}
\end{cases}
$$

3. if $i = (c, H_2, 1)$, then

$$
b_{ij} = \begin{cases}
(n - c - H_2)\lambda, & \text{when } j = (c, H_2 + 1, 2); \\
c v^{(1)}, & \text{when } j = (c - 1, H_2, 1); \\
-[(n - c - H_2)\lambda + c v^{(1)}], & \text{when } j = i; \\
0, & \text{otherwise;}
\end{cases}
$$

4. if $[(i_1 = 0, \ldots, c - 1) \wedge (i_2 = H_1 + 1) \wedge (i_3 = 2)]$, then

$$b_{ij} = \begin{cases} (n - H_1 - 1 - i_1)\lambda, & \text{when } j = (i_1 + 1, H_1 + 1, 2); \\ (H_1 + 1)\mu, & \text{when } j = (i_1 + 1, H_1, 1); \\ i_1 v^{(2)}, & \text{when } j = (i_1 - 1, H_1 + 1, 2) \\ -[(n - H_1 - 1 - i_1)\lambda + (H_1 + 1)\mu + i_1 v^{(2)}], & \text{when } j = i; \\ 0, & \text{otherwise.} \end{cases}$$

Let us define $\pi_{ij}^{(r)}(H)$, $i = 0, 1, \ldots, c$, $r = 1, 2$, $j = 0, 1, \ldots, n - c$ stationary probabilities of the system:

$$\pi_{ij}^{(r)}(H) = \lim_{t \to \infty} P\{C(H, t) = i, N(H, t) = j, R(H, t) = r\}. \tag{4}$$

Consider the following definitions:

$$A_{rj} = \left\| a_{ik}^j(r) \right\|_{i,k=0}^{c-1}, \quad \text{where } a_{ik}^j(r) = \begin{cases} -(n + 1 - i - j)\lambda, & k = i - 1, \\ (n - i - j)\lambda + j\mu + iv^{(r)}, & k = i, \\ -(i + 1)v^{(r)}, & k = i + 1, \\ 0, & \text{otherwise;} \end{cases}$$

when $i \neq 0, c - 1$. When $i = 0$

$$a_{0k}^j(r) = \begin{cases} (n - j)\lambda + j\mu, & k = 0, \\ -v^{(r)}, & k = 1, \\ 0, & \text{otherwise,} \end{cases}$$

and when $i = c - 1$

$$a_{c-1k}^j(r) = \begin{cases} -(n - c + 2 - j)\lambda, & k = c - 2, \\ (n - c + 1 - j)\lambda + j\mu + (c - 1)v^{(r)}, & k = c - 1, \\ 0, & \text{otherwise,} \end{cases}$$

$$B_{rj} = \left\| b_{ik}^j(r) \right\|_{i,k=0}^{c-1}, \quad \text{Where } b_{ik}^j(r) = \begin{cases} (j + 1)\mu, & k = i - 1, \\ 0, & \text{otherwise;} \end{cases}$$

when $i \neq 0, c - 1$. When $i = 0$, $b_{ik}^j(r) = 0$, $k = 0, 1, \ldots, c - 1$, and when $i = c - 1$

$$b_{c-1k}^j(r) = \begin{cases} \dfrac{c(j+1)\mu v^{(r)}}{(n-c-j)\lambda}, & k \neq c - 2, \\ \dfrac{(j+1)\mu[(n-c-j)\lambda + cv^{(r)}]}{(n-c-j)\lambda}, & k = c - 2, \end{cases}$$

$$C_{rj} = \left\| c_{ik}^j(r) \right\|_{i,k=0}^{c-1}, \quad \text{where } c_{ik}^j(r) = \begin{cases} \dfrac{(H_1 + 1)c\mu v^{(r)}}{(n-c-j)\lambda}, & i = c - 1, \\ 0, & \text{otherwise;} \end{cases}$$

$$D = \|d_{ik}\|_{i,k=0}^{c-1}, \quad \text{where } d_{ik} = \begin{cases} 1, & k = 0, \ i = 0, \\ a_{i-1k}^{n-c}(2), & \text{otherwise}; \end{cases}$$

$$h(j) = \begin{cases} 0, & j > H_2 \text{ or } H_1 = H_2, \\ 1, & \text{otherwise}; \end{cases}$$

$$\Phi_j^{(1)} = \left(\prod_{i=j}^{H_1} A_{1i}^{-1} B_{1i} \right) \left[h(j) \left(\prod_{i=j\vee H_1 + 1}^{H_2 - 1} A_{1i}^{-1} B_{1i} \right) A_{1H_2}^{-1} C_{1H_2} + \sum_{k=j\vee H_1 + 1}^{H_2 - 1} \left(\prod_{i=j\vee H_1 + 1}^{k-1} A_{1i}^{-1} B_{1i} \right) A_{1k}^{-1} C_{1k} \right.$$

$$\left. + \left(\text{sgn}(H_1 + 1 - j) \right)^+ E \right] \left[E + \sum_{k=H_1+1}^{H_2} \left(\prod_{i=H_1+1}^{k-1} A_{2i}^{-1} B_{2i} \right) A_{2k}^{-1} C_{2k} \right]^{-1} \left[\prod_{i=H_1+1}^{n-c-1} A_{2i}^{-1} B_{2i} \right] D^{-1} e_0(c),$$

$$\Phi_j^{(2)} = \left(E - \left[\sum_{k=j}^{H_2} \left(\prod_{i=j}^{k-1} A_{2i}^{-1} B_{2i} \right) A_{2k}^{-1} C_{2k} \right] \left[E + \sum_{k=H_1+1}^{H_2} \left(\prod_{i=H_1+1}^{k-1} A_{2i}^{-1} B_{2i} \right) A_{2k}^{-1} C_{2k} \right]^{-1} \prod_{i=H_1+1}^{j-1} A_{2i}^{-1} B_{2i} \right)$$

$$\times \left(\prod_{i=j}^{n-c-1} A_{2i}^{-1} B_{2i} \right) D^{-1} e_0(c).$$

Hereinafter let us omit index H to shorten the formulas. For $\pi_{ij}^{(r)}(H) = \pi_{ij}^{(r)}$ the result similar to Theorem 1 comes true.

Theorem 2. *Stationary probabilities of the system controlled by hysteresis policy can be found in the following form:*

$$\pi_j^{(1)} = \Phi_j^{(1)} \pi_{0n-c}^{(2)}, \ j = 0, \ldots, H_2,$$

$$\pi_j^{(2)} = \Phi_j^{(2)} \pi_{0n-c}^{(2)}, \ j = H_1 + 1, \ldots, n - c,$$

$$\pi_{cj}^{(1)} = \frac{\mu}{(n-c-j)\lambda} 1(c)^T \left[(j+1)\Phi_{j+1}^{(1)} + (\text{sgn}(j - H_1 + 1))^+ (H_1 + 1)\Phi_{H_1+1}^{(2)} \right] \pi_{0n-c}^{(2)},$$
$$j = 0, \ldots, H_2,$$

$$\pi_{cj}^{(2)} = \frac{\mu}{(n-c-j)\lambda} 1(c)^T \left[(j+1)\Phi_{j+1}^{(2)} - (\text{sgn}(H_2 + 1 - j))^+ (H_1 + 1)\Phi_{H_1+1}^{(2)} \right] \pi_{0n-c}^{(2)},$$
$$j = H_1 + 1, \ldots, n - c - 1,$$

$$\pi_{cn-c}^{(2)} = \frac{[\lambda + (n-c)\mu + (c-1)v^{(2)}]e_{c-1}^T(c) - 2\lambda e_{c-2}^T(c)}{cv^{(2)}} D^{-1} e_0(c) \pi_{0n-c}^{(2)}$$

where

$$\pi_j^{(r)} = \left(\pi_{0j}^{(r)} \quad \pi_{1j}^{(r)} \quad \cdots \quad \pi_{c-1j}^{(r)} \right)^T,$$

$$\pi_{0n-c}^{(2)} = \left\{ \sum_{j=0}^{H_2} \frac{(n-c+1-j)\lambda+j\mu}{(n-c+1-j)\lambda} 1(c)^T \Phi_j^{(1)} + \sum_{j=H_1+1}^{n-c} \frac{(n-c+1-j)\lambda+j\mu}{(n-c+1-j)\lambda} 1(c)^T \Phi_j^{(2)} \right.$$

$$\left. + \frac{\left[\lambda+(n-c)\mu+(c-1)v^{(2)}\right]e_{c-1}^T(c) - 2\lambda e_{c-2}^T(c)}{cv^{(2)}} D^{-1} e_0(c) \right\}^{-1}.$$

Theorem 2 gives the analytical representation of the finitesource retrial queue stationary probabilities in case of hysteresis control policy $(H_1, H_2]$. These results are similar to the results of Theorem 1 and in case when $H_1 = H_2$ turn into the same formulas, because the threshold control policy is a partial case of hysteresis policy.

The optimization problem is similar to the previously defined problem for the system with threshold control.

$$C_1 S_1(H_1, H_2) + C_2 S_2(H_1, H_2) - C_3 S_3(H_1, H_2) - C_4 S_4(H_1, H_2) \to \max,$$
$$H_1 = \{-1, 0, 1, \ldots, n-c-1\}, \ H_2 = \{0, 1, \ldots, n-c-1\}, \tag{5}$$
$$H_1 \leq H_2,$$

The problem is to find such thresholds H_1 and H_2, that maximize the total income from the given system.

Boundary functionals $S_i(H)$, $i = 1, \ldots, 4$ can be expressed in terms of stationary probabilities of the system:

$$S_1(H_1, H_2) = v^{(1)} \sum_{j=0}^{H_2} \sum_{i=1}^{c} i\pi_{ij}^{(1)},$$

$$S_2(H_1, H_2) = v^{(2)} \sum_{j=H_1+1}^{n-c} \sum_{i=1}^{c} i\pi_{ij}^{(2)}, S_3(H_1, H_2) = \lambda \sum_{j=0}^{n-c-1} (n-c-j)\left(\pi_{cj}^{(1)} + \pi_{cj}^{(2)}\right),$$

$$S_4(H_1, H_2) = \lambda(n-c-H_2)\pi_{cH_2}^{(1)} + (H_1+1)\mu \sum_{i=0}^{c-1} \pi_{iH_1+1}^{(2)}.$$

The solution of the problem can be found by using the aforementioned results for the stationary probabilities.

4 Numerical Results

To illustrate the application of the obtained results and to show the effect of different types of control policies, let us consider a finitesource retrial queue with the following parameters: $c = 4$, $n = 60$, $\lambda = 0.13$, $v^{(1)} = 5$, $v^{(2)} = 7$, $\mu = 1$. The coefficients of multi-criteria problem are $C_1 = 50$, $C_2 = 10$, $C_3 = 100$, $C_4 = 50$.

For the given system we have obtained solutions of the problems (3) and (5) using computer program, that is based on the results of Theorems 1 and 2. Let us compare total income from the uncontrolled system with the total income from the systems, that are controlled by optimal threshold and hysteresis strategies. The uncontrolled system, that operates in the first mode all the time produces 609.72 of total income, whereas the system, that operates only in the second mode gives the value of *180.18*. The solution of problem (3) gives an optimal threshold policy $H = 12$, that allows to obtain the total income equal to *684.57*. An optimal hysteresis policy, that is a solution of problem (5), is $H_1 = 0$, $H_2 = 50$ with the value of total income equal to *748.37*. Comprehensive results are given in the Table 1.

Table 1. Total income from the system under different control policies.

	Total income	%
Uncontrolled system (first mode only)	609.72	338.40
Uncontrolled system (second mode only)	180.18	100.00
Threshold control policy	684.57	379.90
Hysteresis control policy	748.37	415.40

5 Conclusion

In this paper we have presented research of controlled finitesource retrial queues in stationary regime. Two types of control policies were discussed: threshold and hysteresis. For both types of service control we have obtained stationary probabilities through the model parameters in closed form. These representations coincide with known results. Obtained results allow to provide further detailed analysis of the model, calculate its performance measures, solve optimization problems. We have stated and presented algorithms for finding a solution of multi-criteria optimization problem of total income maximization. These results allow to find an optimal solution in class of threshold and hysteresis control policies.

References

1. Das, I., Lobiyal, D.K., Katti, C.P.: Queuing effect on multipath routing in mobile ad hoc networks. Int. J. Inf. Eng. Electron. Bus. (IJIEEB) **8**(1), 62–68 (2016). https://doi.org/10.5815/ijieeb.2016.01.07
2. Tran-Gia, P., Mandjes, M.: Modeling of customer retrial phenomenon in cellular mobile networks. University of Wurzburg, Institute of Computer Science Report 142 (1996)
3. Saxena, D., Chauhan, R.K., Kait, R.: Dynamic fair priority optimization task scheduling algorithm in cloud computing: concepts and implementations. Int. J. Comput. Netw. Inf. Secur. (IJCNIS) **8**(2), 41–48 (2016). https://doi.org/10.5815/ijcnis.2016.02.05
4. Rao, A.S.: Improving the serviceability of a prepaid autorickshaw counter using queuing model: an optimization approach. Int. J. Inf. Technol. Comput. Sci. (IJITCS) **9**(12), 19–27 (2017). https://doi.org/10.5815/ijitcs.2017.12.03
5. Adebiyi, R.F.O., Abubilal, K.A., Tekanyi, A.M.S., Adebiyi, B.H.: Management of vehicular traffic system using artificial bee colony algorithm. Int. J. Image Graph. Sig. Process. (IJIGSP) **9**(11), 18–28 (2017). https://doi.org/10.5815/ijigsp.2017.11.03
6. Artalejo, J.R., Gómez-Corral, A.: Retrial Queuing Systems: A Computational Approach. Springer, Heidelberg (2008)
7. Falin, G.I., Templeton, J.G.C.: Retrial Queues. Chapman and Hall, London (1997)

8. Gomez Corral, A.: Analysis of a single-server retrial queue with quasi-random input and nonpreemptive priority. Comput. Math. Appl. **43**, 767–782 (2002)
9. Falin, G.I., Gomez Corral, A.: On a bivariate Markov process arising in the theory of single-server retrial queues. Stat. Neerl. **54**, 67–78 (2000)
10. Bolch, G., Roszik, J., Sztrik, J., Wuechner, P.: Modeling finite-source retrial queueing systems with unreliable heterogeneous servers and different service policies using MOSEL. HAS-DFG (2005)
11. Roszik, J., Sztrik, J.: Performance analysis of finite-source retrial queueing systems with heterogeneous non-reliable servers and different service policies. Technical report 6 (2004)
12. Walrand, J.: An Introduction to Queueing Networks. Prentice Hall, Englewood Cliffs (1988)
13. Dudin, A., Klimenok, V.: Optimization of dynamic control of input load in node of informational computing network. Autom. Technol. **3**, 25–31 (1991)

Deobfuscation of Computer Virus Malware Code with Value State Dependence Graph

Ivan Dychka[1] , Ihor Tereikovskyi[1(✉)] , Liudmyla Tereikovska[2] ,
Volodymyr Pogorelov[1] , and Shynar Mussiraliyeva[3]

[1] National Technical University of Ukraine "Igor Sikorsky Kyiv Polytechnic Institute",
Kyiv, Ukraine
dychka@scs.ntu-kpi.kiev.ua, terejkowski@ukr.net,
volodymyr.pogorelov@gmail.com
[2] Kyiv National University of Construction and Architecture, Kyiv, Ukraine
tereikovskal@ukr.net
[3] Al-Farabi Kazakh National University, Almaty, Kazakhstan
mussiraliyevash@gmail.com

Abstract. This paper deals with improvement of malware protection efficiency. The analysis of applied scientific research on malware protection development has shown that improvement of the methods for deobfuscation of program code being analyzed is one of the main means of increasing efficiency of malware recognition. This paper demonstrates that the main drawback of the modern-day deobfuscation methods is that they are insufficiently adapted to the formalized presentation of the functional semantics of programs being tested. Based on the research results, we suggest that theoretical solutions which have been tried out in program code optimization procedures may be used for code deobfuscation. In the course of the study, we have developed a program code deobfuscation procedure utilizing a value state dependence graph. Utilization of the developed procedure was found to enable presentation of the functional semantics of the programs being tested in a graph form. As the result, identification of malware based on its execution semantics became possible. The paper shows that further research should focus on the development of a method for comparison of the value state dependence graph of the program being tested with corresponding graphs of security software and malware.

Keywords: Deobfuscation · Value state dependence graph · Malware
Code optimization

1 Introduction

The results of research in the field of computer system security [12, 15, 16] indicate that malware protection (MW) has been one of the most important and relevant problems in the field of data security over the last decade. The need to enhance malware protection is confirmed by a great number of well-known examples of computer system infection, which leads not only to the loss of functionality but also to the unauthorised use of the

© Springer International Publishing AG, part of Springer Nature 2019
Z. Hu et al. (Eds.): ICCSEEA 2018, AISC 754, pp. 370–379, 2019.
https://doi.org/10.1007/978-3-319-91008-6_37

infected systems. For example, virus-infected computer systems can send spam-messages without authorisation or participate in distributed DDoS attacks. The threat becomes even greater in the light of mainstreamification of web-oriented social networks which require the installation of potentially dangerous specialised software on the client computer to use them. Another aggravating factor is the possibility of computer system infection during a scheduled software update.

Along with that, the analysis of applied scientific research in the field of protection systems development [1–4, 12–15] shows that the main way of increasing their efficiency is to increase malware recognition accuracy. For this, most antivirus protection suites employ cutting-edge solutions in the field of data mining. However, the use of these solutions is seriously complicated because the modern malware is created with extensive use of different techniques of program code distortion, making it impossible to form an input data set for the recognition system. One of the most common distortion techniques is program code obfuscation. By obfuscation, we mean translation of program code to a form, which preserves its functionality but complicates its analysis, understanding of the operation algorithm and modification in the event of decompilation. Therefore, this paper deals with the problematic of obfuscated program recognition for malware recognition.

2   Analysis of Literature Sources in the Field of Research

According to [1–4, 9, 10, 12–15], obfuscation is one of the most common techniques of program code protection in legitimate software and is used to prevent its illegal copying. Thus, the existence of obfuscated program code itself is not a sufficient indication of malware. Consequently, it is necessary to develop a deobfuscation procedure which must be executed before the program code is submitted to the recognition system. The main task of this procedure is to translate program code so that it can be examined, and its functionality analysed. It is thereby concluded that it is reasonable to adapt the deobfuscation procedure to the common techniques of program code obfuscation: minimisation, meshing, and sealing. It is also determined that the following techniques are mainly used for program code obfuscation in web-oriented software: replacement of carriage return characters with line feed characters, replacement of multiple space characters with one space character, replacement of multiple line feed characters with a line feed character, replacement of comments with a line feed character or space characters, declaration of a set of used variables, call of undefined functions in conditional statements with false conditions, encryption of names of variables and functions, division of encrypted program code into visible and hidden parts, JavaScript script packaging into CSS.

The program code obfuscated using the method of executive process meshing, was found to be the most difficult to analyse. The one obfuscated with other methods is quite easy to interpret.

In addition, [1, 13] present the analysis of the main functionality of existing software designed for obfuscation/deobfuscation of web-oriented program code. It is determined that limited functionality of available deobfuscation instruments is primarily due to the imperfection of their mathematical support.

Talking about [1–4, 10, 12–15], one can claim that the result of use of the declared deobfuscation methods fails to reflect the objective of the obfuscated program execution. In other words, it does not reflect the formalised description of the program code execution sequence, which in its turn substantially complicated the analysis of its functional semantics. Consequently, the recognition of destructive properties indicative of malware becomes more complicated too.

[3] shows that obfuscation procedures used to hide malicious code elements employ the same techniques as those designed to protect program code against illegal copying.

Based on the analysis performed, we can claim that the main drawback of the deobfuscation methods available is that they are insufficiently adapted to the formalised presentation of functional semantics of the programs tested. In addition, a specific analogy is pointed out between the procedure of program code deobfuscation and the well-studied procedure for translation of high-level program code into executable code [11]. This suggests a possibility of correcting the mentioned drawback of the well-known deobfuscation methods by means of integrating theoretical solutions used in translator development into them. One of such solutions involves the presentation of program code as a value state dependence graph, which enables formalisation of program execution semantics.

Therefore, the objective of this study is to develop a program code deobfuscation approach utilising a value state dependence graph.

3   Formalization of Obfuscation Procedure

The logic of obfuscation procedure is to exclude most of the obvious connections from the program code, i.e. to transform the code so as to make investigation and modification of the obfuscated program more complicated and expensive than construction of a new algorithm [1–4]. At the same time, obfuscation procedure must be performed automatically, at a minimum estimated cost.

To provide an accurate definition of obfuscation process, we need to introduce the following terms: initial program code $PR1$, transformation process, $TR()$ and the set of algorithms $PR2_1 \ldots PR2_n$ arising as a result of transformation:

$$PR1 \Rightarrow TR(PR2) = \{PR2_1, PR2_2, \ldots PR2_n\}.$$

In this case, transformation function defines the obfuscation procedure if the following requirements are met: program code $PR2_1 \ldots PR2_n$ runs in the same way as program code PR1, program code $PR2_1 \ldots PR2_n$ is substantially different from program code PR1, application of the available reverse engineering algorithms on the program code $PR2_1 \ldots PR2_n$ fails, application of the available algorithms for program code $PR2_1 \ldots PR2_n$ detransformation into program code PR1 fails, each transformation procedure application on program code PR1 generates new program code $PR2_1 \ldots PR2_n$ with unpredictable structure specifics.

Let us consider the use of the procedure developed and formalise the main types of obfuscation algorithms. We should note that such algorithms are classified into two main groups according to [4, 5]. General (abstract) obfuscation algorithms are those, which are not associated with the specifics of programming language and can be applied even to the assembler code. It is considered more efficient to build the obfuscator based on the abstract algorithm of the procedure which uses all advantages of the specific software code [1–4].

Of abstract obfuscation algorithms, the Collberg's algorithm is the most generic one. While studying the types of obfuscation algorithms, it is reasonable to start with this general scheme and then analyse the methods which can be used during its application. Execution of Collberg's algorithm can be conventionally divided into four main stages (Fig. 1):

- Loading of program code elements $PR1$.
- Loading of libraries.
- Cyclic execution of transformation procedure $TR()$ by isolating a code segment, which is repeated until the required level is reached or system resource is exceeded.
- Program code generation $PR2_n$.

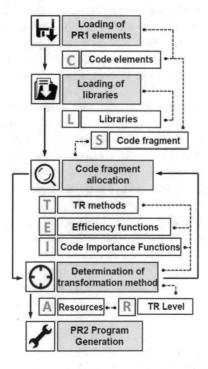

Fig. 1. Pattern of program code obfuscation based on Collberg's algorithm

The input of Collberg's algorithm thus includes:

- Source program code elements *PR*1.
- Standard libraries used in the program code *PR*1.
- Methods of program code transformation.
- The segment of the program code *PR*1 subject to transformation.
- A set of functions that define efficiency of the transformation methods.
- A set of functions defining the importance of code segment S.
- Maximum acceptable cost of system resources that can be used for obfuscation.
- A parameter indicating the required level of program code segment obfuscation.
- Collberg's algorithm is a general pattern of the obfuscation process, while specialised algorithms are defined by obfuscation methods, which can be classified as follows: lexical obfuscation, data obfuscation, control flow obfuscation.

Lexical obfuscation is the simplest type of software protection. It involves code restructuring by deletion or replacement of the comments, deletion of the offsets that are helpful for visual scanning of code, replacement of the identifier (variable, array, function, procedure) names with random character sequences, as well as algorithm block repositioning.

Lexical obfuscation enables the transformation of the program code into a form a programmer cannot analyse quickly and at a moderate cost of hardware resources. However, this method grants extremely low resistance against deobfuscation algorithms.

Data obfuscation, which involves the transformation of data structures, belongs to the group of more complex methods. Obfuscation methods can be divided into three subgroups:

The description of obfuscation subgroups shows that this group of methods requires much more hardware resources, but is more resistant to deobfuscation.

Control flow obfuscation is to obfuscate the sequence of program code execution. Algorithms of this method are based on the use of opaque predicates, i.e. predicates P() the results of which are unknown. In this case, a predicate that always returns "true" is designated as P(t), a predicate that always returns "false" is designated as P(f), and a predicate that can return either is designated as P(t, f).

Opaque predicates can be divided into: local, global, interprocedural.

The efficiency of control flow obfuscation algorithms primarily depends on opaque predicates, which must be sufficiently resistant and flexible in use. In terms of hardware requirements, other important parameters are the time of predicate execution and the number of operations performed during its use. Predicate functions, which aim to increase resistance to the static analysis-based deobfuscation algorithms, should be very similar to software functions.

The control flow obfuscation also includes the methods for computational obfuscation. The most efficient algorithm for computational obfuscation is known as the algorithm of cycle condition extension. Just like in the previous case, it is based on an opaque predicate that simulates influence on the number of cycle code executions. Another efficient pattern is the algorithm of library call elimination. If the software uses standard library functions, the operation principle of these program elements will be known, which can help in reverse engineering. Therefore, names of functions from standard libraries are also transformed in the course of obfuscation. One variety of this approach is to use a proprietary version of libraries (built through the transformation of standard libraries) in software. This technique does not affect the program execution time but increases program size significantly.

4 Deobfuscation Procedure Utilizing a Value State Dependence Graph

Having analysed the available methods of computer virus code deobfuscation, we can claim that deobfuscation procedure is in many ways similar to the procedure of program code optimisation because it often involves incorporation of unnecessary operations and code structure distortion, that does not affect the functionality of the program but hinder the investigation of its operation algorithms. Like deobfuscation, optimisation is aimed at eliminating unnecessary nodes, therefore both can be assigned to the same type of processes on the technical level.

As an internal representation for deobfuscation process we propose to use the Value State Dependence Graph. This graph does not use assignments; the control flow is used only to determine the corresponding operation values, and dependencies are explicit, as well as the conditions for their existence.

In the terms of graph theory, a value state dependence graph ($VSDG$) can be defined as an oriented designated hierarchical graph $G(T, E, l, S, S_0, S_\infty)$, which consists of functional elements. These elements include the following:

- Transitions T are the nodes that correspond to operations.
- Places S are the nodes that correspond to the results of operations.
- Edges E the are operation result dependencies.
- Labelling function l corresponds to each branching operation.
- Arguments S_0 indicate the places wherein the function input arguments are located.
- Results S_∞ indicate the places wherein function output is located.

Each place and each graph edge are typeable by value or state. Edge type is defined by endpoints: The state edge is an edge with a state place being its end-point, and the value edge is the one with a value place being its end-point. Transitions represent VSDG operations effected by the labelling function via the corresponding operator. Transition T's input I_T is a place linked to the branch with an edge. A transition may be considered a place consumer.

In a similar way, a place is called transition T's output O_T is a place with an edge leading from the transition thereto. In this case, a transition may be considered a place producer. A set of transition T's inputs is called transition operands or simply inputs, while a set of transition T's outputs is called transition results or outputs OS_T.

While constructing a VSDG for deobfuscation of potentially malicious software (SW) and code optimisation, the following requirements must be fulfilled:

- Acyclicity: VSDG must not use graph theoretical codes.
- Node arity: each place must have a unique producer (i.e. a distinct edge $E \in T \times S$ must exist).
- Linear use of states: states must act as consumers not more than once.

It is important to note, that VSDG edges must be of the same type, and nodes are described by the following set of simultaneous equations:

$$IS_N = IS_T \wedge OS_N = OS_T.$$

Input nodes are subject to additional conditions:

$$IS_T = \varnothing \wedge OS_T = S_0.$$

Similarly, the following conditions are true for output nodes:

$$OS_T = \varnothing \wedge IS_T = S_\infty.$$

Nodes, which are used in VSDG, can be divided into three types: calculation nodes, γ-nodes, complex nodes.

Calculation nodes simulate simple low-level operations. In turn, they can be subdivided into the following types: value nodes (contain input and output values without additional action), constant nodes (similar to value nodes, but don't have inputs).

State nodes have mixed inputs and outputs and represent operations as additional actions, such as load or store.

γ-nodes are used to express conditional behaviour in VSDG; they perform multiplexing between two sets of operands t and f, which act as predicate functions, based on input predicate p. Operands of both sets, as well as the result of a γ - node execution, shall be of the same type to perform this operation. Characteristically, γ- nodes are the only type of nodes in VSDG that demonstrate the inconsistent behaviour.

Complex nodes are also called regions. A region contains a distinct graph G' and can be substituted with this graph. Characteristically, this graph can contain its own regions; therefore regions, being a separate type of nodes in VSDG, form hierarchic structures.

During code deobfuscation and optimisation, regions may transfer between external and internal regions under certain conditions. However, in this case nesting the property should be kept in mind. A nesting property places a restriction on edges: they must connect nodes only within one region or with a child region. A separate type of complex nodes is θ-nodes. θ-nodes are used on VSDG only for cycles simulation.

Let us consider the example of VSDG application for a program on Fig. 2.

This basic example demonstrates (Fig. 3), how state and value nodes, as well as state and value edges, are used in a graph; state edges are denoted by dashed lines. Figure 3 shows that if the order of value nodes (Product and Sum) is maintained automatically, then state nodes are organised by means of state edges.

```
int32_t fcalc ( int32_t x, int32_t y, int32_t z )
{
z = z + x * y;
return z;
}
```

Fig. 2. Listing of a program code that calculates a mathematical expression

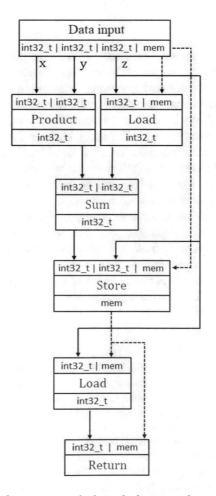

Fig. 3. VSDG of a program code that calculates a mathematical expression

It should be pointed out, that a VSDG, to a certain extent, reflects semantic properties of the program being tested, which are related to the use of computational resources of the computer system. Owing to this capability, application of VSDG is promising for semantic analysis of obfuscated software based on the comparison of the tested program's graph with the corresponding graphs of malware and security software. The development of a relevant comparison method will be tackled in further studies.

Experiments that verify possibility of using the proposed procedure for deobfuscating of JavaScript malware have been conducted. Utilizing VSDG to detect malicious software is reduced to finding on it the states corresponding to attempts to access the program to: the system registry, the file system, network connections, windows, processes, system events, and browser plug-ins. Since in the graph of Fig. 3 such states are absent, it corresponds to a secure software.

5   Conclusion

The analysis of applied scientific research on malware protection development has shown that improvement of the methods for deobfuscation of program code being analysed is one of the main means of increasing efficiency of malware recognition. This paper demonstrates that the main drawback of the modern-day deobfuscation methods is that they are insufficiently adapted to the formalised presentation of the functional semantics of programs being tested. An analogy between the procedures for deobfuscation and program code optimisation has also been identified. Based on the research results, we suggest that theoretical solutions which have been tried out in program code optimisation procedures may be used for code deobfuscation. In the course of the study, we have developed a program code deobfuscation procedure utilising a value state dependence graph. Utilisation of the developed procedure was found to enable presentation of the functional semantics of the programs being tested in a graph form. As the result, identification of malware based on its execution semantics became possible. The paper shows that further research should focus on the development of a method for comparison of the value state dependence graph of the program being tested with corresponding graphs of security software and malware.

References

1. Yadegari, B.N.: Automatic deobfuscation and reverse engineering of obfuscated code. Ph.D. thesis, The University of Arizona, Tucson, USA, 22 September 2016, 200 p (2016)
2. Xu, W., Zhang, F., Zhu, S.: The power of obfuscation techniques in malicious JavaScript code: a measurement study. In: 7th International Conference. Malicious and Unwanted Software, 2012, 8 p (2002). 10.1109
3. Ming, J., Xin, Z., Lan, P., et al.: Impeding behaviour-based malware analysis via replacement attacks to malware specifications. Springer [Electronic], September 2017, pp. 1–13 (2017). https://link.springer.com/article/10.1007/s11416-016-0281-3
4. Robertson, C.: PDF Obfuscation, A Primer [Electronic], SANS Institute Reading Room site, No. 1, 2012, pp. 1–38 (2012). https://www.bing.com
5. Singh, A.: Identifying Malicious Code Through Reverse Engineering. Springer, New York (2009). 196 p
6. Udupa, Sh.K., Debray, S.K., Madou, M.: Deobfuscation: reverse engineering obfuscated code. In: 2005 12th Working Conference on Reverse Engineering (WCRE 2005), No. 13, pp. 1–10 (2005)

7. Lawrence, A.C.: Optimising Compilation with the Value State Dependence Graph. University of Cambridge, Great Britain, 183 p. (Cambridge CB3 0FD) (2008)
8. Nico, R.: Utilising the Value State Dependence Graph for Haskell. University of Gothenburg, Göteborg, Sweden, 68 p (2012)
9. Zhengbing, H., Dychka, I.A., Onai, M., Bartkoviak, A.: The analysis and investigation of multiplicative inverse searching methods in the ring of integers modulo M. Intell. Syst. Appl. **11**, 9–18 (2016)
10. Zhengbing, H., Tereykovskiy, I., Tereykovska, L., Pogorelov, V.: Determination of structural parameters of multilayer perceptron designed to estimate parameters of technical systems. Intell. Syst. Appl. **10**, 57–62 (2017)
11. Pogorelov, V.V., Marchenko, O.I.: Ohlyad vnutrishnikh form predstavlennya prohramy dlya translyatsiyi z protsedurnykh mov prohramuvannya u funktsional'ni movy [Review of internal program presentation forms for translation from procedural programming languages to functional languages]. Sci. Mag. (23), 85–92 (2016). Computer Integrated Technologies: Education, Science, Production
12. Kushnarev, M.V.: Metody i modeli raspoznavaniya vredonosnyh programm na osnove iskusstvennyh immunnyh sistem [Methods and models of malware recognition based on artificial immune systems], Thesis of Candidate of Technical Sciences, Specialty 05.13.23 – Atrificial intelligence systems and tools, Kharkiv, Ukraine, 164 p (2016)
13. Unhul, V.V.: Analysis and development of methods of scripts deobfuscation to identify threats to information computer sustainability. Int. Sci. Mag. **2**(6), 19–27 (2016)
14. Petrov, S.A.: Building adaptive security system based on multi-agent system. In: Materials of the Second International Research and Practice Conference, Westwood, Canada, vol. 2, pp. 196–201 (2013)
15. Hu, Z., Gnatyuk, S., Koval, O., Gnatyuk, V., Bondarovets, S.: Anomaly detection system in secure cloud computing environment. Int. J. Comput. Netw. Inf. Secur. (IJCNIS) **9**(4), 10–21 (2017). https://doi.org/10.5815/ijcnis.2017.04.02
16. Hu, Z., Gnatyuk, V., Sydorenko, V., Odarchenko, R., Gnatyuk, S.: Method for cyberincidents network-centric monitoring in critical information infrastructure. Int. J. Comput. Netw. Inf. Secur. (IJCNIS) **9**(6), 30–43 (2017). https://doi.org/10.5815/ijcnis.2017.06.04

Optimization of Processor Devices Based on the Maximum Indicators of Self-correction

Y. Klyatchenko[(⊠)] [iD], G. Tarasenko, O. Tarasenko-Klyatchenko,
V. Tarasenko, and O. Teslenko

National Technical University of Ukraine "Igor Sikorsky Kyiv Polytechnic Institute",
37, Prosp. Peremohy, Kyiv 03056, Ukraine
k_yaroslav@scs.ntu-kpi.kiev.ua

Abstract. It is proposed to characterize the state of digital processing devices'
outputs with a system of probabilities, where for any binary processing results
the probability of getting the same real result is set, if it is theoretically possible.
Since the outputs of some devices in their compositions are the inputs of other
devices, than such a probability system is universal, that allows taking into
account faults of hardware means. For the original inputs the models of probabilistic information converters (data distortion on the initial inputs system) are
used, on the outputs of which the indicated probabilities system exists, allowing
for distortion of the input data only. Heuristic algorithm for optimal encoding (as
per the maximal possible self-correction) of digital machines (as the models of
processing devices) is proposed for their structural synthesis. The obtained results
enable reaching optimize results on the design stage, that are directed to increase
veracity of computer systems processor devices' functioning.

Keywords: Self-correction · Probability of correct operation
Boolean functions

1 Introduction

In the process of operation of control and monitoring systems of the different objects,
especially of the crucial use objects, the substantial indicator of their efficiency is the
veracity of controlling influences and obtained data. In the computer engineering it is
commonly believed [1], that with input data distortion or with faults the veracity of result
is not guaranteed. At the same time, computer devices can contain latent properties for
correction of input data distortions and failures – the phenomenon of self-correction
(The authors accentuate the term "self-correction" and deliberately avoid the term "auto-
correction", since the latter has an associate connection to the term "automatically", the
property mentioned above is determined by peculiar features of the functions implemented without any automatism). Probability increase of the correct devices operation'
in conditions of the input data distortions and failures with the aide of self-correction,
determines the absolute value of self-correction effect [2]. Ensuring maximal possible
values of self-correction effect on the design stage allows increasing the control quality
and objects monitoring. To carry out the according researches and obtained results

© Springer International Publishing AG, part of Springer Nature 2019
Z. Hu et al. (Eds.): ICCSEEA 2018, AISC 754, pp. 380–390, 2019.
https://doi.org/10.1007/978-3-319-91008-6_38

comparison it is required to make values calculation for self-correction effect. According to [3–7] in the computer engineering it is common that a necessity appears to implement multiple-output combinational schemes upon fractionally-determined Boolean functions, particularly at the stage of encoding of the alphabets' letters of inputs, states and outputs, upon the structural synthesis of digital machines. The goal of the work is to create methods to reduce the number of calculations of self-correction effect' values and the method for optimization of project solutions based on self-correction parameter to use them at the stage of computer systems development.

2   Self-correction Indices Calculation

2.1   Actual Implementations Usage

One of the directions to reducing the volume of calculations of self-correction indices involves using the fact that multiple-digits processing devices are implemented using devices of less-digits capacity. In the work [3] the data distortion' probabilities system is proposed, that is adequate both for input data and results. That has become the basement for the technique creation that enables substantial reduction of computational complexity for self-correction properties' estimates under the condition of corresponding random values, for example upon the following distributing decomposition of the Boolean functions: $f(x_1, x_2, \ldots, x_n) = f_1(x_1, x_2, \ldots, x_t, f_2(x_{t+1}, x_{t+2}, \ldots, x_n))$.

In general case the distributing decomposition is determined the following way:

$$f(x_1, x_2, \ldots, x_n) = f_1(x_1, x_2, \ldots, x_t, f_{21}(x_{t+1}, x_{t+2}, \ldots, x_n), \ldots, f_{2r}(x_{t+1}, x_{t+2}, \ldots, x_n)). \tag{1}$$

The values of $f_{21}(x_{t+1}, x_{t+2}, \ldots, x_n), \ldots, f_{2r}(x_{t+1}, x_{t+2}, \ldots, x_n)$ function are not independent, that does not allow using the indicated technique immediately to calculate the self-correction properties of the f_1 function. The task is to create more universal technique that would allow possibility to calculate the self-correction indicators for variables, separate functions and its corteges.

We will call the cortege of the F functions a normalized set of functions, that depend   on   the   same   variables: $F(x_{t+1}, x_{t+2}, \ldots, x_n) = \langle f_{21}(x_{t+1},$ $x_{t+2}, \ldots, x_n), \ldots, f_{2r}(x_{t+1}, x_{t+2}, \ldots, x_n)\rangle$. Since the cortege functions as random values are not mutually independent, then the probability of the function' cortege value in general cannot be determined using the product of separate functions' probabilities.

For the probability of the functions' cortege it is proposed to use two-dimensional array JF, where the span of each dimension does not exceed 2^r. Assume e as a value of the function' cortege, obtained in the result of data distortions, w is a correct (faultless) value, $(e, w \in \{0, 1, \ldots, 2^r - 1\})$. Then the element of $JF[e, w]$ array is a sum of input data probabilities, when in result of data distortion effect the functions' cortege value F will be equal e, and not w. In case when $e = w$ the probability of self-correction will be located in the corresponding array element.

The proposed array of probabilities is easily determined for a separate independent variable. Indeed, accordingly [3] we have $p_{gi}^0 + p_{gi}^1 = \left(g_{0i}p_i^0 + g_{ci}^0 p_i^0 + g_{ei}^0 p_i^1\right)$

$+\left(g_{0i}p_i^1 + g_{ci}^1 p_i^1 + g_{ei}^1 p_i^0\right)$, where p_{gi}^0, p_{gi}^1 are probabilities of 0 and 1 values on the i-input in the result of distortions; g_{0i} is the probability of any distortions absence; g_{ci}^0 is the sum of the determined distortions, upon which the null-value x_i does not change (incoming self-correction); g_{ei}^0 is the sum of determined distortions probability upon which the singular x_i value change to the null-value; g_{ci}^1 is the sum of probabilities, upon which the singular x_i value does not change (incoming self-correction); g_{ei}^1 is the sum of determined distortions probability, upon which the x_i null value change to the singular. Then $Jx_i[0,0] = g_{0i}p_i^0 + g_{ci}^0 p^0$, $Jx_i[0,1] = g_{ei}^0 p_i^1$, $Jx_i[1,0] = g_{ei}^1 p_i^0$, $Jx_i[1,1] = g_{0i}p_i^1 + g_{ci}^1 p_i^1$.

The proposed technique for probabilities calculation is more general, in comparison with technique proposed in [3], since it has the sense for the functions' cortege, for the input independent data and for the separate Boolean function. For instance, for the function $f(x_1, x_2, \ldots, x_n)$ we have: $Jf[0, 0]$ is probability of correct 0-value, $Jf[1, 1]$ is probability of correct 1-value, $Jf[0, 1]$ is probability of false 0-value, $Jf[1, 0]$ is probability of false 1-value.

Let's examine the self-correction indicators for the $f(x_1, x_2, \ldots, x_n)$ function under conditions of determined incoming distortions regardless of the superposition (1). As in given in [3], we will determine the value of $M(f)$ of all the corteges (organized variables values' sets) $A_j = <a_{1j}, a_{2j}, \ldots, a_{nj}>, (j = 0, 1, \ldots, 2^n - 1)$. Assume $C_l = <c_{1l}, c_{2l}, \ldots, c_{nl}>, (l = 0, 1, \ldots, 2^n - 1)$ is the cortege of variables values, that consists of the A_j cortege as the result of distortions effect. The probability $P[C_j, A_j]$ of the A_j cortege to the C_l cortege is determined as follows:

$$P[C_l, A_j] = \prod_{i=1}^{n} Jx_i[c_{il}, a_{ij}]. \qquad (2)$$

Let us make up a set of $H_{(1,n)}$ products. Since any A_j cortege can transform to any cortege from $M(f)$ as the result of distortions, then the quantity of $H_{(1,n)}$ equals. Let us divide the $H_{(1,n)}$ set into the following subsets (classes):

- Subset $K_w^0(f)$, that contains products where $C_l = A_j$, $af(A_j) = 0$;
- Subset $K_w^1(f)$, that contains products where $C_l = A_j$, $af(A_j) = 1$;
- Subset $K_c^0(f)$, that contains products where $C_l \neq A_j$, $af(C_l) = f(A_j) = 0$;
- Subset $K_c^1(f)$, that contains products where $C_l \neq A_j$, $af(C_l) = f(A_j) = 1$;
- Subset $K_e^0(f)$, that contains products where $C_l \neq A_j$, $af(C_l) = 0, f(A_j) = 1$;
- Subset $K_e^1(f)$, that contains products where $C_l \neq A_j$, $af(C_l) = 1, f(A_j) = 0$.

Adding the products' values of the $K_w^0(f)$ and $K_c^0(f)$ we shall determine the value of $Jf[0,0]$ as the possibility of the correct 0 value. Adding the products' values of the $K_w^1(f)$ and $K_c^1(f)$ we shall determine the value of $Jf[1, 1]$ as the possibility of the correct 1 value. Adding the products' values of the $K_e^0(f)$ we shall determine the value of $Jf[0,1]$ as the possibility of the false null-value of the f function. Adding the products' values of the $K_e^1(f)$ we shall determine the value of $Jf[0,1]$ as the possibility of the false single-value of the f function. Notice that self-correction possibility is determined by the products' sum of $K_c^0(f)$ and $K_c^1(f)$ classes.

Further we examine GF-array formation for the functions cortege of $F(x_{t+1}, x_{t+2}, \ldots, x_n)$. Let's mark $M(F)$ as a set of values corteges of $x_{t+1}, x_{t+2}, \ldots, x_n$ arguments, $Bs = <b(t+1), s, b(t+2), s, \ldots, bn, s>$ from $M(F), (s = 0, 1, \ldots, 2^{n-t} - 1)$. $D_u = <d_{(t+1),u}, d_{(t+2),u}, \ldots, d_{n,u}> (u = 0, 1, \ldots, 2^{n-t} - 1)$, cortege from $M(F)$, that is formed from the B_s in the result of distortions effect. The probability of $P[D_u, B_s]$ distortion of B_s cortege to the D_u cortege is determined as follows:

$$P[D_u, B_s] = \prod_{i=t+1}^{n} Jx_i[d_{iu}, b_{is}]. \tag{3}$$

Let's generate a $H_{(t+1,n)}$ products set (3). Assume e and w any possible values of the F functions cortege. Let's decompose the set $H_{(t+1,n)}$ to subset $K(e, w)$ each contains all products, for which $e = F(D_u), w = F(B_s)$ is valid. Then $JF[e,w]$ is element of the JF array will contain the sum of all products of the $K(e, w)$ subset.

Let's examine the formation of Jf_1 array for the $f_1(x_1, x_2, \ldots, x_t, F)$ function. Let's mark $A_{\alpha,\gamma} = <a_{1\alpha}, a_{2\alpha}, \ldots, a_{t\alpha}, w_\gamma>$, $(\alpha = 0, 1, \ldots, 2^t - 1, \gamma = 0, 1, \ldots, 2^r - 1$ as the cortege of variables of $x_1, x_2, \ldots, x_t$ values and values of the F functions' cortege, $C_{\beta,\delta} = <c_{1\beta}, c_{2\beta}, \ldots, c_{t\beta}, e_\delta> (\beta = 0, 1, \ldots, 2^t - 1, \delta = \gamma = 0, 1, \ldots, 2^r - 1)$ are the values' cortege and the values of functions' cortege F that is formed from $A_{\alpha,\gamma}$ as the result of distortions effect. The probability $P[C_{\beta,\delta}, A_{\alpha,\gamma}]$ of cortege $A_{\alpha,\gamma}$ distortion to $C_{\beta,\delta}$ cortege is determined as follows

$$P[C_{\beta,\delta}, A_{\alpha,\gamma}] = (\prod_{i=1}^{t} Jx_i[c_{i\beta}, a_{i\alpha}])(JF[e_\delta w_\gamma]). \tag{4}$$

Let's generate the $H_{(1,t+r)}$ set of products (4). The quantity of elements of $H_{(1,t+r)}$ equals $4^t \times 2^{r+1}$. Let's decompose the $H_{(1,t+r)}$ set into the following subsets (classes):

- Subset $K_w^0(f)$, that contains products where $C_{\beta,\delta} = A_{\alpha,\gamma}, af_1(A_{\alpha,\gamma}) = 0$;
- Subset $K_w^1(f_1)$, that contains products where $C_{\beta,\delta} = A_{\alpha,\gamma}, af_1(A_{\alpha,\gamma}) = 1$;
- Subset $K_c^0(f_1)$, that contains products where $C_{\beta,\delta} \neq A_{\alpha,\gamma}, af_1(C_{\beta,\delta}) = f_1(A_{\alpha,\gamma}) = 0$;
- Subset $K_c^1(f_1)$, that contains products where $C_{\beta,\delta} \neq A_{\alpha,\gamma}, af_1(C_{\beta,\delta}) = f_1(A_{\alpha,\gamma}) = 1$;
- Subset $K_e^0(f_1)$, that contains products where $C_{\beta,\delta} \neq A_{\alpha,\gamma}, af_1(C_{\beta,\delta}) = 0, f_1(A_{\alpha,\gamma}) = 1$;
- Subset $K_e^1(f_1)$, that contains products where $C_{\beta,\delta} \neq A_{\alpha,\gamma}, af_1(C_{\beta,\delta}) = 1, f_1(A_{\alpha,\gamma}) = 0$.

By the adding products' values of the $K_w^0(f_1)$ and $K_c^0(f_1)$ classes we shall determine $Jf_1[0,0]$ value as the probability of correct value of 0. Adding the products' values of the

$K_w^1(f_1)$ and $K_c^1(f_1)$ we shall determine the value of $Jf_1[1, 1]$ as the possibility of the correct 1 value. Adding the products' values of the $K_e^0(f_1)$ we shall determine the value of $Jf_1[1,0]$ as the possibility of the false null-value of the f_1 function. Adding the products' values of the $K_e^1(f)$ we shall determine the value of $Jf_1[1,0]$ as the possibility of the false 1-value of the f_1 function. Notice that self-correction possibility is determined by the products' sum of $K_c^0(f_1)$ and $K_c^1(f_1)$ classes.

Assertion. The Jf and Jf_1 arrays are identical.

Proving. Assume $\hat{A} = <a_1, a_2, \ldots, a_t>$ as a random cortege of variables' values $x_1, x_2, \ldots, x_t$, $\hat{C} = <c_1, c_2, \ldots, c_t>$ is some distortion of the $\hat{A}$ cortege. $P[\hat{C}, \hat{A}] = Jx_1[c_1, a_1] \times Jx_2[c_2, a_2] \times \ldots \times Jx_t[c_t, a_t]$ is the probability of such distortion. According to the distribution rule in the field of real numbers, by the grouping of corresponding products (2) from the set $H_{(1,n)}$, we shall create subset of $Q(\hat{C}, \hat{A} \, e, w) = P[\hat{C}, \hat{A}] \times K(e, w)$. When forming classes of $H_{(1,n)}$ set any products from the $Q(\hat{C}, \hat{A} \, e, w)$ set can be used independently. At the same time, when forming classes of $H_{(1,t+r)}$ set one product is used ($P[\hat{C}, \hat{A}] \times JF[e, w]$). Consequently, to prove Jf and Jf_1 identical it is sufficient to ascertain that upon any $\hat{C}, \hat{A}, e$ and w one of the following correlations are preserved – $Q(\hat{C}, \hat{A}, e, w) \subset K_w^0 \, (f) \cup K_c^0(f)$, or $Q(\hat{C}, \hat{A}, e, w) \subset K_w^1(f) \cup K_c^1(f)$, or $Q(\hat{C}, \hat{A}, e, w) \subset K_e^0(f)$, or $Q(\hat{C}, \hat{A}, e, w) \subset K_e^0(f)$.

Let's mark $B_0 = <b_{(t+1),0}, b_{(t+2),0}, \ldots, b_{n,0}>$, $B_1 = <b_{(t+1),1}, b_{(t+2),1}, \ldots, b_{n,1}>$ are random corteges from $M(F)$, $D_0 = <d_{(t+1),0}, d_{(t+2),0}, \ldots, d_{n,0}>$ and $D_1 = <d_{(t+1),1}>$ and $D_1 = <d_{(t+1),1}, d_{(t+2),1}, \ldots, d_{n,1}>$ are distorted corteges B_0 and B_1 accordingly. Using concatenation we shall create the $A_0 = \hat{A} \,||\, B_0$, $A_1 = \hat{A} \,||\, B_1$, $C_0 = \hat{C} \,||\, D_0$, $C_1 = \hat{C} \,||\, D_1$ corteges. Let's determine probability of corteges' transformation:

$$P[D_0, B_0] = Jx_{t+1}[d_{(t+1),0}, b_{(t+1),0}] \times \ldots \times Jx_n[d_{n0}, b_{n0}]. \tag{5}$$

$$P[D_1, B_1] = Jx_{t+1}[d_{(t+1),1}, b_{(t+1),1}] \times \ldots \times Jx_n[d_{n1}, b_{n1}]. \tag{6}$$

$$P[C_0, A_0] = Jx_1[c_1, a_1] \times \ldots \times Jx_t[c_t, a_t] \times Jx_{t+1}[d_{(t+1),0}, b_{(t+1),0}] \times \ldots \times Jx_n[d_{n0}, b_{n0}]. \tag{7}$$

$$P[C_1, A_1] = Jx_1[c_1, a_1] \times \ldots \times Jx_t[c_t, a_t] \times Jx_{t+1}[d_{(t+1),1}, b_{(t+1),1}] \times \ldots \times Jx_n[d_{n1}, b_{n1}]. \tag{8}$$

Let's mark $w_0 = F(B_0), e_0 = F(D_0), w_1 = F(B_1), e_1 = F(D_1)$.

Assume $w_0 = w_1 = w$ and $e_0 = e_1 = e$, so the products of (5) and (6) belong to the same subset of $K(e, w)$. Considering (1) for any corteges that are used alongside forming products from $Q(\hat{C}, \hat{A}, e, w)$ we have

$$f(A_0) = f_1(\hat{A}, w) = f(A_1). \tag{9}$$

$$f(C_0) = f_1(\hat{C}, e) = f(C_1). \tag{10}$$

Let's examine the case when $e \neq w$. Then $F(B_0) \neq F(D_0)$, $F(B_1) \neq F(D_1)$, i.e. $B_0 \neq D_0$, $B_1 \neq D_1$, and thus $A_0 \neq C_0$, $A_1 \neq C_1$ independently of equality or inequality of the $\hat{A}$ and $\hat{C}$ corteges. From (9) and (10) it comes out that the products (7) and (8) belong to the same sets separation class of $H_{(1,n)}$. Indeed, if $f(A_0) = f(C_0)$, then this is one of the classes of $K_c^0(f)$ or $K_c^1(f)$ self-correction, if $f(A_0) \neq f(C_0)$, then this is one of the distorted $f - K_e^0(f)$ or $K_e^1(f)$ functions classes.

Let's examine the case when $e = w$ (case of the F functions' cortege self-correction). In this case two options are possible.

Option 1.

$\hat{A} \neq \hat{C}$, then $A_0 \neq C_0$, $A_1 \neq C_1$ independently of the correlation between B_0 and D_0 and between B_1 and D_1. As before, the products (7) and (8) belong to the same sets separation class of $H_{(1,n)}$ or the class of self-correction or the class of distorted values of function f.

Option 2.

$\hat{A} = \hat{C}$, $B_0 \neq D_0$, $B_1 = D_1$. In this case $A_0 \neq C_0$, $A_1 = C_1$. We have $f(A_0) = f_1(\hat{A}, w) = = f(A_1)$, $f(C_0) = f_1(\hat{C}, w) = f(C_1)$.

Since $\hat{A} = \hat{C}$, than $f(A_1) = f(C_1) = f(A_0) = f(C_0)$, then the product (7) will belong to one of the $K_c^0(f)$ or $K_c^1(f)$ classes, and the product (8) is to one of the $K_w^0(f)$ or $K_w^1(f)$ classes, i.e. $Q(\hat{C}, \hat{A}, e, w) \subset K_w^0(f) \cup K_c^0(f)$, or $Q(\hat{C}, \hat{A}, e, w) \subset K_w^1(f) \cup K_c^1(f)$.

Option 3.

$\hat{A} = \hat{C}$, $B_0 = D_0$, $B_1 \neq D_1$, $A_0 = C_0$, $A_1 \neq C_1$. This option is identical to the Option 2.

Option 4.

$\hat{A} = \hat{C}$, $B_0 \neq D_0$, $B_1 \neq D_1$, $A_0 \neq C_0$, $A_1 \neq C_1$. This option is identical to the Option 3.

Application of the outlined technique can substantially reduce the complexity of the self-correction properties of the Boolean functions. For instance, given $n = 8$, and different values of t and r, the acceleration by several times in calculations is being observed, given the $n = 16$ - several thousand times.

2.2   Using the Properties of Boolean Functions

For a random Boolean function of $f(x_1, x_2, \dots, x_n)$, according to the previous, we shall determine the set $M(f)$ of all the $A_j = <a_{1j}, a_{2j}, \dots, a_{nj}>$, $j = 0, 1, \dots, 2^n - 1$ corteges, determine two-dimensional array $GX_{(1,n)}[C, A]$, where any element $GX_{(1,n)}[C_l, A_j]$ will contain the product value of $P[C_j, A_j]$ (2) and two-dimensional array $Jf[0..1, 0..1]$.

In practice it is believed that the indubitably correct function' values are determined as a probability of false argument function values absence. We shall mark this probability as $P(f)$, it is equal to the sum of diagonal elements of $GX_{(1,n)}[A, C]$ array. In fact, the indubitably correct function values are calculated as the sum of $P_c(f) = Gf[0, 0] + Gf[1, 1]$. We shall mark the difference between the given values as

$P_a(f)$ is the possibility of self-correction. In the given context the complexity of self-correction values calculation equals to the complexity of all $GX[C, A]$ array elements calculation, i.e. $4^n(n-1)$ floating-point multiplication operations.

We shall split the $f(x_1, x_2, \ldots, x_n)$ function variables manifold into k groups. We shall mark $m_i (i = 1, 2, \ldots, k)$ as variables quantity in the group, $m_1 + m_2 + \ldots + m_k = n$. We shall calculate the products in every variables group separately, representing the $4^{mi} \times (m_i - 1)$ multiplication operations for the i-group. Using the products multiplications against each other, we shall receive the $H_{(1,n)}$ products set. The total quantity of operations will be equal to:

$$4^{m_1} \times (m_1 - 1) + 4^{m_2}(m_2 - 1) + \ldots + 4^{m_k} \times (m_k - 1) + 4^n \times (k - 1) \le 4^n \times (n - 1). \tag{11}$$

The maximal value $(4^n \times (n - 1))$ of the correlation' left side (11) is achieved upon $m = 1$, i.e. when the separation into groups is absent. The minimal value is achieved upon $k = 2$ and $m_i = n/2$ and equals to $2 \times 4^{n/2} + 4^n$. Herewith, the multiplication operations quantity diminishes approximately $(n - 1)$ times. For the $f(x_1, x_2, \ldots, x_n)$ function we shall select the following option to separate arguments into two groups $(x_1, x_2, \ldots, x_m)$ and $(x_{m+1}, x_{m+2}, \ldots, x_n)$. We shall create the $GX_{(1,m)}$ array of all the products of (2) kind for the first group of values and $GX_{(m+1,n)}$ array for the second group. According to (11), multiplication operations quantity will be $4^m \times (m - 1) + 4^{n-m} \times (m - n - 1) + 4^n$.

Further reduction of this quantity can be achieved taking into account the properties of $f(x_1, x_2, \ldots, x_n)$ function. Using the enumeration of all the possible variables of the first group we shall create 2^m derived $f(d_1, d_2, \ldots, d_m, x_{m+1}, x_{m+2}, \ldots, x_n)$ functions. Assume s as the quantity of different derived $(1 \le s \le 2^m)$ functions. Using r let's mark the quantity of different derived $f(x_1, x_2, \ldots, x_m, d_{m+1}, d_{m+2}, \ldots, d_n)$. Then, the computational complexity will be:

$$4^m \times (m - 1) + 4^{n-m} \times (m - n - 1) + s^2 \times r^2. \tag{12}$$

For instance,

$$f(0, 0, \ldots, 0, x_{m+1}, x_{m+2}, \ldots, x_n) = f(0, 0, \ldots, 1, x_{m+1}, x_{m+2}, \ldots, x_n). \tag{13}$$

From the $GX_{(m+1,n)}$ we shall select a random product of $Gx_{m+1}[c_{m+1}, a_{m+1}] \times \ldots \times Gx_{m+1}[c_n, a_n]$. Creating the $H_{(1,n)}$ we will have the two following products:

$$(Gx_1[0, 0] \times \ldots \times Gx_m[0, 0]) \times (Gx_{m+1}[c_{m+1}, a_{m+1}] \times \ldots \times Gx_{m+1}[c_n, a_n]). \tag{14}$$

$$(Gx_1[0, 0] \times \ldots \times Gx_m[0, 1]) \times (Gx_{m+1}[c_{m+1}, a_{m+1}] \times \ldots \times Gx_{m+1}[c_n, a_n]). \tag{15}$$

Since the condition (13) is met upon any $x_{m+1}, x_{m+2}, \ldots, x_n$ variables values ten the products (14) will belong to the same $H_{(1,n)}$ set class and will be used in the corresponding products' sum to determine the Jf array. The products located in line and column of the $x_1, x_2, \ldots, x_m$ variables with matching derived functions will be located in the same separation class of the $H_{(1,n)}$ set (see Table 1 for the details).

Table 1. Uniting the columns on non-distorted, lines on distorted variables values.

C\A	00..00	...	$<a_1 \ldots a_m>$	...	11..11
00..00	P[00..00,00..00] + P[00..00,00..01] + P[00..01,00..00] + P[00..01,00..01]	...	P[00..00,a_1..a_m] + P[00..01,a_1..a_m]		P[00..00,11..11] + P[00..01,11..11]
...	...	...	...	...	...
$<c_1 \ldots c_m>$	P[c_1..c_m,00..00] + P[c_1..c_m,00..01]	...	P[c_1..c_m, a_1..a_m]	...	P[c_1..c_m, 11..11]
...	...	...	...	...	...
11..11	P[11..11,00..00] + P[11..11,00..01]	...	P[11..11, a_1..a_m]	...	P[11..11,11..11]

The examples given are valid for any $x_1, x_2, \ldots, x_m$ variables values, that result in identical derived functions. The same way, a proof can be made for the second group of variables and derived functions. The final proof of validity (3) is based on the distributive law and symmetry of addition and multiplication operations.

$$\delta = \frac{4^n(n-1)}{4^m \times (m-1) + 4^{n-m} \times (m-n-1) + s^2 \times r^2}. \tag{16}$$

The correlation (16) shows many times the multiplication operation of the technique proposed differs from the standard one.

3 Methods for Devices Optimization

3.1 Optimization of Pre-definitions

Quite often in the engineering practice the Boolean functions of $F(x_1, x_2, \ldots, x_n)$ cortege are not determined on all the codes of $M(F)$ set of argument values. Implementing the combinational scheme, the Boolean functions are pre-defined in order to optimize scheme based on some parameters. For the parameter, we shall choose the self-correction parameter.

We shall mark $W(F) \subset M(F)$ as a set of codes, on which the functioning of combinational scheme is pre-defined under the technical requirements, $E(F) = M(F)/W(F)$. Assume in the result of distortions from the $W(F)$ set' codes the $M(F)$ codes will be created with some probability. We shall determine the $<D_u, B_s, P_{DB}, j>$ corteges, where $B_s \in W(F)_r$, D_u is the result of B_s code distortion, P_{DB} is a probability of such a distortion, $j = F(B_s)$ are codes on the combinational scheme outputs. In the real practice the probability of codes distortion from $W(F)$ to the codes from $W(F)$ can be unequal.

We shall mark Z as a set of $<D_u, B_s, P_{DB}, j>$ corteges, implying that any B_s code from $W(F)$ with one or another probability can be distorted into any code from $M(F)$. For the pre-determination of Boolean functions of the $F(x_1, x_2, \ldots, x_n)$ cortege, in order to ensure the maximal self-correction indicator, the following algorithm is proposed:

Algorithm 1.

(1) Delete the corteges containing $D_u \in W(F)$ from Z-set. According to technical requirements, the D_u distortions are not sensitive, if $D_u \in K_v$ (self-correction), or the functions' cortege forms the false value. In the result such removal a Z_1 set will be formed, where in every cortege $D_u \in E(F)$.

(2) Split the Z_1 set into $Z_1(D_u)$ subsets where in the corteges the values of D_u codes match.

(3) In every set $Z_1(D_u)$ find the cortege with maximal value of P_{DB} and pre-determine the functions' cortege on a D_u code with a V code value of this cortege.

Given the distortion probabilities, a received pre-determination provides the maximal self-correction value. Indeed, a D_u code can be formed in the result of several codes distortion from $W(F)$. Choice of the most probable distortion provides the bigger probability value of self-correction.

3.2   Optimization with Structural Synthesis of Machines

Usually in engineering practice the technical requirements for implementation of combinational scheme are set as $y = f(x)$ function, where $x \in A = \{a_1, a_2, \ldots, a_w\}$, $y \in B = \{b_1, b_2, \ldots, b_u\}$ are finite sets of random objects. At the same time, the problem of optimal encoding by the binary codes of elements of A and B sets on different criteria is being solved. Let's examine encoding optimization based on criteria of self-correction indicators. We shall separate the A set into classes of K_v non-sensitivity, where any a_{i1}, a_{i2} of elements from A belong to K_v, if $f(a_{i1}) = f(a_{i2}) = b_v (v = 1, 2, \ldots, u)$. It is obvious, that the encoding of B-set symbols does not affect indicators of self-correction. To determine optimal elements encoding of the A-set it is sufficient to calculate probabilities sum of mutual distortions within the classes given and choose encoding with maximal sum. Totally, there are $N = (2^n(2^n - 1)(2^n - 2) \ldots (2^n - w + 1)$ encodings possible, where n is codes' number of digits. For example, even with $n = 6$ and $w = 50 N > 10^{78}$. That is why the following heuristic algorithm is proposed. Assume that for the input data an array $GX(1, n)[C, A]$, where any element $GX(1, n)[C_l, A_j]$ $(l, j \in \{0, 1, \ldots, 2^n - 1\})$ contain value of the $P[C_l, A_j]$ product (2). Based on this array, a GXT auxiliary array is created, that contains C_l, A_j codes and value of $P[C_l, A_j] + P[A_j, C_l]$ sum with $A_j \neq C_l$. All elements of the GXT array are listed as non-marked.

Algorithm 2.

(1) Choose the foremost K_v class with the biggest value of $\#K_v$.

(2) Create a TK_v array from all combinations of two elements of the class. All elements of the TK_v array consider as non-marked.

(3) If all elements of the TK_v array array are marked, then go to point 1.

(4) Choose the foremost pair (i.e. a_{i1}, a_{i2}) from the TK_v array.

(5) Choose from the GXT array the foremost non-marked element with the biggest value of probabilities sum. Assign $a_{i1} = C_l$, $a_{i2} = A_j$, where the value (C_l, A_j) correspond to the sum chosen. Assign $c = C_l$, $a = A_j$. Mark up the chosen element in the GXT array. Mark up the $(a_{i1}$, $a_{i2})$ pair in the GXT array.

(6) Choose form the TK_v array the foremost non-marked pair, that contains a_{i1} or a_{i2} elements. Choose from the GXT structures array the foremost non-marked element with maximal sum under condition that one of values of one of the codes of the chosen sum equals to already determined (c or a). For the second element of pair, for instance (a_{i3}) the second from codes chosen is assigned. Mark up pairs $(a_{i1}$, $a_{i3})$ and $(a_{i2}$, $a_{i3})$ of the TK_v array and the corresponding structures of the GXT array. Repeat Point 6 until all pairs containing a_{i1} or a_{i2} elements will not be marked up.

(7) If the K_v class contains only one element, then assign it any code from non-used previously.

(8) If there are classes non-worked out, go to Point 1.

The example examined does not guarantee optimal solution, but provides the result close enough to the optimal. In case when $\#A$ is not the power of 2, for the pre-determination of Boolean functions it is necessary to use Algorithm 1.

4 Conclusions

The Statement proven enables the practical possibility to calculate self-correction parameters considering the real devices structure, in which always the multiple-digits devices are implemented as the composition of lesser digits-quantity' devices. In separate cases a method can be used that is based on the Boolean functions properties, and not on the actual implementation.

The obtained results are a basis for analysis and experimental checkup of the devices optimization methods based on self-correction parameter upon condition of incomplete determinacy of Boolean functions or when to determine Boolean functions it is required to carry out the encoding stage of input alphabet elements. According to point 3 of the Algorithm 1 and points 5 and 6 of the Algorithm 2, upon the equality of probabilities, a choice can be ambiguous. This enables for carrying out the optimization based on other parameters.

References

1. Klyatchenko, Y., Tarasenko, V., Tarasenko-Klyatchenko, O., Teslenko, O.: Reliability evaluation method of functioning logic networks in the distortion of input data determined. Radioelectron. Inform. **5**, 165–169 (2014)
2. Klyatchenko, Y.: The reliability determination of the hardware devices on FPGAs functioning under conditions of input logic signals distortion. Inf. Technol. Comput. Eng. **3**, 9–12 (2015)
3. Klyatchenko, Y., Tarasenko, V., Tarasenko-Klyatchenko, O., Teslenko, O.: The probability of correct operation for logic networks in condition of input signals distortion. Comput.-Integr. Technol.: Educ. Sci. Prod. **8**, 47–52 (2012)
4. Afshord, S.T., Pottosin, Y.: Improved decomposition for a system of completely specified boolean functions. Int. J. Inf. Technol. Comput. Sci. (IJITCS) **6**(1), 25–32 (2014). https://doi.org/10.5815/ijitcs.2014.01.03
5. Wason, R., Soni, A.K., Rafiq, M.Q.: Estimating software reliability by monitoring software execution through opcode. Int. J. Inf. Technol. Comput. Sci. (IJITCS) **7**(9), 23–30 (2015). https://doi.org/10.5815/ijitcs.2015.09.04
6. Vijayakumari, C.K., Mythili, P., James, R.K.: A simplified efficient technique for the design of combinational logic circuits. Int. J. Intell. Syst. Appl. (IJISA) **7**(9), 42–48 (2015). https://doi.org/10.5815/ijisa.2015.09.06
7. Georgiev, D., Tentov, A.: FSM circuits design for approximate string matching in hardware based network intrusion detection systems. Int. J. Inf. Technol. Comput. Sci. (IJITCS) **6**(1), 68–75 (2014). https://doi.org/10.5815/ijitcs.2014.01.08

Simulation of Multithreaded Algorithms Using Petri-Object Models

Inna V. Stetsenko$^{(\boxtimes)}$ and Oleksandra Dyfuchyna

Igor Sikorsky Kyiv Polytechnic Institute, 37 Prospect Peremogy, Kiev 03056, Ukraine
`stiv.inna@gmail.com, sashadif@gmail.com`

Abstract. Multithreaded programming used for the development of faster algorithms is a very effective method. However, the designing, testing and debugging of nontrivial programs are not easy and need to be improved. Stochastic behavior of threads entails their conflicts and in some cases the unpredictable result of the program. Stochastic Petri nets are widely used for the investigation of concurrent processes in many areas: manufacturing, computer systems, workflow management. In this research stochastic multichannel Petri net is considered as a tool for multithreaded programs modeling. The correspondence between main instructions of multithreaded program and fragments of stochastic Petri net is discovered. Petri-object model's formalization and software are used for complicated models' constructions. This approach allows duplicating objects with the same dynamics and aggregating them in model. Models that present the concurrent functioning of multithreaded Java programs are considered. Model's verification indicates its accuracy. The results of experimental research of these models show a strong impact the values of time delay.

Keywords: Multithreaded programming · Stochastic Petri net
Simulation algorithm

1 Introduction

Modern information technologies need fast performance of algorithms that can be achieved using parallel programming. Even simple applications use parallelism in graphics or animation processes. Scientists investigate numerous variants of parallelizing the same algorithm which distinguished by used programming tools and computing resources [24]. The main tools for development of parallel programs are described in [1]. The libraries of parallel programming, which exist in different programming languages, contain standard means which implement the asynchronous and synchronous execution of computing processes. However, the development, testing and debugging of parallel programs is difficult [2]. Furthermore, the debugging of multithreaded programs is even more difficult because of the non-deterministic order of instructions executed by threads. Hence, Lee concludes that "non-trivial multi-threaded programs are incomprehensible to humans" and they should be more predictable and understandable [3].

In general the designing of multithreaded algorithms is a difficult task when use the non-deterministic execution. Special mechanisms of multithreaded execution give the

© Springer International Publishing AG, part of Springer Nature 2019
Z. Hu et al. (Eds.): ICCSEEA 2018, AISC 754, pp. 391–401, 2019.
https://doi.org/10.1007/978-3-319-91008-6_39

possibilities for asynchronies execution. In concern with stochastic behavior of multi-threaded programs their presentation by the tools of discrete event simulation should be considered. Key concepts and algorithms of simulation grounded on system's events description can be found in [4]. Another approach based on formal description by stochastic Petri net is more preferable for complicated dynamic systems [5].

Basics of using Petri nets for systems modeling contains in [6]. The automated transformation of business processes into Petri nets based on the UML description of processes is investigated in [7]. UML diagrams transformation to hierarchical colored Petri nets and the development of Automatic Software Performance Tool are considered in [8]. The use of Petri nets to assess structural properties of multi-agent system is discussed in [22].

Using basic Petri nets for modeling multithreaded applications is outlined in [9]. The development of tool for visualizing of multithreaded java-program using basic Petri nets is presented in [10]. Although the use of basic Petri nets allows presenting separate mechanisms of multithreaded programs, it cannot reproduce the details of parallel algorithm functioning. Only the logic of program is modeled in this manner. However the execution in time is different from execution in steps. As a consequence, to achieve the most accurate reproduction of complicated interaction between threads in one program, simulation modeling should be used. The main purpose of this research is the development of dynamic models of multithreaded program with the use of which the investigation of the concurrent program's functioning can be done.

The functioning of parallel programs strongly depends on computing resources used in program performance. Time performances of different program instructions are distinguish for different operating systems and this fact may cause distinguish functioning of concurrent program. As presented in this research, it depends on ratio of times execution per program instructions as well. Numerous performance issues are dealing with the use of parallel computing resources can be found in [23].

In this research we use stochastic Petri net that allows us to investigate the execution of programs with taking into account the time performance of the actions of threads and to increase the accuracy of constructed model. We can investigate the impact of the time of the actions on algorithm correctness.

Since we are talking about the use of Petri nets for debugging complicated programs, increasing the number of events should not lead to a large increase in both the effort to construct models and the time for its exploration. Formalism and software of Petri-object simulation are the best suited to the constructing of such models.

The implementation of parallel algorithm's simulation is grounded on Petri-object model formalization that allows completing the model of complicated multithreaded algorithm. The Petri-object simulation software is characterized by faster model's performance and construction in comparison with ordinary stochastic Petri net.

2   Multithreaded Programming and Its Problems

Software developers can take advantages of the additional computing resource since the advent of multicore architecture. Instead, additional efforts are required to restructuring software toward parallel execution [11]. The exhaustive description of perils of multi-threading hiding in Java programs can be found in [12].

A parallel program is a collection of simultaneously executing processes. A process is associated with computing subtask described by sequence of programmed instructions. Process runs in an operating system that manages the sharing resources among the community of processes. A thread is an elementary process which is called sometimes lightweight process. Any process consists of one or more threads.

A thread executes the code if resources (cores, shared data) capture is happens. Limit of the resources cause a conflict of threads running. Multithreaded program performance is the concurrently execution of multiple threads.

Thread operations are performed in asynchronous or synchronous manner. Asynchronous performance means that the thread can be executed independently of other threads (except main thread). The synchronous performance means waiting the signal from other threads to continue the performance. Coordination of threads running is needed for example when the shared data access happens. Synchronization defects causing bugs often occur in multithreaded programs [13].

The main problems of parallel program are deadlock and data inconsistency. Deadlock occurs when two or more threads are waiting for signal that never can be received. For example, thread A waits for signal from thread B but the last one is waiting for signal from thread A. Therefore, they fell into a state of mutual expectation of each other. Deadlock has fine description by the means of ordinary Petri net. A well-known example of five philosophers is described by Peterson [6]. Special class of ordinary Petri net called Gadara net is proposed in [14] for a deadlock investigation in multithreaded program.

Data inconsistency means the computing error when more than one thread does concurrent modifying of shared data. The unfolding method of data inconsistency error's detecting in multithreaded programs using specific Petri nets with data (PD-net) is proposed in [15].

In conclusion, the behavior of concurrency algorithm strongly depend on resources and the ratio of the time of instructions executions. Therefore the accuracy model describing such behavior may be constructed using stochastic (not ordinary) Petri net.

3   Petri-Object Models and Simulation

Petri-object simulation is grounded on the stochastic multichannel Petri net and the object-oriented technology. Mathematical description of stochastic Petri net with multichannel transitions is presented in [16]. The features of this Petri net functioning are following. A state S of stochastic Petri net is defined by sets M and E. The set M characterizes the state of places as the values of quantity of tokens in them. The set E characterizes the state of transitions as the values of moments of tokens output. In contrast of other known Petri nets the time delay of transition can be determined by the nonnegative random (with given distribution) or determine value (including zero). While ordinary Petri net perform the tokens input and output as one action [17], timed Petri net needs dividing tokens input and output into two different actions [18]. In every time the slot is corresponded to the nearest event the tokens output and in the same moment the tokens input are performed. Multichannel transition means that tokens input should be repeated until the firing condition is true. A conflict between transitions happens when more than one transition has true condition of firing. Conflict resolution implements the choice of transitions with the biggest priority and then, in the case of equal priority, implements randomly choice from them with taking into account the given probabilities of firing.

The state equations of stochastic Petri net contain the equation that describes the time promotion to the nearest event and the equations that describe the state transformations in accordance with this event [16]:

$$t_n = \min E(t_{n-1}), t_n \geq t_{n-1}, n = 2, 3 \ldots \tag{1}$$

$$S(t_1) = (D^-)^m(S(t_0)), \tag{2}$$

$$S(t_n) = (D^-)^m(D^+(S(t_0))) \tag{3}$$

where the transformations D^+ and $(D^-)^m$ describe tokens output and repeated tokens input correspondently using predicate logic and algebraic expressions.

The main disadvantage of stochastic Petri net is a large number of elements, needed to describe the complex systems and as a consequence much increasing the time of model construction. The Petri-object approach firstly proposed in [19]. Petri-object is the object (in object-oriented technology's terms) that inherits the class PetriSim, whose main field contains description of own dynamics in terms of stochastic multichannel Petri net. In contrast of high level Petri nets, object-oriented Petri nets, net in nets, the Petri-object approach allows to construct model as aggregation (in terms of OO technology) of Petri-objects and to compile the dynamics of model from the objects with given dynamics (in terms of stochastic multichannel Petri net). Moreover, the dynamics of Petri-object model is described by stochastic multichannel Petri net composed of the object's nets. Petri-object simulation algorithm and software is presented in [20].

4   The Main Mechanisms of Multithreading and Their Description in Terms of Stochastic Multichannel Petri Net

4.1   Petri Net Fragments

Considering the main methods of multithreaded algorithm described in Java tutorials [21] we associate them with Petri net fragments. All transitions have such parameters: non negative time delay given as a random value with known distribution or as a determined value, the value of priority and the value of probability. The transition which has higher value of priority is depicted in next figures with bigger width. Note that all transitions that perform program instructions use the core resources (Fig. 1). However, for the simplified presentation those resources will not be shown for every transition, their presence is implied by default.

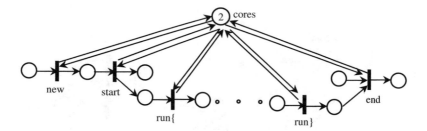

Fig. 1. Implementation of thread's creating, starting and ending by stochastic Petri net.

The creating, starting and ending of thread are initiated by main process the following program code:

```
public static void main(String[] args) {
        Thread thread = new Thread(new Runnable());
        thread.start();
}
```

This code has the equivalent presentation by Petri net that contains branching after event "start", when the new thread is starting (Fig. 1). When the creating thread performance is fulfilled, the event "end" of a thread, which creates another thread, occurs.

The locking is widely used for synchronization and described by following program instructions:

```
private final Lock lock = new ReentrantLock();
try{
    lock = lock.tryLock();
    ...// synchronized actions
} finally {
    lock.unlock();
}
```

Put in line the states "lock" and "unlock" of the object lock. The capture of lock object occurs when it is in state "unlock", and the release of lock object occurs when it is in state "lock" (Fig. 2).

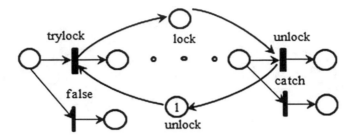

Fig. 2. Implementation of thread's locking by stochastic Petri net.

Guarded block is another mechanism of synchronization that allows setting one of the threads in a waiting state until some condition will not be true. Another thread should run the notify method and send signal to the first thread to check the waiting condition again. Guarded block is always performed within lock block. The program instructions are following:

```
public synchronized void method() throws InterruptedException {
    while (!condition()) {
        wait();
    }
    ...// some actions
    notifyAll();
}
```

Put in line the states of signal the places "signal from another thread" and "signal to another thread" (Fig. 3). Note that alternative event "wait" has higher value of priority. When the threads number in program is more than two the token arrives to place "signal from another thread" from any thread and the token in place "signal to another thread" can be used by any thread.

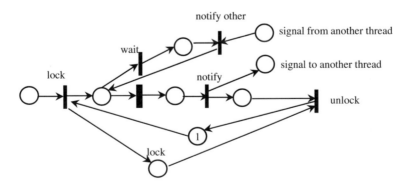

Fig. 3. Implementation of thread's guarded block by stochastic Petri net.

If shared data access is used by more than one thread it can cause a memory consistency error. For example, even a simple increase method for shared data can be executed wrongly. A lock should be used for correct functioning. Appropriate fragment of Petri net contains consequently events "read", "modify" and "write". If lock is not used, another thread can modify shared data value in the same time (Fig. 4).

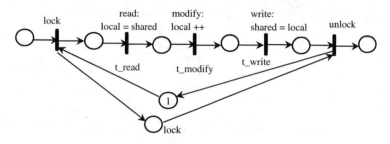

Fig. 4. Implementation of thread's shared data access by stochastic Petri net.

4.2 Petri-Object Model Development for Parallel Program

The main steps of multithreaded program's model construction are: (1) for each class R that implements Runnable interface the stochastic Petri net N that presents the instructions of its run-method should be developed, (2) for each R-class of the program the class S that inherited PetriSim with the developed Petri net N should be developed, (3) for each object O of the class R the object of appropriated class S should be developed (Petri-object), (4) the connections between Petri-objects should be given, (5) the list of Petri-objects is created, (6) the Petri-object model is created with the list of Petri-objects as an argument of its constructor. Thus, each object of Runnable class is put in line with object of PetriSim class. Then we can replicate objects of this class with different parameters using the constructor of corresponded class.

Let us consider the construction of model that simulates the running of producer and consumer threads, given in [21], which are using shared bounded buffer. The producer thread puts object to buffer and uses guarded block to wait while buffer has any free places. The consumer thread has similar dynamics but it takes object from buffer and uses guarded block to wait while buffer has any objects. The methods that created both the producer and consumer threads should be constructed using the collections of appropriated places, transitions and arcs. The connections between threads are created by the common places "signal to another thread", "signal from other thread", "buffers free places", "buffers occupied places". Figure 5 depicts the net of Petri-object Producer and particularly the net elements of Petri-object Consumer which are connected with common places. If the Petri-object may be replicated the model can be extended by adding Producer and Consumer Petri-objects with given parameters to the list of PetriSim objects of model.

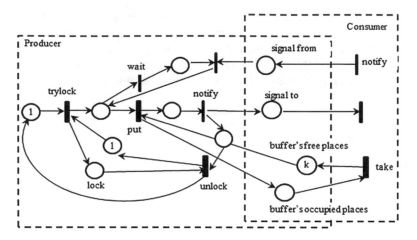

Fig. 5. The net of Petri-object Producer in connection with the net of Petri-object Consumer.

5   Experimentation

5.1   Concurrent Running Model

The deadlock illustrated by friends bowing one to another, then the conflict may be resolved using locks. When one of friends finds the other busied by bowing, he counts faults of tries (Fig. 6). The model is constructed in accordance with the program code, which is given in [21]. The registered values of success and faults quantity are in Table 1. The results show a strong dependence on time delays.

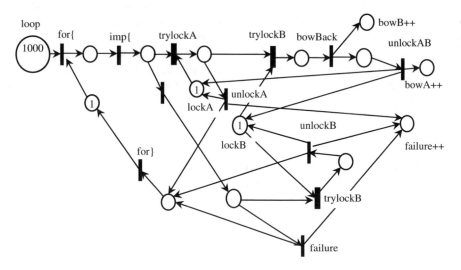

Fig. 6. The net of Petri-object Friend.

The experiment researching the impact of transitions time delays is conducted. The time delay of transition "for{" was set to value d, the other were set to $d \cdot r$, where r is a ratio of transitions time delays. When friend can't bow he will get a failure caused by concurrent running of threads. Hence, exploration of the value of failure relative frequency f characterizes the frequency of threads conflict occurrence. The results of the experiment indicate a strong dependence of the values f and r (Fig. 7). Indeed, when the time delay in the transition "for{" is much greater than the delays in other transitions, it means that the time spent on actions much less than interval between them. Consequently, the probability of simultaneous actions of threads decreases. The comparison of model's result and running program's result is given in Table 1.

Table 1. The comparison of model's result and running of program.

Number of threads	Multithreaded program	Simulation model (d = 100, r = 0.01)	Error
2	0.981450	0.978650	0.29%
4	0.959042	0.963417	0.46%

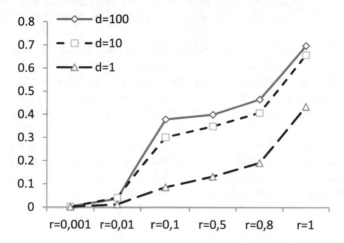

Fig. 7. Model's experimental research of the threads conflict dependence on values of time delays.

5.2   Data Consistency Model

The main problem of shared data access is the data inconsistency when more than one thread tries to concurrently modify shared data. The investigation of this problem is done using a simple example of asynchronous running of counter [21]. If in Petri net presented in Fig. 4 positions "lock" and "unlock" are deleted the Petri net for investigation of asynchronous running of Counter threads will be obtained. Arithmetic operations of counter is done by the method of PetriSim class which is running before transformation D^+ of net and doing operation with data fields of Petri-object. The value of calculation accuracy characterizes the conflict of threads. For example, if k threads do n increments then expected value of counter is kn. However, due to concurrency running of asynchronous threads the value of counter is c. Calculation accuracy is defined by the value $(kn - c)/(kn)100\%$. The simulation results and running of program are presented in Table 2.

Table 2. The comparison of model's result and running of program.

Number of threads	Multithreaded program	Simulation model $(d = 100, r = 1.00)$	Error
2	49.173	48.460	1.45%
10	45.336	48.460	6.89%

6 Conclusion

In this research the method of transforming multithreaded program code into Petri-object models is considered as well as the ways of investigation of such models. The main mechanisms of multithreading such as thread's start, lock, guarded block, shared data access are considered and placed in appropriate fragments of stochastic multichannel Petri net.

A stochastic Petri net represents the behavior of a multithreaded program depending on time and gives the opportunity to explore the work of the program, depending on the ratio of the time characteristics of elementary actions. Thus, it was found that when the time performance of the actions of threads is taken into account, the accuracy of model increases. As a consequence, the impact of the time of the actions on algorithm correctness was investigated.

Petri-object model is a suitable manner for parallel program debugging and testing. The hard question is construction finding real time values for program instructions execution and it needs additional investigation. However, the useful results can be obtains even when only the ratios between time delays are known.

References

1. Lea, D.: Concurrent Programming in Java: Design Principles and Patterns, 2nd edn. Addison Wesley, Boston (1999)
2. Goetz, B., Peierls, T., Bloch, J., Bowbeer, J., Holmes, D., Lea, D.: Java Concurrency in Practice. Addison-Wesley, Boston (2006)
3. EECS at UC Berkeley: The Problem with Threads (E.A.Lee). http://www.eecs.berkeley.edu/Pubs/TechRpts/2006/EECS-2006-1.html. Accessed 24 Nov 2017
4. Law, A.: Simulation Modeling and Analysis. McGraw-Hill International, New York (2014)
5. Haas, P.: Stochastic Petri Nets: Modelling, Stability, Simulation. Springer, New York (2002)
6. Peterson, J.: Petri Nets Theory and the Modelling of Systems. Prentice-Hall, New Jersey (1981)
7. Mahdi, K., Elmansouri, R., Chaoui, A.: On transforming business patterns to labeled petri nets using graph grammars. Int. J. Inf. Technol. Comput. Sci. (IJITCS) **5**(2), 15–27 (2013). https://doi.org/10.5815/ijitcs.2013.02.02

8. Zhu, L., Wang, W.: UML diagrams to hierarchical colored petri nets: an automatic software performance tool. Procedia Eng. **29**, 2687–2692 (2012)

9. Kavi, K., Moshtaghi, A., Chen, D.: Modeling multithreaded applications using petri nets. Int. J. Parallel Prog. **30**(5), 353–371 (2002)

10. Katayama, T., Kitano, S., Kita, Y., Yamaba, H., Okazaki, N.: Proposal of a supporting method for debugging to reproduce java multi-threaded programs by petri-net. J. Robot. Netw. Artif. **1**(3), 207–211 (2014)

11. Rauber, T., Runger, G.: Parallel Programming: for Multicore and Clusters Systems, vol. 2. Springer, Heidelberg (2012)

12. Holub, A.: Taming Java Threads. Apress, Berkeley (2000)

13. Von Praun, C.: Detecting synchronization defects in multi-threaded object-oriented programs, Ph.D thesis, Swiss Federal Institute of Technology, Zurich (2004)

14. Liao, H., Wang, Y., Cho, H., Stanley, J., Kelly, T., Lafortune, S., Mahlke, S., Reveliotis, S.: Concurrency bugs in multithreaded software: modeling and analyzing using petri nets. Discrete Event Dyn. Syst. **23**(2), 157–195 (2013)

15. Xiang, D., Liu, G., Yan, C., Jiang, C.: Detecting data inconsistency based on the unfolding technique of petri nets. IEEE Trans. Industr. Inf. **13**(6), 2995–3005 (2017)

16. Stetsenko, I.V.: State equations of stochastic timed petri nets with informational relations. Cybern. Syst. Anal. **48**(5), 784–797 (2012)

17. Murata, T.: Petri nets: properties, analysis and applications. Proc. IEEE **77**(4), 541–580 (1989)

18. Zaitsev, D., Sleptsov, A.: State equations and equivalent transformations of timed petri nets. Cybern. Syst. Anal. **33**(5), 659–672 (1997)

19. Stetsenko, I.V.: Systems formal description in the form of petri-object models. Visnyk NTUU "KPI" Inf. Oper. Comput. Sci. **53**, 74–81 (2011)

20. Stetsenko, I.V., Dorosh, V., Dyfuchyn, A.: Petri-object simulation: software package and complexity. In: The 8th IEEE International Conference on Intelligent Data Acquisition and Advanced Computing Systems: Technology and Applications (IDAACS 2015), pp. 381–385. IEEE, Piscataway (2015)

21. The Java Tutorials. Lesson: Concurrency. https://docs.oracle.com/javase/tutorial/essential/concurrency/index.html. Accessed 24 Nov 2017

22. Pujari, S., Mukhopadhyay, S.: Petri net: a tool for modeling and analyze multi-agent oriented systems. Int. J. Intell. Syst. Appl. (IJISA) **4**(10), 103–112 (2012)

23. Singh, R.: An Optimized task duplication based scheduling in parallel system. Int. J. Intel. Syst. Appl. **8**(8), 26–37 (2016). https://doi.org/10.5815/ijisa.2016.08.04

24. Sharma, K., Girotra, S.: Parallel bat algorithm using mapreduce model. Int. J. Inf. Technol. Comput. Sci. (IJITCS) **9**(11), 72–78 (2017). https://doi.org/10.5815/ijitcs.2017.11.08

On-line Robust Fuzzy Clustering for Anomalies Detection

Yevgeniy Bodyanskiy[1] (ORCID) and Oleksii Didyk[2(✉)] (ORCID)

[1] Kharkiv National University of Radio Electronics, Kharkiv, Ukraine
yevgeniy.bodyanskiy@nure.ua
[2] Kherson National Technical University, Kherson, Ukraine
olexii.didyk@kntu.net.ua

Abstract. Widly-used fuzzy c-means algorithm (FCM) has been utilized, with much success, in a variety of applications. The algorithm is known as an objective function based fuzzy clustering technique that extends the use of classical k-means method to fuzzy partitions. However, one of the most important drawbacks of this method is its sensitivity to noise and outliers in data since the objective function is the sum of squared distance. New robust fuzzy clustering algorithm (RFC) for exploring of signals of different nature taking into account the presence of noise with unknown density distributions and anomalous outliers in the data being analyzed is presented in this paper. By rejection of the Euclidean distance in the objective function the insensibility to the noise and outliers in the data was archived. Our approach introduces a robust probabilistic clustering procedure and is based on a modified objective function.

Keywords: Robust fuzzy clustering · Fuzzy c-means · Anomalies detection

1 Introduction

Fuzzy clustering techniques have been applied in many engineering fields such as pattern recognition, system modeling, communication, image processing, data mining, etc. [1–4]. The best known approach to studying fuzzy clustering is the method of fuzzy c-means (FCM), proposed by Dunn [5] and Bezdek [6] and generalized by other authors. Careful review of relevant works can be found in [7, 8] and almost all analytic clustering procedures are derived from this approach.

The algorithms based on this approach go under the same name objective function clustering procedures and are effective in situation when the clusters overlap but only under the assumption that the clusters are compact, i.e. they do not have anomalous outliers.

Robust objective function-based algorithm that allows us to solve clustering task at presence of high level of noise in the data in the same time without a priory statistical assumptions neither about the data distribution nor the type of outliers is introduced in this paper.

In the literature, there are a number of techniques that have the ability to tolerate outliers, including fuzzy noise clustering [9, 10], robust fuzzy clustering techniques with M-estimators [11, 12] and others [13–18] outliers resistant approaches.

The aim of the proposed algorithm is its use for anomaly detection in signals of different nature such as biological signals or signals of state of complex technical systems taking into account the existence of anomalous outliers in the data.

2   Robust Objective Function-Based Fuzzy Clustering Algorithm

Probabilistic fuzzy clustering algorithms form subclass of clustering procedures based on objective function and are designed to solve the clustering problem using the optimization of a certain predetermined clustering criterion [6, 14].

The source information for these algorithms is the data set of N n-dimensional attribute vectors $X = \{x(1), x(2), \ldots, x(N)\}, x(k) \in R^n, k = 1, 2, \ldots, N$. The output of the algorithms is the partition of the source data into m clusters with some degree of membership $w_j(k)$ of the k-th feature vector to the j-th cluster $j = 1, 2, \ldots, m$.

These algorithms can be seen as a strategy for minimizing the following objective function for pre-standardized feature vectors (all feature vectors belong to the unit hypercube $[0, 1]^n$)

$$J\left(w_j(k), c_j\right) = \sum_{k=1}^{N} \sum_{j=1}^{m} w_j^{\beta}(k) D\left(x_k, c_j\right) \tag{1}$$

under the constraints

$$\sum_{j=1}^{m} w_j(k) = 1, \qquad k = 1, 2, \ldots, N \tag{2}$$

and

$$0 < \sum_{k=1}^{N} w_j(k) \le N, j = 1, 2, \ldots, m \tag{3}$$

where $w_j \in [0, 1]$ is the degree of membership of the vector $x(k)$ to the j-th cluster, c_j is the prototype (centroid) of the j-th cluster, β is the fuzzifier, non-negative parameter (it is not subject of the optimization process and has to be chosen in advance. A typical choice is $\beta = 2$); $D\left(x(k), c_j\right)$ is the distance between data vector $x(k)$ and cluster prototype c_j in adopted metric.

Usually as a distance function, Minkowsky L^p metric is chosen

$$D\left(x(k), c_j\right) = \left(\sum_{i=1}^{n} \left|x_i(k) - c_{ji}\right|^p \right)^{\frac{1}{p}}, p \ge 1 \tag{4}$$

where $x_i(k)$, c_{ji} is the i-th component of $(n \times 1)$-vectors $x(k)$, c_j accordingly. Given $p = \beta = 2$ we obtain a standard FCM algorithm

$$w_j(k) = \frac{\left\| x(k) - c_j \right\|^{-2}}{\sum_{l=1}^{m} \left\| x(k) - c_l \right\|^{-2}}, \tag{5}$$

$$c_j = \frac{\sum_{k=1}^{N} w_j^2(k) x(k)}{\sum_{k=1}^{N} w_j^2(k)}. \tag{6}$$

A result of clustering process is the $(N \times m)$ matrix $W = w_j(k)$ which is called the fuzzy partition matrix. Since the elements of matrix W are considered as probabilities of the hypothesis of the data vectors membership to certain clusters, all procedures, generated by minimization (1) under the constraints (2), (3) are called probabilistic clustering algorithms.

The estimates connected with the quadratic objective functions (1) with $\beta = p = 2$ are optimal only in the cases when the data belong to the bounded variance distribution class. The most well-known representative of this class is Gaussian distribution. The variation of parameter β allows us to improve the robust properties of the clustering procedures; however the estimation quality is determined by the distribution of the data. For example, the estimates corresponding to $p = 1$ are optimal for Laplasian distribution, but in order to obtain them we have to make a lot of computations.

Approximate normal distribution class seems to be a reasonable compromise [19–21]. This class is a mixture of Gaussian density and a distribution of some arbitrary density contaminating the Gaussian distribution with outliers. The optimal objective function in that case is quadratic-linear.

It should be pointed out that if the data and disturbance distributions are known a priory the corresponding objective function can always be found ensured the optimal processing of the data [19, 22, 23] by means of weakening of the influence of the outliers contained in the data.

The problem become complicated if the nature of the statistic characteristic of the data is unknown or if the data don't have a statistic nature.

In that case as a basis of the objective function we propose using overturned Gaussian density

$$f_i(x_i, c_i) = 1 - \exp\left(-\frac{\left(x_i(k) - c_i \right)^2}{\sigma_i^2} \right) \tag{7}$$

(here σ_i is a width parameter) shown in Fig. 1.

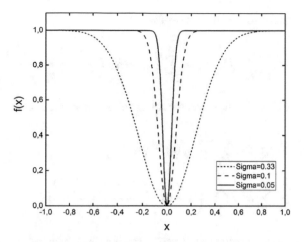

Fig. 1. Overtuned Gaussian with $c_i = 0$.

Influence function [24] for (7) can be written in the form

$$\varphi_i\left(x_i\right) = \frac{df_i\left(x_i\right)}{dx_i} = \frac{2}{\sigma_i^2}x_i\left(\exp-\frac{x_i^2}{\sigma_i^2}\right) \tag{8}$$

and its view is shown in Fig. 2. It is easy to see that the influence of the data out of the range $\pm\dfrac{\sigma_i}{\sqrt{2}}$ is suppressed and the function (7) just does not "see" far-situated observations because the (8) is converging to zero.

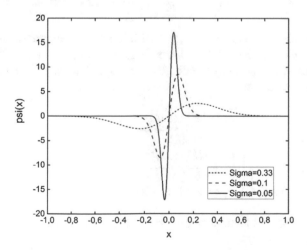

Fig. 2. Influence function of overturned Gaussian.

Using as a distance following construction

$$D^R\left(x(k),c_j\right) = \sum_{i=1}^{n} f_i\left(x_i(k),c_j\right) = \sum_{i=1}^{n}\left(1 - \exp\left(-\frac{\left(x_i(k) - c_{ij}\right)^2}{\sigma_i^2}\right)\right),\tag{9}$$

we can present an objective function for robust clustering in the form

$$J\left(w_j(k),c_j\right) = \sum_{k=1}^{N}\sum_{j=1}^{m} w_j^{\beta}(k) D^R\left(x(k),c_j\right)$$

$$= \sum_{k=1}^{N}\sum_{j=1}^{m} w_j^{\beta}(k) \sum_{i=1}^{n}\left(1 - \exp\left(-\frac{\left(x_i(k) - c_{ij}\right)^2}{\sigma_i^2}\right)\right).\tag{10}$$

It is necessary to notice that this function cannot be considered as metric because it does not satisfy the triangle inequality. Thus we can speak about the distance only in the e neighborhood of a minimum of this function ($e_i = c_{ij} \pm \dfrac{\sigma_i}{\sqrt{2}}$) and as far as moving away from it we can operate such the definition as semi-metric [25, 26] and approximate distance [27, 28] in the qualitative sense.

Let us introduce Lagrangian function

$$L\left(w_j(k),c_j,\lambda(k)\right) = \sum_{k=1}^{N}\sum_{j=1}^{m} w_j^{\beta}(k) D^R\left(x(k),c_j\right)$$

$$= \sum_{k=1}^{N}\sum_{j=1}^{m} w_j^{\beta}(k) \sum_{i=1}^{n}\left(1 - \exp\left(-\frac{\left(x_i(k) - c_{ij}\right)^2}{\sigma_i^2}\right)\right) + \sum_{k=1}^{N}\lambda(k)\left(\sum_{j=1}^{m} w_j(k) - 1\right)\tag{11}$$

where λ is an undetermined Lagrange multiplier that guarantees carrying-out the constraints (2) and (3). The saddle point of Lagrange function (11) could be found by solving the following system of Kuhn-Tucker equations with the help of Arrow-Hurwitz-Uzawa procedure

$$\begin{cases} \dfrac{\partial L\left(w_j(k),c_j,\lambda(k)\right)}{\partial w_j(k)} = 0, \\[2mm] \dfrac{\partial L\left(w_j(k),c_j,\lambda(k)\right)}{\partial \lambda(k)} = 0, \\[2mm] \nabla_{c_j} L\left(w_j(k),c_j,\lambda(k)\right) = 0. \end{cases}\tag{12}$$

Solving of the first two equations of the system (12) leads to the well-known results

$$
\begin{cases}
w_j(k) = \dfrac{\left(D\bigl(x(k), c_j\bigr)\right)^{\frac{1}{1-\beta}}}{\sum_{l=1}^{m}\left(D\bigl(x(k), c_l\bigr)\right)^{\frac{1}{1-\beta}}}, \\[4mm]
\lambda(k) = -\left(\displaystyle\sum_{l=1}^{m}\left(D\bigl(x(k), c_l\bigr)\right)^{\frac{1}{1-\beta}}\right)^{1-\beta}
\end{cases}
\tag{13}
$$

but the third one

$$
\nabla_{c_j} L\bigl(w_j(k), c_j, \lambda(k)\bigr) = \sum_{k=1}^{N} w_j^{\beta}(k)\nabla_{c_j} D^{R}\bigl(x(k), c_j\bigr) = 0
\tag{14}
$$

has no obvious analytical solution.

Solving of the third part of the system (12) could be obtained using local modification of Lagrangian function

$$
L_k\bigl(w_j(k), c_j, \lambda_k\bigr) = \sum_{j=1}^{m} w_j^{\beta}(k)D^{R}\bigl(x_k, c_j\bigr) + \lambda_k\left(\sum_{j=1}^{m} w_j(k) - 1\right)
\tag{15}
$$

and Arrow-Hurwitz-Uzawa procedure of the saddle point searching. As a result we obtain

$$
\begin{cases}
w_j(k) = \dfrac{\left(D\bigl(x(k), c_j\bigr)\right)^{\frac{1}{1-\beta}}}{\sum_{l=1}^{m}\left(D\bigl(x(k), c_l\bigr)\right)^{\frac{1}{1-\beta}}}, \\[4mm]
c_{ji}(k+1) = c_{ji}(k) - \eta(k)\dfrac{\partial L\bigl(w_j(k), c_j, \lambda(k)\bigr)}{\partial c_{ji}} \\[4mm]
\quad = c_{ji}(k) + 2\eta(k)\sigma_i^{-2}\bigl(x_i(k) - c_{ji}(k)\bigr)\cdot \exp\left(-\dfrac{\bigl(x_i(k) - c_{ji}(k)\bigr)^2}{\sigma_i^2}\right)
\end{cases}
\tag{16}
$$

where $c_{ji}(k)$ is the i-th component of the j-th prototype vector which is calculated at the k-th step, $\eta(k)$ is a learning rate parameter.

The process of the calculation starts from the initialization of W matrix (as usual with random values). On the basis of the matrix, the initial set of prototype $c_j(0)$ is calculated for the calculation of a new matrix $W(1)$. Then in batch mode all parameters like $c_j(1), W(2), \ldots, W(k), c_j(k), W_{k+1}$ are calculated until the difference $\|W(k+1) - W(k)\|$ becomes less than some apriori specified threshold Δ.

Here we should mention since the used objective function is non-convex one the algorithm convergence essentially depends on initial conditions [24, 29] which in that case can be found through the standard fuzzy c-means.

3   Conclusion and Future Direction

It is clear that new class of applications of anomaly detection demand the development of new techniques for robust fuzzy clustering which are able to handle noisy, uncertain, vague and incomplete information. In this paper, we have presented new robust probabilistic objective-function-based algorithm.

The method we have presented in this paper is not yet mature, i.e. it is still in the stage of its development and should not be considered as an ultimate solution. We have not taken into account peculiar properties of signals of different nature such as biological signals, signals of state of complex technical systems, or computer networks traffic. Furthermore, we have had to simplify of the problem domain in order to try out our ideas through all different steps. In the future, the method needs to be improved and diversified to improve further results.

References

1. Delen, D.: Real-World Data Mining: Applied Business Analytics and Decision Making. Pearson FT Press, New Jersey (2015)
2. Aggarwal, C.C.: A Data Mining: The Textbook. Springer, New York (2015)
3. Larose, D.T.: Discovering Knowledge in Data: An Introduction to Data Mining. Wiley, Hoboken (2014)
4. Yang, M.-S., Chang-Chien, S.-J., Hung, W.-L.: An unsupervised clustering algorithm for data on the unit hypersphere. Appl. Soft Comput. **42**, 290–313 (2016)
5. Dunn, J.C.: A fuzzy relative of the ISODATA process and its Use in detecting compact well-separated clusters. J. Cybern. **3**(3), 32–57 (1973)
6. Bezdek, J.C.: Pattern Recognition with Fuzzy Objective Function Algorithm. Plenum Press, New York (1981)
7. Bezdek, J.C., Keller, J., Krisnapuram, R., Pal, N.R.: Fuzzy Models and Algorithms for Pattern Recognition and Image Processing. Springer, Boston (1999)
8. Xu, R., Wunsch II, D.: Survey of clustering algorithms. IEEE Trans. Neural Netw. **16**(3), 645–678 (2005)
9. Davé, R.N.: Characterization and detection of noise in clustering. Patt. Recogn. Lett. **12**(11), 657–664 (1991)
10. Krishnapuram, R., Joshi, A., Nasraoui, O., Yi, L.: Low-complexity fuzzy relational clustering algorithms for Web mining. IEEE Trans. Fuzzy Syst. **9**(4), 595–607 (2001)
11. Bodyanskiy, Y.: Computational intelligence techniques for data analysis. In: Proceedings of the LIT 2005, vol. P-72, pp. 15–36. Gesellschaft für Informatik, Bonn (2005)
12. Bodyanskiy, Y., Gorshkov, Y., Kokshenov, I., Kolodyazhniy, V.: Robust recursive fuzzy clustering algorithms. In: Proceedings of the East West Fuzzy Colloqium 2005, pp. 301–308. HS Zittau/Görlitz (2005)
13. Tsuda, K., Senda, S., Minoh, M., Ikeda, K.: Sequential fuzzy cluster extraction and its robustness against noise. Syst. Comp. Jpn. **28**(6), 10–17 (1997)

14. Höppner, F., Klawonn, F., Kruse, R., Runkler, T.: Fuzzy Cluster Analysis: Methods for Classification, Data Analysis and Image Recognition. Wiley, Chichester (1999)
15. Georgieva, O., Klawonn, F.: A clustering algorithm for identification of single clusters in large data sets. In: Proceedings of the East West Fuzzy Colloquium 2004, pp. 118–125. HS Zittau/Görlitz (2004)
16. Butkiewicz, B.S.: Robust fuzzy clustering with fuzzy data. In: Szczepaniak, P.S., Kacprzyk, J., Niewiadomski, A. (eds.) Advances in Web Intelligence, vol. 3528, pp. 76–82. Springer, Heidelberg (2005)
17. Bodyanskiy, Y., Kokshenev, I., Gorshkov, Y., Kolodyazhniy, V.: Outlier resistant recursive fuzzy clustering algorithms. In: International Conference 9th Fuzzy Days in Dortmund: Computational Intelligence, Theory and Applications, pp. 647–652. Dortmund (2006)
18. Gorshkov, Y., Kokshenev, I., Bodyanskiy, Y., Kolodyazhniy, V., Shylo, O.: Robust recursive fuzzy clustering-based segmentation of biological time series. In: Proceedings of the 2006 International Symposium on Evolving Fuzzy Systems (EFS 2006), pp. 101–105 (2006)
19. Tsypkin, Y.Z.: Foundations of the Information Theory of Identification. Science, Moscow (1984). (in Russian)
20. Hu, Z., Bodyanskiy, Y.V., Tyshchenko, O.K., Samitova, V.O.: Fuzzy clustering data given on the ordinal scale based on membership and likelihood functions sharing. Int. J. Intell. Syst. Appl. (IJISA) 9(2), 1–9 (2017). https://doi.org/10.5815/ijisa.2017.02.01
21. Hu, Ż., Bodyanskiy, Y.V., Tyshchenko, O.K., Samitova, V.O.: Fuzzy clustering data given in the ordinal scale. Int. J. Intell. Syst. Appl. (IJISA) 9(1), 67–74 (2017). https://doi.org/10.5815/ijisa.2017.01.07
22. Hu, Z., Bodyanskiy, Y.V., Tyshchenko, O.K., Samitova, V.O.: Possibilistic fuzzy clustering for categorical data arrays based on frequency prototypes and dissimilarity measures. Int. J. Intell. Syst. Appl. (IJISA) 9(5), 55–61 (2017). https://doi.org/10.5815/ijisa.2017.05.07
23. Hu, Z., Bodyanskiy, Y.V., Tyshchenko, O.K., Tkachov, V.M.: Fuzzy clustering data arrays with omitted observations. Int. J. Intell. Syst. Appl. (IJISA) 9(6), 24–32 (2017). https://doi.org/10.5815/ijisa.2017.06.03
24. Zhang, Z.: Parameter estimation techniques: a tutorial with application to conic fitting. Image Vis. Comput. 15(1), 59–76 (1997)
25. Galvin, F., Shore, S.D.: Distance functions and topologies. Am. Math. Mon. 98(7), 620 (1991)
26. Bodyanskiy, Y., Vynokurova, O., Savvo, V., Tverdokhlib, T., Mulesa, P.: Hybrid clustering-classification neural network in the medical diagnostics of the reactive arthritis. Int. J. Intell. Syst. Appl. (IJISA) 8(8), 1–9 (2016). https://doi.org/10.5815/ijisa.2016.08.01
27. Coppola, C., Pacelli, T.: Approximate distances, pointless geometry and incomplete information. Fuzzy Sets Syst. 157(17), 2371–2383 (2006)
28. Perova, I., Pliss, I.: Deep hybrid system of computational intelligence with architecture adaptation for medical fuzzy diagnostics. Int. J. Intell. Syst. Appl. (IJISA) 9(7), 12–21 (2017). https://doi.org/10.5815/ijisa.2017.07.02
29. Li, S.Z.: Markov Random Field Modeling in Computer Vision. Springer, Tokyo (1995)

Data Stream Clustering in Conditions of an Unknown Amount of Classes

Polina Zhernova[✉], Anastasiya Deyneko, Zhanna Deyneko,
Irina Pliss, and Volodymyr Ahafonov

Kharkiv National University of Radioelectronics, Nauka Ave, 14, Kharkiv, Ukraine
polina.zhernova@gmail.com

Abstract. An on-line modified X-means method is proposed for solving data stream clustering tasks in conditions when an amount of clusters is apriori unknown. This approach is based on an ensemble of clustering neural networks that contains the self-organizing maps by T. Kohonen. Each clustering neural network consists of a different number of neurons where an amount of clusters is connected to a quality of the clustering process. All ensemble's members process information which is fed sequentially to the system in a parallel mode. The effectiveness of the clustering process is determined using the Caliński-Harabasz index. The self-learning algorithm uses a similarity measure of a special type. A main feature of the proposed method is an absence of the competition step, i.e. neuron-winner is not determined. A number of experiments has been held in order to investigate the proposed system's properties. Experimental results have confirmed the fact that the system under consideration could be used for solving a wide range of Data Mining tasks when data sets are processed in an on-line mode. The proposed ensemble system provides computational simplicity, and data sets are processed faster due to the possibility of parallel tuning.

Keywords: Clustering · X-means method · Ensemble of neural networks
Self-organization map · Self-learning · Kohonen neural network
Similarity measure

1 Introduction

Data stream clustering is an important part of Data Mining. Many approaches to its solution have been developed [1, 2]. Processing of large information volumes requires, first of all, a high speed and simple numerical implementation of clustering algorithms. One of the most popular procedures is the K- means method due to its simplicity, clarity of results and possibilities for their explicit interpretation [3, 4]. This method refers to the algorithms based on calculation of centroids-prototypes. In the frame of this approach, an initial data set (possibly growing) $X = \{x(1), x(2), \ldots, x(k), \ldots x(N) \ldots\} \subset R^n$, $x(k) = \left(x_1(k), \ldots, x_i(k), \ldots, x_u(k)\right)^T$, $k = 1, 2 \ldots N \ldots$ is partitioned into m clusters where their number m is defined a priori or chosen empirically.

For formal finding an amount of clusters, the X-means method has been developed [5, 6]. This method is based on the statistical analysis of the data distribution in the initial

© Springer International Publishing AG, part of Springer Nature 2019
Z. Hu et al. (Eds.): ICCSEA 2018, AISC 754, pp. 410–418, 2019.
https://doi.org/10.1007/978-3-319-91008-6_41

X data set. If a number of clusters m is chosen correctly during the K-means method's working then obtained results coincide completely with the X-means results [7, 8].

In the last years as a result of intensive development of Data Stream Mining [9], the necessity to solve on-line clustering tasks has emerged. When observations are fed to processing sequentially, a volume of a data set N is not limited and grows with time, and an index k defines a current discrete time moment. In this situation, the standard K-means method is not effective, but T. Kohonen's clustering neural network (the self-organizing map – SOM) can be successfully used [10], because these networks solve the task in an on-line mode. Furthermore, the obtained result completely coincides to the K-means method due to using a common self-learning clustering criterion based on the Euclidean metrics [11]. At the same time, a problem of choosing m remains open. Inclusion of additional "dead" neurons to the network does not usually solve it, and an on-line use of the X-means method in its traditional form is principally impossible [12, 13].

An alternative idea to using the standard X-means method can be an idea of using clustering ensembles. In this case, we form an ensemble based on a set of SOM^m that are connected in a parallel mode with their inputs. Every ensemble is apriori oriented at a various number of possible clusters, $m = 2, 3, \ldots, M$. Thus, the first clustering network in the ensemble suggests that $m = 2$, i.e. the Kohonen layer contains only two neurons with synaptic weights w_1^2 and w_2^2 for a centroid. The second element of the ensemble contains three neurons with synaptic weights w_1^3, w_2^3, w_3^3 [14], and the last SOM^m of the ensemble works under a premise that a number of possible clusters is equal to M, i.e. contains M neurons as adaptive linear associators.

The good of this paper is development of an on-line Data Stream clustering method in conditions of an unknown amount of classes when data are sequentially processed.

2   An Algorithm of Neural Networks' Tuning in the Ensemble

For learning of each separate SOM^m, both the standard Kohonen self-learning WTA- and WTM-rules and their variations can be used. The self-learning task of the m-th Kohonen network which contains m neurons with synaptic weights $\{w_1^m, w_2^m, \ldots w_m^m\} \subset R^n$, is under consideration in this paper.

The algorithm of synaptic weights' tuning is based on the competitive self-learning which is implemented in three basic stages (competition, cooperation, and synaptic adaptation) [15]. It starts with an analysis of an input vector $x(k)$ which enters from a receptive layer to all neurons of the Kohonen layer. For each neuron, a distance is calculated by the expression

$$D\left(x(k), w_j^m(k-1)\right) = \left\|x(k) - w_j^m(k-1)\right\|, j = 1, 2, \ldots m,$$

and if input signals are previously normalized by transformation

$$\tilde{x}(k) = \frac{x(k)}{\|x(k)\|} \tag{1}$$

so that $\|\tilde{x}(k)\| = 1$, and the Euclidean metrics is used as a distance, so a scalar product of $\tilde{x}(k), w_j^m(k-1)$ can be used as a similarity measure

$$sim\left(\tilde{x}(k), w_j^m(k-1)\right) = \tilde{x}^T(k)w_j^m(k-1) = \cos\left(\tilde{x}(k), w_j^m(k-1)\right). \tag{2}$$

Then the "nearest" to the input vector-observation neuron is defined as a winner

$$sim\left(\tilde{x}(k), w^{m^*}(k-1)\right) = \max_j sim\left(\tilde{x}(k), w_j^m(k-1)\right),$$

After temporarily skipping the cooperation process, the winner's synaptic weights can be defined using recurrent procedure

$$w_j^m(k) = \begin{cases} w_j^m(k-1) + \eta(k)\left(\tilde{x}(k) - w_j^m(k-1)\right), & \text{if } w_j^m(k-1) = w^{m^*}(k-1), \\ w_j^m(k-1) \text{ otherwise.} \end{cases} \tag{3}$$

Thus, this procedure implements the "Winner Takes All" rule (WTA) [16, 17], and a vector of the winner's synaptic weights $w^{m^*}(k-1)$ is "winded" to the input vector-observation for the distance which is defined by a learning rate parameter

$$0 < \eta(k) < 1.$$

The learning rate parameter $\eta(k)$ is usually tuned by empiric consideration. A general recommendation is that it must monotonically decrease during the self-learning process. Otherwise, the learning rate parameter can be tuned by a condition

$$\eta(k) = r^{-1}(k), \ r(k) = \alpha r(k-1) + \|x(k)\|^2, \ 0 \le \alpha \le 1$$

or

$$r(k) = \alpha r(k-1) + 1, \ 0 \le \alpha \le 1 \tag{4}$$

for the inputs that are normalized according to the condition (1).

It is clear that if $\alpha = 1, \eta(k) = k^{-1}$, it satisfies requirements of stochastic approximation.

The cooperation step is an important part of the Kohonen neural network tuning when the neuron-winner $w^{m^*}(k-1)$ defines a local area of the topological neighborhood where both it and its closest environment become excited. All neurons which are similar to the winner are more excited than "neighbors" at farther distances [18]. This area is described by a neighborhood function $\varphi(j, l), l = 1, 2, \ldots, m$, that depends on the distance $D\left(w^{m^*}(k-1), w_l^m(k-1)\right) = D\left(w_j^m(k-1), w_l^m(k-1)\right)$, between the winner and any other neuron $w_l^m(k-1)$ in the Kohonen layer. Generally, $\varphi(j, l)$ is a kernel function symmetrical relatively to maximum in a point with $D\left(w_j^m(k-1), w_j^m(k-1)\right)$ and accepting unit value $\varphi(j, l) = 1$ in it. If the distance $D\left(w_j^m(k-1), w_l^m(k-1)\right)$ is increased

[19], then this function is monotonously decreased. Usually the Gaussian is used as a neighborhood function

$$\varphi(j, l) = \exp\left(-\frac{\left\|w_l^m(k-1) - w^{m*}(k-1)\right\|^2}{2\sigma^2}\right).$$

The use of the neighborhood function leads to the algorithm of self-learning

$$w_l^m(k) = w_l^m(k-1) + \eta(k)\varphi(j, l)\big(\tilde{x}(k) - w_l^m(k-1)\big)\forall l = 1, 2, \dots, m, \tag{5}$$

that implements the "Winner Takes More" rule (WTM), and at $l = j$ this algorithm corresponds to the procedure (3).

In fact, competitive and defining the neuron-winner steps could be omitted. In this case, input vector-observation becomes the winner [20], and as a neighborhood function similarity measure (2) is used.

In this situation, the algorithm of self-learning for the m-th ensemble's member takes the form

$$\begin{aligned}
w_l^m(k) &= w_l^m(k-1) + \eta(k)\big[\cos\big(\tilde{x}(k), w_l^m(k-1)\big)\big]_+ \big(\tilde{x}(k) - w_l^m(k-1)\big) \\
&= w_l^m(k-1) + \eta(k)\big[\tilde{x}^T(k)w_l^m(k-1)\big]_+ \big(\tilde{x}(k) - w_l^m(k-1)\big) \\
&= w_l^m(k-1) + \eta(k)\big[y_l^m(k)\big]_+ \big(\tilde{x}(k) - w_l^m(k-1)\big).
\end{aligned} \tag{6}$$

As it is seen in the Eq. (6), an output signal of the neural network $y_l^m(k)$ is used as the neighborhood function. It is also interesting to note that together with "pulling" synaptic weights vector to the input vector-observation $\tilde{x}(k)$ "pushing" at $y_l^m(k) < 0$ can also take place [21]. Thus, not only "winners" are defined during the competitive step, but also and neurons-loosers.

It is clear that the procedure (6) is more simple for calculations and has an obvious physical sense than the standard algorithms (3) and (5) due to excluding the competitive step.

3   Determination of the Clusters' Number

During the ensemble processing, a quality of clustering is permanently evaluated with the help of the Caliński-Harabasz estimation either in its standard form or its on-line modification. In general, this criterion has the following form

$$CH(m) = \frac{1}{m-1} TrS_B^m \left(\frac{1}{N-m} TrS_w^m\right)^{-1} \tag{7}$$

where

$S_B^m = \frac{1}{N} \sum_{j=1}^{m} N_j^m \big(w_j^m - \overline{w}^m\big)\big(w_j - \overline{w}^m\big)^T$ is an intercluster distance matrix for m clusters;

$\overline{w}^m = \dfrac{1}{N} \sum\limits_{j=1}^{m} N_j^m w_j^m$ is a center of a data array;

N_j^m is a number of observations related to the j-th cluster, $j = 1, 2, \ldots m$;

$S_w^m = \dfrac{1}{N} \sum\limits_{j=1}^{m} \sum\limits_{k=1}^{N} u_j(k)\Big(x(k) - w_j^m\Big)\Big(x(k) - w_j^m\Big)^T$ is an intra cluster distance matrix for m clusters;

$u_j = \begin{cases} 1, & \text{if } x(k) \text{ belongs to } j\text{ - th cluster,} \\ 0 - \text{otherwise} \end{cases}$ is a crisp membership function of the k-th observation to the j-th cluster.

Upon rewriting the expression for TrS_B^m as

$$TrS_B^m = \frac{1}{N} \sum_{j=1}^{m} N_j^m \left\| w_j^m - \overline{w}^m \right\|^2,$$

and TrS_w^m is

$$TrS_w^m = \frac{1}{N} \sum_{j=1}^{m} u_j(k)\left\| x(k) - w_j^m \right\|^2;$$

the criterion (7) can be represented as

$$CH(m) = \frac{\dfrac{1}{m-1} \sum\limits_{j=1}^{m} N_j^m \left\| w_j^m - \overline{w}^m \right\|^2}{\dfrac{1}{N-m} \sum\limits_{j=1}^{m} \sum\limits_{k=1}^{N} u_j(k)\left\| x(k) - w_j^m \right\|^2}, \tag{8}$$

which is more convenient for calculations.

When the data are processed sequentially in an on-line mode, it's necessary to organize calculation of the criterion (8) on a sliding window s ($s = 1, 2, \ldots N$) [22]. In this case, $CH(m)$ can be written for a current time moment k as

$$CH(m, k) = \frac{\dfrac{1}{m-1} \sum\limits_{j=1}^{m} N_j^m(\tau) \left\| w_j^m(\tau) - \overline{w}^m(\tau) \right\|^2}{\dfrac{1}{N-m} \sum\limits_{j=1}^{m} \sum\limits_{k=1}^{N} u_j(\tau)\left\| x(\tau) - w_j^m(\tau) \right\|^2}$$

where

$$\overline{w}^m(\tau) = \frac{1}{s} \sum_{\tau=k-s+1}^{k} x(\tau).$$

The index m that ensures maximum in a value $CH(m)$ is taken as an amount of optimal clusters in the sample m^*, i.e.

$$CH(m^*) = \max_{m} \{CH(2), \; CH(3), \; \dots, \; CH(M)\}.$$

For finding $CH(m^*)$, the multilayered ANN-comparator [23] can be used that is formed by L-neurons with a linear activation function and Z-neurons with a threshold activation function and constant synaptic weights.

The ensemble that is organized according to the principle of parallel operating SOM can be cumbersome for calculations, as it is quite difficult to define a number of M-1 elements in this ensemble apriori. It would be much more convenient to organize the ensemble as a growing cascade architecture instead of parallel operating neural networks. In this case, the work starts with the only $SOM^{m\,min}$ where $m\;min$ is an apriori evaluation of a minimally possible number of clusters in the analyzed data stream [24]. If evaluation of the clustering quality with the criterion is not satisfactory, $SOM^{m\,min+1}$ shall be included to the ensemble and on-line data processing is implemented with two neural networks. The clusters' number is increased until the necessary clustering quality is reached.

The proposed procedure of the ensemble on-line clustering based on T. Kohonen's neural networks is basically an adaptive variant of the X-means method oriented on data streams processing. It is relatively simple for numerical implementation. Also, it allows to solve the task of crisp clustering under conditions of apriori unknown or changing clusters' amount.

4   Experimental Results

The efficiency of the proposed ensemble of on-line clustering T. Kohonen's neural networks was investigated on data sets from the UCI repository [25] in two experiments. For the experimental series, two data sets were taken. The data set "Wine" consists of 178 observations that are divided into 3 classes where every observation has 13 features. The data set "Ionosphere" consists of 351 observations that are divided into 2 classes where every observation has 34 features [26]. These clustering results of the proposed ensemble were estimated by the well-known Caliński-Harabasz criterion. The results for a series of 50 experiments are shown in Table 1.

Table 1. The clustering results

The data set "Wine"	
A number of neurons in the network	*The Caliński-Harabasz criterion*
An ensemble of 2 NNs	69.5233
An ensemble of 3 NNs	70.94
An ensemble of 4 NNs	56.2019
The data set "Ionosphere"	
A number of neurons in the network	*The Caliński-Harabasz criterion*
An ensemble of 2 NNs	122.6472
An ensemble of 3 NNs	86.1654
An ensemble of 4 NNs	74.7092

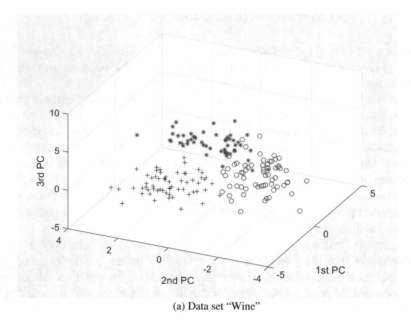

(a) Data set "Wine"

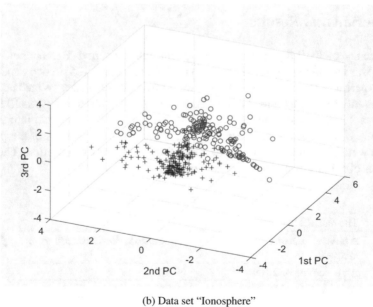

(b) Data set "Ionosphere"

Fig. 1. Visualization of the proposed clustering method

For visualization, all the taken data sets were compressed by the PCA (principal component analysis) method in three principal components. Visualization results of the proposed ensemble of on-line clustering T. Kohonen's neural networks are shown in (Fig. 1).

5 Conclusion

In this paper, the on-line modified version of the X-means method was offered for solving the data stream clustering task in conditions when an amount of clusters is apriori unknown. It is based on the ensemble of parallelly connected clustering Kohonen neural networks that contains a various number of neurons. An optimal amount of clusters is defined by a quantity of neurons in the network with the best clustering quality. The introduced modified X-means method can be used for Data Stream Mining and Dynamic Data Mining tasks.

References

1. Gan, G., Ma, C., Wu, J.: Data Clustering: Theory, Algorithms and Application. SIAM, Philadelphia (2007)
2. Xu, R., Wunsch, D.C.: Clustering. Computational Intelligence. IEEE Press/Wiley, Hoboken (2009)
3. Hu, Z., Bodyanskiy, Y.V., Tyshchenko, O.K., Tkachov, V.M.: Fuzzy clustering data arrays with omitted observations. Int. J. Intell. Syst. Appl. (IJISA) 9(6), 24–32 (2017). https://doi.org/10.5815/ijisa.2017.06.03
4. Zhengbing, H., Bodyanskiy, Y.V., Tyshchenko, O.K., Samitova, V.O.: Possibilistic fuzzy clustering for categorical data arrays based on frequency prototypes and dissimilarity measures. Int. J. Intell. Syst. Appl. (IJISA) 9(5), 55–61 (2017). https://doi.org/10.5815/ijisa.2017.05.07
5. Pelleg, D., Moor, A.: X-means: extending K-means with efficient estimation of the number of clusters. In: Proceedings of 17th International Conference on Machine Learning, pp. 727–730. Morgan Kaufmann, San Francisco (2000)
6. Ishioka, T.: An expansion of X-means for automatically determining the optimal number of clusters. In: Proceedings of 4th IASTED International Conference on Computational Intelligence, pp. 91–96. Calgary, Alberta (2005)
7. Zhengbing, H., Bodyanskiy, Y.V., Tyshchenko, O.K., Samitova, V.O.: Fuzzy clustering data given on the ordinal scale based on membership and likelihood functions sharing. Int. J. Intell. Syst. Appl. (IJISA) 9(2), 1–9 (2017). https://doi.org/10.5815/ijisa.2017.02.01
8. Hu, Z., Bodyanskiy, Y.V., Tyshchenko, O.K., Samitova, V.O.: Fuzzy clustering data given in the ordinal scale. Int. J. Intell. Syst. Appl. (IJISA), 9(1), 67–74 (2017). https://doi.org/10.5815/ijisa.2017.01.07
9. Bifet, A.: Adaptive Stream Mining: Pattern Learning and Mining from Evolving Data Streams. IOS Press, Amsterdam (2010)
10. Kohonen, T.: Self-Organizing Maps. Springer, Heidelberg (1995)
11. Perova, I., Pliss, I.: Deep hybrid system of computational intelligence with architecture adaptation for medical fuzzy diagnostics. Int. J. Intell. Syst. Appl. (IJISA) 9(7), 12–21 (2017). https://doi.org/10.5815/ijisa.2017.07.02
12. Strehl, A., Ghosh, J.: Cluster ensembles – a knowledge reuse framework for combining multiple partitions. J. Mach. Learn. Res. 3, 583–617 (2002)

13. Topchy, A., Jain, A.K., Punch, W.: Clustering ensembles: models of consensus and weak partitions. IEEE Trans. Pattern Anal. Mach. Intell. **27**, 1866–1881 (2005)
14. Alizadeh, H., Minaei-Bidgoli, B., Parvin, H.: To improve the quality of cluster ensembles by selecting a subset of base clusters. J. Exp. Theor. Artif. Intell. **26**, 127–150 (2013)
15. Charkhabi, M., Dhot, T., Mojarad, S.A.: Cluster ensembles, majority vote, voter eligibility and privileged voters. Int. J. Mach. Learn. Comput. **4**, 275–278 (2014)
16. Bodyanskiy, Y.: Computational intelligence techniques for data analysis. Lecture Notes in Informatics. GI, Bonn (2005)
17. Bodyanskiy, Y., Rudenko, O.: Artificial Neural Networks: Architecture, Learning, Application. TELETEKH, Kharkiv (2004)
18. Bodyanskiy, Y., Peleshko, D., Vinokurova, O., Mashtalir, S., Ivanov, Y.: Analyzing and Processing of Data Stream using Computational Intelligence. Lvivska Polytehnika Publishing, Lviv (2016)
19. Murphy, P.M., Aha, D.: UCI Repository of machine learning databases. Department of Information and Computer Science. University of California, CA (1994). http://www.ics.uci.edu/mlearn/MLRepository.html
20. Bodyanskiy, Y.V., Deineko, A.A., Kutsenko, Y.V.: On-line kernel clustering based on the general regression neural network and T. Kohonen's self-organizing map. Autom. Control Comput. Sci. **51**(1), 55–62 (2017)
21. Bodyanskiy, Y., Deineko, A., Kutsenko, Y.: Sequential fuzzy clustering based on neuro-fuzzy approach. Radioelectronics Inform. Control **3**(38), 30–39 (2016)
22. Zakharian, S., Ladevig-Riebler, P., Tores, S.: Neuronale Netze für Ingenieure: Arbeits und Übungsbuch für regelungs-technische Anwendungen. Vieweg, Braunschweig (1998)
23. Perova, I., Pliss, G., Churyumov, G., Eze, F.M., Mahmoud, S.M.K.: Neo-fuzzy approach for medical diagnostics tasks in online-mode. In: 1th IEEE International Conference on Data Stream Mining and Processing (DSMP), pp. 34–38 (2016)
24. Bodyanskiy, Y., Deineko, A., Kutsenko, Y., Zayika O.: Data streams fast EM-fuzzy clustering based on Kohonen's self-learning. In: 1st IEEE International Conference on Data Stream Mining and Processing (DSMP), pp. 309–313 (2016)
25. Frank, A., Asuncion, A.: UCI Machine Learning Repository, University of California, School of Information and Computer Science, Irvine, CA (2013). http://archive.ics.uci.edu/ml
26. Deineko, A., Kutsenko, Y., Pliss, I., Shalamov, M.: Kernel evolving neural networks for sequential principal component analysis and its adaptive learning algorithm. In: International Scientific and Technical Conference Computer Science and Information Technologies (CSIT 2011), Lviv, pp. 107–110 (2015)

Stegoalgorithm Resistant to Compression

M. A. Kozina[1(✉)], A. B. Kozin[2], and O. B. Papkovskaya[1]

[1] Odessa National Politechhic University, Odessa, Ukraine
mashaK1989@rambler.ru
[2] National University "Odessa Academy of Law", Odessa, Ukraine
kozindre@rambler.ru

Abstract. In this paper, we propose a choice of parameters for constructing an efficient stegoalgorithm that is resistant to compression. In order to preserve the reliability of perception of formed stegoimage, we analyze changes of second singular value number for different compression coefficients Q = {70, 80, 90, 100}. Experimental results show that we can choose second singular value number as one of the parameters. In the paper, authors proposed the efficient stegoalgorithm resistant to compression with different quality of compression. It was added a condition that allows to decode the information if the block is close to the maximum border of the range of brightness values 255. It was done the rate of reliability of perception and normalized cross-correlation coefficient.

Keywords: Stegoalgorithm · Singular value decomposition
Resistant to compression · Additional information · Watermark

1 Introduction

During the past decade, it has blossomed research in data hiding with great commercial interests. Illegal change or falsification, destruction or disclosure of a certain part of information in information management systems inflict serious material and moral damage on many entities (the state, legal entities and individuals) that participate in the processes of automated information interaction. Additional information/watermarking was initially perceived to be the answer to protect the content by the music and motion picture industries to blatant and numerous violations of copyrighted material such as music and movies [1]. It was simple and attractive the notion that embedding information directly into the data could help identify violators, check the integrity and authentication container/information, because of it must be firstly resistant to compression and then to other violators.

One of the most common attacks to date for systems of hidden data transmission - a compression attack aimed at stegomessage refers to attacks against an embedded message, the purpose of which is to destroy additional information. The popularity of this attack is associated with the widespread use of formats with losses for storing and forwarding information, which does not attract attention to its addressees, not only talking about hidden data transmission, especially open information, for example in

© Springer International Publishing AG, part of Springer Nature 2019
Z. Hu et al. (Eds.): ICCSEEA 2018, AISC 754, pp. 419–428, 2019.
https://doi.org/10.1007/978-3-319-91008-6_42

social networks, which is an open question for today. In this regard, today work in this area is relevant.

A lot of freeware is available in the Internet for embedding secret messages. As well as there are interested in analyzing image data that may contain secret information more bandwidth is available for efficient sending image data, digital forensics personnel areas. Thus, actual today creating a stegoalgorithm of coding and decoding additional information/watermark, resistant to compression attack.

In steganography a secret message is inserted into a cover file. Most current steganography algorithms insert data in the spatial or transform domains; common transforms include the discrete cosine transform (DCT) [2], the discrete Fourier transform (DFT) [8], and discrete wavelet transform (DWT) [3, 4]. Not less effective steganography techniques have been produced for embedding secret information message based on singular value decomposition (SVD) [3, 6, 7, 9] in scientific articles last decades [1, 10]. These techniques embed message in either right singular vectors, left singular vectors, singular values or combinations of all approaches in spatial domain or transform domain with satisfactory performance to various attacks.

2   Literature and Theoretical Review

In the paper [3] presents a new technique for copyright protection of images using integer wavelet transform (IWT), singular value decomposition (SVD) and Arnold transform. The embedding is done by modifying singular values of the IWT coefficients of the selected sub-image. The use of Arnold transform and SVD increases security and robustness against geometric and several signals processing attacks, while IWT provides computational efficiency. The experimental results show that the proposed technique is more imperceptible and achieves higher security and robustness against various signal processing, such as filtering, compression, noise addition, histogram equalization and motion blur and geometrical: cropping, resizing, rotation attacks. Proposed technique has both advantages and disadvantages. Disadvantages are consider only six images, correlation coefficient of extracted copyright mark from the proposed scheme without attack do not give 1 as seen from experimental part.

The paper [4] focuses on a compression technique known as block truncation coding as it helps in reducing the size of the image so that it takes less space in memory and easy to transmit. Authors in the presented scientific work increase value of PSNR, but it is not clear how the value is considered because of different image sizes, as shown in graphs.

In the paper [6] authors proposed a steganography technique that computes the SVD of submatrices of the image, the proposed method never assigns negative values in the singular values to embed secret message unlike the other existing methods. Experiment tested only for six images, and showed bad estimating secret message from Lena image for compression attack with saving reliability of perception formed stegoimage.

In the scientific work [6] sufficient conditions for the stability of stegomethods and stegoalgorithms to compression, are not dependent on any region of the digital container image in a spatial or frequency domain. The obtaining of these conditions were the result of the further development of a general approach to the analysis of the state and technology of the functioning of information systems, based on matrix analysis and perturbation theory, and its adaptation to the solution of steganography problems. In accordance with the obtained sufficient conditions, the stability to compression of the stegoalgorithm is ensured in case when the process of embedding information into a container is carried out in such way that when the result is formally represented as a set of perturbations of singular numbers and/or singular vectors of blocks of the container matrix obtained as a result of the standard partition, these sets contained perturbations of the maximal singular numbers and/or singular vectors corresponding to the maximal singular numbers locks. Based on the described sufficient conditions obtained earlier, the authors of the works [7] developed a steganographic algorithm that is stable to attack of single and double compression, has a small computational complexity: polynomial degree 2. The characteristics of the developed algorithm do not depend on the format of the container image used. The author [7] suggest resistance to compression steganography algorithm, however, do not provide the reliability of perception formed stegotransform, which is one of the important parameters in the development of efficiency stegomethods/stegoalgoritms proposed. For images without loss of quality (TIF format) the correlation coefficient they did not receive value NC = 1, It is said about the high accumulation of computational errors even without perturbations. However, neither the first, nor the improved stegoalgorithm proposed by the authors of the works does not ensure the reliability of perception and does not give a single correlation coefficient NC = 1 when they say about a lossless format JPEG, which signal about major perturbations of the pixel matrix of the image during stegotransformation. Another modern steganographic algorithm [8] proposed by the authors positions itself as effective, fulfills all the requirements for constructing an effective stegoalgorithm, but problem of compression attack stand open.

The authors of [11] proposed the theoretical development of the efficient steganographic method using the singular decomposition of the container matrix, which solves the disadvantages of the studies in the subject area, which requires the specification of some parameters, for the construction of steganographic algorithms standing to compression attack.

Basic theorems and terms from linear algebra were taken from the Handbook of Linear Algebra [8].

$$A = U\Sigma V^T \tag{1}$$

- Σ is a diagonal matrix with real, nonnegative diagonal entries $\sigma_1, \sigma_2, \ldots, \sigma_n$ such that singular values $\sigma_1 \geq \sigma_2 \geq \cdots \geq \sigma_n$. (SV);
- U and V^T must be real and orthogonal, which means $V^T V = U^T U = I$, the identity matrix; alternately a matrix is orthogonal if its column vectors are pairwise orthogonal unit vectors;
- singular values of A are the diagonal entries of Σ, $\sigma_1, \sigma_2, \ldots, \sigma_n$. These are of the form $\sqrt{\lambda}$ where λ is an is an eigenvalue of AA^T. These eigenvalues λ are always real and nonnegative, and are uniquely determined;

- singular vectors of A are the columns of U and V, respectively called the left singular vectors and right singular vectors. Singular vectors for U will be pairwise orthogonal.

The 'compact' SVD for tall-rectangular matrices, like I, is generated in Matlab by:

$$[U, S, V] = svd(I);$$

In scientific work [11] is described a stegomethod, we call it method M1, where it is necessary to detail and analysis some of the parameters for constructing the effective stego algorithm.

Embedding
1. Matrix image container B is divided into blocks in a standard way B in general size s × s. Each block is used for embedding 1 bit additional information.
2. For each block b of the matrix B:
   2.1   Construction singular value decomposition as shown in formula 1.
   2.2   If the bit of AI
          $$p=0$$
          then correct maximum value of SV
          $$\sigma_1 = roundn(\sigma_1, K) + 1/4 \cdot T$$
          else
          $$\sigma_1 = roundn(\sigma_1, K) + 3/4 \cdot T$$
      The result of the correction is $\bar{\sigma}_1$
3. Return to the block image using inverse SVD
   $$b = U \bar{\Sigma} V^T,$$
   $$\text{where } \bar{\Sigma} = diag(\bar{\sigma}_1, \sigma_2, \dots \sigma_8)$$

Extractions
1. Matrix image container $\bar{B}$ is divided into blocks in a standard way size s × s.
2. For each block $\bar{b}$ of the matrix $\bar{B}$:
   2.1   Construction singular value decomposition, in general way
          $$\bar{b} = \bar{U} \bar{\Sigma} \bar{V}^T,$$
   2.2   if $(\sigma_1 - roundn(\sigma_1, K)) < 1/2 \cdot T$
          $$\bar{p} = 0$$
          else
          $$\bar{p} = 1.$$

3   Analysis Parameters of Stegomethod

From M1 we see that it is necessary to select three parameters s and T. In accordance with the general approach to analyzing the state and technology of functioning of an arbitrary information system based on perturbation theory and matrix analysis, the process of stegotransformation, regardless of the method and region of embedding information, the result of active attack actions, in particular, compression, can formally be presented as a set of perturbations of singular numbers and/or singular vectors of the corresponding container for the matrix (s) [5].

It is proposed to investigate the changes in the second singular number after compression with different compression ratios. We should consider the compression coefficients, which are often choosen when are analyzed algorithms in steganography, Q = {70, 80, 90, 100}, due to the fact that they are less likely to lead to the reliability of perception. But it is considered Q = 70 often leads in visual artifacts on images, so less often involved as an attack on stegoimages. At the same time, Q = 100 does not mean lossless compression. This partitioning is used in the compression algorithm JPEG.

We took more than 150 images for each parameter for experimental test. These images are differ in contrast, sharpness, format (with/without losses) and other parameters that are important for testing stegoalgorithms.

The size of the block for dividing the matrix of the digital image began with a standard size of 8 × 8. Split occurs into disjoint blocks. One more reason that this kind of partitioning is considered is that this partitioning is used in the compression algorithm JPEG. We represent the changes in the second singular number after compression with different quality factors as shown Fig. 1.

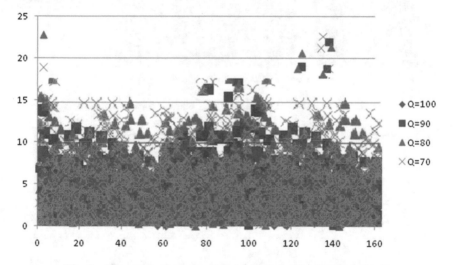

Fig. 1. Changes in the second singular number for the block 8 × 8

Analyzing the results, it can be concluded that the changes that occurred with the second singular number at different compression ratios when selecting the block for partitioning the 8 × 8 container matrix, do not exceed 23 gradations.

Further, the choice of the size block for dividing the matrix of the digital image fell on the minimum block size 2 × 2. In the work [8, 10] it was shown good results in solving the problems of steganography. But the theoretical basis for the development of stegoalgorithms that are stable to compression, based on the construction of a singular value decomposition of the matrix, presented by the authors in the scientific paper [5] absolutely don't work for the decomposition block 2 × 2. Such a partition would increase the throughput, but the results indicate the need for additional analysis of the properties of obtaining singular decomposition for embedding information to the blocks, size 2×2.

However, for size 3×3, not less popular size in steganography it is used in the development of stegoalgorithms using the analysis of pixels from the circle of the center pixel. It will increase the throughput in 7 times, compared to the standard size 8×8 of the container matrix, the changes in the second singular number are shown in Fig. 2. Here, as we can see, the maximum changes that occurred with the second maximum singular number does not exceed 26 gradations.

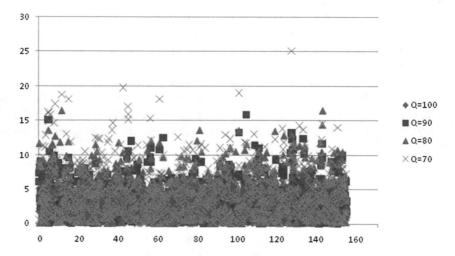

Fig. 2. Changes in the second singular number for the block 3×3

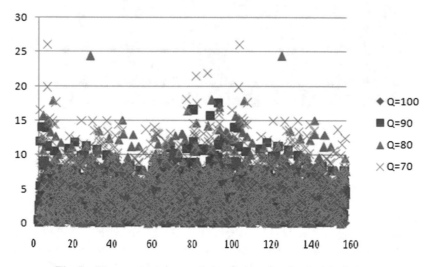

Fig. 3. Changes in the second singular number for the block 4×4

In addition, an analysis of the 4 × 4 size was performed, the results of the second singular number changes for the block of such a partition are shown in Fig. 3. Size block 4 × 4, was chosen for practice experiment from that point of view that on the one hand to increase bandwidth compared with standard partition 8 × 8, on the other hand to select an even partition size block, but less than standard one. We can see, the maximum changes that occurred with the second maximum singular number from singular value decomposition does not exceed 27 gradations.

Thus, it can be seen from the presented analysis (Figs. 1, 2 and 3, Table 1) that regardless of the choice of the size of the decomposition block of the change of the second singular number, for the coefficients of the analyzed compression with qualities from the set Q = {70,80,90,100}, the changes of the second singular number do not exceed 27 gradations.

Table 1. Max Changes in the second singular number for the block different size

Max	70	80	90	100
8 × 8	22	23	21	13
3 × 3	27	24	18	14
4 × 4	22	23	21	13

Further, it is necessary to estimate the maximum possible singular number for blocks of different sizes. To estimate the possibility of rounding by the proposed stegomethod, and also to implement an additional condition that can take into account the fall outside the range of the pixel value. This option will be obtained if all the pixel values are equal to 255. We calculate the following values of the maximum singular numbers: for the block 8 × 8 - 2040, 4 × 4 - 1020, 3 × 3 - 765.

Falling out of the range is suggested to be solved by adding checking the follow condition, where M is the maximum possible singular number for the block s × s and K – number for rounding of the maximum possible singular number:

$$if(\sigma_1 >= roundn(M, K) \; AND \; p = 1) \tag{2}$$

$$\sigma_1 = roundn(\sigma_1, K) - 1/4 \cdot T;$$

4   Proposed Stegoalgorithm

The embedding of additional information in the image will occur, as shown on Fig. 4. The result of the analysis of the maximum change in the second singular number σ_2 indicates the possibility of adopting the parameter $T = \sigma_2$, so that changes in σ_1 in the stegoalgorithm do not lead to a violation of the reliability of perception.

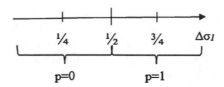

Fig. 4. Proposed changes of the first singular number

The proposed steganographic algorithm, based on the analysis of the parameters, is presented on Fig. 5.

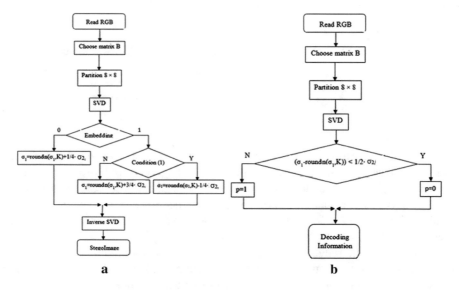

Fig. 5. Proposed stegoalgorithm. a – Embedding, b – Decoding

Experimental results of the proposed stegoalgorithms are presented on Tables 2 and 3. Algorithm is realized in Matlab. The performance is evaluated using on Peak Signal to Noise Ratio (PSNR) and Normalized Cross-Correlation (NC) for images without attacks.

Table 2. The average rate of reliability of perception PSNR for set of images

Q	TIF	100	90	80	70
PSNR	47	46,9	46,5	43	42,6

Table 3. The average rate of NC for set of images

Q	TIF	100	90	80	70
NC	1	0,98	0,91	0,89	0,81

Computational results confirmed the effectiveness of the proposed stegoalgorithm. The rate of reliability of perception gives good result. It is considered that the reliability of perception is not violated if PSNR > 37 Db, the rate of reliability of perception of proposed stegoalgorithms is PSNR > 42,6 Db.

It is considered that the coefficient of normalized cross-correlation is good if NC > 0,7. The coefficient of normalized cross-correlation of proposed stegoalgorithms is NC > 0,81. If the image is saved without loss of quality NC = 1. This result signal about zero perturbations of the pixel matrix of the image during stegotransformation, and that only an attack will make changes in the pixel matrix of the image.

5   Conclusions

In this scientific work, was done the literature review. It was identified disadvantages of the developed stegoalgorithms resistant to compression attack.

It was proposed to study SVD in stegomethod and proposed the changes in the second singular value number after compression with different compression ratios. Received quantitative changes in the second singular number for blocks of a partition of different dimension.

In the paper, authors proposed the efficient stegoalgorithm resistant to compression with different quality of compression $Q = \{70, 80, 90, 100\}$. It was proposed the additional condition that allows to correctly decoding the information if the block is close to the maximum border of the range of brightness values 255. This additional condition helps to get NC = 1, when saving the image without loss of quality (TIFF).

The rate of reliability of perception of proposed stegoalgorithms is PSNR > 42Db. The coefficient of normalized cross-correlation of proposed stegoalgorithms is NC > 0,81.

References

1. Dixit, A., Dixit, R.: A review on digital image watermarking techniques. Int. J. Image Graph. Sig. Process. (IJIGSP) **9**(4), 56–66 (2017). https://doi.org/10.5815/ijigsp.2017.04.07
2. Das, S., Banerjee, M., Chaudhuri, A.: An improved DCT based image watermarking robust against JPEG compression and other attacks. Int. J. Image Graph. Sig. Process. (IJIGSP) **9**(9), 40–50 (2017). https://doi.org/10.5815/ijigsp.2017.09.05
3. Singh, S., Siddiqui, T.J.: Copyright protection for digital images using singular value decomposition and integer wavelet transform. Int. J. Comput. Netw. Inf. Secur. (IJCNIS) **8**(4), 14–21 (2016). https://doi.org/10.5815/ijcnis.2016.04.02
4. Mander, K., Jindal, H.: An improved image compression-decompression technique using block truncation and wavelets. Int. J. Image Graph. Sig. Process. (IJIGSP) **9**(8), 17–29 (2017). https://doi.org/10.5815/ijigsp.2017.08.03
5. Chanu, Y.J., Singh, K.M., Tuithung, T.: A robust steganographic method based on singular value decomposition. Int. J. Inf. Comput. Technol. **4**(7), 717–726 (2014)
6. Kobozeva, A.A., Melnyk, M.A.: Formalnyie usloviya obespecheniya ustoychivosti steganometoda k szhatiyu. Suchasna spetsialna tehnika **4**, 60–69 (2012). (in Russian)

7. Kobozeva, A.A., Melnyk, M.A.: Formalnyie usloviya obespecheniya ustoychivosti steganometoda k szhatiyu i ih realizaciya v novom steganoalgoritme, problems of regional energetic. Electron. J. Acad. Sci. Repub. Moldova **21**(1), 93–102 (2013). (in Russian)
8. Kobozeva, A.A., Kozina, M.A.: Steganography method to provide the integrity and authenticity of data transmitted, problems of regional energetic. Electron. J. Acad. Sci. Repub. Moldova **26**(3), 93–106 (2014)
9. Kozina, M.O., Njike Amougou, S.M.: Steganography method of embedding information with singular value decomposition. In: Legal, Regulatory and Metrological Support Information Security System in Ukraine, No. 2, pp. 56–61 (2016)
10. Kozin, A., Papkovskaya, O., Kozina, M.: Steganography method using Hartley transform. In: XIII International Conference Modern Problem of Radio Engineering, Telecommunications, and Computer Science (TCSET 2016), pp. 473–475 (2016)
11. Hogben, L., Brualdi, R., Greenbaum, A.: Handbook of Linear Algebra, 1904 p. (2006)

NP-Hard Scheduling Problems in Planning Process Automation in Discrete Systems of Certain Classes

Alexander Anatolievich Pavlov[(✉)] [iD], Elena Borisovna Misura[iD],
Oleg Valentinovich Melnikov[iD], and Iryna Pavlovna Mukha[iD]

National Technical University of Ukraine "Igor Sikorsky Kyiv Polytechnic Institute",
37, Prospekt Peremohy, Kyiv 03056, Ukraine
pavlov.fiot@gmail.com, elena_misura@ukr.net,
oleg.v.melnikov72@gmail.com, mip.kpi@gmail.com

Abstract. In this paper, we consider an intractable problem of total tardiness of tasks minimization on single machine. The problem has a broad applications solutions during planning process automation in systems in various spheres of human activity. We investigate the solutions obtained by the exact algorithm for this problem earlier developed by M.Z. Zgurovsky and A.A. Pavlov. We propose an efficient approximation algorithm with $O(n^2)$ complexity with estimate of the maximum possible deviation from optimum. We calculate the estimate separately for each problem instance. Based on this estimate, we construct an efficient estimate of the deviation from the optimum for solutions obtained by any heuristic algorithms. Our statistical studies have revealed the conditions under which our approximation algorithm statistically significantly yields a solution within 1–2% deviation from the optimum, presumably for any problem size. This makes possible obtaining efficient approximate solutions for real practical size problems that cannot be solved with known exact methods.

Keywords: Planning · Process automation · Scheduling
Combinatorial optimization · Heuristics · Exact algorithm · Estimate of deviation
Total tardiness

1 Introduction

Problems arising during the planning process automation in systems of various nature are, as a rule, NP-hard problems of the scheduling theory. All known exact methods for their solving have exponential complexity and are substantially based on branch and bounds methods and dynamic programming. Therefore, an important research area is still the construction of heuristic algorithms and approximation algorithms with an estimate of deviation of the solution from the optimum. Many state-of-the-art scheduling methods, as well as planning systems, are described in the book [1]. Models of planning process automation are researched in [2–10]. Some new scheduling methods, with the use of cloud computing, are considered in a recent review [11].

In this work, we research the known intractable problem of minimizing the total tardiness of tasks on a single machine. The importance of this problem cannot be

© Springer International Publishing AG, part of Springer Nature 2019
Z. Hu et al. (Eds.): ICCSEEA 2018, AISC 754, pp. 429–436, 2019.
https://doi.org/10.1007/978-3-319-91008-6_43

overemphasized, it is used in the automation of planning process in a variety of practical areas: in planning discrete-type productions (in particular, small-scale productions, construction, aircraft and shipbuilding industries), in business process automation, in project management systems, etc. For example, this problem is used during discrete productions planning both as a separate problem and as a basis for algorithms creation for more complex models (in multi-stage systems, in systems with parallel machines etc.). In [2], the algorithm for this problem solving is used as a part of the algorithm for single machine total earliness and tardiness minimization problem.

The Problem Statement. Given a set of independent tasks $J = \{j_1, j_2, \ldots, j_n\}$, each task consists of one operation. For each task $j \in J$, we know its processing time l_j and due date d_j. All tasks are available at the same time zero. Interruption in their processing is not allowed. We need to build a schedule for single machine that minimizes the total tardiness of tasks:

$$f = \sum_{j=1}^{n} \max\left(0, C_j - d_j\right)$$

where C_j is the completion time of a task j.

This problem is studied already for about 50 years. Known state-of-the-art exact algorithms can solve instances with up to 500 tasks in size and are based on methods of dynamic programming, branch and bounds and their modifications [12–14]. Not lesser attention of scientists is attracted to the creation of approximation and heuristic algorithms. A detailed review of known methods (exact and heuristic) for this problem is given in [15]. Estimates of the deviation from optimal solutions of known heuristic (including genetic) algorithms are almost always obtained experimentally. Analytical evaluations are very few (e.g., [16]).

Exact exponential algorithm published in [2] allows to solve instances with up to 1,000 tasks and is based on permutations of tasks. In this paper, we propose an approximation algorithm with an estimate of the maximum possible deviation of the obtained solution from the optimum. We calculate the estimate separately for each individual problem instance. Our result is based on research of properties of sequences obtained by the exact algorithm from [2]. On the basis of the estimate, we show an efficient estimate construction for arbitrary heuristic algorithms of the problem solving.

2 Theoretical Background

We now excerpt from [2] some definitions necessary for our results presentation. Let $j_{[g]}$ denote the number of a task occupying a position g in a schedule. Let $\overline{p, q}$ mean the interval of integer numbers from p to q: $\overline{p, q} = p, p+1, \ldots, q$.

Definition 1. $d_{j_{[g]}} - C_{j_{[g]}} > 0$ is called the time reserve of a task $j_{[g]}$.

Definition 2. A task $j_{[g]}$ is called tardy in some sequence if $d_{j_{[g]}} < C_{j_{[g]}}$.

Definition 3. A permutation of a task $j_{[g]}$ is its move to a later position $k > g$ with shift of the tasks in positions $\overline{g+1, k}$ by one position to the left.

Definition 4. The ordered sequence σ^{ord} is a sequence of tasks $j \in J$ satisfying the order: $\forall j, i, j < i: l_j < l_i$; if $l_j = l_i$, then $d_j \leq d_i$.

Definition 5. A free permutation of a non-tardy task $j_{[k]}$ in the sequence σ^{ord} is its permutation into such $\underline{\text{later position}}$ $q > k$ that $C_{j_{[q]}} \leq d_{j_{[k]}} < C_{j_{[q+1]}}$ if there is at least one tardy task in positions $\overline{k+1, q}$.

So, the position q is the maximum position where the task $j_{[k]}$ remains non-tardy. It is clear that the functional value decreases after free permutations. A free permutation from position k to position q decreases the objective function value by $\sum_{i \in Q} \min\left(l_{j_{[i]}}, C_{j_{[i]}} - d_{j_{[i]}}\right)$ where $Q = \left\{ j_{[i]} \,\middle|\, C_{j_{[i]}} > d_{j_{[i]}}, \quad i = \overline{k+1, q} \right\}$.

Definition 6. Sequence σ^{fp} is the sequence obtained after all free permutations in the ordered sequence σ^{ord}.

Definition 7. A tardy task $j_{[g]}$ in the sequence σ^{fp} is called competing task if for at least one non-tardy preceding task $j_{[l]}$ in the sequence we have $d_{j_{[l]}} > d_{j_{[g]}}$.

The algorithm published in [2] is the series of similar iterations. The number of iterations is determined by the number of competing tasks in the sequence σ^{fp}. During each iteration, we search for a possibility to use the time reserves of preceding tasks for each next competing task of the sequence σ^{fp} and build an optimal schedule for a subsequence of tasks bounded with the current competing task. We can eliminate some tasks from the number of competing tasks during the solving process. The functional value on the entire set of tasks may decrease or remain the same after each iteration. This allows to build efficient heuristic and approximation algorithms on the basis of the algorithm from [2].

3   Approximation Algorithm for the Problem Solving

1. Arrange all tasks in non-decreasing order of l_j. Tasks that have equal processing times, are arranged in non-decreasing order of due dates. Denote the obtained sequence by σ^{ord}.
2. Perform free permutations in the sequence σ^{ord}:
   2.1. Find the next unconsidered non-tardy task $j_{[k]}$ that has a maximal due date.
   2.2. Find such position $q > k$ where $C_{j_{[q]}} \leq d_{j_{[k]}} < C_{j_{[q+1]}}$.
   2.3. If we had at least one tardy task within positions $\overline{k+1, q}$, permute the task $j_{[k]}$ into position q. Go to Step 2.4.
   2.4. If all non-tardy tasks are considered, then denote the current sequence as σ^{fp}, go to Step 3. Otherwise, go to Step 2.1.

3. The sequence σ^{fp} is the approximate solution for the problem. Determine the set of competing tasks in the sequence σ^{fp}: find all tardy tasks $j_{[g]}$ that follow at least one non-tardy task $j_{[l]}$, $d_{j_{[l]}} > d_{j_{[g]}}$. Determine the estimate of deviation of the obtained solution from the optimum for the current problem instance (Sect. 4).

4   Estimate of Deviation for the Sequence σ^{fp}

According to [2], we can decrease the functional value only if there are non-tardy tasks with time reserves. Let R denote the set of such tasks, Z be the set of tardy tasks and K be the set of competing tasks in the sequence σ^{fp}.

Statement 1. The maximum possible deviation of the functional value in the sequence σ^{fp} from an optimal solution due to use of time reserve of a task $j_{[r]} \in R$, $d_{j_{[r]}} > C_{j_{[r]}}$, $d_{j_{[r]}} > d_{j_{[l]}}$ where $j_{[l]} \in K$, is determined as

$$\Delta f_{\max R}\left(j_{[r]}\right) = \frac{d_{j_{[r]}} - C_{j_{[r]}}}{l_{\min}} l_{j_{[r]}}, j_{[r]} \in R \tag{1}$$

where $l_{\min} = \min_{j_{[l]} \in K} l_{j_{[l]}}$.

Proof. Suppose there is a task $j_{[r]} \in R$ in the sequence σ^{fp} followed by tardy competing tasks $j_{[i]} \in K$. We reach the maximum possible decrease in the functional value by free permutation of the task $j_{[r]}$ to a later position. Let us create a fictitious competing tardy task with the processing time $l_{\min}$. We determine the decrease in the functional value as a result of the free permutation for the case when the free permutation interval is occupied by only fictitious tasks:

$$\Delta f_{\max R}\left(\sigma^{fp}\right) = \sum_{i \in Z} \min\left(l_{j_{[r]}}, C_{j_{[i]}} - d_{j_{[i]}}\right)$$

(Definition 5). Suppose that $C_{j_{[i]}} - d_{j_{[i]}} > l_{j_{[r]}}$. Then

$$\Delta f_{\max R}\left(\sigma^{fp}\right) = \sum_{i \in Z} l_{j_{[r]}}.$$

We obtain the expression (1) where $\left(d_{j_{[r]}} - C_{j_{[r]}}\right) \big/ l_{\min}$ is the number of fictitious tasks within the free permutation interval. ☐

If there are several tasks with time reserves in the sequence σ^{fp}, then we calculate the estimate of the deviation using each reserve and determine the sum:

$$\Delta f_{\max R}\left(\sigma^{fp}\right) = \sum_{j_{[r]} \in R} \Delta f_{\max R}\left(j_{[r]}\right).$$

So, based on existing reserves in the sequence σ^{fp}, the estimate of the maximum deviation from an optimal functional value is:

$$\Delta f_{\max R}\left(\sigma^{fp}\right) = \sum_{j_{[r]} \in R} \frac{d_{j_{[r]}} - C_{j_{[r]}}}{l_{\min}} l_{j_{[r]}}. \tag{2}$$

We determine the final estimate as follows:

$$\Delta f_{\max}\left(\sigma^{fp}\right) = \min\left(\Delta f_{\max R}\left(\sigma^{fp}\right), \Delta f_{\max K}\left(\sigma^{fp}\right)\right) \tag{3}$$

where $\Delta f_{\max K}\left(\sigma^{fp}\right) = \sum_{j_{[i]} \in K} C_{j_{[i]}} - d_{j_{[i]}}.$

5   Analysis of the Estimate

Let n_K denote the number of competing tasks in the sequence σ^{fp} and n_R denote the number of tasks with time reserves. The estimate (2) depends on the amount of reserves and is effective for a large n_K and, accordingly, a small n_R. If n_K is small, the value of Ex. (2) can be greater than the total tardiness of competing tasks (because we use in an optimal schedule only a part of reserves that we used to calculate the estimate). Therefore, we determine the resulting estimate by the formula (3).

The algorithm from [2] has exponential complexity (in worst case it is $O(e^{n_K + n_R})$ where $n_K > 0$, $n_R > 0$, because in cases $n_K = 0$ or $n_R = 0$ the sequence σ^{fp} is already optimal). But actual complexity of the problem solving is affected by many factors: the tightness of due dates, the dispersion of processing times, the presence of non-competing tardy tasks, the amount of reserves and the presence of tasks with reserves that have a lesser due date than that of competing tasks. All these factors make cutoffs of unpromising permutations possible during the algorithm execution. This allowed us to solve instances of such large dimensions that could not even theoretically be solved by existing methods. We can determine the possibility of fast exact solving for large n just after the sequence σ^{fp} construction. The condition is a small n_K or a small n_R, with the exception of special complicated cases when the parameters of the tasks differ insignificantly.

If n_K is small, we achieve an optimal solution by the exact algorithm in a short time, because a small number of iterations is executed. Similarly, we reach the exact solution shortly if n_K is large but the number of tasks with reserves is small. The reason is exhaustion of time reserves, therefore we eliminate most of the competing tasks from the set of competing tasks after a small number of iterations. As a result, we obtain an optimal solution of the problem with a functional value f_{opt}. This value is the exact lower bound for a functional value f_{heur} of the solution by an arbitrary heuristic algorithm. Then, the deviation from the optimum is $f_{heur} - f_{opt}$.

If we cannot obtain the exact solution in a short time, then we choose the following value to evaluate the solution obtained by an arbitrary heuristic algorithm:

$$f_{heur} - \left[f\left(\sigma^{fp}\right) - \min\left(\Delta f_{\max R}\left(\sigma^{fp}\right), \Delta f_{\max K}\left(\sigma^{fp}\right)\right)\right]. \tag{4}$$

6   Statistical Research and Recommendations

We have done statistical studies of the approximation algorithm efficiency. We compared the solutions obtained by the approximation algorithm for the known problem instances with the optimal solutions obtained by known exact methods. Then we determined the percentage of deviation of the solution from the optimum.

In Table 1, we give actual deviations as a percentage of the functional value in the sequence σ^{fp} from an optimal solution for all benchmark instances from [13] that had a known optimal solution. There were 200 such instances for dimensions 100...200; 199 instances for dimension 300; 193 for 400 tasks problems, and 176 instances for dimension 500. Authors of the article [13] generated 10 instances for each value of two parameters: the due dates factor R and the tardiness factor T (Fisher's method of generation [17]). The complexity of each instance depends on it's parameters.

Table 1. The average percentage Δ_{avg} of deviation of the functional value in the sequence σ^{fp} from the optimum, depending on the percentage of the number of competing tasks, for all benchmark instances from [13]

$n_{K\%}$, %	n					
	100	150	200	300	400	500
0	0.00	0.00	0.00	0.00	0.00	0.00
(0, 5]	90.16	87.08	89.90	90.20	94.78	93.63
(5, 10]	52.36	53.95	51.65	53.55	52.06	53.64
(10, 15]	35.77	35.14	35.19	31.29	29.88	23.55
(15, 20]	15.04	16.23	14.38	13.88	11.86	12.02
(20, 25]	11.10	10.40	9.12	8.79	9.00	9.13
(25, 30]	5.61	5.75	5.65	5.62	4.86	2.37
(30, 35]	5.01	5.26	5.23	5.79	–	–
(35, 40]	1.51	1.81	1.29	1.23	1.16	1.19
(40, 45]	1.59	0.96	1.01	0.93	0.72	0.59
(45, 50]	1.38	1.18	1.34	1.15	0.89	0.25
(50, 55]	0.77	0.25	0.47	0.31	0.24	0.19
(56, 60]	0.34	0.24	–	–	–	–
Total average	21.28	21.45	20.30	20.26	19.84	21.69

We calculated the percentage of deviation from the optimum Δ_{dev} for each instance as follows:

$$\Delta_{dev} = \frac{f(\sigma^{fp}) - f_{opt}}{f(\sigma^{fp})} \times 100\%.$$

Then we calculated the percentage of the number of competing tasks n_K in the sequence σ^{fp} of the total number of tasks: $n_{K\%} = (n_K/n) \times 100\%$. We grouped instances according to the percentage of the number of competing tasks $n_{K\%}$ and averaged the percentage of deviation from the optimum. In Table 1, we show the percentage of the number of competing tasks as intervals: notation $(0, 5]$ means $0 < n_{K\%} \leq 5$. The table cells contain the average percentage of deviation from the optimum Δ_{avg} in the sequence σ^{fp} for all instances corresponding to the given dimension and the given interval of $n_{K\%}$. The dash in a cell means we did not have any instance for the given dimension and given interval of $n_{K\%}$. In each group of instances, the maximum percentage of deviation from the optimum is only a little different from Δ_{avg}. The last row of the table contains the average percentage of deviation for all instances of the given dimension.

It is clear from the above table that the percentage of deviation of the functional value in the sequence σ^{fp} from the optimum decreases with the growth of the number of competing tasks. If n_K is small (less than 10), then for any dimension from 100 to 500 the percentage of deviation is high (greater than 50%). But we can obtain an exact solution by the algorithm from [2] in a short time for such instances. With the growth of n_K, the value of Δ_{avg} decreases. For example, when the number of competing tasks is from 25 to 35, the average percentage of deviation from optimum is about 5–6%. If the number of competing tasks is from 35 to 50, it falls to about 1–2%. You can similarly find such ranges corresponding the desired percentage of deviation for other dimensions.

The consequence of this analysis is the following recommendation. If $n_K \geq 0.35n$, then we recommend to use the sequence σ^{fp} as an approximate solution. The complexity of its construction is $O(n^2)$, and this solution would be statistically significant and have the average percentage of deviation from the optimum of no more than 1–2%.

Heuristic. This statistical regularity depends on n a little and remains valid for large n $(n \geq 1,000)$ for $n_K \geq 0.35n$.

The rationale for the heuristic is based on the fact that the structure of the sequence σ^{fp} does not depend on the dimension of the problem.

For real practical problems of dimension $n > 1,000$ (that cannot be solved to optimality by known methods), in case if $n_K \geq 0.35n$, we statistically significantly obtain an a solution with an average percentage of deviation from the optimum of no more than 1–2% by the proposed approximation algorithm with $O(n^2)$ complexity.

7   Conclusion

We proposed an efficient polynomial approximation algorithm to solve the known NP-hard problem of minimizing the total tardiness on one machine which has wide applications in the automation of planning processes in various areas of human activity. We constructed an estimate of deviation from an optimal functional value for the solutions obtained for any problem instance. On the basis of this estimate, we build an estimate of deviation from the optimum for any heuristic algorithms. We have studied the dependence between the deviation of the approximate solution from the optimum and the number of competing tasks in a sequence σ^{fp}. The research showed that for problems of dimension $n \leq 1,000$ when the number of competing tasks is more than 35% of the entire number of tasks, we statistically significantly obtain a solution within 1–2% deviation from the optimum by the approximation algorithm with $O(n^2)$ complexity. According to the proposed heuristic, the same dependence is true for real practical dimensions instances $(n > 1,000)$ that cannot be solved by any of the existing methods. This allows us to obtain an efficient solution of the problem of almost any dimension by means of the proposed approximation algorithm.

References

1. Pinedo, M.L.: Scheduling: Theory, Algorithms, and Systems. Springer, Cham (2016). https://doi.org/10.1007/978-3-319-26580-3
2. Zgurovsky, M.Z., Pavlov, A.A.: Prinyatie resheniy v setevyh sistemah s ogranichennymi resursami: Monograph. Naukova dumka, Kiev (2010). (in Russian)
3. Alhumrani, S.A., Qureshi, R.J.: Novel approach to solve resource constrained project scheduling problem (RCPSP). Int. J. Mod. Educ. Comput. Sci. (IJMECS) 8(9), 60–68 (2016). https://doi.org/10.5815/ijmecs.2016.09.08
4. Vollmann, T.E., Berry, W.L., Whybark, D.C.: Manufacturing Planning and Control Systems. McGraw-Hill, Boston (2005)
5. Popinako, D.: ERP-rynok glazami veterana. SK-Press, Kiev, PC Week-Ukr (2011). (in Russian)
6. Zagidullin, R.R.: Upravlenie mashinostroitelnym proizvodstvom s pomoschyu sistem MES, APS, ERP. TNT, Staryi Oskol (2011). (in Russian)
7. Zagidullin, R.R.: Planirovanie mashinostroitelnogo proizvodstva. TNT, Staryi Oskol (2017). (in Russian)
8. Swayamsiddha, S., Parija, S., Sahu, P.K., Singh, S.S.: Optimal reporting cell planning with binary differential evolution algorithm for location management problem. Int. J. Intell. Syst. Appl. (IJISA) 9(4), 23–31 (2017). https://doi.org/10.5815/ijisa.2017.04.03
9. Soudi, S.: Distribution system planning with distributed generations considering benefits and costs. Int. J. Mod. Educ. Comput. Sci. (IJMECS) 5(9), 45–52 (2013). https://doi.org/10.5815/ijmecs.2013.09.07
10. Garg, R., Singh, A.K.: Enhancing the discrete particle swarm optimization based workflow grid scheduling using hierarchical structure. Int. J. Comput. Netw. Inf. Secur. (IJCNIS) 5(6), 18–26 (2013). https://doi.org/10.5815/ijcnis.2013.06.03
11. Soltani, N., Soleimani, B., Barekatain, B.: Heuristic algorithms for task scheduling in cloud computing: a survey. Int. J. Comput. Netw. Inf. Secur. (IJCNIS) 9(8), 16–22 (2017). https://doi.org/10.5815/ijcnis.2017.08.03
12. Szwarc, W., Grosso, A., Della Croce, F.: Algorithmic paradoxes of the single machine total tardiness problem. J. Sched. 4(2), 93–104 (2001). https://doi.org/10.1002/jos.69
13. Tansel, B.C., Kara, B.Y., Sabuncuoglu, I.: An efficient algorithm for the single machine total tardiness problem. IIE Trans. 33(8), 661–674 (2001). https://doi.org/10.1080/07408170108936862
14. Garraffa, M., Shang, L., Della Croce, F., T'Kindt, V.: An exact exponential branch-and-merge algorithm for the single machine total tardiness problem (2017). Preprint, HAL. https://hal.archives-ouvertes.fr/hal-01477835
15. Koulamas, C.: The single-machine total tardiness scheduling problem: review and extensions. Eur. J. Oper. Res. 202, 1–7 (2010). https://doi.org/10.1016/j.ejor.2009.04.007
16. Chang, S., Matsuo, H., Tang, G.: Worst-case analysis of local search heuristics for the one-machine total tardiness problem. Nav. Res. Logist. 37(1), 111–121 (1990). https://doi.org/10.1002/1520-6750(199002)37:1<111::AID-NAV3220370107>3.0.CO;2-V
17. Fisher, M.L.: A dual algorithm for the one machine scheduling problem. Math. Progr. 11(1), 229–251 (1976). https://doi.org/10.1007/BF01580393

Statistic Properties and Cryptographic Resistance of Pseudorandom Bit Sequence Generators

V. Maksymovych, E. Nyemkova[✉], and M. Shevchuk

Institute of Computer Technologies, Automation and Metrology,
Lviv Polytechnic National University, Lviv, Ukraine
volodymyr.maksymovych@gmail.com, cyberlbi12@gmail.com,
shevchuk.mykola93@gmail.com
http://www.lp.edu.ua/education

Abstract. Generators of pseudorandom sequences are widely used in practice. Generators of pseudorandom bit sequences occupy a special place among them; they are necessary for solving a number of important tasks, for example, for strong cryptography. The impossibility of predicting the following values of pseudorandom sequences is one of the basic requirements for such generators. Otherwise, these generators cannot be used to protect of information. It is generally accepted that if the stochastic sequence is stationary, then the prediction of such sequence is impossible. Our research shows that there are invariants for specific pseudorandom sequences that can be used to this prediction.

The article is devoted to the method of prediction of pseudorandom bit sequences. The values of the autocorrelation coefficients for some lags are used. Good results are obtained for software-implemented stationary stochastic sequences.

Keywords: Autocorrelation · Pseudorandom bit sequence · Statistical portrait
Cryptographically strong generator

1 Introduction

Generators of pseudorandom bit sequences (GPRBS) are widely used by modern science in various systems. In the field of information security, pseudorandom numbers are used both in technical and cryptographic tools. It is known that the characteristics of the security systems depend on the characteristics of their subsystems, which are determined not only by the used algorithms, but also by qualitative indicators of the used pseudorandom sequences. Security of cryptosystems is focused on the keys, the whole cryptosystem becomes vulnerable when unreliable process of generating keys is using. Therefore, the problem is the construction of high-quality, reliable GPRBS.

Nowadays the number of industries with GPRBS increased due to massive use of electronics and electronic computing. The main areas of use of GPRBS show in Fig. 1.

© Springer International Publishing AG, part of Springer Nature 2019
Z. Hu et al. (Eds.): ICCSEEA 2018, AISC 754, pp. 437–446, 2019.
https://doi.org/10.1007/978-3-319-91008-6_44

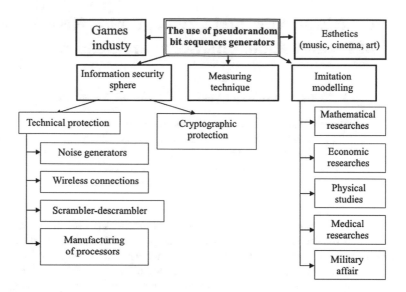

Fig. 1. The main areas of use of GPRBS.

In recent years, more and more applied computer programs use GPRBS to generate the necessary random data due to their simple implementing on the all types of computer systems.

GPRBS are an integral part of any system of information protection and they are used to solve many problems, series of publications show those implementations [2, 11, 13, 14, 16, 17]. GPRBS are also major components of the device for technical protection of information. Most often GPRBS are used in systems "scrambler - descrambler" and as noise generators. Typically, these systems are constructed with using GPRBS which are based on shift registers, because they have high performance.

In modern telecommunication systems GPRBS are widely used. In the first place, the interest in GPRBS in the field of telecommunications is due to the fact that they are the basis for the formation of broadband (noisy) signals.

The main directions of the use of GPRBS in telecommunication systems are presented in publications [2, 12, 17].

GPRBS are often used in imitation modelling. For example, the same program is runned on many processors with different streams of random numbers. Results from each processor are statistically independent and then these results are averaged. Deterministic algorithms for generating pseudorandom numbers are needed in many cases. For example, there is required that the sequence of pseudorandom numbers can be repeated once more for the research of influence of different parameters of process.

Klapper and Goresky [5] used GPRBS in simulation of mutations in biological researches. GPRBS are also used in dosimetry to generate pulse flows that obey the Poisson distribution law.

Another important area of GPRBS application is their use to control the quality of printed circuit boards, individual modules, or entire devices [11, 14]. Kitsos et al. [4] proposed several new interesting research methods for analog circuits and electronic modules. Nas and Van Berkel [12] used the technique of pseudo-random testing as an input exciter (pseudorandom samples).

It is necessary that GPRBS should be of sufficient quality and reliability. For example, game electronic machine requires a sequence that could not be foreseen if the previous value is known. Otherwise the game system would fail, if the player cans determine the following move in play from the analysis of the previous situation. There is similar situation with the encoding of messages. GPRBS must be cryptographically strong generator.

2 Pseudorandom Bit Sequences Generators

Let us consider the statistical properties of several generators, such as pseudorandom bit sequence generators (PSRBG), based on linear feedback shift registers (LFSR) with linear cycles, also feedback with carry shift register (FCSR).

2.1 Pseudorandom Bit Sequences Generators Based on LFSR

The research of LFSR generators was carried out by many scientists [1–4, 6, 7, 11, 12, 14, 17]. It is known that some pseudorandom bit sequences generators (PSRBG) based on linear feedback shift registers with linear loops (LFSR) are not cryptographically strong generators [2, 17]. Despite this, they are often used as building blocks of more complicated devices such as the devises of stream cipher (e.g. algorithms - A3, A5, A8, PIKE, SEAL, RC4) or a set of sequences for encryption keys as Schneier wrote [17]. Ndaw et al. [14] used this type of generators in communication systems such as CDMA (Code Division Multiple Access). This type of generators also are used in the mobile communication systems, navigation systems, in the spread spectrum of communication systems, particularly in the Bluetooth technology, GSM, radar systems and some radio modems. Their main advantages are high speed and simplicity in hardware implementation.

Statistical characteristics of PSRBG based on LFSR are considered in many publications [2, 3, 6, 7]. However, the minimum number of structural elements LFSR in devices based on them is not finally determined. We don't know, when we can say that statistical characteristics of the output bit sequence are satisfactory.

We assume that the statistical characteristics PSRBG are satisfactory if the output bits sequence passes all NIST tests [15]. When complex generators are compared one should take into account the recurrence of the built output sequence (adaptability), complexity of construction and implementing on programmable logic circuits (FPGAs) and the maximum possible length of the encryption key. But main attention is focused to the study of statistical characteristics PSRBG with different parameters.

We should emphasize that the cryptostability of such generators, without cryptanalysis, is not guaranteed. As known, statistical security is necessary, but it is not sufficient condition [8–10, 15].

Simplified block diagram of the PSRBG based on LFSR is shown in Fig. 2.

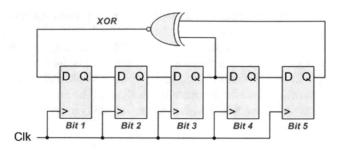

Fig. 2. Simplified block diagram of the PSRBG based on LFSR.

The GPRBS based on LFSR was used for polynomial $F(x) = x^{135} + x^{11} + 1$. The generator is implemented in software in Java and the repetition period was greater than 109. The analysis of this generator by the NIST tests showed that its statistical characteristics do not meet the requirements of randomness, Fig. 3.

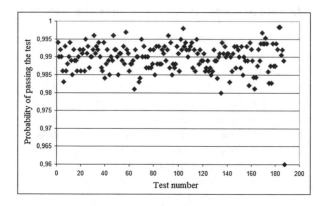

Fig. 3. The statistical portrait of the GPRBS based on LFSR ($F(x) = x^{135} + x^{11} + 1$).

2.2 Generators of Pseudorandom Bit Sequences Based on the Feedback with Carry Shift Register

The transfer register, or FCSR, is similar to the LFSR. In two cases there is a shift register and a feedback function. The difference is that the FCSR has a transfer register instead of executing XOR over all the bits of the batch sequence, as shown in Fig. 4. These bits are added with each other and with the contents of the transfer register. The result, divided by 2, becomes the new content of the transfer register.

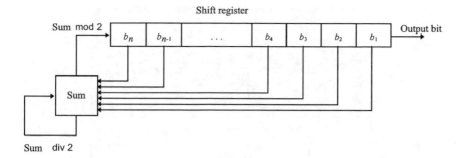

Fig. 4. Feedback with carry shift register.

The idea of using FCSR in cryptography is nominated by Klapper and Gorecky in 1994 [5].

So, the GPRBS based on FCSR with polynomials $F(x) = x^{32} + x^8 + x^3 + x^2$ was used for researches, this generator is implemented in software in Java, the repetition period was greater than 10^9.

Also the generator was used in the next researches, which is shown in Fig. 5.

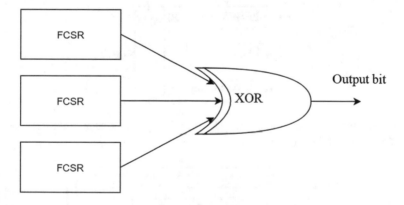

Fig. 5. GPRBS based on XOR of three FCSRs.

The characteristics of this GPRBS are based on XOR of three FCSRs:

$$F(x) = \begin{cases} x^{32} + x^6 + x^3 + x^2 \\ x^{32} + x^7 + x^5 + x^2 \\ x^{32} + x^8 + x^3 + x^2 \end{cases}$$

Statistical portrait of this GPRBS is shown in Fig. 6.

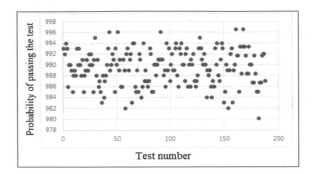

Fig. 6. Statistical portrait of GPRBS based on XOR of three FCSRs.

2.3 Generators of Pseudorandom Bit Sequences Based on the Jiffy Generator

Simplified block diagram of the Jiffy generator is shown in Fig. 7. It consists of three registers LFSR1–LFSR3 and multiplexer MUX [2]. The Fig. 8 show statistical portrait of GPRBS based Jiffy generator.

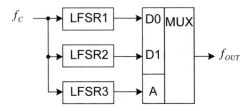

Fig. 7. Simplified block diagram of the Jiffy generator LFSR1($x^{20} + x^3 + 1$), LFSR2($x^{25} + x^3 + 1$), LFSR3($x^{35} + x^2 + 1$).

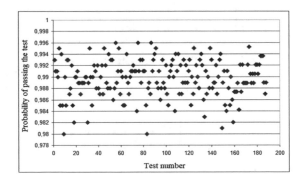

Fig. 8. Statistical portrait of GPRBS based on Jiffy generator.

In Fig. 8 we can see good statistical characteristics of GPRBS based on Jiffy generator, comparatively the previous GPRBS based on LFSR, as show in Fig. 3.

3   Invariants of Pseudorandom Sequences

The pseudorandom sequences $x_i(j)$ of each generator j will be different. Calculations show that for noise-like pseudorandom sequences the autocorrelation functions change insignificantly, they retain own form. The autocorrelation functions for the pseudorandom sequences under study are presented in Figs. 9, 10 and 11. The third subsequence was used to determine how quickly the autocorrelation functions are change. The lag for the second subsequence is equal 1.The lag for the third subsequence is equal 10.

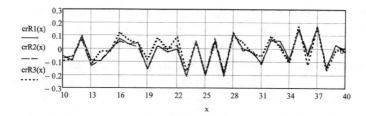

Fig. 9. Autocorrelation functions of subsequences of a pseudorandom sequence $\{rnd(1) - 0.5\}$.

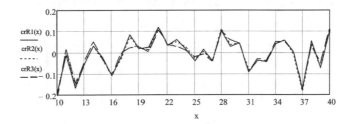

Fig. 10. Autocorrelation functions of subsequences of a pseudorandom bit sequence of FCSR generator.

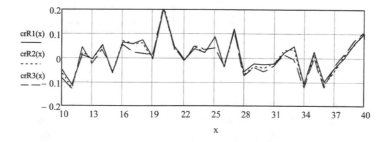

Fig. 11. Autocorrelation functions of subsequences of a pseudo-random bit sequence of LFSR generator.

The function $lcorr(x, x)$ was used to calculate the autocorrelation. The result represents 100 values for each subsequence. The procedure of linear interpolation $linterp\ (i, y(i), x)$ was applied to the values of autocorrelation for convenience of comparison.

Two characteristic features are observed. Firstly, the form of the autocorrelation function for different subsequences of each pseudorandom sequence remains practically constant. Secondly, there are samples for which the autocorrelation function of different subsequences is the same with great accuracy. For example, there are samples number 25 and 28 in the Fig. 9; there are samples number 30 and 35 in the Fig. 10. In the Fig. 11, there are the samples number 15 and 26.

It is possible with great accuracy to perform prediction of the sample number 101 for the first subsequence using the first and second characteristic features. The next method is proposed for this prediction. The sample with number 101 for the first subsequence is the sample number 100 for the second subsequence. The value of this sample $R2_{100} = y$ can take any value from the range of possible $-0.5 \ldots + 0.5$ with increment 0.01 (for example), $y_i = -0.5 + 0.01 \cdot (i - 1), i = 1 \ldots 101$.

For each value y_i, the autocorrelation functions $A3(y_i)$ are calculated and the values $A3(y_i)_l$ for lag l are selected. Each of them is compared with the value of the autocorrelation function $crR1_l$ with the lag l from the first subsequence. The value y_i, which is the solution of the equation $A3(y_i)_l = crR1_l$, determines one of the hundred numbers i and y_i. It should be noted that $crR1_l = crR2_l$. The graphical solution of the equation $A3(y_i)_l = crR1_l$ is shown in Fig. 12. The function $root(A3(y)_l - crR1_l)$ was used to determine y_i analytically.

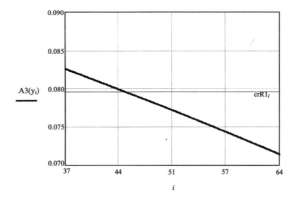

Fig. 12. Graphical solution of the equation to determine the predicted value of the sample.

The forecast was made for a pseudorandom sequence $\{rnd(1) - 0.5\}$. The first forecast value coincided with the true one with great accuracy. In general, the calculations were carried out for the next 10 samples; all the calculated values practically coincided with the true ones. The results of the forecast are presented in Table 1.

It is important to note that the magnitude of the error $\delta_i = Predicted\ r_i - True\ r_i$ does not increase when prediction number enhance.

Table 1. Results of forecasting the values of 10 samples of the sequence $\{rnd(1) - 0.5\}$.

Number i	Predicted r_i	True r_i	δ_i
1	−0,0633244	−0,0633169	$-7,5 \cdot 10^{-6}$
2	0,0778651	0,0778666	$-1,5 \cdot 10^{-6}$
3	0,1287013	0,1286670	$34,3 \cdot 10^{-6}$
4	0,0041477	0,0041493	$-1,5 \cdot 10^{-6}$
5	0,1957697	0,1957680	$1,7 \cdot 10^{-6}$
6	−0,3100367	−0,3100483	$11,6 \cdot 10^{-6}$
7	−0,3216296	−0,3216249	$-4,7 \cdot 10^{-6}$
8	−0,0425355	−0,0425416	$6,0 \cdot 10^{-6}$
9	−0,4024815	−0,4024773	$-4,2 \cdot 10^{-6}$
10	−0,4056022	−0,4055958	$-6,4 \cdot 10^{-6}$

The prediction of the sequence $R1 = \{rnd(1) - 0.5\}$ with the help of the built-in function of the Mathcad "*predict*" shows the absolute inapplicability of this function to the considered sequence. For example, the use of *predict* $(R1, 99, 1)$ gives a value −0.138 instead value −0.063. The peculiarity of the described method is that not all autocorrelation coefficients are used, but only those that are almost equal for neighboring subsequences.

4 Conclusion

The GPRBS based on LFSR and on FCSR were studied in detail. It was found that the repetition period for them is not less than 10^9. The generators were tested by NIST tests; statistical portraits of generators are presented.

The bit sequences of these generators have good statistical properties. As a consequence, their autocorrelation functions retain their form; this was confirmed by calculations in the Mathcad program. Each generator has individual form of autocorrelation function. For some lags the values of the autocorrelation coefficients remain constant. This allowed us to calculate the value of next sample of the pseudorandom sequence. The prediction was performed for the built-in pseudorandom sequence $\{rnd(1) - 0.5\}$. The predicted values of samples almost coincided with the true ones. The method can be used to test of the cryptographically strength of GPRBS.

References

1. Chethan, J., Lakkannavar, M.: Design of low power test pattern generator using low transition LFSR for high fault coverage analysis. Int. J. Inf. Eng. Electron. Bus. (IJIEEB) **5**(2), 15–21 (2013). https://doi.org/10.5815/ijieeb.2013.02.03
2. Ivanov, M.A., Chugunkov, I.V.: Cryptographic methods of information protection in computer systems and networks: a teaching manual. NIUA MIFI, Moscow (in Russian) (2012)
3. Jhansirani, A., Harikishore, K., Basha, F., et al.: Fault tolerance in bit swapping LFSR using FPGA architecture. Int. J. Eng. Res. Appl. **2**(1), 1080–1087 (2012)

4. Kitsos, O., Sklava, P., Zervas, N., et al.: A reconfigurable linear feedback shift register (LFSR) for the bluetooth system. ICECS (2001). https://doi.org/10.1109/ICECS.2001.957640
5. Klapper, A., Goresky, M.: 2-Adic shift registers. In: Fast Software Encryption, Cambridge Security Workshop Proceedings, pp. 174–178. Springer (1994)
6. Maksymovych, V., Shevchuk, M., Mandrona, M.: Research pseudorandom bit sequence generators based on LFSR. J. Autom. Measur. Control **852**, 29–34 (2016). (in Ukraine)
7. Maksymovych, V., Mandrona, M.: Investigation of the statistical characteristics of the modified fibonacci generators. J. Autom. Inf. Sci. (2014). https://doi.org/10.1615/JAutomatInfScien.v46.i12.60
8. Maksymovych, V., Mandrona, M.: Comparative analysis of pseudorandom bit sequence generators. J. Autom. Inf. Sci. (2017). https://doi.org/10.1615/JAutomatInfScien.v49.i3.90
9. Maksymovych, V., Mandrona, M., Garasimchuk, O., Kostiv, Yu.: A study of the characteristics of the fibonacci modified additive generator with a delay. J. Autom. Inf. Sci. (2016). https://doi.org/10.1615/JAutomatInfScien.v48.i11.70
10. Ali, M.A., Ali, E., Habib, M.A., Nadim, M., Kusaka, T., Nogami, Y.: Pseudo random ternary sequence and its autocorrelation property over finite field. Int. J. Comput. Netw. Inf. Secur. (IJCNIS) **9**(9), 54–63 (2017). https://doi.org/10.5815/ijcnis.2017.09.07
11. Milovanovic, E., Stojcev, M., Milovanovic, I., et al.: Concurrent generation of pseudo random numbers with LFSR of fibonacci and galois type. Comput. Inform. **34**, 941–958 (2015)
12. Mondal, A., Pujari, S.: A novel approach of image based steganography using pseudorandom sequence generator function and DCT coefficients. Int. J. Comput. Netw. Inf. Secur. (IJCNIS) **7**(3), 42–49 (2015). https://doi.org/10.5815/ijcnis.2015.03.06
13. Nas, R.J., Van Berkel, C.H.: High throughput, low set-up time, reconfigurable linear feedback shift registers. In: International Conference on Computer Design (2010). https://doi.org/10.1109/iccd.2010.5647572
14. Ndaw, B.A., Sow, D., Sanghare, M.: Construction of the maximum period linear feedback shift registers (LFSR) (Primitive Polynomials and Linear Recurring Relations). Br. J. Math. Comput. Sci. (2015). https://doi.org/10.9734/BJMCS/2015/19442
15. NIST SP 800-22. A Statistic Test Suite for Random and Pseudorandom Number Generators for Cryptographic Application. DIALOG: http://csrc.nist.gov/publications/niatpubs/SP800-22rev1a.pdf. Accessed Apr 2000
16. Nanda, S.K., Tripathy, D.P., Nayak, S.K., Mohapatra, S.: Prediction of rainfall in india using artificial neural network (ANN) models. Int. J. Intell. Syst. Appl. (IJISA) **5**(12), 1–22 (2013). https://doi.org/10.5815/ijisa.2013.12.01
17. Schneier, B.: Applied cryptography, protocols, and algorithms for source code in C. Triumph, Moscow (2002). (in Russia)

Organization of Network Data Centers Based on Software-Defined Networking

Yurii Kulakov[✉], Sergii Kopychko, and Victoria Gromova

National Technical University of Ukraine "Igor Sikorsky Kyiv Polytechnic Institute",
37 Peremohy ave., Kyiv 03056, Ukraine
ya.kulakov@gmail.com, kopychko.sn@gmail.com,
vikvikgrom@gmail.com

Abstract. In this paper, we propose and substantiate organization of modern data center networks (DCN) with Fat Tree topology based on software-configurable network technology (SDN). In order to increase efficiency of traffic engineering in DCN with Fat Tree topology, we propose to construct a set of disjoint paths. We also propose an improved deep search algorithm that enables us to reduce time needed for constructing such a set due to taking into consideration self-similarity of DCN topology. Main difference of this algorithm, compared to known analogies, lies in possibility to construct a maximal set of optimal disjoint paths. Applying SDN technology enabled us to increase efficiency of functioning of large-scale DCNs. Central SDN controller possesses full information about paths and trees that generated them. This enables us to optimize the paths according to specified metrics in the process of their construction. We show the application of the proposed method with an example of constructing a set of disjoint paths in DCN.

Keywords: Data center network · Software-defined networking
Multipath routing · Streaming algorithm · Counter flow method

1 Introduction

Modern computer networks are characterized by a large dimensionality and a variety of equipment, including mobile communications and mobile access points. In this regard, the process of managing this type of networks and building data on their networked centers, clusters distribution and the use of cloud technologies is becoming more complicated. To solve these problems, software-configurable network technology (SDN) [1–3] is now widely used. This technology enables us to increase efficiency of network equipment, reduce operating costs, increase manageability and network security basing on multi-path routing [4]. The technology enables us to manage the network at the software level, thus expanding the functionality of managing the transmission of information on the network. Figure 1 shows the structure of software-configurable networks.

© Springer International Publishing AG, part of Springer Nature 2019
Z. Hu et al. (Eds.): ICCSEEA 2018, AISC 754, pp. 447–455, 2019.
https://doi.org/10.1007/978-3-319-91008-6_45

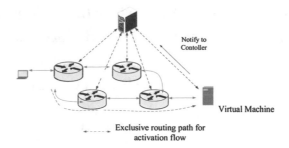

Fig. 1. Structure of software-configurable networks

Currently, due to the increasing demand for processing power and the amount of information, data center networks (DCNs) become relevant. Functionality DCN is the high-speed processing and storage of large amounts of data. DCNs should provide high data throughput and reliability. This, in turn, places high demands on the design of traffic engineering (TE) in such systems. In this regard, it is currently important to develop ways to organize the operation of DCNs and TE, considering the features and benefits of SDN [5–7].

When building modern DCNs, the Fat Tree [8] topology is mainly used, which is one of the most common topologies for building distributed systems aimed at solving high-performance tasks. The Fat Tree topology is a tree whose leaves are the computing devices, and the nodes are the switches. In this case, for higher-level switches, the bandwidth of the channels is greater, i.e. connections with other vertices are "thicker." Therefore, this topology was called Fat Tree.

When using the Fat Tree topology, DCN forms a four-level structure [9]. The top level consists of the core Switches, the aggregation switches are located one level below, the next level has the edge switches, to which the hosts are connected (Fig. 2).

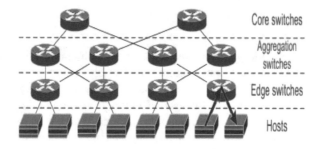

Fig. 2. Fat Tree topology

In turn, for DCN that consists of hundreds and thousands of large-scale servers, the Fat Tree topology, based on the principles of self-similarity, is transformed into the so-called Dragonfly network [10]. Considering the fractal nature of the Dragonfly network topology, it is possible to simplify the process of traffic design as well as the routing process. The main requirements for routing in DCN are reliability, minimal delay, and fault tolerance. These criteria are more satisfied with the methods of multipath routing, which are widely used in DCN [11–13]. One of the main disadvantages of combinatorial algorithms for multi-path routing is their considerable time complexity, which typically has a nonlinear character.

Therefore, the goal of this paper is to create a modified method of constructing a set of disjoint paths that takes SDN technology into account. The method should provide uniform load distribution among data transfer channels and minimal access time to network resources.

2 Modified Access Method Based in Multipath Routing

For Fat Tree topology, the depth-first search (DFS) is effective. It is based on recursive traversing of the vertices of the graph. During the traversing, the nodes are marked as visited and unvisited. At the beginning, all nodes are unvisited. The node that receives the control packet is marked as visited and analyzes information about adjacent nodes. Then, according to a certain rule, this node transmits the packet to one of the unmarked nodes adjacent to it. If this node does not have unmarked adjacent nodes, then it passes the control packet back.

The drawbacks of the basic algorithm are that it does not take into account:

1. The capacity of communication channels.
2. Features of the network topology.

There are three main strategies for selecting channels:

- Worst-Fit: select the channel with the highest available bandwidth;
- First-Fit: selects a channel with any available bandwidth that meets the requirements;
- Best-Fit: select the channel with the available bandwidth, which meets the requirements best.

It should be noted that when creating multiple paths in DCN using the DFS method, the Worst-Fit strategy is the most effective. Further efficiency enhancement of the DFS method in the formation of multiple paths is connected to considering the peculiarity of the hierarchical organization of DCN and the bandwidth of communication channels between nodes of the network.

The DFS method uses a centralized routing method, in which the central controller contains all the information needed to form routes. In addition to the network topology, the central controller contains information about the capacity of communication channels, which is used to select a communication channel at a lower level or to return to a higher level in the tree. After the formation of the next path, the permissible bandwidth of communication channels belonging to the selected path is reduced by a given value.

The first step of the algorithm determines the level of topology, at which the sender and recipient nodes can be unified. This is done on the basis that the central controller contains information about the network topology. If the sender and receiver nodes are connected to the same edge switch, it directly connects these nodes. Otherwise, the path passes through several switches, the number of which depends on the relative location of the nodes.

Determining the connection level of nodes makes it easier to perform routing. In this case, unnecessary transitions between levels are eliminated.

To accelerate the transition between levels in the modified algorithm, a global variable next that indicates the direction of traversal of the nodes of the network is introduced. With next = 1, the move must be done up the tree. When next = -1, the path down the tree should be selected. When passing through the level, at which the sender and receiver nodes can be linked, the value of the next variable becomes -1. This indicates that the traversing has reached a local maximum and one needs to move down the tree.

If, as a result of traversing, an edge switch that is not connected to the receiver is reached, the variable next takes the value next = 1 to return to the previous level. The use of the variable next is expedient when implementing the DFS algorithm, since it enables us to consider the features of the hierarchical topology.

Therefore, the modified DFS algorithm consists of the following sequence of operations:

1. Begin.
2. Determine the connection level of the path.
3. Set next to 1.
4. Set the sender node to the current node.
5. Add the current node to the nodes of the path.
6. If the current node is not a final switch, then go to step 9.
7. If the current node is the recipient's destination switch, then proceed to step 14
8. Set the previous node to the current one.
9. If the upper level is reached, then set next to −1.
10. Select the next node among adjacent nodes.
11. If it is impossible to select the next node, then go to step 8.
12. Set the selected node to the current one.
13. Go to step 5.
14. Add the recipient node to the path.
15. End.

To analyze the effectiveness of the proposed method for the formation of multiple paths, a special modeling system was developed. Figure 3 shows the example of the formation of paths between 1 and 5 DCN nodes. The length of the ith path (L_i) is determined by the number of transitions between the initial and final vertices and is equal to $N - 1$, where N is the total number of vertices of the given path.

As a result of the algorithm, four paths are formed between the vertices n_1 and n_5:

1. n1, s0, s3, s1, s7, s4, n5; L1 = 6.
2. n1, s0, s1, s0, s5, s4, n5; L2 = 6.
3. n1, s0, s3, s1, s7, s6, s5, s4, n5; L4 = 8.
4. n1, s0, s1, s0, s5, s6, s0, s4, n5; L4 = 8.

In this case, the mean path length is $L_{mean} = 7$.

Below is a pseudocode of the modified path generation algorithm in DCN:

```
{// G: network, a: source, b: destination, d: demand
1 H=necessary-layer-to-connect(G, a, b);
2 path={
3 u=a; // temp variable indicating current location
4 next=1; // search direction flag, 1: upstream, -1:
downstream
5 return SEARCH(u, path, next);
};
SEARCH(u, path, next) {
1 path=path+u;
2 if(u=b) return true;
3 if ( layer-of(u)=H) next=•1;// reverse search direc-
tion after reaching connecting layer
4 if(next=•1&& layer-of(u)=1) return false;// failure
at bottom layer
5 neighbors={v| layer-of(v)= layer-of(u)+next,and
available bandwidth of link(u, v)•d};
6 found=false;
7 while (neighbors!=0&&found=false) {
8 v=worst-fit(neighbors); neighbors=neighbors\{v};
9 found=SEARCH(v,path,next);
10 };
11 return found;}
```

A distinctive feature of software-configurable networks is that the organization and management of the network is carried out at the software level with the help of virtual switches [1].

The SDN concept is based on the OpenFlow open standard, which defines the flow principles of managing network traffic. Considering the ability to configure the route of each individual traffic at the level of software-configurable SDN switches, in [12], the implementation of multithreaded traffic routing in a software-configurable network is proposed.

In comparison with known algorithms, streaming algorithms for multipath routing are characterized by a minimum time complexity.

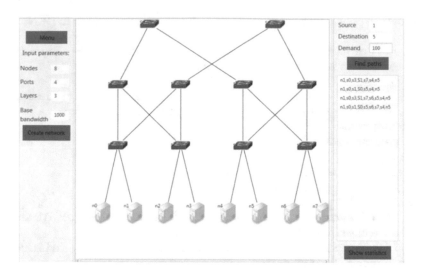

Fig. 3. DCN paths forming

The presence of centralized management of SDN based on the network controller enables us to form a set of disjoint paths by arranging counter flows between the sender and recipient nodes. The task of finding paths is reduced to finding the points of connection of trees from the initial and final vertices. When routing similarly with the modified wave algorithm, as the next for a certain path, a vertex is selected that has a smaller outer degree. Therefore, the formation of disjoint paths is guaranteed.

Paths trees are being constructed until all disjoint paths between the initial and final vertices are constructed. The algorithm for forming a set of disjoint paths is shown in Fig. 4.

```
begin;
Form the path tree T_i( B_i ,E_i) from the initial vertex
b_i;
Form the path tree T_j( B_j ,E_j) from the initial vertex
b_j;
If B_i • B_j  then go to 2;
Find paths L_k ∈ T_i and L_m ∈ T_j with mutual vertex b_s={
B_k•B_m};
Form the united path L_s =( L_k ∈ T_i )∪ ( L_k ∈ T_i );
If (All trees are formed) then go to 8 else go to 2;
end.
```

Fig. 4. Algorithm for forming a set of disjoint paths

In the centralized formation of a set of independent paths, a central SDN controller has complete information about the paths being formed and trees that generated them, which enables us to optimize them in the process of forming the paths according to the given metrics.

For the considered set of vertices, the transition tree is constructed. The basic condition for traversing this tree is the minimum path length to the vertex. Thus, when choosing between vertices B_5 and B_6 (Fig. 5), the tree continues to be built through the vertex B_6 (Fig. 6). If there is a branch whose path length is less than the length of the current path, then it is necessary to return to the branch with a shorter length. So, in Fig. 6 the path $B_1 - B_2 - B_5$ is less than the current path $B_1 - B_2 - B_6 - B_{10}$. Hence, the tree continues to be constructed from the vertex B_5 until the length of the given path exceeds the length of the other path (Fig. 7).

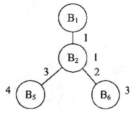

Fig. 5. Start of traversing the transition tree

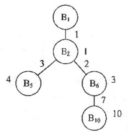

Fig. 6. Continuation of the transition tree through the vertex B_6

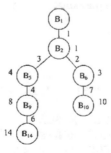

Fig. 7. The transition tree must be continued from vertex B_{10}

Similarly, the tree continues to be constructed from the top of B_{10}, until the length of this path is longer than the length of the other path. As a result, the tree in Fig. 8 is obtained. As we can see, the path with a length of 19: $B_1 - B_2 - B_6 - B_{10} - B_{15} - B_{18} - B_{21}$ is the optimal way for the considered set $\{B_1, B_2, B_5, B_6, B_9, B_{10}, B_{14}, B_{15}, B_{18}, B_{21}\}$.

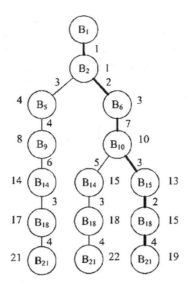

Fig. 8. The generated transition tree for the set of vertices {B1, B2, B5, B6, B9, B10, B14, B15, B18, B21}

3 Conclusions

The modified multipath routing method is proposed, which enables us to reduce the time for the formation of multiple routes of access to network resources. The reduction is gained due to the special features of the SDN organization and due to the presence of a central controller in the network.

The proposed method of organizing network data centers based on SDN technology in combination with the formation of multiple paths makes it possible to improve the efficiency of the procedure for the operation of network data centers.

References

1. Isong, B., Kgogo, T., Lugayizi, F.: Trust establishment in SDN: controller and applications. Int. J. Comput. Netw. Inf. Secur. (IJCNIS) 9(7), 20–28 (2017). https://doi.org/10.5815/ijcnis. 2017.07.03
2. Kumar, P., Dutta, R., Dagdi, R., Sooda, K., Naik, A.: A programmable and managed software defined network. Int. J. Comput. Netw. Inf. Secur. (IJCNIS) 12, 11–17 (2017). https://doi.org/ 10.5815/ijcnis.2017.12.02. In MECS http://www.mecs-press.org/. Accessed Dec 2017
3. Sahoo, K.S., Mishra, S.K., Sahoo, S., Sahoo, B.: Software defined network: the next generation internet technology. Int. J. Wirel. Microwave Technol. (IJWMT) 7(2), 13–24 (2017). https://doi.org/10.5815/ijwmt.2017.02.02
4. Moza, M., Kumar, S.: Analyzing multiple routing configuration. Int. J. Comput. Netw. Inf. Secur. (IJCNIS) 5, 48–54 (2016). https://doi.org/10.5815/ijcnis.2016.05.07. In MECS http:// www.mecs-press.org/. Accessed may 2016

5. Agarwal S., Kodialam M., Lakshman T.: Traffic engineering in software defined networks. In: Proceedings of the 32nd IEEE International Conference on Computer Communications, INFOCOM 2013, pp. 2211–2219 (2013)
6. Shu, Z., Wan, J., Lin, J., Wang, S., Li, D., Rho, S., Yang, C.: Traffic engineering in software-defined networking: measurement and management. IEEE Access **4**, 3246–3256 (2016). http://www.ieee.org/publications_standards/publications/rights/index.html
7. Abbasi, M.R., Guleria, A., Devi, M.S.: Traffic engineering in software defined networks: a survey. J. Telecommun. Inf. Technol. **4**, 3–13 (2016)
8. Kim, J., Dally, W.J., Scott S., Abts D.: Technology-driven, highly-scalable dragonfly topology. In: Proceedings of the 35th Annual International Symposium on Computer Architecture. ISCA 2008, pp. 77–88 (2008). http://dx.doi.org/10.1109/ISCA.2008.19
9. Jo, E., Pan, D., Liu, J., Butler L.: A simulation and emulation study of SDN-based multipath routing for fat-tree data center networks. In: Proceedings of the 2014 Winter Simulation Conference (2014). http://dl.acm.org/citation.cfm?id=2694235
10. Fatmi O., Pan D.: Distributed multipath routing for data center networks based on stochastic traffic modeling. In: IEEE 11th International Conference on Networking, Sensing and Control (ICNSC) (2014). http://ieeexplore.ieee.org/xpl/articleDetails.jsp?arnumber=6819683
11. Jung, E.-S., Vishwanath, V., Kettimuthu, R.: Distributed multipath routing algorithm for data center networks. In: 2014 International Workshop on Data Intensive Scalable Computing Systems (DISCS), pp. 49–56 (2014)
12. Luo, M., Zeng, Y., Li, J., Chou, W.: An adaptive multi-path computation framework for centrally controlled networks. Comput. Netw. **83**, 30–44 (2015)
13. Kulakov, Y., Kogan, A.: The method of plurality generation of disjoint paths using horizontal exclusive scheduling. Adv. Sci. J. **10**, 16–18 (2014). https://doi.org/10.15550/ASJ.2014.10. ISSN 2219-746X

Management of Services of a Hyperconverged Infrastructure Using the Coordinator

Oleksandr Rolik[1] ⓘ, Sergii Telenyk[2] ⓘ, and Eduard Zharikov[1]([✉]) ⓘ

[1] Department of Automation and Control in Technical Systems,
National Technical University of Ukraine
"Igor Sikorsky Kyiv Polytechnic Institute", Kyiv, Ukraine
o.rolik@kpi.ua, stelenyk@pk.edu.pl,
zharikov.eduard@acts.kpi.ua
[2] Department of Theoretical Electrical Engineering and Computer Science,
Faculty of Electrical and Computer Engineering,
Cracow University of Technology, Cracow, Poland

Abstract. Modern data centers' providers are gradually moving away from traditional and multi-vendor IT infrastructures to open, standardized and interchangeable solutions that are based on a software defined approach to managing data center resources. The authors analyze the architectural features, requirements, limitations, hardware and software of hyperconverged infrastructures and their advantages in comparison with traditional and converged architectures deployed in data centers. The authors propose to employ two-level coordination schema to manage compute, storage, network and virtualization subsystems of hyperconverged infrastructure along with the self-management algorithms inside these subsystems.

Keywords: Data center · Hyperconverged system · Coordination

1 Introduction

Evolution and transformation of the data center are influenced by such powerful driving factors and technologies as virtualization [1], distributed computing, GRID [2], cloud computing [3], Fog Computing [4], scale-out and warehouse-scale computing [5, 6], Software Defined Networks [7] and Software Defined Data Center [8, 9]. A modern data center is a complex system that is evolved to a level of a cloud data center offering significant improvement in self-service, scalability and manageability [10].

IDC research shows that operators who monitor the operation of a traditional three-tier IT infrastructure in the data center spend more than 70% of the time on monitoring operations, troubleshooting, finding performance bottlenecks, updating system software and drivers, reconfiguring storage subsystems and networking. Thus, enterprises and service providers while modernizing and deploying new data centers are gradually moving away from traditional IT silos and multi-vendor IT infrastructures to open, standardized and interchangeable solutions that are based on a software defined approach [11].

© Springer International Publishing AG, part of Springer Nature 2019
Z. Hu et al. (Eds.): ICCSEEA 2018, AISC 754, pp. 456–467, 2019.
https://doi.org/10.1007/978-3-319-91008-6_46

Currently, cloud computing has become the main paradigm on which modern information services are based. Enterprises face the objective need for hybrid cloud by integrating private and public cloud solutions. According to RightScale 2017 State of the Cloud Report [12], 95% of organizations are running applications or experimenting with infrastructure-as-a-service (IaaS) using the hybrid cloud and top priority for 50% of enterprise central IT teams is a hybrid cloud.

According to IDC [13], by 2020, the heavy workload demands of next-generation applications and new IT architectures in critical business facilities will have forced 55% of enterprises to modernize their datacenter assets through updates to existing facilities and/or the deployment of new facilities. Thus, there is a need for development of new architectures for data center resource organization and management that meet modern challenges.

Hyperconverged Infrastructure (HCI) has recently gained significant popularity as an environment where servers, storage and networking are brought together using virtualization technologies. To address modern challenges in cloud data center management [14–16] hyperconverged infrastructure is widely used in core and edge IT infrastructure. An HCI is a virtual computing infrastructure solution that seamlessly combines several data center services in an appliance form factor and runs different workloads such as virtual desktop infrastructure, big data analysis, enterprise resource planning, and customer relationship management.

An HCI has a fundamentally different architecture and differs from traditional three-tier infrastructure solutions. Hyper-convergence is an architectural model for IT infrastructure that uses virtualization [1] and employs x86-based CPUs, SSD and HDD storages in a single building block. Next, hyperconverged systems are typically connected over IP-based networks (most commonly Ethernet) and are distributed by nature [17]. An HCI now scales to hundreds of nodes and combines web-scale engineering with client grade design to deliver enterprise level cloud IT infrastructure. So, there is a need for development of methods and algorithms to effective management of an HCI subsystems, such as compute, storage and network to deliver cloud services at scale.

This study explores the opportunity of applying two-level coordination schema [18] to management of an HCI subsystems. The authors propose to employ two-level coordination schema to manage compute, storage, network and virtualization subsystems along with the self-management algorithms inside these subsystems. The authors also analyze the architectural features and capabilities of an HCI with the aim of using them in private and hybrid cloud solutions based on modern hardware platforms for the IT infrastructure of the enterprise data center.

2   Traditional and Integrated Architectures of a Data Center

An architecture of data center has undergone significant transformations over the past two decades, from fully proprietary solutions to complex three-tier systems. Currently, this approach to building a data center is predominant, in which the IT infrastructure is built with the involvement of components, modules, programs and licenses from different vendors.

One of the most complex deployment plans involves developing an architecture based on existing specifications, analyzing solutions and selecting vendors, purchasing server hardware, network devices, storage area network (SAN) or network attached storage (NAS) systems, software and licenses, and configuring hardware, software, and management tools. Deploying and hosting such complex IT infrastructure can require significant capital and operational investment and compatibility checks.

Another approach allows to reduce the cost of designing and deploying the data center IT infrastructure. This approach is to use integrated solutions from a specific vendor. Despite the well-known risks of binding to a particular vendor (vendor lock-in or platform lock-in), the enterprise receives a competitive solution for the data center, which the vendor accompanies, updates and eliminates integration errors. An integrated system is a set of tools that includes a certain number of servers, storage systems and network devices that are selected by the vendor (or a group of vendors) in such a way as to effectively implement management functions using specialized software developed for this class of integrated systems. There are three categories of integrated systems [19]: integrated stack system, integrated infrastructure system, hyperconverged integrated system (or HCI).

The integrated stack system is usually based on blade systems and is designed to create an IT infrastructure for a specific software stack, for example, the IBM PureApplication System, Oracle Exadata Database Machine and Teradata, etc. Such a system usually consists of a server, storage and network device in the form of different modules and is characterized by a strong binding to the vendor.

The integrated infrastructure system has today become the most widely used in small and medium-sized data centers, as well as on-premises solutions in the IT infrastructure of enterprises and organizations due to the availability and ease of management. Such systems are called converged systems. Converged systems are presented as blades and as appliance for rack enclosures for flexible deployments. Examples of converged systems are VCE Vblock, HP ConvergedSystem, CISCO UCS, Lenovo Converged System and others. The main advantage of convergent systems is minimization of downtime of the entire system when updating the system software, firmware, and drivers using one update in the form of roll-up. Adaptation of hardware for certain software and workloads in converged systems leads to increased productivity of applications. Other advantages of convergent systems are the ability to scale-up the hardware and the use of a vendor's support.

Hyperconverged system (HCS) is built on the basis of web-scale architectural model. Hyperconverged system is distributed by nature and consists of building blocks represented by nodes, each of which includes computational, storage and network resources. As a computational node the x86 CPU architecture is used in most cases. The storage resources are represented by server-side NVMe PCI or DIMM storage, SSD and HDD. The network communication is represented by several interfaces with 10 Gb and 40 Gb Ethernet.

Using such building blocks, it is possible to scale-out by building a high-performance cluster that allows to deploy, effectively use and scale a wide class of applications, including virtual desktop infrastructure, large data services including MapReduce workload, SQL and NoSQL databases, block and file storages and unified communications. Along with scale-out option the HCS can be also scaled up.

The main difference from the convergent systems among other important differences is the elimination of the centralized SAN or NAS storages. This approach significantly reduces the capital and operational costs of deploying, managing and scaling storage subsystem. In reality, the CPU capacity of storage controllers may reach maximum while new servers will be deployed in the data center, thus the CPU of storage controllers will become a bottleneck as more VMs will be deployed. Another problem of traditional storage network is scalability of controllers and their management.

Significant results in the development of the hyperconverged systems have been achieved by Nutanix, VMware, HPE, CISCO, Microsoft, Scale Computing. Most integrated systems use Intel and AMD x86 processors, as well as RISC, Power, SPARC, ARM and Intel Atom architectures are presented in some appliances.

3   Technological Aspects of Hyperconverged Architecture

As a result of research in the field of creating high-performance and scalable systems, over the last 15 years, the leaders of public cloud services such as Microsoft, Google, and Amazon have deployed a web-scale technology to build an IT infrastructure. Each cloud service is characterized by such indicators as the use of a certain (often homogeneous) platform, a very large number of clients, the management of large amounts of data compared to enterprise data. Despite of this fact, the ideas and technologies used in the web-scale approach can be applied along with virtualization solution to heterogeneous platforms in an enterprise data center. Thus, the ideas of web-scaling found their application in hyperconverged systems.

HCS uses the principles and technologies of distributed, highly productive and software-defined systems. Web-based scaling is based on such systems and technologies as:

- Cassandra, distributed NoSQL database management system;
- HDFS or other distributed file system;
- MapReduce, a software model for implementing distributed parallel computing;
- Hadoop, a software framework for storing and processing objects in a distributed computing environment;
- Paxos, a family of protocols for organizing the interaction of nodes in a distributed system to ensure their consistent interaction and data integrity;
- Zookeeper, a service used in clustered distributed systems for naming objects and their synchronization;
- REST-based software interface for automation of processes in the IT infrastructure.

Typically, HCS vendors are currently offering their software stack, which uses combinations of various control and orchestration software and technologies, including proprietary and opensource frameworks and architectures. This vector of HCS development reduces the capabilities of the enterprise to the integration of tools and hardware. Thus, there is a need to develop open architectures with orchestration and integration tools based on a software-defined approach and embracing all levels of interaction in a cloud data center.

It should be noted, that there is a very important result of the deployment of HCS in the life cycle of the enterprise IT infrastructure. Hyper converged approach in theory allows a simultaneous use of equipment of various generations, as well as ensuring possibility of seamless integration of new nodes operating based on the latest hardware architectures. Thus, HCS ensures the evolution of the IT infrastructure by allowing new resources to be added to meet new challenges as well as decommissioning obsolete equipment without affecting the service delivery process.

As a result of analysis of the range of enterprise services and workloads, the authors can conclude that the full transition to the HCS is constrained by the following factors:

– the enterprises use different teams for administration of compute, network, storage and virtualized services;
– many line of business applications of the enterprises that implement business processes are not designed for distributed environments;
– the enterprise technologies are still focused on the use of the centralized storages;
– low availability of new architecture application development to transition to DevOps management operations.

HCI is fully software-defined approach based on virtualization of network objects and functions, on the virtualization of the storage system and on the virtualization of the computing resources of a group of physical servers. High degree of HCS scalability allows to increase the computing and storage resources with the necessary increment, which is caused by increasing workload, by the current and prospective customer requirements, and by the introduction of new services. Linear growth of capacity and performance with guaranteed quality of service is a very important advantage of HCI. Furthermore, unplanned downtime is reduced almost twice [20].

Unlike the traditional data center architectures where virtual machines were in a shared storage, HCS uses distributed storage when each node of the system performs both storage functions and computational functions. The hyperconverged infrastructure with m nodes is shown in Fig. 1. Each node may be composed of one or more physical servers. Each physical server has a local storage composed of NVMe PCI, SSD and HDD disks. That storage is represented to cluster storage pool as a software-defined, fault-tolerant, shared storage with local caching on each physical node. The implementation of the data storage and cluster management functions is performed either in a modified hypervisor (for example, as in Scale Computing solution), or as a Controller VM (for example, as in Nutanix solution). This provides redundancy, high data reading and writing performance [21], and the fault tolerance by distributing data on cluster nodes in the form of 2 or 3 copies of the same data. Despite the fact that the useful disk space is reduced, this architecture by default provides fault tolerance in case of failure or maintenance of one node in the three-node cluster.

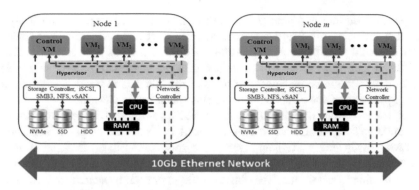

Fig. 1. Hyperconverged infrastructure

The HCI cluster can be created starting from 2, 3 or 4 nodes, depending on the vendor of the hyperconverged solution and the starting functionality of the HCI cluster. Networking of HCI cluster nodes is performed using two common approaches: using a dedicated network for the functioning of a hyperconverged cluster (Private/Backplane Network, Out-Of-Band) or using a single network infrastructure (Public Network).

4   HCI Hardware

In this paper, several HCI configurations are being considered. Each of the considered configurations is characterized by the minimum and maximum number of nodes (servers or appliances) in the cluster, by the functionality of management software, by the requirements for the network topology, by the number of disks and memory size relating to the CPU in the node, by the number of CPU cores per node. The HCS (appliance) may consist of homogenous or heterogenous nodes. At the same time, the nodes are either provided by the manufacturer or the customer can purchase and configure the servers themselves. All vendors of hyperconverged systems provide fully software-defined platforms with support for backup, deduplication, disk balancing, disaster recovery, compression, and erasure coding.

Nutanix is one of the leaders in the development of hyperconverged systems. It offers a hardware and software platform based on the different hypervisors such as Hyper-V, ESXi, Acropolis, and KVM to deploy IT infrastructure for any enterprise or web-scale applications at any scale. The proposed solution allows to begin cluster deployment from three nodes and increase the performance (scale-out) flexibly with 1-node increments, depending on the customer needs, up to 20 or 80 cores per node. Nutanix does not use the dedicated Backplane Network for inter-node communication to manage the cluster. It uses a standard telecommunications network with a 10 Gb or 40 Gb ethernet topology. A 2U node with 80 cores is used [22]. IDC study [20] shows that HCS Nutanix platform allows to reduce the annual IT infrastructure costs by 30.6%, while TSS declines by 58.3% over the five-year period.

vSAN ReadyNode [23] is positioned as an intelligent, economical and reliable integrated solution that integrates computing resources, virtualization, network, storage and management tools into an appliance to create a hyperconverged infrastructure. Two nodes are used to deploy a cluster, further expansion takes place in 1-node increment. Each node contains 12 to 24 processor cores, up to 384 GB RAM, up to 12 TB combined HDD and two SSD, 10 GbE port for network connection. One cluster can scale to 64 nodes and provide up to 120 virtual machines per node to the client. The solution is based on the software products vCenter Server, vSphere and specialized software.

Dell EMC VxRail [24] is a fully integrated, pre-installed, tested and configured hyperconverged IT infrastructure that is based on VMware vSphere software and Dell EMC hardware. VxRail appliance featuring kernel-layer integration between VMware vSAN and the vSphere hypervisor. VxRail is designed to build a Software-Defined Data Center. The cluster deployment can be started from three nodes and it can be scaled up to 64 nodes in steps of one node. One node can be configured using different hardware: processor (s) with number of cores up to 44, RAM from 64 GB to 512 GB, SSD storage from 3 TB to 6 TB, two 10 GbE ports, and two 1 GbE ports.

The HC3 virtualization platform from Scale Computing is positioned as a datacenter in a box and allows to compose the cluster using three nodes with the same hardware configuration within one of the levels (HC1000, HC2000, HC5000). The maximum number of nodes in the cluster is 8, but it is possible to adapt the cluster to work with a larger number of nodes with the help of Scale Computing consultants. The HC3 is designed without a single point of failure. The private network (1 GbE or 10 GbE) is used to manage the cluster. Each node of the HC3 system has a size of 1U [25].

5   Two-Level Model of Management System for Hyperconverged Infrastructure

The authors propose to solve management tasks in HCS using HCI management system (HCIMS) shown in Fig. 2. The HCIMS is an evolution of the two-level IT infrastructure management system proposed in [18].

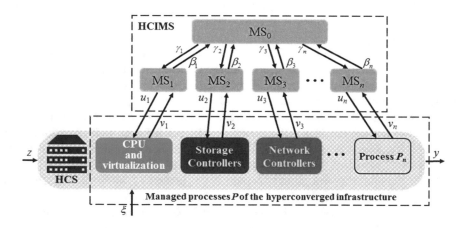

Fig. 2. The HCI management system model

The position of the management subsystems (MS) displays the hierarchical structure of the HCIMS. The model consists of the upstream management subsystem (MS_0) namely the coordinator, n second-level management subsystems (MS_1, ..., MS_n) and the managed processes P that exist in the HCS. The processes P are represented by management processes of each HCS subsystem such as storage management process, network management process, virtualization management process and other.

The interaction of management subsystems along the vertical is performed as follows. Commands and impacts $\gamma_1, \ldots, \gamma_n$ transmitted from MS_0 to MS_1, ..., MS_n, are coordinating impacts. Commands and impacts $(u_1, \ldots, u_n)$ from MS_1, ..., MS_n to processes P are control impacts. Feedback signals $v_1, \ldots, v_n$ are transmitted from the processes P to MS_1, ..., MS_n. Feedback signals $\beta_1, \ldots, \beta_n$. are transmitted from the management subsystems to the coordinator. The description of a two-level HCIMS can be implemented by means of terminal variables such as inputs and outputs.

The controllable subsystems P are impacted by management signals u from MS_1, ..., MS_n, $u \in U$, where U is the set of control actions. The controllable subsystems P are also affected by input signals z, $z \in Z$, where Z is the set of user requests, and by the signals ξ, $\xi \in \Xi$, where Ξ denotes disturbing impacts. Disturbing impacts include faults in the HCI, functional failures in the elements and subsystems of HCS, overloaded state of subsystem. The output of the processes P is y, $y \in Y$, where Y is the set of outputs of subsystems such as monitoring data.

The processes P can be represented in the form of the following mapping on the basis of the cartesian product $P : U \times Z \times \Xi \to Y$.

The set U of management signals impacting on the processes P can be represented as the cartesian product of n sets [26] $U = U_1 \times U_2 \times \ldots \times U_n$.

In this case each management subsystem of MS_1, ..., MS_n has the authority to select individual component from $u_1, \ldots, u_n$ to have a direct impact on the processes P.

Each i-th management subsystem MS_i, $i = \overline{1, n}$ receive two inputs: a coordinating impact $\gamma_i \in \Gamma$ from the MS_0 and feedback signal v_i in the form of monitoring data. The management output of MS_i, is the impact u_i, chosen by MS_i from the set U_i. Suppose that each MS_i realizes the mapping C_i so that $C_i : \Gamma \times V_i \to U_i$, where V_i is a set of monitoring data v_i coming to HCIMS from the subsystems of HCS, $v_i \in V_i$. Monitoring data V_i, $i = \overline{1, n}$, are feedback signals inside i-th HCS subsystem.

Feedback signals v_i, coming to the input of MS_i, are obtained from subsystems of HCS as a result of monitoring. Naturally, these data functionally depend on the control signals u, inputs z, and disturbing impacts. This dependence can be represented by the mapping such that [26]

$$f_i : U \times Z \times \Xi \times Y \to V_i. \tag{1}$$

The management subsystem MS_0 is the coordinator that generates coordinating signals $\gamma_i \in \Gamma$, and the signal from the i-th output MS_0 going only to the input of the i-th lower-lying management subsystem MS_i. The coordinator generates a signal based on the analysis of data coming to its inputs from the MS_i, and representing feedback signals about the state and functioning of the HCS correspondent subsystems. In this case, it can be assumed that in the coordinator the mapping C_0 is realized such that

$$C_0 : B \rightarrow \Gamma,$$

where B is a set of information signals β realizing the feedback. And $\beta = (\beta_1, \ldots, \beta_n)$ is the aggregate of feedback signals β_i coming to the coordinator from subsystems MS_i.

Similarly to (1), the feedback signal β received at MS_0 carries information about the state of all the underlying subsystems, so it is determined by the mapping

$$f_0 : \Gamma \times V \times U \rightarrow B,$$

where $V = V_1 \times \ldots \times V_n$. Thus, B is a function of the coordinating signals γ_i, the feedback signals $v = (v_1, \ldots, v_n)$ arriving at the MS_i, and control impacts $u = (u_1, \ldots, u_n)$.

On the model shown in Fig. 2, the interaction between the management subsystems MS_1, ..., MS_n is not explicitly shown, nor is the direct influence of MS_0 on the functioning of the HCS that can take place in real HCI.

In accordance with [26], the coordination is based on the impact on the management subsystems MS_i, which forces them to act in concert, subjecting the MS actions to a unified policy aimed at achieving the global goal of the HCS, in spite of the fact that this goal may conflict with the local goals of the subsystems. The MS_0 performs coordination and overcomes the contradictions between the local goals of subsystems MS_i.

The success of the coordinator activities for the organization of concerted actions of MS_i is evaluated by how successfully the global goal of HCS management is achieved. Achieving the goal by the coordinator can be considered as a solution to the task, which can be formalized as a decision-making problem that consists in evaluating the effectiveness of coordination process. Since this task is defined with respect to all management subsystems including the processes P, so it is determined as a global task [26].

For two-level systems, coordination must be ensured with respect to the problem solved by MS_0, and be ensured with respect to the global problem. The first one means that the signals from MS_0 have a coordinating effect on the tasks solved by MS_i, and the second one means that the coordinator is able to influence MS_i, so that their combined effect on the processes P is directed at solving the global problem.

Successful functioning of the HCIMS that corresponds to the two-level model can be ensured when the objectives of the subsystems are consistent with each other and are consistent with the global goal of the system [26]. Three types of goals are singled out in a two-level system: the global goal, the goal of the coordinator MS_0 and the goals of the managing subsystems MS_i. The need for compatibility of goals results from the following features. The process P is directly affected only by the MS_i, therefore the global goal can only be achieved indirectly through the actions of MS_i, which must be coordinated with respect to the global goal, as well as the coordinator's goal.

The global goal is to improve the efficiency of services. It goes beyond the immediate activities of the two- level system, shown in Fig. 2, and none of the subsystems of MS_i, is focused on achieving a global goal or solving a global task. The global task can be solved only by joint actions of all management subsystems MS_i.

Global Goal. Given the fact that HCS are deployed to improve the efficiency of production and tenant services, the global goal of HCIMS can be defined as ensuring the maximum quality of IT services with minimal costs while ensuring Service Level Agreement (SLA). The maximum quality of IT services is denoted as max $\mathfrak{Q}$.

The maximum quality of the IT services delivered by HCI will be achieved when

$$\max \mathfrak{Q} \Leftrightarrow \max Q_j, \forall j = \overline{1, K} \Leftrightarrow \max q_{kj},$$
$$\forall j = \overline{1, K}, \ \forall k = \overline{1, M_j},$$

where Q_j, $j = \overline{1, K}$ is a quality of j-th service; q_{kj}, $k = \overline{1, M_j}$ is a value of the k-th quality index of the j-th service.

To achieve the goal of IT infrastructure management in total, it is necessary to continuously scale up and scale out the hardware resources of the IT infrastructure, which is unacceptable primarily from the economic point of view. On the other hand, increasing the economic efficiency of the service delivery requires reducing the total cost ownership, i.e., actions aimed at achieving min $\mathfrak{C}$. Maintaining the quality of services at this level is the main task of the coordinator.

The Goal of the Coordinator. The coordinator's goal is to maintain the quality of services $\mathfrak{Q}$ at an agreed level with minimal costs $\mathfrak{C}$ for the IT resources involved. The goal of the coordinator can be formalized as follows

$$\mathfrak{Q} = \text{const}|_{\min \mathfrak{C}}. \tag{2}$$

Expression (2) means that the coordinator of all possible impacts will choose those that require a minimum cost of implementation.

The requirement to maintain an agreed level of services covers all services and individual service quality indicators:

$$\mathfrak{Q} = \text{const} \Leftrightarrow Q_j = \text{const}, \forall j = \overline{1, K} \Leftrightarrow q_{kj}$$
$$= \text{const}, \forall k = \overline{1, M_j}, \ \forall j = \overline{1, K}.$$

It is necessary to define the following circumstance. The main way to improve the quality of the j-th service is to allocate additional resources to applications supporting the j-th service. If the level of the j-th service exceeds the setpoint, the resources allocated to the corresponding applications are reduced, as required by the criterion min $\mathfrak{C}$. At the same time, the last server that provides j-th service cannot be turned off, despite the fact that the quality of this service is still higher than required, as this will lead to a complete termination of the provision of the service. Thus, there will always be some fixed minimum of costs, after which further cost reduction will be impossible.

Local Goals. The goal of local management is to maintain the specified values of the functioning parameters of the HCS with minimal costs, i.e.

$$q_{kj} = \text{const}|_{\text{min}\mathfrak{C}}, \forall k = \overline{1, M_j}, \ \forall j = \overline{1, K}.$$

In the HCIMS model shown in Fig. 2, management subsystems MS_i, can have their own differing goals of functioning.

6 Conclusion

Currently, to implement HCI in the enterprise data center, it is necessary to adapt existing software systems and services to realize the possibility of distributed data processing. Despite this, for some HCS it is necessary to develop service-aware levels of abstraction, in order to ensure the operation and management of a wide class of software oriented not only to distributed data processing, but also to existing enterprise applications.

As a result of the analysis of hyperconverged architectures, the authors can conclude that many vendors of hyperconverged systems use proprietary equipment and software, which leads to a strong binding to the vendor. Thus, the integration of systems from different manufacturers is not foreseen, except for the use of additional software with REST-based API.

It is possible to deploy a private cloud on small clusters with several tens to several hundred physical servers using a hyperconverged infrastructure. For large clusters that have thousands of physical servers, the problems of deploying a hyperconverged system implementing effective network interconnection and organizing a distributed data storage and processing are being resolved.

The authors propose to employ two-level coordination schema to manage compute, storage, network and virtualization subsystems of a hyperconverged infrastructure along with the self-management algorithms inside these subsystems.

References

1. Barham, P., Dragovic, B., Fraser, K., Hand, S., Harris, T., Ho, A., Neugebauer, R., Pratt, I., Warfield, A.: Xen and the art of virtualization. In: Proceedings of the 19th ACM Symposium on Operating Systems Principles (2003)
2. Foster, I., Zhao, Y., Raicu, I., Lu, S.: Cloud computing and grid computing 360-degree compared. In: Grid Computing Environments Workshop, GCE 2008, pp. 1–10 (2008)
3. Mell, P., Grance, T.: The NIST definition of cloud computing. National Institute of Standards and Technology, Special Publication 800-145, Gaithersburg, USA (2011)
4. Agarwal, S., Yadav, S., Yadav, A.K.: An efficient architecture and algorithm for resource provisioning in fog computing. Int. J. Inf. Eng. Electron. Bus. (IJIEEB) **8**(1), 48–61 (2016). https://doi.org/10.5815/ijieeb.2016.01.06
5. Barroso, L.A., Clidaras, J., Hölzle, U.: The datacenter as a computer: an introduction to the design of warehouse-scale machines. Synth. Lect. Comput. Archit. **8**(3), 1–154 (2013)

6. Barroso, L.: Warehouse-scale Computers. Invited talk at the USENIX Annual Technical Conference, Santa Clara, CA (2007)
7. Open Network Foundation: Software-defined networking: The new norm for networks. https://www.opennetworking.org/images/stories/downloads/sdn-resources/white-papers/wp-sdn-newnorm.pdf. Accessed 13 Dec 2017
8. DMTF DSP-IS0501: Software Defined Data Center (SDDC) Definition (2015)
9. The Software-Defined Data Center. https://www.vmware.com/solutions/software-defined-datacenter.html. Accessed 13 Dec 2017
10. Jararweh, Y., Al-Ayyoub, M., Benkhelifa, E., Vouk, M., Rindos, A.: Software defined cloud: survey, system and evaluation. Future Gener. Comput. Syst. **58**, 56–74 (2016)
11. Li, C.S., Brech, B.L., Crowder, S., Dias, D.M., Franke, H., Hogstrom, M., Lindquist, D., Pacifici, G., Pappe, S., Rajaraman, B., Rao, J.: Software defined environments: an introduction. IBM J. Res. Dev. **58**(2/3), 1–15 (2014)
12. RightScale 2017 State of the Cloud Report. https://www.rightscale.com/2017-cloud-report. Accessed 13 Dec 2017
13. IDC FutureScape: Worldwide Datacenter 2018 Predictions Description, October 2017
14. Asyabi, E., Sharifi, M.: A new approach for dynamic virtual machine consolidation in cloud data centers. Int. J. Mod. Educ. Comput. Sci. (IJMECS) **7**(4), 61–66 (2015). https://doi.org/10.5815/ijmecs.2015.04.07
15. Jain, S., Sharma, V.: Enhanced load balancing approach to optimize the performance of the cloud service using virtual machine migration. Int. J. Eng. Manuf. (IJEM) **7**(1), 41–48 (2017). https://doi.org/10.5815/ijem.2017.01.04
16. Kaur, A., Kaur, B., Singh, D.: Optimization techniques for resource provisioning and load balancing in cloud environment: a review. Int. J. Inf. Eng. Electron. Bus. (IJIEEB) **9**(1), 28–35 (2017). https://doi.org/10.5815/ijieeb.2017.01.04
17. Rolik, O., Telenyk, S., Zharikov, E.: IoT and cloud computing: the architecture of microcloud-based IoT infrastructure management system. In: Emerging Trends and Applications of the Internet of Things, pp. 198–234. IGI Global, Hershey (2017)
18. Rolik, A.: Service level management of corporate IT infrastructure based on the coordinator. Visnyk NTUU "KPI" Inf. Oper. Comput. Sci. **59**, 98–105 (2013)
19. Butler, A., Dawson, P., Palmer, J., Weiss, G.J., Yamada, K.: Magic Quadrant for Integrated Systems (2016)
20. Matthew, M., Eric, S.: Quantifying the Business Value of Nutanix Solutions. IDC, August 2015
21. Million IOPS in 1 VM – World First for HCI with Nutanix. http://longwhiteclouds.com/2017/11/14/1-million-iops-in-1-vm-world-first-for-hci-with-nutanix/. Accessed 13 Dec 2017
22. Nutanix Products Series. http://www.nutanix.com/products/hardware-platforms/. Accessed 13 Dec 2017
23. Hyper-Converged Infrastructure. https://www.vmware.com/products/hyper-converged-infrastructure.html. Accessed 13 Dec 2017
24. Dell EMC VxRail. https://www.dellemc.com/en-ca/converged-infrastructure/vxrail/index.htm#collapse=. Accessed 13 Dec 2017
25. Scale Computing Hardware Platforms. https://www.scalecomputing.com/products/hardware-platforms/. Accessed 13 Dec 2017
26. Mesarovic, M.D., Macko, D., Takahara, Y.: Theory of Hierarchical, Multilevel Systems. Academic Press, New York (1970)

Detection of MAC Spoofing Attacks in IEEE 802.11 Networks Using Signal Strength from Attackers' Devices

R. Banakh[✉], A. Piskozub, and I. Opirskyy

Lviv Polytechnic National University, Lviv, Ukraine
banakh.ri@gmail.com, azpiskozub@gmail.com, iopirsky@gmail.com

Abstract. The main goal of this project is to improve intrusion detection process in IEEE 802.11 based networks in order to provide conditions for further inter-action between attackers and honeypot. In order to gather metadata from clients' devices, part of Wi-Fi Honeypot as a Service model was applied in the experiment, and for the first time ever. MAC addresses of access points and clients' devices, probe requests, beacons and power of signal were used as basic data for further processing. Gathered metadata was used to detect malicious activities against network which is under defense and its clients. Several modifications of MAC spoofing attack were provided by authors in order to find attacks' fingerprints in Wi-Fi ether. Besides base MAC spoofing attack authors suggested a method which allows to identify modification of MAC spoofing where attacker uses power antenna. Also, the new synchronization method for external elements of honeypot was proposed. It is based on centralized random message generation and allows to avoid detection from attackers' side.

Keywords: IEEE 802.11 · Wi-Fi · Intrusion detection · MAC address
Machine learning

1 Introduction

MAC spoofing is not a new attack. Today there already exists method based on packets timestamps counting which can help to detect it in Wi-Fi ether. The main problem is that usually MAC spoofing attack is detected afterwards during digital forensics, when it is actually too late [1].

When there wasn't any declared authentication method in IEEE 802.11, MAC address filtering played a significant role in Wi-Fi security [2]. MAC address filtering is used even today, but using only this method does not meet modern corporate security requirements as it may be easily bypassed using specialized software [3].

Many scientific studies [4, 5] provide description of platform for building a device which can be used in order to intercept service signals in Wi-Fi ether. Using this type of device may help in building external infrastructure described in scientific paper [6].

© Springer International Publishing AG, part of Springer Nature 2019
Z. Hu et al. (Eds.): ICCSEEA 2018, AISC 754, pp. 468–477, 2019.
https://doi.org/10.1007/978-3-319-91008-6_47

Concept of cloud computing infrastructure for Wi-Fi security was also described. Practical realization of this model can help to detect almost all known attacks in Wi-Fi networks including MAC spoofing.

Scientific studies [7] provide suggested approach for realization a reconfigurable antenna which may be used in detection of power antennas not inherent for regular client devices but used by attackers in order to intercept information from wireless networks at large distances.

2 MAC Address in Wi-Fi Specifics

MAC address is integral part of IEEE 802.1x networks. A lot of different mechanisms in IEEE 802.1x use MAC address for different reasons. In this section, we described benefits of MAC address in IEEE 802.11 networks.

2.1 Convenient Access to Wireless Network

By default, each Wi-Fi client saves data about any network it was connected to. It means that client will be reconnected to this network as soon as there is access point (AP) zone-coverage area. This functionality makes users' life easier since they don't need to connect to network each time. But a bunch of metadata might be received and processed by attackers. This type of traffic is available for viewing on data link layer of OSI model.

If client's device is connected to some network it will actively transmit data to AP and receive answers from AP. Of course, in most cases this data is encrypted and encoded but such administrative data as MAC addresses of APs and clients, current APs channels, authentication methods that are set in APs and of course service set identifiers (SSID) are available for viewing if one's wireless card is set to monitoring mode. In case when client is not connected to any AP, data about SSIDs to which client was connected before, might be available as well, as it continuously searches AP to connect to [3].

Another method which provides convenient access to network is called wireless distribution system (WDS). WDS is a technology which helps extend Wi-Fi zone coverage by combining all APs under single Wi-Fi network. This mechanism provides communication between APs without any wired connection.

This communication method unlike other solutions saves clients' MAC addresses on main base station.

2.2 Where Computers Store Data About Wi-Fi Networks

Windows 7 stores data about APs that it was connected to in system register. They are available at HKEY_LOCAL_MACHINE\SOFTWARE\Microsoft\WindowsNT \CurrentVersion\NetworkList\Signatures\Unmanaged. Inside this key might be found sub-keys with such default list of values as:

- DefaultGatewayMac
- Description
- DnsSuffix

- FirstNetwork
- ProfileGuid
- Source

The same information in Debian Linux is stored in directory which is located in /etc/ NetworkManager/system-connections. Info about each connection is stored in separate file. Inside these files different parameters for flexible configuration are available. Similarly, to Linux in MacOS this data is stored in Library/Preferences/SystemConfiguration/com.apple.airport.preferences.plist.

3   MAC Address and Security Issues

MAC addresses provide a lot of benefits, but along with benefits we often get drawbacks. In this section, we described attacks based on MAC address juggling from the attackers' side.

3.1   Metadata for Client-Side and AP-Side Attacks

It is natural for attackers to use administrative data and different additional metadata to perform attacks against clients and whole infrastructure.

In case attacker knows MAC address of client and SSID of AP which client is connected to, s/he may craft AP with similar MAC address and SSID but without any authentication method even if some has been used in legitimate AP. Incensement of signal strength can help attacker lure client away from legitimate AP to its clone. This attack is called Evil Tween. In order to reinforce the client-side attack, attacker may provide DoS attack aimed at legitimate AP.

Another modification of Evil Tween's attack attacker creates clone of AP and redirects all clients' requests to phishing page which asks to enter password in order to confirm client's authenticity. Once client entered password attacker will immediately receive it and service which provides evil tween will be stopped.

If client's device is not connected to any AP it will actively search networks that it was connected before. On data link level this process might be found by attacker. Using data from client's device frames attacker might create fake APs with the same SSIDs.

While client is connected to attacker's AP their traffic might be intercepted and modified. If attacker lures clients out to fake access point then intercepted data will be related just to clients. Otherwise, if user entered password being on phishing web page then attacker will receive access to wireless network and possibly, it may be a key from other segments of whole corporate network.

3.2   Why Security Issues Related to MAC Address Cannot Be Fixed Immediately

Rapid development of any technology often brings security issues that are not obvious on early stages. After technology's core gets a lot of features that are well used by end users fixing of security issues may cause problems with some features workability. Such approach may bring some inconvenience for manufacturers and end users.

IEEE 802.11 based networks allows their users to connect with APs automatically any time once they were authenticated. Let's imagine this functionality will be disabled. First of all, clients will need to reconnect each time. WDS mechanism will be useless as moving from one AP to another user would need to require entering password each time.

4 Clients' Recognition Using MAC Address

There are several solutions of clients positioning identification that are applied by manufacturers in their devices:

- Pattern recognition
- Recognition by access point which is used by client
- Triangulation
- Angulation or positioning with definition of signals' angle

It is not enough to use just one sensor or AP to identify whether client is legitimate or this is attacker who has sniffed MAC addresses and changed actual MAC address of his/her wireless network adapter to legitimate one. In order to find positioning of some device in buildings with single floor, number of such devices should be equal to at least three.

4.1 Initial Synchronization Between Sensors

In order to avoid detection in Wireless Honeypot as a Service (WHaaS) model sensors and honeypot do not communicate between each other via Wi-Fi ether but send sniffed data to processing node which is deployed in cloud. Processing node is virtual machine which makes decisions after receiving data from external elements [6].

Before sensors identify clients' positioning and behavior they should know positioning of neighboring sensors. The main problem is that while sensors are in monitoring mode they cannot see each other as they do not send any signals, but just listen to the

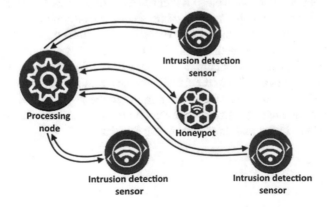

Fig. 1. Data exchange between external elements and processing node in WHaaS model

ether. Figure 1 shows process of data exchange between external elements of WHaaS and WHaaS processing node.

In such a way sensors may send some frame in the ether that may be caught by neighbors. Using model of honeypot for Wi-Fi which is described in [6], it is obvious that after sensors are placed in needed places, they should exchange some administrative data right after user pushes some conditional button. Furthermore, this frame must not be crafted by some template in order to avoid becoming signature for attacker who is aware of honeypots in wireless networks and of some pattern during synchronizing.

Considering the above, authors are suggesting method of synchronization that is based on packet crafting on processing node side. When processing server receives request from user it prepares data which will be used for neighbors' recognition. For example, such data might be represented by beacons that are used by AP to notify clients about its presence or by clients' probes that are used for finding APs that clients were connected to. Next, processing node transmits this data to each external element. Each external element finds its data in received list with instructions and process of neighbors' recognition begins. External elements scan traffic and alternately send their frames to the ether. If element receives data about neighbors in ether it should immediately transmit this data back to processing node. In the end of synchronization process end user can receive map with elements' locations and confirm it if everything is in order [8].

If chosen method of external elements is based on AP beacons transmission then additional disguise should be applied as authenticity of AP might be identified using its MAC address. Manufacturers' MAC address data base might be used to avoid suspicion from the attackers' side. First 24 bits of MAC address may tell attacker about its manufacturer. Default SSID may be used along with MAC.

In order to identify neighbors, it is enough to intercept just one frame, but it might be a marker for attackers. So, another one value that should be generated randomly should be number of frames that will be transmitted to the ether [9].

4.2   Test Intrusion Detection Sensors' Infrastructure Based on WHaaS Model

Let's imagine that object that should be monitored has the form of a perfect square. In order to reach better value of coverage it will be better to place sensors equidistant from each other. For this type of object the following scheme can be applied. Figure 2 [4, 5, 10].

Truth table presented in Table 1 fairly describes the model on Fig. 2.

Intrusion detection infrastructure which is shown in Fig. 2 is a simple case to illustrate its work and shows how it might help in learning process. For example, Fig. 2 in Fig. 3 show signal's strength variation of regular client's device with basic antenna. As we can see, this device has good signal strength in relation to the sensor #7. Also, each sensor from infrastructure can recognize this device. This may indicate that such device is located somewhere in the middle of current infrastructure [11, 12].

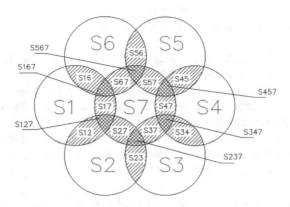

Fig. 2. Example of possible location of intrusion detection sensors' in IEEE 802.11 based network

Table 1. Truth table for scheme depicted in Fig. 2

Sensor's ID	S1	S2	S3	S4	S5	S6	S7
S1		+	-	-	-	+	+
S2	+		+	-	-	-	+
S3	-	+		+	-	-	+
S4	-	-	+		+	-	+
S5	-	-	-	+		+	+
S6	+	-	-	-	+		+
S7	+	+	+	+	+	+	

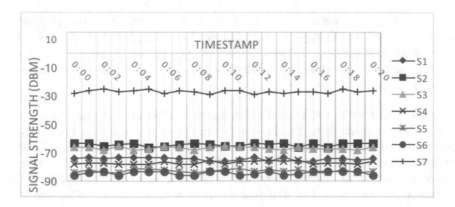

Fig. 3. Signal's strength variation of legitimate Wi-Fi client device presented in time context

5 Metadata Gathering with WHaaS Model

There is one method which can help to identify MAC spoofing of APs in Wi-Fi networks. Such method is based on highly synchronized internal AP clock. For that reason, the access points are constantly exchanging timestamps for synchronization in their beacon frames. In general, attackers attempting an attack called Evil Tween do not try to synchronize the timestamps properly, and you can detect that slip. But detection of such attack becomes impossible in case attacker synchronizes timestamp in their device with AP which is going to be cloned. Also, this method does not work if we need to detect MAC spoofing against clients' devices. But such problem might be partially solved using machine learning.

Each Wi-Fi device has its unique daily road map which depends on its owner. In another case device may statically stays in the same place, for example it might be router. Therefore, different approaches should be applied to clients' devices and to APs.

5.1 Gathering and Analysis of APs' Regular Power of Signal

Quality and stability of signal depends on a large number of external factors. Those include signals from neighbors' APs that work on the same Wi-Fi channel, different electrical devices such as microwave oven, interior walls and furniture. Considering this, power of signal definitely cannot be stable.

In attack called Evil Tween attacker provides AP which is similar to legitimate one. For regular user attacker's AP will not be suspicious. But there is a high probability that attacker's AP works in different place. It means that power of signal will also be different what can be easily detected by intrusion detection sensor.

Position of such type of devices is mostly stable. It means that learning of APs' signal strength is easy to implement as neural networks will not require often retraining.

5.2 Gathering and Analyzing Regular Signal Power of Clients' Devices

Unlike learning APs' behavior, learning of clients' behavior is much more complicated as clients' Wi-Fi devices appear in coverage zone and disappear from it together with their owners. Authors suggest scope of factors that may be used in clients' Wi-Fi devices behavior learning for the purpose of further intrusion detection during which attacker is using MAC spoofing:

Determining of Time When Clients' Devices Appear and Disappear. It is very important to know clients' usual behavior, for example daily time when its devices appear and disappear from coverage zone. For example, let's describe some typical behavior of conditional user: s/he come to office from Monday to Friday at 9AM, at 1PM leaves office and comes back at 2PM, stays there to 6PM and leaves office till next morning. Of course, human cannot come at the same time each day to the same place and follow the same behavior due to various external factors. However, using such information we can easily detect abnormal behavior such as client's appearance on Saturday night in coverage zone which can be considered as possible intrusion.

Detecting of Narrow-Band or Powerful Antenna. The following might indicate possible attack using directional antenna: all or almost all sensors recognize signal from some client's device; the greatest power of signal is detected by sensors that are located on some outer edge of infrastructure; power gradually decreases in the edge opposite direction. Figure 2 in Fig. 4 demonstrate signal's strength variation of potential attacker's device with directional antenna.

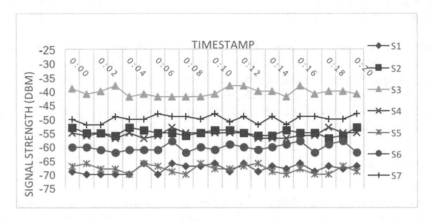

Fig. 4. Signal's strength variation of attacker's Wi-Fi client device (with directional antenna)

The Same MAC Address in Coverage Zone. If another device with the same MAC address appears in coverage zone it might point to the attack. Let's use identification infrastructure from Fig. 2 to provide an example of MAC spoofing detection (Fig. 5).

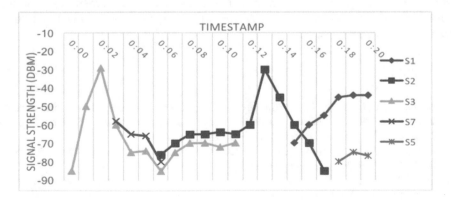

Fig. 5. Detection of MAC spoofing in Wi-Fi ether with intrusion detection infrastructure.

Sensor #3 was the first to identify client as evidenced by a change in signal strength from its device (in Fig. 5 it is time range between 0:00 and 0:03). Subsequently, the motion starts towards the sensor #7 which is adjacent to sensor #3 testifying the natural movement around the room (in Fig. 5 it is time range between 0:03 and 0:06). Then besides the sensors #3 and #7 client's device becomes identified by sensor #2. Client changes its location from sensor #2 and sensor #1 at 0:17, sensors #1 and #5 that are not adjacent detect signals from client with the same MAC address.

Detecting That Client's Device Appeared in Place Where It Definitely Must Not Be. If office is huge then probably there are place where employee must not be. Detection of client's device in not usual place for it (place which is covered by some sensor) can testify either about attacker's presence who is located in convenient place or about violation from the staff side [7].

Detection of Spoofing by Clients' Probe Requests. Each Wi-Fi client device which already was connected to some AP will search it by sending special probe requests. It means that each client's device will have its own list of APs that it should send probe requests. Anyway, even if attacker change MAC address on their device it will send different probe requests or will not send them as in most cases attackers use one time OS to perform an attack.

6 Conclusion

At first time model WHaaS had been deployed for experimental purposes. Model allowed gathering metadata from Wi-Fi ether about APs and clients' devices. In order to find fingerprints of MAC Spoofing both behavior of legitimate users and attackers behavior were simulated. Except of Evil Tween and MAC spoofing attacks WHaaS model allowed to get fingerprint of modified MAC spoofing where conditional attacker used powerful narrow-band antenna.

In future, data about regular signal power from users of IEEE 802.11 based devices will allow to arrange learning process using neural networks. Further usage of deep learning algorithms may help to identify such attacks as MAC spoofing and Evil Tween even at the preparation stage.

References

1. Medjadba, Y., Sahraoui, S.: Intrusion detection system to overcome a novel form of replay attack (data replay) in wireless sensor networks. Int. J. Comput. Netw. Inf. Secur. (IJCNIS) **8**(7), 50–60 (2016). https://doi.org/10.5815/ijcnis.2016.07.07
2. Sobh, T.S.: Wi-Fi networks security and accessing control. Int. J. Comput. Netw. Inf. Secur. (IJCNIS) **5**(7), 9–20 (2013). https://doi.org/10.5815/ijcnis.2013.07.02
3. Gaur, T., Sharma, D.: A secure and efficient client-side encryption scheme in cloud computing. Int. Wirel. Microwave Technol. (IJWMT) **6**(1), 23–33 (2016). https://doi.org/10.5815/ijwmt.2016.01.03
4. Banakh, R., Stefinko, Y.: Single-board workstation as a component of honeypot in Wi-Fi networks. In: Proceedings of the 1st International Conference on Information Security in Modern Society, Lviv, Ukraine, 26 November 2015, pp. 6–7 (2015)
5. Banakh, R., Piskozub, A., Stefinko, Y.: External elements of honeypot for wireless network. In: Modern Proceedings of the XIIIth International Conference TCSET 2016, Lviv-Slavsko, Ukraine, pp. 480–482 (2016)

6. Banakh, R.: Wi-Fi honeypot as a service: conception of business model. In: Proceedings of VI International Conference on "Engineer of XXI Century", Poland, Bielsko-Biała, 02 December 2016, pp. 59–64 (2016)
7. Shah, I.A., Hayat, S., Khan, I., Alam, I., Ullah, S., Afridi, A.: A compact, tri-band and 9-shape reconfigurable antenna for WiFi, WiMAX and WLAN applications. Int. J. Wirel. Microwave Technol. (IJWMT) 6(5), 45–53 (2016). https://doi.org/10.5815/ijwmt.2016.05.05
8. Varadharajan, V., Tupakula, U.: Security as a service model for cloud environment. IEEE Trans. Netw. Serv. Manag. 11(1), 60–75 (2014)
9. Banakh, R., Piskozub, A., Stefinko, Y.: Concept of secured cloud infrastructure using honeypots. Autom. Measur. Control 821, 74–78 (2015)
10. Tao, Z., Nath, B., Lonie, A.: An optimal sensor architecture for Wi-Fi intrusion detection. Int. J. Comput. Sci. Netw. Secur. (IJCSNS) 8(2), 10–19 (2008)
11. Sruthi B, M., Jayanthy, S.: Development of cloud based incubator monitoring system using raspberry Pi. Int. J. Educ. Manag. Eng. (IJEME) 7(5), 35–44 (2017). https://doi.org/10.5815/ijeme.2017.05.04
12. Stefinko, Y., Piskozub, A., Banakh, R.: Concept and model for cloud infrastructure protection by using containers. In: Proceedings of VIII International Conference on Computer Science and Engineering, Ukraine, Lviv, pp. 81–82 (2016)

Steganographic Method of Bitwise Information Hiding in Point-Defined Curves of Vector Images

Oleksiy Kinzeryavyy[1]([✉]), Iryna Kinzeriava[2], Alexander Olenyuk[3], and Krzysztof Sulkowsky[4]

[1] National Aviation University, Kyiv, Ukraine
oleksiykinzeryavyy@gmail.com
[2] Podilsky Special Education and Rehabilitation Socioeconomic College,
Kamyanets-Podilsky, Ukraine
i.kinzeryava@gmail.com
[3] State Agrarian and Engineering University in Podilya,
Kamyanets-Podilsky, Ukraine
unicorn.ua@gmail.com
[4] Państwowa Wyższa Szkoła Zawodowa, Nowym Sącz, Poland

Abstract. In this paper the authors propose steganographic method of bitwise information hiding, which allows embedding information in vector images by splitting point-defined curves into segments. Due to the invariance property of the curves of this type (B-splines, NURB curves, Bezier curves, Hermite curves), the proposed method will provide resistance to active attacks based on affine transformations. On the basis of the proposed method and the properties of the Bezier curves, StegoBIT algorithm was realized. This algorithm allows to embed information in the Bezier curves of the third degree and provides resistance to active attacks based on affine transformations. An experimental study of the stability of proposed algorithm for affine transformations was carried out. 30 arbitrary SVG images were selected for the experiment. Their structural configuration contained parameters for constructing Bezier curves. The information of different sizes was hidden in the curves, by way of its gradual division into visually identical sets of segments. The affine transformations such as transfer, rotation, almost rotation, biasing for the abscissa and ordinate axis and proportional and disproportional scaling was gradually carried out with the obtained steganocontainer. The obtained results of the experiment demonstrate the effectiveness and stability of the proposed StegoBIT algorithm to various transformations that are based on affine transformations.

Keywords: Steganography · Bitwise method of data hiding
StegoBIT algorithm · Vector images · Affine transformation

© Springer International Publishing AG, part of Springer Nature 2019
Z. Hu et al. (Eds.): ICCSEEA 2018, AISC 754, pp. 478–486, 2019.
https://doi.org/10.1007/978-3-319-91008-6_48

1 Introduction

The development of the global Internet network and its dissemination among the planet's population contributes to increasing of the information amount that is being transmitted, processed, stored and destroyed. Using the capabilities and resources of the Internet it can be organized a backup channel, for example, with diplomatic institutions located in the territory of foreign countries. The secrecy of the transmission of information by such a channel will be provided by steganographic protection means. The main difference of steganography from other methods of protecting information is the opportunity to conceal the existence of a secret message in another, not attracting attention to an object – a container that is then openly transported to the addressee [1, 8].

Taking into account the fact that with the images which are sending over the Internet, active filters can be used. They imperceptibly modify the images and thus destroy the hidden message. Such attacks should include different kind of transformations based on affine transformations. Resistance to affine transformations can be ensured by the using of vector images as the containers, which, by their properties and principles of construction, allow to build images of sufficiently high quality.

However, active filters that can visually imperceptibly modify the image and permanently destroy the hidden message can be applied to the image-containers during their transmission. Such attacks include overlays of different transformations, among which the most common are affine transformations [2, 3, 5, 10].

Studies have shown steganographic methods that use raster and fractal images as containers do not provide resistance to affine transformations [2, 6]. While methods of hiding in vector images, thanks to the properties of vector graphics, can successfully resist them. In this regard, the actual task is the development of new steganographic methods of data hiding in vector images, which are resistant to affine transformations and the changed nature of which will not lead to noticeable deviations of objects in the container.

The purpose of the work is to increase the stability of the steganographic information protection against active attacks based on affine transformations.

2 Mechanism of Information Hiding in Vector Images

The main element in vector graphics is a line that has some properties: shape, color, thickness, etc. Each line is shown by using an analytical formula with a certain number of parameters that are necessary for its representation. A large class of curves, which differ in varying degrees of smoothness, can be constructed in a set of points. Such curves are called point-defined. They include broken line and various spline curves (Bezier curves, B-splines, Hermite interpolation curves, and others). Some series of curves of this class possesses the property of affine-invariance, with the help of which resistance to attacks, based on affine transformation, can be provided.

The curves are defined by using a set of points (vertices) $\mathbf{P} = \{P_i\}$, $i = \overline{0, n}$, n is the degree of the curve, $n + 1$ is the number of points that are sequentially connected to form a broken line, which is called supporting, and the vertices of which are the supporting ones. The following curves are set according to the following general formula [9]:

$$R(t) = \sum_{i=0}^{n} f_i(t) P_i$$

where $f_i(t)$ are some functional coefficients, t is parameter for constructing a curve, $t \in [a, b]$.

Construction of any point-defined curve is carried out using a parameter t, $t \in [a, b]$, each value of which strictly determines the position of points on the curve [4, 7, 11]. If you fix a certain step Δt of the parameter changing t, you can make a transition between its values $t_{i+1} = t_i + \Delta t$ within a given interval $[a, b]$. Using the value t_i you can hide the information in the curve by splitting them into a set of segments at certain points, which will then replace the original curves in the vector image.

3   The Method of Bitwise Information Hiding in Point-Defined Curves of Vector Images

Steganographic method of bitwise information hiding in vector images, based on the above-described idea, was developed. Regarding to this, hiding of information occurs as follows:

Step 1.   The secret message $\mathbf{a} = \{a_i\}$, $a_i \in \{0, 1\}$, $i = \overline{1, h}$, a_i is the secret message bit, h is the number of message bits $\mathbf{a}$, is divided into parts $\mathbf{a} = \{a_1^1, \ldots, a_{V_3}^1, a_1^2, \ldots, a_{V_3}^2, \ldots, a_1^j, \ldots, a_{V_3}^j\}$, $\mathbf{a}^j = \{a_1^j, \ldots, a_{V_3}^j\}$, $a_i^j \in \{0, 1\}$, $i = \overline{1, V_3}$, $j = \overline{1, m}$, $m = h/V_3$ where m is the number of curves of the degree V_1 with the minimum allowable distance between the reference points V_2 from the set of vector images in which parts of the message are hidden.

Step 2.   For each sequence $\mathbf{a}^j$ the steganokey Δt^j, $j = \overline{1, m}$, of the step change of the parameter t is defined, where each one $\Delta t^j < 1/V_3$.

Step 3.   Hiding of each sequence a_i^j, $i = \overline{1, V_3}$, in the curve D_j, $j = \overline{1, m}$ is performed by splitting it into a sequence of segments $D_j^* = D_{V_1}^0 \cup D_{V_1}^1 \cup \ldots \cup D_{V_1}^w$ where w is the index of the sequence of the curve segments D_j^*, $w \in N$ (before the hiding $w = 0$ and $D_{V_1}^0 = D_j$). The partition is performed by a given parameter t_i ($t_i = t_{i-1} + \Delta t^j$ where t_0 is arbitrary initial value $0 \le t_0 < 1 - V_3 \cdot \Delta t^j$):

3.1. By a bit hiding $a_i^j = 0$ ($a_i^j = 1$ depend on the choice of the bit value at which the division of curves occurs) at a given point of the partition t_i the curve does not divide, but the transition to the next bit a_{i+1}^j occurs.

3.2. By a bit hiding $a_i^j = 1$ ($a_i^j = 0$ depend on the choice of the bit value by which the division of curves occurs) at a given breakpoint t_i the curve $D_{V_1}^w$ divide into two segments $D_{V_1}^w$ and $D_{V_1}^{w+1}$. The coordinates of the reference points of the resulting segments are calculated with the chosen accuracy V_4. Subsequent insertion of the next bit a_{i+1}^j occurs with the next value t_{i+1} in the resulting segment $D_{V_1}^{w+1}$. Each division of a curve leads to an increasing of the number of segments by one ($w = w + 1$).

3.3. After hiding of the sequence a^j in the curve D_j the resulting sequence of segments D_j^* is written to the steganocontainer instead of the curve D_j.

Parameters V_1, V_2, V_3, V_4 are ancillary parameters that influence on the choice of container and the process of embedding information in selected curves of the image. Given parameters are presented in the work [8] in more detailed.

4 Algorithm of Bitwise Information Hiding in Bezier Curves of Vector Images

StegoBIT algorithms, based on the method that is described above and the properties of Bezier curves, are implemented. This algorithm allows you to embed information in the Bezier curves of the third degree. Pseudocodes of the procedure for embedding information according to these algorithms are presented in Fig. 1.

The operation *SplitMessage*(x, y) involves the bitwise splitting of the incoming message x into parts by the dimension y of the bits. The operation *SelectImage*(x, y, z) performs the selection of permissible containers from the set x that is necessary for embedding parts y of the hidden message, in the structure of which there are Bezier curves of the z degree. The operation *SelectCurves*(x, y, z, h) determines from the plurality of images x the number of Bezier curves of the z degree, required to integrate y parts of the hidden message that correspond to the minimum permissible distance h between the reference points.

Input: Secret message $a = \{a_i\}$, $a_i \in \{0,1\}$, $i = \overline{1,h}$, $h \in N$;

set of vector images (containers) $I = \{I_b\}$, $b \in N$;

set of steganokeys $T = \{\Delta t^u\}$, $\Delta t^u \in (0,1)$, $u \in N$;

set of parameters $V = \{V_i\}$, $i = \overline{1,4}$, $V_i \in N$, $V_1 = 3$;

the value of the bit in which the division of curves of Bezier will occur Q, $Q \in \{0,1\}$.

Output: A set of steganoconteiners $S = \{S_x\}$, $x \in N$.

1. $\{a^j\} = SplitMessage(a, V_3)$, $a^j = \{a_1^j, ..., a_{V_3}^j\}$, $a_i^j \in \{0,1\}$, $i = \overline{1, V_3}$, $j = \overline{1, m}$, $m = \lceil h/V_3 \rceil$;

2. $S = SelectImage(I, m, V_1)$, $S = \{S_x\}$, $x \in N$, $x \le m$, $m = \lceil h/V_3 \rceil$;

3. $D = SelectCurves(S, m, V_1, V_2)$, $D = \{D_j\}$, $j = \overline{1, m}$, $m = \lceil h/V_3 \rceil$;

2. $For\ (j = 1;\ j \le m;\ j++)$

2.1. $\Delta t^j = SelectKey(T, V_3)$, $\Delta t^j < 1/V_3$;

2.2. $w = 0$; $t = 0$; $D_{V_1}^w = D_j$;

2.3. $For\ (i = 1;\ i \le V_3;\ i++)$

2.3.1. $t = t + \Delta t^j$;

2.3.2. $if\ (a_i^j == Q)$

2.3.2.1. $\{P_e\} = GetPoints(D_{V_1}^w)$, $e = \overline{0,3}$;

2.3.2.2. $P_0^1 = (1-t) \cdot P_0 + t \cdot P_1$;

2.3.2.3. $P_1^1 = (1-t) \cdot P_1 + t \cdot P_2$;

2.3.2.4. $P_2^1 = (1-t) \cdot P_2 + t \cdot P_3$;

2.3.2.5. $P_0^2 = (1-t) \cdot P_0^1 + t \cdot P_1^1$;

2.3.2.6. $P_1^2 = (1-t) \cdot P_1^1 + t \cdot P_2^1$;

2.3.2.7. $P_0^3 = (1-t) \cdot P_0^2 + t \cdot P_1^2$;

2.3.2.8. $D_{V_1}^w = CreateCurve(P_0, P_0^1, P_0^2, P_0^3)$;

2.3.2.9. $D_{V_1}^{w+1} = CreateCurve(P_0^3, P_1^2, P_2^1, P_3)$;

2.3.2.10. $w = w + 1$;

2.4. $D_j = ReplaceCurve(S, D_j, \{D_{V_1}^i\})$, $i = \overline{0, w}$;

3. $Return\ S$;

Fig. 1. The pseudo-code of the StegoBIT algorithm.

5   Experimental Study of the Stability of the Stegobit Algorithm to Attacks that Are Based on Affine Transformations

Stability verifications of the proposed algorithm were carried out by embedding information of different sizes in 30 vector images of SVG format by using the following parameters [8]: the parameter $V_1 = 3$ that defines the degree of Bezier curves; the parameter $V_2 \geq 1$ that determines the possible distance between the reference points of the curve; the parameter V_3 that determines the amount of embedded information in one curve which will vary from 40 to 80 bytes (with a change of 20 bytes); the parameter V_4 that define the accuracy of the coordinates of the reference points which will vary from 10^{-5} to 10^{-6} bytes (with the change step 10^{-m}, $m \in \{5,6\}$).

We describe the imposition of affine transformations with vectorial objects of the steganocontainer using the following formula:

$$S_k = F\left(S_{k-1}, \alpha_{1,1}^k, \alpha_{1,2}^k, \alpha_{2,1}^k, \alpha_{2,2}^k, \beta_1^k, \beta_2^k, \varepsilon_1^k, \varepsilon_2^k, \varepsilon_3^k, \varepsilon_4^k, \varepsilon_5^k, \varepsilon_6^k\right), \ k = k+1 \qquad (1)$$

where F is the affine transformation, S_0 is the initial steganocontainer, $\alpha_{1,1}^k$, $\alpha_{1,2}^k$, $\alpha_{2,1}^k$, $\alpha_{2,2}^k$, β_1^k, β_2^k, ε_1^k, ε_2^k , ε_3^k, ε_4^k, ε_5^k, $\varepsilon_6^k \in R$ are the coefficients of affine transformations, $k \in N$ is the number of repetitive iterations imposed by the affine transformation to the steganocontainer. Depending on the selected coefficients $\alpha_{1,1}^k$, $\alpha_{1,2}^k$, $\alpha_{2,1}^k$, $\alpha_{2,2}^k$, β_1^k, β_2^k, ε_1^k, ε_2^k, ε_3^k, ε_4^k, ε_5^k, ε_6^k the transformation of the transfer, rotation, almost rotation, biasing and scaling was performed.

The implementation of the transfer transformation according to (1) was carried out to each steganocontainer by the following coefficients: $\alpha_{1,1}^k = \alpha_{1,2}^k = \alpha_{2,1}^k = \alpha_{2,2}^k = 0$, $\beta_1^k \in [-500, 500]$, $\beta_2^k \in [-500, 500]$, $\varepsilon_1^k = \varepsilon_2^k = \varepsilon_3^k = \varepsilon_4^k = \varepsilon_5^k = \varepsilon_6^k = 0$, $k = \overline{1, 100}$. The average results from the amount of lost information by applying this conversion are shown in Fig. 2.

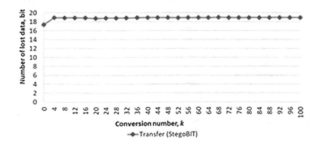

Fig. 2. The results of extracting information from steganocontainer after performing transfer conversions.

Realization of the transfer, rotation (a) and almost rotation (b) around the point $(0, 0, 1)$ according to (1), was carried out to each steganocontainer with the following coefficients: (a) $\alpha_{1,1}^k = \cos\theta$, $\alpha_{1,2}^k = -\sin\theta$, $\alpha_{2,1}^k = \sin\theta$, $\alpha_{2,2}^k = -\cos\theta$, $\theta = 1$, $\beta_1^k = \beta_2^k = 0$, $\varepsilon_1^k = \varepsilon_2^k = \varepsilon_3^k = \varepsilon_4^k = \varepsilon_5^k = \varepsilon_6^k = 0$, $k = \overline{1, 360}$; (b) $\alpha_{1,1}^k = \cos\theta$, $\alpha_{1,2}^k = -\sin\theta$, $\alpha_{2,1}^k = \sin\theta$, $\alpha_{2,2}^k = -\cos\theta$, $\theta = 1$, $\beta_1^k = \beta_2^k = 0$, $\varepsilon_1^k = \varepsilon_4^k = \varepsilon_6^k = -0,0001$, $\varepsilon_2^k = \varepsilon_3^k = \varepsilon_5^k = 0,0001$, $k = \overline{1, 360}$. The average results from the amount of lost information with applying transformation data are shown in Fig. 3.

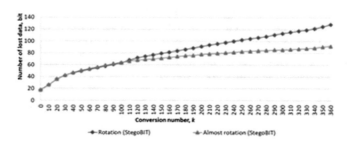

Fig. 3. The results of extracting information from steganocontainers after rotation and almost rotation.

The realization of the offset biasing for the abscissa (a) and ordinate (b) axis in accordance with (1) was performed for each steganocontainer with the following coefficients: (a) $\alpha_{1,1}^k = \alpha_{2,2}^k = 1$, $\alpha_{1,2}^k = 0,01$, $\alpha_{2,1}^k = \beta_1^k = \beta_2^k = 0$, $\varepsilon_1^k = \varepsilon_2^k = \varepsilon_3^k = \varepsilon_4^k = \varepsilon_5^k = \varepsilon_6^k = 0$, $k = \overline{1, 100}$; (b) $\alpha_{1,1}^k = \alpha_{2,2}^k = 1$, $\alpha_{1,2}^k = \beta_1^k = \beta_2^k = 0$, $\alpha_{2,1}^k = 0,01$, $\varepsilon_1^k = \varepsilon_2^k = \varepsilon_3^k = 0$, $\varepsilon_4^k = \varepsilon_5^k = \varepsilon_6^k = 0$, $k = \overline{1, 100}$. The average results from the amount of lost information by applying transformation data are shown in Fig. 4.

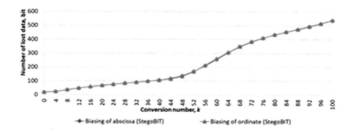

Fig. 4. The results of extracting information from steganocontainers after performing biasing offsets.

The imposition of the proportional (a) and disproportionate scaling over the abscissa (b) and ordinate (c) axis according to (1) was performed by compressing/stretching each steganocontainer according to the following coefficients: (a) compression –

$\alpha^k_{1,1} = \alpha^k_{2,2} = 0,99, \quad \alpha^k_{1,2} = \alpha^k_{2,1} = \beta^k_1 = \beta^k_2 = 0, \quad \varepsilon^k_1 = \varepsilon^k_2 = \varepsilon^k_3 = \varepsilon^k_4 = \varepsilon^k_5 = \varepsilon^k_6 = 0,$
$k = \overline{1,99};$ expansion $- \alpha^k_{1,1} = \alpha^k_{2,2} = 1,01, \alpha^k_{1,2} = \alpha^k_{2,1} = \beta^k_1 = \beta^k_2 = 0, \varepsilon^k_1 = \varepsilon^k_2 = \varepsilon^k_3 =$
$\varepsilon^k_4 = \varepsilon^k_5 = \varepsilon^k_6 = 0, \quad k = \overline{1,100};$ (b) compression $- \alpha^k_{1,1} = 0,99, \quad \alpha^k_{1,2} = \alpha^k_{2,1} = \beta^k_1 =$
$\beta^k_2 = 0, \alpha^k_{2,2} = 1, \varepsilon^k_1 = \varepsilon^k_2 = \varepsilon^k_3 = \varepsilon^k_4 = \varepsilon^k_5 = \varepsilon^k_6 = 0, k = \overline{1,99};$ expansion $- \alpha^k_{1,1} =$
$1,01, \quad \alpha^k_{1,2} = \alpha^k_{2,1} = \beta^k_1 = \beta^k_2 = 0, \quad \alpha^k_{2,2} = 1, \quad \varepsilon^k_1 = \varepsilon^k_2 = \varepsilon^k_3 = \varepsilon^k_4 = \varepsilon^k_5 = \varepsilon^k_6 = 0,$
$k = \overline{1,100};$ (c) compression $- \alpha^k_{1,1} = 1, \alpha^k_{1,2} = \alpha^k_{2,1} = \beta^k_1 = \beta^k_2 = 0, \alpha^k_{2,2} = 0,99,$
$\varepsilon^k_1 = \varepsilon^k_2 = \varepsilon^k_3 = \varepsilon^k_4 = \varepsilon^k_5 = \varepsilon^k_6 = 0, \quad k = \overline{1,99};$ expansion $- \alpha^k_{1,1} = 1, \alpha^k_{1,2} = \alpha^k_{2,1} =$
$\beta^k_1 = \beta^k_2 = 0, \alpha^k_{2,2} = 1,01, \varepsilon^k_1 = \varepsilon^k_2 = \varepsilon^k_3 = \varepsilon^k_4 = \varepsilon^k_5 = \varepsilon^k_6 = 0, k = \overline{1,100}.$ The average
results from the amount of lost information with compressing and extending vector
images are shown in Figs. 5 and 6 respectively.

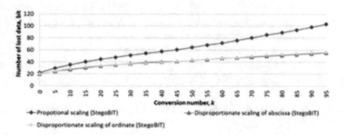

Fig. 5. The results of extracting information from compressed steganocontainers after
performing scaling conversions.

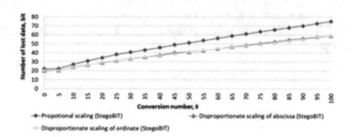

Fig. 6. The results of extracting information from extended steganocontainers after performing
scaling conversions.

6   Conclusions

According to the results of the study, it can be concluded that the developed StegoBIT
algorithm provides reliable resistance to attacks based on affine transformations. But,
with applying rotation, almost rotation, biasing, and scaling, the error of rounding off
the new coordinates of the reference points of vector objects is formed, which affects
the correctness of the information extraction. By reducing the amount of hidden

information in one Bezier curve with the optimal choice of hiding parameters it is possible to provide greater stability to affine transformations of the proposed algorithm.

References

1. Agranovsky, A.V., Derevyanin, P.N., Hady, R.A.: Basics of computer steganography, Radio and connection, Moscow, Russia (2003)
2. Azhbaev, T.G., Azhmukhamedov, I.M.: The analysis of resistance of modern steganos algorithms, pp. 56–61 (2008)
3. Bakhrushina, G.I.: Modeling geometric attacks based on affine transformations. Sci. Notes PNU **4**(4), 1291–1297 (2013)
4. Danik, Yu., Hryschuk, R., Gnatyuk, S.: Synergistic effects of information and cybernetic interaction in civil aviation. Aviation **20**(3), 137–144 (2016)
5. Hu, Z., Gnatyuk, S., Koval, O., Gnatyuk, V., Bondarovets, S.: Anomaly detection system in secure cloud computing environment. Int. J. Comput. Netw. Inf. Secur. (IJCNIS) **9**(4), 10–21 (2017). https://doi.org/10.5815/ijcnis.2017.04.02
6. Hu, Z., Gnatyuk, V., Sydorenko, V., Odarchenko, R., Gnatyuk, S.: Method for cyberincidents network-centric monitoring in critical information infrastructure. Int. J. Comput. Netw. Inf. Secur. (IJCNIS) **9**(6), 30–43 (2017). https://doi.org/10.5815/ijcnis. 2017.06.04
7. Golovanov, N.N.: Geometric Modeling. Publishing House of Physicomathematical Literature, Moscow (2002)
8. Kinzeryavyy, O.M.: Steganographic methods of hiding data in vector images that are resisted to active attacks based on affine transformations, National Aviation University, Kyiv (2015)
9. Matsenko, V.G.: Computer Graphics, Ruta, Chernovtsy, Ukraine (2002)
10. Niu, X., Shao, C., Wang, X.: A survey of digital vector map watermarking. Int. J. Innov. Comput. Inf. Control ICIC Int. **2**(6), 1301–1316 (2006)
11. Rodjers, D., Adams, J.: Mathematical Foundations of Computer Graphics, Mir, Moscow, Russia (2001)

Self-learning Procedures for a Kernel Fuzzy Clustering System

Zhengbing Hu[1] (ORCID), Yevgeniy Bodyanskiy[2] (ORCID),
and Oleksii K. Tyshchenko[2,3](✉) (ORCID)

[1] Central China Normal University, 152 Louyu Road, Wuhan 430079, China
hzb@mail.ccnu.edu.cn
[2] Kharkiv National University of Radio Electronics,
14 Nauky Ave, Kharkiv 61166, Ukraine
yevgeniy.bodyanskiy@nure.ua, lehatish@gmail.com
[3] Institute for Research and Applications of Fuzzy Modeling, CE
IT4Innovations, University of Ostrava, 30. dubna 22,
701 03 Ostrava, Czech Republic

Abstract. The paper exemplifies several self-learning methods through the prism of diverse objective functions used for training a kernel fuzzy clustering system. A self-learning process for synaptic weights is implemented in terms of the competitive learning concept and the probabilistic fuzzy clustering approach. The main feature of the introduced fuzzy clustering system is its capability to cluster data in an online way under conditions when clusters are rather likely to be of an arbitrary shape (which cannot usually be separated in a linear manner) and to be mutually intersecting. Generally speaking, the offered system's topology is mainly based on both the fuzzy clustering neural network by Kohonen and the general regression neural network. When it comes to training this hybrid system, it is grounded on both the lazy and optimization-based learning concepts.

Keywords: Self-learning procedure · Fuzzy clustering
Computational intelligence · Adaptive neuro-fuzzy system · Objective function

1 Introduction

Powerful tools of Computational Intelligence (such as neuro-fuzzy systems and artificial neural networks [1–4]) are widely practiced these days for the clustering scope which is one of the most challenging objects in Data Mining [5–8]. An established research objective for the clustering problem [9–18] implies that formed clusters are linearly separable and are exemplified by a convex shape. These requisitions are rarely carried out in most of real-world goals, and clusters tend to overlap. In this case, the fuzzy clustering technique [19–23] should be applied. To get over the entanglements coming from linearly inseparable clusters of an arbitrary shape, the kernel fuzzy clustering tools have been developed [24–38]. Basically, kernel fuzzy clustering procedures are associated with applying the burdens of radial-basis function neural networks and support

© Springer International Publishing AG, part of Springer Nature 2019
Z. Hu et al. (Eds.): ICCSEEA 2018, AISC 754, pp. 487–497, 2019.
https://doi.org/10.1007/978-3-319-91008-6_49

vector machines [4] as well as some conventional fuzzy clustering procedures like the fuzzy C-means method [19] and the possibilistic C-means method [39–44].

The mentioned concepts of data clustering actually reckon that the whole arrayed data $X = \{x(1), x(2), \ldots, x(k), \ldots, x(N)\} \subset R^n$, $x(k) = (x_1(k), \ldots, x_i(k), \ldots, x_n(k))^T$ is given beforehand, and it is not going to be changed while analyzing the data. But what happens if the data comes in a stream [45–48] (sequentially in an online manner)? In this way, to be sure, the specified methods are barely effective and even useless sometimes. Alternatively, if an initial data array is set preliminarily but its size N and its dimensionality n are high enough, one may stumble over some essential computational problems while bringing into action the typical procedures for kernel fuzzy clustering.

In this context, the paper's novelty underlies in developing an online kernel fuzzy clustering system as well as adaptive algorithms of its self-learning. The introduced learning methods give an option of processing data streams in a sequential way and detecting overlapping clusters of a derived form.

2   A Structure of the Proposed Fuzzy Clustering System

A structure of the system contains generally several layers of data processing.

Initial information enters the first hidden layer as a sequence of n−dimensional vectors $x(k)$. The initial data is centered with respect to a current mean value and normalized to the hypercube via elementary relations either for an online case

$$\bar{x}(k) = \bar{x}(k - 1) + \frac{1}{k}(x(k) - \bar{x}(k - 1)) \tag{1}$$

or for a batch mode

$$\bar{x}(k) = \frac{\sum_{p=1}^{k} x(p)}{k}, \bar{\bar{x}}(k) = x(k) - \bar{x}(k), t_i(k) = 2\frac{\bar{\bar{x}}_i(k) - \bar{\bar{x}}_{i\,\min}}{\bar{\bar{x}}_{i\,\max} - \bar{\bar{x}}_{i\,\min}} - 1, \tag{2}$$
$$\forall i = 1, 2, \ldots, n.$$

Consequently, the data arriving to the second hidden layer meets the condition

$$\left(\sum_{k=1}^{N} t(k)\right) N^{-1} = 0, -1 \leq t_i(k) \leq 1.$$

The preprocessed vectors $t(k)$ are driven up to the layer of radial basis functions. Dimensionality of the input data space R^n is raised here to a feature space [31] of some higher dimensionality R^h, $h > n$.

In a general way, one may exploit typical Gaussian functions

$$G(t(k), t(p)) = \exp\left(-\frac{\|t(k) - t(p)\|^2}{2\sigma^2}\right)$$

to enhance dimensionality of the space. These functions make it possible to guarantee linear discriminability [4] for the clusters being formed.

To appraise a distance between two vectors $t(k)$ and $t(p)$, one may employ a dissimilarity measure [37] which is based on the distance kernel trick

$$d_{DM}^2(t(k),\, t(p)) = G(t(k), t(k)) + G(t(p), t(p)) - 2G(t(k), t(p))$$
$$= 2(1 - G(t(k), t(p))) \tag{3}$$

instead of the conventional Euclidean metric

$$d_{EM}^2(t(k),\, t(p)) = \|t(k) - t(p)\|^2 \tag{4}$$

which is frequently used in data clustering.

It's also possible to apply a similarity measure

$$d_{SM}^2(t(k),\, t(p)) = G(t(k),\, t(p))$$

in the space of the enhanced dimensionality R^h.

The most complex factor while developing the second layer is an option of an amount and centers' coordinates of the radial basis functions. It's always easier to use a concept «neurons in data points» [49] to set a number of the radial basis functions equal to an amount of N observations with centers that fall in with the observations' coordinates $t(k) \; \forall \, k = 1, 2, \ldots, N$ just similarly to how it happens in support vector machines [50, 51] or general regression neural networks [52–54]. But when it comes to processing data streams (a quantity of observations is constantly growing), this approach doesn't give any result.

A more beneficial idea is to use the «sliding window» which means that only a part of the whole sample is used that contains h recent observations $t(k)$, $t(k - 1)$,…, $t(k - h + 1)$. Besides that, a value of h can be changing within data processing.

As a result of increasing the dimensionality of the incoming space, a feature $(h \times 1)-$ vector based on the radial basis functions is generated

$$\phi(k, h) = (\phi(t(k)),\; \phi(t(k-1)), \ldots, \; \phi(t(k-q)), \ldots, \; \phi(t(k-h+1)))^T$$

where

$$\phi(t(k-q)) = \exp\left(-\frac{\|t(k) - t(k-q)\|^2}{2\sigma^2}\right).$$

Apart from the radial basis functions, this layer also consists of $(h - 1)n$ time delay elements z^{-1} externalizing $t(k - 1)$, $t(k - 2)$,…, $t(k - h + 1)$, and a block for N normalization that reshapes the feature vector in a such a way that

$$\|\tilde{\phi}(k, h)\|_2 = 1, \tilde{\phi}(k, h) = \phi(k, h)\|\phi(k, h)\|^{-1}.$$

The feature vector $\tilde{\phi}(k,h)$ enters later the third hidden layer which is technically the self-organizing map by Kohonen [3]. A tuning procedure for prototypes $v_j^k(k)$, $j = 1, 2, \ldots, m$ of configured clusters is performed in an online competitive manner.

Finally, the output layer performs an estimation of membership levels for every feature vector $\tilde{\phi}(k,h)$ to each cluster to have been formed.

3　Self-learning Procedures for the Clustering System

A self-learning process for synaptic weights is implemented in the third hidden layer on the ground of the competitive learning paradigm [3] and fuzzy clustering in terms of the probabilistic approach by Bezdek [19].

According to this method, the clustering process is carried out by minimizing the objective function

$$J\left(u_j(k),\ v_j^k\right) = \sum_{k=1}^{N} \sum_{j=1}^{m} u_j^\omega(k) \left\| \tilde{\phi}(k,h) - v_j^k \right\|^2 \tag{5}$$

under the constraints

$$\sum_{j=1}^{m} u_j(k) = 1, 0 \le \sum_{k=1}^{N} u_j(k) \le N \tag{6}$$

where $u_j(k) \in [0,\ 1]$, ω stands for a non-negative fuzzifier that specifies how blur borders between clusters are.

The conventional non-linear programming technique is utilized in terms of undetermined multipliers by Lagrange. It is trivial enough to arrive at some well-known result within the scope of solving a system of equations by Karush-Kuhn-Tucker:

$$\begin{cases} v_j^k = \dfrac{\sum\limits_{k=1}^{N} u_j^\omega(k)\tilde{\phi}(k,h)}{\sum\limits_{k=1}^{N} u_j^\omega(k)}, \\[4mm] u_j(k) = \dfrac{\left(\|\tilde{\phi}(k,h)-v_j^k\|^2\right)^{\frac{1}{1-\omega}}}{\sum\limits_{l=1}^{m}\left(\|\tilde{\phi}(k,h)-v_l^k\|^2\right)^{\frac{1}{1-\omega}}}. \end{cases} \tag{7}$$

That having been said, one may use either the traditional metric (3) or the dissimilarity measure (4) for the second ratio.

By putting the fuzzifier's value $\omega = 2$ in the expression (7), the frontmost FCM algorithm is gained

$$
\begin{cases}
v_j^k = \dfrac{\displaystyle\sum_{k=1}^{N} u_j^2(k)\tilde{\phi}(k,h)}{\displaystyle\sum_{k=1}^{N} u_j^2(k)}, \\[4mm]
u_j(k) = \dfrac{\left\|\tilde{\phi}(k,h)-v_j^k\right\|^{-2}}{\displaystyle\sum_{l=1}^{m} \left\|\tilde{\phi}(k,h)-v_l^k\right\|^{-2}}.
\end{cases}
\tag{8}
$$

As can be seen, the procedures (7) and (8) embody the batch way of data processing. Multiple recurrent techniques of fuzzy clustering [55–57] could be applied to online data processing. This sort of tools underlies the non-linear programming techniques by Arrow-Hurwicz-Uzawa:

$$
\begin{cases}
u_j(k+1) = \dfrac{\left(\left\|\tilde{\phi}(k+1,h)-v_j^k(k)\right\|^2\right)^{\frac{2}{1-\omega}}}{\displaystyle\sum_{l=1}^{m}\left(\left\|\tilde{\phi}(k+1,h)-v_l^k(k)\right\|^2\right)^{\frac{2}{1-\omega}}}, \\[4mm]
v_j^k(k+1) = v_j^k(k) + \eta(k)u_j^{\omega}(k+1)\left(\tilde{\phi}(k+1,h) - v_j^k(k)\right)
\end{cases}
\tag{9}
$$

where a parameter $\eta(k) > 0$ defines a convergence speed of the expedient (9).

A quick look at the second ratio in the approach (9) shows that a multiplier $u_j^{\omega}(k+1)$ suits a membership function in the WTM self-learning rule [3]. The operation (9) matches a recurrent view of the K-means crisp clustering when $\omega = 1$; for the case $\omega = 0$, the standard WTA self-learning rule comes out

$$
v_j^k(k+1) = v_j^k(k) + \eta(k)\left(\tilde{\phi}(k+1,h) - v_j^k(k)\right),
\tag{10}
$$

which ultimately checks with the recurrent scheme for estimation of the mean value (1). The Eq. (10) minimizes the objective function

$$
J\left(v_j^k\right) = \sum_{k}\left\|\tilde{\phi}(k,h) - v_j^k\right\|^2
$$

with a rate parameter $\eta(k) = \frac{1}{k+1}$, which means that the formula (10) acquires the form of

$$
v_j^k(k+1) = v_j^k(k) + \frac{1}{k+1}\left(\tilde{\phi}(k+1,h) - v_j^k(k)\right).
\tag{11}
$$

By considering the statement (11), the clustering procedure (9) conclusively takes on the form of

$$
\begin{cases}
u_j(k+1) = \dfrac{\left\|\tilde{\phi}(k+1,h)-v_j^k(k)\right\|^{\frac{2}{1-\omega}}}{\sum\limits_{l=1}^{m}\left\|\tilde{\phi}(k+1,h)-v_l^k(k)\right\|^{\frac{2}{1-\omega}}}, \\[4mm]
v_j^k(k+1) = v_j^k(k) + \dfrac{u_j^\omega(k+1)}{k+1}\left(\tilde{\phi}(k+1,h)-v_j^k(k)\right)
\end{cases}
\tag{12}
$$

and it is structurally close to fuzzy competitive self-learning algorithms [58, 59].

Except that, although the probabilistic fuzzy clustering techniques (7), (8), (9), (12) are quite effective in the solution of multiple real-world challenges, they bear some intrinsic weaknesses. To begin with, they are affected by the Concentration of Norms effect if the dimensionality h for processed signals is high enough, and, in the second place, these techniques are not completely fuzzy all in all, since their fuzzifier's value $1 < \omega < \infty$ is crisp in actual practice. This value is chosen empirically in most cases and may manipulate the final results in a meaningful way.

In view of this, instead of exploiting the objective function (5), Klawonn and Höppner [60] offered the criterion

$$
J\left(u_j(k),\ v_j^k\right) = \sum_{k=1}^{N}\sum_{j=1}^{m}\left(au_j^2(k)+(1-a)u_j(k)\right)\left\|\tilde{\phi}(k,h)-v_j^k\right\|^2
\tag{13}
$$

with a crisp fuzzifier ω with the constraints (6) where a tunable parameter $0 < a \le 1$ defines a sort of a resulting solution.

Describing the function by Lagrange

$$
\begin{aligned}
L\left(u_j(k), v_j^k, \lambda(k)\right) &= \sum_{k=1}^{N}\sum_{j=1}^{m}\left(au_j^2(k)+(1-a)u_j(k)\right)\left\|\tilde{\phi}(k,h)-v_j\right\|^2 \\
&+ \sum_{k=1}^{N}\lambda(k)\left(\sum_{j=1}^{m}u_j(k)-1\right)
\end{aligned}
\tag{14}
$$

(where $\lambda(k)$ denotes undefined multipliers by Lagrange) and solving a system of equations by Karush-Kuhn-Tucker, the final result comes out like

$$
\begin{cases}
u_j(k) = \dfrac{a-1}{2a} + \left(1+m\dfrac{1-a}{2a}\right)\dfrac{\left\|\tilde{\phi}(k,h)-v_j^k\right\|^{-2}}{\sum\limits_{l=1}^{m}\left\|\tilde{\phi}(k,h)-v_l^k\right\|^{-2}}, \\[4mm]
v_j^k = \dfrac{\sum\limits_{k=1}^{N}\left(au_j^2(k)+(1-a)u_j(k)\right)\tilde{\phi}(k,h)}{\sum\limits_{k=1}^{N}\left(au_j^2(k)+(1-a)u_j(k)\right)}, \\[4mm]
\lambda(k) = -\dfrac{1+m\frac{1-a}{2a}}{\sum\limits_{l=1}^{m}\left(2a\left\|\tilde{\phi}(k,h)-v_l^k\right\|^2\right)^{-1}}.
\end{cases}
\tag{15}
$$

As is obvious, the statement (15) converts to the FCM algorithm (8) when $a = 1$; reduction of a assigns the result obtained some crisper nature.

The procedure of non-linear programming by Arrow-Hurwicz-Uzawa being applied to the function (14) accounts for the recurrent scheme of fuzzy clustering according to the criterion (13) [61]:

$$
\begin{cases}
u_j(k+1) = \frac{a-1}{2a} + \left(1 + m\frac{1-a}{2a}\right)\dfrac{\left\|\tilde{\phi}(k+1,h)-v_j^k(k)\right\|^{-2}}{\displaystyle\sum_{l=1}^{m}\left\|\tilde{\phi}(k+1,h)-v_l^k(k)\right\|^{-2}}, \\
v_j^k(k+1) = v_j^k(k) + \eta(k)\left(au_j^2(k+1) + (1-a)u_j(k+1)\right) * \left(\tilde{\phi}(k+1,h) - v_j^k(k)\right)
\end{cases}
\tag{16}
$$

where the approach (16) providing $a = 1$ is congruent with the expression (9) assuming $\omega = 2$. It can be marked easily that the second ratio (16) is also the WTM self-learning rule.

4 Experimental Examples

To test the performance capability of the developed kernel fuzzy clustering system and its self-learning mechanisms, two well-known classification data sets (Wisconsin Diagnostic Breast Cancer; Pima Indians Diabetes) were chosen for this reason.

The first data set contained information about the cancer of a lacteal gland to have been previously exposed. The set contains 569 observations split into 2 classes where each object is described by 30 features.

The second data set exemplifies diabetes and contains 768 observations split into 2 classes where each object is described by 8 features.

All the data described were initially normalized to the hypercube in the interval $[-1, 1]$ and centered relatively to the mean value.

To accomplish our comparison, the fuzzy C-means by J. Bezdek with a fuzzifier $\omega = 2$ and the self-organizing map by T. Kohonen were selected.

Since there are labels for the correct classification results, the clustering quality (accuracy) was estimated as some percentage with respect to a reference object. Table 1 demonstrates results (min, max, and avg) for a set of 50 experiments.

Table 1. Comparison of clustering quality.

The methods applied	Wisconsin Diagnostic Breast Cancer			Indians Diabetes		
	Avg	Max	Min	Avg	Max	Min
Fuzzy C-means	87	92	82	89	96	83
Self-organizing map	78	88	73	81	89	72
Kernel fuzzy clustering system	91	94	86	92	95	86

It is impressive to see that the offered system outperforms other competitors in both cases. But the difference between FCM and the offered method is really close. Summarizing the results, it can be concluded that the clustering accuracy is improving if the data sample keeps on growing.

5   Conclusion

The paper described a topology of the multilayer kernel clustering system and several self-learning procedures used for its training. The key aspect of the offered clustering system is its capability to restore overlapping clusters of an arbitrary shape in the context of feeding data in an online way. The system doesn't give any troubles to be implemented in a simple sequence of steps. The kernel fuzzy clustering system is actually a hybrid combination of the fuzzy SOM architecture and the general regression neural network. The learning process of the introduced system also joints their basic training paradigms correspondingly.

Acknowledgment. This scientific work was financially supported by self-determined research funds of CCNU from the colleges' basic research and operation of MOE (CCNU16A02015). The third author also acknowledges the support of the Visegrad Scholarship Program—EaP #51700967 funded by the International Visegrad Fund (IVF).

References

1. Kruse, R., Borgelt, C., Klawonn, F., Moewes, C., Steinbrecher, M., Held, P.: Computational Intelligence. A Methodological Introduction. Springer, Berlin (2013)
2. Mumford, C.L., Jain, L.C.: Computational Intelligence. Springer, Berlin (2009)
3. Kohonen, T.: Self-Organizing Maps. Springer, Berlin (1995)
4. Haykin, S.: Neural Networks and Learning Machines. Prentice Hall, Upper Saddle River (2009)
5. Delen, D.: Real-World Data Mining: Applied Business Analytics and Decision Making. Pearson FT Press, Upper Saddle River (2015)
6. Aggarwal, C.C.: A Data Mining: The Textbook. Springer, Heidelberg (2015)
7. Larose, D.T.: Discovering Knowledge in Data: An Introduction to Data Mining. Wiley, Hoboken (2014)
8. Hastie, T., Tibshirani, R., Friedman, J.: The Elements of Statistical Learning. Data Mining, Inference, and Prediction. Springer Science & Business Media, LLC, New York (2009)
9. Aggarwal, C.C., Reddy, C.K.: Data Clustering: Algorithms and Applications. CRC Press, Boca Raton (2014)
10. Gosain, A., Dahiya, S.: Performance analysis of various fuzzy clustering algorithms: a review. Procedia Comput. Sci. **79**, 100–111 (2016)
11. Xu, R., Wunsch, D.C.: Clustering. IEEE Press Series on Computational Intelligence. Wiley, Hoboken (2009)
12. Yang, M.-S., Chang-Chien, S.-J., Hung, W.-L.: An unsupervised clustering algorithm for data on the unit hypersphere. Appl. Soft Comput. **42**, 290–313 (2016)

13. Babichev, S., Taif, M.A., Lytvynenko, V.: Inductive model of data clustering based on the agglomerative hierarchical algorithm. In: The 2016 IEEE First International Conference on Data Stream Mining and Processing (DSMP), pp. 19–22, Lviv (2016)
14. Babichev, S., Lytvynenko, V., Oypenko, V.: Implementation of the objective clustering inductive technology based on DBSCAN clustering algorithm. In: The XII International Scientific and Technical Conference "Computer Sciences and Information Technologies" (CSIT), pp. 479–484, Lviv (2017)
15. Ivanov, Y., Peleshko, D., Makoveychuk, O., Izonin, I., Malets, I., Lotoshunska, N, Batyuk, D.: Adaptive moving object segmentation algorithms in cluttered environments. In: 2015 15th International Conference the Experience of Designing and Application of CAD Systems in Microelectronics (CADSM), pp. 97–99, Lviv (2015)
16. Babichev, S., Lytvynenko, V., Korobchynskyi, M., Taif, M.: Objective clustering inductive technology of gene expression sequences features. Commun. Comput. Inf. Sci. **716**, 359–372 (2016)
17. Izonin, I., Tkachenko, R., Peleshko, D., Rak T., Batyuk, D.: Learning-based image super-resolution using weight coefficients of synaptic connections. In: 2015 Xth International Scientific and Technical Conference "Computer Sciences and Information Technologies" (CSIT), pp. 25–29, Lviv (2015)
18. Babichev, S.A., Kornelyuk, A.I., Lytvynenko, V.I., Osypenko, V.: Computational analysis of microarray gene expression profiles of lung cancer. Biopolym. Cell **32**(1), 70–79 (2016)
19. Bezdek, J.C.: Pattern Recognition with Fuzzy Objective Function Algorithms. Plenum Press, New York (1981)
20. Hu, Z., Bodyanskiy, Y.V., Tyshchenko, O.K., Samitova, V.O.: Fuzzy clustering data given on the ordinal scale based on membership and likelihood functions sharing. Int. J. Intell. Syst. Appl. (IJISA) **9**(2), 1–9 (2017). https://doi.org/10.5815/ijisa.2017.02.01
21. Hu, Z., Bodyanskiy, Y.V., Tyshchenko, O.K., Samitova, V.O.: Fuzzy clustering data given in the ordinal scale. Int. J. Intell. Syst. Appl. (IJISA) **9**(1), 67–74 (2017). https://doi.org/10.5815/ijisa.2017.01.07
22. Bodyanskiy, Y., Vynokurova, O., Savvo, V., Tverdokhlib, T., Mulesa, P.: Hybrid clustering-classification neural network in the medical diagnostics of the reactive arthritis. Int. J. Intell. Syst. Appl. (IJISA) **8**(8), 1–9 (2016). https://doi.org/10.5815/ijisa.2016.08.01
23. Hu, Z., Bodyanskiy, Y.V., Tyshchenko, O.K., Tkachov, V.M.: Fuzzy clustering data arrays with omitted observations. Int. J. Intell. Syst. Appl. (IJISA) **9**(6), 24–32 (2017). https://doi.org/10.5815/ijisa.2017.06.03
24. Kung, S.Y.: Kernel Methods and Machine Learning. University Press, Cambridge (2014)
25. Czarnowski, I., Jędrzejowicz, P.: Kernel-based fuzzy c-means clustering algorithm for rbf network initialization. In: Czarnowski, I., Caballero, A.M., Howlett, R.J., Jain, L.C. (eds.) Intelligent Decision Technologies 2016: Proceedings of the 8th KES International Conference on Intelligent Decision Technologies (KES-IDT 2016), Part I, pp. 337–347. Springer International Publishing, Cham (2016)
26. Saikumar, T., Neenu Preetam, I.: Optimized kernel fuzzy c means (OKFCM) clustering algorithm on level set method for noisy images. In: 2013 IEEE International Conference on Computational Intelligence and Computing Research (ICCIC), pp. 880–884. IEEE (2013)
27. Zhang, Z., Havens, T.C.: Scalable approximation of kernel fuzzy c-means. In: 2013 IEEE International Conference on Big Data, pp. 161–168. IEEE (2013)
28. Girolami, M.: Mercer kernel based clustering in feature space. IEEE Trans. Neural Netw. **13**(3), 780–784 (2002)
29. Kim, D.-W., Lee, K., Lee, K.H.: Evaluation of the performance of clustering algorithms in kernel-based feature space. Pattern Recogn. **35**, 2267–2278 (2002)

30. MacDonald, D., Fyfe, C.: Clustering in data space and feature space. In: Proceedings of European Symposium on Artificial Neural Networks, ESANN 2002, pp. 137–142, Bruges (2002)
31. Camastra, F., Verri, A.: A novel kernel method for clustering. IEEE Trans. Pattern Anal. Mach. Intell. **5**, 801–805 (2005)
32. Zhang, D.-Q., Chen, S.-C.: Fuzzy clustering using kernel method. In: Proceedings of International Conference on Control and Automation, ICCA 2002, pp. 162–163 (2002)
33. Zhang, D.-Q., Chen, S.-C.: Kernel based fuzzy and possibilistic c-means clustering. In: Proceedings of International Conference on Artificial Neural Networks, ICANN 2003, pp. 122–125 (2003)
34. Miyamoto, S., Mizutani, K.: Fuzzy multiset space and c-means clustering using kernels with application to information retrieval. In: Lecture Notes on Artificial Intelligence, vol. 2715, pp. 387–395. Springer, Heidelberg (2003)
35. Zhang, D.-Q., Chen, S.-C.: A novel kernelized fuzzy c-means algorithm with application in medical image segmentation. Artif. Intell. Med. **32**(1), 37–50 (2004)
36. Du, W., Inoue, K., Urahama, K.: Robust kernel fuzzy clustering. In: Lecture Notes on Artificial Intelligence, vol. 3613, pp. 454–461. Springer, Heidelberg (2005)
37. Filippone, M., Camastra, F., Masulli, F., Rovetta, S.: A survey of kernel and spectral methods for clustering. Pattern Recogn. **41**, 176–190 (2008)
38. Havens, T.S., Bezdek, J.C., Palaniswami, M.: Incremental kernel fuzzy c-means. In: Madani, K., et al. (eds.) Computational Intelligence, SCI, vol. 399, pp. 3–18. Springer, Heidelberg (2012)
39. Krishnapuram, R., Keller, J.M.: A possibilistic approach to clustering. IEEE Trans. Fuzzy Syst. **1**(2), 98–110 (1993)
40. Krishnapuram, R., Keller, J.M.: The possibilistic c-means algorithm: insights and recommendations. IEEE Trans. Fuzzy Syst. **4**(3), 385–393 (1996)
41. Bodyanskiy, Y., Tyshchenko, O., Kopaliani, D.: An evolving neuro-fuzzy system for online fuzzy clustering. In: Proceedings of the International Conference on Computer Sciences and Information Technologies, CSIT 2015, pp. 158–161, Lviv (2015)
42. Hu, Z., Bodyanskiy, Y.V., Tyshchenko, O.K., Samitova, V.O.: Possibilistic fuzzy clustering for categorical data arrays based on frequency prototypes and dissimilarity measures. Int. J. Intell. Syst. Appl. (IJISA) **9**(5), 55–61 (2017). https://doi.org/10.5815/ijisa.2017.05.07
43. Hu, Z., Bodyanskiy, Y.V., Tyshchenko, O.K.: A deep cascade neuro-fuzzy system for high-dimensional online fuzzy clustering. In: Proceedings of the 2016 IEEE 1st International Conference on Data Stream Mining and Processing, DSMP 2016, pp. 318–322, Lviv (2016)
44. Hu, Z., Bodyanskiy, Y.V., Tyshchenko, O.K.: A cascade deep neuro-fuzzy system for high-dimensional online possibilistic fuzzy clustering. In: Proceedings of the 11th International Scientific and Technical Conference "Computer Sciences and Information Technologies", CSIT 2016, pp. 119–122, Lviv (2016)
45. Aggarwal, C.C.: Data Streams: Models and Algorithms. Springer Science & Business Media, Berlin (2007)
46. Bifet, A.: Adaptive Stream Mining: Pattern Learning and Mining from Evolving Data Streams. IOS Press, Amsterdam (2010)
47. Bodyanskiy, Y.V., Tyshchenko, O.K., Kopaliani, D.S.: An evolving connectionist system for data stream fuzzy clustering and its online learning. Neurocomputing **262**, 41–56 (2017)
48. Hu, Z., Bodyanskiy, Y.V., Tyshchenko, O.K., Boiko, O.O.: A neuro-fuzzy Kohonen network for data stream possibilistic clustering and its online self-learning procedure. Appl. Soft Comput. J. (2017, in Press). https://doi.org/10.1016/j.asoc.2017.09.042

49. Zahirniak, D., Chapman, R., Rogers, S., Suter, B., Kabrisky, M., Piati, V.: Pattern recognition using radial basis function network. In: Proceedings of 6th Annual Aerospace Application of Artificial Intelligence Conference, pp. 249–260, Dayton, OH (1990)
50. Vapnik, V.: The Nature of Statistical Learning Theory. Springer, New York (1995)
51. Cortes, C., Vapnik, V.: Support vector networks. Mach. Learn. **20**, 273–297 (1995)
52. Specht, D.F.: A general regression neural network. IEEE Trans. Neural Netw. **2**, 568–576 (1991)
53. Rooki, R.: Application of general regression neural network (GRNN) for indirect measuring pressure loss of Herschel-Bulkley drilling fluids in oil drilling. Measurement **85**, 184–191 (2016)
54. Alilou, V.K., Yaghmaee, F.: Application of GRNN neural network in non-texture image inpainting and restoration. Pattern Recogn. Lett. **62**, 24–31 (2015)
55. Bodyanskiy, Y., Kolodyazhniy, V., Stephan, A.: Recursive fuzzy clustering algorithms. In: Proceedings of 10th East West Fuzzy Colloqium, pp. 276–283, Zittau/Görlitz (2002)
56. Bodyanskiy, Y.: Computational intelligence techniques for data analysis. In: Lecture Notes in Informatics, vol. P.72, pp. 15–36, Bonn, GI (2005)
57. Gorshkov, Y., Kolodyazhniy, V., Bodyanskiy, Y.: New recursive learning algorithm for fuzzy Kohonen clustering network. In: Proceedings of 17th International Workshop on Nonlinear Dynamics of Electronic Systems, pp. 58–61, Rapperswil, Switzerland (2009)
58. Park, D.C., Dagher, I.: Gradient based fuzzy c-means (GBFCM) algorithm. In: Proceedings of IEEE International Conference on Neural Networks, pp. 1626–1631 (1984)
59. Chung, F.-L., Lee, T.: Unsupervised fuzzy competitive learning with monotonically decreasing fuzziness. In: Proceedings of 1993 International Joint Conference on Neural Networks, pp. 2929–2932 (1993)
60. Klawonn, F., Höppner, F.: What is fuzzy about fuzzy clustering? Understanding and improving the concept of the fuzzifier. In: Lecture Notes in Computer Science, vol. 2811, pp. 254–264. Springer, Heidelberg (2003)
61. Kolchygin, B., Bodyanskiy, Y.: Adaptive fuzzy clustering with a variable fuzzifier. Cybern. Syst. Anal. **49**(3), 176–181 (2013)

Information Technology of Data Protection on the Basis of Combined Access Methods

Andrey Kupin[(✉)] ⓘ, Yurii Kumchenko ⓘ, Ivan Muzyka ⓘ, and Dennis Kuznetsov ⓘ

Kryvyi Rih National University, 11, Vitaliya Matusevycha str., Kryvyi Rih 50027, Ukraine
kupin.andrew@gmail.com, y.kumchenko@gmail.com,
musicvano@gmail.com, kuznetsov.dennis.1706@gmail.com

Abstract. The main task of the article is to develop information technology (IT) for data protection based on combined access methods. The need to create a reliable IT for data protection is conditioned by an active increase in confidential information and unauthorized access. The article presents the existing static and dynamic biometric access methods, the evaluation of biometric technologies is reviewed: market segmentation, access errors and a general table of characteristics. A combined access method based on Acuity Market Intelligence and International Biometric Group data is proposed, which includes a combination of voice and face - a multimodal method. The article contains the calculation of the work accuracy by using the characteristic curves: DET (Detection error trade-off), which establish the relationship between FRR errors (False Rejection Rate) and FAR (False Acceptance Rate) and identify the advantages of a multimodal biometric personnel identification system comparing the unimodal one. Also, the mathematical model of IT for data protection has been developed. The proposed scheme of information links is developed for the IT for data protection based on combined access methods.

Keywords: Information technology · Data protection · Biometrics

1 Introduction

The need to create a reliable information technology for data protection is conditioned by an active increase in confidential information and unauthorized access. The analysis of modern access control systems indicates an obvious movement towards combined information security methods [1, 2]. It should also be noted that according to the International Biometric Group and Acuity Market Intelligence, biometrics is the most promising technology for confirming the identity when using an ATM, smartphone, or computer.

The scientific task of the article is to develop information technology for data protection based on combined access methods. The task is topical, since its solution will provide a significant simplification of data protection, as well as increase the reliability of access control systems.

The research methods are based on the integrated use of fundamental provisions, theoretical studies are based on the system approach, application of linear programming

© Springer International Publishing AG, part of Springer Nature 2019
Z. Hu et al. (Eds.): ICCSEEA 2018, AISC 754, pp. 498–506, 2019.
https://doi.org/10.1007/978-3-319-91008-6_50

methods, spectral analysis methods, boundary delineation techniques, computer graphics techniques, the description of information links and the methodology of functional modelling.

2   Biometric Access Methods

For biometric access, the characteristics inherent in a person are used, which are divided into static and dynamic [3]. Static ones are related to unique physical characteristics (papillary finger pattern, palm geometry and/or hand vein patterns, retina, iris, facial geometry, face thermogram, etc.). Dynamic characteristics are related to the features of the person's performance of any actions (voice, tempo typing on the keyboard, handwriting, gait, heart rate, etc.).

Let us consider the evaluation of biometric technologies: market segmentation Fig. 1, access errors Fig. 2 and general characteristics table Fig. 3 [4].

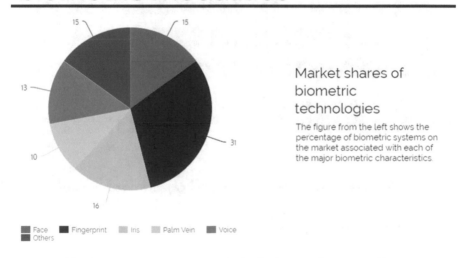

Fig. 1.   Assessment of biometric technologies - market segmentation.

Based on the analysis of modern access technologies, it is proposed to use a multimodal biometric system, which consists of two characteristics: face and voice. Deciding on the user's access to protected data is a logical scheme that takes into account the results of all modules of the identification system.

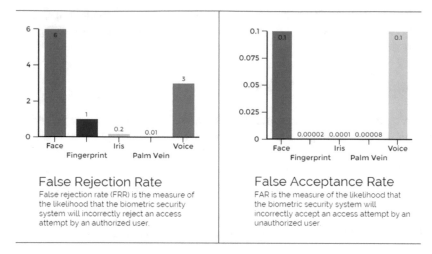

Fig. 2. Assessment of biometric technologies - access errors FRR and FAR.

Modality	Accuracy	Ease to Use	User Acceptance
Face	Low	High	High
Fingerprint	High	Medium	Low
Iris	High	Medium	Medium
Palm Vein	High	High	Medium
Voice	Medium	High	High

Comparison of Biometric Modalities

It's important to realize that there is not one biometric modality which is best for all conditions and implementations. The above figure will help you to decide the right type of biometric system.

www.bayometric.com | sales@bayometric.com | +1 (408) 940-3955

Fig. 3. Assessment of biometric technologies - general characteristics table.

3 Multimodal Biometric System

The proposed multimodal biometric system is a multifactor identification of personnel, which consists of two main statically-dynamic components:

(a) access to a person's face;
(b) voice access.

A multimodal solution is a generalization of the results obtained in the course of facial and voice identification. The result of processing these characteristics are the mathematical probabilities of the similarity of the face and voice received at the input of the audio/video stream, with the reference samples of the user. Based on these values, the complex probability of access is calculated.

To assess the accuracy of any **biometric** system, it is customary to use the DET (Detection error trade-off) curve, which establishes the relationship between FRR and FAR errors [5]. For a multimodal solution, we obtain the following DET curves Fig. 4.

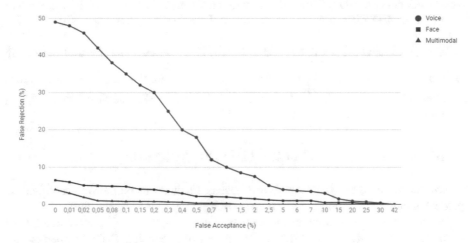

Fig. 4. Estimation of the accuracy of the biometric system (DET curves).

Table 1. Relationship between FRR and FAR errors of unimodal biometric systems and the developed multimodal one.

No.	Biometric characteristics					
	Unimodal				Multimodal	
	Voice		Face		Voice and face	
	FRR, %	FAR, %	FRR, %	FAR, %	FRR, %	FAR, %
1.	48	0.01	6.5	0.01	3	0.01
2.	46	0.02	5.1	0.02	1.9	0.02
3.	42	0.05	5	0.05	1.8	0.03
4.	35	0.1	4.8	0.1	1	0.05
5.	30	0.2	4	0.2	0.9	0.07
6.	20	0.4	3	0.4	0.8	0.1
7.	18	0.5	2.2	0.5	0.8	0.2
8.	10	1	2	1	0.7	0.3
9.	7.5	2	1.5	2	0.6	0.4
10.	4	5	1	5	0.5	0.45
11.	3	10	0.48	10	0.35	0.5
12.	1.5	15	0.48	15	0.3	0.8
13.	0.9	20	0.48	20	0.25	1
14.	0.2	37	0.3	30	0.15	1.1
15.	0.02	38	0.02	42	0.02	1.2

Consequently, Table 1 shows that if using a unimodal or multimodal system in an organization with a staff number of: $N = \sqrt{\dfrac{1}{0.0001}} = 100$ (persons), then the voice system will not admit 48% (FRR) of staff with an the access right, the facial system will not admit 6.5% (FRR); as far as the multimodal system the percentage will make 3% (FRR).

When analyzing the last line of Table 1 (15), it can be seen that if a staff number in an organization is $N = \sqrt{\dfrac{1}{0.01}} = 10$ (persons), then the multimodal system is 33 times more reliable than unimodal systems: the voice is 38% (FAR), the face is 42% (FAR), the multimodal one (voice and face) - 1.2% (FAR).

4   Mathematical Model of the IT Data Protection

To protect the data based on combined access methods in biometric identification systems, a number of biometric parameters should be provided by the staff.

Suppose we have n different human parameters P_1, P_2, ...,P_n and m number of personnel L_1, L_2, ...,L_m. Table 2 shows the number of parameters P_i for one person L_j.

Table 2.   Output data for the mathematical model of IT data protection.

Staff, m	Biometrical parameters of a person, n				Minimum standard for the access
	P_1	P_2	...	P_n	
L_1	x_{11}	x_{12}	...	x_{1n}	d_1
L_2	x_{21}	x_{22}	...	x_{2n}	d_2
...	...	...		...	...
L_m	x_{m1}	x_{m2}	...	x_{mn}	d_k
The system cost	c_1	c_2	...	c_r	

The task is as follows: it is necessary to organize staff access to the data so that the minimum rate for access d_k, is met, which is set for each individual, depending on the level of security, and the cost c_r of such a system was minimal.

1. X the number of biometric parameters of a person.
2. Restriction system:

$$\begin{cases} x_{11} + x_{12} + \ldots + x_{1n} \geq d_1, \\ x_{21} + x_{22} + \ldots + x_{2n} \geq d_2, \\ \ldots \ldots \ldots \ldots \ldots \ldots \ldots \ldots \ldots \\ x_{m1} + x_{m2} + \ldots + x_{mn} \geq d_k, \end{cases} \tag{1}$$

$$x_{ij} \geq 0 \text{ where } i,j = \overline{1,n} \ d_i \geq 0 \text{ where } i = \overline{1,k}.$$

3. The minimum of the objective function:

$$F(X) = c_1x_1 + c_2x_2 + \ldots + c_rx_n \to \min. \tag{2}$$

So, we have introduced determinants for unknown X's in the problem and fixed the constraints for them: $x_{ij} \geq 0$ where $i, j = \overline{1, n}$ where $i = \overline{1, k}$. A system of constraints (1) of the problem with respect to the minimum norm for access d_k and a target function with an established extremum $F(X) = c_1x_1 + c_2x_2 + \ldots + c_rx_n \to \min$ is drawn out.

The above model belongs to the problems of linear programming, so we will calculate it in a mathematical package. Based on the calculation, the minimum 4 values of the function in access technology using a combination of voice, face and password were obtained. The selected combination of biometric parameters and the password corresponds to the established extremum of the objective function (2).

5   Scheme of Information Links of the Developed IT Data Protection

Let us describe the scheme of information communications using the modern American standard NIST Special Publication 800-183 (July 2016) by the National Institute of Standards and Technology [6]. This document was released in the SP 800 series, Computer Security, which implies its direct relevance to information security. The author of NIST SP 800-183 is Jeffrey Voas, who has been known since the early 1990s because of his works on the theory of software evaluation and testing.

NIST SP 800-183 offers a unique approach in describing information links. Five types of primitives are used for this Fig. 5: (1) Sensor, (2) Aggregator, (3) Communication Channel, (4) External Utility (eUtility), (5) Decision Trigger.

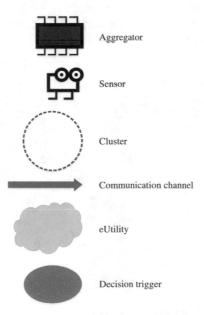

Fig. 5.   The primitives of the standard NIST SP 800-183.

Primitive № 1: Sensor

"Sensor" is a sensor designed to measure physical parameters (temperature, humidity, pressure, acceleration, etc.).

Primitive № 2: Aggregator

"Sensor" transmits information to "Aggregator", which is a software implementation of functions (possibly using artificial intelligence) that turns the output (raw) data into intermediate aggregate data. For "Aggregator", the concept of Actors is introduced for processing the two types of data: "Cluster" & "Weight". "Cluster" refers to the virtual dynamic Cluster of Sensors, which is organized and changed depending on the approach to data aggregation. "Weight" refers to the weighting factor (also, possibly, dynamic) that is used to process data using "Aggregator".

Primitive № 3: Communication Channel

Communication Channel is a virtual or physical communication medium that combines all other primitives.

Primitive № 4: eUtility (External Utility)

eUtility means any hardware device, program or service that is the platform for executing "Aggregator". In the future it is supposed to concretize this primitive, having allocated several categories.

Primitive № 5: Decision Trigger

"Decision Trigger" generates the final result necessary to fulfil the objective function of a particular system [7].

The information communication scheme of the IT data protection consists of the following components Fig. 6: (1) U - user (identification object), (2) C - sensor cluster, (3) S - sensor, (4) A - software implementation of functions, (5) eU - hardware device, program or service, (6) virtual and physical media, (7) the shaper of the final result. To describe a small IT, the number of components is sufficient, and with increasing the dimension, you can apply hierarchical structures.

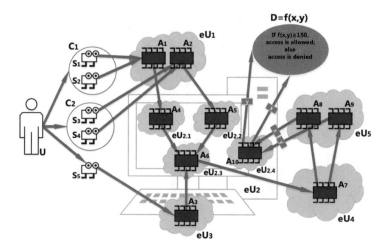

Fig. 6. The scheme of information communications of the developed IT data protection.

User U in non-contact type informs two biometric characteristics (face and voice) to the two clusters C1 and C2, each of which consists of two sensors: S1, S2 - ordinary and infrared cameras; S3, S4 - two microphones. U contacts physically the S5 sensor to enter a password.

The sensors S1, S2, S3, S4 transmit the information flow from U to the aggregators A1 and A2, which are a composite multimodal hardware device eU1. The device eU1 is physically connected by two communication lines (USB and 3.5 mm audio cable) with the eU2 computer.

The eU2 computer contains four programs eU2.1, eU2.2, eU2.3, eU2.4 and a connected eU3 keyboard, which is the platform for the aggregator A3; eU2.1 pre-processes images and forms a reference face pattern [8]; eU2.2 pre-processes the sound and forms the reference voice sample; eU2.3 receives two information streams with preprocessed signals from eU2.1, eU2.2 and a stream from eU3 for local encryption of reference samples using CryptoJS.

The encrypted samples are transmitted to the eU4 database. The A7 database aggregator is connected to the two aggregators A8 and A9 of the cloud service eU5 (Sky Biometry - Cloud Based Biometrics API as a Service, Microsoft Cognitive Services, Google Speech), which serves as a comparative method.

The web browser eU2.4 (Google Chrome) receives information from eU5 as two streams of information x and y, and transmits the data of the final result to the former to perform the objective function of the system $D = f(x, y)$. At the output we will get the result about permission or prohibition of access to confidential information of a particular user U.

6   Conclusion

The use of IT data protection based on combined access methods with the use of multimodal biometric parameters for information protection has significant advantages. Thanks to a combination of methods that take into account several biometric characteristics, it is possible to increase the security of information resources from unauthorized access in general. A mathematical model is developed in the form of a linear programming task with IT data protection limitations to reduce the cost of the system. A description of the information communication scheme developed by IT data protection based on a complex of biometric parameters is developed using the modern American standard NIST Special Publication 800-183 by the National Institute of Standards and Technology. When assessing the accuracy of the IT protection of data, it was found that if a single-modal or multimodal system is used in an organization with 100 employees, the system using voice will not admit 48% (FRR) of personnel having an access right, the facial system will not admit 6.5% (FRR); as far as the multimodal system the percentage will make 3% (FRR), and if there are 10 people in the organization, the multimodal system is 33 times more reliable than the unimodal system: the voice is 38% (FAR), the face is 42% (FAR), the multimodal one (voice and face) - 1.2% (FAR).

References

1. Tran, L.B., Le, T.H.: Person authentication using relevance vector machine (RVM) for face and fingerprint. Int. J. Mod. Educ. Comput. Sci. (IJMECS) **7**(5), 8–15 (2015). https://doi.org/10.5815/ijmecs.2015.05.02
2. Barde, S., Zadgaonkar, A.S., Sinha, G.R.: PCA based multimodal biometrics using ear and face modalities. Int. J. Inf. Technol. Comput. Sci. (IJITCS) **6**(5), 43–49 (2014). https://doi.org/10.5815/ijitcs.2014.05.06
3. Shankar, S., Udupi, V.R., Gavas, R.D.: Biometric verification, security concerns and related issues - a comprehensive study. Int. J. Inf. Technol. Comput. Sci. (IJITCS) **8**(4), 42–51 (2016). https://doi.org/10.5815/ijitcs.2016.04.06
4. Thakkar, D.: Top Five Biometrics: Face, Fingerprint, Iris, Palm and Voice. https://www.bayometric.com/biometrics-face-finger-iris-palm-voice. Accessed 17 Nov 2017
5. Malik, J., Girdhar,D., Dahiya, R., Sainarayanan, G.: Reference Threshold Calculation for Biometric Authentication. Int. J. Image Graph. Signal Process. (IJIGSP) **6**(2), 46–53 (2014). https://doi.org/10.5815/ijigsp.2014.02.06
6. Voas, J.: NIST Special Publication 800-183. Networks of 'Things'. http://nvlpubs.nist.gov/nistpubs/SpecialPublications/NIST.SP.800-183.pdf. Accessed 17 Nov 2017
7. Skliar, V.: NIST recommends: building blocks for describing IoT. https://habrahabr.ru/post/314956. Accessed 17 Nov 2017
8. Kupin, A., Kumchenko, Y.: Improved algorithm for creating a template for the information technology of biometric identification. Metall. Mining Ind. **7**(4), 7–10 (2015)

Neural Network Algorithm for Accuracy Control in Modelling of Structures with Changing Characteristics

Dmitriy Zelentsov 🆔 and Olga Denysiuk(✉) 🆔

Ukrainian State University of Chemical Technology, Gagarina av. 8, Dnipro 49005, Ukraine
dgzelentsov@gmail.com, denolga91@rambler.ru

Abstract. The paper is devoted to the creation of neural network algorithm for control of the accuracy of durability calculation in problems of modelling the behavior of structures with changing characteristics, in particular, corroding trusses. In contrast to known methods, the approach suggested in this paper takes into account the change of forces in elements of the truss over time. First, the analysis of corrosive wear models is given to choose the most suitable one for further research. Then, the mathematical statement of durability determination problem is given. The paper describes the approach to the approximation of the numerical solution error for the durability problem. To build an approximating function, artificial neural networks are used. The architecture of these networks and the procedure of their training are described further in the paper. At the end, the results of numerical experiments, which prove the correctness of the chosen approach, are given. The developed algorithm can be especially effective in solving optimization problems with constraints on the durability of a structure. The same approach can also be generalized to other classes of structures.

Keywords: Artificial neural networks · Accuracy control
Systems of differential equations · Corrosion · Trusses

1 Introduction

During exploitation of metal structures used, in particular, in the chemical industry, there is a noticeable deterioration in their performance as a result of corrosive wear caused by the impact of highly aggressive working media. Corrosive wear leads to the possibility of premature failure of structures. Therefore, ensuring the necessary accuracy of durability prediction for a structure subjected to corrosion is a problem of vital importance. The problem is connected with the problem of optimal design of corroding structures. The proposed paper is devoted to one of the approaches to increasing the calculation accuracy for durability of corroding trusses.

It is proposed in current paper to use artificial neural networks (ANNs) in modeling of corrosive deformation process. ANNs are widely used, for example, to estimate parameters of this process (rate of corrosion) [1]. In [2] the use of artificial neural networks was proposed to determine the numerical solution parameters for systems of differential equations (SDE), which provide the required calculation accuracy for all

© Springer International Publishing AG, part of Springer Nature 2019
Z. Hu et al. (Eds.): ICCSEEA 2018, AISC 754, pp. 507–516, 2019.
https://doi.org/10.1007/978-3-319-91008-6_51

points of the solution space. However, the drawback of these works was the fact that the training of ANNs had not taken into account change in internal forces in the rod elements. This fact caused some difference between the reference error in constraint functions' computation and the error actually obtained while solving the SDE with the numerical solution parameter obtained using ANN.

In this paper, it is proposed to use an ANN to approximate the dependence between the error of numerical solution of the SDE, design parameters and parameters of an aggressive medium, taking into account the change of internal forces in truss elements. This approach helps to improve the accuracy of structural durability calculation to reduce the computational costs.

## 2	Mathematical Statement of Durability Determination Problem

The behavior of a structure in an aggressive medium is modeled by a numerical solution of the Cauchy problem for a system of differential equations (SDE) describing the process of corrosion in structural elements [3]:

$$\frac{d\delta_i}{dt} = v_0\left[1 + k\sigma_i(\bar{\delta})\right]; \delta_i|_{t=0} = 0; \quad i = \overline{1, N} \tag{1}$$

where δ_i is the depth of corrosion damage in i-th element; v_0 is the rate of corrosion in the absence of stress; k is the coefficient of stress state impact on the corrosion rate; σ_i is stress is i-th element; N is the number of elements in a truss.

Calculation of constraint functions (the durability of structure) involves a numerical solution of the system (1). The numerical solution parameters for the SDE are input parameters and usually do not change in the process of solving the optimization problem. As a consequence, the calculation error of constraint functions will be different for all points in the solution space of the optimization problem. In this case, the error of the obtained result cannot be predicted and may be higher than the allowed value.

The presence of feedback in the mathematical calculation models caused by the impact of corrosion on the stresses in the structural elements, and a significant increase in the number of parameters that determine the geometric dimensions of the structure, lead to a significant increase in computational costs. This makes the problem of efficiency and accuracy of numerical algorithms especially important for problems of this class.

In this paper, as an object of study, we consider trusses, elements of which are rectilinear rods. Depending on the design topology, boundary conditions and loading conditions, the forces in the elements are either tensile or compressive. Due to corrosive wear, the geometric characteristics of rod cross-sections (areas and moments of inertia) decrease. The state of a structure is considered safe as long as stress in any rod does not exceed the maximum permissible value, for example, yield stress or critical stress of stability loss (in compressed rods):

$$[\sigma] - \sigma_i(\bar{x}, t) \geq 0;$$

$$\sigma_j * (\bar{x}, t) - \sigma_j(\bar{x}, t) \geq 0; i \in \overline{1, N}; j \in J \tag{2}$$

where $[\sigma]$ is yield stress, $\sigma_j *$ is critical stress of stability loss in j-th element, $\bar{x}$ is the vector of design parameters, J is the set of compressed elements.

The moment when any of the inequalities (2) becomes an equality is a value of durability of the structure. The task of this paper is to develop a computational algorithm for determination of structural durability with minimal computational costs and acceptable accuracy for the entire set of points in the solution space, with constant parameters of the numerical solution of SDE (1).

In most of the known works, one-step Runge-Kutta methods (for example, the Euler method) were used to solve SDE (1). The disadvantages of these methods, in addition to their low efficiency, are fully described in [3].

In this paper, it is proposed to use a modified algorithm of the Euler method with a variable step over the argument to solve SDE (1). It is proposed to set the increment of function $\Delta\sigma_S = const$, and the corresponding value of the increment of argument is determined by the formula, which derivation is given in [3]:

$$\Delta t_S = \frac{\Delta\delta_S}{v_0} - \frac{2kQ}{v_0 d} \ln\left\{\frac{(2a \cdot \Delta\delta_S + b - d)(b + d)}{(2a \cdot \Delta\delta_S + b + d)(b - d)}\right\}. \tag{3}$$

In (3) a is the coefficient of section shape; s is the number of time interval; Q is the value of internal force; $b = -P_{S-1}$; $c = A_{S-1} + kQ$; $d = \sqrt{b^2 - 4ac}$; A_{S-1}, P_{S-1} are area and perimeter of a cross-section at the $(s - 1)$-th interval of time. The increment of depth of corrosion corresponding to the increment of stress is the solution of the equation:

$$A_{S-1} - P_{S-1} \cdot \Delta\delta_S + a \cdot \Delta\delta_S^2 = \frac{Q}{\sigma_{S-1} + \Delta\sigma_S}. \tag{4}$$

A parameter of the computational procedure is a number of equidistant nodal points on interval $[\sigma_0; \sigma *]$. In this case, the condition for the existence of numerical solution is satisfied for the entire domain of SDE parameters.

The algorithm described above allows us to obtain an approximate value of the durability of a structure. The error of this solution will obviously be different for all points of the solution space. An asymptotically exact solution of this problem can be obtained by increasing the number of nodal points, which leads to an increase in computational costs. In optimization problems, this approach is inefficient since the durability value (the constraint function) is calculated at each iteration of the search for an optimal project.

On the other hand, it is possible to get the exact value of durability knowing the approximate solution and its error. The idea of an algorithm proposed by the authors is to approximate the function of error of numerical solution using artificial neural networks.

3   Approximation of the Error of Numerical Solution Using Artificial Neural Networks

To construct the function of numerical solution error for the durability problem, it is necessary to know the reference solution of the problem, as well as the list of arguments on which the error of its numerical solution depends.

Let us dwell in more detail on the procedure for obtaining a reference solution of the durability problem.

If the internal forces in truss elements were constant ($Q = const$), then it is possible to solve the durability problem analytically:

$$t_i *= t_i - \frac{2kQ_i}{v_0 d_i} \ln \left\{ \frac{\left(P_i + d_i - 2a_i\delta_i\right)\left(P_i - d_i\right)}{\left(P_i - d_i - 2a_i\delta_i\right)\left(P_i + d_i\right)} \right\};$$
$$d_i = \sqrt{\left|4a_i\left(A_i + kQ_i\right) - P_i^2\right|}.$$
(5)

In reality, however, the internal forces in the elements of corroding truss change over time. If the rule for change of internal forces in the element $Q(t)$ is known, then it would be easy to obtain a reference (asymptotically exact) solution. However, this function is unknown.

Proceeding from the approximate solution of the durability problem for each of truss elements, it is possible to determine the number of the least durable element. Its failure time will determine the overall durability of a structure.

Solving numerically SDE (1) using the modified Euler method described above, it is possible to obtain information about the internal forces $Q(t_k)$ in a truss, where t_k are nodal points on the time interval. The data can be used to construct the function $Q = Q(t)$, for example, in the form of a polynomial $Q(t) = P_n(t)$. The degree of the polynomial will determine the number of node points t_k of the finite-difference template for the numerical solution of SDE.

Using a polynomial of third degree $Q(t) = P_3(t)$ to describe this dependence, we obtain a function of the form:

$$Q(t) = Q_0(1 + \alpha_1 t + \alpha_2 t^2 + \alpha_3 t^3).$$
(6)

To determine the unknown coefficients of the polynomial $\alpha_1, \alpha_2, \alpha_3$, it is necessary to know the values of the internal forces at four points, including the value Q_0 corresponding to the initial state of a structure.

Since the rod is in a uniaxial stress state, the stresses in it are determined by the formula:

$$\sigma(t, \delta) = \frac{Q(t)}{A(\delta(t))}.$$
(7)

The dependence $A = A(\delta(t))$ can be represented by the form

$$A(\delta(t)) = A_0 - P_0\delta(t) + a\delta^2(t). \tag{8}$$

To obtain a reference solution, it is sufficient to solve numerically a differential equation of the form

$$\frac{d\delta}{dt} = v_0(1 + k \cdot \frac{Q_0(1 + \alpha_1 t + \alpha_2 t^2 + \alpha_3 t^3)}{A(\delta)}). \tag{9}$$

Substituting the dependence (8), we get:

$$\frac{d\delta}{dt} = v_0(1 + k \cdot \frac{Q_0(1 + \alpha_1 t + \alpha_2 t^2 + \alpha_3 t^3)}{A_0 - P_0\delta(t) + s\delta^2(t)}). \tag{10}$$

The reference (asymptotically exact) solution of a given differential equation can be obtained numerically.

The error in the numerical solution will depend on the geometric characteristics of the rod, the initial stress o_0, the parameters of an aggressive medium, and the coefficients of the polynomial (6).

4   Architecture and Training of Neural Networks

To determine the error in the numerical solution of the differential equations, a neural network is used to approximate the dependence of the form:

$$F(\bar{x}, \bar{y}, \bar{\alpha}, \varepsilon *) = 0. \tag{11}$$

The choice of the architecture of neural network approximating the functional dependence (11) will be determined, first of all, by the number of parameters on which the error of solution will depend. In this section, an analysis of influence of various factors on the architecture of the neural network is proposed.

Durability represents the value of the time $t *$ at which the depth of corrosion reaches the limit value: $\delta(t *) = \delta *$.

The limit value is determined from the constraints on strength, continuity of section and stability:

$$\frac{Q}{A(\delta_1)} = [\sigma] \tag{12}$$

$$\delta_2 = \eta \cdot D \tag{13}$$

$$Q = \frac{\pi^2 EI_{\min}(\delta_3)}{L^2} \tag{14}$$

In (13) D is the thickness of the I-beam or structural channel; η is the coefficient (<0.5).

In case of the first two constraints, the solution of SDE will be affected by the area A^0 and perimeter P^0 of the cross-section, the corrosion rate v, the value of initial stress σ^0, and the coefficient of stress state impact on the corrosion rate k. With an active stability constraint, the minimum moment of inertia of the cross-section J^0 and the length of the element L that determines the value of critical stress of stability loss σ_{crit} are also important factors.

Considering these factors, the vector of design parameters can be represented in the form: $\bar{x} = \left(A^0, \; P^0, \; J^0, \sigma^0, \; \sigma_{crit} \right)$. The vector of parameters of an aggressive medium contains the following elements: $\bar{y} = (v_0, \; k)$.

The area, perimeter and moment of inertia of cross-section are determined through its geometric dimensions. Minimum cross-section moment of inertia depends, in addition to the dimensions, on the shape of a cross-section. For the same numerical values of the cross-sectional dimensions, the moment of inertia for different types of cross-sections will be different. The rules for determining other characteristics of a cross-section will also be different for each type of profile. Therefore, the matrices of synaptic weights must be obtained separately for each type of cross-section.

The number of factors influencing the error of solution of the problem depends on the type of active constraints determining the moment of destruction of a structural element. With active constraints on strength and continuity of a cross-section, the significant parameters for constructing the approximating function are A^0, P^0, σ^0, k. In case of active stability constraints, these parameters are A^0, P^0, J^0, σ^0, $\sigma *$, k. Matrices of synaptic weights must be obtained for each of the types of active constraints.

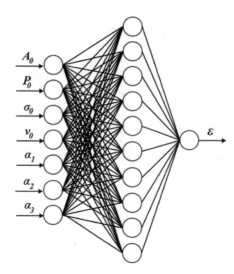

Fig. 1. Architecture of the neural network with active constraints on strength (continuity of cross-section)

The architecture of the neural network with active constraints on strength or continuity of cross-section, created taking into account the factors discussed above, is shown in Fig. 1. In the case of stability constraints, the additional input parameter is the initial critical stress of stability loss σ *.

According to Hecht-Nilsen theorem, any continuous real-valued mapping $f:[0,\ 1]^n \to R^m$ can be approximated to any degree of accuracy by a feedforward network with n input nodes, $2n + 1$ hidden units, and m output nodes [4]. The number of hidden units for the network shown in Fig. 1, sufficient to solve the problem, was chosen on the basis of the numerical experiments.

The training set consisted of the initial area and perimeter of the rod element, the value of initial stress, the value of critical stress (for active stability constraints), the corrosion rate, the coefficients of internal force variation function and the desired network output (error of numerical solution of the problem). Such sets were obtained for every profile type. Subsequently, artificial neural networks were trained using the backpropagation algorithm, described in detail, for example, in [5].

Let us assume that the neural network is trained, i.e. there is a function of the dependence between a SDE numerical solution error and rod parameters, the parameter of an aggressive medium, and the coefficients of a polynomial describing the change of internal forces in a rod over time. In general, the algorithm for solving the durability problem can be represented as follows:

Step 1.   An approximate value of durability t_{approx} is obtained using the numerical-analytical algorithm. The number of an element determining the durability of a structure is found.

Step 2.   A polynomial approximating the change in internal forces is constructed for this rod.

Step 3.   Using an artificial neural network, the error of the approximate solution of the durability problem is determined.

Step 4.   Using an approximate solution of the durability problem and its error, a more accurate solution is obtained:

$$t *= t_{approx}(1 - \varepsilon). \tag{15}$$

5   Numerical Experiment

To illustrate the proposed method, we are going to consider a statically indeterminate 5-rod truss as a model design (Fig. 2). The design parameters and parameters of aggressive medium were the following: $L = 100$ cm; $P = 200$ kN; $[\sigma] = 240$ MPa; $v_0 = 0.1$ cm/year; $k = 0.003$ MPa^{-1}.

The cross-sections of the elements correspond to standard angle profiles: (1) – $160 \times 100 \times 10$; (2) and (3) – $100 \times 63 \times 8$; (4) is $110 \times 110 \times 8$ and (5) is $180 \times 110 \times 12$.

To evaluate the error in the determination of corroding structure durability – which was obtained using the developed algorithm – a reference solution is required. To obtain it, the Euler method with recalculation is used. An error in the numerical solution of the problem will depend, first of all, not on the distance between the nodes of a time grid, but on the number of these nodes n. Therefore, for correct assignment of Δt it is necessary to have at least an approximate value of the durability of a structure.

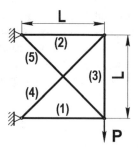

Fig. 2. A calculation scheme of the truss

This value $\tilde{t}$ can be obtained using formula (5), ignoring a change in forces in elements of a structure. The error of this solution can be very significant and will depend only on the degree of change in internal forces. Nevertheless, the approximate value of durability will allow us to choose the parameter $\Delta t = \dfrac{t}{n}$ reasonably and obtain an asymptotically exact solution.

The value $\tilde{t}$ is defined as $\tilde{t} = \min\{\tilde{t}_1; \tilde{t}_2; \ldots; \tilde{t}_5\}$, where $\tilde{t}_1; \tilde{t}_2; \ldots; \tilde{t}_5$ are the values of rod elements durability. In this case, this value was determined by the durability of the fourth element and amounted to 2.512 years.

The value of durability of the structure obtained at 200 nodal points $t_{ref} = 2.67845$ years was used as a reference value, as numerical experiments show no significant change in the result with further decrease in the number of nodal points.

Some results of calculations obtained in the process of solving the problem are presented in Table 1.

Table 1. The change in internal forces and stresses in the elements of the truss

t, years	Q_i, kN (σ_i, MPa)				
	(1)	(2)	(3)	(4)	(5)
0.0	144.69 (57.9)	55.16 (44.5)	55.14 (44.5)	77.90 (45.9)	205.01 (61.5)
1.0	148.71 (79.2)	51.10 (59.0)	51.07 (59.0)	72.13 (60.7)	210.82 (80.0)
2.0	156.52 (128.9)	43.18 (89.5)	43.14 (89.4)	60.89 (91.9)	222.17 (116.8)
2.5	164.77 (195.1)	34.80 (124.9)	34.77 (124.8)	49.03 (128.3)	234.19 (155.5)
2.66	168.79 (233.9)	30.71 (143.4)	30.69 (143.2)	43.24 (147.6)	240.08 (173.8)

As follows from the data in the table, in the elements (2), (3) and (4) the internal forces decrease, and in the elements (1) and (5) they increase. The durability of the structure will be determined by the first element.

To approximate the force function $Q(t)$, it is necessary to determine force values at four points. The SDE was solved using four nodal points. As a result, the following values of coefficients of the polynomial (6) were obtained: $\alpha_1 = 0.01671$, $\alpha_2 = -0.04849$, $\alpha_3 = 0.06833$. The obtained value of durability was 2.90448 years.

At the next stage of the problem solution, the error in the numerical solution was determined using the previously trained ANN, and the value of durability was corrected using this error. The error of the obtained solution of the durability determination problem, according to the result of work of the neural network, was $\varepsilon_{net} = 8.05\%$. Using the approximate solution of the durability problem and its error, a more accurate solution $t* = 2.67067$ was obtained on the basis of formula (15). The relative error of this solution compared to the reference one was $\varepsilon = 0.29\%$. In the process of solving the durability problem, the problem of the stress state was solved five times, i.e. the computational costs decreased by more than 40 times.

The developed algorithm was also tested for other designs. Table 2 shows the results of solving the durability problem for a five-rod truss with elements of the same type.

Table 2. The results of solving the durability problem

Number of a problem	t_{et}, years	t_{approx}, years	ε_{net}, %	$t*$, years	ε, %
1	2.32173	2.20845	5.71	2.33454	0.55
2	2.59140	2.73454	6.01	2.57019	0.81
3	1.57270	1.50232	4.95	1.57668	0.25
4	1.66593	1.69516	2.03	1.66075	0.31
5	1.83823	1.87321	1.75	1.84042	0.11

In the first column of Table 2 the numbers of problems correspond to the next designs:

- problem 1 – all elements – angle beam 125 × 125 × 9;
- problem 2 – all elements – angle beam 140 × 90 × 10;
- problem 3 – all elements – I-beam 160 × 81;
- problem 4 – all elements – channel beam 180 × 70;
- problem 5 – all elements – channel beam 200 × 76.

It follows from the given data that the algorithm's efficiency does not depend on the shape and size of cross-sections of rod elements. The error in the value relative to the reference solution of the problem is determined only by the error of approximating polynomial and the training error of a neural network. Both these errors can be reduced to acceptable values, i.e. they are disposable.

6 Conclusions

With the help of the algorithm, presented in this paper, it is possible to determine the durability of a structure with a sufficiently high accuracy. At the same time, computational costs are significantly lower than for other known methods. This method can be especially effective for optimization problems with constraints on the structural durability [6–9].

The results of problem solution demonstrate that the synthesis of methods of computational intelligence (artificial neural networks) with known numerical methods is a promising approach that can be generalized to other classes of structures.

References

1. Oladipo, B.A., Ajide, O.O., Monyei, C.G.: Corrosion assessment of some buried metal pipes using neural network algorithm. Int. J. Eng. Manuf. (IJEM) **7**(6), 27–42 (2017). https://doi.org/10.5815/ijem.2017.06.03
2. Korotka, L.I.: Use of neural networks in numerical solution of some systems of differential equations. East. Eur. J. Enterp. Technol. **3/4**(51), 24–27 (2011)
3. Zelentsov, D.G., Lyashenko, O.A., Naumenko, N.Y.: Information support for calculation of corroding objects. In: Mathematical Models and Conception of System Design. Dnipropetrovsk, USUCT (2012)
4. Du, K.-L., Swamy, M.N.S.: Neural Networks and Statistical Learning. Enjoyor Labs, Enjoyor Inc., China (2014). https://doi.org/10.1007/978-1-4471-5571-3
5. Callan, R.: The Essence of Neural Networks. Prentice Hall Europe, London (1999)
6. Webb, D., Alobaidi, W., Sandgern, E.: Structural design via genetic optimization. Modern Mech. Eng. **7**(3), 73–90 (2017)
7. Assimi, H., Jamali, A.: Sizing and topology optimization of truss structures using genetic programming. Swarm Evolut. Comput. **37**, 90–103 (2017)
8. Dey, S., Roy, T.K.: Optimized solution of two bar truss design using intuitionistic fuzzy optimization technique. Int. J. Inf. Eng. Electron. Bus. (IJIEEB) **6**(4), 45–51 (2014). https://doi.org/10.5815/ijieeb.2014.04.07
9. Dey, S., Roy, T.K.: Multi-objective structural optimization using fuzzy and intuitionistic fuzzy optimization technique. Int. J. Intell. Syst. Appl. (IJISA) **7**(5), 57–65 (2015). https://doi.org/10.5815/ijisa.2015.05.08

The Study of Visual Self-adaptive Controlled MeanShift Algorithm

P. H. Wu[1], G. Q. Hu[1], and D. Wang[2(✉)]

[1] South China University of Technology, Guangzhou, China
pianhui@hotmail.com, gqhu@scut.edc.cn
[2] Guangzhou Huashang Vocational College, Guangzhou, China
wd000111@hotmail.com

Abstract. This paper proposed a self-adaptive visual control system which is controlled by human eyes, the visual image tracking algorithm utilized by this system is also introduced in this paper. Through eye-gaze detection and electrical device control corresponding, it will automatically respond to the provided interface. This paper mainly introduces helmets and remote vision-based eye-gaze tracking algorithms; the algorithm has good performance in aspects of usability and adaptability.

Keywords: Eye tracking · Eye-gaze · Helmet · MeanShift algorithm

1 Introduction

A survey of 2009 revealed that there were millions of patients suffered by ALS (Amyotrophic Lateral Sclerosis), MND (Motor Neurone Disease), SMA (Spinal Muscular Atrophy), TS (Tourette Syndrome), stroke and traumatic brain injury and etc. Those diseases made human brains degraded or even dysfunction, a patient's ability to move is decreased. To help those patients, many research institutions and hospitals have started research work about vision-based eye-gaze tracking system. As a result of research work, prof. David Wooding (visual science center, University of Derby) demonstrated an eye-gaze tracking experiment in British art gallery in October, 2000. In which, eye tracking device was provided by ASL (Applied Science-Laboratories) of the U.S. Intelligently processing images captured by a camera to achieve the same affect that a human brain processes images captured by eyes is one of important subjects in electrical and computer science fields, and it has been developed to members of neighboring scientific research fields [1]. This paper mainly researches human psychological activity through human eye movement, combining with MeanShift algorithm, this visual tracking system is faster and more effective [2]. Kernel-based object tracking theory proposed by Comaniciu in 2003 is simple to compute and has certain self-adaptivity in aspects of target's noise and scale change [3].

© Springer International Publishing AG, part of Springer Nature 2019
Z. Hu et al. (Eds.): ICCSEEA 2018, AISC 754, pp. 517–528, 2019.
https://doi.org/10.1007/978-3-319-91008-6_52

2 Control System of Eye Gaze

Eye Gaze Communication System developed by LG technology company of the United States utilized eye tracking device to help those disables, their eyes are used to operate keyboard instead of their hands. An operator watches a key or a subject displayed on the screen to give an order of 'press it' or 'select it', the computer will take the order and perform the corresponding action, this is one of functions of eye tracking device. To increase usability of the device, it needs to design and add a user-centric control component. A user's eyes gaze a three-dimensional environment control device, three-dimensional control device is a special customized device which can provide selection function of simple responding interface. At present, as user's selection menu, input control of a screen operation system is implemented by eye gazing, interface control drive responding is controlled by eye gazing, input control is usually one-way control, such a method can effectively reduce operating errors, and allows more detailed control options, this can increase the environment of conscious and self-adaptive control. The self-adaptive algorithm is based on MeanShift algorithm.

HEGTS (Head-mounted Eye Gaze Test System) is deigned to achieve better experimental environment for eye-gazing control. HEGTS consists of 5 components: eye tracking device, scenario monitor, controllable electrical device, screen with operation menu, Wi-Fi controller.

Eye movement tracking determines the position and gazing direction of user's eyeball (pupil center), which are the calculating result of corneal reflection data collected by a sensor that is watching user's eye. To ensure eye track of high precision, the system employs high-speed and high-resolution camera in a three-dimensional space, a computer calculates collected eye images. The system has achieved precise positioning for eyeball and automatic track for eye gazing position. High precision for target tracking is ensured by head movement compensation algorithm. The system broke the limit of cap and head care, a user wearing glasses can use it, even there is no need to reposition target when the user come back and use the system again. This extended the scope of psychology and visual research. The experiment is taken place in the natural state, which reflects real psychological state, it can continuously determine user's eyeball movement and rationally analyze scenario of eye gazing under any condition. Besides, it can identify and locate the controllable electrical device [4].

There are two eye track devices containing in the system, the first device provides the whole movement track of user's head and eye. In contrast, the second device can easily detect user's movement state because it is fixed outside of the helmet. The sensor of external equipment or electrical device can fast and effectively detect service condition of every time. Gaze tracking system technical process shown in the flow diagram is shown below (Fig. 1).

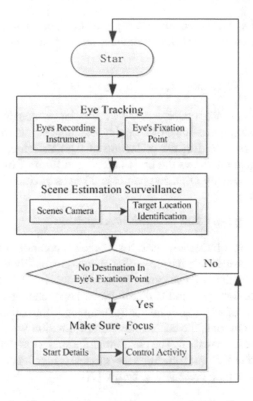

Fig. 1. Gaze tracking system response technique flowchart

The system decides whether take any action for a fixed location where eyes are gazing. Thus, the system is especially suitable to select specific device and control it. Compared with other eye tracking algorithms, MeanShift algorithm is a pattern matching algorithm with high efficiency. Simple theory and better robustness make MeanShift algorithm has been applied to some subject tracking technology of sensitive real-time [5]. However, there are a few defects of MeanShift algorithm, it sensitively dependent on the contrast between subject and background. The defects are possible to be overcome if three-dimensional environment is applied to the system.

3   The Physical Level of MeanShift Algorithm

Formula (1) presents: in an a-dimensional space $\mathbb{R}^d$, sample point x_i (i = 1…n,) for x point, the basic form definition of MeanShift vector value is:

$$M_h(x) \equiv \frac{1}{k} \sum_{x_i \in S_h} (x_i - x) \tag{1}$$

In Formula (1), k presents the number of points dropped to S_h area, S_h is the area of high dimensional area with radio h, it is the collection of y point which meet the following request.

$$S_h(x) = \{y:(y-x)^T(y-x) \le h^2\} \tag{2}$$

Formula (2) presents the offset vector to the point, which is the offset value of summarizing and taking the average of dropping k sample points to S_h area, S_h is a high dimensional area of radio h, it is the collection of y points which meet the following requirement. The equation shows the basic calculation form of eye-gazing algorithm, the area of maximum density of gazing points is taken as gazing direction.

3.1   Target Tracking Based on MeanShift Algorithm

MeanShift algorithm can be widely used for image segmentation and target tracking, Dorin Comaniciu proposed to define the similarity measure by Bhattacharyya coefficient [6]. Bhattacharyya coefficient, also called Papier distance, is usually used to measure the similarity of continuous probability distribution of two relative subjects. The process of target identification of MeanShift image tracking algorithm utilizes calculated mean vector to do repeat iterations, the central position of function window is updated until the judgment condition is satisfied. The position of the subject and pixel distribution are two factors to describe this subject. Suppose subject center located at x_0, the feature model of this subject can be presented as below [7]:

$$\hat{q}_u = C \sum_{i=1}^{n} k\left(\left\|x_i^*\right\|^2\right) \delta\left[b(x_i^*) - u\right] \tag{3}$$

Formula (3) $u \in \{1, \cdots m\}$ is set to capture probability, k represents the contents of the unit, in function b, object requires pixel reflection b is limited in the scope. In alternative area y, subject position feature vector can be described as below:

$$\hat{p}(y) = C_h \sum_{i=1}^{n} k\left(\left\|\frac{y - x_i}{h}\right\|^2\right) \delta\left[b(x_i) - u\right] \tag{4}$$

In the Eqs. (3) and (4), δ is Leopold Kronecker symbol (Kronecker Delta) and C is the normalization constant, so

$$\sum_{u=1}^{m} p(y) = 1 \tag{5}$$

In formula (5), subject tracking can be simplified to seek the best y value, make $\hat{p}u(y)$ similar to $\hat{q}u(y)$. The similarity of $\hat{p}u(y)$ to $\hat{q}u(y)$ can be measured by Bhattacharyya coefficient [7].

$$\hat{\rho}(y) \equiv \rho\left[p(y), q_u\right] \tag{6}$$

The γ_{i+1} in formula (7) is variation degree of related coefficient:

$$\gamma_{i+1} = \frac{\rho[p(y_{i+1}), q_u]}{\rho[p(y_i), q_u]} \tag{7}$$

Get the maximum Bhattacharyya coefficient.

$$\hat{\rho}(y) \equiv \rho[p(y), q_u] + \frac{1}{2} \sum_{u=1}^{m} p_u(y) \sqrt{\frac{q_u}{p_u(y_0)}} \tag{8}$$

The function selection of Bhattacharyya coefficient is between 1 and 1, $\hat{p}_u$ for (y), the bigger value, the more matching degree that selected target matches to standard target, which suggests the possibility of canter y being tracked target is much higher [8].

MeanShift tracking algorithm uses histograms of kernel functions to model the target, in the process of computer implementation, it needs to determine feature region of the target, and then implements match tracking [9]. MeanShift algorithm needs less parameters, its robustness is better and can fast implement mode calculation. It therefore has good real-time performance [8].

Kernel function histogram is weak in describing target's feature, if the target moves out of template's range, the system will not achieve satisfied tracking result due to the lack of space information. Because the MeanShift algorithm is a kind of ladder-type algorithms, it is prone to occur regional maximum value and slow convergence speed when the target tracking speed becomes fast. The target movement state is in fact disordered (i.e. scale rotate), the window width of simplifying kernel function calculation keeps constant, it is not well adaptive to track the target under constant change. The system cannot solve the problem that the target is covered. To overcome the weaknesses of the MeanShift algorithm mentioned above, many researchers have done a lot of research work and have proposed corresponding improvement methods, some of proposed methods have been proven effective by related experiments.

In summary, the MeanShift algorithm is suitable for static probability distribution, computing color histogram in HSV space is a good tracking algorithm.

3.2 Improved Method and Application

Many researchers have contributed their research work to improve the MeanShift algorithm because it has good characteristics including faster operation, good robustness and resistance against reforming and covering target. This paper mainly study image tracking method based on the MeanShift algorithm and proposes some improvement.

When weighted gray histogram of the target model area is established, the image pixels of tracked target affect the distribution of target's gray histogram. When target's model area contains less background area pixels, traditional MeanShift tracking algorithm can achieve batter tracking performance; when target's model area contains more background image pixels or background is constantly changing, traditional method is prone to occur best matching position errors between candidate regional histogram and target model area histogram [6]. For tracking the target of fast moving, it is prone to loss the target.

In contrast, if certain feature value is the weight of the smaller in the background gray histogram, this gray feature value will be given the weight of the larger when the target model and background model are established. This method can effectively reduce the background pixels impact on the target tracking. It improves tracking precision of MeanShift algorithm, the implementation steps are introduced below:

Formula (9) calculates current weight coefficient of each pixel [5]:

$$w_i = \sum_{u=1}^{m} \sqrt{\frac{q_u}{p_u(y_0)}} \delta[b(y_i) - u] \tag{9}$$

Formula (10) calculates the new position of the target [5]:

$$y_i = \frac{\sum_{i=1}^{n} x_i w_i g\left(\left\|\frac{y_0 - y_i}{h}\right\|^2\right)}{\sum_{i=1}^{n} w_i g\left(\left\|\frac{y_0 - y_i}{h}\right\|^2\right)} \tag{10}$$

$g(x) = -k'(x)$

How to determine whether it continues to calculate. It firstly determines whether $\|y_i - y_0\| < \varepsilon$ is true. If not, the calculation will stop, the position information of tracked target will be obtained; If it is true, use y_i to replace y_0 and return the first step of target positioning process, it will continue to seek candidate target position which can meet judgment condition. Because photoelectric tracking system has a high claim for real-time capability, the maximum number of iterations set practically should be n [9]. MeanShift is in nature a self-adaptive method with gradient ascent to search peak value. It is shown below, if data collection $\{x_i, i = 1, \dots n\}$ complies with probability density function $f(x)$, define a_m initial point x [10], MeanShift algorithm will trace the move of the first step, it will be finally constricted to the first peak point [11].

MeanShift replaces neighboring value which distance is K-nearst what self-adaptive bandwidth Bayesian is being used [14]. This algorithm can be applied to single or multiple values (multi-modal merge), it is suitable to segment different images. Shown as formula (1), it can get Mh(x) if the sample point of Sh is found. Thus, once K-nearst value is used, the system can find all x related values when Mh is being calculated. This algorithm is normalized probability density gradient of Mh(x), it can aggregate to get the stable point of probability density function [16]. It can get better resolution and can assess the performance of image edge quality that MeanShift adds K-nearst. In the mean time, the experiment uses Kalman filter and gets effective outcomes.

4   Experiment Results and Analysis

Experimental system adopts eye-ball movement recorder, which the tester carries. In this device, tester's head, the recorder and live camera and in constant condition. In practice, it is hard to precisely locate previous position when the system tracks user's head position every time. Employing ASL makes such a calibration much easier to perform.

Sight line movement of the tester is detected by infrared ray detection technique, the system extracts and calculates changes of both eye pupils and eye gazing time in a certain scope, image formation is made on tester's corneal facet by infrared ray source. Corneal reflection center (Pirkinje) is referenced point (two-dimensional vector between Pirkinje point and pupil center is called corneal reflection vector P-CR), convert P-CR vector to screen coordinate vector X_s, it is shown as formula (11) [14]:

$$X_s = A \cdot X_e \tag{11}$$

Gazing point in formula (11) is screen coordinate system. $X_e = (x_e, y_e)^T$ Represents the coordinate vector P-CR of camera (eye) image coordinate system. Use matrix A as mapping relation between image coordinate and screen coordinate, the mapping relation can be expressed by the following quadratic polynomial (12) [15]:

$$
\begin{aligned}
X_s &= a_0 + a_1 x_e + a_2 y_e + a_3 x_e y_e + a_4 x_e^2 + a_5 y_e^2 \\
Y_s &= a_6 + a_7 x_e + a_8 y_e + a_9 x_e y_e + a_{10} x_e^2 + a_{11} y_e^2
\end{aligned}
\tag{12}
$$

In Fig. 2, there are 9 coordinate vector of gazing point in screen coordinate system, which is responded by 9 P-CR vectors in eye coordinate, this can get 18 overdetermined equations and multiple corresponding coefficients ($a_0 a_1 \cdots a_{11}$).

Fig. 2. Gaze tracking system of hardware

This experiment implements a 3-dimensional coordinate rebuild of pixels; the target picture shown in Fig. 2 which is captured by a 10-million-pixel camera. Through a series of technical processing such as image noise reduction, image expansion and erosion, image edge detection processing technique, there are 9 points captured by the coordinate in the image, the tester can see them on the screen, use 260×240 as the record of each point, 2 groups of data measured are output result of eye movement tracking controller.

Calibration time is usually 30 s, to complete a successful calibration, it firstly needs to real-time track the scenario where the eye is gazing, and then combines the target saved in the computer in advance with the real coordinate in the scenario where the eye is gazing.

Thus, the target which is needed to display enhancing information that can be obtained. Based on the current target which the eye is gazing, virtual enhancing information which matches to real scenario is obtained, the computer outputs real eye-gazing point in real-time 3-dimensional special environment. The theory is shown in Fig. 3.

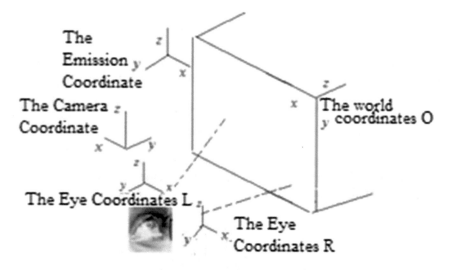

Fig. 3. 3D coordinate system for **gaze** tracking system

Screen coordinate O: the coordinate system on computer screen The original point O is defined at right-upper corner of the screen, set the line starting from the original point to the left as X axis, set the line starting from the original point vertical down as Y axis, set Z axis perpendicular to the surface XOY and meet the rules of right-handed coordinate system. The sensor coordinate of position tracking system is S, the coordinate system of the launcher is B, eye coordinate system is separately set as L and R.

Put the red line of the launcher point at any region on the board, and set (x_i, y_i) as initial position, the vector is P_i. When the red point moves to another position (x_{i+1}, y_{i+1}), the vector of the next prediction point is P_{ij}. Prediction formula (14) is shown as below:

$$d_{ij} = \sqrt{\left(x_i - x_{i+1}\right)^2 + \left(y_j - y_{j+1}\right)^2} \tag{13}$$

$$P_{ij} = \sqrt{\sum_{k=1}^{n} f_k (p_{ik} - p_{jk})^2} \tag{14}$$

d_{ij} and p_{ij} are movement vectors of pixels, f_k is the frequency of weighted pixel value. The difference distance between d_{ij} and p_{ij} is called the compensation value, if the value equals to 0, then D equals to 0, their position is the central position, use formula (15) to express D:

$$D = d_{ij} - p_{ij} \pm off \cdot set \tag{15}$$

Detected prediction red point (D value) must be greater than or equal to zero, which means the red point roughly locates at prediction point, it needs the position through setting offset value to precisely specify. These pixels can be aggregated by the MeanShift algorithm, use formula (16) to express:

$$PC_{ij} = \begin{cases} 1, & if\ D_{ij} \geq 0 \\ 0, & otherwise \end{cases} \tag{16}$$

Experimental steps:

Firstly, it needs to trace the position in real time that the user is gazing in a real scenario, interested area and specific target of the user can be analyzed by determining the real coordinate of gazing point in the scenario and the scenario pictures captured by the camera, it then can obtain the object which is required to display enhancing information.

(1) The system performs target mode identification for the images captured by the camera, the images are preprocessed by a filter, in this experiment system, most of collected pictures are disturbed by many factors, it therefore needs to do smoothing processing, during the process, the image quality is reduced, the image becomes blurred. Kalman filter is employed to use for this experiment to improve image display quality [14].

(2) The system captures images to identify the target's mode. It firstly uses a filter to preprocess, many images captured in this experiment are interfered by some factors, thus, it must do smooth process for input images. During the smooth process, image's quality is reduced to be not clear. To achieve better image's quality, this experiment adopts Kalman filter get rid of image's noise [18].

(3) To predict the position of next move of eye gazing, the next frame takes previous frame as prediction standard, picture 7 shows the prediction position of infrared light point (x_{i+1}, y_{i+1}), (x_{i+2}, y_{i+2}) and calculate their error value.

(4) Figure 4 shows calibrated data coordinates (red cross) seen by the eyes, surrounding black points represents the error coordinate obtained after calibrating results. In the experiment, the samples of 2 data points are selected randomly to calculate the distance threshold value t and the amount of points of ellipse parameters. Thus, it can get known of the model distance between two points and probability distribution, such a method is called ellipse fitting, the fitting inspects the target from 2

aspects, one is the target's edge, the other one is the options of determining threshold value in the center of ellipse.

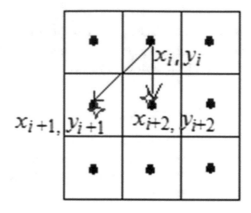

Fig. 4. Infrared light prediction calculation

(5) In Fig. 5, Red point is set as the real coordinate of the calibrated point, blue point is set as the calibrated point coordinate obtained by using calibrated results to calculate. The basic principle of the inspection is to use the vector of matrix X(m × n) to describe, in which there is a non-linear parameters (a1,···,ap) in n basis function, both linear parameters and non-linear parameters are obtained from the following formula (17).

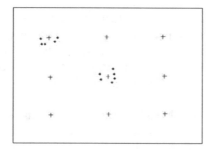

Fig. 5. The calibration data from the eyes

$$y(t) \approx \beta_1 \phi_1(t, a) + \cdots \beta_n \phi_n(t, a) \tag{17}$$

The method is to find reflection coefficient between the image coordinate of sight line recorder and screen coordinate, which can finally get the system vector of visual recorder coordinate. $(X_i, Y_i)^T$ expresses eye-gazing point vector in the screen coordinate system. Optimized function R is shown below:

$$R^2 = \sum_{i=1}^{n} \left[X_i - (a_0 + a_i x_i + a_2 y_i + a_3 x_i y_i + a_4 x_i^2 + a_5 y_i^2) \right] \tag{18}$$

The residual value R^2 tends to zero, it separately seeks partial derivative for a_j and make the value of a_j equal to 0.

$$\frac{\partial R^2}{\partial a_j} = -2 \sum_{i=1}^{n} \left[X_i - (a_0 + a_i x_i + a_2 y_i + a_3 x_i y_i + a_4 x_i^2 + a_5 y_i^2) \right] = 0 \tag{19}$$

Sight line recorder monitors screen target point and collects 5 groups of data, in which a_j $(j = 1, 2, 3, 4, 5)$. The error is calculated by formula (20), the width of the monitor display pixels is 0.27 mm.

$$Error = \frac{1}{N} \sum_{i=1}^{N} 0.27 \times \left(\sqrt{(X_i + x_i)^2 + (Y_i + x_i)^2} \right) \tag{20}$$

(X_i, Y_i) is the real coordinate of the target point, (x_i, y_i) is the target coordinate obtained by calibrated calculation, N is the number of either calibration point group or tracing date. Through a series of calculations, the calibration error of sight line recorder is 3.7254 mm, the average error of real-time tracking is 8.7871 mm.

5   Conclusion

This paper introduced HEGTS system as experimental environment, MeanShift algorithm and its improvement method are proposed for eye gazing tracking system.

MeanShift algorithm adopts the weighted probability distribution $(q = \{q_u\})$ of feature value to target modeling, and then matches tracking algorithm, MeanShift algorithm uses less parameters and simplifies the tracking actions to find the best similarity between $p_u(y)$ and $\hat{q}_u(x)$, the algorithm implements ellipse fitting for feature point and utilizes comparison method to inspect such a fitting by target feature model, which greatly increases precision.

To reduce dependency of contrast of the target and the background, it employs background weighted method for target model and candidate model and changes gray value to reduce the impact of the target tracking, such a method enables the system to be stable, precise and has good capability of resisting disturbance. Thus, in the area of visual self-adaptive control technology, this system has good application future, it provides an effective way for bionics study. Currently, visual self-adaptive control system is useful in aspects of assisting brain injured patients, industrial automated control and etc. there is a lot of research work in the future.

Compared to normal object tracking algorithms, the object tracking system proposed in this paper combines with Kalman filter to calculate and predict position of eye gazing. Therefore, tracking precision, time and performance achieved by this system is much better than others. There is board application prospect with great value for this object tracking system.

References

1. Wooding, D.S., Mugglestone, M.D., Purdy, K.J., Gale, A.G.: Eye movements of large populations: I. implementation and performance of an autonomous public eye tracker. Behav. Res. Methods Instrum. Comput. **34**(4), 509–517 (2002)
2. Comaniciu, D., Meer, P.: MeanShift: a robust approach toward feature space analysis. IEEE Trans. Pattern Anal. Mach. Intell. **24**(5), 603–619 (2002)
3. Comaniciu, D., Ramesh, V., Meer, P.: Kernel-based object tracking. IEEE Trans. Pattern Anal. Mach. Intell. **25**(5), 564–575 (2003)
4. Javed, A., Aslam, Z.: An intelligent alarm based visual eye tracking algorithm for cheating free examination system. Int. J. Intell. Syst. Appl. (IJISA) **5**(10), 86–92 (2013). https://doi.org/10.5815/ijisa.2013.10.11
5. Comaniciu, D., Ramesh, V., Meer, P.: Real-time tracking of non-rigid objects using MeanShift, pp. 142–149 (2000)
6. Ning, J., Zhang, L., Zhang, D., Wu, C.: Robust mean-shift tracking with corrected background-weighted histogram. Comput. Vis. IET **6**(1), 62–69 (2012)
7. Kailath, T.: The divergence and bhattacharyya distance measures in signal selection. IEEE Trans. Commun. Technol. **15**, 52–60 (1967)
8. Lipton, A.J., Fujiyoshi, H., Patil, R.S.: Moving target classification and tracking from real-time video. In: IEEE Workshop on Applications of Computer Vision, Princeton, pp. 8–14 (1998)
9. Pu, X., Zhou, Z.: A more robust MeanShift tracker on joint color-CLTP histogram. Int. J. Image Graph. Signal Process. (IJIGSP) **4**(12), 34–42 (2012). https://doi.org/10.5815/ijigsp.2012.12.05
10. McKenna, S.J., Raja, Y., Gong, S.: Tracking color objects using adaptive mixture models. Image Vis. Comput. **17**, 223–229 (1999)
11. Paragios, N., Deriche, R.: Geodesic active regions for motion estimation and tracking. In: IEEE International Conference on Computer Vision, Kerkyra, Greece, pp. 688–674 (1999)
12. Fitzgibbon, A.W., Pilu, M., Fisher, R.B.: Direct least squares fitting of ellipses. IEEE Trans. PAMI **21**, 476–480 (1999)
13. Snekha, S.C., Birok, R., et al.: Real time object tracking using different MeanShift techniques–a review. Int. J. Soft Comput. Eng. (IJSCE) **3**(3), 98–102 (2013). ISSN 2231-2307
14. Jatoth, R.K., Gopisetty, S., Hussain, M.: Performance analysis of Alpha Beta filter, Kalman filter and MeanShift for object tracking in video sequences. Int. J. Image Graph. Signal Process. (IJIGSP) **7**(3), 24–30 (2015). https://doi.org/10.5815/ijigsp.2015.03.04
15. Kim, P., Chang, H., Song, D., et al.: Fast support vector data description using k-means clustering. In: Advances in Neural Networks, pp. 506–514 (2007)
16. Dong, Y., Jae, K., Bang Rae, L., et al.: Non-contact eye gaze tracking system by mapping of corneal reflections. In: Proceeding of the Fifth IEEE International Conference on Automatic Face and Gesture Recognition, vol. 5, pp. 94–99 (2002)
17. Jahangir Alam, S.M.: Based on Arithmetic Study of Image Processing and Recognition for Mosquito Detecting and Position Tracking. Xiamen University (2014)
18. Mallikarjuna Rao, G., Satyanarayana, C.: Object tracking system using approximate median filter, Kalman filter and dynamic template matching. Int. J. Intell. Syst. Appl. (IJISA) **6**(5), 83–89 (2014). https://doi.org/10.5815/ijisa.2014.05.09

Computer Science for Medicine and Biology

Method for Research of the Human Static Equilibrium Function

Yurii Onykiienko[(✉)] [iD]

National Aviation University, Kyiv 03058, Ukraine
yurii.onykiienko@gmail.com

Abstract. The paper deals with problems of determination of the informative indices, which characterize the function of the human balance (human static equilibrium). The kefalographic method for research of the human static equilibrium is suggested. The kefalographic plant for this method implementation was modified. The informative indices, which characterize the space dynamic range and features of the human body oscillations relative to the axis z, were determined. Such indices represent the coefficients, which characterize changes of the sampling mathematical expectation $K_{\tilde{m}_r}$, variance $K_{\tilde{D}_r}$, skewness $K_{\tilde{a}_r}$ and kurtosis $K_{\tilde{e}_r}$ for the vector projection of central position the human body.

Keywords: Kefalography · Informative indices · Statistic characteristics
Human static equilibrium · Distribution law · Human extreme activity

1 Introduction

Specialists working in extreme situations may have problems with the state of health depending on the kind of their professional activity. This is connected with the low human ability to adaptation, which had not been timely detected at the stage of the professional selection. Usually, such specialists are the most sensitive to the negative influence of environment. Moreover, they are able to the disadaptive and desynchronous disorders, which are accompanied by symptoms of depression, stress, local and generalized hypoxia, oxidative stress and so on.

Diagnostics of the human dysfunctional states is extraordinarily complicated by absence of the specific criteria and signs, which may be registered by means of the traditional biomedical methods and means. Imperfection of the methods for the human psychophysiological state assessment increases the possibility of unforeseen dysfunction development and loss of working capacity. This may lead to disruption of the scheduled work or reduction of its efficiency.

So, search of the additional criteria for assessment of the psychophysiological state of specialists working in extreme situations at stages of professional selection and professional activity is of great importance. Now, the biotechnical systems for assessment of the human psychophysiological state and early diagnosis of the adaptation possibilities of extreme activity specialists are not available both in Ukraine and in the world. Taking this situation into consideration, the method of determination of the information indices, which characterize the function of the human static equilibrium,

© Springer International Publishing AG, part of Springer Nature 2019
Z. Hu et al. (Eds.): ICCSEA 2018, AISC 754, pp. 531–539, 2019.
https://doi.org/10.1007/978-3-319-91008-6_53

was developed. Basic results of this development are given in the paper. The function of the human static equilibrium represents an additional criterion of assessment of the psychophysiological state of specialists working in extreme situations.

2   Method of Research of Human Static Equilibrium

Problems of assessment of static equilibrium function are discussed in many papers of both domestic [1–3] and foreign [4–12] researchers. Methods represented in these papers (for example, stabilometry) are directed for usage in the clinical conditions and require using of the specialized equipment in addition to the high qualification of a researcher. It is sufficiently difficult to use such equipment in conditions of extreme situations.

The method of research of the static equilibrium function represented in the paper is based on the improved modification of the kefalograph plant. The structural scheme of this plant is represented in Fig. 1 [13, 14].

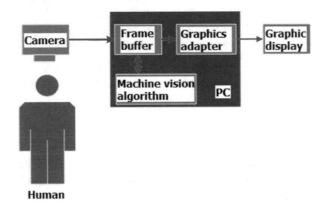

Fig. 1. The structural scheme of the modernized kefalographic plant.

In contrast to construction of available kefalograph plants [1–3], this modification differs by absence of the mechanical connection between a man to be tested and an apparatus for his motion registration. This gives the possibility to avoid distortions in the resulting measurements and to improve the measuring accuracy due to usage of the camera for the human motion registration.

The method of research by means of the modernized plant includes following steps. The little black cap made of the textile fabric is dressed on the head of the specialist to be tested. The white marker, which represents a 10×10 mm square, is lying outside of the cap. A tested man stands with his outstretched arms (the Romberg pose) at the special platform [14].

The video camera is attached to the stand (auxanometer) at the given fixed distance (25 cm) from the head in the vertical direction. The necessary distance is determined in an experimental way.

Research of the marker oscillations is carried out in the plane relative to the central position of the body of a tested man during 2 min in two modes with open and closed eyes respectively. An image obtained by means of the camera is transmitted to a computer, which carries out processing of measured information in the real-time mode. As a result of this processing, the vector projection of the marker displacement relative to the central position of the body of a tested specialist is represented at the graphic display.

This projection looks like a two-dimensional drawing (kefalogram) of the individual oscillations due to the resultant activity of the static and statokinetic reflexes. Examples of kefalograms, which characterize the function of a human equilibrium in modes with open and closed eyes changing in time, are represented in Figs. 2 and 3.

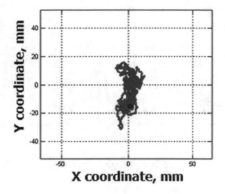

Fig. 2. The vector projection of the marker displacement relative to the central position of the body of a tested specialist without pathology of the vestibular apparatus.

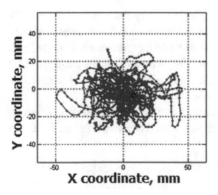

Fig. 3. The vector projection of the marker displacement relative to the central position of the body of a tested specialist with pathology of the vestibular apparatus.

The quantitative indices, which characterize the function of the static equilibrium of a tested man, may be determined by means of further processing of kefalograms. This allows to assess the state of health of the extreme activity specialists.

3   Method of Static Equilibrium Function Assessment

In fact the suggested method represents an approach to the analysis of the vector projection of the marker displacement r (radius) relative to the axes OX and OY during a specified research time (see Fig. 4).

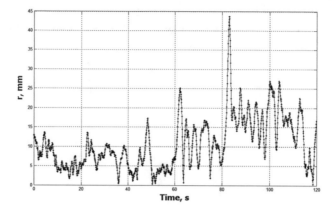

Fig. 4. The graph of changing of the vector projection r (radius) of the marker displacement in time.

Determination of the information indices, which characterize the function of the static equilibrium of a tested specialist, includes the following stages:

- calculation of the sampling (point) parameters of the distribution of the displacement vector projection r for both test modes (with open and closed eyes);
- determination of the confidence intervals for calculated sampling (point) parameters of the distribution of the displacement vector projection for both test modes;
- choice of the theoretical and empiric laws of the distribution of the displacement vector projection for both test modes;
- calculation of indices, which characterize the determined distribution law;
- calculation of coefficients, which characterize the function of the static equilibrium of a tested specialist.

Some of above mentioned stages it is expedient to analyze in more details. It is proposed to analyze the quality of the researched system by means of the distribution sample parameters (point estimates) such as the sampling mathematical expectation $\tilde{m}_r$, variance $\tilde{D}_r$, root-mean-square deviation $\tilde{\sigma}_r$, skewness $\tilde{a}_r$ and kurtosis $\tilde{e}_r$, which may be calculated by means of the expressions [15–19].

$$\tilde{m}_r = \frac{1}{n} \sum_{i=1}^{n} r_i, \tag{1}$$

$$\tilde{D}_r = \frac{1}{n-1} \sum_{i=1}^{n} \left(r_i - \tilde{m}_r\right)^2, \tag{2}$$

$$\tilde{\sigma}_r = \sqrt{\tilde{D}_r}, \tag{3}$$

$$\tilde{a}_r = \frac{1}{(n-1)\tilde{\sigma}_r^3} \sum_{i=1}^{n} \left(r_i - \tilde{m}_r\right)^3, \tag{4}$$

$$\ddot{e}_r = \frac{1}{(n-1)\tilde{\sigma}_r^4} \sum_{i=1}^{n} \left(r_i - \tilde{m}_r\right)^4 - 3. \tag{5}$$

Parameters obtained by the formulas (1)–(5) are calculated separately for a tested man with open and closed eyes (two modes). They are denoted $\tilde{m}_{rv}, \tilde{D}_{rv}, \tilde{a}_{rv}, \tilde{e}_{rv}, \tilde{m}_{rz}, \tilde{D}_{rz}, \tilde{a}_{rz}, \tilde{e}_{rz}$ respectively. The given modes are bounded by the time domains at the kefalograms. The first minute corresponds to the mode with open eyes and the second – to the mode with closed eyes respectively. It should be noticed, that the skewness and kurtosis are the most important parameters among above mentioned ones. They characterize a level of displacement in the sagittal and frontal planes and rate of the recovery of the initial state (returning to the central axis of a human body).

The informative indices, which characterize the function of a static equilibrium of the body of a tested specialist, are of great importance for the suggested method. Such indices represent the coefficients, which characterize changes of the sampling mathematical expectation $K_{\tilde{m}_r}$, variance $K_{\tilde{D}_r}$, skewness $K_{\tilde{a}_r}$ and kurtosis $K_{\tilde{e}_r}$ of the vector projection of the marker displacement. These informative indices are calculated as ratios of the sampling (point) parameters of the distribution of the displacement vector projection r in the mode of testing with closed eyes to sampling (point) parameters of the distribution of the displacement vector projection r in the mode of testing with open eyes by the formulas:

$$K_{\tilde{m}_r} = \frac{\tilde{m}_{rz}}{\tilde{m}_{rv}}, \tag{6}$$

$$K_{\tilde{D}_r} = \frac{\tilde{D}_{rz}}{\tilde{D}_{rv}}, \tag{7}$$

$$K_{\tilde{a}_r} = \frac{\tilde{a}_{rz}}{\tilde{a}_{rv}}, \tag{8}$$

$$K_{\tilde{e}_r} = \frac{\tilde{e}_{rz}}{\tilde{e}_{rv}}. \tag{9}$$

The informative indices determined by the expressions (6)–(9) characterize the function of a static equilibrium of the body of a tested specialist as a whole.

4   Obtained Results

The developed method and kefalographic plant were used for the research of the static equilibrium of the specialists worked at the Ukrainian Antarctic station "Academician Vernadsky".

The normalized values of the informative indices and confidence intervals of the possible changes under action of the extremal environment were determined based on the transatlantic experiment carried out during five years.

These normalized values and appropriate confidence intervals are represented in Tables 1, 2 and 3.

Table 1. Distribution normalized values and confidence intervals for test mode "Open Eyes"

Levels of the possibilities of the human organism to adaptation	Parameters of marker oscillations			
	Expectation $\tilde{m}_{rv}$	Variance $\tilde{D}_{rv}$	Skewness $\tilde{a}_{rv}$	Kurtosis $\tilde{e}_{rv}$
Super-strong	$2{,}17 < 2{,}21 < 2{,}25$	$1{,}32 < 1{,}41 < 1{,}51$	$-0{,}12 < -0{,}05 < -0{,}02$	$-0{,}15 < -0{,}1 < -0{,}07$
			$0{,}02 < 0{,}05 < 0{,}12$	$0{,}07 < 0{,}1 < 0{,}15$
Strong	$5{,}25 < 5{,}36 < 5{,}47$	$6{,}7 < 7{,}17 < 7{,}67$	$-0{,}69 < -0{,}3 < -0{,}13$	$-0{,}82 < -0{,}54 < -0{,}36$
			$0{,}13 < 0{,}3 < 0{,}69$	$0{,}36 < 0{,}54 < 0{,}82$
Middle	$8{,}34 < 8{,}51 < 8{,}68$	$12{,}1 < 12{,}9 < 13{,}8$	$-1{,}61 < -0{,}7 < -0{,}3$	$-1{,}96 < -1{,}3 < -0{,}86$
			$0{,}3 < 0{,}7 < 1{,}61$	$0{,}86 < 1{,}3 < 1{,}96$
Weak	$11{,}86 < 12{,}1 < 12{,}34$	$17{,}9 < 19{,}2 < 20{,}6$	$-2{,}99 < -1{,}3 < -0{,}57$	$-3{,}1 < -2{,}05 < -1{,}36$
			$0{,}57 < 1{,}3 < 2{,}99$	$1{,}36 < 2{,}05 < 3{,}1$

Table 2. Distribution normalized values and confidence intervals for test mode "Closed Eyes"

Levels of the possibilities of the human organism to adaptation	Parameters of marker oscillations			
	Expectation $\tilde{m}_{rv}$	Variance $\tilde{D}_{rv}$	Skewness $\tilde{a}_{rv}$	Kurtosis $\tilde{e}_{rv}$
Super-strong	$2{,}36 < 2{,}41 < 2{,}46$	$1{,}21 < 1{,}3 < 1{,}39$	$-0{,}23 < -0{,}1 < -0{,}04$	$-0{,}45 < -0{,}3 < -0{,}2$
			$0{,}04 < 0{,}1 < 0{,}23$	$0{,}2 < 0{,}3 < 0{,}45$
Strong	$6{,}00 < 6{,}12 < 6{,}24$	$11{,}59 < 12{,}4 < 13{,}27$	$-1{,}04 < -0{,}45 < -0{,}2$	$-1{,}1 < -0{,}73 < -0{,}48$
			$0{,}2 < 0{,}45 < 1{,}04$	$0{,}48 < 0{,}73 < 1{,}1$
Middle	$9{,}64 < 9{,}83 < 10{,}03$	$21{,}96 < 23{,}5 < 25{,}15$	$-2{,}14 < -0{,}93 < -0{,}40$	$-2{,}36 < -1{,}56 < -1{,}03$
			$0{,}40 < 0{,}93 < 2{,}14$	$1{,}03 < 1{,}56 < 2{,}36$
Weak	$13{,}9 < 14{,}2 < 14{,}5$	$36{,}1 < 38{,}6 < 41{,}3$	$-2{,}97 < -1{,}29 < -0{,}56$	$-3{,}2 < -2{,}12 < -1{,}4$
			$0{,}56 < 1{,}29 < 2{,}97$	$1{,}4 < 2{,}12 < 3{,}2$

Obtained normalized values were divided conditionally on four levels depending on adaptation capabilities of organism of extreme activity specialists. Such levels as weak, middle, strong, super-strong were formed respectively. The best execution of the production function may be implemented by specialists, which belong to the strong and super-strong levels.

It should be noticed, that the skewness and kurtosis may be both left-side and right-side. Therefore, these indices in Tables 1, 2 and 3 are represented in two columns, where the upper rows correspond to the left-side skewness and kurtosis and the lower rows – to right-side ones respectively.

Table 3. Distribution normalized values and confidence intervals for test modes "Open Eyes" and "Closed Eyes"

Levels of the possibilities of the human organism to adaptation	Informative indices of static equilibrium			
	Expectation $K_{\bar{m}_r}$	Variance $K_{\bar{D}_r}$	Skewness $K_{\bar{a}_r}$	Kurtosis $K_{\bar{e}_r}$
Super-strong	0,48 < 0,49 < 0,5	0,23 < 0,25 < 0,27	0,46 < 1,05 < 2,42	0,71 < 1,07 < 1,62
Strong	0,85 < 0,87 < 0,89	1,94 < 2,08 < 2,23	0,48 < 1,1 < 2,53	1,21 < 1,83 < 2,76
Middle	1,27 < 1,3 < 1,33	5,52 < 5,91 < 6,32	0,52 < 1,2 < 2,76	1,66 < 2,5 < 3,78
Weak	1,79 < 1,83 < 1,87	8,61 < 9,21 < 9,85	0,57 < 1,32 < 3,04	2,03 < 3,07 < 4,64

Moreover, approximations of the possible laws of the distribution of the displacement vector projection values r including:

- distribution of extreme values;
- γ - distribution;
- Weibull distribution;
- logistic distribution

were determined by means of Pirson χ^2 (chi-square) and Kolmogorov-Smirnov criteria.

If distribution of the displacement vector projection values r corresponds to any above mentioned distributions, the physiological norm from the point of view of the human psychophysiological state takes place. As for other possible distributions of the displacement vector projection r (for example, the normal distribution law, Rayleigh law and others), it should be noticed, that they were observed only in the cases of some illness or worsening state of health of the extreme activity specialists.

5 Conclusions

The developed method of static equilibrium function assessment allows determining of the function of the human static equilibrium and the informative indices, which characterize the space dynamic range and features of the human body oscillations relative to the axis z. Both these indices and determined normalized values given in Tables 1, 2 and 3 may be used during the process of the professional selection of the extreme activity specialists.

The suggested criteria were firstly used for selection of the polar explorers worked at the Antarctic station "Academician Vernadsky". Their application has decreased a time necessary for selection of the polar explores with improved possibilities for extreme activity.

References

1. Bazarov, V.G.: Clinical Vestibulometry. Zdorovie, Kyiv (1988). (in Russian)
2. Barilyak, P.A., Kitsera, A.E., Borisov, A.V.: Improvement of the kephalography method. J. Aural Nasal Guttural Dis. **6**, 66–67 (1981). (in Russian)
3. Gazhe, P.M., Veber, B.: Post-Urology. Regulation and Disequilibrium of Human Body. SPbMAPO, St-Peterburg (2008). (In Russian).
4. Rather, N.N., Patel, C.O., Khan, S.A.: Using deep learning towards biomedical knowledge discovery. Int. J. Math. Sci. Comput. (IJMSC) **3**(2), 1–10 (2017). https://doi.org/10.5815/ijmsc.2017.02.01
5. Chaudhry, H., Bukiet, B., Ji, Z., Findley, T.: Measurement of balance in computer posturography: comparison of methods. a brief review. J. Bodyw. Mov. Ther. **15**(1), 82–91 (2011)
6. Chiari, L., Rocchi, L., Cappello, A.: Stabilometric parameters are affected by anthropometry and foot placement. Clin. Biomech. **17**(9), 666–677 (2002)
7. Kaur, S., Malhotra, J.: On statistical behavioral investigations of body movements of human body area channel. Int. J. Comput. Netw. Inf. Secur. (IJCNIS) **8**(10), 29–36 (2016). https://doi.org/10.5815/ijcnis.2016.10.04
8. Maatar, D., Fournier, R., Naitali, A., Lachiri, Z.: Human balance and stability behavior analysis using spatial and temporal stabilometric parameters. Int. J. Image Graph. Signal Process. (IJIGSP) **5**(6), 33–42 (2013). https://doi.org/10.5815/ijigsp.2013.06.05
9. Visser, J.E., Carpenter, M.G., van der Kooij, H., Bloem, B.R.: The clinical utility of posturography. Clin. Neurophysiol. **119**(11), 2424–2436 (2008)
10. Maatar, D., Lachiri, Z., Fournier, R., Nait-Ali, A.: Stabilogram mPCA decomposition and effects analysis of several entries on the postural stability. Int. J. Image Graph. Signal Process. (IJIGSP) **4**(5), 21–30 (2012). https://doi.org/10.5815/ijigsp.2012.05.03
11. Tossavainen, T., Toppila, E., Pyykkö, I., Forsman, P.M., Juhola, M., Starck, J.: Virtual reality in posturography. IEEE Trans. Inf Technol. Biomed. **10**(2), 282–292 (2006)
12. Prieto, T.E., Myklebust, J.B., Hoffmann, R.G., Lovett, E.G., Myklebust, B.M.: Measures of postural steadiness: differences between healthy young and elderly adults. IEEE Trans. Biomed. Eng. **43**(9), 956–966 (1996)
13. Onykiienko, Y.Y., Kuzovik, V.D.: Hardware-software complex of estimation of operator psychophysiological state. In: Newest Scientific-Educational Achievements in Transport Medicine, pp. 109–111 (2011). (In Russian)
14. Onykiienko, Y.Y., Kuzovik, V.D.: Hardware-software system of evaluation criteria of the additional information CNS. In: Proceedings of Sixth World Congress "Aviation in the XXIst Century" "Safety in Aviation and Space Technologies", Kyiv, 23–25 September, pp. 1.7.23–1.7.25 (2014)
15. Sharma, G.: Performance analysis of image processing algorithms using matlab for biomedical applications. Int. J. Eng. Manuf. (IJEM) **7**(3), 8–19 (2017). https://doi.org/10.5815/ijem.2017.03.02
16. Volodarsky, E.T., Koshova, L.O.: Statistical Data Processing, vol. 308. National Aviation University, Kyiv. (in Ukrainian)

17. Goswami, S., Chakraborty, S., Saha, H.N.: An univariate feature elimination strategy for clustering based on metafeatures. Int. J. Intell. Syst. Appl. (IJISA) **9**(10), 20–30 (2017). https://doi.org/10.5815/ijisa.2017.10.03
18. Mandel, J.: The Statistical Analysis of Experimental Data. Courier Corporation, Mineola (2012)
19. Johnson, R.A., et al.: Applied Multivariate Statistical Analysis. Prentice hall, Englewood Cliffs (1992)

Levitating Orbitron: Grid Computing

Stanislav S. Zub[1(✉)] 🆔, N. I. Lyashko[2] 🆔, S. I. Lyashko[1] 🆔,
and Andrii Yu. Cherniavskyi[3] 🆔

[1] Taras Shevchenko Kyiv National University, 2 Glushkov ave., corp. 6, 05100 Kyiv, Ukraine
stah@univ.kiev.ua
[2] V.M. Glushkov Institute of Cybernetics, 40 Glushkov ave., 03187 Kyiv, Ukraine
[3] National Aerospace University "Kharkiv Aviation Institute",
Chkalova str. 17, 61070 Kharkiv, Ukraine

Abstract. Mathematical model of interaction for magnetic symmetric top (i.e. a rigid body and magnetic dipole simultaneously) in external magnetic field under uniform gravitational field is presented. Numerical modeling of the top dynamics, i.e. spinning and rotating around the axis of symmetry in axially-symmetric magnetic field is proposed. Investigation of the dynamics in some neighborhood of a given relative equilibrium for physically reasonable parameters of the system was required to generate a set of random trajectories (Monte-Carlo simulation) with small variations of parameters or initial conditions. More than 1000 of trajectories with 100 turns for each have been tested using grid computing on Grid-clusters of Ukrainian Academic Grid. The motion was limited in certain region for the trajectories with disturbed initial conditions and parameters within 1%. Executed analysis shows the possibility of stable motions and levitation in some neighborhood of a given relative equilibrium. It corresponds to the long trajectories observed in a physical experiment.

Keywords: Orbitron magnetic levitation · Symmetric top · Grid computation

1 Introduction

It is proved [1] that quasi-periodic motions of the magnetic symmetric top in specially constructed system without gravity force exist. The system is named Orbitron. Its configuration is optimal to investigate stability in magnetic system without gravity. Latest investigations show that the original Orbitron is not suitable for magnetic levitation, because in this very case it is difficult to reach compensation of gravity force by a magnetic force and simultaneously provide stability conditions. The problem can be solved by combining magnetic field of the Orbitron and the simplest magnetic field for gravity force compensation. So, the suitable system consists of two magnetic poles that are the same as for the Orbitron and supplementary axially symmetric magnetic field (with respect to z-axis) that is linearly depends on spatial coordinates for gravity force compensation.

It is necessary to mention here the experimental work of Kozorez [2] who was the first to have demonstrated quasi-periodic motions for small permanent magnet in such a magnetic field. In the experiment the magnet performs more than 1000 revolutions (i.e. ~6 min of flight). But successful launches were too rare because they were not based

© Springer International Publishing AG, part of Springer Nature 2019
Z. Hu et al. (Eds.): ICCSEA 2018, AISC 754, pp. 540–544, 2019.
https://doi.org/10.1007/978-3-319-91008-6_54

on an adequate mathematical model. The required initial conditions for the experiment were reproduced accidentally. It makes a sense to mention that neither natural experiment nor numerical simulation were able to provide the evidence of system stability, but could provide the arguments in favor of its existence.

2 Mathematical Model

2.1 Magnetic Field of the Orbitron

There are two magnetic poles $\pm Q$ on the axis z that are placed at a distance $\mp h$.

So, the magnetic field of the Orbitron $\mathbf{B}^O$ has the form[1]:

$$\mathbf{B}^O(\mathbf{r}) = \sum_{\varepsilon=\pm 1} \mathbf{B}_\varepsilon(\mathbf{r}), \ \mathbf{B}_\varepsilon = -\frac{\mu_0}{4\pi} \varepsilon \cdot Q \frac{\mathbf{r} - \varepsilon h \mathbf{e}_z}{|\mathbf{r} - \varepsilon h \mathbf{e}_z|^3} \tag{1}$$

where $\mathbf{B}^O$ field has axial symmetry with respect to the z-axis.

2.2 The Axially Symmetrical Field

$\mathbf{B}^L$ linearly depends on coordinates and used for gravity compensation. Decomposition of the field is shown in [3].

So, the total magnetic field $\mathbf{B}(\mathbf{x})$ that acts on the dipole is a sum of the fields:

$$\mathbf{B}(\mathbf{x}) = \mathbf{B}^O(\mathbf{x}) + \mathbf{B}^L(\mathbf{x}). \tag{2}$$

2.3 Symmetric Top

Let's suggest that magnetic dipole is a small rigid body with axial symmetry and its magnetic moment is also directed along the axis of symmetry, i.e. symmetries of the magnetic and mass distributions coincide.

3 Mathematical Model

We have the Poisson brackets [4]:

$$\begin{cases} \{x_i, p_j\} = \delta_{ij}, & \{v_i, v_k\} = 0, \\ \{\pi_i, v_j\} = \varepsilon_{ijl} v_l, & \{\pi_i, \pi_j\} = \varepsilon_{ijl} \pi_l \end{cases} \tag{3}$$

and Hamiltonian of the system is given by the expression

$$h = \frac{1}{2M}\mathbf{p}^2 + \frac{1}{2}\alpha\pi^2 + \frac{1}{2}\beta\langle\pi, \nu\rangle^2 - m\langle\mathbf{B}(\mathbf{x}), \nu\rangle + Mg\mathbf{x} \cdot \mathbf{e}_3 \tag{4}$$

---

[1] Throughout the article all arithmetical vectors are designated by bold symbols.

where x_i are coordinates of the mass center of the rigid body, A is the matrix of rotation for transformation from space coordinates (in the inertial frame) to the coordinates, connected with the body (in the body frame), p_i are components of momentum of the body, π_i are components of the angular momentum in the inertial frame. M is dipole mass, g is acceleration of gravity, m is quantity of the magnetic moment, $\alpha = 1/I_\perp$, $I_\perp = I_1 = I_2$ are moments of inertia of a symmetric top, $\beta = 1/I_3 - 1/I_1$, $\nu = A\mathbf{e}_3$, $\mathbf{e}_3 = (0, 0, 1)$ arithmetic vector (column), and thus, ν is the third column of the matrix A.

Applying $\dot{f} = \{f, h\}$ to the basic dynamic variables we got motion equations [4].

4   Estimation of Physical Parameters for Relative Equilibrium

Real parameters of relative equilibrium for the numerical simulation have been taken from the [4]. The magnetic poles and tablet (the magnetic dipole) were made of $Nd - Fe - B$ and have the following physical characteristics: $\rho = 7.4 \cdot 10^3 (\text{kg/m}^3)$ - density of the material and $B_r = 0.25(\text{T})$ - residual induction. Then the magnetic "charge" of the Orbitron's poles, required to ensure the stability is equal to $Q = 351.5625(\text{A} \cdot \text{m})$, and the distance between the poles is $L = 2h = 0.1(\text{m})$. Linear field for compensation of the gravity force is characterized by two parameters of our model: $B' = 0.357(\text{T/m})$, $B_0 = 2.985(\text{T})$. Movable magnet has a disk shape (i.e. tablet) of a diameter $d = 0.014(\text{m})$ and height $l = 0.0068(\text{m})$. Its parameters are as follows: magnetic moment $m = 0.18375(\text{A} \cdot \text{m}^2)$ with mass $M = 0.0068(\text{kg})$ and $\alpha = 0.9594 \cdot 10^7 (\text{kg}^{-1} \cdot \text{m}^{-2})$. For these parameters of our model in the point of the researched relative equilibrium total magnetic field will be $B_1 = -0.178(\text{T})$, $B_2 = 2.989(\text{T})$. So for the orbit with $r_0 = 1.5(\text{m})$ and $h = 0.075(\text{m})$ we have the relative equilibrium. These values seem to be quite reasonable for experimental implementation.

5   Estimation of Physical Parameters for Relative Equilibrium

Dimensionless variables are introduced by formulas

$$
\begin{cases}
\mathbf{x} = h\vec{x}, \mathbf{p} = Mh\sqrt{\dfrac{g}{h}}\vec{v}, \pi = Mh^2\sqrt{\dfrac{g}{h}}\vec{\pi}, \\[2mm]
t = \tau\sqrt{\dfrac{h}{g}} \to \omega = \varpi\sqrt{\dfrac{g}{h}} \to \xi = \varpi\sqrt{\rho}, \\[2mm]
I_\perp = Mh^2 i_\perp, \mathbf{B} = hB'\vec{b}, Q = \dfrac{4\pi}{\mu_0}h^3 B'q.
\end{cases}
\tag{5}
$$

Initial condition of relative equilibrium in dimensionless variables:

$$
\begin{cases}
\vec{x}_0 = r_0\vec{e}_1, \vec{p}_0 = r_0\omega\vec{e}_2, \vec{\pi}_0 = i_\perp\omega(\vec{e}_3 - v_z\vec{v}_0) + c_2\vec{v}_0, \\
\vec{v}_0 = v_z\vec{e}_3 + v_r\vec{e}_1, \quad v_r^2 + v_z^2 = 1.
\end{cases}
\tag{6}
$$

For magnetic dipole the motion equations in dimensionless variables have a form:

$$
\begin{cases}
\dot{\vec{x}} = \vec{p}, \\
\dot{\vec{p}} = \kappa^{-1} \cdot \nabla \langle \vec{v}, \vec{b} \rangle - \vec{e}_3, \\
\dot{\vec{v}} = i_\perp^{-1} \cdot \vec{\pi} \times \vec{v}, \\
\dot{\vec{\pi}} = \kappa^{-1} \cdot \vec{v} \times \vec{b},
\end{cases}
\tag{7}
$$

$$
\kappa = Mg / mB', \quad b_{i,j}^O = \sum_{\varepsilon = \pm 1} b_{i,j}^\varepsilon,
$$

$$
b_{i,j}^\varepsilon = \frac{\varepsilon \cdot q}{|\vec{x} - \varepsilon \cdot \vec{e}_z|^3} \left(\delta_{ij} - \frac{3(x_i - \varepsilon \cdot \delta_{i3})(x_j - \varepsilon \cdot \delta_{j3})}{|\vec{x} - \varepsilon \cdot \vec{e}_z|^2} \right)
$$

Starting from the point (6) we calculated initial values and parameters of our model for variation. Random variations were simulated in some neighborhood of a given relative equilibrium so that deviation from it was less than 1% for the corresponding dynamical variable. Totally 10^6 tests were performed with Grid computations [4–7].

Monte Carlo method, applied to test stability of the trajectories (100 turns each), didn't destroy the hypothesis of stability. Obviously, in case of unstable motion the magnetic dipole falls down or flies away during the first turn of the orbit. For the analysis we have created the Poincare sections in the neighborhood of relative equilibrium for physically reasonable parameters of the system.

6   Discussion

Most of the modern systems of magnetic levitation cannot be stable without automatic stabilization since physical stability in them occurs in the vertical direction only. The physical stability can be achieved on the base of the effects related with superconductivity or by the reason of dynamic effects. Levitron is a rare example of the second class systems. As an example of the other stable dynamic magnetic system we proposed Orbitron [1] where investigated system was purely magnetic, i.e. without homogeneous gravity field. It is also interesting to note that this study sheds light on the stability of the orbital motion of the magnet in the experiment conducted in Glushkov Institute of Cybernetics (Kiev) in 1970.

It is also important to note that Levitron and Orbitron are essentially different systems. Levitron stability is observed near the axis of symmetry of the system as contrasted to the investigated system and Orbitron where stability takes place at a sufficient distance from the axis of symmetry [1, 4]. As in case of Orbitron we have analytically proved the stability of the investigated system, but the proof concerns only small deviations from the relative equilibrium, while for practical purposes it's important to know the stability factor and other parameters of quasi orbits for the finite deviations. These parameters are important for planning of a real physical experiment. Mathematical model proposed in this paper describes a physical experiment (see [2], pp. 124–126) with magnet orbital flight but neglect the energy dissipation.

7   Conclusions

Dimensionless system model suitable for analysis and visualization has been obtained. Extensive sessions of quasi-orbits generation and numerical simulation in some neighborhood of a given relative equilibrium have been carried out. Grid computing allows visualizing expected trajectories to obtain statistically reliable estimates. The proposed computational model and numerical results are presented for the first time and reflect the current state of researches for the above mentioned system.

References

1. Zub, S.S.: Stable orbital motion of magnetic dipole in the field of permanent magnets. Physica D **275**, 67–73 (2014)
2. Kozorez, V.V.: Dynamic Systems of Free Magnetically Interacting Bodies. Naukova dumka, Kyiv (1981)
3. Simon, M.D., Heflinger, L.O., Geim, A.K.: Diamagnetically stabilized magnet levitation. Am. J. Phys. **6**(69), 702–713 (2001)
4. Zub, S.S.: Magnetic levitation in orbitron system. Probl. At. Sci. Technol. **5**(93), 31–34 (2014)
5. Agarwal, M., Srivastava, G.M.S.: Cloud computing: a paradigm shift in the way of computing. Int. J. Mod. Educ. Comput. Sci. (IJMECS) **9**(12), 38–48 (2017). https://doi.org/10.5815/ijmecs.2017.12.05
6. Fotohi, R., Effatparvar, M.: A cluster based job scheduling algorithm for grid computing. Int. J. Inf. Technol. Comput. Sci. (IJITCS) **5**(12), 70–77 (2013). https://doi.org/10.5815/ijitcs.2013.12.09
7. Thomas, M.V., Dhole, A., Chandrasekaran, K.: Single sign-on in cloud federation using CloudSim. Int. J. Comput. Netw. Inf. Secur. (IJCNIS) **7**(6), 50–58 (2015). https://doi.org/10.5815/ijcnis.2015.06.06

A Fuzzy Model for Gene Expression Profiles Reducing Based on the Complex Use of Statistical Criteria and Shannon Entropy

Sergii Babichev[1,2(✉)], Volodymyr Lytvynenko[1], Aleksandr Gozhyj[3],
Maksym Korobchynskyi[4], and Mariia Voronenko[1]

[1] Kherson National Technical University, Kherson 73008, Ukraine
sergii.babichev@ujep.cz, immun56@gmail.com, mary_voronenko@i.ua
[2] Jan Evangelista Purkyne University in Usti nad Labem, 400 96 Usti nad Labem, Czech Republic
[3] Petro Mohyla Black Sea National University, 68 Desantnikov str., 10, Nikolaev 54003, Ukraine
[4] Military-Diplomatic Academy named Eugene Bereznyak, Kiev, Ukraine
alex.gozhyj@gmail.com

Abstract. The paper presents the technology of gene expression profiles reducing based on the complex use of fuzzy logic methods, statistical criteria and Shannon entropy. Simulation of the reducing process has been performed with the use of gene expression profiles of lung cancer patients. The variance and the average absolute value were changed within the defined range from the minimum to the maximum value, and Shannon entropy from the maximum to the minimum value during the simulation process. 311 gene expression profiles from 7129 were removed as non-informativity during simulation process. The structural block diagram of the step-by-step data processing in order to remove non-informativity gene expression profiles has been proposed as the simulation results.

Keywords: Fuzzy logic · Gene expression profiles · Reducing · Statistical criteria
Shannon entropy

1 Introduction

One of the current directions of modern bioinformatics is reconstruction and simulation of gene regulatory network on the basis of DNA-microchip experiments or technique of RNA molecules sequencing [15]. Implementation of this process involves the preliminary processing of experimental data in order to increase their informativity and decrease the dimension of feature space by properly performed processes: filtration, reducing and grouping of gene expression profiles of the investigated genes by their mutual correlation. In the case of DNA-microchip experiments, the initial data includes a complex noise component, which appears at the stages of DNA microchip creation and reading information from it.

Moreover, the diverse nature of biological processes in an organism that are not related to the identified disease also contributes to the specificity of the noise component. The peculiarity of the gene expression profiles is also the large dimension of the feature space, which complicates the process of gene regulatory network reconstructing [13].

© Springer International Publishing AG, part of Springer Nature 2019
Z. Hu et al. (Eds.): ICCSEEA 2018, AISC 754, pp. 545–554, 2019.
https://doi.org/10.1007/978-3-319-91008-6_55

Preprocessing of the studied data involves the reducing of gene expression profiles, which are not informative in terms of statistical and entropy criteria. It is assumed that if variance or average absolute value of gene expression profiles less than the corresponding boundary values, or if Shannon entropy of the corresponding gene expression profiles is greater than the boundary value, then these profiles are not informative and they can be removed without significant loss of useful information. But at the same time there is a problem of determining the boundary values of the appropriate criteria, which allow us to divide objectively the gene expression profiles into informative and non-informative ones. In addition, the use of these criteria independently of each other is not objective because the non-informativeness of the corresponding gene according to one of the criteria does not mean that this gene is not informative according to other criteria. The complex use of all criteria for determining the level of gene expression profiles informativity is rational in this case. A gene, identified as non-informative based on the use of all criteria can be removed from the data without significant loss of useful information.

In this paper this problem is solved with the use of fuzzy logic technique [14], which involves possibility for complex estimation of various criteria influence on integral parameter, which determines the level of the corresponding gene informativity. A large number of works are devoted to the issue of the fuzzy-system use for various purposes in various fields of research [6–12]. However, it should be noted that the author's research were mainly focused on data processing of small dimension data in order to create fuzzy systems for direction and control the functioning of the system. The problem of decrease of features space dimension by clustering the gene expression profiles was considered in works [2–4]. However, the author's research were focused at grouping genes in accordance with the corresponding affinity functions. The questions of genes reduction according to the relevant criteria in these works were not considered.

The aim of this work is development of technology for reducing gene expression profiles based on the complex use of fuzzy logic methods and statistical and entropy criteria, which allow us to estimate the level of the studied genes informativity.

2   Stages of Fuzzy Inference Process

During fuzzy inference process the values of the input variables of the model are transformed into the values of the output variables using fuzzy rules, which are formed by experts in this subject field. The block diagram of fuzzy inference process is shown in Fig. 1. The stage of knowledge base formation involves:

- formation of sets of input $X = \{x_1, x_2, \dots, x_n\}$ and output $D = \{d_1, d_2, \dots, d_m\}$ variables;
- formation of the basic term-set with the corresponding membership functions for each term: $A = \{a_1, a_2, \dots, a_i\}$;
- formation a set of fuzzy rules agreed with the variables used:

$$\bigcup_{k=1}^{m} \left[\bigcap_{i=1}^{n} \left(x_i = a_i^k \right), \ for \ \omega_k \right] \rightarrow D = d_k \tag{1}$$

where $k = 1,\ldots, m$ is the quantity of logical statements, $i = 1,\ldots, n$ is the quantity of terms used, ω_k represents the weight of the k-th statement. The fuzzification stage involves establishing the correspondence between the specific values of the input variables of fuzzy inference system and the values of the membership function of the term corresponding to the given variable. For this, the membership functions defined on the input variables are applied to their actual values. In other words, the values of the membership functions $\mu^{a_i^k}\left(x_i \right)$ versus the variables x_i for the corresponding terms a_i^k are determined. The fuzzification stage is completed when all values of the membership functions $b_i = \mu\left(a_i \right)$ for all rules which are included in this knowledge base of the fuzzy inference system are found.

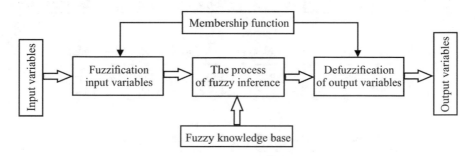

Fig. 1. Structural block diagram of fuzzy inference process

The process of fuzzy inference includes the following steps: aggregation of prerequisites, activation, accumulation of fuzzy rules inferences and defuzzification. Depending on the method of the fuzzy logic process using the classical fuzzy model can be implemented on the basis of the following algorithms: Mamdani, Sugeno, Larsen and Tsukamoto. The choice of model type is determined by the nature of the data used. In the case of creating a fuzzy model for evaluating the quality of gene expression profile on the basis of statistical and entropy criteria, the values of the input functions do not require scaling, all membership functions can be isotropic, resulting membership function can be represented as a simple set and the implementation of the defuzzification process does not require the use of a special functional. In this case, it is reasonable the use of Mamdani fuzzy inference algorithm.

3 Simulation Process of Gene Expression Profiles Reducing

The data of patients with lung cancer E-GEOD-68571 of Array Express database [5], which were pre-processed according to the method presented in [1], was used during simulation process. The initial data contained gene expression profiles of 96 objects, each of which was characterized by an expression vector of 7129 genes. The data was presented in the form of a matrix in size of (96 × 7129). Tables 1, 2 and 3 show the variation intervals of statistical values of variance, average absolute value, and Shannon entropy for 7129 gene expression profiles of the studied data. Analysis of the data from Tables 1, 2 and 3 shows that most gene expression profiles have small average values of expression and variance and large values of Shannon entropy.

Table 1. Variations of gene expression profiles variance

Minimum	1 Quartile	Median	Mean	3 Quartile	Maximum
0,03	2,31	15,46	693756,1	1813,5	124462000

Table 2. Variations of gene expression profiles average absolute values

Minimum	1 Quartile	Median	Mean	3 Quartile	Maximum
0,695	3,48	9,16	334,63	29,48	25755,45

Table 3. Variations of gene expression profiles Shannon entropy values

Minimum	1 Quartile	Median	Mean	3 Quartile	Maximum
0,06	2,15	2,81	2,47	3,08	3,47

Table 4. Formalisation of the input variables to linguistic terms

Input variables	Range	Terms of linguistic variables
Variance (Vr)	0–15	mean = 1.5, sd = 1.2 - Low
		mean = 6, sd = 1.2 - Median (Md)
		mean = 9, sd = 2 - High (Hg)
Average absolute value (Abs)	0–9	mean = 1.5, sd = 1.3 - Low
		mean = 4.5, sd = 1.3 - Median (Md)
		mean = 7.5, sd = 2 - High (Hg)
Shannon entropy (Entr)	2.8–3.5	mean = 2.87, sd = 0.07 - Low
		mean = 3.02, sd = 0.07 - Median (Md)
		mean = 3.3, sd = 0.11 - High (Hg)

We assume that the gene is not informative if the variance and the average absolute value of the appropriate gene expression profile are in the range from the minimum to the value corresponding to 25% quartile of the distribution of this value, and the value of Shannon entropy is in the range from the maximum to 75% quartile. Formalization of the input variables (variance, average absolute value and Shannon entropy) that determines the level of gene expression profile informativity is presented in Table 4. The range of change of the variance and the average absolute value of gene expression profiles from the minimum to the median of the corresponding intervals, and in the case of Shannon entropy from the maximum to the median have been used during setting up the system. Gaussian functions for input variables and triangular functions for the output variable have been used as a membership function. The value of the output variable varied in the range from 0 to 1, and this range was divided into five equal sections, each of which corresponded to the level of gene expression profile informativity (Very Low, Low, Median, High, Very High). Table 4 shows the parameters of the corresponding functions for the linguistic terms of the input parameters. Figure 2 shows the fuzzy membership functions of the input and output variables, which were created in accordance with the values of the Table 4. The result of the obtained fuzzy system operation is presented in Fig. 3.

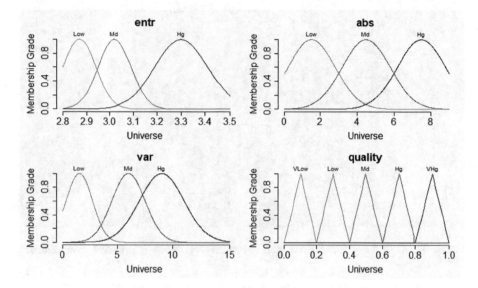

Fig. 2. Fuzzy membership functions for the input and the output variables

The variance and the average absolute values were changed within the defined range from the minimum to the maximum value (Fig. 3a), and Shannon entropy from the maximum to the minimum value (Fig. 3b) during simulation process. The step was chosen so that the appropriate interval was divided into 50 s. The gene expression profile was identified as non-informative if the value of the output variable (Quality) was less than the boundary between the low and the median values. In our case this value equals 0.4. The chart of the output parameter (Quality) versus the step of the input variables values changing is shown in Fig. 3c. The analysis of the charts allows us to conclude that during the process of step-by-step increase of the variance and the average absolute value and appropriate decrease of Shannon entropy the quality of gene expression profiles, which determines the level of their informativity, is gradually increasing.

This parameter reaches the boundary value on 23-th step. The membership function of the final fuzzy subset for the output variable corresponding to 0.4 boundary value is shown in Fig. 3d. The input variables at this step are the following: the variance −6.75; the average absolute value −4.37; Shannon entropy −3.16. The simulation results showed that if var ≤ 6.72, abs ≤ 4.37 and entr ≥ 3.16, then 311 gene expression profiles from 7129 are non-informative. The size of the output data is changed from (96×7129) to (96×6818). The simulation results allow us to propose the technology of gene expression profiles reducing based on the complex use of fuzzy logic methods, statistical criteria and Shannon entropy.

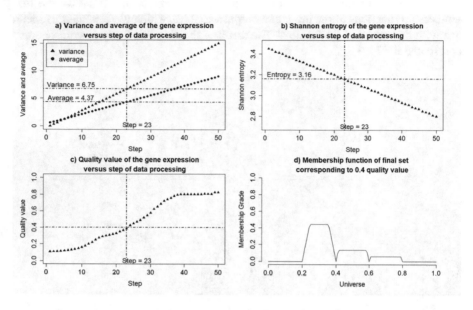

Fig. 3. Result of fuzzy system parameters determination for non-informative genes reducing

4 Architecture of the Technology of Gene Expression Profiles Reducing

The architecture of the technology of gene expression profiles reducing in the form of a block diagram of stepwise information processing is presented in Fig. 4. Practical implementation of this technology involves the following stages:

Stage I. Preparation of data for analysis and processing.

1. Calculation of the variance, the average absolute value and Shannon entropy for the expression profiles of the studied genes. Formation of data in the form of corresponding vectors: $var = \{var_1, var_2, \ldots, var_m\}$, $abs = \{abs_1, abs_2, \ldots, abs_m\}$, $entr = \{entr_1, entr_2, \ldots, entr_m\}$.
2. Statistical analysis of the obtained vectors, determining the range from var_{min}, abs_{min} and $entr_{max}$ to the median values of the appropriate values, fixing the values corresponding to the first quartile for the variance and the average absolute value of the genes expressions and the third quartile for Shannon entropy.

Stage II. Setup a fuzzy logic inference system for the studied data.

3. Formation of the basic term-set for input variables (variance, average absolute expression, Shannon entropy), and the output variable that determines the level of informativity of gene expression profiles QL (Quality).
4. Formation of the fuzzy rules, which agreed with the input variables and the output parameter.
5. Determination of the boundary value of the output parameter QL_{lim}, which allows the gene expression profiles to divide into informative and non-informative. Determine the step of the input variables changing within a given range.

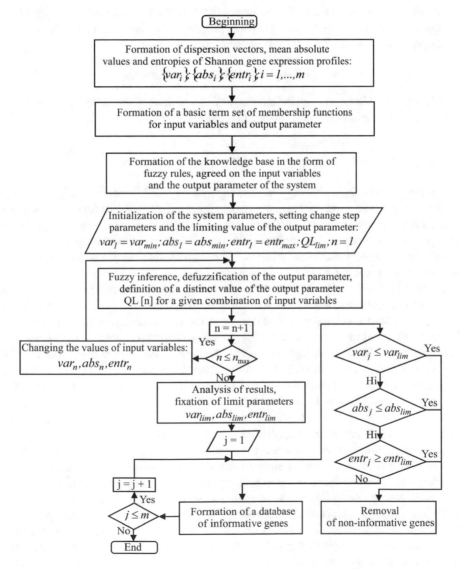

Fig. 4. Structural block diagram of the technology of gene expression profile reducing

Stage III. Calculation and analysis of the output parameter values, which determine the level of gene expression profile informativity.

6. Calculation of the output parameter QL for each combination of the input variables values corresponding to the appropriate gene expression profile. The result is formed as a vector: $QL = \{QL_1, QL_2, \dots, QL_m\}$.
7. Analysis of the results. Determine the values of the input variables that correspond to the boundary value of the output parameter.

Stage IV. Formation of the informative gene expression profiles database.

8. A stepwise comparison of the variance, the average absolute value and Shannon entropy values of the gene expression profiles with the boundary values of the appropriate criteria. If the following conditions are true:

$$var \leq \text{var}_{\text{lim}}; abs \leq abs_{\text{lim}}; entr \leq entr_{\text{lim}}$$

then this gene is allocated from the data as non-informative. Otherwise, the gene profile is recognized as informative for further analysis.

5   Conclusions

The paper presents the technology of gene expression profiles reducing, which is based on the complex use of fuzzy logic methods, statistical criteria and Shan non entropy. The knowledge base, which allows us to automate the process of determining the boundary parameters for estimating the level of the studied genes informativity is determined. Simulation of the reducing process has been performed with the use of gene expression profiles of lung cancer patients. The variance and the average absolute value were changed within the defined range from the minimum to the maximum value, and Shannon entropy from the maximum to the minimum value during the simulation process. The step was chosen so that the given interval is divided into 50 s. Gaussian functions for input variables and triangular functions for output variable have been used as membership functions for input and output variables. The value of the output variable was varied in the range from 0 to 1, and this range is divided into five equal sections, each of which corresponds to the level of informativity of the gene expression profile (Very Low, Low, Median, High, Very High). The gene expression profile is considered as non-informative if the value of the output variable (Quality) was less than the difference between low and average quality of the examined gene expression profiles. The value of this bounder parameter was determined as 0.4 during simulation process. Analysis of the simulation results has showed that during stepwise increase of the average absolute value and variance of appropriate gene expression profiles and decrease of Shannon entropy, the quality of gene expression profiles, which determines the level of their informativity, is gradually increasing. The boundary value of this parameter reaches at the 23-th step. The values of input variables at this step are: the variance -6.75; the average absolute value -4.37; Shannon entropy -3.16. The simulation results also have showed that if var ≤ 6.72, abs ≤ 4.37 and entr ≥ 3.16, then 311 gene expression profiles from 7129 are non-informative. The size of the studied dataset was changed in this case from (96×7129) to (96×6818).

The structural block diagram of a step-by-step processes of data formation, setup of the fuzzy logic inference system and the following data processing of data in order to remove non-informative genes according to the criteria: variance, average absolute value and Shannon entropy have been proposed. Prospects for further authors research are practical implementation of this technology within the framework of the hybrid model of gene expression profiles preprocessing at the early stage of the gene regulatory network reconstruction.

References

1. Babichev, S.A., Kornelyuk, A.I., Lytvynenko, V.I., Osypenko, V.: Computational analysis of microarray gene expression profiles of lung cancer. Biopolym. Cell **32**(1), 70–79 (2016). http://biopolymers.org.ua/content/32/1/070/
2. Babichev, S., Lytvynenko, V., Korobchynskyi, M., Taif, M.: Objective clustering inductive technology of gene expression sequences features. Commun. Comput. Inf. Sci. **716**, 359–372 (2016). https://doi.org/10.1007/978-3-319-58274-0_29
3. Babichev, S., Lytvynenko, V., Skvor, J., Fiser, J.: Model of the objective clustering inductive technology of gene expression profiles based on SOTA and DBSCAN clustering algorithms. Adv. Intell. Syst. Comput. **689**, 21–39 (2018). https://doi.org/10.1007/978-3-319-70581-1_2
4. Babichev, S., Taif, M.A., Lytvynenko, V., Osypenko, V.: Criterial analysis of gene expression sequences to create the objective clustering inductive technology. In: Proceeding of the 2017 IEEE 37th International Conference on Electronics and Nanotechnology, ELNANO 2017, pp. 244–248 (2017). http://ieeexplore.ieee.org/document/7939756/
5. Beer, D., Kardia, S.: Gene-expression profiles predict survival of patients with lung adenocarcinoma. Nat. Med. **8**(8), 816–824 (2012). http://www.nature.com/nm/journal/v8/n8/full/nm733.html
6. Bodyanskiy, Y., Dolotov, A., Vynokurova, O.: Evolving spiking wavelet-neuro-fuzzy self-learning system. Appl. Soft Comput. J. **4**(8), 252–258 (2014). https://doi.org/10.1016/j.asoc.2013.05.020
7. Bodyanskiy, Y., Vynokurova, O., Pliss, I., Peleshko, D., Rashkevych, Y.: Hybrid generalized additive wavelet-neuro-fuzzy-system and its adaptive learning. Adv. Intell. Syst. Comput. **470**, 51–61 (2016). https://doi.org/10.1007/978-3-319-39639-2_5
8. Hu, Z., Bodyanskiy, Y.V., Tyshchenko, O.K., Samitova, V.O.: Fuzzy clustering data given in the ordinal scale. Int. J. Intell. Syst. Appl. (IJISA) **9**(1), 67–74 (2017). https://doi.org/10.5815/ijisa.2017.01.07
9. Hu, Z., Bodyanskiy, Y.V., Tyshchenko, O.K., Samitova, V.O.: Fuzzy clustering data given on the ordinal scale based on membership and likelihood functions sharing. Int. J. Intell. Syst. Appl. (IJISA) **9**(2), 1–9 (2017). https://doi.org/10.5815/ijisa.2017.02.01
10. Hu, Z., Bodyanskiy, Y.V., Tyshchenko, O.K., Samitova, V.O.: Possibilistic fuzzy clustering for categorical data arrays based on frequency prototypes and dissimilarity measures. Int. J. Intell. Syst. Appl. (IJISA) **9**(5), 55–61 (2017). https://doi.org/10.5815/ijisa.2017.05.07
11. Hu, Z., Bodyanskiy, Y.V., Tyshchenko, O.K., Tkachov, V.M.: Fuzzy clustering data arrays with omitted observations. Int. J. Intell. Syst. Appl. (IJISA) **9**(6), 24–32 (2017). https://doi.org/10.5815/ijisa.2017.06.03

12. Kondratenko, Y., Korobko, O., Kozlov, O.: Synthesis and optimization of fuzzy controller for thermoacoustic plant. Stud. Fuzziness Soft Comput. **342**, 453–457 (2016). https://doi.org/ 10.1007/978-3-319-32229-2_31
13. Yaghoobi, H., Haghipour, S., Hamzeiy, H., Asadi-Khiavi, M.: A review of modeling techniques for genetic regulatory networks. J. Med. Sig. Sens. **2**(1), 61–70 (2012). https:// www.ncbi.nlm.nih.gov/pmc/articles/PMC3592506/
14. Zadeh, L.: Fuzzy logic = computing with words. IEEE Trans. Fuzzy Syst. **4**(2), 103–111 (1996). http://ieeexplore.ieee.org/document/493904/
15. Zak, D., Vadigepalli, R., Gonye, E., Doyle, F., Schwaber, J., Ogunnaike, B.: Unconventional systems analysis problem in molecular biology: a case study in gene regulatory network modeling. Comput. Chem. Eng. **29**(3), 547–563 (2005). http://www.sciencedirect.com/ science/article/pii/S0098135404002443

Quality of Symptom-Based Diagnosis of Rotavirus Infection Based on Mathematical Modeling

Serhii O. Soloviov[1,2(✉)] [ID], Mohamad S. Hakim[3,4] [ID], Hera Nirwati[4] [ID], Abu T. Aman[4] [ID],
Yati Soenarto[5] [ID], Qiuwei Pan[3] [ID], Iryna V. Dzyublyk[2] [ID], and Tatiana I. Andreeva[6,7] [ID]

[1] Department of Applied Mathematics, National Technical University of Ukraine "Igor Sikorsky
Kyiv Polytechnic Institute", Kyiv, Ukraine
solovyov.nmape@gmail.com
[2] Department of Virology, Shupyk National Medical Academy of Postgraduate Education,
Kyiv, Ukraine
[3] Department of Gastroenterology and Hepatology, Erasmus MC-University Medical Center,
Rotterdam, The Netherlands
[4] Department of Microbiology, Faculty of Medicine, Universitas Gadjah Mada,
Yogyakarta, Indonesia
[5] Department of Child Health, Faculty of Medicine, Universitas Gadjah Mada,
Yogyakarta, Indonesia
[6] Department of Public Health, Babeş-Bolyai University, Cluj-Napoca, Romania
[7] Alcohol and Drug Information Center, Kyiv, Ukraine

Abstract. Rotavirus is the leading cause of severe childhood gastroenteritis worldwide. The laboratory diagnosis requires testing of fecal specimens with commercial assays that often are not available in low resource settings. Therefore, estimation of rotavirus presence based on clinical symptoms is expected to improve the disease management without laboratory verification.

We aimed to develop and compare different mathematical approaches to model-based evaluation of expected rotavirus presence in patients with similar clinical symptoms. Two clinical datasets were used to develop clinical evaluation models of rotavirus presence or absence based on Bayesian network (BN), linear and nonlinear regression.

The developed models produced different levels of reliability. BN compared with regression models showed better rotavirus detection results according to optimal cut-off points. Such approach is viable to help physicians refer patient to the group with suspected rotavirus infection to avoid unnecessary antibiotic treatment and to prevent rotavirus infection spread in a hospital ward.

Keywords: Rotavirus infection · Symptoms · Bayesian network · Regression

1 Introduction

Acute gastroenteritis (diarrhea) remains a significant global health problem, particularly disturbing in the developing countries [1–3]. The range of pathogens causing

© Springer International Publishing AG, part of Springer Nature 2019
Z. Hu et al. (Eds.): ICCSEEA 2018, AISC 754, pp. 555–566, 2019.
https://doi.org/10.1007/978-3-319-91008-6_56

gastroenteritis varies widely and includes pathogenic and opportunistic bacteria, viruses and parasites. Among viruses that may cause an acute gastroenteritis, rotavirus plays a dominating role, followed by norovirus and sapovirus [4].

In Asia, rotavirus is responsible for estimated 145,000 deaths per year, with the highest burden found in India, Pakistan and Indonesia [3]. In Indonesia, several diarrhea outbreaks due to rotavirus have been reported [5, 6] with rotaviruses being responsible for about 50–60% of acute diarrhea cases in children under 5 years of age [7–9] and rotavirus-positive children more likely suffering severe clinical symptoms such as vomiting and dehydration [7, 9]. Rotavirus is not routinely tested in patients with gastro-enteritis because the results do not alter clinical management. However, the knowledge of the laboratory diagnosis can prevent the excessive use of antibiotics in many cases [10]. Mathematical modeling typically provides a retrospective analysis of existing data [11] and can be used for developing decision-making algorithms. Such model-based decision-making algorithms, developed for clinical and diagnostic solutions, are called clinical and diagnostic decision-support systems (CDDSS) [12] and are used in medical practice for management of patients with various pathologies [13–18].

We did not find any CDDSS for management of patients with rotavirus infection while its timely detection is needed to prevent the spread of virus in the community or in a hospital ward. Due to limited availability of resources and diagnostic tests, rotavirus is not routinely tested in clinical settings, especially in developing countries such as Indonesia. Therefore, development of high-quality symptom-based evaluation models of rotavirus infection becomes extremely relevant in settings like in Indonesia.

2 Objectives

The aim of the study was to build and compare mathematical models for evaluation of rotavirus presence in patients based on clinical data. Here we used Bayesian networks (BN), linear and nonlinear regression for evaluation.

3 Materials and Methods

Two clinical datasets were used. The first dataset contains data of 105 children aged <5 years hospitalized due to acute diarrhea in Kodya Yogyakarta General Hospital, Yogya-karta, Indonesia from February to August 2009.

The second dataset comprises data of 1962 children aged <5 years hospitalized from January to December 2006 due to acute diarrhea at 6 teaching hospitals in Indonesia [7]. The participating hospitals were: (1) Muhammad Hussein Hospital (Palembang, South Sumatera), (2) Cipto Mangunkusumo Hospital (Jakarta, Capital City Special Territory), (3) Hasan Sadikin Hospital (Bandung, West Java), (4) Sardjito Hospital (Yogyakarta, Yogyakarta Special Territory Province), (5) Sanglah Hospital (Denpasar, Bali), and (6) Mataram Hospital (Mataram, West Nusa Tenggara).

Acute diarrhea in both datasets was defined as ≥3 loose stools within 24 h for the duration of <2 weeks. A standardized clinical form was completed including in forma-tion on the date of admission, age and sex of the patients, clinical symptoms and final laboratory-confirmed diagnosis.

3.1 Statistical Analysis of Retrospective Clinical Data

Both clinical datasets were randomly divided into a training (70%) and a test subsets (30%). Each training subset was used to parametrize mathematical models and each smaller subset was used for checking their model quality. The collected variables repre-sent demographics (age, sex, weight, height), clinical (e.g. vomiting, temperature, dehy-dration, etc.) and laboratory (rotavirus presence or absence) data for each patient in both datasets. The next step was the development of mathematical models describing asso-ciations between demographic and clinical data as potential explanatory variables, and laboratory data as the outcome variable, for their use in the development of clinical and diagnostic decision support system (DSS).

3.2 Bayesian Network

BNs for prediction of rotavirus presence were developed based on the described varia-bles of both clinical datasets, reflecting the relationships between clinical and laboratory variables [19–23]. Free license software Hugin Expert A/S Lite 8.4 (http://www.hugin.com) was used for development, training, and testing of each BN (Figs. 1 and 2).

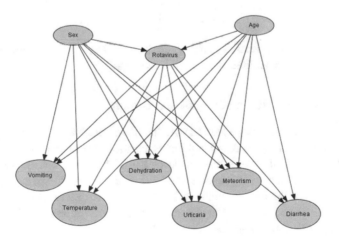

Fig. 1. BN for dataset 1, developed in Hugin Expert A/S Lite 8.4

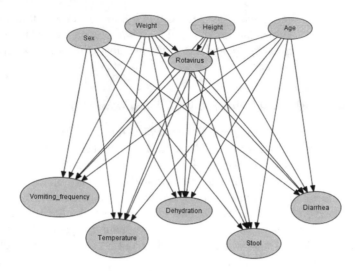

Fig. 2. BN for dataset 2, developed in Hugin Expert A/S Lite 8.4

3.3 Linear and Nonlinear Regression

Model parameters of linear (formula 1) and nonlinear regression (formula 2) were estimated with the method of least squares [24, 25].

$$f_n = a_i \cdot x_{in} + z \tag{1}$$

$$f_n = a_i \cdot b_i^{x_{in}} + z \tag{2}$$

where a_i and b_i are coefficients, x_{in} are values of variables, i is the number of variables, n is the number of observations and z is an intercept.

3.4 ROC-Curve as Criteria of Evaluation Quality

Quality of developed models was tested with the use of receiver operating characteristic curve, (ROC curve), a graphical plot that illustrates the performance of a binary classifier system as its discrimination threshold is varied [26]. The curve was created by plotting the true positive rate (sensitivity) against the false positive rate (1 - specificity) at various threshold settings of the model outcome. These thresholds are model output cut-points above which the result is assumed to signal about rotavirus presence. Pairs of sensitivity and specificity at various cut-points were used to plot a ROC-curve.

4   Results

4.1   Sample Characteristics

Case distribution by clinical and laboratory variables for both clinical datasets are presented in Figs. 3 and 4.

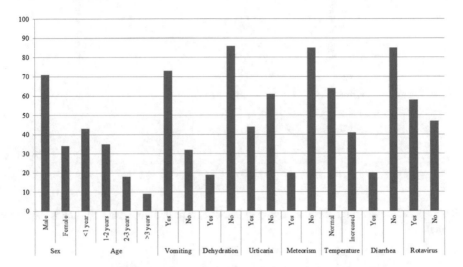

Fig. 3. Case distribution by variables (Dataset 1).

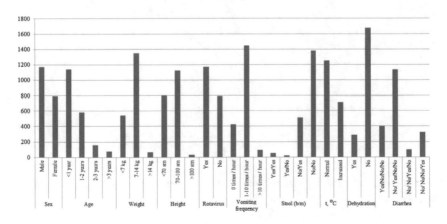

Fig. 4. Case distribution by variables (Dataset 2).

4.2   Evaluation Based on Bayesian Network

Probabilities calculated for each input variable were used to calculate the conditional probability of rotavirus presence based on the above-described BNs (Figs. 1 and 2) in Hugin Expert A/S Lite 8.4. Received probabilities of rotavirus presence: [0.1; 1] for the

Dataset 1 and [0; 1] for Dataset 2, were compared with real values of rotavirus presence. The variation of outcome probabilities thresholds produced different values of sensitivity and specificity and ROC-curves with estimated optimum points (Figs. 5 and 6). Both BNs produced level of sensitivity and specificity above 0.5.

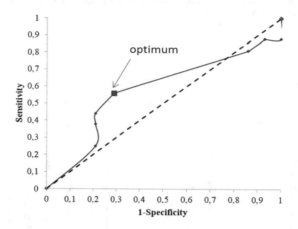

Fig. 5. ROC-curve for the Bayesian network (dataset 1). The optimum (cut-off) point is (Se; Sp) = (0,56; 0,71).

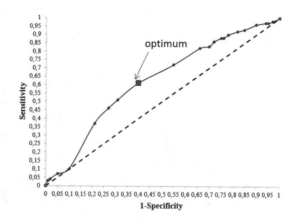

Fig. 6. ROC-curve for the Bayesian network (dataset 2). The optimum (cut-off) point is (Se; Sp) = (0,61; 0,60).

4.3   Evaluation Based on Linear and Nonlinear Regression

Presence or absence of rotavirus was regressed on other variables. Coefficients were found as mean values of several iterations, based on training datasets. Regression models validation and respective ROC-curves development showed that both linear and nonlinear regression approaches produced better specificity (>0.5) than sensitivity (<0.5) with the use of dataset 1 (small dataset), and better sensitivity (>0.5) than specificity (<0.5) with the use of dataset 2 (big dataset) (Figs. 7, 8, 9, 10, 11, 12, 13 and 14), but not balancing sensitivity and specificity as BNs with both datasets. Each model produced only a few discrete outcome values, so the number of thresholds differed in each case.

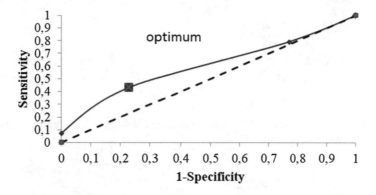

Fig. 7. ROC-curve for linear regression without intercept (dataset 1). The optimum (cut-off) point is (Se; Sp) = (0,43; 0,77).

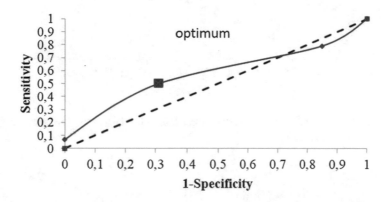

Fig. 8. ROC-curve for linear regression with an intercept (dataset 1). The optimum point is (Se; Sp) = (0,50; 0,69).

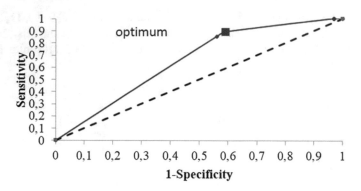

Fig. 9. ROC-curve for linear regression without intercept (dataset 2). The optimum (cut-off) point is (Se; Sp) = (0,89; 0,41).

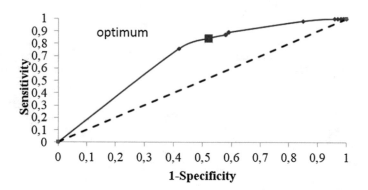

Fig. 10. ROC-curve for linear regression without intercept (dataset 2). The optimum (cut-off) point is (Se; Sp) = (0,84; 0,48).

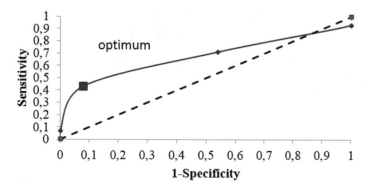

Fig. 11. ROC-curve for nonlinear regression without intercept (dataset 1). The optimum (cut-off) point is (Se; Sp) = (0,43; 0,92).

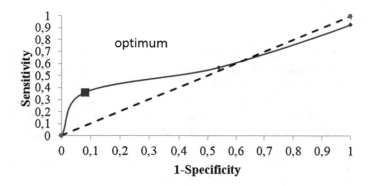

Fig. 12. ROC-curve for linear regression with an intercept (dataset 1). The optimum (cut-off) point is (Se; Sp) = (0,36; 0,92).

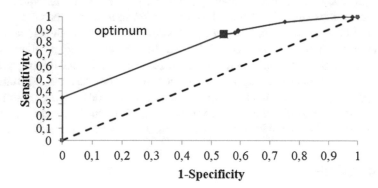

Fig. 13. ROC-curve for nonlinear regression without intercept (dataset 2). The optimum (cut-off) point is (Se; Sp) = (0,86; 0,46).

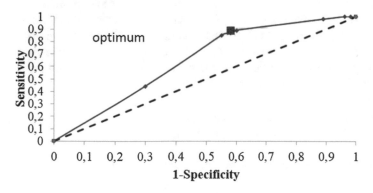

Fig. 14. ROC-curve for nonlinear regression with an intercept (dataset 2). The optimum (cut-off) point is (Se; Sp) = (0,89; 0,42).

5   Discussion

We considered the use of multiple analytical models for evaluation of expected rotavirus gastroenteritis in children. The models were based on relationships between certain clinical symptoms and rotavirus presence. We found that these models produce different levels of reliability. From this perspective, we found that BN-based modeling was a better choice to achieve this goal, producing a better balance of sensitivity and specificity. This can be explained by the fact, that BN development is based on conditional probabilities of each variable, so taking into account relationships between variables improves prediction. On the other hand, regression models function like "black box", producing coefficients that cannot be clearly explained.

This report bears limitations typical for modeling studies. An additional limitation is related to the use of only two mathematical approaches for evaluation of possible rotavirus presence. Sure, several other approaches, such as neural networks, machine learning, and others could be also used for evaluation in further research and are to be explored in future. Also, in our study, we used variables relating to demographics, clinical symptoms and etiological agent of disease. In the future studies, other important variables, e.g. hygiene measures, could be included in the model, and this might improve the outcome. The proposed modeling approaches can be used as a basis of future diagnostic support system that would help physicians in the developing countries with limited resources to predict the presence or absence of rotavirus infection and make a consequent rational decision in treatment or control measures of symptomatic patients.

6   Conclusions

The control of rotavirus spread needs its timely detection. Two available clinical datasets were used to develop several clinical models for evaluation of expected rotavirus presence or absence based on such approaches as BN, linear and nonlinear regression. In our work, we found that the use of BN compared with regression models showed better results according to optimal cut-off points of appropriate ROC-curves. Such approach could help physicians refer their patients to the group with suspected rotavirus infection to avoid unnecessary antibiotic treatment and to develop strategies for preventing nosocomial rotavirus spread in a hospital ward.

Acknowledgements. The authors thank the Indonesia Endowment Fund for Education (LPDP) for funding Ph.D. fellowship to Mohamad S. Hakim.

Conflict of Interest. The authors declare to have no conflict of interest.

References

1. GBD 2015 Mortality and Causes of Death Collaborators: Global, regional and national age-sex specific all-cause and cause-specific mortality for 240 causes of death, 1990–2013: a systematic analysis for the Global Burden of Disease Study 2013. Lancet **385**(9963), 117–171 (2015)
2. Tate, J.E., et al.: 2008 estimate of worldwide rotavirus-associated mortality in children younger than 5 years before the introduction of universal rotavirus vaccination programmes: a systematic review and meta-analysis. Lancet Infect. Dis. **12**(2), 136–141 (2012)
3. Kawai, K., et al.: Burden of rotavirus gastroenteritis and distribution of rotavirus strains in Asia: a systematic review. Vaccine **30**(7), 1244–1254 (2012)
4. Lanata, C.F., et al.: Global causes of diarrheal disease mortality in children <5 years of age: a systematic review. PLoS One **8**(9), e72788 (2013)
5. Pratiwi, E., Setiawaty, V., Putranto, R.H.: Molecular characteristics of rotavirus isolated from a diarrhea outbreak in October 2008 in Bintuni Bay, Papua, Indonesia. Virology (Auckl) **5**, 11–14 (2014)
6. Corwin, A.L., et al.: A large outbreak of probable rotavirus in Nusa Tenggara Timur Indonesia. Am. J. Trop. Med. Hyg. **72**(4), 488–494 (2005)
7. Soenarto, Y., et al.: Burden of severe rotavirus diarrhea in Indonesia. J. Infect. Dis. **200**(Suppl. 1), S188–S194 (2009)
8. Nirwati, H., et al.: Detection of group a rotavirus strains circulating among children with acute diarrhea in Indonesia. Springerplus **5**, 97 (2016)
9. Wilopo, S.A., et al.: Rotavirus surveillance to determine disease burden and epidemiology in Java, Indonesia, August 2001 through April 2004. Vaccine **27**(Suppl. 5), F61–F66 (2009)
10. Parashar, U.D., Nelson, E.A., Kang, G.: Diagnosis, management, and prevention of rotavirus gastroenteritis in children. BMJ **347**, f7204 (2013)
11. Oliveira, J.: A shotgun wedding: business decision support meets clinical decision support. J. Healthc. Inf. Manag. **16**(4), 28–33 (2002)
12. DeGruy, K.B.: Healthcare applications of knowledge discovery in databases. J. Healthc. Inf. Manag. **14**(2), 59–69 (2000)
13. Hardin, J.M., Chhieng, D.C.: Data mining and clinical decision support systems. In: Clinical Decision Support Systems, pp. 44–63. Springer (2007)
14. Perreault, L.E., Metzger, J.B.: A pragmatic framework for understanding clinical decision support. J. Healthc. Inf. Manage. **13**, 5–22 (1999)
15. Khabbaz, R.F., et al.: Challenges of infectious diseases in the USA. Lancet **384**(9937), 53–63 (2014)
16. Tleyjeh, I.M., Nada, H., Baddour, L.M.: VisualDx: decision-support software for the diagnosis and management of dermatologic disorders. Clin. Infect. Dis. **43**(9), 1177–1184 (2006)
17. Aminu, E.F., Ogbonnia, E.O., Shehu, I.S.: A predictive symptoms-based system using support vector machines to enhanced classification accuracy of malaria and typhoid coinfection. Int. J. Mathe. Sci. Comput. (IJMSC) **2**(4), 54–66 (2016). https://doi.org/10.5815/ijmsc.2016.04.06
18. Abumelha, M., Hashbal, A., Nadeem, F., Aljohani, N.: Development of infection control surveillance system for intensive care unit: data requirements and guidelines. Int. J. Intell. Syst. Appl. (IJISA) **8**(6), 19–26 (2016). https://doi.org/10.5815/ijisa.2016.06.03
19. Li, Z.N., et al.: Novel multiplex assay platforms to detect influenza a hemagglutinin subtype specific antibody responses for high-throughput and in-field applications. Influenza Other Respir. Viruses (2017)

20. Sesen, M.B., et al.: Bayesian networks for clinical decision support in lung cancer care. PLoS One **8**(12), e82349 (2013)
21. Ewings, S.M., et al.: A Bayesian network for modelling blood glucose concentration and exercise in type 1 diabetes. Stat. Methods Med. Res. **24**(3), 342–372 (2015)
22. Dai, M., et al.: Mutation of the 2nd sialic acid-binding site resulting in reduced neuraminidase activity preceded emergence of H7N9 influenza a virus. J. Virol. (2017)
23. Mani, K., Kalpana, P.: An efficient feature selection based on bayes theorem, self information and sequential forward selection. Int. J. Inf. Eng. Electron. Bus. (IJIEEB) **8**(6), 46–54 (2016). https://doi.org/10.5815/ijieeb.2016.06.06
24. Abente, E.J., et al.: A highly pathogenic avian-derived influenza virus H5N1 with 2009 pandemic H1N1 internal genes demonstrates increased replication and transmission in pigs. J. Gen. Virol. **98**(1), 18–30 (2017)
25. Ullah, Z., Fayaz, M., Iqbal, A.: Critical analysis of data mining techniques on medical data. Int. J. Mod. Educ. Computer Science (IJMECS) **8**(2), 42–48 (2016). https://doi.org/10.5815/ijmecs.2016.02.05
26. Kedzierski, L., et al.: Suppressor of cytokine signaling (SOCS)5 ameliorates influenza infection via inhibition of EGFR signaling. Elife (2017)

Optimization of Convolutional Neural Network Structure for Biometric Authentication by Face Geometry

Zhengbing Hu[1], Igor Tereykovskiy[2]([✉]) [ID], Yury Zorin[2] [ID],
Lyudmila Tereykovska[3] [ID], and Alibiyeva Zhibek[4]

[1] School of Educational Information Technology, Central China Normal University,
Wuhan, China
hzb@mail.ccnu.edu.cn

[2] National Technical University of Ukraine "Igor Sikorsky Kyiv Polytechnic Institute",
Kyiv, Ukraine
terejkowski@ukr.net, yzorin@gmail.com

[3] Kyiv National University of Construction and Architecture, Kyiv, Ukraine
tereikovskal@ukr.net

[4] Kazakh National Research Technical University named after K.I. Satpayev,
Almaty, Republic of Kazakhstan
alibieva_j@mail.ru

Abstract. The article presents development of the methodology of using a convolutional neural network for biometric authentication based on the analysis of the user face geometry. The need to create a method of the structural parameters of convolutional neural network adaptation to the expected conditions of its use in a biometric authentication system is postulated. It is proposed to adapt the convolutional neural network structural parameters based on the maximum similarity to the process of recognizing a human face image by an average user considering peculiar properties of computer input and display. A group of principles for optimization methods is formulated by combining this assumption with the generally accepted concept of a convolutional neural network constructing. The number of convolution layers should be equal to the number of the person image recognition levels by an average user. The number of feature maps in the n-th convolutional layer should be equal to the number of features at the n-th recognition level. The feature map in the n-th layer, corresponding to the j-th recognition feature, is associated only with those feature maps of the previous layer that are used to build the specified figure. The size of the convolution kernel for the n-th convolutional layer should be equal to the size of the recognizable feature on the n-th hierarchical level. Based on these principles, a method of the structural parameters optimization of a convolutional neural network has been developed. Advisability of these principles use has been proved experimentally.

Keywords: Biometric authentication · Neural network model
Convolutional neural network · Optimization · Recognition · Facial geometry

© Springer International Publishing AG, part of Springer Nature 2019
Z. Hu et al. (Eds.): ICCSEEA 2018, AISC 754, pp. 567–577, 2019.
https://doi.org/10.1007/978-3-319-91008-6_57

1   Introduction

Currently one of the most significant trends in the information security systems evolution is active introduction of the user biometric authentication tools. The reason for it is increase of the of confidential information flow, the proven fundamental drawbacks of classical user authentication systems, as well as objective requirements to ensure the privacy of access control systems. In addition, increased attention to biometric authentication technologies is explained by existence of a wide range of commercial and social applications where automatic human authentication is very useful [14–17].

Analysis of modern biometric authentication systems shows that in most of them geometric parameters of hands or ears, 'keyboard handwriting', geometric parameters of user's handwritten symbols, pattern of the blood vessels on hands or surface of the ocular fundus, voice, fingerprints, and person's face geometric parameters are used as a biometric key. The advantages of the latter include the possibility of hidden control, reliability and low cost of reading devices. A broad perspective of user biometric authentication systems based on the recognition of the geometry of his face is confirmed both by the fairly wide spread of these means of authentication and by a large number of relevant theoretical and practical studies [1, 3, 9–14]. At the same time the range of their application is substantially limited by insufficient recognition accuracy, a significant development period, and low adaptability to certain features of information systems, which predetermines the importance of research in this field.

2   Literature Review and Problem Postulation

The analysis of scientific and practical works on the user biometric authentication by his face image allows us to state that at present two main classes of corresponding devices are used [3, 7, 14, 15, 18]. They analyze two- and three-dimensional images of a person's face respectively. Two-dimensional analyzers of images find both the characteristic anthropometric points in the person's portrait and the distances between these points. The transition to a more complex three-dimensional analysis of the facial geometry makes it possible to significantly increase the volume of obtained biometric information. A literature review on neural network methods of human recognition by the face image is presented in [9, 10]. It is shown that the main problems being solved when recognizing a person by the face image with neural networks are the following: splitting the feature space into regions corresponding to the individual person's face; extraction of key characteristics; compression and reconstruction of images; topologically ordered transformation of space; recognition with considering the space topology. In this article the architectures of neural networks of various types are presented and it is shown how they are used to recognize people. A method of flexible comparison on graphs is proposed in [1] as an algorithm of graphic images recognition corresponding to the face image. A mathematical description of the method is presented, and the advisability of using a convolutional neural network is denoted. An approach to preprocessing a face image in order to remove noise components, scaling, normalization and centering is presented in [8]. A neural network model of the multilayer perceptron was used for image recognition. The methodology of constructing neural network models for face recognition in real time is presented in [2] and the need to optimize the parameters of the neural network model is denoted.

A technique of network training proposed in [12] uses alternation of learning epochs with and without symbols distortion. A method of selection and modification of the learning rate is proposed. A convolutional neural network used is different from the classical one by slightly modified parameters of convolutional layers and by the absence of subsample layers. An alternate compression of the feature maps in order to isolate the features of the higher hierarchy level is ensured by displacement of the convolution kernel with step 2. At the same time it was proved experimentally that, despite the simplification of the network structure, the recognition accuracy deteriorated insignificantly.

It was suggested to optimize the structure of a convolutional neural network for image recognition by a self-configuring evolutionary algorithm [6]. At the same time it is pointed out that, because of the high resource consumption, the proposed approach has a very limited scope of use.

The use of convolutional neural networks for the recognition of handwritten symbols is presented in [5]. Unlike the convolutional neural network, which structure is shown in Fig. 1, a radial basis function network was used as a prototype of the last layers. It is stated that the most significant shortcoming of the convolutional network is its low learning rate. It is suggested to level this feature by parallelizing the learning algorithm. A neural network system for searching, selecting and recognizing faces in images was presented in [11]. Although it declares the development of its own neural network model, in fact, the convolutional neural network for recognition of handwritten symbols is used which parameters are described in [12, 13].

As it stated in [4, 10, 13, 18], in order to improve the quality of recognition and acceleration of convergence of a convolutional neural network, it is necessary: to make the network structure more complicated by adding additional convolutional and subsampling layers, to use the rectified linear unit as an activation function and to parallelize the learning process of the network inside each layer.

Therefore, it can be concluded that the shortcomings of modern neural network biometric authentication tools based on the analysis of the two-dimensional image of the user's face are related to the shortcomings of the methodological basis for using convolutional neural networks for recognition. First of all, the issues of determining the optimal network structural parameters are not solved in modern methodology. This negatively affects the accuracy of users' face recognition which in turn can lead to unauthorized access to the information in protected system.

3 Purposes and Objectives of the Study

The main goal of the study is to develop a method of determination the optimal structural parameters of convolutional neural network for user recognition by a two-dimensional image of his face.

4   Determination of Structural Parameters Optimization Model

A typical structure of a convolutional neural network for image recognition is shown in Fig. 1 [5, 6, 11, 12]. The input parameters of such a neural network model correspond to individual pixels. Therefore, the number of input parameters is equal to the size of the image. The number of output neurons is equal to the number of recognizable images and the number of hidden neurons is selected experimentally based on maximum recognition accuracy.

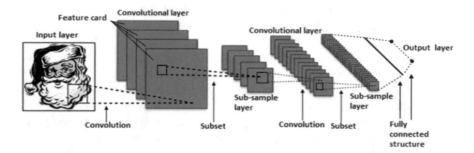

Fig. 1. Typical structure of a convolutional neural network

A somewhat simplified convolutional neural network is also known which structure is distinguished by the absence of subsample layers. In this case the alternate compression of the feature maps in order to isolate the features of the hierarchy higher level is ensured by displacement of the convolution kernel with step 2.

Based on the analysis of these structures and taking into consideration the results of theoretical works [4, 8, 10, 14] it was determined that the main structural parameters of this type neural network model are the following: the number L_{in} of input neurons, the number L_{out} of output neurons, the number L_f of neurons in a fully connected layer, the number K_{ls} of convolution layers, the number $K_{h,k}$ of feature cards in each convolutional layer, the number K_{ld} of subsampling layers, the size $(b \times b)_k$ of the convolution kernel for each k-th convolutional layer, displacement d_k of the receptive field for each k-th convolution procedure, the size $(a \times a)_k$ of the feature map for each k-th convolutional layer, structure of links between adjacent convolution/subsample layers.

Considering the need to minimize the recognition error, the model for optimizing the structural parameters of a convolutional neural network can be represented by

$$\Delta\left(L_{in}, L_{ls}, L_{out}, K_{h,k}, b_k, K_{ls}, |Q_{i,i+1}|_{K_{ls}}\right) \to \min \tag{1}$$

Where Δ is a recognition error, $|Q_{i,i+1}|_{K_{ls}}$ is a vector consisting of matrices that define connections between adjacent hidden layers. The absence of other convolutional neural network structural parameters in (1) is explained by the fact that they are derived from the components of the expression above.

5   Formulation of Approach and Principles of Structural Parameters Optimization

The optimization of above mentioned structural parameters is based on the following approach. In systems of biometric authentication the process of recognition by a convolutional neural network of a two-dimensional image of user's face should be as close as possible to its biological prototype. By the biological prototype we mean the process of recognition by an average user a person's face geometric parameters displayed on a computer monitor with average characteristics.

Combination of the approach formulated above with the generally accepted concept of a convolutional neural network functioning and particular properties of the biometric user authentication system based on the analysis of two-dimensional face geometry allowed to formulate the following group of principles determining the directions of optimization.

Principle 1. The number of convolutional layers should be equal to the number of recognition levels of a two-dimensional face image by average user.

Principle 2. The number of feature maps in the n-th convolutional layer should be equal to the number of features at the n-th recognition level.

Principle 3. The feature map in the n-th layer, corresponding to the j-th recognition feature (the j-th recognizable geometric figure) is associated only with those feature maps of the previous layer that are used to build the specified figure.

Principle 4. The size of the convolution kernel for the n-th convolutional layer should be equal to the size of the recognizable features on the n-th hierarchical level.

The recommendations [7, 9] allow applying the following restrictions on the input and preliminary processing of a two-dimensional face image in a biometric authentication system:

The authentication system uses a video camera with a resolution of 1920×1080 pixels. At present video cameras with such characteristics are quite widespread in modern universal computer systems.

The image to be recognized is in black and white. This restriction simplifies the preliminary processing of graphic information before it is fed into the recognition system. Thus, the face image has a binary representation.

The maximum size of the recognized image is equal to the resolution of the camera while the minimum size is 30×30 pixels.

Before submission to the convolutional neural network, the image undergoes preliminary processing which consists of scaling, centering, and trimming.

6   Method of Structural Parameters Optimization

Combination of generally accepted methodology for neural network recognition systems constructing [3, 4, 8, 12, 14] and proposed optimization principles with certain restrictions on input and preliminary image processing in the biometric authentication system, made it possible to propose an optimization method the implementation of which consists of 7 steps.

Step 1. Identify the recognizable users set which elements will correspond to the output neurons of the network. Thus, the number of output neurons is equal to the set cardinality.

Step 2. Determine the number of input parameters of the network. To do this, calculate the size of images to be recognized. It is necessary to take into consideration the size of users' face images in available databases used for training the network. If the size of the images of available training samples is much different from the ones to be recognized, they should be scaled.

Step 3. Based on the first and second principles, and using the geometric characteristics of the recognized images standards, determine the number of convolution layers and the number of feature maps in each convolutional layer.

Step 4. Based on the third principle, determine the structure of the links between neighboring feature maps.

Step 5. Based on the requirements to accuracy and recognition resource consuming, determine the presence and parameters of subsampling layers.

Step 6. Based on the fourth principle and considering the need to convolve the image to the feature vector, determine the kernel size and convolution displacement step for each recognition level.

Step 7. Based on the method of determining the hidden neurons number in a multilayer perceptron [10], determine the number of neurons in a fully connected layer.

Let us consider the application of the proposed method on example of authentication of 15 users whose face frontal images are represented in the Yale Faces database. It is assumed that video camera captures simultaneously the face of only one person. Before submission to the neural network, an image is preprocessed by isolating the face contours with following normalizing, centering and trimming the contour to a size of 29×29. The size of the whole image is 33×33. To isolate the face contour classical Viola-Jones algorithm is used [3, 5]. Given the number of authenticated users and the size of the recognized image, we get $L_{out} = 15$ and $L_{in} = 1089$.

As a result of the analysis of the frontal projection of human faces, an assertion has been made about the possibility of recognizing them on the basis of combinations of elementary figures, which dimensions in most cases range from 10% to 60% of the entire image. Such figures include open and closed triangles, ellipses, and vertical and horizontal line segments.

Since the convolutional neural network is stable only to small affine transformations, differently oriented and scaled figures can be considered as different features. For all five elementary figures, the number of possible transformations is assumed to be 16. Therefore, the number of significant features in the upper level of the recognition hierarchy is 80 which corresponds to the number of feature maps in the last convolutional layer. Also, as a result of analyzing the geometric parameters of mentioned above elementary figures, it is determined that for their displaying it is enough to have 4 segments that are oriented in different ways (horizontally, vertically, and at an angle $\pm45°$). Thus, two layers of convolution ($K_{ls} = 2$) are sufficient to isolate significant features. In this case, the first layer of the convolution will contain 4 feature maps, and in the second layer will contain 80 ($K_{h,1} = 4$, $K_{h,2} = 80$). At the same time, it was determined experimentally that at a resolution of 1920×1080 pixels, the minimum distinguishable length of the segments separated by the first convolutional layer is 5 pixels. Thus, the size of the convolution kernel of the first convolutional layer is $(b \times b)_1 = 5 \times 5$ pixels. In accordance with [15], for such convolution kernel the displacement should be equal to 1 pixel ($d_1 = 1$). With such parameters, the size of the feature maps of the first convolutional layer is $(a \times a)_1 = 29 \times 29$.

Note that the size of selected segments is minimally distinguishable. Because of this, there is no need to scale the feature maps of the first convolutional layer. We also take into account that described restrictions on the input of geometry of face into the system of biometric authentication assume that practically the entire recognition area is used. For this reason it is not necessary for the convolutional neural network to perform the scaling of elementary figures recognized by the second convolutional layer. Therefore, it is expedient to use the network structure without layers of subsampling ($K_{ld} = 0$). The rationality of such a solution is confirmed by the results of an experimental analysis of a similar convolutional neural network [12].

Taking into consideration the size of the first convolutional layer feature maps and the expediency of separating them from new features which are approximately three times as large as the previous ones, we obtain the size of the convolution kernel of the second convolutional layer as $(b \times b)_2 = 15 \times 15$. With a displacement at 1 pixel, the size of the feature maps of the second convolution layer is equal to $(a \times a)_2 = 15 \times 15$, which determines the size of the third convolution kernel as $(b \times b)_3 = 15 \times 15$.

For simple implementation a fully connected structure of the links between first and second hidden layers is chosen. In the future the structure can be modified by taking into account the peculiarities of recognition of closed and open figures (features of the second level of the hierarchy). Thus, all elements of the matrix (1) are equal to 1.

The number of hidden neurons of a fully connected layer is calculated on the assumption of minimum sufficiency defined by the Hecht-Nielsen theorem $L_f = 190$. The structure of a convolutional neural network adapted to the problem of face recognition is shown in Fig. 2.

To verify obtained results we have developed computer program modeling the system for human face recognition by convolutional neural network. The system have been trained and tested on samples from Yale Faces database. In computer experiments the accuracy of test samples recognition depending on the number of feature maps in the first and second convolutional layers has been calculated. The values of $K_{h,1}$ were 3, 4, 5 and the values of $K_{h,2}$ were 20, 40, 60, 80, 100.

The experiment results are shown in Figs. 2, 3 and in Table 1. As it follows from the obtained results, an increase of the number of learning epochs in excess of 100 does not lead to a guaranteed increase of recognition accuracy. Therefore, in order to minimize the computational resources, the most effective network is that which parameters will ensure maximum recognition accuracy at 100 learning epochs. In our case it is a network with $K_{h,1} = 4$, $K_{h,2} = 80$ (Fig. 4).

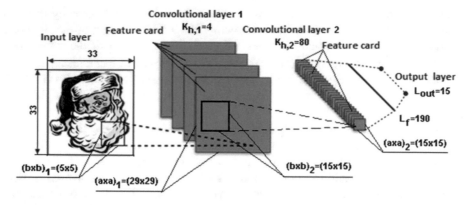

Fig. 2. The structure of an adapted convolutional neural network

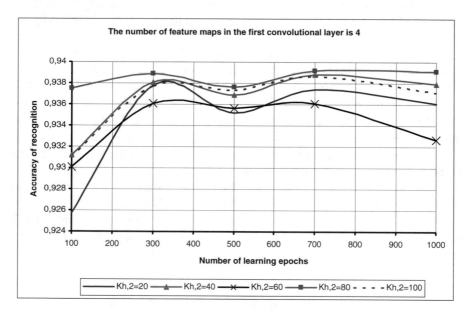

Fig. 3. Dependence of recognition accuracy on the number of learning epoch with $K_{h,1} = 4$ and different values of $K_{h,2}$

Table 1 Presents maximum recognition accuracy for each investigated combination of $K_{h,1}$ & $K_{h,2}$. The number of learning epochs is 100.

It is worth to note that the experimental values of the feature maps number for which the network error is minimal correspond to the values obtained using the proposed adaptation method.

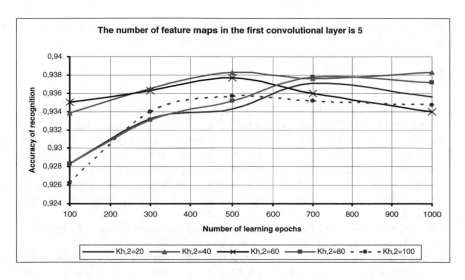

Fig. 4. Dependence of recognition accuracy on the number of learning epoch with $K_{h,1} = 5$ and different values of $K_{h,2}$

Table 1. Maximum recognition accuracy for different combinations of $K_{h,1}$ & $K_{h,2}$.

Number of feature maps in 1st convolutional layer	Number of feature maps in 2nd convolutional layer	Recognition accuracy
3	40	0,9132
4	80	0,9375
5	60	0,9354

7   Conclusions

An approach to structural parameters optimization of a convolutional neural network for face recognition in biometric authentication systems is proposed. It is assumed that the optimization is based on conditions of its maximum similarity to the process of face recognition by an expert.

Based on the proposed approach and using the generally accepted concept of the convolutional neural network functioning, additional principles for optimizing its structure have been formulated. Based on the proposed principles, a method for optimizing the structural parameters of a convolutional neural network for face recognition in a biometric authentication system was developed.

Computer experiments on face recognition by an optimized convolutional neural network in conditions close to the biometric authentication system demonstrated satisfactory recognition accuracy and confirmed the prospects of the proposed solutions.

Acknowledgment. This scientific work was financially supported by self-determined research funds of CCNU from the colleges' basic research and operation of MOE (CCNU16A02015).

References

1. Arsentyev, D.A., Biryukova, T.S.: Method of flexible comparison on graphs as algorithm of images recognition. Bull. Ivan Fedorov MGUP **6**, 74–75 (2015). (in Russian)
2. Bryliuk, D., Starovoitov, V.: Application of recirculation neural network and principal component analysis for face recognition. In: The 2nd International Conference on Neural Networks and Artificial Intelligence, BSUIR, Minsk, pp. 136–142 (2001)
3. Chirchi, V.R.E., Waghmare, L.M.: Iris biometric authentication used for security systems. Int. J. Image Graph. Sig. Process. (IJIGSP) **6**(9), 54–60 (2014). https://doi.org/10.5815/ijigsp.2014.09.07
4. Connaughton, R., Bowyer, K.W., Flynn, P.J.: Fusion of face and iris biometrics. In: Handbook of Iris Recognition, pp. 219–237. Springer, London (2013)
5. Dyomin, A.A.: Adaptive processing of the calligraphic information presented in the form of hand-written symbols. Ph.D. dissertation, Moscow, p. 182 (2014). (in Russian)
6. Fedotov, D.V., Popov, V.A.: Optimisation of convolutional neural network structure with self-configuring evolutionary algorithm in one identification problem. Vestnik SibGAU **16**(4), 857–862 (2013). (in Russian)
7. Hurshudov, A.A.: Development of the system of recognition of visual images in data stream. Ph.D. dissertation, Krasnodar, p. 130 (2015). (in Russian)
8. Ibragimov, V.V., Arsentyev, D.A.: Algorithms and methods of person recognition in modern information technologies. Bull. Ivan Fedorov MGUP **1**, 37–41 (2015). (in Russian)
9. Korchenko, A., Tereykovsky, I., Karpinsky, N., Tynymbayev, S.: Neural network models, methods and means of safety parameters assessment of the Internet focused information systems. In: Our Format, p. 275 (2016). (in Russian)
10. Narendira Kumar, V.K., Srinivasan, B.: New biometric approaches for improved person identification using facial detection. Int. J. Image Graph. Sig. Process. (IJIGSP) **4**(8), 43–49 (2012). https://doi.org/10.5815/ijigsp.2012.08.06
11. Zoubida, L., Adjoudj, R.: Integrating face and the both irises for personal authentication. Int. J. Intell. Syst. Appl. (IJISA) **9**(3), 8–17 (2017). https://doi.org/10.5815/ijisa.2017.03.02
12. Mishchenko, V.A.: Algorithm of recognition of graphic images. Bull. Voronezh State Tech. Univ. **5**(12), 103–105 (2009). (in Russian)
13. Ross, A., Jain, A.K.: Fusion techniques in multibiometric systems. In: Hammound, R.I., Abidi, B.R., Abidi, M.A. (eds.) Face Biometrics for Personal Identification, pp. 185–212. Springer, Heidelberg (2007)
14. Soldatova, O.P., Garshin, A.A.: Application of convolutional neural network for recognition of hand-written figures. Comput. Opt. **2**, 252–258 (2010). (in Russian)
15. Hu, Z., Tereykovskiy, I.A., Tereykovska, L.O., Pogorelov, V.V.: Determination of structural parameters of multilayer perceptron designed to estimate parameters of technical systems. Int. J. Intell. Syst. Appl. (IJISA) **9**(10), 57–62 (2017). https://doi.org/10.5815/ijisa.2017.10.07
16. Hu, Z., Gnatyuk, S., Koval, O., Gnatyuk, V., Bondarovets, S.: Anomaly detection system in secure cloud computing environment. International Journal of Computer Network and Information Security (IJCNIS) **9**(4), 10–21 (2017). https://doi.org/10.5815/ijcnis.2017.04.02
17. Hu, Z., Gnatyuk, V., Sydorenko, V., Odarchenko, R., Gnatyuk, S.: Method for cyberincidents network-centric monitoring in critical information infrastructure. Int. J. Comput. Netw. Inf. Secur. (IJCNIS) **9**(6), 30–43 (2017). https://doi.org/10.5815/ijcnis.2017.06.04

18. Hu, Z., Dychka, I.A., Mykola, O., Andrii, B.: The analysis and investigation of multiplicative inverse searching methods in the ring of integers modulo M. Int. J. Intell. Syst. Appl. (IJISA) **8**(11), 9–18 (2016). https://doi.org/10.5815/ijcnis.2017.06.04

19. Umyarov, N.H., Yu, K.G., Fedyaev, O.I.: A software model of convolutional neural network. Information management systems and computer monitoring. In: Proceedings of III Ukrainian Scientific Technical Conference of Students, Graduate Students that Young Scientists, 16–19 April 2012, Donetsk, pp. 343–347 (2012). (in Russian)

Model and Principles for the Implementation of Neural-Like Structures Based on Geometric Data Transformations

Roman Tkachenko and Ivan Izonin[(✉)]

Lviv Polytechnic National University, Lviv, Ukraine
roman.tkachenko@gmail.com, ivanizonin@gmail.com

Abstract. In this paper, the concept of information modeling based on a new model of geometric transformations is considered. This concept ensures the solutions of the following tasks like pattern recognition, predicting, classification, the principal independent components selection, optimization, recovering of lost data or their consolidation and implementing the information protection and privacy methods. Neural-like structures based on the Geometric Transformations Model as universal approximator implement principles of training and self-training and base on an algorithmic or hardware performing variants using the space and time parallelization principles. Geometric Transformations Model uses a single methodological framework for various tasks and fast non-iterative study with predefined number of computation steps, provides repeatability of the training outcomes and the possibility to obtain satisfactory solutions for large and small training samples.

Keywords: Geometric Transformations Model · Neural-like network
Training and self-training algorithms
Basic properties of the geometric transformations model

1 Introduction

The most popular modern neuro-paradigms are demonstrating their effectiveness in the numerous practical applications examples [1–3]. They are constructed on the biological analogies and iterative learning algorithms concepts [2, 4, 5]. This construction principle determines their principal drawbacks and limitations in complex multi-dimensional problems solving, in particular the instability and time delays in the ANN learning process, the inability to accurately display the solutions due to random initialization, the significant difference between the baseline calculation procedures for the ANN learning and applying and the difference between some neuro-paradigms [5, 6].

In this paper, a common Geometric Transformations Model (GTM) is presented. The GTM is a new concept of the information objects modelling that is presented either tabular data or production rules of the logical conclusions and their combinations [7]. It provides an effective solution for a wide range of problems, which detailed describe in [7]. The GTM realizes the training and self-training principles, where GTM's learning

© Springer International Publishing AG, part of Springer Nature 2019
Z. Hu et al. (Eds.): ICCSEEA 2018, AISC 754, pp. 578–587, 2019.
https://doi.org/10.1007/978-3-319-91008-6_58

and use procedures are the same type and based on the use of a common set of general arithmetic operations. Most of the developed GTM neural-like structures were implemented as software models that underwent extensive testing and are used in the real-world information and analytical systems. Hardware version of GTM neural-like structures was created due to the simplicity and uniformity of the algorithms, which, particularly, is used in car audio and video systems.

2 Basic Assumptions of the GTM

2.1 Analogies with the Artificial Neural Networks

A feature of GTM's learning and use algorithms is the same type [7, 8]. The topology of GTM's learning algorithm is presented in the form of some graph, vertices of which correspond to the basic operations of the algorithm: the scalar multiplication of the input signals vector on the weighting coefficients vector and then non-linear transformation from the scalar multiplication. So, the graph vertices can be considered as equivalents of neural elements of the ANN's hidden layer, and model of neural-like structure type described the GTM. This interpretation is productive as when debugging the GTM model as during the construction of appropriate system architectures based on GTM for solving many tasks, including forecasting of time series, the selection of principal components, shrinking data etc. using neural network approaches. At the same time, we note a fundamental difference between GTM and traditional type neural networks. Learning procedure in artificial neural networks is carried out (usually iterative) to set settings for the preselected structure. The structure of GTM formed as a result of training according to given characteristics. So, in a neural network interpretation, GTM is only graphs of corresponding algorithms, which, however, contain an internal dimensional and timing parallelism and can be implemented as hardware. Therefore, the GTM concept provides the implementation of principles of functional modelling, but not structural, as well as structure of the GTM model is completely determined by the structure of tabular data.

2.2 Variant of the GTM's Autoassociative Type Construction

An autoassociative mode is a basic version of the construction and application of the GTM's neural-like structures. An autoassociation means that the initial separation components (attributes) of the implementations-vectors for input and output is not used and the aim of the modelling is to construct an intermediate coordinate system with certain properties, which also provides the inverse transformation from the intermediate to the primary system without loss of information. As an analogue of the unsupervised GTM's neural-like structures mode may be considered two-layer autoassociative type perceptron based on a "bottleneck" [9].

All available attributes of vectors samples are provided to the inputs of the perceptron simultaneously, the same attributes are repeated as output signals of learning vectors of the perceptron. In general, the use of "narrow throat" when the number of neural elements of hidden layer is less than the number of inputs (outputs), the conversion of input vectors to identical output vectors occurs with some error. The output signals of the hidden layer in the neural elements represent the signals of principal components. Due to the use of optimization procedures for learning, the error value of input vectors conversion to identical output vectors is minimized and the output signals of the neural elements of the hidden layer set the optimized representation of the input vectors in the new coordinate system of reduced dimensionality. If the implementation vector of autoassociative neural network includes signals of input variables and callback signals, and the number of neural elements of hidden layer corresponds to the dimension of the ellipsoid scattering inputs space, then the space of callbacks are approximated by hyperspace (space) on the network output, and the values of additional measurements are errors of such approximation.

However, there are significant differences between neural-like autoassociative structures of GTM and multilayer perceptrons when working in autoassociation mode. The principal difference is that the mode of "narrow throat" is not required for the implementation of such structures, therefore, it is possible to accurately (with zero methodological error) display vectors of input signals into output vectors, at the same time to separate all components of information object on the outputs of the neural elements of hidden layer.

Moreover, the first component specifies the direction along which the variance is maximal, the next component is made so that along it maximizes the residual variation, etc. The advantages of this approach over classical algorithms PCA and SVD are: significantly increased the speed of implementation, the possibility of solving large dimension tasks, generalization property, which does not provide for mandatory retraining when new vector data available. Close enough, with using the separation of principal components is also the representation of callbacks based on given inputs (projective task), as there is a correspondence between the coordinates of the ellipsoid points of the inputs scattering and hyperbody of the object that is modelled. Based on a unified approach also to solve the tasks of data consolidation, factor analysis, clustering using self-training, signal filtering and so on.

3   The Process of the GTM Training

The GTM training process is performed by step-by-step geometric transformations in $(n + 1)$-dimensional implementations space, where n is the number of input attributes of the model. The longest axis of the scattering of inputs ellipsoid is determined on the first transformations step and the direction along this axis coincides with the first coordinate of the intermediate coordinate system that is formed within the training process. The next by size ellipsoid axis (the second inputs coordinate) is determined and the residue from the previous approximation step is approximated on the second step and etc. Overall, the training results are the parameters of the intermediate coordinate system and the parameters of elementary approximation surface for each transformation step of supervised training mode. Therefore, there are some differences between GTM in the training and self-training modes.

3.1   Unsupervised GTM Training

The training process is performed by step-by-step geometric transformations in n-dimensional space of realization, where n – the number of attributes of the model. As a result of these transformations, an intermediate coordinate system is formed where the direction of the coordinates coincides with the longest axis of the scattering ellipsoid that is given by the implementation-points training set. A training data sample (matrix) is the basis for the training model where each data sample row corresponds to the ellipsoid point [8]. The random matrix element is defined by $X_{i,j}^{(1)}$ for $i = \overline{1, N}$, $j = \overline{1, n}$, where i is the line number (implementations-vector) and j is the column number (number of properties attribute). The separation of attributes on the input and output is not performed in this mode. For the consistent training process should be performed the following transformations:

(1) Set the initial value of the transformation step number $S = 1$.
(2) Choose the initial basic line of the implementations-matrix $X_{b,j}^{(S)}$ for $j = \overline{1, n}$, where the sum of squares of the elements is the maximum [8]. This is corresponding to the vector with the highest rate. Setting up a number of the re-computation steps by p. 3 and 4, perform the forming of the basic line where the direction of the coordinates coincides with the longest axis of the scattering ellipsoid.
(3) Perform calculations:

$$K_i^{(S)} = \frac{\sum\limits_{j=1}^{n}(X_{i,j}^{(S)} \times X_{b,j}^{(S)})}{\sum\limits_{j=1}^{n}(X_{b,j}^{(S)})^2},$$

(1)

for $i = \overline{1, N}$.
(4) Construct transposed matrix $\left\| X_{j,i}^{T(S)} \right\|$ and perform computation of its elements by the transposing of the implementations-matrix. As a result, the elements of the basic line is modified.

$$X_{b,j}^{(S)} = \frac{\sum\limits_{i=1}^{N}(X_{j,i}^{T(S)} \times K_i^{(S)})}{\sum\limits_{i=1}^{N}(K_i^{(S)})^2},$$

(2)

for $j = \overline{1, n}$.
(5) Subtract the basic line elements from the elements of each line of the training matrix where basic line elements are multiplied by a coefficient calculated by the formula (2).

$$X_{i,j}^{(S+1)} = X_{i,j}^{(S)} - K_i^{(S)} \times X_{b,j}^{(S)} \text{ for } i = \overline{1, N}, j = \overline{1, n}.$$

(3)

(6) Do the next transformation step by performing $S = S + 1$ and going to the p. 2.
(7) As the transformations in accordance with (2) and (3) formulas correspond to the process of Gram-Schmidt orthogonalization, then after $S = n$ step in case columns of the implementations-matrix are not linearly dependent, the following will be obtained [10]:

$$X_{i,j}^{(S+1)} = \|0\|. \tag{4}$$

Consequently, a matrix with zero elements, generally the zero matrix will be formed after $Smax \leq n$ step.

As a result of this transformations is a set of $X_{b,j}^{(S)}$, $K_i^{(S)}$ vectors for $S = \overline{1, S_{max}}$, $i = \overline{1, N}$, $j = \overline{1, n}$.

Corresponding inverse transformation is also truthful

$$X_{i,j}^{(S)} = X_{i,j}^{S+1} + K_i^{(S)} \times X_{b,j}^{(S)}, \tag{5}$$

for $S = \overline{Smax, 1}$, where $X_{i,j}^{smax+1} = \|0\|$.

The GTM training outcomes is a set of vectors of the $X_{b,j}^{(S)}$ basic lines for $S = \overline{1, S_{max}}$ $j = \overline{1, n}$ that define the autoassosiative ANN parameters.

The straightforward and reverse algorithm of transformation the implementations-matrix elements could be easily presented as a graphical interpretation using the topology of GTM's autoassociative neural-like structure (Fig. 1).

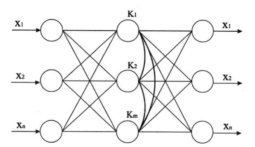

Fig. 1. The GTM's autoassosiative neural-like structure

The converting of a vector of $\overline{X}$ input signals into an appropriate vector of $\overline{K}$ coefficients has corresponded to their transfer from the ANN entrance to outputs of the linear neural elements of the hidden layer. Given these facts, not only projective relations between the layers of neural elements that take place in multilayer perceptron, but also the ordered lateral (side) relations between the hidden layer neurons are added that reflect a dependence of each successive transformation step from the previous one.

The reverse recovery of the initial implementations-matrix elements is displayed by the output part of the neural network. Taking into account the orthogonality and selection of the basic lines principles respectively the variance of a set of vectors along the transformations directions is maximum and the output signals will be given on the neurons outputs of the hidden layer that are consistent with the principal components signals.

One of the principal features of the PCA method [11] is to carry out the exact inverse transformation of the principal components into the vectors of the initial set. The neural network interpretation of the GTM is profuse in terms to transfer the setting tasks by the neurocomputer technique onto the GTM and the formation of new problems, where straightforward and reverse transformations algorithms are shown as the neural-like structures.

The appropriate transformation steps can be also displayed as the weights of corresponding synaptic relations between neural elements that may be used in the hardware implementation of the GTM neural networks. The following type GTM neural structure represents the principle of training without the supervisor mode and can be used to solve the problems of analyzing the principal components, the factor analysis problems and data consolidation before processing. Providing the implementations-vectors onto the network inputs is performed in the mode of a trained ANN according to the (1), (3) and (5) equations, where the values of the basic lines elements are obtained during the training process.

3.2 Supervised GTM Training

The supervised GTM training is specified by that the components of vector data are divided into input and output, where the latter are known only to the elements of the training samples [5]. Thus, the $K_i^{(S)}$ coefficient is required to perform a sequence of geometric transformations that may calculate only for training sample vectors. In this case, it is sufficient to calculate the $K_i^{(S)}$ coefficient from (2), where n is the number of vector input components and represent the desired coefficient as an approximation of the dependency (6) that is defined in the tabular form.

$$K_i^{(S)} = f(\overline{K_i^{(S)}}). \tag{6}$$

The GTM's neural-like structure topology of such a type is shown in Fig. 2, where the linear or nonlinear neural elements with the function of the dependency (6) approximation, could be used in the hidden layer. For the case of linear variant ANN type, the activation function of neural elements is the following:

$$K_i^{(S)} = \alpha \times \overline{K_i^{(S)}}, \tag{7}$$

where the α coefficient is calculated on the base of the least squares criterion [8].

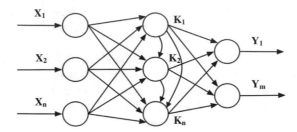

Fig. 2. The topology of the GTM neural-like structure in supervised mode

3.3 The GTM Linear Type Features

Hypersurface responses that they form, are hyperplanes where extra dimension model is determined by the non-linearity, noise components and rounding errors. As follows from the GTM concept (what is confirmed by the experiments), the vectors of the $K_i^{(S)}$ coefficients derived from the GTM almost coincide with the principal components obtained by known PCA [12], SVD [13] methods.

However, the GTM provides several important advantages like a function of the model without its retraining mode, the high speed where the result without error accumulation and significant dimension restrictions are obtained by a pre-defined number of steps. There is no need to apply the procedure of the normal equations or iterative adaptation solving.

Despite the linear structural elements, mixing them based on linear superposition to the isolated neural elements set (by the network outputs) is possible. However, it is devoid of practical meaning, since the signals are shown in the outputs of the hidden layer neural elements and approximately equal to the principal components signals of the implementations vectors and proportional to the principal components of the inputs vectors. These signals may be useful in the data type analysis and in establishing the parameters of the structure where some principal components could be masked [14].

Note that the number of the hidden layer neural elements equal to the number of n inputs and the ANN training or functioning input vectors data will be reset after the nth step of the changes for the linear neural network of the GTM where n inputs are not linearly related. There is a significant positive effect of the linear neural network of the GTM use for the supervised training that is used for the degenerate or almost degenerate problems. The problem solving is implemented automatically by masking the hidden layer neural elements where the variance of output signals is lower than the defined threshold.

3.4 The Neural-Like Structure Based on the GTM of Nonlinear Feature's Type

In generally dependencies (6) are approximated by the simple formula, particularly by the exponential polynomial. Thus, the neural elements of the polynomial activation function are used in the hidden layer of the ANN. It is possible to use more accurate approximation type like a rational Pade polynomials as a result that $K_i^{(S)}$ value functionally dependent on a set of network inputs.

$$K_i^{(S)} = \frac{a_0 + \sum_{j=1}^{n} a_j \times X_j}{1 + \sum_{j=1}^{n} b_j \times X_j}. \tag{8}$$

The number of nonlinear neural elements of the hidden layer can exceed the number of input attributes n in comparison with the ANN linear type and the number of geometric transformations steps while training will be determined by a defined training accuracy parameter. It was designed a non-linear version of the neural-like structure based on the GTM with the synaptic weights polynomial functions of the synaptic relations where can be used both linear and nonlinear neural elements in the hidden layer. One of the drawbacks of this neural network is the need to repeat the training again in case the new vectors of the training set are available.

4   The GTM Neural-Like Computers Applications

The GTM training algorithms and applying sameness simplify the software and hardware implementation tasks of the neural-like computers. In the Fig. 3 a window of the software neural-like computers application is shown. This application is used for solving the task of assessing the cost of housing [15, 16].

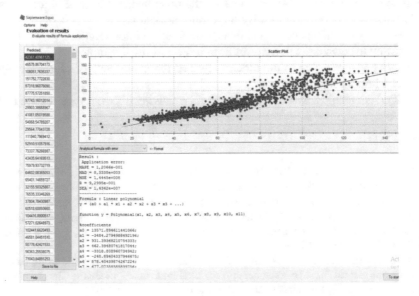

Fig. 3. The window of the GTM's neural-like computers application

The application provides high-speed training and deep analysis of the available data, particularly the principal components selection, visualization, evaluation of accuracy, and also the construction of a model by high power polynomials and rational fractions. It is also implemented the other data processing functions like a completely new method of filling gaps and restoring data in the tables and time series based on the procedure of the implementation point projection on the object's hyperbody. This method essentially excels the existing average substituting, interpolation and extrapolation methods by the accuracy and can be the basis for the adaptive control [17] and autonomous intelligent agents models [18] construction.

The GTM neural-like structures construction is an important way that is provided high accuracy and simplicity of the defuzzification functions and possibility to solve more dimension tasks. The following data processing approaches based on the GTM are effective, versatile and could be used in many other industries.

5 Discussion and Conclusion

In the paper, the concept of the GTM is described. The GTM is a reasonable interpretation of the neural-like network focused on the modern neuro-paradigms deficiencies and limitations elimination including slow and unreliable training that results from initial randomization. The transition from the "black box" to the "grey box" concept, where some results are justified, is also important as a result. The GTM's features like non-iterative training, the solutions step orthogonality and kinship between the learning and use procedures are providing the following advantages:

- high-speed training that is performed by the predefined number of algorithm steps what gives an opportunity to solve large-dimensional tasks;
- repeatability of the outcomes and their mathematical interpretation;
- the ability to solve tasks in terms of both large and small samples;
- the data pre-processing in the hidden layer of the neural network by selecting the signal's principal components;
- a unified geometric transformations model for solving different types of the tasks.

References

1. Hu, Z., Bodyanskiy, Y.V., Tyshchenko, O.K., Boiko, O.O.: An ensemble of adaptive neuro-fuzzy Kohonen networks for online data stream fuzzy clustering. Int. J. Mod. Educ. Comput. Sci. (IJMECS) **8**(5), 12–18 (2016). https://doi.org/10.5815/ijmecs.2016.05.02
2. Hu, Z., Bodyanskiy, Y.V., Tyshchenko, O.K., Boiko, O.O.: An evolving cascade system based on a set of neo-fuzzy nodes. Int. J. Intell. Syst. Appl. (IJISA) **8**(9), 1–7 (2016). https://doi.org/10.5815/ijisa.2016.09.01
3. Hu, Z., Bodyanskiy, Y.V., Tyshchenko, O.K., Samitova, V.O.: Fuzzy clustering data given on the ordinal scale based on membership and likelihood functions sharing. Int. J. Intell. Syst. Appl. (IJISA) **9**(2), 1–9 (2017). https://doi.org/10.5815/ijisa.2017.02.01

4. Bodyanskiy, Y., Vynokurova, O., Savvo, V., Tverdokhlib, T., Mulesa, P.: Hybrid clustering-classification neural network in the medical diagnostics of the reactive arthritis. Int. J. Intell. Syst. Appl. (IJISA) **8**(8), 1–9 (2016). https://doi.org/10.5815/ijisa.2016.08.01
5. Muller, B.: Neural Networks: An Introduction. Springer, Berlin (1991)
6. Bishop, C.M.: Neural Networks for Pattern Recognition. Oxford University Press, Oxford (1999)
7. Verbenko, I., Tkachenko, R.: Gantry and bridge cranes neuro-fuzzy control by using neural-like structures of geometric transformations. Tech. Trans. Autom. Control Issue 3-AC (11) (2013). https://doi.org/10.4467/2353737xct.14.057.3965
8. Tkachenko, R., Tkachenko, P., Izonin, I., Tsymbal, Y.: Learning-based image scaling using neural-like structure of geometric transformation paradigm. In: Hassanien, A., Oliva, D. (eds.) Advances in Soft Computing and Machine Learning in Image Processing. Studies in Computational Intelligence, vol. 730, no. 1, pp. 537–567. Springer, Cham (2018). https://doi.org/10.1007/978-3-319-63754-9_25
9. Zajíc, Z., Machlica, L., Müller, L.: Bottleneck ANN: dealing with small amount of data in shift-MLLR adaptation. In: 2012 IEEE 11th International Conference on Signal Processing, Beijing, pp. 507–510 (2012). https://doi.org/10.1109/icosp.2012.6491536
10. Tsymbal, Y., Tkachenko, R.: A digital watermarking scheme based on autoassociative neural networks of the geometric transformations model. In: 2016 IEEE First International Conference on Data Stream Mining & Processing (DSMP), Lviv, pp. 231–234 (2016). https://doi.org/10.1109/dsmp.2016.7583547
11. Ku, W., Storer, R.H., Georgakis, C.: Disturbance detection and isolation by dynamic principal component analysis. Chemometr. Intell. Lab. Syst. **30**(1), 179–196 (1995). https://doi.org/10.1016/0169-7439(95)00076-3
12. Sharma, A., Paliwal, K.K.: Fast principal component analysis using fixed-point algorithm. Pattern Recogn. Lett. **28**(10), 1151–1155 (2007). https://doi.org/10.1016/j.patrec.2007.01.012
13. Xu, H.: An SVD-like matrix decomposition and its applications. Linear Algebra Appl. **368**, 1–24 (2003). https://doi.org/10.1016/s0024-3795(03)00370-7
14. Dronyuk, I., Fedevych, O.: Traffic flows Ateb-prediction method with fluctuation modeling using Dirac functions. In: Gaj, P., Kwiecień, A., Sawicki, M. (eds.) Computer Networks, CN 2017, Communications in Computer and Information Science, vol. 718. Springer, Cham (2017)
15. Picus, L.O., Adamson, F., Montague, W., Owens, M.A.: New Conceptual Framework for Analyzing the Costs of Performance Assessment. Stanford University, Stanford Center for Opportunity Policy in Education, Stanford (2010)
16. Pawson, H., Milligan, V., Phibbs, P., Rowley, S.: Assessing management costs and tenant outcomes in social housing: developing a framework. AHURI Positioning Paper No.160. Australian Housing and Urban, Melbourne (2014)
17. Ivanov, Y., et al.: Adaptive moving object segmentation algorithms in cluttered environments. In: The Experience of Designing and Application of CAD Systems in Microelectronics, Lviv, pp. 97–99 (2015). https://doi.org/10.1109/cadsm.2015.7230806
18. Kazarian, A., Teslyuk, V., Tsmots, I., Mashevska, M.: Units and structure of automated "smart" house control system using machine learning algorithms. In: 2017 14th International Conference the Experience of Designing and Application of CAD Systems in Microelectronics (CADSM), Lviv, pp. 364–366 (2017). https://doi.org/10.1109/cadsm.2017.7916151

New Symmetries and Fractal-Like Structures in the Genetic Coding System

Sergey Petoukhov$^{(\boxtimes)}$, Elena Petukhova, and Vitaliy Svirin

Mechanical Engineering Research Institute, Russian Academy of Sciences,
M. Kharitonievsky Pereulok, 4, Moscow, Russia
spetoukhov@gmail.com

Abstract. The achievements of molecular genetics and bioinformatics lead to significant changes in technological, medical and many other areas of our lives. This article is devoted to new results of study of structural organization of genetic information in living organisms. A new class of symmetries and fractal-like patterns in long DNA-texts is represented in addition to two Chargaff's parity rules, which played an important role in development of genetics and bioinformatics. Our results provide new approaches for modeling genetic informatics from viewpoints of quantum informatics and theory of dynamic chaos.

Keywords: DNA · Symmetry · Fractal · Probability · Quantum informatics
Cancer

1 Tetra-Group Symmetries in Long DNA-Texts

The achievements of molecular genetics and biotechnology lead to significant changes in our lives. Genetic engineering and related fields provide not only the diagnostic and therapeutic possibilities of medicine that were unthinkable before, but also the emergence of new materials with surprising properties, new approaches to solving problems of nanotechnology, robotics, artificial intelligence systems, etc. Specialists consider projects for the cultivation of finished bodies of cars from chitin or bones. Several DNA strings connected together form a hinge-type mechanism for nanorobots, capable of bending and unbending by a chemical signal. Of particular importance is the knowledge of the principles of noise immunity of the genetic code in connection with the problem of ensuring noise immunity of information systems of control [1–6].

A road to the knowledge of bioinformational patents of living matter for their use in engineering, medicine and education is inextricably linked with the study of hidden regularities of hereditary information recorded in DNA molecules. The species of living organisms are amazingly diverse, but in all organisms genetic information is recorded in DNA and RNA molecules in the form of long texts of four letters: adenine A, cytosine C, guanine G and thymine T (in RNA uracil U is used instead of thymine). This article represents a new class of symmetries and fractal-like relations in long DNA-texts, the discovery of which shows elements of their fractal grammar. The goal of our research is revealing a participation of fractal structures in long DNA-texts.

© Springer International Publishing AG, part of Springer Nature 2019
Z. Hu et al. (Eds.): ICCSEEA 2018, AISC 754, pp. 588–600, 2019.
https://doi.org/10.1007/978-3-319-91008-6_59

DNA molecules are very long. For example, the human genome is a text with several billions of genetic letters A, T, C and G (it is equivalent to a text of thousands of thick books). DNA-texts of different organisms are represented in the GenBank (https://ru.wikipedia.org/wiki/GenBank), which contains hundreds of millions of sequences for more than one hundred thousand organisms. The set of known DNA-texts contains hundreds of billions of letters A, T, C and G.

What rules exist in these basic texts of living organisms? The modern situation is described by the following citation: *"What will we have when these genomic sequences are determined? ... We are in the position of Johann Kepler when he first began looking for patterns in the volumes of data that Tycho Brahe had spent his life accumulating"* [7]. Kepler did not make his own astronomic observations, but he found – in the huge astronomic data from the collection of Tycho Brahe - his Kepler's laws of symmetric planetary movements relative to the Sun. In 100 years after Kepler, thanks to the laws of Kepler, Newton discovered the law of universal gravitation. We have revealed new hidden symmetries in many long texts of single-stranded DNA of several dozen species of organisms, including the complete genomes of some organisms from the GenBank (without exceptions till now).

Below we explain our study but previously we should remind about two Chargaff's parity rules, which are known in genetics long ago. They are important because they point to a kind of "grammar of biology" [8]: a set of hidden rules that govern the structure of DNA. The first Chargaff's parity rule states that in any double-stranded DNA segment, the number of frequencies of adenine A and thymine T are equal, and so are frequencies of cytosine C and guanine G [8, 9]. The rule was an important clue to model the double helix structure of DNA by J. Watson and F. Crick.

The second Chargaff's parity rule states that both %A ≈ %T and %G ≈ %C are approximately valid in single stranded DNA for long nucleotide sequences. Many works of different authors are devoted to confirmations and discussions of this second Chargaff's rule [10–25]. Originally, CSPR is meant to be valid only to mononucleotide frequencies in single stranded DNA. *"But, it occurs that oligonucleotide frequencies follow a generalized Chargaff's second parity rule (GCSPR) where the frequency of an oligonucleotide is approximately equal to its complement reverse oligonucleotide frequency ... This is known in the literature as the Symmetry Principle"* [25, p. 2]. The work [22] shows the implementation of the Symmetry Principle in long DNA-sequences for cases of complementary reverse n-plets with n = 2, 3, 4, 5 at least. In all these works, authors concentrate their attention on the comparison of frequencies (or probabilities) of separate fragments of DNA-texts. By contrast to this, we study not individual probabilities of separate fragments but collective (or total) probabilities of special groups of fragments in long DNA-texts.

Let us explain our approach more detailed. Each of long DNA-sequences (for example, the sequence CAGGTATCGAAT...) can be represented not only in the form of the text of 1-letter words (C-A-G-G-T-A-T-C-G-A-A-T-...) but also in the form of the text of 2-letter words (CA-GG-TA-TC-GA-AT-...) or in the form of the text of 3-letter words (CAG-GTA-TCG-AAT-...) or in the form of the text of n-letter words in a general case. We briefly call such representations "n-letter representations" of DNA-texts. In each of such n-letter representations, we study total probabilities of all members of each of 4 groups of those n-letter words, which have the same letter

(A, T, C or G) on the same position $k \leq n$ inside words (we call such groups "tetra-groups"). Each of four DNA-letters A, T, C and G defines its own group of n-letter words inside a complete tetra-group of each n-representation of DNA-text.

By definition, in an n-letter representation of DNA-text, a total (or collective) probability of a group of n-letter words is the ratio: total quantity of all n-letter words of this group divided by the total quantity of all n-letter words. For example, the 2-letter text with 7 words AT-CT-GG-AG-AA-CA-AC contains 4 words with the letter A at their first position. In this text, the total probability $P_2(A_1)$ of such words is equal to $4/7 = 0,571$. This text contains also 2 words with the letter A at their second position. Correspondingly their total probability $P_2(A_2)$ is equal to $2/7 = 0,286$. In a general case, we denote by the symbol $P_n(A_k)$ a total probability of a group of n-letter words that have the letter A at their position k ($k \leq n$) in a long DNA-text. The similar symbols $P_n(T_k)$, $P_n(C_k)$ and $P_n(G_k)$ denote correspondingly total probabilities of all those n-letter words, which have the letters T, C and G at their position k.

It was unexpectedly for us to discover that all these n-letter representations of a long DNA-text are symmetrically interrelated each other on the basis of approximate equalities of total probabilities of all words with the same letter on the same position inside words (n = 1, 2, 3, 4, 5, …. is not too large). These approximate equalities are symmetrical relations, whom we call "tetra-group symmetries in long DNA-texts". For example, Fig. 1 shows results of our calculation of total probabilities $P_n(A_k)$, $P_n(T_k)$, $P_n(C_k)$ and $P_n(G_k)$ for DNA-text of the human chromosome № 1 that contains 248956422 letters (here the values of probabilities are rounded to the third decimal place; more detailed results are shown in [4]). One can see in Fig. 1 approximate equalities of high level of accuracy inside the set of $P_n(A_k)$, $P_n(T_k)$, $P_n(C_k)$ and $P_n(G_k)$ under different values n = 1, 2, 3, 4, 5 and $k \leq n$.

$P_1(A_1) = P_2(A_1) = P_2(A_2) = P_3(A_1) = P_3(A_2) = P_3(A_3) = P_4(A_1) = P_4(A_2) = P_4(A_3) = P_4(A_4) = P_5(A_1) = P_5(A_2) = P_5(A_3) = P_5(A_4) = P_5(A_5) = 0,291$.
$P_1(T_1) = P_2(T_1) = P_2(T_2) = P_3(T_1) = P_3(T_2) = P_3(T_3) = P_4(T_1) = P_4(T_2) = P_4(T_3) = P_4(T_4) = P_5(T_1) = P_5(T_2) = P_5(T_3) = P_5(T_4) = P_5(T_5) = 0,292$.
$P_3(C_1) = P_5(C_4) = 0,208$; $P_1(C_1) = P_2(C_1) = P_2(C_2) = P_3(C_2) = P_3(C_3) = P_4(C_1) = P_4(C_2) = P_4(C_3) = P_4(C_4) = P_5(C_1) = P_5(C_2) = P_5(C_3) = P_5(C_5) = 0,209$.
$P_1(G_1) = P_2(G_1) = P_2(G_2) = P_3(G_1) = P_3(G_2) = P_3(G_3) = P_4(G_1) = P_4(G_2) = P_4(G_3) = P_4(G_4) = P_5(G_1) = P_5(G_2) = P_5(G_3) = P_5(G_4) = P_5(G_5) = 0,209$.

Fig. 1. Total probabilities $P_n(A_k)$, $P_n(T_k)$, $P_n(C_k)$ and $P_n(G_k)$ in the corresponding tetra-groups of n-letter words (n = 1, 2, 3, 4, 5) in the DNA-text of the following sequence, which contains 248956422 letters: Homo sapiens chromosome 1, GRCh38.p7 Primary Assembly. NCBI Reference Sequence: NC_000001.11; https://www.ncbi.nlm.nih.gov/nuccore/NC_000001.11

We have got similar results about an approximate equality of probabilities $P_n(A_k)$, $P_n(T_k)$, $P_n(C_k)$ and $P_n(G_k)$ (where n = 1, 2, 3, 4, 5) for several dozen long DNA-texts, including the complete genomes of a number of organisms. These results have given the opportunity to formulate the following rules of the tetra-group symmetries in long DNA-texts [4].

The First Rule of Tetra-Group Symmetries in Long DNA-Texts: if a long sequence of single stranded DNA is represented in different forms of texts of n-letter words (n = 1, 2, 3, 4, 5…. is not too large), then - in these texts - probabilities of words with the letter X (X = A, T, C, G) in their position k ≤ n are approximately equal to each other independently on values n.

The Second Rule of Tetra-Group Symmetries in Long DNA-Texts: if a long sequence of single stranded DNA is represented in different forms of texts of n-letter words (n = 1, 2, 3, 4, 5…. is not too large), then - in these texts - probabilities of words with the letter X (X = A, T, C, G) in their position k ≤ n are approximately equal to each other independently on values k.

The Third Rule of Tetra-Group Symmetries in Long DNA-Texts: if a long sequence of those single-stranded DNA, that satisfy the second Chargaff's rule, is represented in different forms of texts of n-letter words (n = 1, 2, 3, 4, 5…. is not too large), then - in these texts - probabilities of words with the complementary letters A and T in their position k are approximately equal to each other. The same is true for probabilities of words with the complementary letters C and G in their position k.

These rules are candidacies for the role of universal rules of long DNA-texts in living bodies. Further research is needed to define a degree of universality of these rules and these cooperative genetic symmetries. These phenomenologic rules can be modelled on the basis of a quantum informational approach [4].

2   Tetra-Group Symmetries in Complete Sets of Chromosomes

Human organisms contain 24 chromosomes: 22 autosomes and 2 sex chromosomes X and Y. These chromosomes are long DNA molecules, the length of texts in which lie in the range from 50 to 250 million letters approximately. Autosomes are numbered from 1 to 22. We have studied tetra-group symmetries of long DNA-texts in each of 24 human chromosomes for cases n = 1, 2, 3, 4, 5. In the result we have obtained not only a confirmation of the described 3 rules of the tetra-group symmetries for all separate chromosomes but also an additional unexpected result concerning the complete set of chromosomes: numeric characteristics of tetra-group symmetries of long DNA-texts of separate chromosomes are approximately equal to each other for all human chromosomes. This result was unexpected since 24 human chromosomes differ greatly by their molecular dimensions, their sequences of letters, kinds and quantities of genes in them, cytogenetic bands (which shows biochemical specificity of different parts of chromosomes), etc. But in relation to values of tetra-group symmetries of their DNA-texts (that is, in relation to their total probabilities $P_n(A_k)$, $P_n(T_k)$, $P_n(C_k)$ and $P_n(G_k)$) 24 human chromosomes are very similar each other [4]. Figure 2 shows the average values of the probabilities $P_n(A_k)$, $P_n(T_k)$, $P_n(C_k)$ and $P_n(G_k)$ for all 24 human chromosomes.

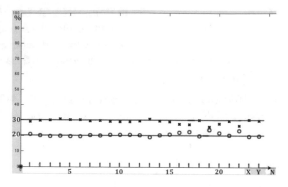

Fig. 2. The graphical representation of the average values of the probabilities $P_n(A_k)$, $P_n(T_k)$, $P_n(C_k)$ and $P_n(G_k)$ for all 24 human chromosomes. The abscissa axis contains numberings N of chromosomes, and the ordinate axis contains average values of these probabilities in percent. The symbol "o" corresponds the average values of $P_n(C_k) \approx P_n(G_k)$, and the symbol "x" corresponds the average values of $P_n(A_k) \approx P_n(T_k)$. Direct lines correspond values 20% and 30%.

One should add that fluctuations of values of probabilities $P_n(A_k)$, $P_n(T_k)$, $P_n(C_k)$ and $P_n(G_k)$ for different n and k (n = 1, 2, 3, 4, 5) in relation to their average values were very small. For example, in the case of the human chromosome № 1, fluctuations of the probabilities were ±0,006% (Fig. 3). For other chromosomes, the fluctuations of the probabilities were of the same order of magnitude.

Average value of $P_n(A_k)$ and fluctuations (%)	Average value of $P_n(T_k)$ and fluctuations (%)	Average value of $P_n(C_k)$ and fluctuations (%)	Average value of $P_n(G_k)$ and fluctuations (%)
29,100±0,005	29,176±0,006	20,850±0,006	20,874±0,005

Fig. 3. Fluctuations of the probabilities $P_n(A_k)$, $P_n(T_k)$, $P_n(C_k)$ and $P_n(G_k)$ under different values n and k in relation to their average values in the case of the DNA-text of human chromosome № 1.

It seems that – in the case of all human chromosomes - the average values of the probabilities of $P_n(A_k) \approx P_n(T_k)$ are concentrated around the value of 30% and the average values of the probabilities of $P_n(C_k) \approx P_n(G_k)$ are concentrated around the value of 20%. In theory of musical harmony, the ratio 30/20 = 3/2 is called "quint" (or "fifth").

We also analyzed tetra-group symmetries of the DNA-texts in the complete sets of chromosomes of a few organisms, which are traditionally used as model organisms in the study of genetics, development and disease: a nematode *Caenorhabditis elegans*, a fruit fly *Drosophila melanogaster*, a plant *Arabidopsis thaliana*. All received results show that the represented tetra-group rules are implemented not only for separate long DNA-texts but also for studied complete sets of chromosomes of eukaryotes. These initial results allow putting forward the hypothesis about existence of the following

general rule of tetra-group symmetries of DNA-texts in complete sets of chromosomes of different organisms: in the complete set of chromosomes of each of eukaryot organisms, characteristics of tetra-group symmetries of the DNA-texts of separate chromosomes are approximately equal to each other for all chromosomes [4].

Further researches are needed to check a degree of universality of this rule.

3   Fractal Genetic Nets, Tetra-Group Symmetries of DNA-Texts and a Fractal Grammar of Biology

Our article [26] has introduced the notion of "fractal genetic nets" (FGN) of texts for revealing hidden regularities in long DNA-texts in a connection with Charaff's thoughts about a "grammar of biology" [8]). Each FGN of texts can contain different fractal genetic trees (FGT). In that work we have represented results testifying in favor of existence of new symmetry principles in long nucleotide sequences in an addition to the known symmetry principle on the basis of the generalized Chargaff's second parity rule.

Below we represent our results about implementation of the rules of tetra-group symmetries of long DNA-texts at different levels of different fractal genetic trees and nets. In line with our article [26], FGT of various types are constructed by the method of sequential positional convolutions of a long DNA-text into a set of ever-shorter texts. Figure 4 explains a construction of FGT of various types by means of an example of FGT for a DNA-text, which is represented as a sequence S_0 of 3-letter words (a sequence of triplets). In each triplet, 0, 1 and 2 numbers its three positions correspondingly. At the first level of the text convolution, an initial long sequence S_0 of triplets is transformed by means of a positional convolution into three new sequences of nucleotides $S_{1/0}$, $S_{1/1}$ and $S_{1/2}$, each of which is 3 times shorter in comparison with the initial sequence S_0 (in this notation of sequences, numerator of the index shows the level of the convolution, and the denominator - the position of the triplets, which is used for the convolution): the sequence $S_{1/0}$ includes one by one all the nucleotides that are in the initial position "0" of triplets of the original sequence S_0; the sequence $S_{1/1}$ includes one by one all the nucleotides that are in the middle position "1" of triplets of the original sequence S_0; the sequence $S_{1/2}$ includes one by one all the nucleotides that are in the last position "2" of triplets of the original sequence S_0. At the final stage of the first level of the positional convolution, each of the sequences of nucleotides $S_{1/0}$, $S_{1/1}$, $S_{1/2}$ is represented as a sequence of triplets, where three positions inside each of triplets are numbered again by 0, 1 and 2. To construct the second level of the convolution, each of the sequences $S_{1/0}$, $S_{1/1}$, $S_{1/2}$ is transformed by means of the same positional convolution into three new sequences: $S_{1/0}$ is convolved into $S_{2/00}$, $S_{2/01}$, $S_{2/02}$; $S_{1/1}$ – into $S_{2/10}$, $S_{2/11}$, $S_{2/12}$; $S_{1/2}$ – into $S_{2/20}$, $S_{2/21}$, $S_{2/22}$. Similarly, the third level and subsequent levels of the convolution are constructed to form a multi-level tree of sequences of triplets called "the fractal genetic tree for the triplet convolution" or briefly "FGT-3". Texts at lower levels of any FGT can be figuratively called "daughter texts" of the original long DNA-text S_0.

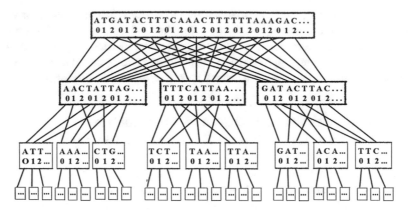

Fig. 4. The scheme of the fractal genetic tree (FGT-3) of a DNA-text, which is represented as a sequence of triplets (from [26]).

This FGT possesses a fractal-like character if the enumeration of positions is only taken into account: each of long sequences of this FGT can be taken as an initial sequence to form a similar genetic net on its basis (Fig. 4). In general case, the FGT can be built not only for triplets, but also for other n-plets (n = 2, 4, 5,...) by means of a repeated positional convolution of each of sequences from the previous level into "n" sequences of the next level of the convolution. This way one can build FGT-2, FGT-4, FGT-5, etc. for n = 2, 3, 4, 5,... correspondingly. A set of these FGT-2, FGT-3, FGT-4, FGT-5, ... forms a net of separate trees; FGN is a set of such separate trees.

For a long DNA-text of any biological organism, one can study implementation of the described rules of tetra-group symmetries in long texts at different levels of the convolution in cases of FGT-2, FGT-3, FGT-4, etc. Our own results of initial study of enough long DNA-texts of different organisms show implementation of these tetra-group rules in all texts at initial levels of the FGT-2, FGT-3 and FGT-4. Moreover values of the probabilities $P_n(A_k)$, $P_n(T_k)$, $P_n(C_k)$ and $P_n(G_k)$ are approximately repeated in all convoluted texts at different initial levels of convolutions in these tested cases of FGT. Figures 5 and 6 illustrate this phenomenologic fact for long texts at initial levels of convolutions in FGT-2 and FGT-3 for human sex chromosomes X and Y, whose DNA-texts contain 156040895 and 57227415 letters correspondingly.

One can see from Figs. 5 and 6 that fluctuation intervals of studied probabilities are very narrow for the set of texts at each of the initial levels of the fractal genetic trees. Moreover, fluctuation intervals for each of separate kinds of probabilities $P_n(A_k)$, $P_n(T_k)$, $P_n(C_k)$ and $P_n(G_k)$ are approximately equal to each other for the sets of texts at all considered levels of the FGT-2 and the FGT-3. Our results can be considered as evidences in favor of a fractal grammar of genetics in line with the Chargaff's problem about a grammar of biology.

	Level 0	Level 1/0	Level 1/1	Level 2/00	Level 2/01	Level 2/10	Level 2/11
$P_n(A_k)\in$	0.3017÷ 0.3019	0.3017÷ 0.3019	0.3016÷ 0.3020	0.3016÷ 0.3017	0.3017÷ 0.3019	0.3015÷ 0.3015	0.3015÷ 0.3019
$P_n(T_k)\in$	0.3028÷ 0.3029	0.3028÷ 0.3028	0.3027÷ 0.3027	0.3025÷ 0.3029	0.3028÷ 0.3029	0.3027÷ 0.3031	0.3027÷ 0.3027
$P_n(C_k)\in$	0.197÷ 0.1971	0.1969÷ 0.1971	0.1969÷ 0.197	0.1967÷ 0.1971	0.1969÷ 0.1971	0.1968÷ 0.197	0.1971÷ 0.1971
$P_n(G_k)\in$	0.1981÷ 0.1982	0.1981÷ 0.1982	0.1981÷ 0.1982	0.1981÷ 0.1982	0.1981÷ 0.1982	0.198÷ 0.1984	0.1981÷ 0.1983

	Level 0	Level 1/0	Level 1/1	Level 1/2	Level 2/00	Level 2/01
$P_n(A_k)\in$	0.3017÷ 0.3019	0.3016÷ 0.3020	0.3017÷ 0.3017	0.3017÷ 0.3019	0.3015÷ 0.3018	0.3016÷ 0.3019
$P_n(T_k)\in$	0.3028÷ 0.3029	0.3027÷ 0.3029	0.3028÷ 0.3029	0.3027÷ 0.3029	0.3024÷ 0.3029	0.3025÷ 0.3028
$P_n(C_k)\in$	0.1970÷ 0.1971	0.1970÷ 0.1970	0.1970÷ 0.1974	0.1967÷ 0.1971	0.1969÷ 0.1971	0.1967÷ 0.1968
$P_n(G_k)\in$	0.1981÷ 0.1982	0.1981÷ 0.1981	0.1980÷ 0.1980	0.1981÷ 0.1982	0.198÷ 0.1981	0.1979÷ 0.1985

	Level 2/02	Level 2/10	Level 2/11	Level 2/12	Level 2/20	Level 2/21	Level 2/22
$P_n(A_k)\in$	0.3015÷ 0.3019	0.3018÷ 0.3019	0.3015÷ 0.3018	0.3014÷ 0.3014	0.3016÷ 0.3019	0.3016÷ 0.3017	0.3015÷ 0.3018
$P_n(T_k)\in$	0.3025÷ 0.3026	0.3025÷ 0.3025	0.3026÷ 0.3030	0.3024÷ 0.3030	0.3025÷ 0.3029	0.3025÷ 0.3030	0.3027÷ 0.3030
$P_n(C_k)\in$	0.1966÷ 0.1973	0.1967÷ 0.1969	0.1968÷ 0.1969	0.1969÷ 0.1974	0.1968÷ 0.1971	0.1966÷ 0.1968	0.1967÷ 0.1971
$P_n(G_k)\in$	0.1980÷ 0.1981	0.1977÷ 0.1987	0.1979÷ 0.1983	0.1978÷ 0.1982	0.1980÷ 0.1981	0.1980÷ 0.1985	0.1978÷ 0.1981

Fig. 5. Tables of fluctuation intervals of probabilities $P_n(A_k)$, $P_n(T_k)$, $P_n(C_k)$ and $P_n(G_k)$ for the set of all texts at each of levels of convolutions in the FGT-2 (upper table) and in the FGT-3 (bottom tables) in the case of the human chromosome X (NCBI Reference Sequence: NC_000023.11).

	Level 0	Level 1/0	Level 1/1	Level 2/00	Level 2/01	Level 2/10	Level 2/11
$P_n(A_k) \in$	0.2983÷ 0.2987	0.2982÷ 0.2982	0.2979÷ 0.2987	0.2980÷ 0.2989	0.2981÷ 0.2981	0.2981÷ 0.2987	0.2976÷ 0.2992
$P_n(T_k) \in$	0.3009÷ 0.3012	0.3008÷ 0.3021	0.3007÷ 0.3014	0.3009÷ 0.3013	0.3005÷ 0.3020	0.3005÷ 0.3010	0.3005÷ 0.3009
$P_n(C_k) \in$	0.1998÷ 0.1998	0.1997÷ 0.1997	0.1995÷ 0.1995	0.1996÷ 0.1999	0.1997÷ 0.1997	0.1996÷ 0.2001	0.1995÷ 0.1995
$P_n(G_k) \in$	0.1998÷ 0.2003	0.1994÷ 0.2001	0.1999÷ 0.2004	0.1995÷ 0.1998	0.1994÷ 0.2002	0.1995÷ 0.2002	0.1996÷ 0.2003

	Level 0	Level 1/0	Level 1/1	Level 1/2	Level 2/00	Level 2/01
$P_n(A_k) \in$	0.2983÷ 0.2987	0.2981÷ 0.2988	0.2984÷ 0.2989	0.298÷ 0.2982	0.2981÷ 0.2988	0.2976÷ 0.2984
$P_n(T_k) \in$	0.3009÷ 0.3012	0.3008÷ 0.3012	0.3007÷ 0.3013	0.3008÷ 0.3012	0.300÷ 0.3013	0.3010÷ 0.3019
$P_n(C_k) \in$	0.1998÷ 0.1998	0.1995÷ 0.2004	0.1999÷ 0.2002	0.1999÷ 0.1999	0.199÷ 0.1999	0.1994÷ 0.1994
$P_n(G_k) \in$	0.1998÷ 0.2003	0.1996÷ 0.1996	0.1996÷ 0.1997	0.200÷ 0.2007	0.1996÷ 0.2001	0.1991÷ 0.2003

	Level 2/02	Level 2/10	Level 2/11	Level 2/12	Level 2/20	Level 2/21	Level 2/22
$P_n(A_k) \in$	0.2974÷ 0.2982	0.2978÷ 0.2991	0.2979÷ 0.2979	0.2983÷ 0.2985	0.2975÷ 0.2976	0.2974÷ 0.2990	0.2975÷ 0.2979
$P_n(T_k) \in$	0.3005÷ 0.3023	0.2997÷ 0.3006	0.3002÷ 0.3010	0.3006÷ 0.3008	0.3005÷ 0.3023	0.3005÷ 0.3010	0.3005÷ 0.3006
$P_n(C_k) \in$	0.1994÷ 0.2005	0.1994÷ 0.1996	0.1995÷ 0.2004	0.1996÷ 0.2011	0.1993÷ 0.2003	0.1997÷ 0.2004	0.1993÷ 0.2002
$P_n(G_k) \in$	0.1990÷ 0.1990	0.2000÷ 0.2008	0.1988÷ 0.2007	0.1991÷ 0.1996	0.1997÷ 0.1998	0.1996÷ 0.1996	0.1998÷ 0.2013

Fig. 6. Tables of fluctuation intervals of probabilities $P_n(A_k)$, $P_n(T_k)$, $P_n(C_k)$ and $P_n(G_k)$ for the set of all texts at each of levels of convolutions in the FGT-2 (upper table) and in the FGT-3 (bottom tables) in the case of the human chromosome Y (NCBI Reference Sequence: NC_000024.10).

These results about a fractal grammar of long DNA-texts are additionally interesting by the following reasons:

- Many biological organisms have fractal-like inherited configurations in their bodies. This phenomenon can be considered as a consequence of the fractal-like organization of long DNA texts with the participation of tetra-group symmetries;
- As known, fractals allow a colossal compression of information (https://en.wikipedia.org/wiki/Fractal_compression). It is obvious that an opportunity of information compression is essential for genetic systems. Modern computer science knows a great number of methods of information compression including many methods of fractal compression. Our described results about fractal genetic nets can lead to a discovery of those «genetic» methods of information compression, which are used in genetic systems and in biological bodies in the whole;
- Many authors published their ideas and materials about relations of genomes with fractal structures in different aspects [27–33]. For example, work [33] shown an existence of fractal globule in the three dimensional architecture of whole genomes, where spatial chromosome territories exist and where maximally dense packing is provided on the basis of a special fractal packing, which provides the ability to easily fold and unfold any genomic locus. One should note that, by contrast to the work [33], in our work we study not spatial packing of whole genomes in a form of fractal globules but the quite another thing: we study the fractal organization of long DNA-texts, in particular, in the form of described fractal genetic nets or trees of different kinds (FGT-n, where n = 1, 2, 3, 4,...), which are connected with tetra-group symmetries of these texts (these symmetries and fractals were never studied early).
- Fractals are actively used in study of cancer; some modern data testify that cancer processes are related with fractal patterns and their development [34–38].
- Fractals are connected with theory of dynamic chaos, which has many applications in engineering technologies. We believe that the discovery of fractal-like properties of DNA-texts related with their tetra-group symmetries can lead to new ideas in theoretical and application areas, including problems of artificial intelligence and in-depth study of genetic phenomena for medical and biotechnological tasks [3, 39–44].

4   Conclusions

A special class of symmetries is implemented in long DNA-texts of different organisms. Long DNA-texts are constructed by Nature with using fractal structures. This can be one of reasons of existence of a great number of inherited fractal configurations in biological bodies in their normal and pathologic states including fractal patterns in cancerous and biorythmic structures. Fractal patterns are connected with the theory of dynamic chaos, which has many applications in sciences and technology. A specificity of fractal genetic nets can provoke a further development of the theory of dynamic chaos and its applications.

References

1. Petoukhov, S.V.: Matrix Genetics, Algebras of the Genetic Code, Noise Immunity. RCD, Moscow, Russia (2008). (in Russian)
2. Petoukhov, S.V., He, M.: Symmetrical Analysis Techniques for Genetic Systems and Bioinformatics: Advanced Patterns and Applications. IGI Global, Hershey (2010)
3. Petoukhov, S., Petukhova, E., Hazina, L., Stepanyan, I., Svirin, V., Silova, T.: The genetic coding, united-hypercomplex numbers and artificial intelligence. In: Hu, Z., Petoukhov, S., He, M. (eds.) Advances in Artificial Systems for Medicine and Education. AIMEE 2017. Advances in Intelligent Systems and Computing, vol. 658. Springer, Cham (2017). https://doi.org/10.1007/978-3-319-67349-3_1. Print ISBN 978-3-319-67348-6, Online ISBN 978-3-319-67349-3. https://link.springer.com/search?query=978-3-319-67348-6. Accessed 20 Aug 2017
4. Petoukhov, S.V.: The rules of long DNA-sequences and tetra-groups of oligonucleotides (2017). (https://arxiv.org/abs/1709.04943)
5. Hu, Z.B., Petoukhov, S.V., Petukhova, E.S.: I-Ching, dyadic groups of binary numbers and the geno-logic coding in living bodies. Progress in Biophysics and Molecular Biology (2017, in press). http://authors.elsevier.com/sd/article/S0079610717300949. Accessed 18 Sept 2017
6. Hu, Z.B., Petoukhov, S.V.: Generalized crystallography, the genetic system and biochemical esthetics. Struct. Chem. 28(1), 239–247 (2017). https://doi.org/10.1007/s11224-016-0880-0. http://link.springer.com/journal/11224/28/1/page/2
7. Fickett, J., Burks, C.: Development of a database for nucleotide sequences. In: Waterman, M.S. (ed.) Mathematical Methods in DNA Sequences, pp. 1–34. CRC Press, Florida (1989)
8. Chargaff, E.: Preface to a grammar of biology: a hundred years of nucleic acid research. Science 172, 637–642 (1971)
9. Chargaff, E.: Structure and function of nucleic acids as cell constituents. Fed. Proc. 10, 654–659 (1951)
10. Albrecht-Buehler, G.: Asymptotically increasing compliance of genomes with Chargaff's second parity rules through inversions and inverted transpositions. Proc. Nat. Acad. Sci. USA 103(47), 17828–17833 (2006)
11. Baisnee, P.-F., Hampson, S., Baldi, P.: Why are complementary DNA strands symmetric? Bioinformatics 18(8), 1021–1033 (2002)
12. Bell, S.J., Forsdyke, D.R.: Deviations from Chargaff's second parity rule correlate with direction of transcription. J. Theor. Biol. 197, 63–76 (1999)
13. Chargaff, E.: A fever of reason. Ann. Rev. Biochem. 44, 1–20 (1975)
14. Dong, Q., Cuticchia, A.J.: Compositional symmetries in complete genomes. Bioinformatics 17, 557–559 (2001)
15. Forsdyke, D.R.: A stem-loop "kissing" model for the initiation of recombination and the origin of introns. Mol. Biol. Evol. 12, 949–958 (1995)
16. Forsdyke, D.R.: Symmetry observations in long nucleotide sequences: a commentary on the discovery of Qi and Cuticchia. Bioinf. Lett. 18(1), 215–217 (2002)
17. Forsdyke, D.R.: Evolutionary Bioinformatics. Springer, New York (2006)
18. Forsdyke, D.R., Bell, S.J.: Purine-loading, stem-loops, and Chargaff's second parity rule. Appl. Bioinf. 3, 3–8 (2004)
19. Mitchell, D., Bridge, R.: A test of Chargaff's second rule. BBRC 340, 90–94 (2006)
20. Okamura, K., Wei, J., Scherer, S.: Evolutionary implications of inversions that have caused intra-strand parity in DNA. BMC Genomics 8, 160–166 (2007). http://www.gutenberg.org/files/39713/39713-h/39713-h.htm#Page_264

21. Perez, J.-C.: The "3 genomic numbers" discovery: how our genome single-stranded DNA sequence is "self-designed" as a numerical whole. Appl. Math. **4**, 37–53 (2013). http://dx.doi.org/10.4236/am.2013.410A2004
22. Prabhu, V.V.: Symmetry observation in long nucleotide sequences. Nucleic Acids Res. **21**, 2797–2800 (1993)
23. Rapoport, A.E., Trifonov, E.N.: Compensatory nature of Chargaff's second parity rule. J. Biomol. Struct. Dyn. 1–13 (2012). https://doi.org/10.1080/07391102.2012.736757
24. Sueoka, N.: Intrastrand parity rules of DNA base composition and usage biases of synonymous codons. J. Mol. Evol. **40**, 318–325 (1995)
25. Yamagishi, M., Herai, R.: Chargaff's "Grammar of Biology": New Fractal-like Rules. http://128.84.158.119/abs/1112.1528v1 (2011)
26. Petoukhov, S.V., Svirin, V.I.: Fractal genetic nets and symmetry principles in long nucleotide sequences. Symmetry Cult. Sci. **23**(3–4), 303–322 (2012). http://petoukhov.com/PETOUKHOV_SVIRIN_FGN.pdf
27. Jeffrey, H.J.: Chaos game representation of gene structure. Nucleic Acids Res. **18**(8), 2163–2170 (1990)
28. Peng, C.K., Buldyrev, S.V., Goldberger, A.L., Havlin, S., Sclortino, F., Simons, M., Stanley, H.E.: Long-range correlations in nucleotide sequences. Nature **356**, 168–170 (1992)
29. Peng, C.K., Buldyrev, S.V., Goldberger, A.L., Havlin, S., Sclortino, F., Simons, M., Stanley, H.E.: Fractal landscape analysis of DNA walks. Phys. A **191**(1–4), 25–29 (1992)
30. Pellionis, A.J.: The principle of recursive genome function. Cerebellum **7**, 348–359 (2008). https://doi.org/10.1007/s12311-008-0035-y
31. Pellionisz, A.J., Graham, R., Pellionisz, P.A. Perez, J.C.: Recursive genome function of the cerebellum: geometric unification of neuroscience and genomics, In: Manto, M., Gruol, D.L., Schmahmann, J.D., Koibuchi, N., Rossi, F. (eds.) Handbook of the Cerebellum and Cerebellar Disorders, pp. 1381–1423 (2012)
32. Perez, J.C.: Codon populations in single-stranded whole human genome DNA are fractal and fine-tuned by the Golden Ratio 1.618. Interdiscip. Sci. Comput. Life Sci. **2**, 228–240 (2010). https://doi.org/10.1007/s12539-010-0022-0
33. Lieberman-Aiden, E., van Berkum, N.L., Williams, L., Imakaev, M., Ragoszy, T., Telling, A., Lajoie, B.R., Sabo, P.J., Dorschner, M.O., Sandstrom, R., Bernstein, B., Bender, M.A., Groudine, M., Gnirke, A., Stamatoyannopoulos, J., Mirny, L.A., Lander, E.S., Dekker, J.: Comprehensive mapping of long-range interactions reveals folding principles of the human genome. Science **326**(5950), 289–293 (2009). https://doi.org/10.1126/science.1181369
34. Baish, J.W., Jain, R.K.: Fractals and cancer. Can. Res. **60**, 3683–3688 (2000)
35. Bizzarri, M., Giuliani, A., Cucina, A., Anselmi, F.D., Soto, A.M., Sonnenschein, C.: Fractal analysis in a systems biology approach to cancer. Semin. Cancer Biol. **21**(3), 175–182 (2011). https://doi.org/10.1016/j.semcancer.2011.04.002
36. Lennon, F.E., Cianci, G.C., Cipriani, N.A., Hensing, T.A., Zhang, H.J., Chen, C.-T., Murgu, S.D., Vokes, E.E., Vannier, M.W., Salgia, R.: Lung cancer - a fractal viewpoint. Nat. Rev. Clin. Oncol. **12**(11), 664–675 (2015) https://doi.org/10.1038/nrclinonc.2015.108
37. Dokukin, M.E., Guz, N.V., Woodworth, C.D., Sokolov, I.: Emergence of fractal geometry on the surface of human cervical epithelial cells during progression towards cancer. New J. Phys. **17**(3), 033019 (2015)
38. Perez, J.C.: Sapiens mitochondrial DNA genome circular long range numerical meta structures are highly correlated with cancers and genetic diseases mtDNA mutations. J. Cancer Sci. Ther. **9**, 6 (2017). https://doi.org/10.4172/1948-5956.1000469
39. Abo-Zahhad, M., Ahmed, S.M., Abd-Elrahman, S.A.: Genomic analysis and classification of exon and intron sequences using DNA numerical mapping techniques. Int. J. Inf. Technol. Comput. Sci. (IJITCS) **4**(8), 22–36 (2012)

40. Abo-Zahhad, M., Ahmed, S.M., Abd-Elrahman, S.A.: A novel circular mapping technique for spectral classification of exons and introns in human DNA sequences. Int. J. Inf. Technol. Comput. Sci. (IJITCS) **6**(4), 19–29 (2014). https://doi.org/10.5815/ijitcs.2014.04.02

41. Meher, J.K., Panigrahi, M.R., Dash, G.N., Meher, P.K.: Wavelet based lossless DNA sequence compression for faster detection of eukaryotic protein coding regions. Int. J. Image Graph. Sig. Process. (IJIGSP) **4**(7), 47–53 (2012). https://doi.org/10.5815/ijigsp.2012.07.05

42. Srivastava, P.C., Agrawal, A., Mishra, K.N., Ojha, P.K., Garg, R.: Fingerprints, Iris and DNA features based multimodal systems: a review. Int. J. Inf. Technol. Comput. Sci. (IJITCS) **5**(2), 88–111 (2013). https://doi.org/10.5815/ijitcs.2013.02.10

43. Mousa, H.M.: DNA-genetic encryption technique. Int. J. Comput. Netw. Inf. Secur. (IJCNIS) **8**(7), 1–9 (2016). https://doi.org/10.5815/ijcnis.2016.07.01

44. Hossein, S.M., Roy, S.: A compression & encryption algorithm on DNA sequences using dynamic look up table and modified Huffman techniques. Int. J. Inf. Technol. Comput. Sci. (IJITCS) **5**(10), 39–61 (2013). https://doi.org/10.5815/ijitcs.2013.10.05

Triply Stochastic Cubes Associated with Genetic Code Numerical Mappings

Matthew He[1($\boxtimes$)], Zhengbing Hu[2], and Sergey Petoukhov[3]

[1] Nova Southeastern University, Fort Lauderdale, FL 33314, USA
hem@nova.edu
[2] Central China Normal University, Wuhan, China
hzb@mail.ccnu.edu.cn
[3] Mechanical Engineering Research Institute, Russian Academy of Sciences,
Moscow 101830, Russia
spetoukhov@gmail.com

Abstract. Knowledge about genetic coding systems are useful for computer science, engineering and education. In this paper we derive triply stochastic cubes associated with the triplet genetic code numerical mappings. We also demonstrate the symmetrical patterns between the entries of the cubes and DNA molar concentration accumulation via an arithmetic sequence. The stochastic cubes based on genetic code were derived by using three kinds of chemically determined equivalences. We have shown that at each stage (Nth step) of matrix evolution, hydrogen bonds expansion is triply stochastic and its accumulation is governed by an arithmetic sequence with a common difference of total number of hydrogen bonds of **5N**; the pyrimidines/purines ring expansion is triply stochastic and its accumulation is governed by an arithmetic sequence with a common difference of total number of rings of **3N**; and the amino-mutating absence/present expansion is also triply stochastic and its accumulation is governed by an arithmetic sequence with a common difference of total number of amino-mutating of **1N**. Data about the genetic stochastic matrices/cubes associated with the genetic codes can lead to new understanding of genetic code systems, to new effective algorithms of information processing which has a perspective to be applied for modeling mutual communication among different parts of the genetic ensemble.

Keywords: Genetic code equivalence · DNA numerical mapping
Triply stochastic cubes · Arithmetic sequence

1 Introduction

Knowledge about genetic coding systems are useful for computer science, engineering and education. A mathematical definition of genetic code is a mapping **g** as the following:

© Springer International Publishing AG, part of Springer Nature 2019
Z. Hu et al. (Eds.): ICCSEEA 2018, AISC 754, pp. 601–616, 2019.
https://doi.org/10.1007/978-3-319-91008-6_60

$$\mathbf{g} : \mathbf{V} \to \mathbf{W},$$

where $\mathbf{V} = \{(x_1 x_2 \, x_3): x_i \in \mathbf{R} = \{A, C, G, U\} = \text{RNA base set}\} = \text{the set of codons and}$ $\mathbf{W} = \{\text{Ala, Arg, Asp, ..., Val, UAA, UAG, UGA}\} = \text{the set of amino acids and termi-}$ nation codons. The inheritable information is encoded by the texts from three-alphabetic words - *triplets* compounded on the basis of the alphabet consisted of four characters being the nitrogen bases: A (adenine), C (cytosine), G (guanine), T (thiamine).

The RNA base set $\mathbf{R}$ has 4 elements. Each codon consists of three elements from the set $\mathbf{R}$. A complete list of $4^3 = 64$ triplets (codons) with corresponding amino acids with three (one) letter code and stop codons in the universal genetic code is listed on the Table 1 below:

Table 1. The universal genetic code

		Second Position of Codon				
		U	C	A	G	
First Position	U	UUU Phe [F]	UCU Ser [S]	UAU Tyr [Y]	UGU Cys [C]	U
		UUC Phe [F]	UCC Ser [S]	UAC Tyr [Y]	UGC Cys [C]	C
		UUA Leu [L]	UCA Ser [S]	UAA *Ter* [end]	UGA *Ter* [end]	A
		UUG Leu [L]	UCG Ser [S]	UAG *Ter* [end]	UGG Trp [W]	G
	C	CUU Leu [L]	CCU Pro [P]	CAU His [H]	CGU Arg [R]	U
		CUC Leu [L]	CCC Pro [P]	CAC His [H]	CGC Arg [R]	C
		CUA Leu [L]	CCA Pro [P]	CAA Gln [Q]	CGA Arg [R]	A
		CUG Leu [L]	CCG Pro [P]	CAG Gln [Q]	CGG Arg [R]	G
	A	AUU Ile [I]	ACU Thr [T]	AAU Asn [N]	AGU Ser [S]	U
		AUC Ile [I]	ACC Thr [T]	AAC Asn [N]	AGC Ser [S]	C
		AUA Ile [I]	ACA Thr [T]	AAA Lys [K]	AGA Arg [R]	A
		AUG Met [M]	ACG Thr [T]	AAG Lys [K]	AGG Arg [R]	G
	G	GUU Val [V]	GCU Ala [A]	GAU Asp [D]	GGU Gly [G]	U
		GUC Val [V]	GCC Ala [A]	GAC Asp [D]	GGC Gly [G]	C
		GUA Val [V]	GCA Ala [A]	GAA Glu [E]	GGA Gly [G]	A
		GUG Val [V]	GCG Ala [A]	GAG Glu [E]	GGG Gly [G]	G

Observing this table, it's easy to see that some amino acids are encoded by several different but related base codons or triplets. Three triplets (UAA, UAG, and UGA) are stop codons. These stop codons have no amino acids corresponding to their codes. The remaining 61 codons represent 20 different amino acids. Furthermore, one can see that the number of variants of location of 64 triplets in octet tables is equal 64! Which is approximately 10^{89}. There have been many studies and investigations on formal characterizations of the particular structure of the code, exploring a justification from physico-chemical and/or evolutionary points of view (Knight et al. 1999). Swanson in

(Swanson 1984) proposed a Gray code representation of the genetic code. A representation of the genetic code as a six-dimensional Boolean hypercube was proposed in Jiménéz-Montaño et al. (1994). In (Yang 2003) Yang applied a topological approach to rearranging the Hamiltonian-type graph of the codon map into a polyhedron model. The universal metric properties of the genetic code were defined by means of the nucleotide base representation on the square with vertices U or T = 00, C = 01, G = 10 and A = 11 in Štambuk (2000). Petoukhov (1999, 2001, 2002) introduced the "Biperiodic table of genetic code" as shown below:

This table demonstrates a great symmetrical structure and has led to many discoveries. The stochastic characteristic of the biperiodic table and symmetries in structure of genetic code were recently investigated in He (2003a, b, 2004); He et al. (2004).

In this paper we derive triply stochastic cubes associated with triplet genetic codes by using genetic code equivalences in Sect. 2. By iterating the stochastic cubes, we demonstrate the symmetrical patterns embedded inside the entries of the cube and DNA molar concentration accumulation. We present the iterative cubes associated with genetic code in Sect. 3 and discuss different applications and perspectives of stochastic genetic cubes in Sect. 4.

2   Three Kinds of Attributive Equivalence Mappings

In 1953, Watson and Crick worked out the double helix structure of DNA in 1953 (Watson 1968), the development and advances of genetics in biology have grown rapidly. In 1999, Ian Stewart went beyond the DNA structure and points out that "life is a partnership between genes and mathematics" (Stewart 1999). Between 1999–2001, Petoukhov showed that the "elementary" four-letter alphabet of genetic code comprises three binary sub-alphabets according to three kinds of biochemical attributes equivalence relation (marked by symbol "=") (Petoukhov 1999, 2001).

- 1st kind of equivalence: G = U = 0, A = C = 1, amino-mutating absence/present (0, 1)-combination,
- 2nd kind of equivalence: C = U = 1, A = G = 2, pyrimidines/purines ring-based (1, 2)-combination,
- 3rd kind of equivalence: A = U = 2, C = G = 3, hydrogen bonds-based (2, 3)-combination.

These three kinds of equivalences are also consistent with the early classification in Wittmann (1961). Associating these three kinds of attribute equivalences with basic numbers of 0, 1, 2, and 3 provides strong connections between chemical properties of RNA with three logical control of pair of numbers (0, 1), (1, 2), and (2, 3).

Based on these three attributes equivalences and assignments, three mapping relations from $\Re = \{A, C, G, U\}$ to $\Re = \{0, 1, 2, 3\}$ were defined in He (2004) as follows:

- α: $\{A, C, G, U\} \rightarrow \{0, 1\}$ with $\alpha (G) = \alpha (U) = 0$, $\alpha (A) = \alpha (C) = 1$,
- β: $\{A, C, G, U\} \rightarrow \{1, 2\}$ with $\beta (C) = \beta (U) = 1$, $\beta (A) = \beta (G) = 2$,
- γ: $\{A, C, G, U\} \rightarrow \{2, 3\}$ with $\gamma (A) = \gamma (U) = 2$, $\gamma (C) = \gamma (G) = 3$.

These three mappings were applied to the Table 2 in He (2004) to generate three matrices. The basic symmetrical properties were investigated in He (2004); He et al. (2004); He and Petoukhov (2011); Petoukhov and Petukhova (2016). Here we further investigate the symmetrical structures of matrix generated by these three mappings α, β, and γ. Most recently DNA numerical mapping techniques have applied to genomic analysis, classification of exon and intron sequences, and artificial neural network with some applications (Abo-Zahhad et al. 2012, 2014a, b).

Table 2. Biperiodic table of genetic code by Petoukhov

CCC	CCA	CAC	CAA	ACC	ACA	AAC	AAA
CCU	CCG	CAU	CAG	ACU	ACG	AAU	AAG
CUC	CUA	CGC	CGA	AUC	AUA	AGC	AGA
CUU	CUG	CGU	CGG	AUU	AUG	AGU	AGG
UCC	UCA	UAC	UAA	GCC	GCA	GAC	GAA
UCU	UCG	UAU	UAG	GCU	GCG	GAU	GAG
UUC	UUA	UGC	UGA	GUC	GUA	GGC	GGA
UUU	UUG	UGU	UGG	GUU	GUG	GGU	GGG

3   Triply Stochastic Cubes Associated with Genetic Code

A cube is triply (or 3-way) stochastic if each (frontal, horizontal and lateral) sliced section is doubly stochastic matrix. The term "stochastic matrix" goes back at least to Romanovsky (1931). A square matrix is a stochastic matrix if all entries of the matrix are nonnegative and the sum of the elements in each row (or column) is unity or a constant. If the sum of the elements in each row and column is unity or the same, the matrix is called doubly stochastic. Stochastic matrices have many remarkable properties. The properties of stochastic matrices are mainly spectral theoretic and are motivated by Markov chains. The structures of the stochastic cubes associated with genetic codes demonstrate the stochastic patterns hidden inside the genetic codes.

In this section we represent these three kinds of equivalences from Sect. 2 by square matrices:

- First kind of equivalence: G = U = 0, A = C = 1, amino-mutating absence/present (0, 1)-combination, this equivalence can be represented as a 2×2 matrix with genetic codes and corresponding numerical mutating values of 0 or 1 as follows:

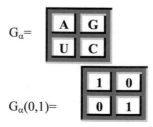

- Second kind of equivalence: $C = U = 1$, $A = G = 2$, pyrimidines/purines ring-based $(1, 2)$-combination, this equivalence can be represented as a 2×2 matrix with genetic codes and corresponding numerical ring values of 1 or 2 as follows:

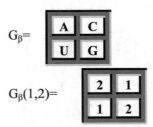

$$G_\beta =$$

$$G_\beta(1,2) =$$

- Third kind of equivalence: $A = U = 2$, $C = G = 3$, hydrogen bonds-based $(2, 3)$-combination, this equivalence can be represented as a 2×2 matrix with genetic codes and corresponding numerical bond values of 2 or 3 as follows:

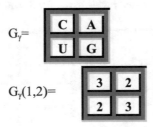

$$G_\gamma =$$

$$G_\gamma(1,2) =$$

These 2×2 matrices provide building blocks for constructing higher dimensions of matrices of 4×4, 8×8, and so on. The construction process is straight forward. We append each entry of the 2×2 matrix to the same 2×2 matrix and form 4 blocks of 2×2 matrices which generate a new 4×4 matrix consisting of 16 new entries. We then append each entry of the 4×4 matrix to the same 4×4 matrix and form 4 blocks of 4×4 matrices which generate a new 8×8 matrix consisting of 64 new entries. This process can be continued on for higher dimension of matrices.

For a purpose of illustration, we demonstrate this process in **a reverse order** of 3rd hydrogen bonds-based γ equivalence, 2nd pyrimidines/purines ring-based β equivalence, and 1st amino-mutating absence/present α equivalence as the γ equivalence has been given much attention in comparison with other two equivalences.

3.1 Stochastic Cubes Associated with Hydrogen Bonds-Based $\gamma(2, 3)$ Equivalence

The corresponding numeric matrices can be constructed by means of standardizing quaternary partition with an operation of **addition** of mapping numerical values of each DNA letter as shown below. We apply the γ-mapping to all 64 codons with the addition operations to each codon in the Table 3a, for example, $C = G = 3$, $A = U = 2$, $CC = 3 + 3 = 6$, $AAA = 2 + 2 + 2 = 6$, $CGG = 3 + 3 + 3 = 9$, we get the following matrices, denoted by $G_{\gamma^2}^+ \Rightarrow G_{\gamma^4}^+ \Rightarrow G_{\gamma^8}^+$ (Table 3b).

Table 3a. Hydrogen bonds-based (2, 3) equivalence matrix

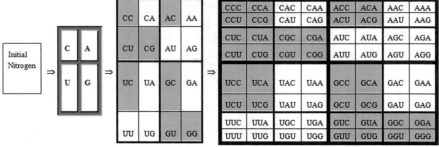

Table 3b. Hydrogen bonds-based (2, 3) equivalence matrix

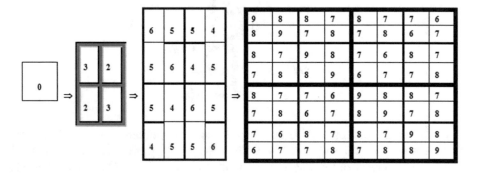

In each stage of this process $(G_{\gamma^2}^+ \Rightarrow G_{\gamma^4}^+ \Rightarrow G_{\gamma^8}^+)$, it generates a series of doubly-stochastic matrices of

1. (2×2) denoted by $G_{\gamma 2}$ with a row/column common sum of 5;
2. (4×4) denoted by $G_{\gamma 4}$ with a row/column common sum of 20;
3. (8×8) denoted by $G_{\gamma 8}$ with a row/column common sum of 60.

From the matrix $G_{\gamma 2}$, we can construct a $2 \times 2 \times 2$ cube $C_{\gamma 2}$ with the following 2 frontal/lateral/horizontal slices of doubly stochastic matrices;

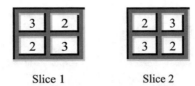

Slice 1   Slice 2

Take the sum of this $2 \times 2 \times 2$ cube from the frontal, lateral, and horizontal directions, respectively, each gives the common sum of 5. This cube $\mathbf{C}_{\gamma 2}$ is a triply stochastic cube. This represents the evolution of hydrogen bonds system of random values over time. As we move on the next stages of this process, we'll see a sequence of triply stochastic cubes will emerge.

From the matrix $\mathbf{G}_{\gamma 4}$, we start with the central 2×2 submatrix to form a $2 \times 2 \times 2$ cube $\mathbf{C}_{\gamma 4}^{(2 \times 2 \times 2)}$ with 2 slices of matrices:

Slice 1   Slice 2

Take the sum of this $2 \times 2 \times 2$ cube from the frontal, lateral, and horizontal directions, respectively, each gives the common sum of **10**. This cube $\mathbf{C}_{\gamma 4}^{(2 \times 2 \times 2)}$ is a triply stochastic cube.

Next we expand the central matrix to all directions (left, right, up, and down) by one matrix dimension to form a $4 \times 4 \times 4$ cube $\mathbf{C}_{\gamma 4}^{(4 \times 4 \times 4)}$ with 4 slices of 4×4 matrices through a permutation of 4 columns.

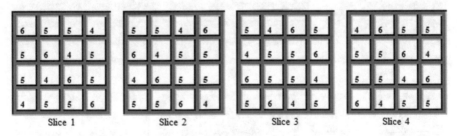

Slice 1   Slice 2   Slice 3   Slice 4

Take the sum of this $4 \times 4 \times 4$ cube from the frontal, lateral, and horizontal directions, respectively, each gives the common sum of **20**. This cube $\mathbf{C}_{\gamma 4}^{(4 \times 4 \times 4)}$ is a triply stochastic cube.

Next we apply the same process to the matrix of $\mathbf{G}_{\gamma 8}$ and form a series of $2 \times 2 \times 2$, $4 \times 4 \times 4$, $6 \times 6 \times 6$, and $8 \times 8 \times 8$ cubes as follows:

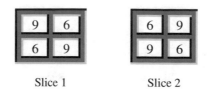

Slice 1   Slice 2

Take the sum of this $2 \times 2 \times 2$ cube $\mathbf{C}_{\gamma^8}^{(2\times2\times2)}$ from the frontal, lateral, and horizontal directions, respectively, each gives the common sum of **15**.

Next we generate $\mathbf{C}_{\gamma^8}^{(4\times4\times4)}$ as follows:

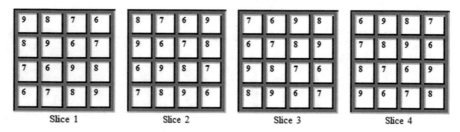

Slice 1   Slice 2   Slice 3   Slice 4

Take the sum of this $4 \times 4 \times 4$ cube $\mathbf{C}_{\gamma^8}^{(4\times4\times4)}$ from the frontal, lateral, and horizontal directions, respectively, each gives the common sum of **30**.

Next we generate $\mathbf{C}_{\gamma^8}^{(6\times6\times6)}$ as follows:

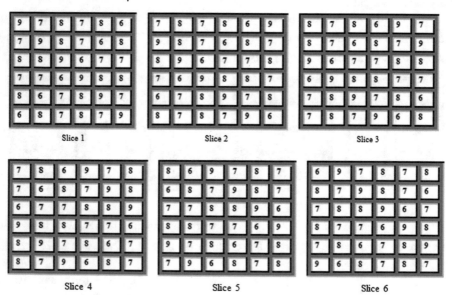

Slice 1   Slice 2   Slice 3

Slice 4   Slice 5   Slice 6

Take the sum of this $6 \times 6 \times 6$ cube $\mathbf{C}_{\gamma 8}^{(6 \times 6 \times 6)}$ from the frontal, lateral, and horizontal directions, respectively, each gives the common sum of **45**. Finally we generate $\mathbf{C}_{\gamma 8}^{(6 \times 6 \times 6)}$ as follows:

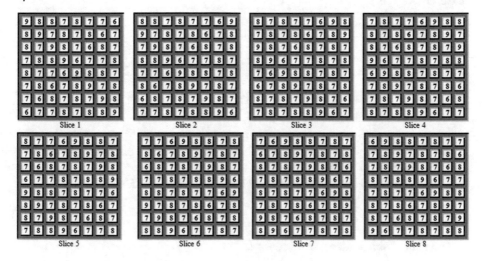

Take the sum of this $8 \times 8 \times 8$ cube $\mathbf{C}_{\gamma 8}^{(8 \times 8 \times 8)}$ from the frontal, lateral, and horizontal directions, respectively, each gives the common sum of **60**.

Here we summarize all the cubes generated by the 3rd hydrogen bonds-based (2, 3) equivalence under addition operation on each genetic code (Table 3c).

Table 3c. Common sums of cube slices of γ equivalence

γ Equivalence	No. of cubes	Common sums of cube slices
Initial stage	0	0
First stage	1	5
Second stage	2	10, 20 (d = 10)
Third stage	4	15, 30, 45, 60 (d = 15)

Next we extend this process to the 4th stage and so on..., for each equivalence. At the Nth stage, we can generate a $2^N \times 2^N$ genetic code-based matrix and a doubly stochastic matrix. From $2^N \times 2^N$ doubly stochastic matrix, we then can generate a series of $(2 \times 2 \times 2), (4 \times 4 \times 4), (6 \times 6 \times 6), \ldots (2k \times 2k \times 2k), \ldots (2^N \times 2^N \times 2^N)$ triply stochastics cubes for $k = 1, 2, 3, \ldots, 2^{N-1}$. We summarize this process by the following Table 3d:

Table 3d. Nth common sums of cube slices of γ equivalence

γ Equivalence	No. of cubes	Common sums of cube slices
Initial stage(2^0)	0	0
1st stage(2^1)	1	5
2nd stage(2^2)	2	10, 20 (d = 10, k = 1, 2)
3rd stage (2^3)	4	15, 30, 45, 60 (d = 15, k = 1, 2, 3, 4)
4th stage (2^4)	8	20, 40, 60, 80,100, 120, 140,160 (d = 20, k = 1, 2, 3,..., 8)
......	...	...
N^{th} stage(2^N)	2^{N-1}	5N, 10N, 15N, ... 5Nk, ..., $5N2^{N-1}$ (d = 5N, k = 1, 2, 3,..., 2^{N-1})

At the Nth stage (N = 1, 2, 3, 4, ...), the common sums of ($2 \times 2 \times 2$), ($4 \times 4 \times 4$), ($6 \times 6 \times 6$), ... ($2k \times 2k \times 2k$), ...($2^N \times 2^N \times 2^N$) stochastic cubes form an arithmetic sequence with the common difference d = 5N,

$$5N, 10N, 15N, \ldots 5Nk, \ldots 5N2^{N-1} \tag{1}$$

The sum of this arithmetic sequence is

$$S_\gamma(2,3) = 5N\left(1+2+3+\ldots+2^{N-1}\right) = 5N\left(1+2^{N-1}\right)2^{N-1}/2$$
$$= \mathbf{5N2^{N-2}\left(1+2^{N-1}\right), N = 1, 2, 3, \ldots}$$

This shows that hydrogen bonds expansion is triply stochastic and its accumulation is governed by this arithmetic sequence (1) with a common difference total number of bonds of **5N**.

3.2   Stochastic Cubes Associated with Pyrimidines/Purines Ring-Based (1, 2) Equivalence

Next we apply the same process to the 2nd pyrimidines/purines ring-based (1, 2) equivalence and get the following (Table 4a):

The corresponding matrices can be constructed by means of standardizing quaternary partition with an operation of **addition** of mapping numerical values of each DNA letter as shown below (Table 4b).

Here we summarize all the cubes generated by the 2nd pyrimidines/purines ring-based (1, 2) equivalence under addition operation on each genetic code (Table 4c).

Similar to γ (2, 3) equivalence, we extend this process to the 4th stage and so on..., for each equivalence. At the Nth stage, we can generate a $2^N \times 2^N$ genetic code-based matrix and a doubly stochastic matrix. From $2^N \times 2^N$ doubly stochastic matrix, we then can generate a series of ($2 \times 2 \times 2$), ($4 \times 4 \times 4$), ($6 \times 6 \times 6$), ... ($2k \times 2k$ 2k), ...($2^N \times 2^N \times 2^N$) triply stochastics cubes for k = 1, 2, 3,..., 2^{N-1}. We summarize this process by the following Table 4d:

Table 4a. Pyrimidines/purines ring-based (1, 2) equivalence matrix

Initial Nitrogen ⇒

A	C
U	G

⇒

AA	AC	CA	CC
AU	AG	CU	CG
UA	UC	GA	GC
UU	UG	GU	GG

⇒

AAA	AAC	ACA	ACC	CAA	CAC	CCA	CCC
AAU	AAG	ACU	ACG	CAU	CAG	CCU	CCG
AUA	AUC	AGA	AGC	CUA	CUC	CGA	CGC
AUU	AUG	AGU	AGG	CUU	CUG	CGU	CGG
UAA	UAC	UCA	UCC	GAA	GAC	GCA	GCC
UAU	UAG	UCU	UCG	GAU	GAG	GCU	GCG
UUA	UUC	UGA	UGC	GUA	GUC	GGA	GGC
UUU	UUG	UGU	UGG	GUU	GUG	GGU	GGG

Table 4b. Pyrimidines/purines ring-based (1, 2) equivalence matrix

0 ⇒

2	1
1	2

⇒

4	3	3	2
3	4	2	3
3	2	4	3
2	3	3	4

⇒

6	5	5	4	5	4	4	3
5	6	4	5	4	5	3	4
5	4	6	5	4	3	5	4
4	5	5	6	3	4	4	5
5	4	4	3	6	5	5	4
4	5	3	4	5	6	4	5
4	3	5	4	5	4	6	5
3	4	4	5	4	5	5	6

At the Nth stage (N = 1, 2, 3, 4, ...), the common sums of $(2 \times 2 \times 2)$, $(4 \times 4 \times 4)$, $(6 \times 6 \times 6)$, ... $(2k \times 2k \times 2k)$, ...$(2^N \times 2^N \times 2^N)$ stochastic cubes form an arithmetic sequence with the common difference d = 3N,

$$3N, 6N, 9N, \ldots 3Nk, \ldots, 3N2^{N-1} \tag{2}$$

The sum of this arithmetic sequence is

$$S_\beta(1, 2) = 3N \left(1 + 2 + 3 + \ldots + 2^{N-1}\right) = 3N \left(1 + 2^{N-1}\right) 2^{N-1}/2$$
$$= 3N2^{N-2}\left(1 + 2^{N-1}\right), \quad N = 1, 2, 3, \ldots$$

This shows that pyrimidines/purines ring expansion is triply stochastic and its accumulation is governed by this arithmetic sequence (2) with a common difference total number of rings of **3N**.

Table 4c. Common sums of cube slices of β equivalence

β (1, 2) Equivalence	No. of cubes	Common sums of cube slices
Initial stage	0	0
First stage	1	3
Second stage	2	6, 12 (d = 6)
Third stage	4	9, 18, 27, 36 (d = 9)

Table 4d. Nth common sums of cube slices of β equivalence

β (1, 2) Equivalence	No. of cubes	Common sums of cube slices
Initial stage(2^0)	0	0
1st stage(2^1)	1	3
2nd stage(2^2)	2	6, 12 (d = 6, k = 1, 2)
3rd stage (2^3)	4	9, 18, 27, 36 (d = 9, k = 1, 2, 3, 4))
4th stage (2^4)	8	12, 24, 36, 48, 60, 72, 84, 96 (d = 12, k = 1, 2, 3, ..., 8)
......	...	...
N^{th} stage(2^N)	2^{N-1}	3N, 6N, 9N, ...3Nk,..,$3N2^{N-1}$ (d = 3N, k = 1, 2, 3,..., 2^{N-1})

3.3 Stochastic Cubes Associated with Amino-Mutating Absence/Present (0, 1)

Now we apply the same process to the 1st amino-mutating absence/present (0, 1) equivalence (Table 5a):

Table 5a. Amino-mutating absence/present (0, 1) equivalence matrix

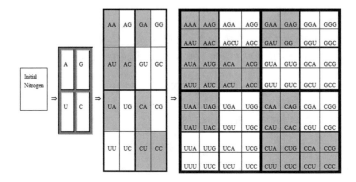

The corresponding matrices can be constructed by means of standardizing quaternary partition with an operation of **addition** of mapping numerical values of each DNA letter as shown below (Table 5b).

Table 5b. Amino-mutating absence/present (0, 1) equivalence matrix

Here we summarize all the cubes generated by the 1st amino-mutating absence/present (0, 1) equivalence under addition operation on each genetic code (Table 5c).

Table 5c. Common sums of cube slices of α equivalence

α (0, 1) Equivalence	No. of cubes	Common sums of cube slices
Initial stage	0	0
First stage	1	1
Second stage	2	2, 4 (d = 2)
Third stage	4	3, 6, 9, 12 (d = 3)

Finally, we extend this process to the 4th stage and so on..., for each equivalence. At the Nth stage, we can generate a $2^N \times 2^N$ genetic code-based matrix and a doubly stochastic matrix. From $2^N \times 2^N$ doubly stochastic matrix, we then can generate a series of $(2 \times 2 \times 2)$, $(4 \times 4 \times 4)$, $(6 \times 6 \times 6)$, ... $(2k \times 2k \times 2k)$, ...$(2^N \times 2^N 2^N)$ triply stochastics cubes for k = 1, 2, 3,..., 2^{N-1}. We summarize this process by the following Table 5d.

At the Nth stage (N = 1, 2, 3, 4, ...), the common sums of $(2 \times 2 \times 2)$, $(4 \times 4 \times 4)$, $(6 \times 6 \times 6)$, ... $(2k \times 2k \times 2k)$,...$(2^N \times 2^N \times 2^N)$ stochastic cubes form an arithmetic sequence with the common difference d = N,

$$N, 2N, 3N, \ldots kN, \ldots, N2^{N-1} \tag{3}$$

Table 5d. Nth common sums of cube slices of α equivalence

α (0, 1) Equivalence	No. of cubes	Common sums of cube slices
Initial stage(2^0)	0	0
1st stage(2^1)	1	1
2nd stage(2^2)	2	2, 4 (d = 2, k = 1, 2)
3rd stage (2^3)	4	3, 6, 9, 12 (d = 3, k = 1, 2, 3, 4)
4th stage (2^4)	8	4, 8, 12, 16, 20, 24, 28, 32 (d = 4, k = 1, 2, 3,..., 8)
......	...	...
Nth stage(2^N)	2^{N-1}	N, 2N, 3N,...,Nk,...N2^{N-1}(d = N, k = 1, 2, 3,..., 2^{N-1})

The sum of this arithmetic sequence is

$$S_\alpha(0, 1) = N\left(1 + 2 + 3 + \ldots + 2^{N-1}\right) = N\left(1 + 2^{N-1}\right)2^{N-1}/2$$
$$= \mathbf{1}N2^{N-2}\left(1 + 2^{N-1}\right), N = 1, 2, 3, \ldots$$

This shows that amino-mutating absence/present expansion is also triply stochastic and its accumulation is governed by this arithmetic sequence (3) with a common difference total number of amino-mutating of **1N**.

4   Applications and Perspectives of Stochastic Genetic Matrices

Our study showed a close relation between genetic code and trilby stochastic cubes by using three kinds of chemically determined genetic equivalences. Why the data about double stochastic matrices and cubes are interesting for bio-information with its tasks of understanding interconnections among different subsystems of the genetic system?

Development of bio-informational sciences is based, in particular, on searching and using mathematical achievements in theory of communication. This theory includes a part devoted to an important class of satellite-switched, time division multiple access systems, which are used in problems of group satellite communication (Brualdi 1988) and which has a perspective to be applied for modeling mutual communication among different parts of the genetic ensemble. As known, the theory of these time division multiple access systems are connected with doubly stochastic matrices (Brualdi 1982, 1988). One can add that modern genetics often uses graph theory, where double stochastic matrices are also applied (Hartfiel and Spellmann 1972; Zhang 2011). Stochastic matrix genetics can be interpreted as a part of algebraic biology on the genetic systems by means of their matrix forms of presentation.

The discovery of connections of the genetic matrices with doubly/triply stochastic matrices leads to many new possible investigations using methods of symmetries, of spectral analysis, etc. Data about the genetic stochastic matrices together with data about algebras of the genetic code can lead to new understanding of genetic code systems, to new effective algorithms of information processing and, perhaps, to new directions in the field of quantum computing. The matrices/cubes are storages of digital data. The matrices/cubes appear in various dimensions with different shapes. Stochastic matrices/cubes motivated by language of probability show up repeatedly in the nature. Many literatures on mathematics and biological systems have merged in recent years (Percus 2002; Pevzner 2000; Higgins and Taylor 2000) to further advance our understanding of life and its evolutions (Kay 2000; Patterson 1999).

References

Abo-Zahhad, M., Ahmed, S.M., Abd-Elrahman, S.A.: Genomic analysis and classification of exon and intron sequences using DNA numerical mapping techniques. Int. J. Inf. Technol. Comput. Sci. (IJITCS) **4**(8), 22–36 (2012). https://doi.org/10.5815/ijitcs.2012.08.03

Abo-Zahhad, M., Ahmed, S.M., Abd-Elrahman, S.A.: A novel circular mapping technique for spectral classification of exons and introns in human DNA sequences. Int. J. Inf. Technol. Comput. Sci. (IJITCS) **6**(4), 19–29 (2014). https://doi.org/10.5815/ijitcs.2014.04.02

Abo-Zahhad, M., Ahmed, S.M., Abd-Elrahman, S.A.: Integrated model of DNA sequence numerical representation and artificial neural network for human donor and acceptor sites prediction. Int. J. Inf. Technol. Comput. Sci. (IJITCS) **6**(8), 51–57 (2014). https://doi.org/10. 5815/ijitcs.2014.08.07

Brualdi, R.A.: Notes on the Birkhoff algorithm for doubly stochastic matrices. Canad. Math. Bull. **25**(2), 191–199 (1982)

Brualdi, R.A.: Some applications of doubly stochastic matrices. Linear Algebra Appl. **107**, 77–100 (1988)

Hartfiel, D.J., Spellmann, J.W.: A role for doubly stochastic matrices in graph theory. Proc. Am. Math. Soc. **36**(2), 389–394 (1972)

He, M.: Genetic code, attributive mappings and stochastic matrices. Bull. Math. Biol. **66**(5), 965–973 (2004)

He, M.: Double helical sequences and doubly stochastic matrices. In: Symmetry: Culture and Science: Symmetries in Genetic Information International Symmetry Foundation, Budapest, 2004, pp. 307–330 (2003a)

He, M.: Symmetry in structure of genetic code. In: Proceedings of the Third All-Russian Interdisciplinary Scientific Conference "Ethics and the Science of Future. Unity in Diversity", 12–14 February 2003, Moscow (2003b)

He, M., Petoukhov, S.V.: Mathematics of Bioinformatics: Theory, Practice, and Applications. Wiley, Hoboken (2011)

He, M., Petoukhov, S.V., Ricci, P.E.: Genetic code, hamming distance and stochastic matrices. Bull. Math. Biol. **66**, 1405–1421 (2004)

Higgins, D., Taylor, W.: Bioinformatics, Oxford University Press, Oxford (2000)

Kay, L.: Who Wrote the Book of Life? A History of the Genetic Code. Stanford University Press, Stanford (2000)

Jimenéz-Montaño, M.A., Mora-Basáñez, C.R., Pöschel, T.: On the hypercube structure of the genetic code. In: Lim, H.A., Cantor, C.A. (eds.) Proceedings of the 3rd International Conference on Bioinformatics and Genome Research, p. 445. World Scientific (1994)

Knight, R.D., Freeland, S.J., Landweber, L.F.: Selection, history and chemistry: the three faces of the genetic code. TIBS **24**, 241–247 (1999)

Percus, J.: Mathematics of Genome Analysis. Cambridge University Press, New York (2002)

Pevzner, P.: Computational Molecular Biology. MIT Press, Cambridge (2000)

Patterson, C.: Evolution, 2nd edn. Cornell University Press, Ithaca (1999)

Petoukhov, S.V.: The Bi-periodic Table of Genetic Code and Number of Protons. MKC (in Russian), Moscow (2001)

Petoukhov, S.V.: Genetic code and the ancient Chinese "Book of Changes". Symmetry Cult. Sci. **10**(3–4), 211–226 (1999)

Petoukhov, S.V.: Binary sub-alphabets of genetic language and problem of unification bases of biological languages. In: IX International Conference "Mathematics, Computer, Education", Russia, Dubna, 28–31 January, 191p. (in Russian) (2002)

Petoukhov, S.V., Petukhova, E.S.: Symmetries in genetic systems and the concept of geno-logical coding. Information **8**(1), 2 (2016). https://doi.org/10.3390/info8010002. Online - uploaded 25 December 2016. http://www.mdpi.com/2078–2489/8/1/2/htm

Romanovsky, V.: Sur les zeros des matrices stocastiques. C. R. Acad. Sci. Paris **192**, 266–269 (1931)

Štambuk, N.: Universal metric properties of the genetic code. Croat. Chem. Acta **73**(4), 1123–1139 (2000)

Stewart I.: Life's Other Secret: The New Mathematics of the Living World, Penguin, New York (1999)

Swanson, R.: A unifying concept for the amino acid code. Bull. Math. Biol. **46**(2), 187–203 (1984)

Watson, J.: The Double Helix. Atheneum, New York (1968)

Wittmann, H.G.: Ansatze zur Entschlusselung des Genetishen Codes. Die Naturwissenschaften **B.48**(24), S.55 (1961)

Yang, C.M.: The Naturally Designed Spherical Symmetry in the Genetic Code (2003). http://arxiv.org/abs/q-bio.BM/0309014

Zhang, X.-D.: Vertex degrees and doubly stochastic graph matrices. J. Graph. Theory **66**, 104–114 (2011)

Mathematical Modeling of Blood Vessel Stenosis and Their Impact on the Blood Vessel Wall Behavior

Igor Tereshchenko and Ivan Zhuk[✉]

National Technical University of Ukraine "Igor Sikorsky Kyiv Polytechnic Institute", Kyiv, Ukraine
tereshchenko.igor@gmail.com, zis96@ukr.net

Abstract. This article presents experimental result and mathematical analysis of the blood vessel stenosis influences on blood vessel wall behavior by the action of blood circulation. Stenosis leads to less vessel wall segment stretching depending on size and stiffness of vessel segment with stenosis. During the research, the MRI data processing approaches were performed to get blood vessel through-time behavior information and simplified model of blood vessel behavior was determined for obtained information processing to detect stenosis automatically. The results determined empirical dependences, which are necessary for the scientific study of blood vessel behavior. Mathematical analysis of research data was also carried out. Research results are compared with expert's opinion about stenosis segment of blood vessel's projection.

Keywords: Stenosis · MRI · Spring model of blood vessel wall
Blood vessel mathematical and physical modeling

1 Introduction

Today cardiovascular disease becomes a hard and keen health problem. Stenosis is one of the reasons for this disease, and, therefore, blood flows through vessels with less velocity.

One of the stenosis detection methods is to investigate the vessel behaviour during blood circulation. The unusual vessel area overstretching is another cardiovascular disease reason, which makes metabolism become worse and gives a possibility to accumulate blood-cells' ends in this area. This thesis is addressed to the analysis of existing approaches of investigations about vessel behaviour with a purpose to create the simplified vessel behaviour model and the vessel's surface reconstruction system, which is based on this simplified model.

2 MRI Data Filtration

All types of filtrations based [6] on deconvolution approaches [1, 2]. Combining different PSF functions acceptable results for two-stage filtration approach were obtained, where

© Springer International Publishing AG, part of Springer Nature 2019
Z. Hu et al. (Eds.): ICCSEEA 2018, AISC 754, pp. 617–626, 2019.
https://doi.org/10.1007/978-3-319-91008-6_61

on the first stage boundary of vessels had become clearer (Fig. 1b) and on the second stage noise reductions were performed (Fig. 1c). Using Gabor filter banks [2] for texture classification this allowed to make the segmentation not only by brightness of pixels but also by the structure of biological tissue. With additional functions this allowed to determine the inner and outer surface of vessels, where its thickness was not less than two pixels.

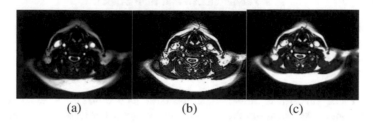

(a)   (b)   (c)

Fig. 1. (a) Original image, (b) sharpen filtration, (c) Gauss filtration.

To achieve more accuracy in problem area techniques of image recognition with the usage of neural network (NN) [3].

Firstly, threshold filtration (Fig. 2b) was carried out to obtain the approximate location of the vessels on the image, after that shape recognition and restoring were carried out (Fig. 3b) as simple contour recognitions techniques based on gradient method [4] made inaccuracies (Fig. 3a) in area and was not enough. Restoring based on building an enveloped curve for the recognized area. For bifurcation recognition, information about upper and lower layer was a need for NN, which helped to trace vessel centre-point evolution in space. If the lower layer contains bifurcation point on upper layers artificial image fitting carried out for areas where merge of vessels was pictured (Fig. 3).

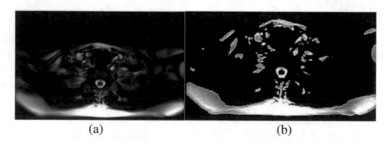

(a)   (b)

Fig. 2. Original image (a), threshold filtration (b).

Fig. 3. NN recognition and restoring.

The proposed technique allowed building more accurate surfaces, which fit properly original MRI images (Fig. 4a).

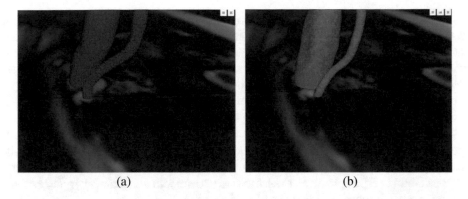

(a)   (b)

Fig. 4. Surface with NN processing (a), surface without correction techniques (b)

For restoring information between layers, techniques of space approximation were used. As this approach introduces some error, especially in the area of a bifurcation, where some algorithm for correction, based on the hypothesis that diameter of vessels does not change too much in 6 millimetres gap between layers, was produced. Areas before and after problem area approximated by cubic splines and bifurcation area are constructed as an approximation of small separating vessel from upper layers. All of 25 series were approximated and obtained additional 60 images for the spatial model building.

3   Vessel' Projection Mathematical Model

Vessel' projection turn is taken into account as follows. We assume, that enveloped curve constructed for initial vessel' projection image, is used to calculate rotation angles of another enveloped curve from.

Calculate centroid of enveloped curve k:

$$c_n^k = \begin{pmatrix} c_{nx}^k \\ c_{ny}^k \end{pmatrix} = \begin{pmatrix} \dfrac{1}{L} \sum_{l=1}^{L} b_{lnx}^k \\ \dfrac{1}{L} \sum_{l=1}^{L} b_{lnx}^k \end{pmatrix}.$$

$$B_n = \left\{ B_n^k \mid 1 \leq k \leq K \right\}, 1 \leq n \leq N,$$

$$B_n^k = \left\{ \overrightarrow{b_{ln}^k} \,\middle|\, 1 \leq l \leq L \right\},$$

$$\vec{b}_{ln}^{k} = \begin{pmatrix} b_{lnx}^{k} \\ b_{lny}^{k} \end{pmatrix},$$

B_n is a denotation of a set of enveloped curves, which correspond to vessel' projection n at different time moments, B_n^k is a denotation of vessel' projection enveloped curve n at time moment k; b_{ln}^k is a denotation of point l of vessel' projection enveloped curve n at time moment k, b_{lnx}^k is a denotation of X-axis value of point l, which belongs to vessel' projection enveloped curve n at time moment k, b_{lny}^k is a denotation of Y-axis value of point l, which belongs to vessel' projection enveloped curve n at time moment k, K is a denotation of images amount or time moments amount, N is a denotation of vessel's projections amount, L is a denotation of enveloped curve points amount, which is given before the vessel's projection contour recognition. After the rotation angle φ^{k*} is found, enveloped curve B_n^k is rotated at an φ^{k*}. Rotated enveloped curve is redenoted and now is considered as B_n^k. In practice, it was necessary to smooth a function $\varphi^*(t_k) = \varphi^{k*}$. An example of such function is shown on Fig. 5.

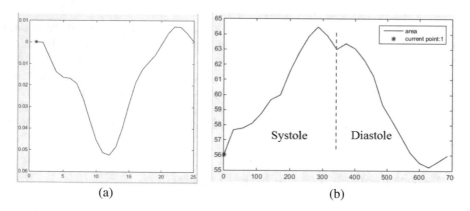

(a)   (b)

Fig. 5. Rotation angle function as a function of time (a) which corresponds to systolic and diastolic behavior of blood flow (b).

For this reason, interpolation [3–5] was applied to this function. Spline interpolation was chosen as an interpolation approach.

Rotation is performed for every enveloped curve, which is made for every vessel's projection image k, where $2 \leq k \leq K$.

Now let's find the set of n-th vessel's projection control points.

Set up an amount of contour control points. If amount of control points L is not equal to L, then new envelope curve construction process with amount of curve points equal to $\tilde{L}$ is performed for every k-th B_n^k.

3.1 Vessel' Projection Mathematical Model

Blood vessel wall can be represented as a set of small particles. Therefore vessel's projection can be represented as successively spring-connected particles (Fig. 6a) [5].

Model with three particles sequentially connected by two springs is chosen as a particles' connections model (Fig. 6b), k_m denotes stiffness of m spring, m = b, s. A mathematical model of spring's behaviour is a Hook's law.

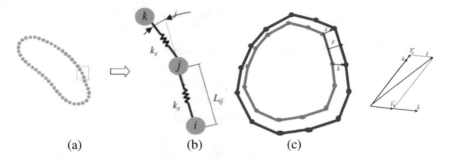

(a)   (b)   (c)

Fig. 6. (a) Vessel's wall (b) projection model, (c) geometric intuition of proposed physical consideration.

Model inputs are next:

$$X_n = \{X_n^k | 1 \le k \le K\}, 1 \le n \le N,$$

$$X_n^k = \left\{ \overrightarrow{x_{jn}^k} \Big| 1 \le j \le J \right\},$$

$$\overrightarrow{x_{jn}^k} = \begin{pmatrix} x_{jn}^k \\ y_{jn}^k \end{pmatrix}.$$

X_n is a denotation of set of n-th vessel's projection images for different time moments, X_n^k is a denotation of n-th vessel's projection image at time moment k, in fact polygon, which is a boundary of vessel's projection, $\overrightarrow{x_{jn}^k}$ denotes a point j, which belongs to polygon k of vessel's projection n, x_{jn}^k is a denotation of X-axis of point j, which belongs to polygon k of vessel's projection n, y_{jn}^k is a denotation of Y-axis of point j, which belongs to polygon k of vessel's projection n, K is a denotation of images amount or time moments amount, N is a denotation of vessel's projections amount, J is a denotation of polygon points amount.

Firstly, springs are not connected between themselves by additional spring. Secondly, assume, that springs do not deflect It corresponds to the fact, that blood uniformly and equally presses on every vessel's area. Therefore, there is a force, which is acting perpendicular to the polygon sections from the polygon inner side. This force appears by inner pressure in the vessel at particular time moment in particular vessel's projection.

The geometric interpretation of given consideration is shown in Fig. 6c.

Therefore, springs' stiffnesses can be calculated in a next way:

find vectors $\begin{pmatrix} F_{ina}^k \\ 0 \end{pmatrix}$ та $\begin{pmatrix} 0 \\ F_{inb}^k \end{pmatrix}$, which is set up in basis {a;b}, in initial basis:

$$\overrightarrow{F^k_{inaxy}} = T\begin{pmatrix} F^k_{ina} \\ 0 \end{pmatrix},$$

$$\overrightarrow{F^k_{inbxy}} = T\begin{pmatrix} 0 \\ F^k_{inb} \end{pmatrix};$$

project the vectors found on spring axis at time moment k:

$$F^k_{inax} = \frac{\overrightarrow{F^k_{inaxy}}(\overrightarrow{x^k_{(i+1)n}} - \overrightarrow{x^k_{in}})}{\left|\overrightarrow{x^k_{(i+1)n}} - \overrightarrow{x^k_{in}}\right|},$$

$$F^k_{inbx} = \frac{\overrightarrow{F^k_{inbxy}}(\overrightarrow{x^k_{(i+1)n}} - \overrightarrow{x^k_{in}})}{\left|\overrightarrow{x^k_{(i+1)n}} - \overrightarrow{x^k_{in}}\right|},$$

where $|\vec{c}| = \sqrt{c_x^2 + c_y^2}$;

find the total force, which stretches the spring:

$$F^k_{in3a\Gamma} = F^k_{inax} + F^k_{inbx};$$

find spring stiffness at time moment k:

$$g^k_{in} = \frac{F^k_{in3a\Gamma}}{\Delta l^k_{in}}$$

where $\Delta l^k_{in} = \left|\left(\overrightarrow{x^{k+1}_{(i+1)n}} - \overrightarrow{x^{k+1}_{in}}\right) - \left(\overrightarrow{x^k_{(i+1)n}} - \overrightarrow{x^k_{in}}\right)\right|$ - subtraction of i-th spring's lengths at time moments k and $k + 1$.

Vessel's projection pressure dependence from vessel's projection square is obtained as follows. Pressure, that acts on arbitrary section at time moment t_i is determined as [9]:

$$P_i = \frac{F^{LM}_i}{S^{LM}_i}$$

where F^{LM}_i is a denotation for a pressure force projection, which act on arbitrary section at time moment t_i on arbitrary line segment LM.

S^{LM}_i is a denotation for square of vessel's wall segment at time moment t_i, which is projected on arbitrary line segment LM:

$$S^{LM}_i = l^{LM}_i * h^{LM}_i$$

where l_i^{LM}, h_i^{LM} is a length and height of vessel's wall arbitrary segment at time moment t_i, which is projected on arbitrary line segment LM.

Since vessel projection is under consideration, the height for all vessel's wall segments is threatened equally ($h_i^{LM} = h$) at any time moment.

Therefore, $F_i^{LM} \sim P_i l_i^{LM}$

Then pressure changing in the vessel, by Poiseuille's law, can be represented as:

$$\Delta P_i = \frac{8\mu\pi\left(S_{i+1} - S_i\right)}{S_i^2\left(t_{i+1} - t_i\right)}$$

where S_i is a vessel's projection square at time moment t_i, μ – blood viscosity.

The pressure in vessel's projection can be represented as:

$$P_i = P_0 + \Delta P_i,$$

where $P_0 = 100$ mmHg. $* 133,3334 = 13333,34$ Pa.

3.2 Pressure Results for Vessel Projection

This correspondence allows the determining pressure at discontinuous time moments by known vessel's projection square. Results received about pressure can be used in simplified vessel's wall model.

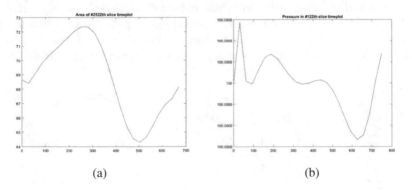

(a)   (b)

Fig. 7. The input data about vessel's projection square as a function of time (a), results about pressure in vessel as a function of time (b).

The input data about contour square as a function of time is shown in Fig. 7a. In practice, results have non-uniform distribution character, but trend-line is close to polynomial (Fig. 7b). Therefore, polynomial regression [7, 8] was applied to received data for further work with them.

3.3 Joining of Single Vessel's Projections

The building of dynamic motion of centerlines in different time steps needed to perform vessels' surface motion. In areas of bifurcation, because of lack of information obtained from MRI image, gaps between vessels connected artificially according to centerlines behaviour [4]. Centerline-i, where i is an index of time step, separated into three parts in bifurcation point: root, main vessel part and subsidiary vessel. Each part is divided into equal segments (red points on Fig. 8).

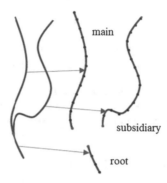

Fig. 8. Centerline division.

All parts of different time step have equal amount of segments that connected by curves to each other. These curves are spatial movement of centerlines (Fig. 9a). Joining of mathematical models of single vessel's projections is performed by putting control points sets (polygon vertices) of these models together (Fig. 9b). Since these sets are ordered and have equal amount of control points, therefore mapping is unambiguous.

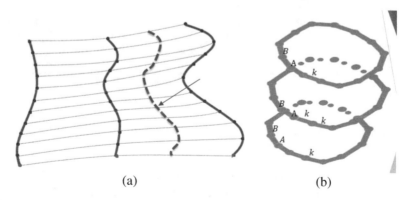

(a)   (b)

Fig. 9. Centerline evolution (a). Mapping (blue) between control points of single contours (b), segment with springs of equal stiffness (orange).

3.4   Stiffnesses Results for Vessel Projection

Stiffnesses data processing is performed as follows. In practice, stiffnesses, which is calculated for different time moments and for one and the same vessel's projection's spring, are not always the same (Fig. 10a). Values' concentrations had been filtered [10, 11] and filtration results became to be considered as the springs' stiffness values (Fig. 10b).

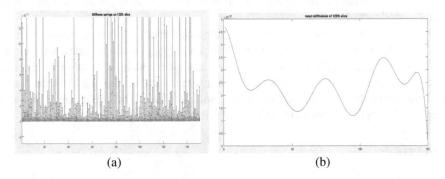

(a)   (b)

Fig. 10. Concentration of the stiffness values of single springs (a), interpolation of the mean stiffness value among vessel's projection springs (b).

Based on results of interpolation [7, 8], for detection of sick vessel's segment by interpolation received there was used next criteria: vessel's projection sick segment is such, that every spring of this segment has a stiffness, which is less than mean stiffness, calculated through all springs.

3.5   Stenosis Detection Results for Vessel Projection

Given changes in time of vessel's projection (Fig. 11a), that had been received from the results of recognition, expert pointed the sick part of vessel's projection (Fig. 11b). The

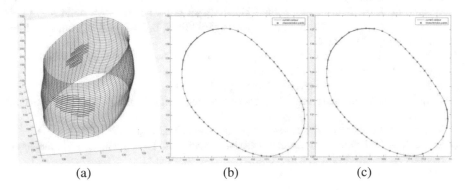

(a)   (b)   (c)

Fig. 11. Changes in time of vessel's projection (a), sick part of vessel's projection, expert opinion (b) and software result (c).

software gave the result (Fig. 11c). The accuracy of vessel's projection sick segment detection during the experiment is 85%.

4 Conclusion

The simplified model showed that the less movable the vessel wall is, the more stiffness vessel wall is. Created vessel's projection model should be used for developing of blood vessel spatial model.

The results of the research show that simplified model correlates with a real-life situation. The accuracy of vessel's projection sick segment detection during the experiment is 85%. Loss of 15% accuracy is gotten because of made filtration. Filtration was necessary because of presented anomalies in stiffness data. Presenting of anomalies can be explained by the simplicity of the model. But from the other side, it can be explained because of recognition precision due to MRI data smoothing.

References

1. Gourav, Sharma, T., Singh, H.: Computational approach to image segmentation analysis. Int. J. Modern Educ. Comput. Sci. (IJMECS) 9(7), 30–37 (2017). https://doi.org/10.5815/ijmecs. 2017.07.04
2. Gupta, S., Porwal, R.: Implementing blind de-convolution with weights on x-ray images for lesser ringing effect. Int. J. Image Graph. Signal Process. (IJIGSP) 8(8), 30–36 (2016). https:// doi.org/10.5815/ijigsp.2016.08.05
3. Bajaj, K., Kaur, N.: Integrated gabor filter and trilateral filter for exudate extraction in fundus images. Int. J. Image Graph. Signal Process. (IJIGSP) 9(1), 10–17 (2017). https://doi.org/ 10.5815/ijigsp.2017.01.02
4. Isah, R.O., Usman, A.D., Tekanyi, A.M.S.: Medical image segmentation through bat-active contour algorithm. Int. J. Intell. Syst. Appl. (IJISA) 9(1), 30–36 (2017). https://doi.org/ 10.5815/ijisa.2017.01.03
5. Suprijadi, Sentosa, M.R.A., Subekti, P., Viridi, S.: Application of computational physics: blood vessel constrictions and medical infuses. In: AIP Conference Proceedings, vol. 1587, no. 1, pp. 3–6 (2013)
6. Wihandika, R.C., Suciati, N.: Retinal blood vessel segmentation with optic disc pixels exclusion. Int. J. Image Graph. Signal Process. (IJIGSP) 5(7), 26–33 (2013). https://doi.org/ 10.5815/ijigsp.2013.07.04
7. Feldman, L.P., Petrenko, A.I., Dmytriieva, O.A:. Chyselni metody v informatytsi. Vydavnycha hrupa BHV, p. 480 (2006)
8. Kalitkin, N.N.: Chislennye metody. Nauka, p. 512 (1978)
9. Kittel', C., Nait, U., Ruderman, M.: Berkleevskii kurs fiziki. Tom 1. Mekhanika. Nauka, p. 479 (1971)
10. Venttsel', A.D.: Kurs teorii sluchainykh protsessov. Fizmatlit, p. 399 (1978)
11. Kobzar', A.I.: Prikladnaya matematicheskaya statistika. Dlya inzhenerov i nauchnykh rabotnikov. FIZMATLIT, p. 816 (2006)

Simulation of Electrical Restitution
in Cardiomyocytes

N. Ivanushkina[1], K. Ivanko[1(✉)], Y. Prokopenko[1], A. Redaelli[2], V. Timofeyev[1],
and R. Visone[2]

[1] Igor Sikorsky Kyiv Polytechnic Institute, Kyiv, Ukraine
koondoo@gmail.com
[2] Politecnico di Milano, Milan, Italy

Abstract. The efforts of many scientists are directed to study of heart electrical instability by experimental methods and mathematical modeling of cardiomyocytes' functional properties. The development of arrhythmias can be caused by cardiac beat-to-beat alternations in action potential duration (APD), concentration of intracellular Ca^{2+} and contraction force. One of the methods for investigation of dangerous arrhythmias genesis is based on the restitution hypothesis.

Motivated by theoretical foundations and experimental research of the arrhythmias, the new approach to stimulation of action potential (AP) alternans in cardiomyocytes due to the heart rate variability was proposed. The main attention was paid to study of electrical restitution dynamics of cardiomyocytes using several pacing protocols. Computational simulation of the action potential and currents for potassium, sodium, calcium ions in cardiomyocytes was carried out using parallel conductance model. Numerical experiments, performed in Matlab environment, allowed us to study electrical properties of cardiomyocytes. The occurrence of APD alternans in areas of electrical restitution curve with a maximum slope is presented.

Keywords: Cardiomyocyte · Action potential · Parallel conductance model
Electrical restitution · Alternans

1 Introduction

Arrhythmias, which are the most common heart diseases, are caused by violations in formation and/or propagation of the action potential (AP) in cardiomyocytes [1]. To investigate such violations of heart electrical activity, experimental methods and methods of mathematical modeling as well as signal processing are performed [2–7].

It is known, that experimental methods on animals are not sufficient in modeling of the human myocardium. Today new experimental methods of modeling, based on the design and manufacture of lab-on-chip, are developed. These devices, which use 3D cell-laden hydrogel constructs and human-induced pluripotent stem cells (hiPSCs), are applied for personalized medicine studies with the individual drug response of patients [8–11].

However, experimental modeling has many problems and limitations (mapping of cardiac conduction, simultaneous examination of membrane potentials and several

© Springer International Publishing AG, part of Springer Nature 2019
Z. Hu et al. (Eds.): ICCSEEA 2018, AISC 754, pp. 627–637, 2019.
https://doi.org/10.1007/978-3-319-91008-6_62

membrane currents, maintenance of required research conditions). To solve such problems, mathematical models are used for the study of the heart functional properties at all levels: cellular, tissue and organ [2, 4, 5] since they allow researchers to carry out investigations of individual phenomena with different assumptions. The dynamics of the cardiomyocytes electrical activity and changes in AP is described as intracellular and extracellular processes of electrical impulses generation and their propagation through the conduction system to cause contraction of the heart.

The study of AP is especially important, because the electrical and mechanical processes in cardiomyocytes can be explained on the base of AP characteristics (shape, duration, generation frequency). Furthermore, the large number of investigations shown that alternans in the action potential duration is a precursor to arrhythmias. Many scientists studied the electrical restitution that is an intrinsic heart property of APD's change according to heart rate (HR) [12–16].

According to the "restitution hypothesis", the relationship between oscillations in myocardium and the slope of restitution curve was established. It is believed, that under the slope of restitution curve greater than 1, small changes in stimulation cycle length lead to large changes in APD and hence dynamic instability due to transformation of excitation waves into multiple wavelets. Therefore, the goal of different studies was to measure restitution curves and estimate their slopes.

The aim of our study is to investigate restitution curves and their maximum slopes by means of computational model for cardiomyocytes, as well as to focus on the simulation of heart rate variability using various stimulation protocols and on the determination of conditions for electrical stimulation in which APD alternans may arise.

2 Computational Modeling

Simulation of action potential in heart cells was carried out by alteration of Hodgkin-Huxley generalized model [1]. This model, developed for nerve tissue, was used by many scientists and improved specifically for cardiomyocytes [2, 16–18]. In the proposed model [19] a branch for calcium ion channels modeling was supplemented to parallel-conductance model. Independent potassium, sodium, calcium ion channels and leakage through the cell membrane are taken into account in the model. The total membrane current is determined by the sum of the contributions of ions of each type. The permeability of the membrane for ions of different types is various.

Transmembrane potential has two components: $V_m(t) = V_{m0} + v_m(t)$, one of which is the resting membrane potential V_{m0} and other is the variable component of membrane potential $v_m(t)$. The resting membrane potential can be determined from the parallel-conductance model as

$$V_{m0} = \frac{-g_{K0}E_K + g_{Na0}E_{Na} + g_{Ca0}E_{Ca} - g_l E_l}{g_{K0} + g_{Na0} + g_{Ca0} + g_l},$$

where g_{K0}, g_{Na0} and g_{Ca0} are conductances for K^+, Na^+ and Ca^{2+} ions at rest; g_l is conductance of leakage. The voltage sources E_K, E_{Na} and E_{Ca} in the parallel-conductance model simulate Nernst potentials for potassium, sodium, calcium; E_l simulates Nernst potential for leakage and other factors.

The variable component of membrane potential of pacemaker cell satisfies the following differential equation

$$\frac{dv_m}{dt} = \frac{1}{C_m}\left(-I_K(u_m, t) - I_{Na}(u_m, t) - I_{Ca}(u_m, t) - I_l\right),$$

where time dependences of K^+, Na^+ and Ca^{2+} currents and current of leakage are presented below

$$
\begin{aligned}
I_K(v_m, t) &= g_K(v_m, t)(V_{m0}+v_m+E_K) \\
I_{Na}(v_m, t) &= g_{Na}(v_m, t)(V_{m0}+v_m-E_{Na}) \\
I_{Ca}(v_m, t) &= g_{Ca}(v_m, t)(V_{m0}+v_m-E_{Ca}) \\
I_l &= g_l(V_{m0}+v_m + E_l)
\end{aligned}
$$

In case of membrane depolarization, conductance of potassium, sodium and calcium channels can be described by the equations:

$$
\begin{aligned}
g_K(u_m, t) &= g_{K_{max}} n^4(v_m, t) \\
g_{Na}(v_m, t) &= g_{Namax} m^3(v_m, t)h(v_m, t) \\
g_{Ca}(v_m, t) &= g_{Camax} d(v_m, t)f(v_m, t)
\end{aligned}
$$

where g_{Kmax}, g_{Namax} and g_{Camax} are membrane conductances for potassium, sodium and calcium ions, when all the channels of these types of ions are in the open state; n is activation function of K^+ channels; m is activation function of Na^+ channels; h is inactivation function of Na^+ channels; d is activation function of Ca^{2+} channels and f is inactivation function of Ca^{2+} channels. Gating variables n, m, h, d and f change from 0 to 1 and define the probability that subunit of ion channel is in the open state. These variables determine the ionic currents of the model and can be found as the solutions of the set of nonlinear ordinary differential equations.

$$\frac{dn}{dt} = \frac{n_\infty - n}{\tau_n}, \quad \frac{dm}{dt} = \frac{m_\infty - m}{\tau_m}, \quad \frac{dh}{dt} = \frac{h_\infty - h}{\tau_h}, \quad \frac{dd}{dt} = \frac{d_\infty - d}{\tau_d}, \quad \frac{df}{dt} = \frac{f_\infty - f}{\tau_f},$$

where

$$n_\infty = \frac{1}{1 + \exp\left(\dfrac{V_{hn} - v_m}{V_{Sn}}\right)}, m_\infty = \frac{1}{1 + \exp\left(\dfrac{V_{hm} - v_m}{V_{Sm}}\right)},$$

$$h_\infty = \frac{1 - h_{min}}{\left(1 + \exp\left(\dfrac{v_m - V_{hh}}{V_{Sh}}\right)\right)^{P_h}} + h_{min}, d_\infty = \frac{1}{1 + \exp\left(\dfrac{V_{hd} - v_m}{V_{Sd}}\right)},$$

$$f_\infty = \frac{1 - f_{min}}{\left(1 + \exp\left(\dfrac{v_m - V_{hf}}{V_{Sf}}\right)\right)^{P_f}} + f_{min},$$

n_∞, m_∞, and d_∞ are the steady-state values of activation function for potassium, sodium and calcium channels respectively; h_∞ and f_∞ are the steady-state values of inactivation function for sodium and calcium channels; h_{min} and f_{min} are minimum inactivation for sodium and calcium channels; V_{hn}, V_{hm}, and V_{hd} are voltages of semi-activation for potassium, sodium and calcium channels; V_{Sn}, V_{Sm}, and V_{Sd} are voltages of activation shape for potassium, sodium and calcium channels; V_{hh} and V_{hf} are voltages of semi-inactivation for sodium and calcium channels; V_{Sh} and V_{Sf} are voltages of inactivation shape for sodium and calcium channels.

Relaxation periods of activation τ_n, τ_m, and τ_d for potassium, sodium and calcium channels were determined as follows

$$\tau_n = \tau_{nmax}\left((1 - r_n)\left(1 + \exp\left(\frac{v_m - V_{htn}}{V_{stn}}\right)\right)^{-P_{\tau n}} + r_n\right),$$

$$\tau_m = \tau_{mmax}\left((1 - r_m)\left(1 + \exp\left(\frac{v_m - V_{htm}}{V_{stm}}\right)\right)^{-P_{\tau m}} + r_m\right),$$

$$\tau_d = \tau_{dmax}\left((1 - r_d)\left(1 + \exp\left(\frac{v_m - V_{htd}}{V_{std}}\right)\right)^{-P_{\tau d}} + r_d\right),$$

τ_{nmax}, τ_{mmax}, and τ_{dmax} are maximum relaxation periods of activation for potassium, sodium and calcium channels; r_n, r_m and r_d are ratios of minimum relaxation period to maximum relaxation period of activation for potassium, sodium and calcium channels; V_{htn}, V_{htm}, and V_{htd} are voltages of semi-relaxation of activation; V_{Stn}, V_{Stm} and V_{Std} are voltages of relaxation shape of activation; $P_{\tau n}$, $P_{\tau m}$, and $P_{\tau d}$ are degrees of relaxation of activation for potassium, sodium and calcium channels respectively.

Relaxation periods of inactivation τ_h and τ_f for sodium and calcium channels were determined as follows

$$\tau_h = \tau_{hmax}\left((1 - r_h)\left(1 + \exp\left(\frac{v_m - V_{hth}}{V_{Sth}}\right)\right)^{-P_{\tau h}} + r_h\right),$$

$$\tau_f = \tau_{fmax}\left((1 - r_f)\left(1 + \exp\left(\frac{v_m - V_{htf}}{V_{Stf}}\right)\right)^{-P_{\tau f}} + r_f\right),$$

$\tau_h$max and $\tau_f$max are maximum relaxation periods of inactivation for sodium and calcium channels; r_h and r_f are ratios of minimum relaxation period to maximum relaxation period of inactivation for sodium and calcium channels; $V_{h\tau h}$ and $V_{h\tau f}$ are voltages of semi-relaxation of inactivation; $V_{S\tau h}$ and $V_{S\tau f}$ are voltages of relaxation shape of inactivation; $P_{\tau h}$ and $P_{\tau f}$ are degrees of relaxation of inactivation for sodium and calcium channels respectively.

The initial conditions are $v_m(0) = 0$ and $n(0) = n_0$; $m(0) = m_0$; $d(0) = d_0$; $h(0) = h_0$; $f(0) = f_0$.

Parameters n_0, m_0, d_0, h_0 and f_0 can be found as n_∞, m_∞, d_∞, h_∞ and f_∞ for $v_m = 0$. Maximal conductance of the membrane for K^+, Na^+ and Ca^{2+} channels is defined as

$$g_K\text{max} = \frac{g_{K0}}{n_0^4}; \quad g_{Na}\text{max} = \frac{g_{Na0}}{m_0^3 h_0}; \quad g_{Ca}\text{max} = \frac{g_{Na0}}{d_0 f_0}.$$

Differential equation defining the variable component of membrane potential and the set of equations for 5 gating variables, as well as their initial conditions form Cauchy problem relatively v_m, n, m, d, h and f. The Cauchy problem was solved by implicit methods of integration [20]. A detailed description of the functions and numerical values for parameters of the proposed model is given in [19].

3 Numerical Experiments

Numerical experiments for modeling of AP in cardiomyocytes were performed in *Matlab* environment. The main attention was paid to the research of cardiomyocytes' electrical restitution using several protocols of stimulation.

It is known that electrical restitution is the dependence of cardiomyocytes' APD from stimulation frequency (Fst) or stimulation cycle length (CL) or, more correctly, from preceding diastolic interval (DI). Usually, the preceding DI is determined as the time interval between two action potentials over a range of cycle lengths. It is established that APD shortens with decreasing CL length and thus with decreasing DI. The phenomenon of electrical restitution takes place because calcium current does not fully recover at short DI, which leads to short APD at short DI.

As a rule, electrical restitution data were measured using the standard (STRT) or dynamic (DYRT) restitution protocol [12]. In STRT protocol the cardiac cells are paced at a constant cycle length using stimuli S1 and another stimulus S2 at variable time intervals from the S1 stimulus. According to the DYRT protocol, cardiomyocytes were stimulated in the physiologically determined frequencies with increasing Fst (decreasing CL) incrementally until APs failed, that means refractory period has been reached. On the base of the stimulation results the restitution curve is constructed and the areas with a maximum slope are defined.

At presented work the investigation of the new approach to stimulation of cardiomyocytes for modeling of APD alternans due to the heart rate variability was made.

This approach is based on successive application of the DYRT protocol with the definition of Fst (CL), when cardiomyocyte failed to capture AP (the maximum capture rate), and new protocol of more detailed stimulation with small changes of Fst (CL) in the range of the detected maximum capture rate.

On the base of the proposed model, the responses (APs, currents and conductances for K^+, Na^+, Ca^{2+} channels) from ventricular cardiomyocytes, stimulated according to DYRT protocol at various Fst, are calculated. Electrical stimulation started at 1 Hz and increased up to 6 Hz with assessment of the maximum capture rate. Numerical experiments showed that ventricular cardiomyocyte failed to capture AP at stimulation frequency Fst = 3.27 Hz (CL = 306 ms). Therefore, the maximum capture rate of cardiomyocyte is achieved under the stimulation frequency Fst = 3.27 Hz.

The transient process for APDs before a 'steady-state' reaching (Fig. 1) was investigated by simulation APs of ventricular cardiomyocytes stimulated with the constant Fst = 3.27 Hz (CL = 306 ms). These date demonstrate that cardiomyocytes produce stimulus-induced APs with changing APDs, as the duration of AP adapts to the changing diastolic interval. The dependence of APDs from preceding DIs reflects physiological rate-adaptation.

According to proposed new protocol, the research of transient process for the appearance and maintenance of AP alternans was made by simulation of AP in ventricular cardiomyocytes, stimulated in the range of the detected maximum capture rate with the decrease and increase of CLs. Figure 2 demonstrates AP alternans, currents and

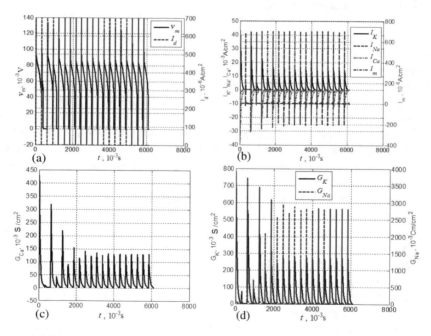

Fig. 1. (a) Adaptation process of APDs for ventricular cardiomyocytes with constant CLs: action potentials; (b) currents of K^+, Na^+, Ca^{2+} ions; (c) conductance of Ca^{2+} channels; (d) conductance of K^+, Na^+ channels.

conductances for K^+, Na^+, Ca^{2+} channels in the range of the detected maximum capture rate Fst = 3.257 Hz (CL = 307 ms) with shortening CLs (at every step CL was shortened by 2 ms). Under lengthening CLs (at every step CL was extended by 2 ms) the rate dependence of APDs, currents and conductances for K^+, Na^+, Ca^{2+} channels is similar to the transient process shown in Fig. 1 (absence of the sustainable alternans).

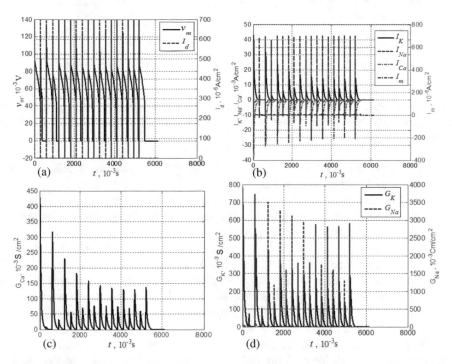

Fig. 2. (a) Cardiomyocyte alternans with shortening CLs: action potentials; (b) currents of K^+, Na^+, Ca^{2+} ions; (c) conductance of Ca^{2+} channels; (d) conductance of K^+, Na^+ channels.

4   Simulation Results and Discussion

The restitution hypothesis has activated a series of experimental and computational studies [2, 16–18], which showed that steepness of electrical restitution may be a key marker in predicting of the onset of alternans in cardiac myocytes and dangerous arrhythmias such as ventricular fibrillation (VF). In addition, there is evidence that the effect of drugs and vagus nerve stimulation, reducing the slope of the restitution curve, prevents the induction of VF [15].

The main attention of the presented study was paid to electrical restitution dynamics and to APD alternans appearance in cardiomyocytes using several pacing protocols. The electrical restitution curves for ventricular cardiomyocytes, calculated for the cases of absence and presence of sustainable alternans, are shown in Fig. 3. The restitution curve (Fig. 3a) was constructed according to adaptation process of APDs (Fig. 1a), and the

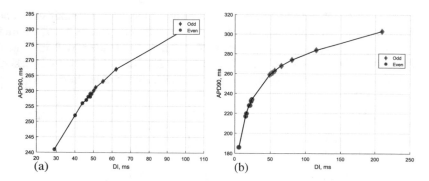

Fig. 3. (a) Electrical restitution curves for ventricular cardiomyocytes: APD90 against DI in the case of the sustainable alternans absence; (b) APD90 against DI in the case of the sustainable alternans presence.

restitution curve (Fig. 3b) was constructed according to the process of maintenance of APDs sustainable alternans (Fig. 2a).

Plotted restitution curves have several phases with the various steepness. The restitution curve in Fig. 3b is steeper than restitution curve in Fig. 3a due to new protocol with shortening CLs and, respectively, DIs.

The clear correlation is visible between changes of APD90 and DI in Fig. 4, which demonstrates dependence of APD90 and DI from sequential number of stimulus in the case of absence and presence of sustainable alternans. Obtained pattern is typical for wave alternans, when one type of wave is observed in even cycles (the waves with a lower amplitude in Fig. 4), and the different type of wave can be seen in odd cycles (the waves with a larger amplitude in Fig. 4).

Using restitution curves, the areas with the maximum slope, which determines arrhythmogenic properties of heart cells, were detected. The results of the research shown that the maximum slope of the curves, based on the refractoriness of cardiomyocytes, corresponds to the range of maximum capture rate.

The simulation of CLs shortening or lengthening reflects changes of the pacing pulses generation under heart rate variability. The decrease of CLs leads to the decrease of diastolic intervals, which take on values from the range with the maximum slope of the restitution curve that causes the appearance of APD sustainable alternans. The performed study demonstrated the effect of several stimulation protocols on APD, but other factors were not examined, such as cardiac memory and conduction velocity [2, 16].

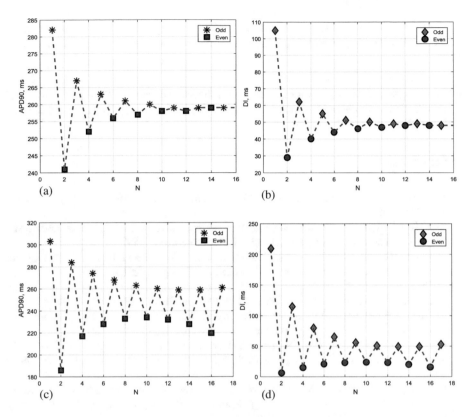

Fig. 4. (a, b) Dependence of APD90 and DI from sequential number of stimulus: the case of the sustainable alternans absence; (c, d) the case of the sustainable alternans presence.

5   Conclusions

The new approach to AP stimulation in cardiomyocytes for modeling of AP alternans due to the heart rate variability was proposed. Computational simulation of cardiomyocytes' restitution properties, performed according to the combined pacing protocol, allowed us to explain the interaction mechanism of intracellular processes in cardiomyocytes and extracellular processes of generation of electrical excitation impulses. On the base of the improved parallel conductance model, the range of diastolic intervals in electrical restitution curve with the maximum slope was determined. The novel pacing protocol with decrease or increase of cycle length in the range of maximum slope of restitution allowed us to reflect the heart rate variability. Due to proposed protocol the onset of AP alternans were shown. Obtained results suggest that the new approach to research of cardiomyocytes' electrical activity unmasked the mechanisms for appearance of alternans, which cannot be seen with current protocols of stimulation. Computational results of cardiomyocytes' electrical restitutions can be useful for methods of mathematical modeling and signal processing [21, 22], as well as can help plan experiments on cardiomyocytes and, especially, design the experimental research on hiPSC-CMs.

Acknowledgement. The study was supported by EU-financed Horizon 2020 project AMMODIT (Approximation Methods for Molecular Modeling and Diagnosis Tools) - Grant Number MSCA-RISE 645672.

References

1. Plonsey, R., Barr, R.: Bioelectricity: A Quantitative Approach, 3rd edn., 528 p. Springer, New York (2007)
2. Tusscher, K.H.W.J.: Alternans and spiral breakup in a human ventricular tissue model. Am. J. Physiol. Heart Circulatory Physiol. 3(291), H1088–H1100 (2006)
3. Coronel, R.: Electrophysiological changes in heart failure and their implications for arrhythmogenesis. Biochem. Biophys. Acta. **1832**, 2432–2441 (2013)
4. O'Hara, T.: Simulation of the undiseased human cardiac ventricular action potential: model formulation and experimental validation. PLoS Comput. Biol. 7, e1002061 (2011)
5. Solid, Z.: Reentry via high-frequency pacing in a mathematical model for human-ventricular cardiac tissue with a localized fibrotic region. Sci. Rep. 5(7), 15350 (2017)
6. Olaniyi, E.O., Oyedotun, O.K., Adnan, K.: Heart diseases diagnosis using neural networks arbitration. Int. J. Intell. Syst. Appl. (IJISA) 7(12), 75–82 (2015). https://doi.org/10.5815/ijisa.2015.12.08
7. Goshvarpour, A., Shamsi, M., Goshvarpour, A.: Spectral and time based assessment of meditative heart rate signals. Int. J. Image Graph. Sig. Process. (IJIGSP) 5(4), 1–10 (2013). https://doi.org/10.5815/ijigsp.2013.04.01
8. Ellis, B.W.: Human iPSC-derived myocardium-on-chip with capillary-like flow for personalized medicine. Biomicrofluidics 2(11), 024105 (2017)
9. Liang, P.: Drug screening using a library of human induced pluripotent stem cell-derived cardiomyocytes reveals disease-specific patterns of cardiotoxicity. Circulation **127**, 1677–1691 (2013)
10. Pavesi, A.: Controlled electromechanical cell stimulation on-a-chip. Sci. Rep. **5**, 1–12 (2015). 11800
11. Marsano, A.: Beating heart on a chip: a novel microfluidic platform to generate functional 3D cardiac microtissues. Lab Chip **16**, 599–610 (2016)
12. Rawan, A.K.: Electrical Restitution and Action Potential Repolarisation Studies in Acutely Isolated Cardiac Ventricular Myocytes. http://hdl.handle.net/2381/38815. Accessed 11 Dec 2017
13. Traxel, S.J.: A novel method to quantify contribution of electrical restitution to alternans of repolarization in cardiac myocytes: a simulation study. FASEB J. 1(23) (2009). Supplement 624.7
14. Shattock, M.J.: Restitution slope is principally determined by steady-state action potential duration. Cardiovasc. Res. 7(113), 817–828 (2017)
15. Ng, G.A.: Autonomic modulation of electrical restitution, alternans and ventricular fibrillation initiation in the isolated heart. Cardiovasc. Res. 4(73), 750–760 (2007)
16. Orini, M.: Interactions between activation and repolarization restitution properties in the intact human heart: in-vivo whole-heart data and mathematical description. PLoS ONE **9**(11), e0161765 (2016)

17. Luo, C.: A dynamic model of the ventricular cardiac action potential: I. simulations of ionic currents and concentration changes. Circ. Res. **6**(74), 1071–1096 (1994)
18. Hund, T.J.: Rate dependence and regulation of action potential and calcium transient in a canine cardiac ventricular cell model. Circulation **20**(110), 3168–3174 (2004)
19. Ivanko, K.: Simulation of action potential in cardiomyocytes. In: Proceedings of 2017 IEEE 37th International Scientific Conference on Electronics and Nanotechnology (ELNANO), pp. 358–362 (2017)
20. Spiteri, R.J.: On the performance of an implicit– explicit Runge-Kutta method in models of cardiac electrical activity. IEEE Trans. Biomed. Eng. **5**(55), 1488–1495 (2008)
21. Ahmad, A.A., Kuta, A.I., Loko, A.Z.: Analysis of abdominal ECG signal for fetal heart rate estimation using adaptive filtering technique. Int. J. Image Graph. Sig. Process. (IJIGSP) **9**(2), 19–26 (2017). https://doi.org/10.5815/ijigsp.2017.02.03
22. Gowri, T., Rajesh Kumar, P.: Muscle and baseline Wander artifact reduction in ECG signal using efficient RLS based adaptive algorithm. Int. J. Intell. Syst. Appl. (IJISA) **8**(5), 41–48 (2016). https://doi.org/10.5815/ijisa.2016.05.06

Deep Learning with Lung Segmentation and Bone Shadow Exclusion Techniques for Chest X-Ray Analysis of Lung Cancer

Yu. Gordienko[1(✉)], Peng Gang[2], Jiang Hui[2], Wei Zeng[2], Yu. Kochura[1], O. Alienin[1], O. Rokovyi[1], and S. Stirenko[1]

[1] National Technical University of Ukraine "Igor Sikorsky Kyiv Polytechnic Institute", Kyiv, Ukraine
yuri.gordienko@gmail.com, iuriy.kochura@gmail.com, sergii.stirenko@gmail.com
[2] Huizhou University, Huizhou City, China
peng@hzu.edu.cn
http://comsys.kpi.ua, http://www.hzu.edu.cn

Abstract. The recent progress of computing, machine learning, and especially deep learning, for image recognition brings a meaningful effect for automatic detection of various diseases from chest X-ray images (CXRs). Here efficiency of lung segmentation and bone shadow exclusion techniques is demonstrated for analysis of 2D CXRs by deep learning approach to help radiologists identify suspicious lesions and nodules in lung cancer patients. Training and validation was performed on the original JSRT dataset (dataset #01), BSE-JSRT dataset, i.e. the same JSRT dataset, but without clavicle and rib shadows (dataset #02), original JSRT dataset after segmentation (dataset #03), and BSE-JSRT dataset after segmentation (dataset #04). The results demonstrate the high efficiency and usefulness of the considered pre-processing techniques in the simplified configuration even. The pre-processed dataset without bones (dataset #02) demonstrates the much better accuracy and loss results in comparison to the other pre-processed datasets after lung segmentation (datasets #02 and #03).

Keywords: Deep learning · Convolutional neural network · Tensorflow · GPU
JSRT · Chest X-ray · Segmentation · Bone shadow exclusion · Lung cancer

1 Introduction

Lung cancer is a significant burden in the world and especially in China, where as more than half of adult men in that country are and heavy air pollution aggravate the consequences of the disease. Lung cancer in the world and China has a very high disease incidence and the current solution consists in early screening which often leads to good outcomes at a relatively low cost. Screening, especially by computed tomography, has been recognized worldwide as an approach to reduce lung cancer mortality [1]. But its relatively low outspread is aggravated by the high cost and limited availability for the most parts of the world. Chest X-ray (CXR) imaging is currently the most popular and the most available

© Springer International Publishing AG, part of Springer Nature 2019
Z. Hu et al. (Eds.): ICCSEEA 2018, AISC 754, pp. 638–647, 2019.
https://doi.org/10.1007/978-3-319-91008-6_63

diagnostic tool for health monitoring and diagnosing many lung diseases, including pneumonia, tuberculosis, cancer, etc. However, detecting marks of these diseases from CXRs is a very complicated procedure, which takes involvement of the expert radiologists. Application of CXRs is postponed by long manual analysis and detection of lung cancer and it is limited also by shortage of experts. For example, in China the annual number of the diagnosed lung cancer cases is huge (>600 thousands), but the number of certified radiologists is low (<80 thousands) for the nation-wide screening of >1.4 billion of citizens.

Meanwhile, the recent disruptive progress of computing, especially computing on the general purpose graphic processing units (GPU) [2, 3], machine learning, and especially deep learning [4], for image recognition brings a meaningful effect. For example, recently, CheXNet model was announced that can automatically detect pneumonia from chest X-rays at a level exceeding practicing radiologists [5]. That is why any automated assistance tools and related machine learning techniques are of great importance for the faster and better identification, classification and segmentation of suspicious regions (like lesions, nodules, etc.) for the subsequent diagnostic.

The main aim of this paper is to demonstrate efficiency of lung segmentation and bone shadow exclusion techniques for analysis of 2D CXRs by deep learning approach to help radiologists identify suspicious regions in lung cancer patients.

2   Problem and Related Work

Several screening approaches are used now to detect suspicious lung cancer lesions. Computed tomography (CT) is especially sensitive to hard-to-detect nodules and enhances radiologists' diagnosis accuracy. X-ray is often thought as an obsolete medical imaging method, but usage of digital technologies and machine learning now revives the significance of X-ray in medical imaging diagnosis. For example, they allow to detect more different kinds of cardiothoracic lesions and especially sensitive to lung nodules on X-ray scans. The success is strengthened by the fast progress in machine learning and GPU computing research for medical data processing. Recently, several promising results were obtained in the field of lung diseases diagnostic by machine learning and, especially, by deep learning approaches. As a result, the current GPU-based platform can process hundreds of high-resolution medical images per second. Availability of the open datasets with CT and CXR images allow data scientists to train, verify, and tune their new algorithms. The release of the image database with and without lung cancer nodules proposed by Japanese Society of Radiological Technology (JSRT) open this way for many research groups around the world [6]. This initiative was supported by other medical and research institutions. Lung Image Database Consortium (LIDC) proposes the image database containing data captured by various modalities (computed tomography—CT, digital radiography—DX, computed radiography—CR) for >1000 patients, and contain >244 000 images with in-plane resolution of the 512×512-pixel sections ranged from 0.542–0.750 mm [7]. The U.S. National Library of Medicine has made two open datasets of CXRs to foster research in computer-aided diagnosis of pulmonary diseases with a special focus on pulmonary tuberculosis [8, 9]. Montgomery County (MC) dataset has been collected in collaboration with the Department of Health and Human Services,

Montgomery County in Maryland, USA. Shenzhen Hospital dataset (SH) was acquired from Shenzhen No. 3 People's Hospital in Shenzhen, China. Both datasets contain normal and abnormal chest X-rays with manifestations of tuberculosis and include associated radiologist readings. Now ChestX-ray14 is the largest publicly available chest X-ray dataset, containing over 100 000 frontal-view X-ray images with 14 different lung diseases [10]. The aforementioned CheXNet model, which is a 121-layer convolutional neural network, was trained on ChestX-ray14 dataset [5]. To increase the accuracy of predictions researchers try to exclude the regions which are not pertinent to lungs or other regions of interest. Such task includes segmenting the left and right lung fields in standard CXRs. Several segmentation approaches were proposed recently, which are based on active shape models, active appearance models, and a multi-resolution pixel classification method. The methods have been tested on JSRT database, in which all objects have been manually segmented by two human observers [11]. Although many new segmentation methods have been proposed for CXR and MRI images [12] in medical image applications, the lung field segmentation remains a challenge.

Additional promising option for improvement of prediction is related with exclusion of some body parts that shadow the lung, for example, ribs and clavicles. As a service to the medical imaging community, Chest Diagnostic System Research Group (Budapest, Hungary) provided the bone shadow eliminated (BSE) version of JSRT dataset (BSE-JSRT) [13]. In other study, the two-step algorithm for eliminating rib shadows in CXRs was based on delineation of the ribs using a novel hybrid self-template approach and then suppression of these delineated ribs using an unsupervised regression model that takes into account the change in proximal thickness of bone in the vertical axis [14]. Below the further details are given for the datasets used in this work.

3   Data and Methods

3.1   Bone Elimination

Below the segmentation techniques were applied to datasets JSRT and BSE-JSRT datasets. JSRT image dataset contains 247 images (Fig. 1a): 154 cases with lung nodules and 93 cases without lung nodules [6]. BSE-JSRT dataset contains 247 images (Fig. 1b) of the JSRT dataset, but without clavicle and rib shadows removed by the special algorithms [13]. The one of the aims of the work was to check the difference between application of the deep learning approach to the original JSRT dataset (below it is mentioned as dataset #01) and BSE-JSRT dataset, i.e. the same JSRT dataset, but without clavicle and rib shadows (dataset #02).

3.2   Lung Segmentation

To perform lung segmentation, i.e. to cut the left and right lung fields from the heart and other parts in standard CXRs, the UNet-based convolutional neural network (CNN) was applied. The UNet has CNN architecture for fast and precise segmentation of images, which demonstrated high accuracy on several challenges dedicated to segmentation of neuronal structures in electron microscopic stacks, detection of caries in bitewing radiography, and cell tracking from transmitted light microscopy [15]. Recently, this approach was successfully used for segmenting lungs on CXRs from MC and JSRT datasets with usage of manually prepared masks [16]. The additional aim of the work was to check the effect of lung segmentation for application of the deep learning approach to the original JSRT dataset after segmentation (below it is mentioned as dataset #03) and the same BSE-JSRT dataset after segmentation (below it is mentioned as dataset #04).

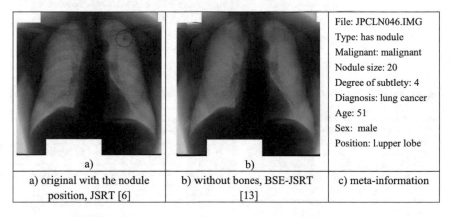

		File: JPCLN046.IMG Type: has nodule Malignant: malignant Nodule size: 20 Degree of subtlety: 4 Diagnosis: lung cancer Age: 51 Sex: male Position: l.upper lobe
a)	b)	
a) original with the nodule position, JSRT [6]	b) without bones, BSE-JSRT [13]	c) meta-information

Fig. 1. Example of the original image (2048 × 2048 pixels) with the cancer nodule from JSRT dataset [6] (a), the correspondent image without bones from BSE-JSRT dataset [13] (b), and the related meta-information (c). The nodule location and region are denoted by the point and circle respectively (a).

3.3 Model, Running Time and Speedup Analysis

The selection of training model was motivated by the reasons of simplicity and short running time to investigate the effect of segmentation and bone elimination only in the simplified configuration without the emphasize on the highest possible accuracy and lowest loss (which will be investigated in the further works). The simple and standard CNN model with only 7 convolution 2D layers was used for training on the original, segmented, and bone eliminated datasets which were mentioned above. The running time (Fig. 2) and speedup analysis (Fig. 3) was performed after numerous tests for images of various sizes and batch sizes in "single-CPU" (1 core of Intel i7), "multi-CPU" (8 cores of Intel i7), and "GPU" (graphic processing unit—NVIDIA Tesla K40c) modes. The obtained results demonstrated the maximal speedup of 3.0 times in multi-CPU mode and up to 9.5 times in GPU mode for the biggest image sizes (1024 × 1024) for the batch size of 8 images. These results allowed to estimate the realistic scenarios for the feasibility tests of segmentation and bone elimination techniques. Finally, the images of sizes 256 × 256 were selected for the previous training and reported here, and the training of the more complicated models on the bigger images are under work now and will be reported elsewhere [17].

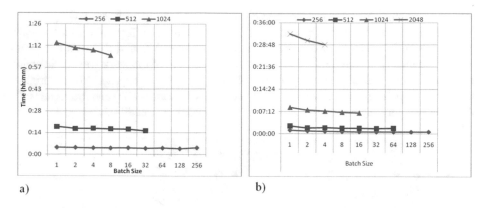

Fig. 2. Running time in CPU (a) and GPU (b) regimes for various batch and images sizes.

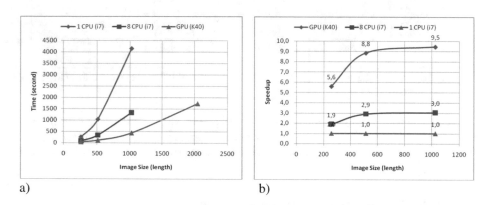

Fig. 3. Speedup of multi-CPU and GPU regimes in comparison to single-CPU mode

4   Results

In this section, the results are presented as to effect of bone elimination and lung segmentation on training with regard to: the original JSRT dataset (below it is mentioned as dataset #01), BSE-JSRT dataset, i.e. the same JSRT dataset, but without clavicle and rib shadows (dataset #02), original JSRT dataset after segmentation (dataset #03), and the same BSE-JSRT dataset after segmentation (dataset #04).

4.1   Segmentation

This segmentation stage was applied to the original images from JSRT dataset (dataset #01, Fig. 4a) to obtain their segmented versions (dataset #03, Fig. 4c) and consisted in the following stages:

- training the UNet-based CNN for lung segmentation (search of lungs borders) on MC dataset with manually prepared masks (lung borders),

- predicting the lung borders in the shape of black-and-white lung masks (Fig. 4b) by means of the trained UNet-based CNN for each of original images from JSRT dataset (Fig. 4a),
- cutting the regions of interest (right and left lungs) (Fig. 4c) from their original images (Fig. 4a).

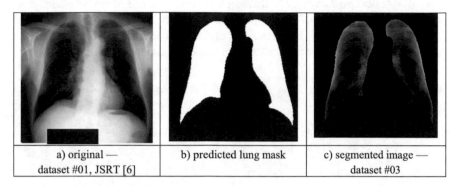

| a) original — dataset #01, JSRT [6] | b) predicted lung mask | c) segmented image — dataset #03 |

Fig. 4. Example of the original image (inversed version of Fig. 1a) (a), the correspondent lung mask predicted by machine learning approach (b), and its segmented version (c).

The similar procedure was applied to the images without bones from BSE-JSRT dataset (dataset #02, Fig. 5a) to obtain their segmented versions without bones (dataset #04, Fig. 5c)

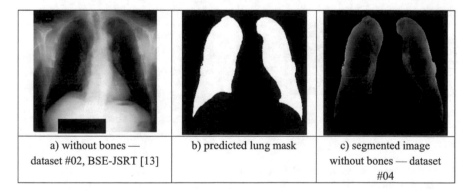

| a) without bones — dataset #02, BSE-JSRT [13] | b) predicted lung mask | c) segmented image without bones — dataset #04 |

Fig. 5. Example of the image without bones (inversed version of Fig. 1b) (a), the correspondent lung mask predicted by machine learning (b), and its segmented version without bones (c).

4.2 Training and Validation

Finally, the simple CNN was trained in GPU mode (NVIDIA Tesla K40c card) by means of Tensorflow machine learning framework [18] with regard to the 4 datasets: original JSRT dataset (dataset #01), original BSE-JSRT dataset, i.e. the same JSRT dataset, but without clavicle and rib shadows (dataset #02), JSRT dataset after segmentation (dataset #03), and BSE-JSRT dataset after segmentation (dataset #04).

The previous results (Fig. 6) clearly demonstrate the drastic difference in training and validation results between raw data, namely, the original JSRT dataset #01, and any of the pre-processed datasets #02, #03, or #04. Despite the original JSRT dataset #01 (red line on Fig. 6) does not show any sign of training at all (because of low image size and negligibly small nodule size), all of the pre-processed datasets #02, #03, or #04 (orange, dark blue, and blue lines on Fig. 6) have tendency to train and demonstrate high training accuracy and training low loss for the late stages in such simplified configuration even.

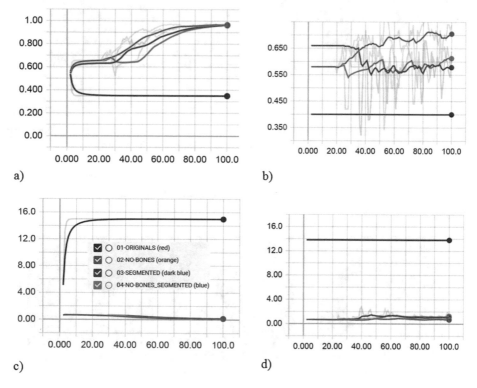

Fig. 6. Training accuracy (a), validation accuracy (b), training loss (c), and validation loss (d) for the original images, JSRT, dataset #01 (line 01), the images without bones, BSE-JSRT, dataset #02 (line 02), the segmented images, dataset #03 (line 03), and the segmented images without bones, dataset #04 (line 04).

The more careful analysis of training and validation curves of the pre-processed datasets #02, #03, or #04 (Fig. 7) allows to note that accuracy (and loss) obtained after training is much higher (lower) than ones observed after validation. This can be considered as the clear manifestation of overtraining with the overestimated accuracy and underestimated loss for the raw image data from the original JSRT dataset #01. Nevertheless, the pre-processed dataset without bones (dataset #02) demonstrates the much better accuracy and loss results in comparison to the other pre-processed datasets after lung segmentation (datasets #02 and #03).

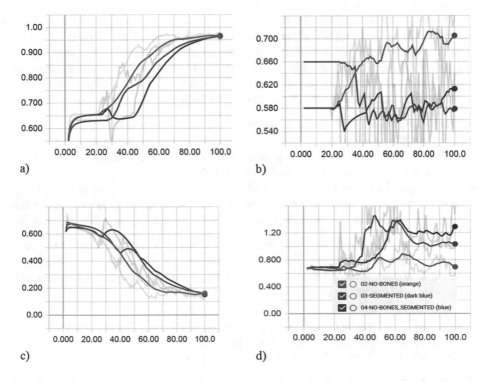

Fig. 7. Training accuracy (a), validation accuracy (b), training loss (c), and validation loss (d) for the images without bones (line 02), the segmented images (line 03), and the segmented images without bones (line 04).

5   Discussion and Conclusions

The results obtained demonstrate the usefulness of pre-processing techniques like bone shadow elimination and segmentation with the clear tendency to train in such simplified configuration even. The overtraining effect with the lower validation accuracy and higher validation loss in comparison to the training accuracy and loss is considered to be related with training to artifacts like the shape of the lungs and lung border pattern.

The further potential for improvement consists in an increase of the investigated datasets:

- the image size from the current 256×256 values up to 1024×1024 (ChestXRay dataset) and 2048×2048 (JSRT dataset);
- the number of images from current 247 (JSRT and BSE-JSRT datasets) up to >1000 (MC dataset) and $>100\,000$ (ChestXRay dataset);
- by data augmentation with regard to lossy and lossless transformations [17].

The much bigger progress can be obtained from integration of many similar datasets from numerous hospitals around the world in the spirit of the open science data, volunteer data collection, data processing and computing [19].

In conclusion, the results demonstrate the high efficiency and usefulness of the considered pre-processing techniques in the simplified configuration even. It should be emphasized that the pre-processed dataset without bones (dataset #02) demonstrates the much better accuracy and loss results in comparison to the other pre-processed datasets after lung segmentation (datasets #02 and #03). That is why the additional reserve of development could be related with improvement of pre-processing algorithms for:

- lung segmentation on the basis of the bigger datasets with masks,
- bone shadow elimination on the basis of the more complicated semantic segmentation techniques applied not only to lungs and body parts outside of them (like heart, arms, etc.), but also to ribs and clavicles inside the lungs,
- training itself by means of increase of size and complexity of the deep learning network from the current miniature size of 7 layers up to >100 layers like in the current most accurate networks like CheXNet used for diagnostics of other diseases [5].

Training of the more complicated models on the bigger images are under work now and will be reported elsewhere [17]. In this connection the fine tuning of datasets and deep learning models should be taken into account, because it can play the crucial role for efficiency of the model used [20–22].

Acknowledgments. The work was partially supported by Huizhou Science and Technology Bureau and Huizhou University (Huizhou, P.R. China) in the framework of Platform Construction for China-Ukraine Hi-Tech Park Project # 2014C050012001.

References

1. National Lung Screening Trial Research Team: Reduced lung-cancer mortality with low-dose computed tomographic screening. N. Engl. J. Med. **365**, 395–409 (2011)
2. Owens, J.D., Luebke, D., Govindaraju, N., Harris, M., Krüger, J., Lefohn, A.E., Purcell, T.J.: A survey of general-purpose computation on graphics hardware. Comput. Graph. Forum **26**(1), 80–113 (2007)
3. Smistad, E., Falch, T.L., Bozorgi, M., Elster, A.C., Lindseth, F.: Medical image segmentation on GPUs – a comprehensive review. Med. Image Anal. **20**(1), 1–18 (2015)
4. LeCun, Y., Bengio, Y., Hinton, G.: Deep learning. Nature **521**(7553), 436–444 (2015)
5. Rajpurkar, P., Irvin, J., Zhu, K., Yang, B., Mehta, H., Duan, T., Ding, D., Bagul, A., Langlotz, C., Shpanskaya, K., Lungren, M.P., Ng, A.Y.: CheXNet: radiologist-level pneumonia detection on chest X-rays with deep learning. arXiv preprint arXiv:1711.05225 (2017)
6. Shiraishi, J., Katsuragawa, S., Ikezoe, J., Matsumoto, T., Kobayashi, T., Komatsu, K., Matsui, M., Fujita, H., Kodera, Y., Doi, K.: Development of a digital image database for chest radiographs with and without a lung nodule: receiver operating characteristic analysis of radiologists' detection of pulmonary nodules. Am. J. Roentgenol. **174**, 71–74 (2000)
7. Armato, S.G., et al.: The lung image database consortium (LIDC) and image database resource initiative (IDRI): a completed reference database of lung nodules on CT scans. Med. Phys. **38**(2), 915–931 (2011)

8. Jaeger, S., Candemir, S., Antani, S., Wáng, Y.X.J., Lu, P.X., Thoma, G.: Two public chest X-ray datasets for computer-aided screening of pulmonary diseases. Quant. Imag. Med. Surg. **4**(6), 475 (2014)

9. Jaeger, S., Karargyris, A., Candemir, S., Folio, L., Siegelman, J., Callaghan, F., Xue, Z., Palaniappan, K., Singh, R.K., Antani, S., Thoma, G., Wang, Y.X.J., Lu, P.X., McDonald, C.J.: Automatic tuberculosis screening using chest radiographs. IEEE Trans. Med. Imag. **33**(2), 233–245 (2014)

10. Wang, X., Peng, Y., Lu, L., Lu, Z., Bagheri, M., Summers, R.M.: Chestx-ray8: hospital-scale chest X-ray database and benchmarks on weakly-supervised classification and localization of common thorax diseases. arXiv preprint arXiv:1705.02315 (2017)

11. Van Ginneken, B., Stegmann, M.B., Loog, M.: Segmentation of anatomical structures in chest radiographs using supervised methods: a comparative study on a public database. Med. Image Anal. **10**(1), 19–40 (2006)

12. Bagherieh, H., Hashemi, A., Pilevar, A.H.: Mass detection in lung CT images using region growing segmentation and decision making based on fuzzy systems. Int. J. Image Graph. Sig. Process. (IJIGSP) **6**(1), 1–8 (2014). https://doi.org/10.5815/ijigsp.2014.01.01

13. Juhász, S., Horváth, Á., Nikházy, L., Horváth, G.: Segmentation of anatomical structures on chest radiographs. In: XII Mediterranean Conference on Medical and Biological Engineering and Computing 2010, pp. 359–362. Springer, Heidelberg (2010)

14. Oğul, H., Oğul, B.B., Ağıldere, A.M., Bayrak, T., Sümer, E.: Eliminating rib shadows in chest radiographic images providing diagnostic assistance. Comput. Meth. Programs Biomed. **127**, 174–184 (2016)

15. Ronneberger, O., Fischer, P., Brox, T.: U-net: convolutional networks for biomedical image segmentation. In: International Conference on Medical Image Computing and Computer-Assisted Intervention, pp. 234–241. Springer, Cham (2015)

16. Pazhitnykh, I., Petsiuk, V.: Lung Segmentation (2D) (2017). https://github.com/imlab-uiip/lung-segmentation-2d

17. Gordienko, Y., Kochura, Y., Alienin, O., Rokovyi, O., Stirenko, S., Gang, P., Hui, J., Zeng, W.: Dimensionality reduction in deep learning for chest x-ray analysis of lung cancer. In: 10th International Conference on Advanced Computational Intelligence, Xiamen, China. arXiv preprint arXiv:1801.06495 (2018)

18. Abadi, M., et al.: TensorFlow: large-scale machine learning on heterogeneous distributed systems. arXiv preprint arXiv:1603.04467 (2016)

19. Gordienko, N., Lodygensky, O., Fedak, G., Gordienko, Y.: Synergy of volunteer measurements and volunteer computing for effective data collecting, processing, simulating and analyzing on a worldwide scale. In: Proceedings of IEEE 38th International Convention on Information and Communication Technology, Electronics and Microelectronics (MIPRO), pp. 193–198 (2015)

20. Gordienko, N., Stirenko, S., Kochura, Yu., Rojbi, A., Alienin, O., Novotarskiy, M., Gordienko, Yu.: Deep learning for fatigue estimation on the basis of multimodal human-machine interactions. In: XXIX IUPAP Conference in Computational Physics. arXiv preprint arXiv:1801.06048 (2017)

21. Kochura, Yu., Stirenko, S., Alienin, O., Novotarskiy, M., Gordienko, Yu.: Performance analysis of open source machine learning frameworks for various parameters in single-threaded and multi-threaded modes. In: Conference on Computer Science and Information Technologies, pp. 243–256. Springer, Cham (2017)

22. Rather, N.N., Patel, C.O., Khan, S.A.: Using deep learning towards biomedical knowledge discovery. Int. J. Math. Sci. Comput. (IJMSC) **3**(2), 1–10 (2017). https://doi.org/10.5815/ijmsc.2017.02.01

RNA Quasi-Orthogonal Block Code

Han Hai ⓘ and Moon Ho Lee(⊠)

Division of Electronics and Information Engineering,
Chonbuk National University, Jeonju, Korea
{hhhtgy,moonho}@jbnu.ac.kr

Abstract. This paper presents a single strand ribonucleic acid (RNA) Kronecker product of double stochastic matrix to a deoxyribose nucleic acid (DNA) double helix based on the block circulant Jacket matrix (BCJM) characteristics which is used to develop a bioinformatics for the molecular communications. The RNA matrix decomposition is the form of the Kronecker product of Hadamard matrices with its pair complementarity. The variants of kernel of the Kronecker families are produced by permutations of the four letters C, A, U, G on positions in the matrix. This decomposition of DNA to RNA leads very clearly to the Kronecker product of the symmetrical genetic matrices. We also analyze DNA quasi-orthogonal matrix.

Keywords: RNA · DNA · Kronecker product · Hadamard/identity matrix
Quasi-orthogonal matrix

1 Introduction

The deoxyribose nucleic acid (DNA) double helix plays an important role in DNA replication, which prepares genetic information to the next generation. The DNA has two helical chains in the same axis. Both chains follow right handed helices, but the sequences of the two chains run in opposite directions [1]. Niels Bohr's pair of complementarity [8] is also utilized in DNA transcription, which generates a ribonucleic acid (RNA) strand from a DNA template. In 1962, H. M. Temin investigated reverse transcription RNA to DNA and he had gotten Nobel Prize in 1975 [2]. Nowadays, nobody doesn't prove it by mathematically. The detection of natural realization of the block circulant Jacket matrices (BCJMs) on the basis of parameters of the molecular genetic systems, which serves to transfer discrete genetic information, shows that known advantages of BCJM [3, 4, 6] can be utilized in bioinformatics. However, we prove mathematically RNA to DNA by the bioinformatics signal processing system. The four letter of the genetic code alphabet has 64 triplets [5, 9, 10]. We can get 6 subset characteristic property of the block circulant matrix.

A complementary strand RNA or DNA may be produced based on nucleobase complementarity. Each position in the sequences of the nucleotide bases will be complementary much like looking in the mirror and seeing the reverse of things such as 0 and 1 i.e., 2 bit case 00 01; 10 11. The complementarity of DNA in double helix prepares to use one strand as a template to build up each other.

© Springer International Publishing AG, part of Springer Nature 2019
Z. Hu et al. (Eds.): ICCSEEA 2018, AISC 754, pp. 648–660, 2019.
https://doi.org/10.1007/978-3-319-91008-6_64

The main contributions of this paper are summarized as follows:

- We prove mathematically a RNA reverse transcription to DNA double helix based on the BCJMs characteristic property.
- The kernels of the Kronecker families of genomatrices are produced by permutations of the four letters C, A, U, G on positions in the matrix. We get 6 subset pattern of the block circulant matrix.

We propose the RNA Yin-Yang model and also analyze the quasi-orthogonal matrix.

2 Block Circulant Jacket Matrix

In this Section, we discuss the block circulant Jacket matrix [3, 4].

Definition 2.1. Let $[C]_N = \begin{pmatrix} C_0 & C_1 \\ C_1 & C_0 \end{pmatrix}$ be 2×2 block matrix of order $N = 2p$. If $[C_0]_p$ and $[C_1]_p$ are $p \times p$ Jacket matrices, then $[C]_N$ is a Jacket matrix if and only if

$$C_0 C_1^{RT} + C_1 C_0^{RT} = [0]_N \tag{1}$$

where RT is reciprocal transpose which mean transpose of element-wise inverse.

Proof. Since C_0 and C_1 are Jacket matrices, we have $C_0 C_0^{RT} = p[I]_p$ and $C_1 C_1^{RT} = p[I]_p$. Note that $[C]_N$ is Jacket matrix if and only if $[C][C]^{RT} = NI_N$. Then C is Jacket matrix if and only if

$$[C][C]^{RT} = \begin{pmatrix} C_0 & C_1 \\ C_1 & C_0 \end{pmatrix} \begin{pmatrix} C_0 & C_1 \\ C_1 & C_0 \end{pmatrix}^{RT} = \begin{pmatrix} 2p[I]_p & C_0 C_1^{RT} + C_1 C_0^{RT} \\ C_0 C_1^{RT} + C_1 C_0^{RT} & 2p[I]_p \end{pmatrix}$$
$$= NI_N.$$

Hence $[C]_N$ is a Jacket matrix if and only if

$$C_0 C_1^{RT} + C_1 C_0^{RT}.$$

By using Definition 2.1, we may construct many BCJMs.

Example 2.1. *Let*

$$C_0 = \begin{pmatrix} 1 & 1 \\ 1 & -1 \end{pmatrix}, C_1 = \begin{pmatrix} a & -a \\ -1/a & -1/a \end{pmatrix}.$$

Since $C_0 C_0^{RT} = 2[I]_2$ and $C_1 C_1^{RT} = 2[I]_2$, C_0 and C_1 are Jacket matrices of order 2. Moreover,

$$C_0 C_1^{RT} + C_1 C_0^{RT} = \begin{pmatrix} 1 & 1 \\ 1 & -1 \end{pmatrix} \begin{pmatrix} a & -a \\ -1/a & -a \end{pmatrix} + \begin{pmatrix} a & -a \\ -1/a & -1/a \end{pmatrix} \cdot \begin{pmatrix} 1 & 1 \\ 1 & -1 \end{pmatrix}$$

$$= \begin{pmatrix} 0 & -2a \\ 2/a & 0 \end{pmatrix} + \begin{pmatrix} 0 & 2a \\ -2/a & 0 \end{pmatrix} = [0]_2 \tag{2}$$

Hence a circulant matrix

$$C_{4=} \begin{pmatrix} C_0 & C_1 \\ C_1 & C_0 \end{pmatrix} = \begin{pmatrix} 1 & 1 & a & -a \\ 1 & -1 & -1/a & -1/a \\ a & -a & 1 & 1 \\ -1/a & -1/a & 1 & -1 \end{pmatrix} \tag{3}$$

$$= \begin{pmatrix} 1 & 1 & 0 & 0 \\ 1 & -1 & 0 & 0 \\ 0 & 0 & 1 & 1 \\ 0 & 0 & 1 & -1 \end{pmatrix} + \begin{pmatrix} 0 & 0 & 1 & -1 \\ 0 & 0 & -1 & -1 \\ 1 & -1 & 0 & 0 \\ -1 & 1 & 0 & 0 \end{pmatrix} = \left(\begin{array}{cc|cc} 1 & 1 & 1 & -1 \\ 1 & -1 & -1 & -1 \\ \hline 1 & -1 & 1 & 1 \\ -1 & -1 & 1 & -1 \end{array} \right)$$

is a BCJM. C_0 and C_1 are the Hadamard matrix.

The size of the circulant submatrices is 2×2, i.e., it has the property that block diagonal cyclic shifts. We can specify to these submatrices as circulant blocks. Then, the BCJM C_4 can be rewritten as

$$\underbrace{C_4 \triangleq I_0 C_0' + I_1 \otimes C_1} \tag{4}$$

where $I_0 = \begin{pmatrix} 1 & 0 \\ 0 & 1 \end{pmatrix}$, $I_1 = \begin{pmatrix} 0 & 1 \\ 1 & 0 \end{pmatrix}$, $C_0' = \begin{pmatrix} 1 & 1 \\ 1 & -1 \end{pmatrix}$, $C_1 = \begin{pmatrix} 1 & -1 \\ -1 & -1 \end{pmatrix}$ and $\otimes$ is the Kronecker product.

Example 2.2. Let

$$H_4 = \begin{pmatrix} -1 & 1 & 1 & 1 \\ 1 & -1 & 1 & 1 \\ 1 & 1 & -1 & 1 \\ 1 & 1 & 1 & -1 \end{pmatrix} = \begin{pmatrix} C_0 & C_1 \\ C_1 & C_0 \end{pmatrix} = I_0 \otimes C_0'' + I_1 \otimes C_1' \tag{5}$$

where $I_0 = \begin{pmatrix} 1 & 0 \\ 0 & 1 \end{pmatrix}$, $I_1 = \begin{pmatrix} 0 & 1 \\ 1 & 0 \end{pmatrix}$, $C_0'' = \begin{pmatrix} -1 & 1 \\ 1 & -1 \end{pmatrix}$, $C_1' = \begin{pmatrix} 1 & 1 \\ 1 & 1 \end{pmatrix}$. It is con-
jectured that there is no other circulant Hadamard (H_4) matrix [4].

3   Analysis RNA and DNA Sparse Matrix

From (4), similar fashion as an example of RNA sequence with the genetic matrix $[C \ U; A \ G]^3$ is given by [5, 6]

$$P^1 = \begin{pmatrix} C & U \\ A & G \end{pmatrix}, P^2 = \begin{pmatrix} C & U \\ A & G \end{pmatrix} \otimes \begin{pmatrix} C & U \\ A & G \end{pmatrix}, P^3 = \begin{pmatrix} C & U \\ A & G \end{pmatrix}^2 \otimes \begin{pmatrix} C & U \\ A & G \end{pmatrix} \quad (6)$$

Where $\otimes$ is the Kronecker product, C is cytosine, U is uracil, A is adenine and G is guanine. By analogy with theory of noise-immunity coding, mosaic gene matrix $[C\ U; A\ G]^3$ can be represented by replacements of 64 triplets with strong roots $(CC, CU, CG, AC, UC, GC, GU, GG)$ and weak roots $(CA, AA, AU, AG, UA, UU, UG, GA)$ by means of numbers +1 and −1 respectively. Thus, $[R]_8$ is the Rademacher singular matrix as shown in Eq. (7).

Where $\otimes$ is the Kronecker product, C is cytosine, U is uracil, A is adenine and G is guanine. By analogy with theory of noise-immunity coding, mosaic gene matrix $[C\ U; A\ G]^3$ can be represented by replacements of 64 triplets with strong roots $(CC, CU, CG, AC, UC, GC, GU, GG)$ and weak roots $(CA, AA, AU, AG, UA, UU, UG, GA)$ by means of numbers +1 and −1 respectively. Thus, $[R]_8$ is the Rademacher singular matrix as shown in Eq. (7).

$$[R]_8 = \begin{pmatrix} 1 & 1 & 1 & 1 & 1 & 1 & -1 & -1 \\ 1 & 1 & 1 & 1 & 1 & 1 & -1 & -1 \\ -1 & -1 & 1 & 1 & -1 & -1 & -1 & -1 \\ -1 & -1 & 1 & 1 & -1 & -1 & -1 & -1 \\ 1 & 1 & -1 & -1 & 1 & 1 & 1 & 1 \\ 1 & 1 & -1 & -1 & 1 & 1 & 1 & 1 \\ -1 & -1 & -1 & -1 & -1 & -1 & 1 & 1 \\ -1 & -1 & -1 & -1 & -1 & -1 & 1 & 1 \end{pmatrix} \quad (7)$$

The Eq. (7) is the DNA double helix as shown the color.
Note that,

$$\underbrace{[R]_8 \triangleq I_0 \otimes C_0 \otimes P_2 + I_1 \otimes C_1 \otimes P_2} \quad (8)$$

where $I_0 = \begin{pmatrix} 1 & 0 \\ 0 & 1 \end{pmatrix}$, $I_1 = \begin{pmatrix} 0 & 1 \\ 1 & 0 \end{pmatrix}$, $C_0 = \begin{pmatrix} 1 & 1 \\ -1 & 1 \end{pmatrix}$, $C_1 = \begin{pmatrix} 1 & -1 \\ -1 & -1 \end{pmatrix}$, and P_2 is the permutation double stochastic matrix as $P_2 = \begin{pmatrix} 1 & 1 \\ 1 & 1 \end{pmatrix}$. The Eq. (7) is a certainly redundancy row repeated, and cancelled the repeated row. From the Rademacher matrix $[R]_8$, then we have

$$[R]_8' = \begin{bmatrix} 1 & 1 & 1 & 1 & 1 & 1 & -1 & -1 \\ -1 & -1 & 1 & 1 & -1 & -1 & -1 & -1 \\ 1 & 1 & -1 & -1 & 1 & 1 & 1 & 1 \\ -1 & -1 & -1 & -1 & -1 & -1 & 1 & 1 \end{bmatrix}. \quad (9)$$

Also, cancelled the repeated column, then (10) is single strand RNA matrix as shown the color.

$$[R]''_4 = \begin{pmatrix} 1 & 1 & 1 & -1 \\ -1 & 1 & -1 & -1 \\ 1 & -1 & 1 & 1 \\ -1 & -1 & -1 & 1 \end{pmatrix} = \begin{pmatrix} C_0 & C_1 \\ C_1 & C_0 \end{pmatrix}. \tag{10}$$

We show that the information is replicated from DNA and transcript to the RNA. Finally, it is translated into protein.

Therefore,

$$R''_4 \triangleq I_0 \otimes C_0 + I_1 \otimes C_1. \tag{11}$$

Thus,

$$R''_4 \otimes P_2 \Rightarrow R_8 = R_{4 \times 2^k}, \text{ where, } k = 1. \ [2] \tag{12}$$

The Eq. (12) is proved the RNA single strand to the DNA double helix; vice versa, perfectly.

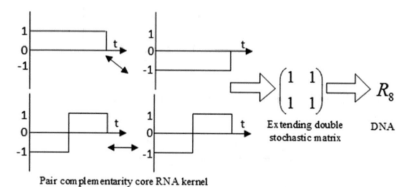

(a) R''_4 RNA to R_8 DNA for Eq. (29).

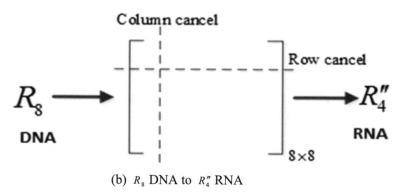

(b) R_8 DNA to R''_4 RNA

Fig. 1. Block Diagram: the construction RNA of 4×4 to 8×8 DNA matrix, vice versa.

These self-replication matrices are the orthogonal Hadamard matrix, but also the symmetric pair complementary of the core RNA as similar Niels Bohr's [8] pair complementarity.

This processes of the RNA can be extended to show in Fig. 1 as RNA to DNA, DNA to RNA, respectively. The complementarity is the basic principle of DNA replication and transcription as it is a property shared between two DNA or RNA sequences, such that when they are aligned antiparallel to each other, the nucleotide bases at each position in the sequences will be complementary as shown in Fig. 1(a) i.e. the contrary is the complementary to each other.

The 4×4 RNA matrix (10) can be used to formulate a sparse matrix decomposition, then we can get

$$
\begin{pmatrix}
1 & 1 & 1 & -1 \\
-1 & 1 & -1 & -1 \\
1 & -1 & 1 & 1 \\
-1 & -1 & -1 & 1
\end{pmatrix}
=
\begin{pmatrix}
1 & 1 & 0 & 0 \\
-1 & 1 & 0 & 0 \\
1 & -1 & 0 & 0 \\
-1 & 1 & 0 & 0
\end{pmatrix}
\begin{pmatrix}
1 & 0 & 1 & 0 \\
0 & 1 & 0 & -1 \\
0 & 0 & 0 & 0 \\
0 & 0 & 0 & 0
\end{pmatrix}
\tag{13}
$$

From (10) and (7), the 4 by 4 RNA and 8 by 8 DNA are implemented Cooley-Tukey fast algorithm and 6 & 16 addition/subtraction butterfly flow graph of the RNA as shown in Fig. 2.

$$
\begin{bmatrix}
1 & 1 & 1 & 1 & 1 & 1 & -1 & -1 \\
1 & 1 & 1 & 1 & 1 & 1 & -1 & -1 \\
-1 & -1 & 1 & 1 & -1 & -1 & -1 & -1 \\
-1 & -1 & 1 & 1 & -1 & -1 & -1 & -1 \\
1 & 1 & -1 & -1 & 1 & 1 & 1 & -1 \\
1 & 1 & -1 & -1 & 1 & 1 & 1 & 1 \\
-1 & -1 & -1 & -1 & -1 & -1 & 1 & 1 \\
-1 & -1 & -1 & -1 & -1 & -1 & 1 & 1
\end{bmatrix}
=
\begin{bmatrix}
1 & 0 & 0 & 0 & 1 & 0 & 0 & 0 \\
1 & 0 & 0 & 0 & 1 & 0 & 0 & 0 \\
0 & 0 & 1 & 0 & 0 & 0 & 1 & 0 \\
0 & 0 & 1 & 0 & 0 & 0 & 1 & 0 \\
0 & 1 & 0 & 0 & 0 & 1 & 0 & 0 \\
0 & 1 & 0 & 0 & 0 & 1 & 0 & 0 \\
0 & 0 & 0 & 1 & 0 & 0 & 0 & 1 \\
0 & 0 & 0 & 1 & 0 & 0 & 0 & 1
\end{bmatrix}
$$

$$
\left(
\begin{bmatrix}
1 & 1 & 0 & 0 & 0 & 0 & 0 & 0 \\
1 & -1 & 0 & 0 & 0 & 0 & 0 & 0 \\
-1 & 1 & 0 & 0 & 0 & 0 & 0 & 0 \\
-1 & -1 & 0 & 0 & 0 & 0 & 0 & 0 \\
0 & 0 & 1 & 1 & 0 & 0 & 0 & 0 \\
0 & 0 & 1 & -1 & 0 & 0 & 0 & 0 \\
0 & 0 & -1 & 1 & 0 & 0 & 0 & 0 \\
0 & 0 & -1 & -1 & 0 & 0 & 0 & 0
\end{bmatrix}
\begin{bmatrix}
1 & 0 & 0 & 0 & 1 & 0 & 0 & 0 \\
0 & 0 & 1 & 0 & 0 & 0 & -1 & 0 \\
0 & 1 & 0 & 0 & 0 & 1 & 0 & 0 \\
0 & 0 & 0 & 1 & 0 & 0 & 0 & -1 \\
0 & 0 & 0 & 0 & 0 & 1 & 0 & 0 \\
0 & 0 & 0 & 0 & 0 & 1 & 0 & 0 \\
0 & 0 & 0 & 1 & 0 & 0 & 0 & 0 \\
0 & 0 & 0 & 1 & 0 & 0 & 0 & 0
\end{bmatrix}
\right)
\tag{14}
$$

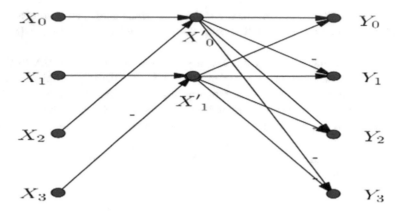

(a) RNA 4 by 4 fast butterfly, 6 addition/subtraction as $(N-1)\log_2 N; N=4$.

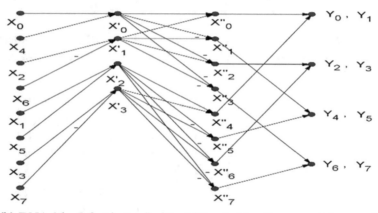

(b) DNA 8 by 8 fast butterfly 16 addition/subtraction as $N/2(1+\log_2 N); N=8$

Fig. 2. RNA 4 by 4 and 8 by 8 butterfly structure.

4   RNA Quasi-Orthogonal Block Code (QO-BC)

We have proved the DNA double helix (8) to the RNA single helix (11). We have

$$R_8 = (I_0 \otimes C_0 + I_1 \otimes C_1) \otimes P_2$$
$$R_4'' = I_0 \otimes C_0 + I_1 \otimes C_1$$

(15)

At first we analyze R_4'' RNA, we can get

$$R_4''^T R_4'' = (I_0 \otimes C_0 + I_1 \otimes C_1)^T (I_0 \otimes C_0 + I_1 \otimes C_1) \tag{16}$$

From (16), we can get the weighted coefficient,

$$[Wc]_4 \triangleq R_4''^T \cdot R_4'' = 4[D]_4 \tag{17}$$

Thus, the formula

$$(A \otimes B + C \otimes D)^T \cdot (A \otimes B + C \otimes D)$$
$$= (AA^T \otimes BB^T)(CC^T \otimes DD^T) \tag{18}$$

We follow the formula (18), then we have

$$[Wc]_4 = (I_0 \otimes C_0 + I_1 \otimes C_1)^T (I_0 \otimes C_0 + I_1 \otimes C_1)$$
$$(I_0 I_0^T \otimes C_0 C_0^T + I_1 I_1^T \otimes C_1 C_1^T) \tag{19}$$

Example 4.1. Let

$$R_4''^T R_4'' = \begin{pmatrix} 1 & 1 & 1 & -1 \\ -1 & 1 & -1 & -1 \\ 1 & -1 & 1 & 1 \\ -1 & -1 & -1 & 1 \end{pmatrix}^T \begin{pmatrix} 1 & 1 & 1 & -1 \\ -1 & 1 & -1 & -1 \\ 1 & -1 & 1 & 1 \\ -1 & -1 & -1 & 1 \end{pmatrix}$$
$$= 4 \begin{pmatrix} 1 & 0 & 1 & 0 \\ 0 & 1 & 0 & -1 \\ 1 & 0 & 1 & 0 \\ 0 & -1 & 0 & 1 \end{pmatrix} = 4[D]_4 \tag{20}$$

Example 4.2. Similar way, we will analyze DNA double helix. We can get the weighted coefficient

$$[Wc]_8 = R_8^T \cdot R_8 = 8 \begin{bmatrix} D_4 & D_4 \\ D_4 & D_4 \end{bmatrix}. \tag{21}$$

CCC 000 x_0	CAC 010 $-x_2$	ACC 100 x_4	AAC 110 $-x_6$	CCA 001 x_1	CAA 011 $-x_3$	ACA 101 x_5	AAA 111 $-x_7$
CUC 010 x_2	CGC 000 x_0	AUC 110 $-x_6$	AGC 100 $-x_4$	CUA 011 x_3	CGA 001 x_1	AUA 111 $-x_7$	AGA 101 $-x_5$
UCC 100 x_4	UAC 110 $-x_6$	GCC 000 x_0	GAC 010 $-x_2$	UCA 101 x_5	UAA 111 $-x_7$	GCA 001 x_1	GAA 011 $-x_3$
UUC 110 $-x_6$	UGC 100 $-x_4$	GUC 010 x_2	GGC 000 x_0	UUA 111 $-x_7$	UGA 101 $-x_5$	GUA 011 x_3	GGA 001 x_1
CCU 001 x_0	CAU 011 $-x_2$	ACU 101 x_4	AAU 111 $-x_6$	CCG 000 x_1	CAG 010 $-x_3$	ACG 100 x_5	AAG 110 $-x_7$
CUU 011 x_2	CGU 001 x_0	AUU 111 $-x_6$	AGU 101 $-x_4$	CUG 010 x_3	CGG 000 x_1	AUG 110 $-x_7$	AGG 100 $-x_5$
UCU 101 x_4	UAU 111 $-x_6$	GCU 001 x_0	GAU 011 $-x_2$	UCG 100 x_5	UAG 110 $-x_7$	GCG 000 x_1	GAG 010 $-x_3$
UUU 111 $-x_6$	UGU 101 $-x_4$	GUU 011 x_2	GGU 001 x_0	UUG 110 $-x_7$	UGG 100 $-x_5$	GUG 010 x_3	GGG 000 x_1

$$(22)$$

The Eq. (21) is the two times double stochastic. The rank of matrix is 4. The Eq. (22) signal constellation comes from $(x_0\, x_2\, x_4\, x_6\, x_1\, x_3\, x_5\, x_7)$ multiply $[R]_8^T$. From (7), the $(C\, U; A\, G)_{123}$ becomes the $(C\, A; U\, G)_{231}$. Therefore, $(C\, A; U\, G)_{231}$ is given by (22) [5]. Examples of genomatrix quasi-orthogonal designs are the 2×2 subset block design from (22), where $x_0, x_2, x_4, x_6, x_1, x_3, x_5, x_7$ are the indeterminate variables. Therefore, we have a symmetric pattern as (21). From (21) and (22), this [ABAB] pattern can be partitioned as

$$\begin{bmatrix} A_{02} & A_{46} & B_{13} & B_{57} \\ A_{46} & A_{02} & B_{57} & B_{13} \\ A_{02} & A_{46} & B_{13} & B_{57} \\ A_{46} & A_{02} & B_{57} & B_{13} \end{bmatrix} \tag{23}$$

Where $A_{02} = \begin{pmatrix} x_0 & -x_2 \\ x_2 & x_0 \end{pmatrix}$, $A_{46} = \begin{pmatrix} x_4 & -x_6 \\ -x_6 & -x_4 \end{pmatrix}$, $B_{13} = \begin{pmatrix} x_1 & -x_3 \\ x_3 & x_1 \end{pmatrix}$, and $B_{57} = \begin{pmatrix} x_5 & -x_7 \\ -x_7 & -x_5 \end{pmatrix}$.

Obviously, the pattern of the eight real genomatrix QO-BC designs are from the four block as [ABAB] pattern. It is desirable to construct the full code rate transmission schemes for any number of transmit antennas. Since full rate codes are bandwidth efficient. Moreover, the measurement of error probability is from calculating the diversity product λ [7]. From (20), we have

$$\lambda = \min_{\{C \neq \tilde{C}\}} \left| \det \left[(C - \tilde{C})^H (C - \tilde{C}) \right] \right|^{1/2n} \tag{24}$$

where n is the number of transmit antennas and $\tilde{C}$ is the error code words from C. Assuming $B_i = (C_i - \bar{C}_j)$, $B_j = (C_i - \bar{C}_j)$, and $\hat{x}_i = x_i - \bar{x}_i, i \in \{0, 2, 4, 6\}$ as elements of error matrix B, the diversity products from the Jafarkhani and its modified 4×4 case,

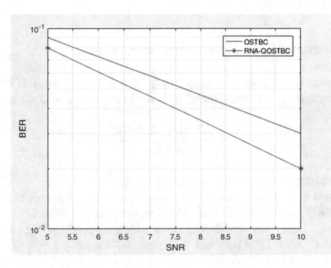

Fig. 3. Comparison of bit error rate versus SNR for the RNA QO-BC and the wireless orthogonal STBC, 4Tx, 1Rx antennas and 2 bits/s/Hz.

$$\lambda_4 = \min_{\{C \neq \tilde{C}\}} \left| \det \left(B_i^H B_i \right) \right|^{1/2n} = \min_{\{C \neq \tilde{C}\}} \left| \det \left(B_j^H B_j \right) \right|^{1/2n}$$

$$= \min_{\{C \neq \tilde{C}\}} \left| \det \begin{bmatrix} \hat{a} & 0 & \hat{b}_j & 0 \\ 0 & \hat{a} & 0 & -\hat{b}_j \\ \hat{b}_j & 0 & \hat{a} & 0 \\ 0 & -\hat{b}_j & 0 & \hat{a} \end{bmatrix} \right|^{1/2n} \tag{25}$$

where

$$\hat{a} = \sum_{i=0,\text{even}}^{6} |\hat{x}_i|^2$$

and

$$\hat{b}_i = \hat{b}_j = \left(\hat{x}_0\hat{x}_6^* + \hat{x}_0^*\hat{x}_6\right) - \left(\hat{x}_2\hat{x}_4^* + \hat{x}_2^*\hat{x}_4\right).$$

Thus, the diversity product (25) [ABAB] is a simple pattern as [ABBA] TBH STBC [8, 11, 12]. The simulation results show that full transmission rate is more important for low signal-to-noise-ratios (SNRs) with high bit error rates (BERs). The RNA quasi-orthogonal performs better than the orthogonal space time block code (OSTBC) at all low SNRs as shown in Fig. 3.

5   RNA Symmetric Complementarity Codon

In this Section, we propose our main results for $24(= 4 \times 4C_2)$ DNA pair class characteristic property. The main kernel of body in the equation is given by

$$\underbrace{\{(I_0 \otimes A) + (I_1 \otimes B)\}}_{MainBodyKernel} \tag{26}$$

A RNA pattern of main kernel is same as the (11).

The national flag of South Korea bears symbols of trigrams and Yin-Yang in the center. The pattern of Fig. 4 is similar pattern as the Korean national flag. We analyze block circulant matrix based on RNA to DNA which are produced by permutations of the four letters C, A, U, G on positions in the matrix. We can get $6(= 4C_2)$ subclasses [C U; A G] from the block circulant case, i.e., $C = G$, $A = U$, then, the following subclass pattern is given by the Eqs. (27–32),

$$I_0 \otimes C_0$$

$$I_1 \otimes C_1$$

Fig. 4. Yin-Yang Tae Goek RNA pattern [8] of block circulant symmetric pair complementary codon   as $I_0 \otimes C_0 + I_1 \otimes C_1$ i.e., $C_0 = \begin{pmatrix} 1 & 1 \\ -1 & 1 \end{pmatrix} = \begin{pmatrix} 1 & 0 \\ 0 & 1 \end{pmatrix} + \begin{pmatrix} 0 & 1 \\ -1 & 0 \end{pmatrix} C_1 = \begin{pmatrix} 1 & -1 \\ -1 & -1 \end{pmatrix} = \begin{pmatrix} 0 & 1 \\ -1 & 0 \end{pmatrix} + \begin{pmatrix} 1 & 0 \\ 0 & -1 \end{pmatrix}.$

$$\begin{bmatrix} A & G \\ C & U \end{bmatrix} = \begin{bmatrix} 1 & 0 \\ 0 & 1 \end{bmatrix} \otimes \begin{bmatrix} -1 & -1 \\ 1 & -1 \end{bmatrix} + \begin{bmatrix} 0 & 1 \\ 1 & 0 \end{bmatrix} \otimes \begin{bmatrix} -1 & 1 \\ 1 & 1 \end{bmatrix}, \tag{27}$$

$$\begin{bmatrix} G & A \\ U & C \end{bmatrix} = \begin{bmatrix} 1 & 0 \\ 0 & 1 \end{bmatrix} \otimes \begin{bmatrix} 1 & -1 \\ 1 & 1 \end{bmatrix} + \begin{bmatrix} 0 & 1 \\ 1 & 0 \end{bmatrix} \otimes \begin{bmatrix} -1 & -1 \\ -1 & 1 \end{bmatrix}, \tag{28}$$

$$\begin{bmatrix} G & U \\ A & C \end{bmatrix} = \begin{bmatrix} 1 & 0 \\ 0 & 1 \end{bmatrix} \otimes \begin{bmatrix} 1 & 1 \\ -1 & 1 \end{bmatrix} + \begin{bmatrix} 0 & 1 \\ 1 & 0 \end{bmatrix} \otimes \begin{bmatrix} -1 & -1 \\ -1 & 1 \end{bmatrix}, \tag{29}$$

$$\begin{bmatrix} U & G \\ C & A \end{bmatrix} = \begin{bmatrix} 1 & 0 \\ 0 & 1 \end{bmatrix} \otimes \begin{bmatrix} -1 & -1 \\ 1 & -1 \end{bmatrix} + \begin{bmatrix} 0 & 1 \\ 1 & 0 \end{bmatrix} \otimes \begin{bmatrix} 1 & 1 \\ 1 & -1 \end{bmatrix}, \tag{30}$$

$$\begin{bmatrix} A & C \\ G & U \end{bmatrix} = \begin{bmatrix} 1 & 0 \\ 0 & 1 \end{bmatrix} \otimes \begin{bmatrix} -1 & -1 \\ -1 & 1 \end{bmatrix} + \begin{bmatrix} 0 & 1 \\ 1 & 0 \end{bmatrix} \otimes \begin{bmatrix} -1 & 1 \\ 1 & 1 \end{bmatrix}, \tag{31}$$

$$\begin{bmatrix} C & A \\ U & G \end{bmatrix} = \begin{bmatrix} 1 & 0 \\ 0 & 1 \end{bmatrix} \otimes \begin{bmatrix} 1 & -1 \\ 1 & 1 \end{bmatrix} + \begin{bmatrix} 0 & 1 \\ 1 & 0 \end{bmatrix} \otimes \begin{bmatrix} 1 & -1 \\ -1 & -1 \end{bmatrix}. \tag{32}$$

Hence, we get 2 the anti-pairs as (A G; C U), (G A; U C) and 4 half pairs as (A C; G U), (G U; A C), (C A; U G), (U G; C A) of complementary RNA.

6 Conclusion

We have presented a simple method of developing reverse transcription RNA to double helix DNA using BCJM. This method provided its simplicity and clarity which it decomposes a DNA matrix in term of sparse matrices having only cancel column and row. We also analyzed RNA Yin-Yang pair symmetric complementary matrix which the contrary is the complementary to each other based on a QO-BC.

Acknowledgments. This work was supported by Ministry of Education Science and Technology (MEST) 2015R1A2A1A05000977, National Research Foundation (NRF), Republic of Korea.

References

1. Watson, J.D., Crick, F.H.C.: Molecular structure of nucleic acids. Nature **171**(4356), 737–738 (1953)
2. Temin, H.M.: Nature of the provirus of rous sarcoma. Nat. Cancer Inst. Monogr. **17**, 557–570 (1964)
3. Lee, M.H., Hou, J.: Fast block inverse Jacket transform. IEEE Signal Process. Lett. **13**(8), 461–464 (2006)
4. Lee, M.H., Hai, H., Zhang, X.D.: MIMO Communication Method and System using the Block Circulant Jacket Matrix, USA Patent 9,356,671, 31 May 2016
5. Petoukhov, S., Matthew, H.: Symmetrical Analysis Techniques for Genetic Systems and Bioinformatics. Wiley, New York (2011)

6. Lee, S.K., Park, D.C., Lee, M.H.: RNA genetic 8 by 8 matrix construction from the block circulant Jacket matrix. In: Symmetric Festival 2016, 18–22 July 2016, Vienna, Austria (2016)
7. Hou, J., Lee, M.H., Park, J.Y.: Matrices analysis of quasi-orthogonal space time block codes. IEEE Commun. Lett. **7**(8), 385–387 (2003)
8. Favrholdt, D. (ed.): Complementarity Beyond Physics. Niels Bohr Collected Works, vol. 1. Elsevier, Amsterdam (1928–1962). ISBN 978-0-444-53286-2
9. Hamdy, M.M.: DNA-genetic encryption technique. Int. J. Comput. Netw. Inf. Secur. (IJCNIS) **8**(7), 1–9 (2016). https://doi.org/10.5815/ijcnis.2016.07.01
10. Khalil, M.I.: A new heuristic approach for DNA sequences alignment. Int. J. Image Graph. Signal Process. (IJIGSP) **7**(12), 18–23 (2015)
11. Ajra, H., Hasan, M.Z., Islam, M.S.: BER analysis of various channel equalization schemes of a QO-STBC encoded OFDM based MIMO CDMA system. Int. J. Comput. Netw. Inf. Secur. (IJCNIS) **6**(3), 30–36 (2014). https://doi.org/10.5815/ijcnis.2014.03.04
12. Siva Kumar Reddy, B., Lakshmi, B.: BER analysis with adaptive modulation coding in MIMO-OFDM for WiMAX using GNU radio. Int. J. Wirel. Microwave Technol. (IJWMT) **4**(4), 20–34 (2014).https://doi.org/10.5815/ijwmt.2014.04.02

Computer Science and Education

Information Technologies for Maintaining of Management Activity of Universities

V. A. Lakhno[1](✉) and V. V. Tretynyk[2]

[1] European University, Kyiv, Ukraine
valss2l@ukr.net
[2] Igor Sikorsky Kyiv Polytechnic Institute, Kyiv, Ukraine
viola.tret@gmail.com

Abstract. A specified method and model for analyzing the organizational management of information flows of a large educational institution is suggested. The described solutions allow to reveal information resources, for systems of support of management decision making and control of development plans. As a data source, information obtained from software complexes is used, in particular, from information complexes and electronic document circulation of educational institutions. The configuration and characteristics of information exchange in the information systems supporting the management decisions of the European University are studied. During the research, approaches to the creation and modernization of effective information systems of large educational institutions have been improved.

It is shown that the suggested models which are implemented in software tools supporting management activities of universities, allow to increase the flexibility and adaptability of existing information and electronic document management systems.

Keywords: Managerial activity · Information exchange
Decision support system · Document circulation · Queuing system

1 Introduction

The modern stage of development of the postindustrial society and, first of all, organizational management, differs with the widespread implementation of information systems (IS) and technologies (IT). To the formed trends of development, one can also include the organization of electronic interaction between various state-owned and private structures, including interdepartmental systems of electronic document circulation (EDC), etc. Another global trend has been the expansion of the functionality of corporate education systems (CES) and social networks [1]. The last allows to build integrated social-information networks to solve a diverse range of problems. In particular, this concerns the sphere of education. At the same time, the technologies X.25, Frame Relay, Ethernet, Fast Ethernet, Gigabit Ethernet, etc. can be used. Development of e-education systems, CES, EDC, etc. are closely related to the expansion of opportunities for remote access to information assets of informational education systems (IES) of public and private universities [1]. Large-scale processes of informatization of organizational structures of many large educational institutions (EI), which began in the 20th century, required the evolution of conventional IS into management support systems [1, 2]. In addition, in recent years, the implementation of intellectual components in the IT has significantly expanded [3].

Despite the undeniable success in this direction, the problem of synthesis methods and models that contributes to the information support of management processes remains urgent. This task is also relevant for such complex objects of informatization as large EI - universities, colleges, etc.

2   References Review

Research in the field of organizational management of the IT infrastructure of large organizations, was carried out by many authors, for example, [4, 5]. In [6–10], methods and models for managing the IT infrastructure of large organizations, in particular, EI, are described. The given methods and models have not yet developed in the format of completed hardware-software solutions, which are to be introduced in everyday practice.

In [11, 12], the functional efficiency of IS and EDC in EI was studied on the basis of complex indicators of service quality [13].

However, to date, there are no completed studies that can provide a joint analysis of application software system data (APS - IS and EDC) in the tasks of supporting the decision-making of management of EI. This, in particular, is due to the disparity in approaches to solving the problems of automatization of information processing in applied IS and their orientation to local functions, supporting business processes of EI. In [14, 15], the potential of using data as the primary source for decision support (decision-making), decisions on organization management, APS was analyzed. The research is not finished.

Analysis of literature sources showed that at present the methodology that considers information resources, in particular, the data of IS and EDC in the tasks of supporting management decisions of large EI, has not been fully investigated [16–18]. The methods of implementing the function of supporting management activities in the IS remain unexplored. The methodology of development and modernization of the SDM systems (SSDM) in the field of management of EI (for example, large universities) also has a poor scientific foundation.

Thus, it is significant to continue research aimed at solving the problems of information support for the management of universities (ISMU). The problem of the further development of methods and algorithms for informational support of management processes with the purpose of increasing the efficiency of EI operation remains relevant.

3   The Aim of the Paper

The aim of the paper is the development of methods and models that allow creating information systems for supporting the management activities of large educational institutions.

To achieve the goal of the work, we need to solve the following tasks:

- clarify the method and model of support for management activities, which are based on information resources of EI;
- to investigate the structure and characteristics of information exchange in ISMU for the improvement of approaches to the creation of effective IS and EDC of EI.

4   Data and Methods

During the refinement of existing methods and models of ISMU, in particular EI, an analysis of the relationship between management functions and applied tasks of IS, EDC, automated EI systems was made with the purpose of finding data sources maintaining management activity.

Let A be a set of tasks (instructions). Assignments are related to the tasks of the functional divisions of the EI. Thus $A = \{A_1, \ldots, A_n\}$ where A_i is an array of data of application software subsystems of the EI (IS, automated systems, SDM, EDC and other i.e. APS). Otherwise $A_n = \{x_{n1}, \ldots, x_{np}\}$ where x_{ni} are the APS reference module. Blocks can be used as data sources in the development of SDM services for the management of EI. The set $D = \{D_1, \ldots, D_m\}$ describes the business processes (operations) of EI. Their support requires the receipt of data from the APS. The coefficients $k_1, \ldots, k_m$ characterize the existence or absence of information for the corresponding operation. The coefficients can take the value 0 or 1 [11].

In Table 1, an example is given of the interconnection of sets A and D for the management option of EI processes (for example, the European University, Kiev). Cells filled with a "*" sign contain coefficients with a value of 1. In the remaining cells, it is 0.

Stating of the problem. Units of EI are defined as objects with many characteristics (OMC), i.e. $O = \{O_1, \ldots, O_n\}$. The activity of subdivisions (objects - O) is evaluated on the basis of the fulfillment of instructions. Instructions are fixed in the database (DB). They are evaluated by a set of criteria $C = \{C_1, \ldots, C_m\}$. Examples of criteria or metrics are shown in Table 2.

The criteria given in Table 2 have an evaluation scale [17, 18, 20, 21], either quantitative or qualitative. The scale of evaluation is indicated by $C_s = \{c_s^{es}\}$, $e_s = 1, \ldots, h_s$, $s = 1, \ldots, m$. Evaluations are sorted from best to worst $c_s^1 \succ c_s^2 \succ \ldots \succ c_s^{hs}$.

The refinement of the method [1, 4, 11] consists in using for each OMC a multiset:

$$O_i = \left\{ \begin{matrix} v_{O_i}\left(c_1^1\right) \circ c_1^1, \ldots, v_{O_i}\left(c_1^{h1}\right) \circ c_1^{h1}; \ldots; \\ v_{O_i}\left(c_m^1\right) \circ c_m^1, \ldots, v_{O_i}\left(c_m^{hm}\right) \circ c_m^{hm} \end{matrix} \right\}, \quad i = 1, \ldots, n \qquad (1)$$

where v_C is a the multiplicity function of the multiset for the OMC that generates the domain $X = C_1 \cup \ldots \cup C_m$, for m criteria scales;

$v_{O_i}(c_s^{es})$ is the number of documents in the IS, which allows to make a conclusion about the criterial evaluation C_s;

$\circ$ is the designation of the fact that in the description of OMC the copies of the characteristic c_s^{es} were previously used.

The problem of regulating OMC is reduced to the ordering of the corresponding multisets. The latter is described as follows:

$$O_i^+ = \{v \circ c_1^1, 0, \ldots, 0; \; v \circ c_2^1, 0, \ldots, 0; \ldots; v \circ c_m^1, 0, \ldots, 0\}, \tag{2}$$

$$O_i^- = \{0, \ldots, v \circ c_1^{h1}; \; 0, \ldots, 0, v \circ c_2^{h2}; \; 0, \ldots, 0, v \circ c_m^{hm}\}. \tag{3}$$

Table 1. Relationship between sets A and D

Actions / Task for University	Business Activity Planning	Making operational decisions	Making recommendations	Monitoring of decision implementation	Evaluation of the implementation processes of decisions	Evaluation of the quality of execution of instructions	Control of plans
	D_1	D_2	D_3	D_4	D_5	D_6	D_7
	k_1	k_2	k_3	k_4	k_5	k_6	k_7
A_1 Documentation management	*	*	*	*	*	*	*
A_2 Human resources in accordance with the requirements of the Ministry	*			*			
A_3 The material-technical supply	*			*			
A_4 Financial resources	*			*		*	*
A_5 Scientific research	*	*			*		
A_6 Remote educational resources	*			*		*	*
A_7 Other business processes	*	*	*	*	*	*	*

Table 2. Example of criteria (metrics) for object estimation within the framework of EI

№	The criteria adopted in the ISMU model	Scale of estimation [11, 17, 18, 20, 21]	
		Tasks (instructions)	
		Fulfilled	Unfulfilled
Example: indicator - "curriculum processing (document)"			
1.	***The state of assignment fullfilment***	Early; on time; violation of the term; other	The deadline is not set; the period is not expired; expired; other
2.	***Prompt acceptance of documents accompanying the assignment***	Accepted on the day of mailing; accepted on the next business day; taken 2–3 days later; taken after 4–7 days	The same
n	...	...	...
Example: indicator - "control of execution of decisions"			
1.	***Execution of target dates***	Removed early from control; withdrawn on time; withdrawn with violation of the term	The deadline is not set; The term has not expired; violation of the term
2.	***Adjusting the deadlines***	Not implemented; Completed once; It was carried out repeatedly	The same
n	...	...	...

In the process of refinement of the method, the problem of selecting the unit with the best and worst object is solved. The following criterion for ordering an object is used: proximity to a conditionally better object in multisets O_i^+, O_i^- i.e.:

$$\delta(C_i) = l^+(C_i)/[l^+(C_i) + l^-(C_i)] \tag{4}$$

where $l^+(C_i) = l(C^+, C^-)$ and $l^-(C_i) = l(C^-, C^+)$.

The best option for an object Q^* is determined by the value min $\delta(C_i)$. As a result, we get ranked objects. Ranking can be performed for the equivalent or unequal in importance criteria (metrics) [19, 22].

The specified model of document circulation of EI was considered as a set of information processes. In quantitative terms, the document flow of a large EI is characterized by tens of thousands of documents per year. The movement of documents into EDC of EI is considered as an open stochastic network serving the workstations of employees.

The average response time for an application is described by the expression:

$$U = \sum_{i=1}^{n} f_i \cdot t_i^{ex} + \sum_{i=1}^{n} f_i \cdot t_i^{se} = \sum_{i=1}^{n} f_i \left(t_i^{ex} + t_i^{se}\right) = \sum_{i=1}^{n} f_i \left(t_i^{ex} + \frac{1}{\mu_i}\right) \tag{5}$$

where f_i is a frequency of receipt to the i-th node of the application; t_i^{ex}, t_i^{se} mean wait and maintenance time for the i-th node, respectively; $\mu_i = 1/t_i^{se}$ is a service intensity, $i = \overline{1, n}$.

Average time of passage of the document in the EDC of EI network was determined as follows:

$$U = \sum_{i=1}^{n} f_i \cdot \left(\frac{\phi_i^2}{\lambda \cdot (1 - \phi_i)} + \frac{1}{\mu_i} \right) \tag{6}$$

where $\phi_i = \lambda_i/t_i^{se}$; $\lambda_i = \rho \cdot f_i$; ρ is the average intensity of the flow of documents arriving at the i-th node of the EI.

The residence time at the i-th node can be defined as follows:

$$T_i = \phi_i/(\lambda_i \cdot (1 - \phi_i)) \tag{7}$$

The resulting expressions (6) and (7) were implemented in the ISMU module of EI (for the European University). A refined model can also be used in ISMU of other major universities, since it makes it possible to identify problem nodes in the EDC network.

5   Experiment

During the investigation as one of the estimated metrics, the value of the data transmission network efficiency index for ISMU of EI was considered (see Table 3 [4, 11, 21, 23–25]). The indicators are presented in conditional scores (from 1 to 5), Fig. 1.

Table 3. Efficiency of data transmission networks in IES of EI

Technology	Cost	Packet delay	Productivity	Relative efficiency, %
X.25	3	1	1	0,3
Ethernet	3	3	3	5,9
Frame Relay	3	3	3	7,4
Fast Ethernet	3	3	3	11,8
40/100 Gb Ethernet	1	3	3	19,7
Gigabit Ethernet	2	3	3	23,5
10 Gb Ethernet	2	3	3	31,4

We can see from Fig. 1, the most efficient technology for the IES of EI is the 10 Gb Ethernet one.

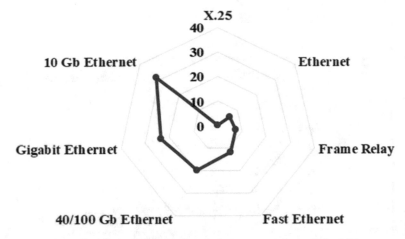

Fig. 1. Efficiency of data transmission networks in IES of EI

As an example, a variant of the European University network, including a central office and several (5–10) units (modeling for Ethernet technologies and 10 Gb Ethernet) is considered.

Suppose, that the input document flow enters node 1 with the following probabilistic parameters: $P_{01} = 1$ is a probability of denial of service of document on node 1; $P_{0i} = 0$ is a probability of denial of document maintenance on nodes $i = \overline{2,10}$. Completed tasks (processed documents) with probability "1" ($P_{i1} = 1$, $i = \overline{2,10}$) arrive at node 1 (central office or dean's office). In the calculations it is assumed that the output from the system occurs with a probability of "0.1".

In Fig. 2 is shown the simulation results (solid lines) and practical tests (dashed lines) of the outcomes for the time documents stay in the node from the following combinations of parameters: $\Lambda = \lambda_i/t_i^{ex} = 0 - 25$ document/h.

In the ideal (normative) case it was assumed that the processing time of the document does not exceed 20–30 min. From the obtained graphs, one can judge the rational value of the document service intensity $\mu_i = 1/t_i^{se}$, which is the $\mu_i = 100 - 150$ document/h for the received initial values. Thus, if the capacity of the terminal (workstation) is about 8–11 documents/h, then a minimum number of such jobs can be determined. For initial data on the European University, this value will be 22–25 workplaces (including remote branches).

Consequently, as an operating influence in the EI "European University" and its workflow system should be an increase in jobs. Or, as an alternative, the development of measures to unite the tasks to be performed by units.

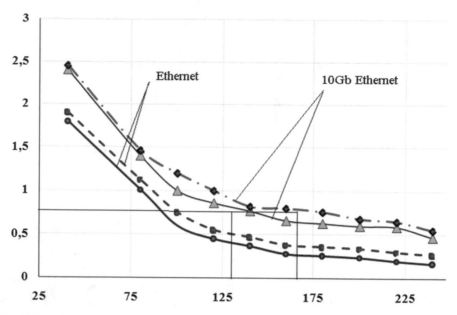

Fig. 2. Results of modeling and validating outcomes for time of documents stay in the node IES of EI

6 Discussion of the Results of SDMS Testing and Perspectives for Further Research

The results obtained during the refinement of the method and model of management support for EI, based on IS and EDC resources, allowed to increase the effectiveness of decision support services and assessment of management processes at a large university by 10–14%. As alternatives, the techniques described in [5, 16, 20, 21] were considered.

The advantage of the proposed approach is the possibility of integrating developed software into existing IS and EDC of EI.

The revealed lack of work is the absence of approbation on a large number of large EI, as far as investigation have only been carried out at the European University.

Further investigations provide for the possibility of approbation and expansion of the use of the proposed program complexes for the organizational management of large organizations in other universities in Ukraine.

In addition, the design of specialized interfaces for various modes of management support is envisaged. In the future, this will ensure that the preferences of employees are reflected in the interaction with used information systems by universities.

In general, based on the conducted investigation, it is possible to state the effectiveness of the proposed refinements to the method and model of support for making managerial decisions and controlling (correcting) plans based on the data of the involved software complexes.

7 Conclusions

The method and model for the analysis of organizational management in a large university were specified. Refinements to the method, which allows to identify the information resources used to support the adoption of managerial decisions by university management were suggested. The data received from the information systems and the electronic document circulation of educational institutions were used as a source.

The configuration and characteristics of the information exchange in information support for the management of universities in a large university were studied. The approaches to the creation and modernization of effective information systems of educational institutions were improved. It was found that the proposed clarifications to the model of support for management activities, allow to increase the flexibility and adaptability of existing information systems and electronic document circulation of educational institutions.

References

1. Leidner, D.E., Jarvenpaa, S.L.: The use of information technology to enhance management school education: a theoretical view. MIS Q. **19**, 265–291 (1995)
2. Galliers, R.D., Leidner, D.E. (eds.): Strategic Information Management: Challenges and Strategies in Managing Information Systems. Routledge, New York (2014)
3. Demirkan, H., Delen, D.: Leveraging the capabilities of service-oriented decision support systems: putting analytics and big data in cloud. Decis. Support Syst. **55**(1), 412–421 (2013)
4. Kerzner, H.: Project Management: A Systems Approach to Planning, Scheduling, and Controlling. Wiley, Hoboken (2013)
5. Peppard, J., Ward, J.: The Strategic Management of Information Systems: Building a Digital Strategy. Wiley, Hoboken (2016)
6. Wilson, S., Liber, O., Johnson, M.W., Beauvoir, P., Sharples, P., Milligan, C.D.: Personal learning environments: challenging the dominant design of educational systems. J. E-Learn. Knowl. Soc. **3**(2), 27–38 (2007)
7. Crone, D.A., Hawken, L.S., Horner, R.H.: Building Positive Behavior Support Systems in Schools: Functional Behavioral Assessment. Guilford Publications (2015)
8. McKelvey, B., Tanriverdi, H., Yoo, Y.: Complexity and information systems research in the emerging digital world. MIS Q. 1–3 (2016)
9. Bakanova, N., Atanasova, T.: Development of the combined method for dataflow system. Inf. Technol. Knowl. **2**(3), 262–266 (2008)
10. Winch, G., Leiringer, R.: Owner project capabilities for infrastructure development: a review and development of the "strong owner" concept. Int. J. Proj. Manage. **34**(2), 271–281 (2016)
11. Bakanova, N.B.: Issledovanie dinamiki deyatelnosti organizatsii na osnove analiza dokumentooborota. Ekonomika. Nalogi. Pravo **1**, 4–8 (2011)
12. Wang, S., Noe, R.A., Wang, Z.M.: Motivating knowledge sharing in knowledge management systems: a quasi–field experiment. J. Manage. **40**(4), 978–1009 (2014)
13. Urbach, N., Ahlemann, F.: Structural equation modeling in information systems research using partial least squares. JITTA J. Inf. Technol. Theory Appl. **11**(2), 5 (2010)

14. Bardhan, I.R., Demirkan, H., Kannan, P.K., Kauffman, R.J., Sougstad, R.: An interdisciplinary perspective on IT services management and service science. J. Manage. Inf. Syst. **26**(4), 13–64 (2010)
15. Kwon, O., Lee, N., Shin, B.: Data quality management, data usage experience and acquisition intention of big data analytics. Int. J. Inf. Manage. **34**(3), 387–394 (2014)
16. Giachetti, R.E.: Design of Enterprise Systems: Theory, Architecture, and Methods. CRC Press, Boca Raton (2016)
17. Dudin, A., Gortsev, A., Nazarov, A., Yakupov, R. (eds.): Information Technologies and Mathematical Modelling-Queueing Theory and Applications: 15th International Scientific Conference, ITMM 2016, Named After AF Terpugov, Katun, Russia, 12–16 September 2016, Proceedings, vol. 638. Springer (2016)
18. Luo, A., Fu, J., Liu, J.: An impact analysis method of business processes evolution in enterprise architecture. In: 2016 International Conference on Progress in Informatics and Computing (PIC), pp. 733–737. IEEE (2016)
19. Lahno, V.: Ensuring of information processes' reliability and security in critical application data processing systems. MEST J. **2**(1), 71–79 (2014)
20. Pearlson, K.E., Saunders, C.S., Galletta, D.F.: Managing and Using Information Systems, Binder Ready Version: A Strategic Approach. Wiley, New York (2016)
21. Wager, K.A., Lee, F.W., Glaser, J.P.: Health Care Information Systems: A Practical Approach for Health Care Management. Wiley (2017)
22. Arezki, S.A., Djamila, H.B., Bouziane, B.C.: AQUAZONE: a spatial decision support system for aquatic zone management. Int. J. Inf. Technol. Comput. Sci. (IJITCS) **7**(4), 1–13 (2015). https://doi.org/10.5815/ijitcs.2015.04.01
23. Tank, D.M.: Enable better and timelier decision-making using real-time business intelligence system. Int. J. Inf. Eng. Electron. Bus. (IJIEEB) **7**(1), 43–48 (2015). https://doi.org/10.5815/ijieeb.2015.01.06
24. Cheowsuwan, T.: The strategic performance measurements in educational organizations by using balance scorecard. Int. J. Modern Educ. Comput. Sci. (IJMECS), **8**(12), 17–22 (2016). https://doi.org/10.5815/ijmecs.2016.12.03
25. Alqahtani, S.S., Alshahri, S., Almaleh, A.I., Nadeem, F.: The implementation of clinical decision support system: a case study in Saudi Arabia. Int. J. Inf. Technol. Comput. Sci. (IJITCS), **8**(8), 23–30 (2016). https://doi.org/10.5815/ijitcs.2016.08.03

Adaptive Expert Systems Development for Cyber Attacks Recognition in Information Educational Systems on the Basis of Signs' Clustering

V. Lakhno[1(✉)], S. Zaitsev[2], Y. Tkach[2], and T. Petrenko[2]

[1] European University, Kyiv, Ukraine
valss21@ukr.net
[2] Chernihiv National University of Technology, Chernihiv, Ukraine

Abstract. The article proposes a new approach to solving the issue of efficiency in systems of cyberattacks intelligent recognition, anomalies and threats for the educational and informational environment of universities and colleges. The solution is based on models and methodology of creating an adaptive expert system capable of self-learning. Unlike the existing ones, the model proposed in the article, takes into account the known statistical and remote parameters of cyberattacks signs' clustering, as well as third-type errors during the machine learning process. It is proposed to evaluate the quality of signs' space partitioning recognition of objects in an adaptive expert system with the use of a modified information performance condition as an evaluation indicator. It is proved that model and application of the method of clustering of signs based on the entropy and information-distance Kullback–Leibler criterion, allows getting the input fuzzy classified educational matrix which is used as an object of study.

Keywords: Adaptive expert system · Recognition · Cyberattack
Anomaly · Clustering of signs

1 Introduction

The entry of humanity into the era of high technology stimulates the further expansion of the capabilities of computing systems, particularly in the field of education. The development of e-education systems and corporate education systems (CES) are closely linked with the expansion of opportunities of remote access to information and education facilities (IEF) of educational institutions (EI) on the basis of advanced technologies [1, 2]. Expanding the range of educational services and functions of CES and EI networks, integrating services with social networks [3], further using of remote access in various forms of education, builds a system of education with critical cybernetic infrastructure.

The active expansion of IEF in many countries of the world is accompanied by the emergence of new threats to cyber security (CS), as evidenced by the increase in the number of incidents involving the protection of information [4].

© Springer International Publishing AG, part of Springer Nature 2019
Z. Hu et al. (Eds.): ICCSEEA 2018, AISC 754, pp. 673–682, 2019.
https://doi.org/10.1007/978-3-319-91008-6_66

One can oppose the increase of numerous and complex destructive effects on IEF by using adaptive intelligence systems for cyber threats and cyber attacks.

Consequently, the researches aimed at the further development of cyber defence models and methods based on the use of adaptive, self-learning systems able for intelligent cyberattack detection, anomalies and threats to the IEF are relevant.

2   Literature Review

The growing interest to CS issue and information security (IS) has caused a burst of research in the field of effective systems development for detecting and preventing cyber threats over the last decade [5, 6]. The analysis of the existing world experience [7, 8] confirms that the extensive approach to solving the tasks of the IS and CS of IEF of EI by means of increasing the finances and measures for the information protection (IP) is often not very effective. A perspective direction of the research has been devoted to the creation of intelligent decision-making support systems (DMSS) [9] and expert systems (ES) [10, 11] in the IS and CS. These studies have not been completed yet.

In [1, 2, 12] the experience of introducing commercial DMSS and ES with IP and CS is analysed. It is noted that commercial systems are of a closed nature, and their acquisition by individual EI involves significant financial costs.

Thus, taken into consideration the controversy in publications [1, 2, 5, 13, 14], the actual task is to improve the models and methods of cyber defence based on the use of adaptive, capable for self-learning systems for the intelligent recognition of cyberattacks, anomalies and threats in the IEF.

3   The Objective of the Research

The objective of the research is to increase the efficiency of the systems of intelligent recognition of cyberattacks, anomalies and threats for IEF of EI on the basis of the creation of a self-learning adaptive expert system (AES) that takes into account the known statistical and remote parameters of clustering of signs of cyberattacks, as well as third-type errors during the machine learning procedures.

4   Methods and Models

The research is a continuation of the works [11, 14], in which, in particular, a structural scheme was suggested and a model of self-learning adaptive expert system for recognition of cyberattacks was presented.

Solving the issue of forming an incoming mathematical description of the AES in the subsystems of the CS of IEF of EI is to create an object used for training – MLMS (i.e., a multidimensional learning matrix of signs) $\left\| lm_{m,i}^{(j)} | m = \overline{1,M}; i = \overline{1,N}, j = \overline{1,n} \right\|$.

To do this: the dictionary of signs for each class of anomalies [7, 11, 15], cyber threats [2] and attacks [14, 15] have been formulated. In the course of the research, the alphabet classes in terms of recognition objects (RO) have been compiled, the minimum size of the representative educational matrix (REM) has been determined, and the standardized tolerances for the signs of recognition of the illegitimate interference in the work of the IEF of EI have been determined. The alphabet of class of anomalies, threats or cyberattacks (recognition objects - RO) for AES $\{lm_m^o\}$ is formed at the first stage by the developer with the involvement of IS specialists. At the second stage of the synthesis of the alphabet, with the help of the AES, the processing of input data with the use of clustering methods continues.

It is accepted that the alphabet of classes $\{CT_m^o | m = \overline{1, M}\}$ and the multidimensional binary educational matrix (MBEM) of the RO are well-known. The matrix describes the m state in which the IEF of EI is located. In this case, the MBEM of RO for the recognition class of CT_m^o will have the following form:

$$
\left\| lm_{m,i}^{(j)} \right\| = \begin{vmatrix} lm_{m,1}^{(1)} & lm_{m,2}^{(1)} & \cdots & lm_{m,1}^{(1)} & \cdots & lm_{m,N}^{(1)} \\ lm_{m,1}^{(2)} & lm_{m,2}^{(2)} & \cdots & lm_{m,1}^{(2)} & \cdots & lm_{m,N}^{(2)} \\ \cdots & \cdots & \cdots & \cdots & \cdots & \cdots \\ lm_{m,1}^{(j)} & lm_{m,2}^{(j)} & \cdots & lm_{m,1}^{(j)} & \cdots & lm_{m,N}^{(j)} \\ \cdots & \cdots & \cdots & \cdots & \cdots & \cdots \\ lm_{m,1}^{(n)} & lm_{m,2}^{(n)} & \cdots & lm_{m,1}^{(n)} & \cdots & lm_{m,N}^{(n)} \end{vmatrix} \tag{1}
$$

In the matrix (1) the following notation was adopted: matrix line - implementation of the "representation" of the RO $\left\{ lm_{m,i}^{(j)} | i = \overline{1, N} \right\}$, N - the number of signs of the RO; column is a stochastic training sample $\left\{ lm_{m,i}^{(j)} | j = \overline{1, n} \right\}$ with volume n. All possible values of each property of a RO are proposed to be encoded in a binary form, where zero corresponds to an indefinite value of the RO property. By combining the data obtained during the preliminary monitoring, into clusters, one can analyse the typical representatives of each cluster and decide whether such evidence is the indication of the attack or not. Then, this decision is transferred to all representatives of the investigated cluster. This approach significantly reduces the amount of efforts necessary for a successful classification of the attack against IEF. The information condition for functional performance (ICFP) capable of self-studying at AES with CS is as follows:

$$
CE_m^* = \max_{IS} CE_m \tag{2}
$$

where CE_m - ICFP procedures for machine training of the AES during the recognition of the class of RO CT_m^0; IS - allowable values of the IEF of EI parameters.

During the training of AES and the formation of the knowledge base (KB), the work of AES is regulated by a specialist in the IS, who, according to the recommendations of AES, forms the management teams $\{CC\{hy_m\} | m = \overline{1, M}\}$. Application of the entropy measure and the Kullback–Leibler criterion [14] have been substantiated for the developed AES as informational measures. The value of the normalized entropy of ICFP, taking into account the priori probability of approving the hypothesis for the recognition of RO, is given as follows:

$$IND = 1 + 0,5 \sum_{l=1}^{2} \sum_{m=1}^{2} p(hy_m/hy_l) \log_2 p(hy_m/hy_l), \tag{3}$$

where $p(hy_l)$ - the priori probability of approving an assumption (hypothesis) hy_l; $p(hy_m/hy_l)$ - a posteriori probability of approving the assumption hy_m, provided that the hypothesis hy_l was adopted; $M = 2$ - the number of considered assumptions in the recognition process.

Considering (2) and (3), the normalized entropy of ICFP study of the ES, which takes into account errors of the 1st and 2nd kind, is as follows:

$$
\begin{aligned}
CE_m^{(ls)} = 1 + 0,5 \cdot \Bigg(& \frac{mis1_m^{(ls)}(cr)}{mis1_m^{(ls)}(cr) + AU_{2,m}^{(ls)}(cr)} \log_2 \frac{mis1_m^{(ls)}(cr)}{mis1_m^{(ls)}(cr) + AU_{2,m}^{(ls)}(cr)} \\
& + \frac{mis2_m^{(ls)}(cr)}{AU_{1,m}^{(ls)}(cr) + mis2_m^{(ls)}(cr)} \log_2 \frac{mis2_m^{(ls)}(cr)}{AU_{1,m}^{(ls)}(cr) + mis2_m^{(ls)}(cr)} \\
& + \frac{AU_{1,m}(cr)}{AU_{1,m}^{(ls)}(cr) + mis2_m^{(ls)}(cr)} \log_2 \frac{AU_{1,m}(cr)}{AU_{1,m}^{(ls)}(cr) + mis2_m^{(ls)}(cr)} \\
& + \frac{AU_{2,m}^{(ls)}(cr)}{mis1_m^{(ls)}(cr) + AU_{2,m}^{(ls)}(cr)} \log_2 \frac{AU_{2,m}^{(ls)}(cr)}{mis1_m^{(ls)}(cr) + AU_{2,m}^{(ls)}(cr)} \\
& + \frac{mis1_m^{(ls)}(cr)}{mis1_m^{(ls)}(cr) + AU_{3,m}^{(ls)}(cr)} \log_2 \frac{mis1_m^{(ls)}(cr)}{mis1_m^{(ls)}(cr) + AU_{3,m}^{(ls)}(cr)} \\
& + \frac{AU_{3,m}^{(ls)}(cr)}{mis3_m^{(ls)}(cr) + AU_{3,m}^{(ls)}(cr)} \log_2 \frac{AU_{3,m}^{(ls)}(cr)}{mis3_m^{(ls)}(cr) + AU_{3,m}^{(ls)}(cr)} \Bigg),
\end{aligned}
\tag{4}
$$

where $AU_{1,m}^{(ls)}(cr)$ - the first validation procedure; $AU_{2,m}^{(ls)}(cr)$ - the second validation procedure; $mis1_m^{(ls)}(cr)$ - the errors of the first kind of making decision for the $ls - 2o$AES training step; $mis2_m^{(ls)}(cr)$ - the mistakes of the second kind of decision making for the $ls - 2o$ AES training step; cr - the radius of the hyper-spherical containers of the cluster.

Ensuring of stable operation for reliable processing of data on the state of the IS of IEF of EI at an any time in the conditions of the influence of cyberattacks is achieved by the implementation of the formula:

$$SO : SS \times CA \rightarrow SS_{res} = \{SS_{res}^i\}, \tag{5}$$

where SS_{res} - the set of permitted states of IEF of EI; $CA = \{CA_0, CA_1, \ldots, CA_N\}$ - the plurality of RO implementation.

The functional, which defines a generalized indicator of the effectiveness of counteracting cyberattacks or threats, takes into account the efficiency of recognition and characterizes the stability of the functioning of the IEF of EI. Taking into account the above mentioned, it is given in the following form:

$$IE = F[(SCA, CE), (SS, T_s, VIL), (CO, CM, ME)],\tag{6}$$

where SCA - the cyberattack scenarios; CE - the criterion of effectiveness of RO recognition; SS - the set of parameters of IEF of EI: T_s - the periods of execution of functional tasks in IEF of EI; VIL - vulnerabilities of IEF of EI; set of parameters of counteraction to cyber-attacks: CO - parameters of regulation of IEF of EI; CM - methods of counteracting cyber-attacks in IEF of EI; ME - the means of prevention, detection, analysis and active counteraction to cyberattacks.

The dependence of the Kullback–Leibler information measure on the parameters of the AES for the variant of application of control commands based on three alternatives has been determined. Alternatives: the first - the main working hypothesis (basic) - hy_{γ_1}: a sign (signs) of RO (RS) and the indicator IE is within the normal state of IEF of EI; the second - the hypothesis hy_{γ_2}: the sign (or signs) rc_i of the RO (RS) and the indicator IE allow us to conclude that the value of the indicator is lower than the norm; the third - the hypothesis hy_{γ_3}: the indicator IE allows us to conclude that the values of the indicator IE are higher than the norm;

For the taken hypotheses, the following result has been obtained:

$$CE_m^{(ls)} = 1/3 \cdot \left\{ \begin{bmatrix} AU_{1,m}^{(ls)} + AU_{2,m}^{(ls)} + AU_{3,m}^{(ls)} \end{bmatrix} - \\ - \begin{bmatrix} mis1_m^{(ls)} + mis2_m^{(ls)} + mis3_m^{(ls)} \end{bmatrix} \right\} \cdot \log_2 \frac{AU_{1,m}^{(ls)} + AU_{2,m}^{(ls)} + AU_{3,m}^{(ls)}}{AU_{1,m}^{(ls)} + AU_{2,m}^{(ls)} + AU_{3,m}^{(ls)}} \cdot \tag{7}$$

where $AU_{1,m}^{(ls)} = p(hy_{\gamma_1}/hy_{\mu_1})$ - the first validation of the hypothesis is based on the conclusions; $AU_{2,m}^{(ls)} = p(hy_{\gamma_2}/hy_{\mu_2})$ - second validation of the hypothesis is based on the comparison of deviations from $\{ca_{K,i}^*\}$; $AU_{3,m}^{(ls)} = p(hy_{\gamma_3}/hy_{\mu_3})$ - the third validation of the hypothesis is based on the results of the processing of the predicate form of the calculation of the number of episodes when it is established that the implementation of the RO does not belong to the container $C_{1,m}^o$, if indeed $\{ct_1^{(j)}\} \in CT_1^o$ the number of episodes when it is established that the implementation of the RO belongs to the container $C_{1,m}^o$, if they actually belong to the class CT_2^o; $mis1_{1,m}^{(ls)} = p(hy_{\gamma_2}/hy_{\mu_1})$ and $mis1_{2,m}^{(ls)} = p(hy_{\gamma_3}/hy_{\mu_1})$ - the number of false positives of the AES in the process of detecting anomalies or cyberattacks respectively; $mis1_{2,m}^{(ls)} = p(hy_{\gamma_3}/hy_{\mu_1})$ and $mis2_{1,m}^{(ls)} = p(hy_{\gamma_1}/hy_{\mu_2})$ - the number of anomalies or cyberattacks not detected during the operation of the AES respectively; $mis3_{1,m}^{(ls)} = p(hy_{\gamma_1}/hy_{\mu_3})$ and $mis3_{2,m}^{(ls)} = p(hy_{\gamma_2}/hy_{\mu_3})$ - the errors of the third kind arise when the model does not take into account some elements of the method and the intellectual technology of learning (MITL) of the ES [11, 14]; hy_{μ_1} - the value of the sign (signs) belongs to the field of admissible deviations (FAD) ca, hy_{μ_2} - the value of the sign (signs) is to the left of the FAD; hy_{μ_3} - the value of the sign (signs) is to the right of the FAD.

Expression (7) takes into account the modified entropy criterion and the Kullbach-Leibler measure. It is a functional of the decision characteristics that are accepted when recognizing the corresponding anomalies or cyberattacks in the IEF of EI.

The correct ultimate rule determines the vector assignment of the implementation parameters of known or unknown cyberattack scenarios SCA_m^{CT} of the m object of the ct - class to one of the known classes of RO $RS_{m_j}^{CT}$ on the j-step of the cyber defence facilities. In accordance with the Bayes criterion, the decisive rule looks like this:

$$P\left(RS_{m_i}^{CT}\right) \cdot P\left(SCA_m^{CT} / RS_{m_i}^{CT}\right) \geq P\left(RS_{m_k}^{CT}\right) \cdot P\left(\overline{SCA_m^{CT}} / RS_{m_k}^{CT}\right), \qquad (8)$$

where $P\left(RS_{m_i}^{CT}\right)$ - the probability of assigning AES of RO to a class of known $RS_{m_i}^{CT}$; $P\left(\overline{SCA_m^{CT}} / RS_{m_i}^{CT}\right)$ - the density of the conditional probability of assigning the AES to a known class $RS_{m_i}^{CT}$; - the probability of assigning AES NPs to a class of known RO; $P\left(\overline{SCA_m^{CT}} / RS_{m_k}^{CT}\right)$ - the density of the conditional probability of assigning AES of detected RO to an unknown class $RS_{m_k}^{CT}$.

Based on the Bayes criterion, the average "price" of the risk of acceptance in the AES of the decision on assigning the vector of parameters of unknown RO to the class $RS_{m_k}^{CT}$ has been determined.

$$PR\left(RUL_i / \overline{SCA_m^{CT}}\right) = \sum_{j=1}^{\gamma} np\left(\frac{RUL_i}{RS_{m_k}^{CT}}\right) \cdot P\frac{RS_{m_k}^{CT}}{SCA_m^{CT}}, \qquad (9)$$

Where RUL_i - a decisive rule, according to which the binary educational vector (BEV) RO $\overline{SCA_m^{CT}}$ determines the belonging of the object to $RS_{m_k}^{CT}$; $np\left(RUL_i / RS_{m_k}^{CT}\right)$ - conditional "price" of the AES decision RUL_i; $P\left(RS_{m_k}^{CT} / \overline{SCA_m^{CT}}\right)$ - conditional probability that the $\overline{SCA_m^{CT}}$ belongs to the AES class $RS_{m_k}^{CT}$.

For the case where the AES performs a comparative analysis of two MBEM, an ultimate rule using the Bayes criterion is given as:

$$\frac{P\left(\overline{SCA_m^{CT}} / RS_{m_1}^{CT}\right)}{P\left(\overline{SCA_m^{CT}} / RS_{m_2}^{CT}\right)} \geq \frac{P(RS_{m_2}^{CT})}{P(RS_{m_1}^{CT})}. \qquad (10)$$

The proposed model, in contrast to the other ones, takes into account the known statistical and deterministic (remote) criteria for the optimization of the procedure of signs' clustering of RO at the previous stage of functioning of the AES capable of studying.

At the first stage, the procedure of partitioning the signs' space (PSS) and the subsequent clustering, for any class of RO CT_m^o, is proposed to be implemented by transforming the PSS into a hyperspherical form. Since the main stage of clustering during the breakdown of PSS into groups is an increase in the radius (cr_m) of the container (CRI) at each learning step, then a recursive expression is used for this:

$$cr_m(ls) = [cr_m(ls - 1) + \xi | cr_m(ls) \in IS_m^{cr}], \qquad (11)$$

where ls - the number of steps to increase the radius of the container C_m^o; ξ - chosen for the selected signs steps of increasing CRI; IS_m^{cr} - acceptable value of the container radius.

The method of the AES training in the IEF of EI is an iterative procedure for searching the global ICFP in the allowable range for determining its function

$$ca^* = \arg \max_{IS_{ca}} \{ \max_{IS_{CE} \cap IS_{cr}} \overline{CE} \}, \qquad (12)$$

where IS_{ca} - the allowable range of values of control deviations ca for the class RO $\{CT_m^o\}$; IS_{CE} - working range of the definition of ICFP $\overline{CE}$; IS_{cr} - the allowable range of the CRT value cr.

At the second stage of the method implementation, the possibility of hyperlipsoid correction of the decisive rules is checked, that allowed to create an adaptive mechanism of self-learning of the cyberattacks recognition system [14].

5   Experiment

According to results [12, 14] the following conclusions are made. The averaged maximal value of ICFP training for AEF is: for attacks of the DoS/DDoS class $\overline{CE} = 3,19$; for attack of Probe class $\overline{CE} = 3,15$; for attacks of R2L class $\overline{CE} = 2,84$; for attacks of U2R class $\overline{CE} = 3,27$; for virus attacks (VA) $\overline{CE} = 2,56$.

During the simulation of the ICEP training of the AEF, it was determined that the quasi-optimal value of the parameter $ca_{n,i}$ of the control system of permissible deviations (CSPD) is equal to CSPD = $8 - 16\%$ with maximum value $CE_{max} = 6,16$.

The results of simulation modelling confirm that the proposed model and method of clustering of RO characteristics, based on the entropy and information-distance criterion of Kullback–Leibler, allow receiving incoming fuzzy classified training matrices for AEF.

The proposed methods and models were implemented in the AEF software complex - "Threats Analyser". Testing of the AES "Threats Analyser" was conducted for the AEF of several universities in Kyiv and Chernihiv. Figure 1 shows the main results obtained during modelling of the indicator CE for the network cyberattacks.

During the study it was found that in the model of "voting" ILTM for representative sets of signs of threats, anomalies and cyberattacks, it is enough to restrict oneself to the construction of representative sets with the length of 5–7 signs. Compared to the reference vectors method, ILTM for a small number of signs of RO (2–4) has a significant advantage of the indicator CE by 25–50%, but is 20–55% inferior to the indicator CE obtained for the hybrid neural network model.

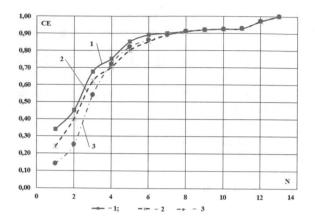

1 – the model of the hybrid neural network;
2 – the model of intellectual learning technology (ILTM) expert system of cyber security of IEF;
3 – the method of reference vectors

Fig. 1. Graph of dependence of the ICFP (IND) training of the AES capable of self-studying on the number of signs (N) used for training

6 Discussion

The comparative analysis was performed on the basis of the data obtained during the AES "Threats Analyser" testing and the data for the Intrusion of (CBB) AIDS Detection System - application-based IDS, the combined IDS & IPS (Intrusion prevention system) solution. The proposed approach to the detection of anomalies, threats and cyberattacks, based on ICFP, can increase the level of detecting network cyberattacks in the CS [7, 11, 14, 15]. Detection of various types of attacks with the help of AES is done with a probability of 77–99%. In addition, the proposed method is not demanding for IS resources and is capable of detecting unknown cyberattacks types in the CS.

A certain disadvantage of the proposed method is the complexity of the formation of the multidimensional training matrix at the first stages of the AES training.

The obtained results allow us to conclude that, unlike the existing methods, the proposed one does not take much time to systematize and transform the threats, anomalies and cyberattacks into the form of the MBEM with their subsequent introduction into the AES.

7 Conclusion

The suggested new approach to solving actual scientific and applied task of improving the intellectual recognition efficiency system of cyberattacks, anomalies and threats to information and educational environment of the educational institutions, is based on developed models and the creation of methodology capable of self - learning expert system that takes into account known statistics and remote parameters of signs' clustering of cyberattacks, as well as third-type errors during the machine learning procedure.

The proposed AES model has been made using procedure of fuzzy clustering of signs of abnormalities or of cyberattacks and the possibility of hyperelipsoid correction of decisive rules. This will allow the creation of adaptive mechanisms for self-learning of the system of intellectual recognition of cyberattacks, anomalies and threats in the IEF of EI.

It is proposed to evaluate the quality of the partitioning the space of objects signs' recognition in the AES using as an evaluating indicator the modified ICFP capable of self-learning recognition system. It is proved that the application of model and method of signs' clustering of RO, based on entropy and information distance criteria of Kullback–Leibler, provides input fuzzy classified training matrix that is used as an object of study, and within smart technologies and teaching methods of AES learning to create correct decisive rules for recognizing cyberattacks in IEF of EI.

References

1. Rezgui, Y., Marks, A.: Information security awareness in higher education: an exploratory study. Comput. Secur. 27(7), 241–253 (2008)
2. Sultan, N.: Cloud computing for education: a new dawn? Int. J. Inf. Manag. 30, 109–116 (2010). http://dx.doi.org/10.1016/j.ijinfomgt.2009.09.004
3. Robles, A.C.M.O.: Evaluating the use of Toondoo for Collaborative E-Learning of Selected pre-service teachers. Int. J. Mod. Educ. Comput. Sci. (IJMECS) 9(11), 25–32 (2017). https://doi.org/10.5815/ijmecs.2017.11.03
4. Schneider, F.B.: Cybersecurity education in universities. IEEE Secur. Priv. 11(4), 3–4 (2013)
5. Conklin, A.: Cyber defense competitions and information security education: an active learning solution for a capstone course. In: 2006 Proceedings of the 39th Annual Hawaii International Conference on System Sciences, HICSS 2006, vol. 9. IEEE (2006)
6. Schuett, M., Rahman, M.: Information Security Synthesis in Online Universities. arXiv preprint arXiv:1111.1771 (2011)
7. Azka, S.R., Geetha, A.: A survey of applications and security issues in software defined networking. Int. J. Comput. Netw. Inf. Secur. (IJCNIS) 9(3), 21–28 (2017). https://doi.org/10.5815/ijcnis.2017.03.03
8. Jalali, M., Siegel, M., Madnick, S.: Decision Making and Biases in Cybersecurity Capability Development: Evidence from a Simulation Game Experiment [Electronic resource] (2017). https://arxiv.org/ftp/arxiv/papers/1707/1707.01031.pdf
9. Gordon, L.A., Loeb, M.P., Zhou, L.: Investing in cybersecurity: insights from the Gordon-Loeb model. J. Inf. Secur. 7(02), 49 (2016). https://doi.org/10.4236/jis.2016.72004
10. Goztepe, K.: Designing fuzzy rule based expert system for cyber security. Int. J. Inf. Secur. Sci. 1(1), 13–19 (2012)
11. Akhmetov, B., Lakhno, V., Boiko, Y., Mishchenko, A.: Designing a decision support system for the weakly formalized problems in the provision of cybersecurity. East.-Eur. J. Enterp. Technol. 1(2(85)), 4–15 (2017)
12. Lakhno, V., Boiko, Y., Mishchenko, A., Kozlovskii, V., Pupchenko, O.: Development of the intelligent decision-making support system to manage cyber protection at the object of informatization. East.-Eur. J. Enterp. Technol. 2/9(86), 53–61 (2017)
13. Keerthi Vasan, K., Arun Raj Kumar, P.: Taxonomy of SSL/TLS attacks. Int. J. Comput. Netw. Inf. Secur. (IJCNIS) 8(2), 15–24 (2016). https://doi.org/10.5815/ijcnis.2016.02.02

14. Lakhno, V., Tkach, Y., Petrenko, T., Zaitsev, S., Bazylevych, V.: Development of adaptive expert system of information security using a procedure of clustering the attributes of anomalies and cyber attacks. East.-Eur. J. Enterp. Technol. **6/9**(84), 32–44 (2016). https://doi.org/10.15587/1729-4061.2016.85600
15. Melese, S.Z., Avadhani, P.S.: Honeypot system for attacks on SSH protocol. Int. J. Comput. Netw. Inf. Secur. (IJCNIS) **8**(9), 19–26 (2016). https://doi.org/10.5815/ijcnis.2016.09.03

Research on the Use of OLAP Technologies in Management Tasks

Daria Yu. Yashchuk[✉] and Bella L. Golub

National University of Life and Environmental Sciences of Ukraine, Kyiv, Ukraine
yashchuk.dasha@gmail.com, bella.golub55@gmail.com

Abstract. The article is dedicated to research of application of OLAP technologies when management. Much attention in the article is paid to creation of the decision-making support system. Such system is set aside for increase of effectiveness of governing, owing to analysis of data collected for a few years. The author demonstrates structure of data warehouse, on which basis analysis of university's processes is kept. The article is important because the study uses the latest information technologies like MS Excel, Power BI, Business Intelligence Development Studio. Their use will allow management getting a system that will enhance the position in competitive environment.

Keywords: Decision support system · Data warehouse · Business Intelligence Information Technology · OLAP · Database · Data mart · Power BI

1 Introduction

Use of information approaches is the main mechanism for managing. In order to achieve strategic goals, improvement of management and decision-making, one needs to create a reliable and efficient system. It will be based on data accumulated over a period of time that includes knowledge and allows decision making by analyzing data accumulated from different sources during some years (Alkhafaji and Sriram 2013).

Technologies are changing fast and IT implementation becomes more popular. There are two factors that influence decision-making within any institution - intuition and rationality are these factors. It is not appropriate to use intuition in process of making strategic decisions. Use of the decision support system will reduce such factor as intuition (Tank 2015).

The purpose of the research is to investigate efficiency of using OLAP technologies in management of Institutions of Higher Education.

2 Overall Structure of the System

Any institution of higher education has a complex structure, which is represented by a large number of subdivisions: administration, enrolment board, educational unit, accounting department, personnel department, faculty and others. Each division performs certain functions, exchanging information that generally ensures functioning

© Springer International Publishing AG, part of Springer Nature 2019
Z. Hu et al. (Eds.): ICCSEEA 2018, AISC 754, pp. 683–691, 2019.
https://doi.org/10.1007/978-3-319-91008-6_67

of the institution. Correlation and execution of functions of each unit, as a rule, are performed automated with the help of data processing system, which allows simplifying work of unit employees. There are many automation systems existing today and providing data flows between higher education institutions.

Leaders are interested in effective and high-quality work. Therefore, there is a problem in obtaining of generalized information from all subdivisions, on which basis it is possible to draw conclusions about life of university and make managerial decisions. Such functions can be implemented as a decision support system (Golub and Yashchuk 2017).

The obtained system will provide problems' solutions for improving efficiency of management with identification of indicators:

- number of entrants (specialty, forms of studying) from separate regions for a certain period of time;
- results of studying of students in terms of areas for a certain time interval;
- disciplines that students pass/fail with excellent marks;
- distribution of students according to the results of progress (specialty, education form, semester).

These factors will help university management solve the following issues:

- carrier-guidance tutorials;
- determination of sequence of disciplines depending on success of students;
- determination of quality rates of studying of students.

The articles (Glazunova and Voloshyna 2016; Yasenova 2007) identified problems on the basis of which indicators of the system were determined.

Decision support system is a system that influences the process of making managerial decisions by collecting and analyzing large amounts of information. Such systems are a tool to help decision makers (Golub and Yashchuk 2017).

In the research, the decision-support system is intended to implement data analysis tasks in order to make effective management decisions of in universities. Figure 1 shows the overall structure of this system.

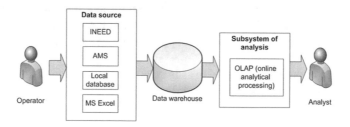

Fig. 1. Overall structure of the decision support system in higher education institutions

The basic elements of system's structure:

- **data sources** can be represented by spreadsheets, local databases, automated management system of Institutions Higher Education (AMS), the Integrated National Education Electronic Database (INEED);
- **data warehouse.** Received information must be integrated into a single data warehouse for decision-making which is based on efficiency index. The DW concept implies allocation of storage structures for rapid data processing and analytical enquiries;
- **analysis subsystem.** OLAP technology is used to analyze data.

3   Database

Microsoft SQL Server Management Studio 2008 tool was chosen to develop relational database (Fig. 2).

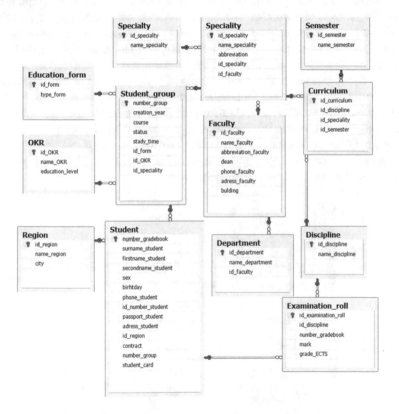

Fig. 2. Database schema

Existing database contains the following information: specialty, speciality, student group, educational degree, education form, student, region, examination roll, faculty, department, semester, curriculum, discipline.

4   Data Warehouse

Data warehouse contains consistent historical data and provides instrumental means for their analysis in order to support strategical decision-making. Information files of data warehouse are based on instant snapshots of same-time information system fixed during long period of time and, possibly, taken from different external sources. OLAP database system, in-depth data analysis and data visualization are used in data warehouses.

Seldom changed data is input into data warehouses. They are designed to perform analytical queries, providing managers with decision-making support.

As one can see from these requirements, domain databases significantly differ from data warehouses according to its structure. Data mining in relational domain DBMS leads to the creation of a number of tables associated with each other. Fact table and dimension table are the main components of data warehouse (Rahman and Khan 2017).

Data Mart – data's structure that provide analytical task solution in particular functioning sphere or subdivision. Data Marts can be considered as small repositories, which are created for maintenance of analytical tasks of definite management subdivisions of an organization. As a matter of fact, data mart contains considerably less information than electronic data warehouse. Data marts can be imagined in the form of logical or physical subdivisions of data warehouses (Alguliyev and Nabibayova 2015).

Data mart (Fig. 3) was designed by means of the basic SQL Server Management Studio. It is intended for saving and analyzing large volumes of data for solving the problem of performance of career-guidance work.

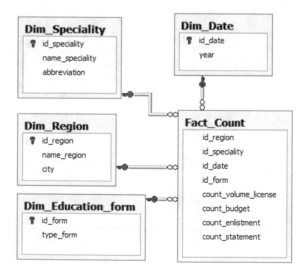

Fig. 3. Data mart 1

Data Mart contains Fact_Count and 4 dimension tables:

- Dim_Speciality;
- Dim_Region;
- Dim_Date;
- Dim_Education_form.

Fact_Count table contains count_volume_license, count_budget, count_enlistment, count_statement, id_speciality, id_form, id_date and id_region.

To solve the task of student success analysis, data mart displayed in the Fig. 4 was designed.

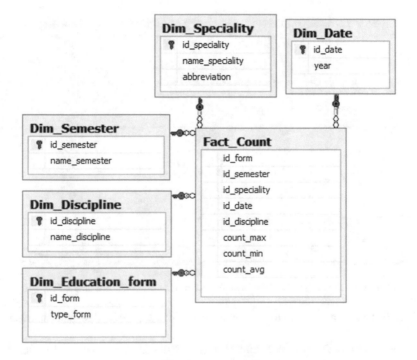

Fig. 4. Data mart 2

Data Mart contains Fact_Count and 5 dimension tables:

- Dim_Speciality;
- Dim_Discipline;
- Dim_Semester;
- Dim_Date;
- Dim_Education_form.

Fact_Count contains count_max, count_min and count_avg in section id_speciality, id_form, id_date, id_semester and id_discipline.

5   Subsystem of Analysis

You can view and analyze data using MS Excel, Power BI, Business Intelligence Development Studio tools. The review of these technologies is described in work by Golub and Trochymenko (2017).

To solve the problem of conducting of carrier-guidance, we use Power BI tool, which serves for improvement of visualization of the results of intellectual analysis. For example, the National University of Life and Environmental Sciences of Ukraine is being analyzed. Histogram of entry to the university during 2012–2017 (computer science faculty) is shown in Fig. 5.

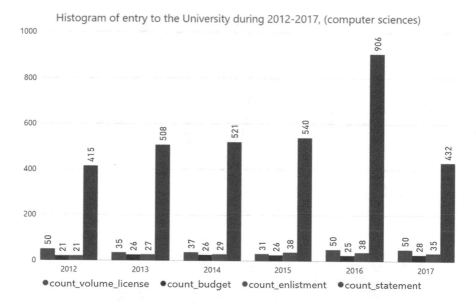

Fig. 5. Histogram of entry to the University during 2012–2017, (computer sciences)

Report of university entrance contest during 2013–2017 is presented in Fig. 6. The report contains the following: the year of entry, abbreviation, the number of allocated licensed and budget places, the number of applications submitted and the number of those who joined the university.

University entrance contest report

year	abbreviation	count_volume_license	count_budget	count_enlistment	count_statement
2013	AKIT	50	40	32	375
	EK	50	25	28	340
	KH	35	26	27	508
	Total	**135**	**91**	**87**	**1223**
2014	AKIT	30	24	24	390
	EK	40	25	25	307
	KH	37	26	29	521
	Total	**107**	**75**	**78**	**1218**
2015	AKIT	27	22	22	368
	EK	50	25	34	457
	IΠ3	25	5	8	167
	KI	25	5	18	175
	KH	31	26	38	540
	Total	**158**	**83**	**120**	**1707**
2016	AKIT	50	25	25	322
	EK	50	30	15	296
	IΠ3	50	10	22	552
	KI	50	10	18	485
	KH	50	25	38	906
	Total	**250**	**100**	**118**	**2561**
2017	AKIT	50	25	30	177
	EK	50	1	25	502
	IΠ3	50	13	31	381
	KI	50	13	30	303
	KH	50	28	35	432
	Total	**250**	**80**	**151**	**1795**

Fig. 6. University entrance contest report

To solve the task of student success analysis, we use MS Excel tool. Histogram of results of studying of the students during 2015 and 2016 in the specialty "Computer Science and Information Technologies" is shown in Fig. 7.

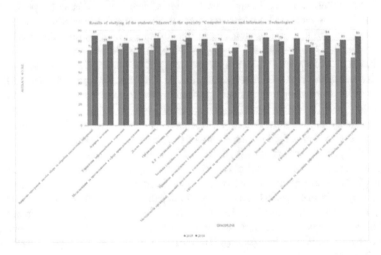

Fig. 7. Histogram of results of studying of the students during 2015 and 2016 in the specialty "Computer Science and Information Technologies"

6 Conclusions

Analyzing obtained visual data, it can be seen that the number of applications has increased by this year significantly each year, as IT specialty has become more popular due to high wages and development of information technologies, but the number of applications has decreased by 2 times during the last year - this may be due to poor economic situation in the country and changing conditions of acceptance of applications.

Also it is necessary to better professional orientation work, because, when the number of applications was increased in comparison with other years, last year the number of students willing to study at the university has been decreased by half.

Today, decision support systems have become one of the main types of systems that provide answers to the exciting questions of management. However, multipurpose systems have not yet been created yet; therefore, there is a need to create such a system that will provide an opportunity for management to carry out data analysis in order to make optimal decisions.

References

Alkhafaji, S., Sriram, B.: Conceptualization and integration of information systems in educational business activities. Int. J. Inf. Eng. Electron. Bus. (IJIEEB) 5(2), 28–33 (2013). https://doi.org/10.5815/ijieeb.2013.02.05

Tank, D.M.: Enable better and timelier decision-making using real-time business intelligence system. Int. J. Inf. Eng. Electron. Bus. (IJIEEB) 7(1), 43–48 (2015). https://doi.org/10.5815/ijieeb.2015.01.06

Yashchuk, D.Yu., Golub, B.L.: Sovremennye informacionnye tehnologii v upravlenii vysshimi uchebnymi zavedenijami (Modern information technologies in the management of higher educational institutions), Sbornik trudov mezhdisciplinarnaja nauchnaja konferencija "Mnogofaktornye podhody k formirovaniju komfortnoj sredy" (Proceedings of International Conference on "Multifactor approaches to the formation of a comfortable environment"), Netanya, Israel, 15 November 2017, pp. 36–40 (2017). (in Russian)

Glazunova, O.G., Voloshyna, T.: Hybrid cloud-oriented educational environment for training future it specialists. In: ICTERI, Kyiv, Ukraine, 21–24 June 2016, pp. 157–167 (2016)

Yasenova, I.S.: Matematychna model' tekhnolohiyi formuvannya navchal'noho planu vidpovidno do kredytno-modul'noyi systemy orhanizatsiyi navchal'noho protsesu (Mathematical model of the technology of formation of the curriculum in accordance with the credit-module system of educational process organization), Avtomatizirovannye sistemy upravlenija i pribory avtomatiki 139, 68–73 (2007). (in Ukrainian)

Golub, B.L., Yashchuk, D.Yu.: Zahal'ni zasady pobudovy systemy pidtrymky pryynyattya rishen' dlya VNZ (General principles for building a decision support system for higher education institutions), Informatsiyni tekhnolohiyi v ekonomitsi tapryrodokorystuvanni 1(1), 29–36 (2017). http://journals.nubip.edu.ua/index.php/Inf/article/view/8926. (in Ukrainian)

Rahman, A., Khan, M.N.A.: An assessment of data mining based CRM techniques for enhancing profitability. Int. J. Educ. Manage. Eng. (IJEME) 7(2), 30–40 (2017). https://doi.org/10.5815/ijeme.2017.02.04

Alguliyev, R.M., Nabibayova, G.C.: The method of measuring the integration degree of countries on the basis of international relations. Int. J. Intell. Syst. Appl. (IJISA) **7**(11), 10–18 (2015). https://doi.org/10.5815/ijisa.2015.11.02

Golub, B.L., Trochymeko, V.: Porivnyal`ny`j analiz instrumental`ny`x zasobiv Microsoft dlya analizu dany`x (Comparative analysis of Microsoft tools for data analysis), Visnyk inzhenernoyi akademiyi Ukrayiny, vol. 1, pp. 61–65 (2017). (in Ukrainian)

Impact of the Textbooks' Graphic Design on the Augmented Reality Applications Tracking Ability

N. Kulishova[(✉)] and N. Suchkova

Kharkov National University of Radioelectronics, Kharkiv 61166, Ukraine
nokuliaux@gmail.com, tallisia77@gmail.com

Abstract. Augmented reality (AR) is very effective in school education. Thus, a number of these applications are growing permanently. In most cases, these applications use school textbooks as target images for the AR technology. In developing a textbook design accompanied by the AR application, it is important to use such elements will ensure a stable tracking property when a gadget is held by a kid. There are also various graphic design elements as addition to texts and illustrations in modern textbooks. It is necessary to study how these elements ensure the tracking stability when conditions of textbook viewing are changing. The use of corner detectors to assess the tracking ability of different graphic elements in a textbook is considered. A comparative analysis of the tracking stability for textbook pages is carried out by means of the Harris-Stephens method, BRISK, FAST, Shi & Tomasi methods (also known as the minimum eigenvalue algorithm) which detect features and form their descriptors for the image. Results for these methods are collated with the tracking ability of targets used for a rating estimation on the basis of the Augmented Reality Platform Qualcomm Vuforia.

Keywords: Augmented reality · Features · Tracking · Corner detectors
Textbook · School education

1 Introduction

The augmented reality is a result of any sensory data inclusion into a perception field to expand and improve the environment spirit. For this reason, the AR technology opens wide opportunities for the school textbooks attractiveness increasing; they now can be made not only informative and cognitive, but also interactive and animated.

A basis of the AR technology is its optical tracking system. The following components are required for this system: a label (a visual identifier for computer models); a camera catches a real-world scene; a software finds and analyzes labels in real-world scene and combines virtual models with images of real objects. In this system, an effect of physical presence in the environment (the surrounding space) is achieved [1].

A main task of the tracking system is to determine the three-dimensional coordinates of a real label from its snapshot obtained by a camera. Devices that interact with the augmented reality have rather limited computational abilities, and this fact complicates

© Springer International Publishing AG, part of Springer Nature 2019
Z. Hu et al. (Eds.): ICCSEEA 2018, AISC 754, pp. 692–701, 2019.
https://doi.org/10.1007/978-3-319-91008-6_68

the image processing and objects contours recognition greatly. However, the main cause of visual distortion and image instability in the tracking is an information loss while target (label) image creating.

A special target image could be chosen as a label where contrast zones combination serves as a natural marker which the augmented reality is attached to. After the marker has been detected in a video stream and its location has been calculated, it becomes possible to build a projection matrix and to site virtual models. This allows superimposing a virtual object on the video stream to achieve the AR effect. The main difficulty is to find a label in a video frame, determine its real-world location and project a virtual model into a device screen in an appropriate way.

In developing AR applications for school education, the most commonly used labels are images of textbook pages. A various elements used for a modern school textbooks design are:

- navigation elements accelerating a search through the desired book chapter – column figures, footers, column lines;
- visual text splitting tools – indents, retracts, piles, initial letters, fonts and color allocations;
- classification elements and marginalia;
- border graphic elements – illustrations, maps, diagrams, and vignettes (Fig. 1).

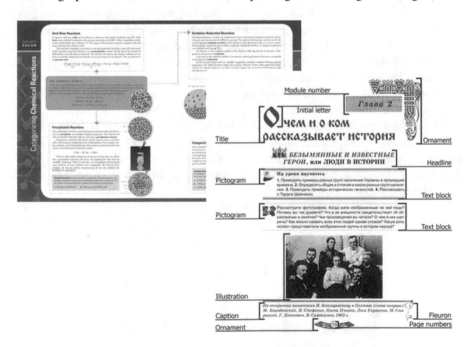

Fig. 1. Modern school textbooks. A design of elements

These elements differ in both structural and geometric characteristics, so they can provide a stable tracking property in AR application in various degrees. School textbooks are used by children in their classes, during breaks, and at home. Lighting conditions and a distance to the textbook are changing and kids often move a book or a gadget in their hands. All these factors worsen the augmented reality label binding in a serious way. However, in school textbooks design it is desirable to use elements that can guarantee the most stable tracking even in such difficult conditions.

The AR technologies require assessing an images/labels quality in terms of detection and tracking. For this reason, different approaches are used; their action is based on the detection such image features as isolated points, corners, curves, contours, and related areas. However, a precise estimation of the image's visibility for AR demands every image downloading to a development platform and evaluating it on this platform. This textbook design process organization leads to time costs increase that can make the developed application too expensive. It is necessary to find a traditional method for an image tracking ability quantitative estimation which helps avoid a development platform access for this assessment and reduce the AR textbook development time.

A purpose of this paper is a comparative analysis of the textbook design elements from the augmented reality tracking stability viewpoint.

2   Feature Detection Standard Algorithms

Characteristic features of different types are used in most computer vision tasks [2]. Among such features are corner points, blobs, straight and curve contours, color and texture differing areas and other image elements and their characteristics. The analysis of the information about the neighborhood selected in the image is necessary to detect characteristics, and for this, the clustering methods are applied. Many approaches for clustering problem solving have been developed; the most promising are the fuzzy methods [3–7]. They are effective under uncertainty, which inevitably accompanies a real time video stream captured by an AR application. After clustering, when features in each frame are found, a transformation between adjacent frames is constructed, establishing a correspondence between the projections of the same real object points on the gadget screen [8]. Thus, tracking consists in constructing of such transformations for all video stream frames.

Corners are widely chosen as characteristic points for image tracking tasks. They are the intersection of two or more edges, and the edge usually define the boundary between different objects and/or parts of the same object in the image (Fig. 2) [9].

Fig. 2. Different types of corner feature points

There are many algorithms for determining singular points for different applications. This article focuses on the corner detectors: Harris, FAST, BRISK, Shi & Tomasi (Minimum eigenvalue algorithm).

The Harris detector action is based on calculating the neighborhood intensity weighted change (sum of squared differences (SSD)) of the moving window in the image. For this change, an autocorrelation matrix and its eigenvalue are calculated. Since the eigenvalues direct computation is a time-consuming task, Harris and Stephen proposed a corner response measure [10]. The Harris detector is invariant to rotations, is partially invariant to affine intensity changes. The detector's disadvantages are the sensitivity to noise and the image scale dependence.

The FAST algorithm considers a circle of 16 pixels around the candidate point P [11]. The point P is the corner if there are N adjacent pixels on the environing circle whose intensities are greater than some threshold value. Experiments have shown that the smallest value of N at which the features begin to be stably detected is nine.

To find features, the BRISK method uses the FAST detector, but with some modifications [12]. To achieve the scale invariance, it chooses the best pixel with the maximum intensity value in the pyramid, which consists of octaves. Octaves are formed as a compression of the original image in 2^i times. The features search in octaves is performed by the FAST detector. To form a descriptor, the area around the singular point is divided into 60 sections and the gradient mean for the internal octaves centers set is calculated. The descriptor consists of a binary string of length 512 with the results of the calculations.

The Shi-Tomasi or Kanade-Tomasi corner detector [13] largely coincides with the Harris detector, but differs in the response measure computation: the algorithm directly calculates the autocorrelation matrix eigenvalues, since it is assumed that this will make the feature search more stable.

3   Feature Detection for Test Images

For the study, three images were taken, representing pages of a textbook with a different components: column figures, footers, column lines, indents, retracts, piles, initial letters, font and color allocations, rubricating elements, illustrations, maps and diagrams, vignettes.

The first image (Fig. 3) has only a text block separated by paragraph indentation. There is also a page number on the page; font style – normal with serifs.

In the second image (Fig. 4) there are several small text blocks, illustration, column, rubric elements, vignettes, font and color highlighting, and initial letter; font styles – normal, bold and italic, without of serifs.

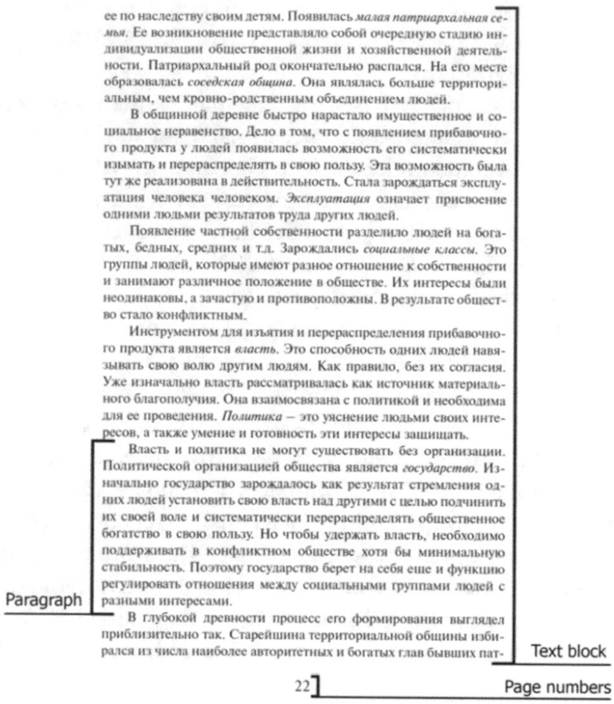

ее по наследству своим детям. Появилась *малая патриархальная семья*. Ее возникновение представляло собой очередную стадию индивидуализации общественной жизни и хозяйственной деятельности. Патриархальный род окончательно распался. На его месте образовалась *соседская община*. Она являлась больше территориальным, чем кровно-родственным объединением людей.

В общинной деревне быстро нарастало имущественное и социальное неравенство. Дело в том, что с появлением прибавочного продукта у людей появилась возможность его систематически изымать и перераспределять в свою пользу. Эта возможность была тут же реализована в действительность. Стала зарождаться эксплуатация человека человеком. *Эксплуатация* означает присвоение одними людьми результатов труда других людей.

Появление частной собственности разделило людей на богатых, бедных, средних и т.д. Зарождались *социальные классы*. Это группы людей, которые имеют разное отношение к собственности и занимают различное положение в обществе. Их интересы были неодинаковы, а зачастую и противоположны. В результате общество стало конфликтным.

Инструментом для изъятия и перераспределения прибавочного продукта является *власть*. Это способность одних людей навязывать свою волю другим людям. Как правило, без их согласия. Уже изначально власть рассматривалась как источник материального благополучия. Она взаимосвязана с политикой и необходима для ее проведения. *Политика* – это уяснение людьми своих интересов, а также умение и готовность эти интересы защищать.

Власть и политика не могут существовать без организации. Политической организацией общества является *государство*. Изначально государство зарождалось как результат стремления одних людей установить свою власть над другими с целью подчинить их своей воле и систематически перераспределять общественное богатство в свою пользу. Но чтобы удержать власть, необходимо поддерживать в конфликтном обществе хотя бы минимальную стабильность. Поэтому государство берет на себя еще и функцию регулировать отношения между социальными группами людей с разными интересами.

В глубокой древности процесс его формирования выглядел приблизительно так. Старейшина территориальной общины избирался из числа наиболее авторитетных и богатых глав бывших пат-

22

Fig. 3. A textbook page that contains only a text block

Глава 2

О чем и о ком рассказывает история

§11 БЕЗЫМЯННЫЕ И ИЗВЕСТНЫЕ ГЕРОИ, или ЛЮДИ В ИСТОРИИ

На уроке научитесь

1. Приводить примеры разных групп населения Украины в прошедшие времена. 2. Определять общее и отличия в жизни разных групп населения. 3. Приводить примеры исторических личностей. 4. Рассказывать о Тарасе Шевченко.

Рассмотрите фотографию. Когда жили изображенные на ней люди? Почему вы так думаете? Что в их внешности свидетельствует об образовании и занятии? Чьи произведения вы читали? О чем в них идет речь? Как можно назвать всех этих людей одним словом? Какую роль играют представители изображенной группы в истории народа?

На открытии памятника И. Котляревскому в Полтаве (слева направо): М. Коцюбинский, В. Стефаник, Олена Пчилка, Леся Украинка, М. Старицкий, Г. Хоткевич, В. Самойленко, 1903 г.

79

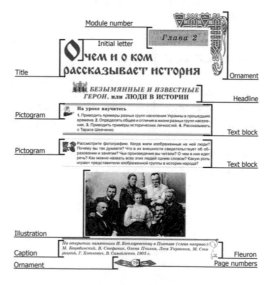

Fig. 4. A textbook page that contains a text and multiple graphic elements

On the third image (Fig. 5) is an illustration – map, occupying 70% of the page area, a text block is separated by indentation and page number: font style – normal with serifs.

Fig. 5. A textbook page that contains a text and a map

Using standard detectors, features on these samples are found. Feature amount is shown in Table 1.

Table 1. An amount of initial features found on originals using standard detectors prior to photographing

	Standard detectors			
	Harris	BRISK	Shi-Tomasi (Minimum eigenvalue algorithm)	FAST
Image 1	13690	59955	23103	33992
Image 2	7184	34143	14873	14013
Image 3	8898	36986	13962	22264

The test environment lighting conditions can significantly affect the label detection and tracking; sufficient illumination ensures good image visibility in the camera field of view. With artificial lighting a series of photos were taken on the distance of 30–57 cm from the image surface to the camera lens with a step of 3 cm (examples of the images after grayscale conversion in Fig. 6 are shown).

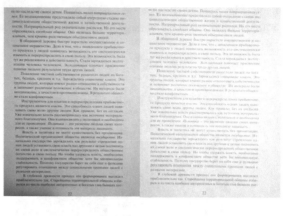

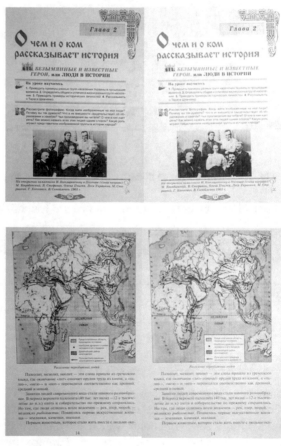

Fig. 6. Textbook pages (photos) that were taken under different lighting conditions

For each position of the camera, scene lighting was changed: images were taken with general uniform illumination combined with a camera flash. General uniform illumination excludes the possibility of the main luminous flux direction changing, and also has excessive light dispersion. Local lighting (stationary or portable) illuminate only the working surfaces and allow selecting only a specific area of the room, which is brightly illuminated by a localized light source. To eliminate these shortcomings combined lighting is designed by local and common light integration.

To estimate the image tracking ability rating, AR SDK Vuforia from Qualcomm was chosen, because it allows you to create applications for Android and iOS. Vuforia SDK is an AR software library offered by Qualcomm, which supports the development of mobile AR applications [14, 15].

The Vuforia Target Manager image rating is an indicator of how reliable the target image within this framework will be, and is a simplified combined assessment. Each image is assigned a rating from 1 to 5, showing the image potential relative to the recognition speed and tracking stability. The highest rating is assigned to an image rich in small details and having a high contrast.

The total features amount for the distorted images was found and the part of results is presented in Tables 2 and 3.

Table 2. An amount of features found by the Harris detector for the image 1

Shooting distance (cm)	LF	Vuforia	L	Vuforia	F	Vuforia	General	Vuforia
30	26252	0	26354	0	25859	0	26836	0
33	20968	0	21708	0	20893	0	21710	0
36	18528	0	18740	0	18061	0	18769	0
39	15839	0	16134	0	15768	0	15760	0
42	14779	0	14580	0	14268	0	14087	0
45	13674	0	13782	0	12896	0	13150	0
48	13036	0	13005	0	11986	0	12248	0
51	12326	0	12290	0	11411	0	11627	0
54	11753	0	11846	0	10658	0	11110	0
57	11298	0	11276	0	5410	0	10523	0

Explaining variables to be mentioned in Tables 2 and 3, LF denotes shooting images in the combined lighting with a flash light; L stands for shooting images in the combined lighting; F determines shooting images in the general lighting with a flash light; General describes shooting images in the general lighting; Vuforia marks a rating assigned to an image in Vuforia Target Manager.

The largest features amount is found using the Shi-Tomasi method, but image was corrupted by digital noise. The number of features found on distorted images using Harris detectors and the Shi-Tomasi algorithm significantly exceeds the number of features in the original image. This is due to the raster nature of printed image, when raster dots are seen from close distance. As the camera moves away from the image surface, the number of detected features decreases, because the camera does not recognize the raster dots and the print quality ceases to play such an important role.

Table 3. An amount of features found by the FAST detector for the image 2

Shooting distance (cm)	LF	Vuforia	L	Vuforia	F	Vuforia	General	Vuforia
30	9720	4	10714	4	9995	4	4006	4
33	5271	5	5776	4	5364	4	3414	5
36	5036	4	5065	5	4254	4	3087	5
39	4803	5	5214	4	4227	5	3304	5
42	4669	5	5022	5	4121	5	3215	5
45	4802	5	4878	5	4065	4	3146	5
48	4745	5	4989	5	4106	5	3212	5
51	4592	5	4771	5	3164	4	3115	5
54	4456	5	4663	5	3606	5	2795	5
57	4517	5	4069	5	1470	4	1671	5

The selected features amount for all images was significantly reduced at a shooting distance of 30–40 cm, and then, as this distance increased, the decrease in the number of signs was less essential. This suggests that textbook pages provide more stable tracking for AR applications, when the gadget is located at a distance of 40–60 cm from them.

Among the considered types of image originals, the largest features amount was detected in image 1, but the same image received the lowest tracking ability rating in Vuforia. This rating was not affected even by a reduction of the shooting distance.

On the other hand, the image 2 has got the highest estimate from Vuforia, in which the least features amount from the proposed images has been found by standard detectors. Thus, the corner detectors do not give complete information about whether this image can guarantee stable tracking in AR application.

The highest rating of the Vuforia was assigned to a page with such elements as a title with the initial letter, ornament, text blocks, pictograms, illustrations. This suggests that when designing a textbook with augmented reality application, it is necessary to use as many graphic elements as possible. The presence of only text blocks on the pages makes tracking extremely unstable and almost completely eliminates the possibility of augmented reality using.

4   Conclusion

The school education is one of the areas where the augmented reality can be a very effective tool. Therefore, it is necessary to develop various approaches for an optimal using of this technology for school textbooks designing. During the research, it was found that the best condition of the stable tracking to a textbook for AR applications is a viewing distance of 46–60 cm between a target and a gadget.

A number of detected corner points by means of the detectors cannot indicate the possible tracking stability: a text block contains the largest number of these points, but a page that contains mostly a text does not provide tracking at all in a practical sense.

A traditional approach to a textbook designing usually assumes an existence of a text and some simple illustrations and does not allow linking the augmented reality to pages. Conversely, the use of graphic, structuring and text navigation elements provides the highest rank of the tracking stability.

References

1. Azuma, R.T.: A survey of augmented reality. Presence Teleoperators Virtual Env. **6**(4), 355–385 (1997). The MIT Press, Cambridge, MA
2. Hu, Z., Bodyanskiy, Y.V., Kulishova, N.Y., Tyshchenko, O.K.: A multidimensional extended neo-fuzzy neuron for facial expression recognition. Int. J. Intell. Syst. Appl. (IJISA) **9**(9), 29–36 (2017). https://doi.org/10.5815/ijisa.2017.09.04
3. Hu, Z., Bodyanskiy, Y.V., Tyshchenko, O.K., Samitova, V.O.: Fuzzy clustering data given in the ordinal scale. Int. J. Intell. Syst. Appl. (IJISA) **9**(1), 67–74 (2017). https://doi.org/10.5815/ijisa.2017.01.07
4. Hu, Z., Bodyanskiy, Y.V., Tyshchenko, O.K., Samitova, V.O.: Fuzzy clustering data given on the ordinal scale based on membership and likelihood functions sharing. Int. J. Intell. Syst. Appl. (IJISA) **9**(2), 1–9 (2017). https://doi.org/10.5815/ijisa.2017.02.01
5. Hu, Z., Bodyanskiy, Y.V., Tyshchenko, O.K., Samitova, V.O.: Possibilistic fuzzy clustering for categorical data arrays based on frequency prototypes and dissimilarity measures. Int. J. Intell. Syst. Appl. (IJISA) **9**(5), 55–61 (2017). https://doi.org/10.5815/ijisa.2017.05.07
6. Hu, Z., Bodyanskiy, Y.V., Tyshchenko, O.K., Tkachov, V.M.: Fuzzy clustering data arrays with omitted observations. Int. J. Intell. Syst. Appl. (IJISA) **9**(6), 24–32 (2017). https://doi.org/10.5815/ijisa.2017.06.03
7. Bodyanskiy, Y.V., Tyshchenko, O.K., Kopaliani, D.S.: An evolving connectionist system for data stream fuzzy clustering and its online learning. Neurocomputing **262**, 41–56 (2017)
8. Hu, Z., Mashtalir, S.V., Tyshchenko, O.K., Stolbovyi, M.I.: Video shots' matching via various length of multidimensional time sequences. Int. J. Intell. Syst. Appl. (IJISA) **9**(11), 10–16 (2017). https://doi.org/10.5815/ijisa.2017.11.02
9. Rodehorst, V., Koschan, A.: Comparison and evaluation of feature point detectors. In: Proceedings of the 5th International Symposium Turkish-German Joint Geodetic Days TGJGD 2006, Berlin (2006)
10. Harris, C., Stephens, M.: A combined corner and edge detector. In: Proceedings of the 4th Alvey Vision Conference, pp. 147–151 (1988)
11. Rosten, E., Drummond, T.: Fusing points and lines for high performance tracking. In: Proceedings of the IEEE International Conference on Computer Vision, vol. 2, pp. 1508–1511 (2005)
12. Leutenegger, S., Chli, M., Siegwart, R: BRISK: Binary Robust Invariant Scalable Keypoints. In: Proceedings of the IEEE International Conference on Computer Vision (ICCV) (2011)
13. Shi, J., Tomasi, C.: Good features to track. In: Proceedings of the IEEE Conference on Computer Vision and Pattern Recognition, pp. 593–600 (1994)
14. Amin, D., Govilkar, S.: Comparative study of augmented reality SDK's. Int. J. Comput. Sci. Appl. (IJCSA) **5**(1), 11–26 (2015)
15. Qualcomm Vuforia Core Samples. https://developer.vuforia.com/downloads/samples. 2 Feb 2015

Information Technologies of Modeling Processes for Preparation of Professionals in Smart Cities

Andrii Bomba[1], Nataliia Kunanets[2], Mariia Nazaruk[2],
Volodymyr Pasichnyk[2], and Nataliia Veretennikova[2(✉)]

[1] Informatics and Applied Mathematics Department, State Humanitarian University,
Rivne, Ukraine
abomba@ukr.net
[2] Information Systems and Networks Department, Lviv Polytechnic National University,
Lviv, Ukraine
nek.lviv@gmail.com, marinazaruk@gmail.com, vpasichnyk@gmail.com,
nataver19@gmail.com

Abstract. It is proposed the training process of qualified specialists in accordance with the needs of a person and the requirements of the labor market in the smart city to be presented in the form of five consecutive functional stages: determination of professional inclinations and abilities; monitoring of the urban labor market; a choice of the future profession; a choice of educational institution; formation of an individual learning trajectory. The model of the data analysis process of career orientation testing with obvious uncertainty and hidden redundancy is presented. For storing and analyzing big data, it is suggested to use data warehouses and the model of data warehouse of the complex assessment of educational institution activities is provided. The diffusion-liked model of the multicomponent knowledge potential dissemination is described and variants of the problem solution of identifying the component parameters of the knowledge potential are described, with the aim of their further usage for the formation of individual learning trajectories. The architecture of the software and algorithmic complex of information and technological support of the processes for specialist training in the smart city is developed.

Keywords: Smart city · Big data · Data warehouse · Knowledge potential
Diffusion-liked model · Career orientation test

1 Introduction

Modern cities are social city-states with a complex infrastructure that needs to be effectively formed, developed, modernized and adapted to the needs of the community. As the results of many studies indicate, one of the most effective innovative concepts for the modern cities is a concept of smart cities.

The purpose of the work is to form an information and technological platform for analyzing the processes of training specialists according to personal characteristics and demands of the labor market in smart cities.

© Springer International Publishing AG, part of Springer Nature 2019
Z. Hu et al. (Eds.): ICCSEEA 2018, AISC 754, pp. 702–712, 2019.
https://doi.org/10.1007/978-3-319-91008-6_69

1.1 Analysis of the Research State

Implementation of the concept of a smart city involves the reorganization of all spheres of its life by creating and implementing modern information and telecommunication technologies, communication engineering and transport networks, computerized control systems, cool and data centers, diagnostic, service and environmental services [1, 2]. Smart city is smart management, smart living, smart people, smart environment, smart economy, and smart mobility [3].

The development of a smart city is not limited to the improvement of hardware, but more attention is paid to the accessibility and the quality of communications, the processes of knowledge dissemination and acquisition, as well as the social infrastructure that contributes to the growth of human and social capital. There is a change both from the angle of view and from the focus of the professional vision: from infrastructure-oriented technologies to citizen-oriented programs [4–6].

Education contributes to the development of creativity and innovative thinking, and increases the competitiveness of urban centers in the global knowledge economy [7]. However, only a few promising systematic studies [8] of smart cities include profiles of educational perspectives. A new transformational paradigm was formed, which IBM experts called as "educational continuum", that includes: life-long learning technologies (critical thinking, information literacy, etc.), data analytics for student and institutional data analysis as well as efficiency indicators (which serve as a basis for improving resource allocation, the creation of new training programs, etc.), personalized learning trajectories (the choice of individual learning opportunities), the use of acquired competencies for economic development and potential growth of the city in general [9–11]. In this paper, the main attention is paid to the study of methods and tools, information technology for modeling the specialist training processes in smart cities.

2 Stages of Specialist Training Taking into Account the Inclinations and Abilities According to the Needs of Smart Cities

The process of qualified personnel training taking into account the person's inclinations and abilities and the requirements of the labor market in a smart city, is a complex, multistep, iterative process that requires consideration of a large number of parameters and prerequisites.

It can be expanded in five consecutive functional steps (Fig. 1).

Step 1. The determination of professional inclinations and abilities is based on the analysis of the accumulated results of the career orientation test by Holland [12].

Step 2. Monitoring of the labor market in order to identify trends in development and changes as well as job vacancies in the city. It is carried out on the basis of reporting informational analytical and statistical materials, which are formed by city employment centers.

Step 3. Choice of the future profession, carried out on the basis of the results of career orientation testing, taking into account the level of knowledge potential and the needs of the city by the qualified specialists of one or another profile.

Step 4. Choice of educational institution. Taking into account the recommendations on the choice of a profession, the selection of educational institutions operating in the city is carried out on the basis of available complete and consistent information about them.

Step 5. Formation of an individual learning trajectory (ILT) is a personal program of student's acquisition of professional competencies corresponding to the chosen specialty, taking into account abilities, interests, motivation, psychodynamic characteristics and level of knowledge potential.

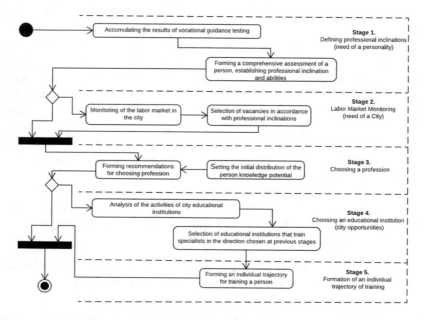

Fig. 1. "Activity diagram" of information and technological support of specialist training

3 Information Technologies for Modeling the Processes of Specialist Training in Smart Cities

3.1 Analysis of Career Orientation Testing with Explicit Uncertainty and Hidden Redundancy

The determination of professional inclinations and abilities of a person is carried out on the basis of analysis of the accumulated results for the career orientation test by Holland [12]. This test is a set of questions answering with the utmost frankness; the recipient undergoes objective testing of emotional and professional preferences.

In order to identify the general relationships that help to make decisions on a profession choice by a person undergoing testing, we suggest improved tools for analyzing the results of career orientation testing.

The model of data analysis process proposed in [13] is used to present the process of determining the person's professional inclinations and abilities in the following form:

$$M = (M_1, M_2, M_3, M_4),$$

where M_1 is a subprocess of a domain forming, M_2 is a subprocess of preliminary data processing, M_3 is a subprocess of dependency detection, M_4 is a subprocess of evaluation and interpretation of the analysis results.

Model of M_1 subprocess for forming the domain description is:

$$M_1 = (X, A, d, \mu(x, a), \eta(x, d)),$$

where X is a number of people involved in the test, A is a set of properties, d is the test results. Functions $\mu(x, a)$, $\eta(x, d)$ will be used to calculate the attribute values of the decision making table.

The set of properties A is divided into two subsets $A = \{A_1, A_2\}$ where A_1 is informative properties, A_2 is test results. For each object, we put the value set of the decision-making table.

The set A_1 of informative attributes consists of 5 attributes:

1. The attribute $a_1 =$ "Name".
2. The attribute $a_2 =$ "Year of birth" with the continuous domain $Va_2 \in N$.
3. The attribute $a_3 =$ "Education" is a domain, whose Va_3 is defined by the rule:

$$V_{a_3} = \begin{cases} 1, & \textit{if } a_3(x) =" \textit{ complete sec ondary}", \\ 2, & \textit{if } a_3(x) =" \textit{ incomplete sec ondary}", \\ 3, & \textit{if } a_3(x) =" \textit{ incomplete higher}", \\ 4, & \textit{if } a_3(x) =" \textit{ basic higher}", \\ 5, & \textit{if } a_3(x) =" \textit{ complete higher}" \end{cases}$$

This attribute splits $\bar{X} = \{X_1, X_2, X_3, X_4, X_5\}$ of the set of X objects where $X_i = \{x \mid x \in X \wedge p(x, a_5) = i\}$.

4. Attribute a4 = "Gender" is a domain, whose Va_4 is defined by the rule:

$$V_{a_4} = \begin{cases} 0, & \textit{if } a_4(x) =" \textit{ male}", \\ 1, & \textit{if } a_4(x) =" \textit{ female}" \end{cases}$$

This attribute splits $\bar{X} = \{X_1, X_2\}$ of the set of X objects where $X_i = \{x \mid x \in X \wedge p(x, a_2) = i - 1\}$.

5. The attribute a_s = "Age" with a continuous domain $Va_5 = [15, 45]$ is discretized by age-division and specified by the function:

$$\rho(x, a_5) = \begin{cases} 1, & \text{if } a_5(x) \in [15, 17], \\ 2, & \text{if } a_5(x) \in [18, 22], \\ 3, & \text{if } a_5(x) \in [23, 30], \\ 4, & \text{if } a_5(x) \in [30, 45], \end{cases}$$

This attribute splits $\bar{X} = \{X_1, X_2, X_3, X_4\}$ of the set of X objects, where $X_i = \{x | \eta(x, a_4) = i\}$.

The set A_2 of test attributes will divide into the following subsets $A_2 = \{A_{21}, A_{22}\}$:

1) A_{21} is "Choice of a profession", the set of which is given by the attribute values $A_{21} = \{a_1, a_1, a_1, a_1, \ldots, a_{42},\}$, where a_1 = "engineer mechanic", a_2 = "lawyer", a_3 = "cook", a_4 = "accountant", a_5 = "choreographer", a_6 = "teacher", a_7 = "speech therapist" and so on.

2) A_{22} is "Professional type", which we set with the attribute values $A_{22} = \{a_1, a_2, a_3, a_4, a_5, a_6,\}$, where a_1 = "realistic", a_2 = "intellectual", a_3 = "social", a_4 = "conventional", a_5 = "entrepreneurial", a_6 = "artistic".

The attributes of the subset A_{22} are the elements of the set d, thus, the decision-making table, created in the subprocess of the domain description, takes on the form:

$$T = (X, A_1 \cup \{d_1, d_2, d_3, d_4, d_5, d_6\},$$

The proposed data structure for attribute sets implies the presence of uncertainties and redundancies that will be used in subsequent analysis procedures.

The subprocess model of the previous data processing with the application of methods of approximate sets (*rough set*) has the form

$$M_2 = (T, Discr(age), EscC(a)),$$

where the $Discr(age)$ function performs a discretization of the continuous values of the age using Boolean reasoning, the $EscC(a)$ function eliminates the non-essential attributes by constructing the reducers using the Johnson algorithm.

The subprocess model for detecting data dependencies takes the form of:

$$M_3 = (T, S, Pat(x)),$$

where the $Pat(x)$ function builds a classifier in the form of classification rule set.

The model of subprocess of evaluation and interpretation is presented as:

$$M_4 = (T, Test(x), F(x), Evl(d)),$$

where the $Test(x)$ function creates a test set of HT objects, on which their values are estimated $F(x)$, $Evl(d)$ is an evaluation for the classification quality.

Detection of excess attributes during the analysis of the decision-making table helps to optimize the process of conducting career orientation tests to determine the person's professional characteristics.

3.2 Information Technology of Big Data for the Analysis of the Educational Institution Activities in the City

During the activity process educational institutions generate big data in the smart city, requiring prompt processing and reliable storage. The processing of such data is hampered by the huge variety of documents by their nature and methods of filing. For the rapid analysis and processing of big data streams, the traditional approaches developed in the coordinates of database concepts and data warehouses cannot be effectively used. The application of technology of big data is natural in these cases.

According to Clifford Lynch's definition, big data in information technology is a set of methods and tools for processing structured and unstructured multi-type, dynamic, big data in order to analyze and use them in decision-making processes [13].

The implementation of the stages of the qualified specialists training according to the person's inclinations and abilities and the requirements of the smart city implies:

- data storing and managing that amounts to tens and hundreds of terabytes. These are data describing the activities of educational institutions of different types and levels of accreditation (for example, there are 72 higher education institutions in Kiev, 21 higher vocational schools, 43 colleges, 260 schools, 33 high schools, 46 gymnasium schools; and there are 29 higher educational institutions in Lviv, 18 higher professional schools, 22 colleges, 99 schools, 17 lyceums, 19 gymnasium, etc.);
- processing of structured data, in particular the database of educational institutions, and unstructured data characterizing the urban labor market, the implementation of procedures for analyzing large amounts of statistical data;
- the analysis of diverse and varied information resources, which are formed on the basis of different sources, using the integration approaches in the form of federalization and consolidation of data.

The training process of qualified professionals in accordance with the person's inclinations and abilities and the requirements of a smart city include the stage "Choice of an educational institution", in the course of which, taking into account the chosen profession, the selection of educational institutions from the set operating in the city is carried out. The decision is made on the basis of complete and consistent information, in particular, about the basic characteristics of the respective educational institutions, their departmental affiliation, forms of ownership, information about the faculties, departments, teaching staff, curricula and educational programs, etc.

3.3   Data Warehouses of Educational Institutions in the Smart City

Information technology which helps to implement the procedures for analyzing and processing big data in the educational environment of smart city is a technology of data warehouses. Data warehouse (DW) is subject-oriented, integrated, time variant, non-destructive set of data, designed to support decision-making [15].

The data warehouse has a number of features that allow you to implement efficient storage and analytical processing of big data and, in many cases, are provided using relational and multidimensional data models [16].

Data warehouse model for comprehensive assessment of the educational institutions activities in the city is presented in the form:

$$SD_{NZ} = \langle DB, rf, RF, rm, RM, func \rangle$$

where *DB* is a set of relationships, their schemes and constraints, which contain information from input databases (databases of educational institutions), *RF* is a scheme of the set of fact relations *rf*, *RM* is a scheme of the set of metadata relations *rm*, *func* is a set of decision-making procedures.

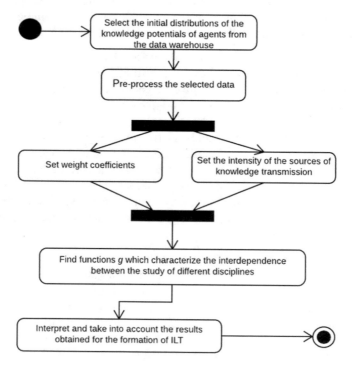

Fig. 2. Diagram of the multidimensional analysis process for functioning the educational institution in a smart city

The projected logical scheme of the multidimensional data warehouse has 104 entities and 785 attributes, which allows to display both detailed and aggregated information about users, activities of secondary, vocational, higher educational institutions, and institutions of postgraduate education, which operate in a particular city or a territorial community.

Getting new solutions involves extracting data from the warehouse by implementing relevant functions in relation to the facts, taking into account the requirements that are put forward for such decision. The set of requirements for the proposed solution depends on the user needs. A link between relations *rf* and *DB* forms a data hypercube, the measure is a set of relations in data warehouse of complex assessment of the educational institution activities of different types and levels of accreditation in the city. In Fig. 2 a diagram of the multidimensional analysis process of the functioning of educational institutions in the city is shown. Data hypercube is analyzed not in all measurements at the same time. Typically, a data sample from a hypercube is made for the specific values of a particular measurement set, and as a rule, one or two measurements are free, and using them the further analysis is carried out.

The process of analyzing data accumulated in the warehouse is considered to consist of four subprocesses, such as the information model formation of a domain, the preliminary data processing, the dependency detection, the evaluation and interpretation of results. Usage and technological realization of well-known methods and data analysis tools [17–22] makes it possible to investigate efficiently accumulated big data and to detect hidden dependencies on them.

3.4 Diffusion-Liked Model of Multicomponent Knowledge Potential Determination

The main factor that is taken into account while forming the individual learning trajectory of a person or an "agent" is to find the distribution of the corresponding knowledge potentials [23]. We believe that the knowledge potential (φ) is a sum of knowledge of an individual (an agent) accumulated during the corresponding life period.

In the case when each k agent is characterized by two knowledge potentials $\varphi_{l,k,m}(l = 1, 2)$. For example, k agent in the moment of time m is inherent with potentials that characterize knowledge of mathematics and language. A model describing the redistribution of these potentials with the possible consideration of an "impact" of one of them on another (obtaining a high knowledge potential in mathematics can affect the reduction of the knowledge potential from the language for the given object, or vice versa, positive interactions), will be presented in the form:

$$\begin{cases} \varphi_{1,k,m+1} = \varphi_{1,k,m} + \sum_{i=1}^{k_j} \alpha_{1,k,i,m}(\varphi_{1,k,m} - \varphi_{1,i,m}) + f_{1,m} + g_{1,m}(\varphi_{1,k,m}, \varphi_{2,k,m}) \\ \varphi_{2,k,m+1} = \varphi_{2,k,m} + \sum_{i=1}^{k_j} \alpha_{2,k,i,m}(\varphi_{2,k,m} - \varphi_{2,j,m}) + f_{2,m} + g_{2,m}(\varphi_{1,k,m}, \varphi_{2,k,m}) \end{cases}$$

where $f_{1,m}$, $f_{2,m}$ are intensities of the knowledge distribution sources, $g_{1,m}(\varphi_{1,k,m}, \varphi_{2,k,m})$, $g_{2,m}(\varphi_{1,k,m}, \varphi_{2,k,m})$ are functions characterizing the interdependence (mutual influence) of the mathematics and language studying in this case. In Fig. 3 (a) the calculation results are given in the case when an increase in the "mathematical component" α leads to an increase in the corresponding component of the knowledge potential of the given agent while preserving the "language component". And in Fig. 3 (b) the results are presented when an increase in the "language component" β leads to a decrease in the "mathematical component" (without affecting the "language").

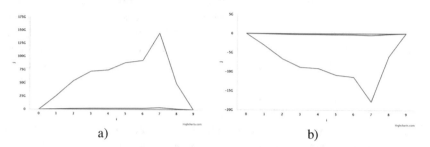

a)   b)

Fig. 3. Redistribution of two-component knowledge potential of an agent

Similarly, in the case when objects (agents) are characterized by many potentials, such as $l = 1, 2, \ldots, l_*$, we have:

$$\varphi_{l,k,m+1} = \varphi_{l,k,m} + \sum_{i=1}^{k_j} \alpha_{l,k,j,m}(\varphi_{l,k,m} - \varphi_{l,j,m}) + f_{l,k,m} + g_{l,k,m}(\varphi_{1,k,m}, \ldots, \varphi_{l_*,k,m})$$

where $\alpha_{l,k,j,m}$ are the coefficients of perception. Identifying such coefficients we can measure the knowledge potentials of all agents at different times, calculating the in tensity values of knowledge distribution sources $f_{l,k,m}$, functions $g_{l,k,m}$ that characterize the interactions, and solving the algebraic equation system:

$$\begin{cases} \sum_{j=1}^{k_j} a_{l,k,j,m}\alpha_{l,k,j,m} = b_{l,k,m} \\ l = \overline{1, l_*}, \quad k = \overline{1, k_*}, \quad m = 1, 2, 3, \ldots \end{cases},$$

where

$$a_{l,k,j,m} = \varphi_{l,k,m} + \varphi_{l,j,m}, b_{l,k,m} = \varphi_{l,k,m+1} + \varphi_{l,k,m} - f_{l,k,m} + g_{l,k,m}(\varphi_{1,k,m}, \ldots, \varphi_{l_*,k,m}),$$

k_* is a number of agents.

Thus, knowing these parameters and the characteristic values of the interactions of various knowledge components of agents, we can find the intensity of the knowledge distribution sources by the formula:

$$f_{l,k,m} = b_{*,l,k,m} - b^*_{l,k,m}$$

where

$$b_{*,l,k,m} = \varphi_{l,k,m+1} - \varphi_{l,k,m} - g_{l,k,m}(\varphi_{1,k,m}, \ldots, \varphi_{l_*,k,m}), b^*_{l,k,m} = \sum_{i=1}^{k_j} \alpha_{l,k,j,m}(\varphi_{l,k,m} - \varphi_{l,j,m}).$$

In general, to identify the parameters of the knowledge potential components, with the purpose of their further usage in creating individual learning trajectories of a person, it can be used the procedure of step-by-step fixations of certain parameters (block iterations).

4 Conclusion

The development of methods and means of information and technological support of the training specialist processes in a smart city involves designing a data warehouse, developing data mining methods and online analytical processing, as well as applying big data technology Data array characterizing the educational institution activities in the city is presented in the form: essence – characteristic – association. It is proposed to use interaction simulating (mutual influence) of the knowledge potential components of agents, as well as on the basis of the analysis of the proposed multi-component vector of knowledge potential proposed by the authors. The architecture of the software and algorithmic complex of information and technological support of the training specialist processes in the smart city has been developed and its program implementation has been implemented.

References

1. Bouskela, M., Casseb, M., Bassi, S., De Luca, C., Facchina, M.: The Road Toward Smart Cities: Migrating from Traditional City Management to the Smart City: Monograph. Inter-American Development Bank, Washington (2016)
2. Kupriyanovsky, V.P., Bulancha, S.A., Chernykh, K.Y., Namiot, D.E.: Smart cities as the "capitals" of the digital economy. Int. J. Open Inf. Technol. **2**, 41–52 (2016). (in Russian)
3. Boulton, A., Brunn, S.D., Devriendt, L.: Cyberinfrastructures and "smart" world cities: physical, human, and soft infrastructures. In: Taylor, P., Derudder, B., Hoyler, M., Witlox, F. (eds.) International Handbook of Globalization and World Cities. Edward Elgar, Cheltenham (2012)
4. Caragliu, A., Del Bo, C., Nijkamp, P.: Smart cities in Europe. J. Urban Technol. **18**(2), 65–82 (2011). https://doi.org/10.1080/10630732
5. Carey, K.: The End of College: Creating the Future of Learning and the University of Everywhere. Riverhead Books, New York (2015)
6. Will universities be responsible for the success of cities? https://www.ecampusnews.com/campus-administration/universities-smart-cities/
7. Plumb, D., Leverman, A., Gray, R.: The learning city in 'planet of slums'. Stud. Contin. Educ. **29**(1), 37–50 (2007)
8. Liu, D., Huang, R., Wosinski, M.: Smart Learning in Smart Cities. LNET. Springer, Singapore (2017). https://doi.org/10.1007/978-981-10-4343-7

9. Haidine, A., Aqqal, A., Ouahmane, H.: Evaluation of communications technologies for smart grid as part of smart cities. In: El Oualkadi, A., Choubani, F., El Moussati, A. (eds.) Proceedings of the Mediterranean Conference on Information and Communication Technologies 2015. Lecture Notes in Electrical Engineering, vol. 381, pp. 277–285. Springer, Cham (2015). https://doi.org/10.1007/978-3-319-30298-0_29

10. Concilio, G., Marsh, J., Molinari, F., Rizzo, F.: Human smart cities: a new vision for redesigning urban community and citizen's life. In: Skulimowski, A., Kacprzyk, J. (eds.) Knowledge, Information and Creativity Support Systems: Recent Trends, Advances and Solutions. Advances in Intelligent Systems and Computing, vol. 364, pp. 269–278. Springer, Cham (2016). https://doi.org/10.1007/978-3-319-19090-7_21

11. Scuotto, V., Ferraris, A., Bresciani, S.: Internet of things: applications and challenges in smart cities: a case study of IBM smart city projects. Bus. Process Manag. J. **22**(2), 357–367 (2016)

12. Holland, J.: Making Vocational Choices: A Theory of Careers. Prentice Hall, Upper Saddle River (1973)

13. Nikolskyi, Y.: Model of data analysis process. Comput. Sci. Inf. Technol. **663**, 108–116 (2010)

14. Clifford, L.: Big data: how do your data grow? Nature **455**, 28–29 (2008). https://doi.org/10.1038/455028a

15. Jacobs, A.: The pathologies of big data. Databases **7**(6), 1–12 (2009)

16. Inmon, W.: Corporate Information Factory, 3rd edn. Wiley, New York (2000)

17. Chai, H., Wu, G., Zhao, Y.: A document-based data warehousing approach for large scale data mining. In: Zu, Q., Hu, B., Elçi, A. (eds.) ICPCA/SWS 2012. LNCS, vol. 7719, pp. 69–81. Springer, Heidelberg (2013). https://doi.org/10.1007/978-3-642-37015-1_7

18. Oracle Data Mining. http://www.oracle.com/technetwork/database/options/advanced-analytics/odm/overview/index.html

19. Dilawar, M.U., Syed, F.A.: Mathematical modeling and analysis of network service failure in datacentre. Int. J. Mod. Educ. Comput. Sci. (IJMECS) **6**(6), 30–36 (2014). https://doi.org/10.5815/ijmecs.2014.06.04

20. Peleshko, D., Rak, T., Izonin, I.: Image superresolution via divergence matrix and automatic detection of crossover. Int. J. Intell. Syst. Appl. (IJISA) **8**(12), 1–8 (2016). https://doi.org/10.5815/ijisa.2016.12.01

21. Kotevski, Z., Tasevska, I.: Evaluating the potentials of educational systems to advance implementing multimedia technologies. Int. J. Mod. Educ. Comput. Sci. (IJMECS) **9**(1), 26–35 (2017). https://doi.org/10.5815/ijmecs.2017.01.03

22. Jian, G.: Reform of database course in police colleges based on working process. Int. J. Mod. Educ. Comput. Sci. (IJMECS) **9**(2), 41–46 (2017). https://doi.org/10.5815/ijmecs.2017.02.05

23. Bomba, A., Nazaruk, M., Kunanets, N., Pasichnyk, V.: Constructing the diffusion-liked model of biocomponent knowledge potential distribution. Int. J. Comput. **16**(2), 74–81 (2017). (in Ukrainian)

The Procedures for the Selection of Knowledge Representation Methods in the "Virtual University" Distance Learning System

Vasyl Kut[1], Nataliia Kunanets[2(✉)], Volodymyr Pasichnik[2],
and Valentyn Tomashevskyi[3]

[1] Information Technologies and Analysts Department, Carpathian University named after
Augustine Voloshin, Uzhhorod, Ukraine
kut.vasilij81@gmail.com
[2] Information Systems and Networks Department, Lviv Polytechnic National University,
Lviv, Ukraine
nek.lviv@gmail.com, vpasichnyk@gmail.com
[3] Automated Systems of Information Processing and Management Department,
National Technical University of Ukraine "Igor Sikorsky Polytechnic Institute", Kyiv, Ukraine
simtom@i.ua

Abstract. The advantages of usage and the main functions of the "Virtual University" distance learning system are analyzed, which provides opportunities for planning the processes for course developing, creating and accounting of arbitrary hierarchy of learning objects, accounting of learning outcomes as well as providing interactive communication (forums, graphical chats, virtual classes, trainings, video broadcasts, webinars, etc.). The basic models of the "Virtual University" system prototype as the educational web-based environment of distance learning are presented. The functional structure of the system and the architecture of software and algorithmic complex, which is implemented on the basis of the GPL-license of the developer tools, are disclosed. The most common ways of knowledge presentation are analyzed as well as its parametrization and expert evaluation of the basic characteristics are carried out. A hierarchy analysis method is used to select the method of presenting knowledge in the system of distance education. Our calculations showed that it is suitable to use the onto-logical representation of knowledge in the "Virtual University" distance learning system.

Keywords: Virtual University · Distance learning system
Hierarchy analysis method · Expert estimation method
Means of knowledge presentation

1 Introduction: General Statement of the Research Question

Modern approaches to the technology of organizing learning process are generated by the rapid development and improvement of computer tools and information science in general. The use of information technology improves the efficiency of knowledge

© Springer International Publishing AG, part of Springer Nature 2019
Z. Hu et al. (Eds.): ICCSEEA 2018, AISC 754, pp. 713–723, 2019.
https://doi.org/10.1007/978-3-319-91008-6_70

acquisition in the remote mode. Distance learning systems are significant component of such learning process that contains informational content from a variety of subject areas. This ensures the formation of virtual learning environments, which undoubtedly generates the need to choose the best ways to present knowledge and improve the processes of its formalization.

One of the main prerequisites for the creation of modern integrated educational Web environments is a rapid introduction of different levels and profiles of remote forms into the educational process which is a set of information technologies that provide the possibility of obtaining knowledge in the interactive mode of the teacher to the student. The problem of the systematic formation of integrated educational information resources (databases and knowledge) which make up the core of such systems, is among the main issues to be solved.

1.1   Literature Review and the Article's Purpose

Domestic and foreign scientists have made a significant contribution to the development of information technologies for learning and education, mainly Zhurovskyi [1], Haydar [2], Pohrebniuk [3], Robey et al. [4], Memon and Meyer [5], Rakhi [6], Haji et al.[7], Karpagam and Saradha [8]. A number of problems with the implementation of information technology in education is analyzed in the publication "Improving Learning with Information Technology: A Report of a Workshop" [9].

The aim of the paper is to substantiate the advantages of using and analyze the main functions of the "Virtual University" distance learning system and implement the procedure for choosing a way of presenting knowledge in the system of distance education using the method of pairwise comparisons.

2   The "Virtual University" – Basic Concept

A systemically formed integrated educational Web environment allows creating the effective mechanisms of development, implementation, accumulation and use of educational information resources in the learning process, which are necessary for lecturers and students of all types of programs at any institution and outside it.

An important element here is to provide support of integration (consolidation) of various distributed, information resources for efficient use by different groups of users employing a unified interface. Such capabilities are realized, in particular, in the developed prototype of modern system of distance learning "Virtual University" and its main functions are formation of conceptual models of subject environments, unification of knowledge resources for educational purposes, construction of effective schedules, planning course development processes, creation and maintenance of an arbitrary hierarchy of learning objects, accounting of learning outcomes, etc. The system is built using the Apache web server and is based on the web-based programming language PHP and the MySQL database management system. This software is freely distributed in accordance to the terms of the GPL license.

The system consists of the following main components:

- information support subsystem for an administrator (dean, head of the department, chair of the department, organization, etc.);
- information support subsystem for a teacher (tutor);
- information support subsystem for a student (employee, worker that is studying);
- communication subsystem (forums, webinars, e-mail, etc.);
- subsystem of a library.

All these allow you to manage from any level of the management hierarchy: a teacher, a head of the department, a dean of the faculty, etc.

2.1 Distance Adaptive Learning System DAOS

In order to implement the proposed concept of the "Virtual University", a distance adaptive learning system DAOS (Distance Adaptive Open System) was developed. It is characterized by multicomponence and is technologically convenient for frequent changes that implement dynamic development procedures. In the course of its development, the methodology of the system approach was used, which allowed taking into account dependencies that exist between the constituents of the system – subsystems and components, taking into account not only the current requirements and opportunities, but also the prospects for its development. Using this context, the following models of components for the distance adaptive learning system were identified: (1) a structural model; (2) a functional model; (3) an information and technological model; (4) an evolutionary model.

The structural model of the system is formed on the basis of the institution structure for which it is developed and planned to be used. The components of the structural model

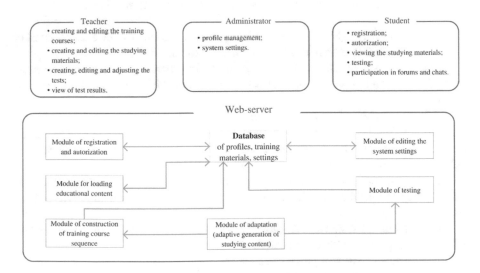

Fig. 1. Functional model of the distance adaptive learning system.

correspond to the units involved in the distance learning process. For example, an admission office, a program coordinator, departments.

The functional model is intended for analysis of the system features which are interconnected with the internal and external elements. The functional model DAOS allows you to submit system components according to their functional characteristics. For example, lectures, tests, practical tasks, etc. The functional model of the distance adaptive learning system is shown in Fig. 1.

The information and technological model reflects information flows in the system and technology used in the process of inspection and technical design of the system.

The evolutionary model reflects the system development processes in time as well as the prospects for its development in the future.

2.2   The Structure of the "Virtual University" Distance Learning System

In Fig. 2 the structure of the "Virtual University" distance learning system is submitted.

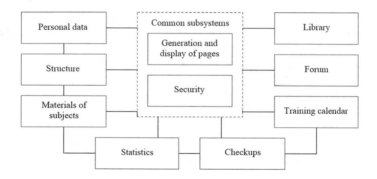

Fig. 2.   Structure of the "Virtual University" distance learning system

The subsystem that generates and displays pages is used to provide a user with interface features for navigation and actions defined by the use rights.

The Security subsystem is used during the login process, provides confidentiality during the registration process and alerts the user about third-party infiltration into the system.

The Personal data subsystem provides an administrator with personal data of users, registration of new users, removal of user records that have completed or dropped out from a training, user data editing, ensures users with access to registration information, such as login and password.

The Structure subsystem defines an organizational structure of an educational institution (or institution where a training takes place), as well as the structure of the training courses.

The Materials subsystem of subjects gives teachers an opportunity to form a teaching and methodological complex for a certain discipline.

The Statistics subsystem provides statistical information about educational process in the context of all user categories, such as:

- information about current student success (for a student);
- student's test results by subject, the number of attempts to pass a test, analysis of test results by section (for a teacher);
- training results for all students and reporting (for an administrator).

The Checkups subsystem provides processes for student knowledge testing and determines their type, list of questions, etc.

The Training calendar subsystem helps to control the timing of testing and to determine the groups of students who should be tested.

The Forum subsystem serves as an effective platform for a group discussion.

The Library subsystem provides various information such as electronic articles, textbooks, methodological materials, video files and facilitates in-depth study of the disciplines.

There are additional service function modules in the system such as Schedule Manager, Publisher, Search, and Analytical module.

The structure of the system for maintaining the educational web environment is presented in Fig. 3. It contains the Information and Controlling Core, Resource Management and Web Portal Subsystems as well as a user-friendly interface.

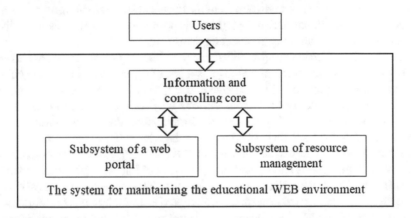

Fig. 3. The structure of the system for maintaining the educational web environment

The Information and Controlling Core is a basic element that implements system functionality and, above all, supports the processes of integrating distributed digital resources in a comprehensive educational Web environment. The Core is connected to the Resource Management Subsystem, Web Portal, and Users by means of the Application Program Interfaces (APIs). The fulfillment of these functions is carried out by the structural component of the Web Portal Subsystem which implements a significant part of the system's instrumental functions, in particular functions of component interoperability.

Web Portal acts as an independent structural element that provides integrated access to existing resources as well as electronic information resources that were created by means of a unified web interface. At the same time, this component also performs representative functions. Web Portal provides informational interaction of the support system of the educational web environment with educational institutions, organizations and other information systems.

The Resource Management subsystem is responsible for content managing the distance education system and maintenance of information manipulation processes (e.g. storage, reading, editing, copying, deletion, sorting, classification, searching, etc.), as well as their sharing and interaction with system users.

3 The Methods of Knowledge Representation in Distance Learning Systems

The hierarchically organized content is usually used in distance learning systems for teaching materials, consisting of modules, which in turn contain topics, chapters, subdivisions, etc. Since distance courses cannot exist independently and they are interconnected with other courses in a particular subject area, it is advisable to consider the set of distance courses. They are related to each other indirectly through certain concepts (definitions, theorems, algorithms, rules, key events and other support points). It should be realized the possibility of transition from the classical structure of the distance course to the so-called transitional, and in the future, it is possible to form not a separate distance course, but a model of an integral object environment.

In the process of creating a specialized distance learning system, you have to choose an effective way of presenting knowledge as a set of syntactic and semantic rules that provide a formal expression of knowledge about the subject area in an understandable form for both a computer and a person. Various ways of presenting knowledge can be used in the formation of such system. Logical models, product rules, frames, semantic networks, ontologies, agent technologies, and others are widespread.

Knowledge presentation using logical models is based on the construction of a formal system with unambiguous and logical interpretation of knowledge, given by four sets: the basic elements (alphabet) T; syntactic rules P, which ensure the accuracy of formula construction; axioms A, assertions that are accepted as true; the rules of output B. Preferably, the logical model of knowledge presentation is based on the formalities of the predicate logic of the first order [10].

The production model consists of a set of products (rules). "If (condition) is A, then (action) is B". The "condition" (antecedent) is formed as a search pattern in the form of a sentence-sample, which promotes an efficient search in the knowledge base, and "action" (consequent) is treated as a process performed after obtaining a relevant search result. Therefore, the implication can be interpreted as the logical flow B from the true A. In other words, the production model implies the presentation of knowledge through a plurality of products where (i) is a label of products by which it is identified in a plurality of products; Q is a domain; R is a condition of product use; $A =>$ is the core of products; interpretation of products depending on the situation ($=>$); N is the condition of the product, which postulates the procedures intended to be executed after the core implementation [11].

Network model (semantic network) of the knowledge representation is based on the graph fuming whose nodes correspond to certain concepts, entities, and arcs correspond to the relationships and connections between them. It is believed that network models $\langle I, C1, C2, \dots, Cn, \Gamma \rangle$ are formed from the set of I information units, which are connected with each other by certain types of connections $C1, C2, \dots, Cn$ [12], G, by means of which links are provided between them. The semantic network is presented as a directed graph where each node is a certain object (concept, core) of a certain domain, which is usually assigned with a specific label (name), and arcs denote the connections between them.

Knowledge presentation using frames is an alternative to productive and logical models and allows forming lineal hierarchies. A frame is a minimal possible description of a certain entity; further reduction of this description leads to the loss of this entity [13] and receives an identifier which is unique for the entire frame system. In other words, the frame is an abstract image of the representation of an object, a concept or a situation. The frame has a certain internal structure consisting of a plurality of elements. The main element of the frame is a slot that is used to store unit knowledge and also gets a name. Slots are formed from spats, which contain data that is the current values of the slots.

Formally, an object using a frame model is described as follows:

Frame name
((Attribute_1, value_1),
(Attribute_2, value_2)
...
(Attribute_n, value_n)).

The hierarchy of concepts is typical for these models.

Agent technologies support a knowledge model about the subject area and are a part of the hybrid intelligence system [14]. Such knowledge can be used to construct a set of texts. The base model, based on agent technologies, is constructed as an interpretation of the perceptual cycle model.

An ontological presentation of knowledge, in fact, is a knowledge base that describes facts that are always believed to be true within a particular community using generally accepted meaning in the dictionary. The concept of ontology defines the doctrine of being, the essence, in contrast to epistemology – the doctrine of knowledge [15]. Further, we will assume that ontologies are knowledge bases of a special type that can be "read" and understood, alienated from the developer or physically shared by their users.

3.1 The Hierarchy Analysis Method in the Selection Procedures for Knowledge Presentation Methods

The choice of the knowledge presentation method in the distance education system will be carried out using the hierarchy analysis method. To do this, we will construct the matrices of pairwise comparisons of the above-described methods of knowledge representation. The data for the matrices is obtained through their expert evaluation.

Each matrix consists of expert assessments of alternative pairs, which are the ways of knowledge presentation in the distance education system. The types of relations, the method of obtaining value, the inheritance of properties, the complexity of assessing the integral knowledge and the efficiency of knowledge processing are chosen as the criteria.

We form the set of criteria based on the results of the analysis of the main characteristics for the chosen methods of knowledge presentation and choice will be made with their help.

For matrices of pairwise comparisons, the following parameters are calculated:

- estimation of the largest eigenvalue calculated by the formula [16] $\lambda_{max} = \sum_{i=1}^{n} w_i s_i$, where w_i is the weight of an alternative with the number i; s_i is the sum of the column elements with the number of the matrix of pairwise comparisons; n is the number of alternatives;

- index of coherence $CI = \dfrac{\lambda_{max} - n}{n - 1}$;

- index of the sequence of relations $CR = \dfrac{CI}{RI}$.

Here $RI = 1,24$ is a random index for $n = 6$, the value of which is the same for all subsequent calculations of the weight of alternatives [17].

For example, we take the results of calculating the weight of alternatives to the methods of knowledge representation (logical models, productive models, frames, semantic networks, ontologies, and agent technologies) by a matrix of pairwise comparisons constructed for the criterion of "types of relations". The matrix of pairwise comparisons for this criterion is given in Table 1.

Table 1. Matrix of pairwise comparisons of alternatives by criterion "types of relations"

Ways	Logical models	Ontologies	Production models	Frames	Semantic networks	Agent technology
Logical models	1.00	0.33	5.00	3.00	7.00	3.00
Ontologies	3.00	1.00	7.00	3.00	9.00	5.00
Production models	0.20	0.14	1.00	0.33	1.00	0.20
Frames	0.33	0.33	3.00	1.00	5.00	1.00
Semantic networks	0.14	0.11	1.00	0.20	1.00	0.14
Agent technology	0.33	0.20	5.00	1.00	7.00	1.00
Sum	5.01	2.12	22.00	8.53	30.00	10.34

The results of these calculations are given in Table 2.

Table 2. Weight alternatives by criterion "types of relations"

Alternatives	Logical models	Ontologies	Production models	Frames	Semantic networks	Agent technology	Sum	Weight alternatives
Logical models	0.200	0.157	0.227	0.352	0.233	0.290	1.459	0.2432
Ontologies	0.599	0.472	0.318	0.352	0.300	0.483	2.524	0.4206
Production models	0.040	0.067	0.045	0.039	0.033	0.019	0.244	0.0407
Frames	0.067	0.157	0.136	0.117	0.167	0.097	0.741	0.1234
Semantic networks	0.029	0.052	0.045	0.023	0.033	0.014	0.197	0.0328
Agent technology	0.067	0.094	0.227	0.117	0.233	0.097	0.835	0.1392

Based on the results of three computational experiments with different constraints, the weight of alternatives is received. In each of the experiments, the ontological approach has the greatest weight, but when the area is expanded in which we search the minimum, its value increases.

So, we come to the conclusion that it is best to use a method based on ontologies to present knowledge in the distance learning systems.

The process of further construction of ontology is organized in the form of learning based on texts from a given subject area, which are organized by the increase in processing complexity. Using this approach, the selection procedures were modeled at the Information Systems and Networks Department of Lviv Polytechnic National University, that are:

- methods of knowledge presentation in the system of library services for people with special needs, in particular, the formation of information resources for users with hearing impairments, vision and locomotive;
- ways of presenting educational content in the information system of the regional educational and consulting center for pupils with special needs in the Transcarpathian region;
- better "cloud manager" for effective storage and use of informational content in the university library.

A significant advantage of the ontological way of knowledge presentation is that it is flexible enough. It adapts easily to new changes that inevitably appear in the process of information service. It works in on-line mode and provides the possibility to obtain adequate data.

Thanks to the presentation of the learning material through the ontological approach, it is possible to organize effective testing and determine the student's knowledge of any distance course or student knowledge of previous distance courses. For this purpose, a matrix of knowledge is constructed (Fig. 4), which contains an assessment of students' knowledge of each topic. It enables to diagnose student's knowledge, determine their readiness to study a particular course, identify "gaps" in student's knowledge and provide them with an opportunity to study the required block of information.

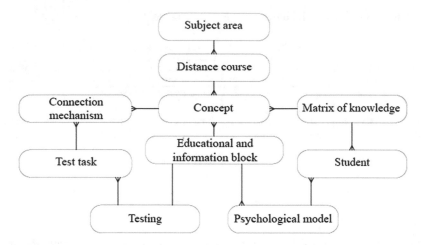

Fig. 4. The general structure of the teaching and information subsystem of the distance learning system

Unlike classic classroom teaching, distance learning allows you to have classes in the most convenient time for the participants of the educational process.

4   Conclusions

The introduction of the "Virtual University" distance learning system has shown the high efficiency of the development, which provides improvements in the mechanism of knowledge control and the average level of student success. Using the system of distance learning in educational institutions and training centers of skill development has allowed reducing the costs of maintaining the system. By creating an entire web-based information resource of learning materials and by defining its structure and component functions we unite both the means of access to learning resources and the resources themselves, and create standard mechanisms for the development, implementation, accumulation and use of information materials in the learning process.

Due to the use of ontological approach and the automated mode of forming the parameters of distance learning, it is possible to ensure the transparency and openness of this process, thereby providing a high-quality performance of tasks, and vastly improving the quality of knowledge acquisition processes.

The use of the hierarchy analysis method allowed the expert assessment procedure to be implemented, which contributed to the effective choice of the method for knowledge presenting in the distance learning system. The results of the research showed that the most convenient way of knowledge presenting in distance learning systems is an ontological approach.

References

1. Zhurovskii, M.: Bologna process: the main principles and ways of structural reform of higher education in Ukraine, Kyiv, p. 544 (2006)
2. Haydar, A.: Information technology and the learning environment in primary schools. Procedia Soc. Behav. Sci. **93**, 695–698 (2013)
3. Pohrebniuk, I.: Model of knowledge estimation at adaptive testing. In: Mathematical and Simulation Modeling Systems: Proceedings of the Sixth Scientific-Practical Conference with International Participation, Chernihiv, pp. 378–382 (2011)
4. Robey, D., Boudreau, M.-C., MRose, G.: Information technology and organizational learning: a review and assessment of research. Account. Manag. Inf. Technol. 10(2), 125–155 (2000)
5. Memon, A.B., Meyer, K.: Why we need dedicated web-based collaboration platforms for inter-organizational connectivity? A research synthesis. Int. J. Inf. Technol. Comput. Sci. **9**(11), 1–11 (2017)
6. Rakhi, G.L.P.: A reliable solution to load balancing with trust based authentication enhanced by virtual machines. Int. J. Inf. Technol. Comput. Sci. (IJITCS) **9**(11), 64–71 (2017). https://doi.org/10.5815/ijitcs.2017.11.07
7. El Haji, E., Azmani, A., El Harzli, M.: Using AHP method for educational and vocational guidance. Int. J. Inf. Technol. Comput. Sci. (IJITCS) **9**(1), 9–17 (2017). https://doi.org/10.5815/ijitcs.2017.01.02
8. Karpagam, K., Saradha, A.: A mobile based intelligent question answering system for education domain. Int. J. Inf. Eng. Electron. Bus. (IJIEEB) **10**(1), 16–23 (2018). https://doi.org/10.5815/ijieeb.2018.01.03
9. Pritchard, G.E.: Improving Learning with Information Technology: Report of a Workshop, p. 84. National Academy Press, Washington (2002)
10. Rogers, R.L.: Mathematical Logic and Formalized Theories: A Survey of Basic Concepts and Results, p. 248. Elsevier (2014)
11. Cadima, E.L: Fish Stock Assessment Manual Rome, p. 161 (2003)
12. Rashid, P.Q.: Semantic network and frame knowledge representation formalisms in artificial intelligence. Submitted to the Institute of Graduate Studies and Research in partial fulfillment of the requirements for the Degree of Master of Science in Applied Mathematics and Computer Science, p. 51 (2015)
13. Antscherl, D.: The Fully Framed Model, HMN Swan Class Sloops 1767-1780, p. 301. Pier Books, New York (2007)
14. Luck, M., McBurney, P., Preist, C.: Agent Technology: Enabling Next Generation Computing: A Roadmap for Agent Based Computing, p. 94. Southampton (2003)
15. Laclavík, M.: Ontology and Agent based Approach for Knowledge Management, Bratyslava (2005)
16. Saaty, T.: Multycriteric Decision Making: The Analytic Hierarchy Process, 300 p. McGraw Hill International, New York (1980)
17. Zgurovsky, M.Z., Pavlov, A.A., Shtankevich, A.S.: Modified hierarchy analysis method. Syst. Res. Inf. Technol. **1**, 7–25 (2010)

Perceptual Computer for Grading Mathematics Tests within Bilingual Education Program

Dan Tavrov$^{(\boxtimes)}$ [ID], Liudmyla Kovalchuk-Khymiuk [ID],
Olena Temnikova [ID], and Nazar-Mykola Kaminskyi [ID]

National Technical University of Ukraine "Igor Sikorsky Kyiv Polytechnic
Institute", 37 Peremohy ave., Kyiv 03056, Ukraine
dan.tavrov@i.ua

Abstract. In this paper, we propose an outline of a perceptual computer for grading mathematical tests written by students studying within the bilingual education program. A generic approach to implementing such a computer is proposed. Concrete implementation is described for the case of teaching mathematics in French. The perceptual computer constructed for this case is tested with real tests written by students of one of Kyiv bilingual schools. Results show that the grades obtained using words are compatible with the grades assigned using conventional numbers, which validates the use of the perceptual computer to reduce subjectivity and uncertainty for a teacher.

Keywords: Perceptual computing · Bilingual education · Type-2 fuzzy set

1 Introduction

There are many definitions of *bilingual education*. The one that is used in this paper states that it is [1] "use of two or more languages as media of instruction in subjects other than the languages themselves." In general, bilingualism can assume two different meanings [2]:

- a bilingual person as an individual who is proficient in more than one language. This approach includes people who are fluent in more than one language but don't actively participate in any multilingual community;
- a bilingual person as someone who participates in multiple language communities. We adhere to this approach in this paper.

Bilingual education and issues that arise with its implementing has been covered in the literature before. E.g., in [3], problems of bilingual education in computer science are addressed. In [4], the task of creating a Myanmar-English bilingual lexicon is considered. In this paper, however, we consider teaching mathematics in French and Ukrainian as the primary subject. Teaching mathematics in bilingual environment has special features, because bilingual students typically follow two common practices [2], namely, switching languages during arithmetic computation and switching languages during conversation (code-switching). This means that the student's excellence hinges on two major competences, mathematical and linguistic one. Following the findings of [5], we split these two competences down into six basic competences that a student must possess:

- *mathematical competences*: ability to explain steps of solution of a problem in a foreign language (competence 1), ability to explain steps of solution of a problem in a native language (competence 2), ability to solve a problem (competence 3);
- *linguistic competences*: understanding of a problem definition (competence 4), ability to explain a problem in a foreign language (competence 5), ability to explain a problem in a native language (competence 6).

The task of assessing levels of competences possessed by each student, let alone defining her final grade, is a highly subjective task that involves not only assessment of correctness of the mathematical solution, but also assessment of language proficiency. The latter is complicated by the fact that [2] such assessment should consider not only proficiency in language per se, but proficiency for using the language to talk or write about particular mathematical topics. This means that it is impossible to come up with a set of tasks for a test that assess each of the above competences separately. Each task is bound to touch upon several, if not all, of the competences at once. Therefore, the task of grading mathematics tests within the framework of bilingual education is challenging and requires building models that can reduce the level of subjectivity.

Some models involving fuzzy mathematics for evaluating students' performance have been proposed in the literature [6, 7]. In this paper, we go one step further and propose to grade mathematics tests using words only.

The goal of this paper is to create a method for grading mathematics tests within the framework of bilingual education that allows the teacher to grade each task using *words*, not exact numbers, and aggregate these words into the final grade by taking into consideration the relative importance of each task.

2 Perceptual Computing Basics

2.1 General Architecture of a Perceptual Computer

Perceptual computing is a type of *computing with words* (CWW) [8], i.e. a methodology, in which objects of computations are words and propositions expressed in a natural language. This basic definition of CWW is very general, so in different contexts, different versions of CWW are used. Perceptual computing is CWW used for making subjective judgments [9], which is the kind of problem outlined in Sect. 1. Therefore, in this paper, we adhere to methodology of perceptual computing for solving the problem of grading math tests.

Any perceptual computer (Per-C) takes *words* as inputs and produces other *words* as outputs, making it possible for a human to interact with the Per-C using a *vocabulary*, i.e. a set of words modeled as *type-2 fuzzy sets* (T2FS). Words are processed in a Per-C using the following three components [10]:

- *encoder*, which transforms words into their T2FS representation. A vocabulary of words with associated T2FS representations is called a *codebook*;
- *CWW engine*, which processes these T2FSs and outputs one or more other T2FSs;
- *decoder*, which maps the output of the CWW engine to a word from the codebook.

2.2   Encoder and Codebook

In Per-C, words are modeled using type-2 fuzzy sets. A *fuzzy set* (*type-1 fuzzy set*) A is a subset of a universal set X that is characterized by a *membership function* $\mu_A(x) : X \rightarrow [0, 1]$ that associates with each value $x \in X$ a *degree of membership* of x in A. A *type-2 fuzzy set* $\tilde{A}$ is characterized by a *type-2 membership function* $\mu_{\tilde{A}}(x, u) : X \times J_x \rightarrow [0, 1]$, where x is called the *primary variable*, u is called the *secondary variable* that has domain $J_x \subseteq [0, 1]$ at each x. This domain is called the *primary membership* of x. The value of $\mu_{\tilde{A}}(x, u)$ is called the *secondary grade* of $\tilde{A}$. For each x, $\mu_{\tilde{A}}(x)$ is the *secondary membership function* of $\tilde{A}$, i.e., for each x, the degree of its membership in $\tilde{A}$ is by itself a type-1 fuzzy set.

In practice, the most widely used type-2 fuzzy sets are *interval type-2 fuzzy sets*, for which secondary grades are $\mu_{\tilde{A}}(x, u) = 1 \; \forall x \in X \; \forall u \in J_x$, and the primary membership is an interval $J_x = \left[\underline{\mu}_{\tilde{A}}(x), \overline{\mu}_{\tilde{A}}(x) \right]$, where $\underline{\mu}_{\tilde{A}}(x)$ is called the *lower membership function* (LMF) of $\tilde{A}$, and $\overline{\mu}_{\tilde{A}}(x)$ is called the *upper membership function* (UMF) of $\tilde{A}$. LMF and UMF bound what is called a *footprint of uncertainty* $FOU(\tilde{A}) = \bigcup_{\forall x \in X} J_x$ of a T2FS, which represents uncertainty about membership grades of x's in $\tilde{A}$. We will use the terms interval T2FS and T2FS interchangeably.

Using T2FSs as models for words in the codebook is justified because [11] a type-1 fuzzy set has a well-defined membership function that is totally certain once all of its parameters are specified. On the other hand, words mean different things to different people, and so are uncertain. Therefore, T2FSs are better suited to model uncertainties involved in interpretation of words.

In this paper, following the advice in [10], the words are represented using *trapezoidal T2FS*, for which LMF and UMF are trapezoidal membership functions:

$$\mu_{trap}(x; a, b, c, d, h) = \begin{cases} (x - a)/(b - a), & a \leq x \leq b \\ h, & b \leq x \leq c \\ (d - x)/(d - c), & c \leq x \leq d \\ 0, & \text{otherwise} \end{cases} \tag{1}$$

Graphical illustration of (1) is given in Fig. 1, where $\underline{\mu}_{\tilde{A}}(x) = \mu_{trap}(x; e, f, g, i, h)$, $\overline{\mu}_{\tilde{A}}(x) = \mu_{trap}(x; a, b, c, d, 1)$. The shaded area represents the FOU for this set.

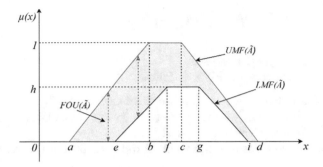

Fig. 1. Trapezoidal interval type-2 fuzzy set

2.3   Linguistic Weighted Average as a CWW Engine

In the literature, several kinds of CWW engines are introduced, e.g. if-then rules or *linguistic weighted averages* (LWA). In this paper, we will use the latter one.

In the most general case, LWA is expressed as

$$Y_{LWA} = \frac{\sum_{i=1}^{n} X_i W_i}{\sum_{i=1}^{n} W_i} \tag{2}$$

where each criterion to be weighted X_i and each weight W_i are T2FSs. To calculate (2), one needs to proceed along the following lines.

First, consider *interval weighted average* (IWA), which is (2) where all criteria and weights are intervals $[a_i, b_i]$ and $[c_i, d_i]$, respectively ($i = 1, \ldots, n$). Then, IWA can be expressed as $Y_{IWA} = [y_l, y_r]$, where

$$y_l = \frac{\sum_{i=1}^{L} a_i d_i + \sum_{i=L+1}^{n} a_i c_i}{\sum_{i=1}^{L} d_i + \sum_{i=L+1}^{n} c_i}, \qquad y_r = \frac{\sum_{i=1}^{R} b_i c_i + \sum_{i=R+1}^{n} b_i d_i}{\sum_{i=1}^{R} c_i + \sum_{i=R+1}^{n} d_i} \tag{3}$$

where L and R are *switch points* that can be found, e.g., using the enhanced Karnik-Mendel algorithms (EKM) described in [12].

If in (2), criteria and weights are modeled using type-1 fuzzy sets, we obtain a *fuzzy weighted average* (FWA), which can be computed using the α-*cut decomposition theorem* [13], according to which each fuzzy set A can be represented as $A = \bigcup_{\alpha \in [0,1]} {}_{\alpha}A$, where $\mu_{{}_{\alpha}A}(x) = \alpha \cdot \mu_{{}_{\alpha}A}(x)$, ${}^{\alpha}A = \{x | \mu_A(x) \geq \alpha\}$ is the α-*cut* of A.

One fundamental principle in fuzzy mathematics is *extension principle*, which states that any function $f : X \to Y$ induces a function $f : F(X) \to F(Y)$, where $F(X)$ is the set of all fuzzy subsets of X, such that $\mu_{f(A)}(y) = \sup_{x|y=f(x)} \mu_A(x)$. Applying this principle to α-cut decomposition, one gets *functional α-cut decomposition theorem* [14], according to which,

$$\alpha f(X_1, \ldots, X_k) = f(\alpha X_1, \ldots, \alpha X_k) \tag{4}$$

where $X_1, \ldots, X_k$ are type-1 fuzzy sets.

Using (4), we can compute FWA as follows:

1. Compute, for each $\alpha \in \{\alpha_1, \ldots, \alpha_m\}$, α-cuts of each X_i and W_i.
2. For each $\alpha \in \{\alpha_1, \ldots, \alpha_m\}$, compute $\alpha Y_{FWA} = [\alpha y_l, \alpha y_r]$ as an IWA of appropriate α-cuts, using (3) in application to αX_i and αW_i.
3. Connect all left coordinates $(\alpha y_l, \alpha)$ and all right coordinates $(\alpha y_r, \alpha)$ to get Y_{FWA}.

In practice, for trapezoidal T2FS, $m = 2$ is sufficient, because of piecewise linear nature of the membership function.

2.4   Decoder in the Perceptual Computer

The output of the Per-C can have several forms [10]: a word from a codebook most similar to the output of the CWW engine, a rank of several competing alternatives, or a class, into which the output of the CWW engine should be mapped. As mentioned before, in this paper, the output of the Per-C is one of the four classes ("A," "B," "C," or "D") representing overall performance of a given student in a given math test. Therefore, the decoder used in this paper is a classifier based on a Vlachos and Sergiadis's IT2 FS subsethood measure [15], which can be written as [10]

$$ss_{VS}\left(\tilde{A}, \tilde{B}\right) = \frac{\sum_{i=1}^{N} \min\left(\underline{\mu}_{\tilde{A}}(x_i), \underline{\mu}_{\tilde{B}}(x_i)\right) + \sum_{i=1}^{N} \min\left(\overline{\mu}_{\tilde{A}}(x_i), \overline{\mu}_{\tilde{B}}(x_i)\right)}{\sum_{i=1}^{N} \underline{\mu}_{\tilde{A}}(x_i) + \sum_{i=1}^{N} \overline{\mu}_{\tilde{A}}(x_i)}$$

where x_i $(i = 1, \ldots, N)$ are equally spaced points in X.

2.5   Decision Making in a Perceptual Computer for Grading Math Tests

In Fig. 2, the decision making process in the Per-C for grading math tests is given. It is hierarchical and distributed, i.e. it is [10] a "decision making [...] made by a single individual [...] based on aggregating independently made recommendations about an object from other individuals, groups or organizations." For our task, by independent recommendations we understand grades received by a student for completing a number of tasks, Task_1 to Task_N. These grades are expressed in terms of words from the codebook. Let us call this part of the codebook *grade subcodebook*.

There are three hierarchical levels in the Per-C, which involve computing:

- level of each competence, obtained as an LWA of student's grades for each task;
- level of two major competences, mathematical and linguistic ones, obtained as an LWA of student's levels of individual competences;
- overall grade, obtained as an LWA of student's level of two major competences.

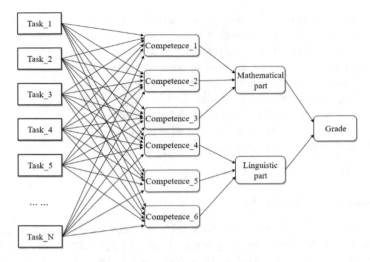

Fig. 2. Hierarchical and distributed decision making in the perceptual computer for grading math tests

Weights for each of the above LWAs are expressed as the words from the codebook and should be assigned by an expert to each connecting arc in Fig. 2 before using the Per-C. Let us call this part of the codebook *weight subcodebook*.

3   Application of the Perceptual Computing

3.1   Structure of a Sample Test

For validating the Per-C with the above codebook and decision making process, we decided to consider bilingual teaching math in French and Ukrainian. The following test was used (translations of original French task formulations are given in brackets):

1. Vrai ou faux (donner l'exemple) [True or False (show an example)] *(2 points)*:
   a. Si la somme de deux nombres relatifs est positive, alors les deux nombres sont positifs [If the sum of two integers is positive, then both numbers are positive]
   b. Si deux nombres non nuls ont des signes contraires, alors ils sont opposes [If two non-zero numbers are of opposite signs, then they are opposite]
2. Complète avec les mots convenables [Complete with correct words] *(2 points)*:
   a. Le nombre 0 est à la fois ___ et ___ [Zero is ___ and ___ at the same time]
   b. Cas de deux nombres négatifs le plus petit celui qui a la ___ ___ ___
   c. [Out of two negative numbers, the lowest is the one that has ___ ___ ___]
3. Calcule [Calculate] *(2 points)*:

   a. $\frac{-5,4\cdot3,9\cdot(-0,02)}{0,42\cdot(-0,18)\cdot(-2,6)}$ =   b. $-1\frac{2}{9} : (-0,25 \cdot 1\frac{2}{9})$ =

4. Mettre les parenthèses oubliées pour corriger l'égalité [Insert parentheses to obtain an identity] *(1 point)*: $D = 5 + 2 \times 3 + 2 + 4 - 1 = 18$

5. Connaissant la valeur du produit ab, calculer l'expression A
   [Given ab, calculate A] *(1 point)*: $A = 4 \times (-a) \times 2 \times b$, $a = 0,5$
6. Calculer pour 1,5 et −9 [Calculate for 1.5 and −9] *(2 points)*:
   a. la somme de leurs opposes [sum of their opposites]
   b. la somme de leurs inverses [sum of their inverses]
   c. l'opposé de leur somme [opposite of their sum]
   d. l'inverse de leur produit [inverse of their product]
   e. le produit de leurs inverses [product of their inverses]
7. Résous [Solve the equations] *(2 points)*:

   a. $|3x - 1| = 5$ b. $0,02 + 0,5y = -0,18$

3.2 Constructing the Codebook

According to 2.5, the codebook in the Per-C presented here consists of grade and weight subcodebooks. Words in the grade subcodebook are defined on the continuous scale from 1 to 12 (the standard grading scale in Ukrainian secondary educational institutions), whereas words in the weight subcodebook are defined on the interval [0,1], which is sufficient because rescaling the weights doesn't influence the LWA.

For both subcodebooks, words were selected that can cover the whole scale and are numerous enough to give the user of the Per-C flexibility in choosing linguistic grades and weights. As such, the grade subcodebook initially included 10 linguistic grades, *perfect* (P), *excellent* (EX) *very good* (VG), *good* (G), *sufficient* (S), *satisfactorily* (SA), *bad* (BA), *unsatisfactorily* (U), *very bad* (VB), *awful* (AW). The weight subcodebook initially included 5 linguistic weights, *not influential* (NI), *weakly influential* (WI), *moderately influential* (MI), *influential* (I), and *highly influential* (HI).

T2FS models for each word were created using the *interval approach* (IA) [16]. Its idea is to collect numeric intervals from a group of experts that, in their opinion, correspond to each word, and to aggregate answers into one T2FS for each word. Due to space limitations, we will not describe the algorithm of IA. For each word, we asked 20 experts (professional teachers in Secondary bilingual school #20 of Kyiv, Ukraine, as well as faculty of the Applied Mathematics Department of the Igor Sikorsky Kyiv Polytechnic Institute) to provide intervals for each word. Results of processing these intervals using software from [17] are given in Table 1 up to two significant digits. For each word, means (m) and standard deviations (σ) of left (l) and right (r) ends of the intervals are given, along with the final model in accordance with Fig. 1.

After having obtained results from Table 1, it was decided to discard words EX and U, because they are redundant (either subsume or are subsumed by other words). Therefore, the final codebook consists of 8 grade words and 5 weight words.

3.3 Assignment of Weights

In Table 2, weights are shown that correspond to arcs in Fig. 2 that connect each task with each competence (represent relative importance of each task for assessing level of each competence). In Table 3, weights are shown that correspond to other arcs in Fig. 2 (represent relative importance of each competence). Weights in both tables were assigned by the second author based on her expertise as a bilingual teacher.

Table 1. Interval statistics for words from the codebook, and their type-2 fuzzy set models

Word	Interval Statistics				Parameters of Trapezoidal T2FS								
	m_l	σ_l	m_r	σ_r	a	b	c	d	e	f	g	i	h
P	10.88	0.35	12.00	0.00	9.37	11.82	12.00	12.00	10.68	11.91	12.00	12.00	1.00
EX	9.77	0.60	11.46	0.52	7.05	10.72	12.00	12.00	9.68	11.83	12.00	12.00	1.00
VG	8.80	0.42	10.20	0.42	7.59	9.00	10.00	11.41	8.79	9.50	9.50	10.21	0.59
G	7.75	0.46	9.25	0.46	6.59	8.00	9.00	10.41	7.79	8.50	8.50	9.21	0.59
S	6.67	0.52	8.33	0.53	5.38	7.00	8.00	9.62	6.79	7.50	7.50	8.21	0.59
SA	4.63	0.52	6.00	0.00	3.59	5.00	5.50	6.41	4.79	5.32	5.32	6.21	0.74
BA	2.67	0.52	4.25	0.61	1.59	3.00	4.00	5.41	2.90	3.35	3.35	3.60	0.41
U	2.44	0.98	4.25	0.46	0.38	2.50	4.25	5.41	3.19	3.58	3.58	4.21	0.37
VB	1.63	0.74	3.38	0.52	0.59	2.00	3.00	4.41	1.79	2.50	2.50	3.21	0.59
AW	0.83	0.41	2.17	0.41	0.59	1.50	2.00	3.41	0.79	1.68	1.68	2.21	0.74
NI	0.00	0.00	0.26	0.13	0.00	0.00	0.05	0.66	0.00	0.00	0.02	0.26	1.00
WI	0.07	0.04	0.27	0.07	0.01	0.15	0.25	0.46	0.08	0.18	0.18	0.22	0.59
MI	0.31	0.08	0.56	0.05	0.12	0.38	0.50	0.68	0.38	0.44	0.44	0.52	0.53
I	0.58	0.06	0.84	0.09	0.45	0.63	0.75	0.96	0.62	0.70	0.70	0.80	0.59
HI	0.73	0.08	1.00	0.01	0.47	0.96	1.00	1.00	0.74	0.98	1.00	1.00	1.00

Table 2. Weights representing relative importance of each task for assessing level of each competence

Task #	Competences					
	1	2	3	4	5	6
1	HI	MI	I	HI	I	NI
2	MI	MI	I	HI	I	WI
3	WI	MI	I	HI	NI	MI
4	WI	MI	I	I	NI	MI
5	MI	MI	I	I	NI	WI
6	I	MI	HI	HI	MI	WI
7	MI	MI	HI	HI	WI	WI

Table 3. Weights representing relative importance of each competence

Competence	1	2	3	4	5	6	Math.	Ling.
Weight	I	MI	HI	HI	I	MI	HI	I

3.4   Comparison of Perceptual Computer and Conventional Grading

Table 4 shows the results of grading a set of tests of the above structure written by students of the 7th grade of the Secondary bilingual school #20 of Kyiv, where the second author of this paper teaches math in French. Each test was graded in two modes:

Table 4. Grades for the test written by students of one of the Kyiv bilingual schools. Grades that don't match are marked with an asterisk

#	Point Grade for Each Task								Word Grade for Each Task							
	1	2	3	4	5	6	7	Tot.	1	2	3	4	5	6	7	Tot.
1	1	1	0.75	1	1	2	2	9	S	S	BA	VG	VG	VG	P	B
2	2	2	0.75	1	1	2	1.75	11*	P	VG	BA	VG	VG	VG	VG	B*
3	2	2	1.75	1	1	0.5	1.5	10*	VG	VG	G	VG	VG	BA	G	B*
4	0	0	0	1	0	0.75	1.5	3*	BA	BA	VB	VG	VB	SA	G	C*
5	1	1	2	0	0	1	1.5	7*	S	SA	VG	BA	VB	SA	G	C*
6	0	1	2	1	0.75	0.75	2	8	BA	SA	VG	VG	S	SA	P	B
7	1	1	1.75	1	1	0.25	2	8	S	SA	G	VG	VG	BA	P	B
8	2	1	1	1	1	0.25	0.75	8	VG	SA	S	VG	VG	BA	BA	B
9	2	1	2	1	0	0.25	2	8	VG	SA	VG	VG	VB	BA	P	B
10	1	1	2	1	0	0.5	2	8	S	SA	VG	VG	VB	G	P	B
11	2	1	2	1	0	0.25	2	8	VG	SA	VG	VG	VB	BA	P	B
12	2	1	1.5	1	1	1.5	2	9	VG	SA	S	VG	VG	G	P	B
13	1	2	0	1	0	1	0.5	6	S	VG	VB	VG	VB	SA	BA	C
14	1	1.75	2	0	1	0.75	1	8	S	G	VG	BA	VG	SA	SA	B
15	1	0	0	0	1	0	1.5	4	S	VB	VB	BA	VG	VB	S	C
16	1	1.5	1	0	0.5	0.25	1.75	6	S	G	SA	BA	SA	BA	VG	C
17	1	1	2	1	1	0	1	7	S	SA	VG	VG	VG	VB	SA	B
18	1	1	2	1	1	1.25	1.25	9	S	SA	VG	VG	VG	S	SA	B
19	1	1	2	1	0	0	1.5	7*	S	SA	VG	VG	VB	VB	S	C*
20	1	0	2	1	0.5	0	0	5	S	VB	VG	VG	SA	VB	BA	C
21	0	2	2	1	0.75	0	2	8	BA	VG	VG	VG	S	VB	P	B
22	1	2	2	1	0	0.25	1.5	8	S	VG	VG	VG	VB	BA	S	B
23	1	2	1	1	1	1.25	1.5	9	S	VG	SA	VG	VG	S	S	B
24	1	1	2	1	1	0.25	1.5	8	S	SA	VG	VG	VG	BA	S	B
25	1	2	1	1	1	1.25	1.5	9	S	VG	SA	VG	VG	S	S	B
26	1	1	1	1	0	0	2	6	S	SA	SA	VG	VB	VB	P	C
27	1	0	2	1	1	0.25	1.25	7*	S	VB	VG	VG	VG	BA	SA	C*
28	1	0	1	1	0	0	0.5	4	S	VB	SA	VG	VB	VB	BA	C

- using conventional point system by assigning a number that summarizes the student's performance in a given task;
- using words from the codebook of the Per-C given above.

The "Tot." column for point grades was calculated as a rounded sum of single points assigned to each task. The "Tot." column for word grades was obtained using the Per-C, in which classes in a decoder were modeled as type-1 fuzzy sets with the following membership functions: $\mu_D(x) = \mu_{trap}(x; 1, 3, 3, 1, 1)$, $\mu_C(x) = \mu_{trap}(x; 4, 6, 6, 4, 1)$, $\mu_B(x) = \mu_{trap}(x; 7, 9, 9, 7, 1)$, $\mu_A(x) = \mu_{trap}(x; 10, 12, 12, 10, 1)$.

In the majority of cases (78.57%), the Per-C produces the output that is consistent with conventional point grades. Differences in rows 2, 3, 4, 5, 19, and 27 could be attributed either to incorrect specification of the Per-C (weights assigned to different tasks don't represent actual influence on the final grade), or to mistakes in assigning point grades, which is precisely why the Per-C was designed in the first place.

4   Conclusions and Further Research

In this paper, we outlined the first steps in building a perceptual computer to aid in grading mathematics tests within the bilingual education. These are just the first steps, and much needs to be done to improve the overall performance of the Per-C, including:

- automating the process of assigning weights to different tasks and competences by analyzing the point grades assigned by the teacher without eliciting them explicitly;
- expanding the dataset of intervals for higher accuracy of T2FSs in the codebook;
- applying the Per-C to a wider selection of tests, graded by different teachers, to assess their degree of satisfaction with it.

References

1. Skutnabb-Kangas, T., McCarty, T.L.: Key concepts in bilingual education: Ideological, historical, epistemological, and empirical foundations. In: Hornberger, N. (ed.) Encyclopedia of Language and Education, pp. 1466–1482. Springer, New York (2008)
2. Moschkovich, J.N.: Bilingual/multilingual issues in learning mathematics. In: Lerman, S. (ed.) Encyclopedia of Mathematics Education, pp. 57–61. Springer, Dordrecht (2014)
3. Xingle, F., Zhaoyun, S., Yan, C., Yupu, B.: The exploration and research on bilingual education of computer discipline. Int. J. Educ. Manage. Eng. (IJEME) 2(10), 52–58 (2012). https://doi.org/10.5815/ijeme.2012.10.09
4. Phyue, S.L.: Development of Myanmar-English Bilingual WordNet like Lexicon. Int. J. Inf. Technol. Comput. Sci. (IJITCS) 6(10), 28–35 (2014). https://doi.org/10.5815/ijitcs.2014.10.04
5. Kovalchuk-Khymiuk, L.O., Tavrov, D.Y.: The fuzzy inference system for assessing of students' performance during the bilingual teaching of mathematics. In: System Analysis and Information Technologies: Materials of the 17th International Scientific and Technical Conference, SAIT 2015, pp. 245–247 (2015). (in Ukrainian)

6. Mitra, M., Das, A.: A fuzzy logic approach to assess web learner's joint skills. Int. J. Mod. Educ. Comput. Sci. (IJMECS) **7**(9), 14–21 (2015). https://doi.org/10.5815/ijmecs.2015.09.02

7. Liu, S., Chen, P.: Research on fuzzy comprehensive evaluation in practice teaching assessment of computer majors. Int. J. Modern Educ. Comput. Sci. (IJMECS) **7**(11), 12–19 (2015). https://doi.org/10.5815/ijmecs.2015.11.02

8. Zadeh, L.A.: From computing with numbers to computing with words—from manipulation of measurements to manipulation of perceptions. IEEE Trans. Circ. Syst. I Fundam. Theory Appl. **46**(1), 105–119 (1999)

9. Mendel, J.M.: The perceptual computer: an architecture for computing with words. In: Proceedings of Modeling with Words Workshop in the Proceedings of FUZZ-IEEE 2001, pp. 35–38 (2001)

10. Mendel, J.M., Wu, D.: Perceptual Computing. Aiding People in Making Subjective Judgments. Wiley, Hoboken (2010)

11. Mendel, J.M.: Fuzzy sets for words: a new beginning. In: Proceedings of FUZZ-IEEE 2003, St. Louis, MO, pp. 37–42 (2003)

12. Wu, D., Mendel, J.M.: Enhanced Karnik-Mendel algorithms. IEEE Trans. Fuzzy Syst. **17**(4), 923–934 (2009)

13. Klir, G.J., Yuan, B.: Fuzzy Sets and Fuzzy Logic, Theory and Applications. Prentice Hall, Upper Saddle River (1995)

14. Wu, D., Mendel, J.M.: Aggregation using the linguistic weighted average and interval type-2 fuzzy sets. IEEE Trans. Fuzzy Syst. **15**(6), 1145–1161 (2007)

15. Vlachos, I., Sergiadis, G.: Subsethood, entropy, and cardinality for interval-valued fuzzy sets —an algebraic derivation. Fuzzy Sets Syst. **158**, 1384–1396 (2007)

16. Liu, F., Mendel, J.M.: Encoding words into interval type-2 fuzzy sets using an interval approach. IEEE Trans. Fuzzy Syst. **16**, 1503–1521 (2008)

17. Type-2 Fuzzy Logic Software. http://sipi.usc.edu/~mendel/publications/software/software. zip. Accessed 29 Nov 2017

Modelling Nonlinear Nonstationary Processes in Macroeconomy and Finances

P. Bidyuk[1] , T. Prosyankina-Zharova[2](✉) , and O. Terentiev[2]

[1] Institute for Applied System Analysis NTUU "Igor Sikorsky
Kyiv Polytechnic Institute", Kyiv, Ukraine
pbidyuke_00@ukr.net
[2] Institute of Telecommunications and Global Information Space of the National
Academy of Sciences of Ukraine, Kyiv, Ukraine
t.pruman@gmail.com, o.terentiev@gmail.com

Abstract. Modern decision support systems need the methods of predictive modelling, which would allow create models of systems with given parameters, such that they are capable of being promptly subjected to changes and additions. It could be used to deal with uncertainties of different types, to maximize the automation of the process of constructing predictive models and improve the quality of forecasts estimated. This article is devoted to the study and solving the problem of modeling and forecasting nonlinear nonstationary processes in economy and finances using the methodology proposed based on systemic approach to model structure and parameter estimation. We present preliminary data processing techniques necessary for eliminating possible uncertainties, application of data correlation analysis for model structure estimation, and a set of model parameter estimation methods providing a possibility for computing unbiased estimates of parameters. Proposed methodology can be applied in decision support systems used in finance and economy spheres under conditions of various uncertainties and risks that usually take place in modeling and forecasting using statistical data. In the present paper we will describe in short the usage of the methodology of adaptive modelling and give a couple of examples presenting new results of its application for forecasting behavior of several economic and financial processes.

Keywords: Nonlinear nonstationary process · Uncertainties
Modelling · Forecasting · Macroeconomy · Finances

1 Introduction

Most of the process taking place in economy and finances today are nonlinear and nonstationary or piecewise linear and stationary. Practically all the process analyzed exhibit trends of various order and/or their variance is not constant. The trends are stochastic or deterministic dependently on the set of specific influence factors to them, and heteroscedasticity is practically inherent to all financial process related to price and return forming, exchange rates etc. [1–4]. Mathematical modeling and forecasting of the processes of the economy or financial dynamics based on the use of statistical and/or experimental data usually need to consider various kinds of uncertainties related

© Springer International Publishing AG, part of Springer Nature 2019
Z. Hu et al. (Eds.): ICCSEEA 2018, AISC 754, pp. 735–745, 2019.
https://doi.org/10.1007/978-3-319-91008-6_72

to statistical data, structure of the process (and its model) under study, parametric uncertainty, and uncertainties relevant to forecasts estimates. To identify and take into consideration the uncertainties in relevant computational algorithms, and improve this way quality of intermediate and the final results (processes evolution forecasts and the decisions based upon them) it is necessary to analyze the reasons for the uncertainties to appear, the consequences of their influence and to construct appropriate computational algorithms for solving multiple related specific problems. Development and application of the methodologies possessing the necessary features mentioned is an important task that is being solved nowadays by many researchers [5–7].

Existing methodologies developed for studying nonlinear nonstationary process (NNP) and constructing mathematical models in various research fields using statistical procedures and state space model representation. Another approach to development models for NNP is based upon intellectual data analysis (IDA) techniques such as artificial neural networks, the group method for data handling (GMDH) [6], and Bayesian networks (both static and dynamic). On the other side these methodologies need refinement so that to produce better results regarding models adequacy and quality of the forecasts based upon them. The refinement may touch preliminary data processing algorithms, aiming to improvement of statistical characteristics of data, model structure and parameters estimation procedures, as well as the forecast estimates. A very important point is hiring of appropriate sets of statistical quality criteria necessary to monitor all stages of computations: data quality analysis, model adequacy estimation, and determining the forecasts quality. Finally the quality criteria should analyze alternative decisions (alternatives) based on the forecasts generated. For example, in a case of computing the alternatives of optimization type very often is used popular quadratic criterion taking into consideration input control energy (or other equivalent of input control variables) and deviations of the controlled system states from prescribed trajectories. Many other quality criteria are available or could be constructed for specific application if necessary [6–17].

In this paper the analysis and development of requirements to the preliminary data processing algorithms (preparing of data to model constructing) are performed; development of the software system architecture for model constructing, process evolution forecasting of various dynamic processes in economy and finances including nonlinear and nonstationary ones; identification of some uncertainties relevant to model structure and parameters estimating, and selection of mathematical techniques for minimizing influence of the uncertainties identified; illustration of the software developed application to solving selected problems of modeling and forecasting using appropriate statistical data.

2   The Main Material

2.1   The Modern Applied Software Systems

Modern applied software systems for modeling and forecasting (ASSMF) [17–20] are rather complex multifunctional (very often possibly distributed) highly developed computing systems with hierarchical architecture that corresponds to the nature of decision making by a human being. To make functionality of the ASSMF maximum useful and convenient for users of different levels (like engineering personnel and managerial staff) they should satisfy some specific requirements: availability of model adaptation features making possible models (structure and parameters) adaptation to new data, and possible changes in the modes of functioning of the system under study; application of optimization techniques (where possible and necessary) aiming to obtain optimal state and parameter estimates, and optimized forecast estimates as well; identification of possible uncertainties and availability of computational techniques directed towards elimination or minimization of negative influence of the uncertainties detected; availability of several sets of statistical quality criteria for estimating quality of data, models, forecasts, and decision alternatives, accordingly etc.

Appropriate satisfaction of all the requirements to development of ASSMF provides good possibilities for effective practical usage of the system developed and for enhancing its general behavioral effect for specific applications in the area of model constructing with statistical data and forecasting of the relevant processes evolution for a given time horizon [6, 20].

To find "the best" model structure it is recommended to apply adaptive estimation schemes that provide automatic search in a definite selected range of model structure parameters (type of distribution, model order, time lags, and nonlinearities) [3, 16, 20]. The adaptive estimation schemes also help to cope with the model parameters uncertainties. New data are used to compute repeatedly model parameter estimates that correspond to possible changes in the object under study. In the cases when model is nonlinear alternative parameter estimation techniques could be hired to compute alternative (though admissible) sets of parameters and to select the most suitable of them using statistical quality criteria. The use of a specific adaptation scheme depends on a volume and quality of collected data, specific problem statement, requirements to forecast estimates, etc. The method proposed is based on the ideologically different techniques of modelling and risk forecasting what creates a convenient basis for combination of various approaches to achieve the best results. The method could be used successfully for solving practical problems of dynamic processes forecasting and risk estimation in decision support systems.

The problem of testing the processes for linearity (nonlinearity) is considered in many studies [7–17]. Usually combination of several tests helps to detect existing nonlinearity and to select appropriate model structure. To detect nonlinearity the following test was applied:

- construct linear regression model for dependent variable $y(k)$ and right hand side (RHS) vector $w(k)$ using Least Squares method (LS) [9] (1):

$$y(k) = \beta^T w(k) + u(k), \tag{1}$$

- where $w(k) = [1, y(k-1), ..., y(k-p); x_1(k), ..., x_l(k)]^T$ is vector of measurement for dependent variable and regressors; $v(k) = [u(k-1), ..., u(k-q)]^T$ is vector of random variables; and $u(k) = g(\beta, \theta, w(k), v(k))\varepsilon(k); \varepsilon(k)$ is martingale process with the following statistical characteristics: $E[\varepsilon(k)|\mathbf{I}(k)] = 0$, $\text{cov}[\varepsilon(k)|\mathbf{I}(k)] = \sigma_\varepsilon^2; I(k) = \{y(k-j), j > 0; x(k-i), i \geq 0\}$ is available observation information;

- compute residuals of the model $\tilde{u}(k)$ and the sum of squared errors SSR_0 for the model constructed;
- construct regression model for $\tilde{u}(k)$ with regressors $w(k)$, and compute sum of squared errors for the model SSR_1;
- compute test statistics (2):

$$F(m, \ N-n-m) \ = \ \frac{(SSR_0 \ - \ SSR_1)/m}{SSR_1/(N-n-m)} \tag{2}$$

where $n = l + p + 1$; m is dimension of parameter vector θ; the value computed has F – distribution with $\theta = 0$. The use of F – statistics instead of χ^2 test is recommended for short samples.

Some nonlinear models result from studying econometric time series. For example, nonlinear regression of the following type was used to describe gross domestic product (GDP) and tax income:

$$y_1(k) \ = \ a_0 + a_1 y_1(k-1) + b_{12} \exp(y_2(k)) + a_2 x_1(k) x_2(k) + \varepsilon_1(k),$$

$$y_2(k) \ = \ c_0 + c_1 y_2(k-1) + b_{21} \exp(y_1(k)) + c_2 x_1(k) x_2(k) + \varepsilon_2(k),$$

where $y_1(k)$ is logarithm of GDP; $y_2(k)$ is logarithm of tax income; $x_1(k)$ – internal investments; $x_2(k)$ external investments. Another useful structure is generalized bilinear model:

$$y(k) \ = \ a_0 + \sum_{i=1}^{p} a_i y(k-i) + \sum_{j=1}^{q} b_j v(k-j) + \sum_{i=1}^{m} \sum_{j=1}^{s} c_{i,j} y(k-i) v(k-j) + \varepsilon(k),$$

Where p, q, m and S are positive numbers. The model can also be represented in state space form where the states are presented in the form of a product of former innovations and vectors of random coefficients [10].

Some models of nonlinear describe specific financial or economic processes. They can take into consideration possibilities for development optimization of the systems under study using appropriate cost functional or utility function. One of possible model structures is as follows (3):

$$y(k) \ = \ \min(\beta^T \mathbf{w}(k), \ \theta^T \mathbf{w}(k)) + \varepsilon(k) \tag{3}$$

where estimate of dependent variable $\hat{y}(k)$ is determined as a smaller one of two possible outcomes computed via alternative functions: $\beta^T w(k)$ or $\theta^T w(k)$. If "min" in the model (3) is replaced by another variable $z(k-d)$, that could be an element of vector $w(k)$, though not equal one, then we get so called switching model (4):

$$y(k) \ = \ \beta^T \mathbf{w}(k) + \theta^T \mathbf{w}(k) F(z(k-d)) + \varepsilon(k), \tag{4}$$

where

$$F(z(k-d)) = \begin{cases} 0, \text{ якщо } z(k-d) \le c; \\ 1, \text{ якщо } z(k-d) > c; \end{cases}$$

c is some threshold value that is used for switching from one model to another; $d = 0$, $1, 2, \ldots$ is discrete delay time.

A scalar version of the model is called threshold autoregression with two modes. It can be generalized to the set of possible functioning modes using the function of the type (5):

$$F(z(k-d)) = \frac{1}{1 + \exp\left[-\gamma\left(z(k-d) - c\right)\right]}, \quad \gamma > 0 \tag{5}$$

This is a model of logistic smooth transition regression (LSTR). The function F can also be used in the form of probability density function (PDF). In a scalar case the model will correspond to the exponential smooth transition autoregressive (ESTAR) model.

A convenient approach to modeling nonlinear processes is based on the models that contain linear and nonlinear components, or flexible models (6):

$$y(k) = \beta^T \mathbf{z}(k) + \sum_{i=1}^{p} \alpha_i \, \phi_i(\theta_i^T \mathbf{z}(k)) + \varepsilon(k), \tag{6}$$

where $z(k)$ is a vector of time delayed values of dependent variable $y(k)$, as well as former and current values of the vector of explaining variables $x(k)$ plus shift constant. The first component of the model is linear, and $\varphi_i(x)$ is a set of functions that could include the following components:

- power function $\varphi_i(x) \equiv x^i$, where variable x can be delayed in time value of y or some other variable;
- trigonometric function $\varphi_i(x) = \sin x$ or $\varphi_i(x) = \cos x$;
- Equation (7) can be expanded with quadratic function $z^T(k)Az(k)$, that will result in a flexible functional form;
- $\varphi_i(x) = \varphi(x)$, $\forall i$, where $\varphi(x)$ is a link function, for example appropriate PDF or logistic function of the following type:

$$\varphi(x) = \frac{1}{1 + \exp(-x)};$$

- $\varphi(x)$ can also be represented by nonparametric function.

Some general class of nonlinear models is given in the form (7):

$$\mathbf{y}(k) = \sum_{j=1}^{p} \phi_j(\mathbf{x}(k-1)) \, \mathbf{y}(k-j) + \mu(\mathbf{x}(k-1)) + \varepsilon(k), \tag{7}$$

Where $y(k)$ is $[n \times 1]$ stochastic vector of dependent variables; $x(k) = [y(k), y(k-1), ..., y(k-n+1)]$ is a vector of state variables dynamics of which is described by the model:

$$\mathbf{x}(k) = h(\mathbf{x}(k-1)) + \mathbf{F}(\mathbf{x}(k-1))\,\mathbf{x}(k-1) + v(k). \tag{8}$$

Equation (7) can also include the moving average members. It means that in this case to describe selected process we use two models simultaneously what may result in some difficulties with the model structure estimation.

Equation (8) is a state space model that can be supplemented with the measurement equation. The elements of matrix $\mathbf{F}(\cdot)$ are linear functions or low order polynomials. The models (7, 8) can also contain the members that reflect availability of long memory what takes place very often when we study ecological, financial and economic processes.

In the process of constructing forecasting we build several candidates and select the best one of them with a set of model adequacy statistics. The following techniques are used to fight structural uncertainties: improvement of model order (NAR(p) or NAR-MAX(p, q)) applying adaptive approach in a loop to modeling and automatic search for the "best" structure using complex statistical adequacy criteria mentioned above; adaptive estimation of delay time (lag) and the type of data distribution with its parameters; describing detected process nonlinearities with alternative analytical forms with subsequent estimation of model adequacy and quality of the forecasts generated.

A wide subclass of models is created today by the models describing dynamics of conditional variance for heteroscedastic process (HP). HP are nonlinear by definition as far as variance description is based on quadratic variables and functions. Popularity of variance is explained by its wide possibilities for practical applications such as following: it is a parameter used in stock trading systems for supporting decisions regarding operations with various stocks; variance characterizes evolution of prices for many market goods; there is no technical or medical diagnostic system that does not use variance as a parameter incorporated in decision making rules. The simplest mathematical model of conditional variance is autoregressive conditionally heteroscedastic (ARCH) equation of the form (9):

$$E_k[\hat{\varepsilon}^2(k+1)] = \alpha_0 + \alpha_1 \hat{\varepsilon}^2(k-1) + \alpha_2 \hat{\varepsilon}^2(k-2) + \ldots + \alpha_q \hat{\varepsilon}^2(k-q), \tag{9}$$

where $\hat{\varepsilon}(k)$ is a stochastic part of equation describing HP under study. It can be estimated by hiring low order autoregressive equation (such as AR(1) or AR(2)) for description of goal variable $y(x)$ in LHS; $E_k(\cdot)$ is a symbol of conditional mathematical expectation computed for specific moment of time k. Usually Eq. (9) does not allow compute acceptable results of short variance forecasting.

The structure of Eq. (9) was improved by introducing another variable into right hand side (RHS) as follows (10):

$$h(k) = \alpha_0 + \sum_{i=1}^{q} \alpha_i \, \varepsilon^2(k-i) + \sum_{i=1}^{p} \beta_i \, h(k-i), \tag{10}$$

where sample conditional variance $h_s(k)$ is computed as follows:

$$h_S(k) = \frac{1}{w-1} \sum_{i=k-\frac{w-1}{2}}^{k+\frac{w-1}{2}} [y(i) - \bar{y}_S(i)]^2, \quad k = 2, 3, \ldots, N,$$

where $\bar{y}_S(k)$ is sample mean; w is size of moving window for computing conditional variance, which is usually selected as odd number for convenience. Equation (10) is generalized ARCH (GARCH) which is usually much more efficient for describing and short term forecasting conditional variance.

Very good results of short term forecasting can be achieved with the exponential GARCH (EGARCH) that has the following structure:

$$\log[h(k)] = \alpha_0 + \sum_{i=1}^{p} \alpha_i \frac{|\varepsilon(k-i)|}{\sqrt{h(k-i)}} + \sum_{i=1}^{p} \beta_i \frac{\varepsilon(k-i)}{\sqrt{h(k-i)}} + \sum_{i=1}^{q} \gamma_i \log[h(k-i)] + \upsilon(k).$$

This equation contains so called "standard" part that takes into account the innovations $\varepsilon(k)$, and another part that takes into consideration sign of the innovation. The values of $\varepsilon(k)$ are normalized by volatility what leads to reduction of possible high values, and the logarithm function is applied for smoothing the volatility.

3 Results and Discussion

The developed procedure for linear and nonlinear model constructing using statistical data includes the steps given below.

1. Preliminary processing of data and expert estimates with application of data quality criteria such as missing values counters, parameters of information content (computing of variance, and number of derivatives for approximating polynomials), power of samples. Filtering of data and imputation of missing values where necessary; estimation of non-measurable components.
2. Application of statistical tests aiming to discovering nonlinearity and nonstationarity; correlation data analysis giving the grounds for estimation of models structure.
3. Estimation of candidate models structure and their parameters. To reduce the influence of possible parametric uncertainties three parameter estimation techniques are applied: least squares (LS), non-linear least squares (NLS), maximum likelihood (ML), and Markov chain Monte Carlo (MCMC) procedures.
4. Application of model adequacy statistics and selection of the "best" models. If model quality is not acceptable we return to the step 2 to get more information regarding the model structure and repeat the procedures of structure and parameter estimation.
5. The model(s) selected is used for computing forecasts that are analyzed with another set of quality criteria. Among them are mean absolute percentage error and Theil coefficient.

6. The final step is practical application of the model constructed. If the model is not satisfactory for practical usage the process of model constructing is repeated with extra statistical data.

Very often uncertainties of model parameter estimates such as bias and inconsistency result from low informative data, or data do not correspond to selected type of distribution, what is required for correct application of parameter estimation method. Such situation may also take place in a case of multicollinearity of independent variables and substantial influence of process nonlinearity that for some reason has not been taken into account when model structure was estimated. When power (size) of data sample is not satisfactory for model construction it can be expanded by special techniques or Monte Carlo simulation is hired, or special model constructing techniques, such as GMDH, are applied. GMDH produces very often results of acceptable quality with short samples. If data does not correspond to normal distribution, then ML technique could be used or appropriate MCMC procedures [8]. The last techniques can be applied with quite acceptable computational expenses when the number of parameters is not large.

Example 1. Consider the problem of modeling return $y(k)$ for a selected stock on the basis of monthly data including 300 observations. According to partial autocorrelation function computed the model of the process should include lags $1 - 3$. Thus the bilinear model selected may look as follows:

$$
\begin{aligned}
y(k) &= \mu + a_1 y(k-1) + a_2 y(k-2) + a_3 y(k-3) + (1 + \beta_1 v(k-1) \\
&+ \beta_2 v(k-2) + \beta_3 v(k-3)) \cdot \varepsilon(k)
\end{aligned}
$$

Where $v(k)$ is moving average process; it was suggested that $\{\varepsilon(k)\} \sim N(0,1)$. The model parameters were estimated with conditional maximum likelihood:

$$
y(k) = 0,0117 + 0,173\, y(k-1) + 0,115\, y(k-2) - 0,089\, y(k-3)
$$

$$
+ 0,077 \cdot (1,0 + 0,383\, v(k-1) + 0,103\, v(k-2) - 0,551\, v(k-3)) \cdot \varepsilon(k)
$$

Adequacy of the model is rather high: $R^2 = 0,89$; DW $= 1,92$, with mean absolute percentage error for one-step-ahead prediction on test sample: $MAPE = 5,2\%$. All parameter estimates are statistically significant at the confidence level of 5%.

The model for random process $\hat{\varepsilon}(k)$ from the last equation is as follows:

$$
\hat{\varepsilon}(k) = \frac{y(k) - 0,0117 - 0,173\, y(k-1) + - 0,115\, y(k-2) + 0,089\, y(k-3)}{0,077 \cdot (1,0 + 0,383\, v(k-1) + 0,103\, v(k-2) - 0,651\, v(k-3))},
$$

where $\hat{\varepsilon}(k) = 0$ for $k \leq 3$. The sample autocorrelation function for the process $\hat{\varepsilon}(k)$ shows that it does not contain statistically significant correlations.

Example 2. Models were developed for the bitcoin exchange rate using statistical data for the period of 10.01.2017 and 15.10.17; the measurements were taken every two weeks. The adaptive models are in the form of autoregression equations, and extended autoregression (ARX).

Partial autocorrelation function (PACF) are as follows: PACF(1) = 0.807; PACF(2) = −0.019; PACF(3) = 0.107; PACF(4) = −0.019; PACF(5) = 0.188; PACF(6) = −0.161; PACF(7) = −0.015; PACF(8) = −0.030; PACF(9) = −0.024; PACF(10) = −0.054. It can be seen that lagged values have influence up to the sixth lag.

The models constructed are as follows:

AR(1):   Price(t) = 36.19836 + 1.083651· Price(t − 1).

ARX(1):   Price(t) = −31484.05508 + 0.7050509055·   Price(t − 1) + 1453.535691· Log_Hash(t).

AR(3):   Price(t) = 50.96291164 + 1.609236977· Price(t − 1) − 1.437068· Price (t − 2) + 0.927195· Price(t − 3).

ARX(3):   Price(t) = −31626.76748 + 1.378278663· Price(t − 1) − 1.3039654· Price (t − 2) + 0.6249846446 · Price(t − 3) + 1458.369272· Log_Hash(t)

AR(6):   Price(t) = 73.62380553 + 1.245017642·   Price(t − 1) − 1.066082835· Price(t − 2) + 0.6064832892·   Price(t − 3) + 0.02360764846·   Price (t − 4) + 0.2026933078· Price(t − 5) + 0.2156083914· Price(t − 6)

ARX(6):   Price(t) = 29875.40335 + 1.109829578·   Price(t − 1) − 0.9588075814· Price(t − 2) + 0.2520366147· Price(t − 3) + 0.2324871042· Price(t − 4) − 0.03031657673· Price(t − 5) + 0.234847922· Price(t − 6) + 1375.430553· Log_Hash(t),

where Price - is actual bitcoin price.

Log_Hash – decimal logarithm for complexity of the whole net.

Mean absolute percentage error was estimated at Table 1:

Table 1 Models constructed and one-step forecasts for bitcoin

Model	R^2	DW	RMSE	MAPE
AR(1)	0,9392	1,29	341	9,19
ARX(1)	0,9518	1,21	304	10,37
AR(3)	0,9763	2,38	208	8,33
ARX(3)	0,9819	2,35	182	7,66
AR(6)	0,9778	2,07	191	5,84
ARX(6)	0,9812	2,02	176	6,24

Statistical characteristics of the "best" model: ARX(6): R^2 = 0,9812; DW = 2,02, RMSE = 176, MAPE = 6,24

4   Conclusions

The general methodology was proposed for developing automatized software system for mathematical modeling and forecasting nonlinear nonstationary economic and financial processes using statistical data. A short review of models for nonlinear nonstationary process was presented which shows growing popularity of the model structures considered. The examples of the software system application show that it can be used successfully for solving practical problems of mathematical model building and forecasts estimation. The system proposed could be used for support of decision making in multiple areas of human activities including strategy development for the state government, various financial institutions, industrial and agricultural enterprises, investment companies etc. Further extension of the system functions is planned with new forecasting techniques based on probabilistic models, neural networks, and fuzzy sets and rules.

References

1. Trofymchuk, O., et al.: Probabilistic and statistical uncertainty Decision Support Systems. Visnyk Lviv Polytech. Natl Univ. **826**, 237–248 (2015)
2. Sallam, E., Medhat, T., Ghanem, A., Ali, M.E.: Handling numerical missing values via rough sets. Int. J. Math. Sci. Comput. (IJMSC) **3**(2), 22–36 (2017). https://doi.org/10.5815/ijmsc.2017.02.03
3. Diebold, F.X.: Forecasting in Economics, Business, Finance and Beyond. University of Pennsylvania, Philadelphia (2015)
4. Hansen, B.E.: Econometrics. University of Wisconsin, Madison (2017)
5. Tsay, R.S.: Analysis of Financial Time Series. Wiley, New York (2010)
6. Dovgij, S.O., Trofymchuk, O.M., Bidyuk, P.I.: DSS Based on Statistical and Probabilistic Procedures. Logos, Kyiv (2014)
7. Shah, Y.A., Mir, I.A., Rathea, U.M.: Quantum mechanics analysis: modeling and simulation of some simple systems. Int. J. Math. Sci. Comput. (IJMSC) **2**(1), 23–40 (2016). https://doi.org/10.5815/ijmsc.2016.01.03
8. Ramsey, J.B.: Tests for specification errors in classical linear least squares regression analysis. J. Roy. Stat. Soc. B **31**, 350–371 (1969)
9. Terasvirta, T., Tjostheim, D., Granger, C.W.J.: Aspects of modeling nonlinear time series. In: Engle, R.F., McFadden, D.L. (eds.) Handbook of Econometrics, vol. 4, pp. 2919–2957 (1994)
10. Tsay, R.S.: Nonlinearity tests for time series. Biometrika **73**, 461–466 (1986)
11. Tjostheim, D.: Some doubly stochastic time series models. J. Time Ser. Anal. **7**, 51–72 (1986)
12. Tjostheim, D.: Nonlinear time series: a selective review. Scand. J. Stat. **21**(2), 97–130 (1994)
13. Krishna, G.V.: Prediction of rainfall using unsupervised model based approach using K-means algorithm. Int. J. Math. Sci. Comput. (IJMSC) **1**(1), 11–20 (2015). https://doi.org/10.5815/ijmsc.2015.01.02
14. Khashei, M., Montazeri, M.A., Bijari, M.: Comparison of four interval ARIMA-base time series methods for exchange rate forecasting. Int. J. Math. Sci. Comput. (IJMSC) **1**(1), 21–34 (2015). https://doi.org/10.5815/ijmsc.2015.01.03
15. Stensholt, B.K., Tjostheim, D.: Multiple bilinear time series models. J. Time Ser. Anal. **8**, 221–233 (1987)
16. Tsay, R.S.: Testing and modeling threshold autoregressive processes. J. Am. Stat. Assoc. **84**, 231–240 (1989)

17. Aminu, E.F., Ogbonnia, E.O., Shehu, I.S.: A predictive symptoms-based system using support vector machines to enhanced classification accuracy of Malaria and Typhoid coinfection. Int. J. Math. Sci. Comput. (IJMSC) **2**(4), 54–66 (2016). https://doi.org/10.5815/ijmsc.2016.04.06

18. Anderson, W.N., Kleindorfer, G.B., Kleindorfer, P.R., Woodroofe, M.B.: Consistent estimates of the parameters of a linear system. Ann. Math. Stat. **40**(6), 2064–2075 (1969)

19. Yavin, Y.A.: Discrete Kalman filter for a class of nonlinear stochastic systems. Int. J. Syst. Sci. **3**(11), 1233–1246 (1982)

20. Eeckhoudt, L., Gollier, C., Schlesinger, H.: Economic and Financial Decisions Under Uncertainty. Princeton University Press, Princeton (2004)

Information and Technology Support of Inclusive Education in Ukraine

Tetiana Shestakevych$^{(\boxtimes)}$ ⓘ, Volodymyr Pasichnyk,
and Nataliia Kunanets ⓘ

Lviv Polytechnic National University, S. Bandery Str., 12, Lviv, Ukraine
tetiana.v.shestakevych@lpnu.ua

Abstract. In modern understanding of education of people with special needs, the inclusive education considers being the beneficial way of socialization. With the growing number of children, who are suggested to be educated inclusively, the use of information technologies will enable a new level of support for all participants of such education – the children, their parents, as well as a wide range of specialists who work with people with special needs.

In this paper, we propose the model of information and technological support of inclusive education, built according to the specialty of such process in Ukraine. We present a model of data warehouse, designed to process data of complex psychophysical assessment of persons with special needs. The paper presents the key aspects and main formal criteria of evaluation of effects, caused by the implementation of complex information technology of inclusive educational support. These criteria reflect the impact on social, scientific and technological effects from implementation of information technologies of every stage of inclusive education support.

Keywords: Information and technology support · Inclusive education
Data warehouse · Cognitive modeling

1 Introduction

At the United Nations World Summit on society informatization, as one of the key issues was stated the lack of quality education with information and technology support [1]. Actual requirements in the field of education (since the beginning of the 2000s), were the individualization of education, the organization of systematic knowledge control, the consideration of the psychophysical features of each child, etc. These requirements have become acuter with the acceptance of the concept of training children who have special needs, in mass education institutions.

The format of training of people with special needs, which takes into account the specific needs of a particular person, arose and began to be used by educators for more than three decades ago – as an effective way of socializing for individuals with peculiarities of psychophysical development [2]. The involvement of a person with peculiarities of psychophysical development into the study is known as inclusion. The integration of children with peculiarities of development into mass educational institutions is a global trend, legally fixed in Canada, Cyprus, Denmark, Iceland, India, Malta,

© Springer International Publishing AG, part of Springer Nature 2019
Z. Hu et al. (Eds.): ICCSEEA 2018, AISC 754, pp. 746–758, 2019.
https://doi.org/10.1007/978-3-319-91008-6_73

the Netherlands, the USA, etc. There are a large number of scientific studies that take into account the various specific characteristics of inclusive education [3, 4], intelligence is also conducted by the UNESCO Commission, the UN Organization, the European Agency for the Development of Education for Persons with Special Needs. In 2008, with the support of the Canadian International Development Agency (CIDA) and partnerships between Canadian and Ukrainian public and state organizations, the Canadian-Ukrainian project "Inclusive education for children with special needs in Ukraine" was launched.

Traditionally, the support of the education of people with special needs is carried out according to technological, pedagogical, psychological, medical, sport and social directions. The process of inclusive education basically consists of four consecutive stages, the implementation of each stage means to perform certain educational tasks related to the organization and support of training of people with special needs (conceptually presented by the UML activity diagram, Fig. 1). The development and implementation of modern comprehensive information and technological (IT) support of all stages of inclusive education (IE), taking into account the national specifics of such a process, promotes complete and better access to education and social integration of people with special needs. IT support for IE is a system of interconnected information technologies, designed to reduce the labor-capacity of tasks for the organization and monitoring of inclusive education processes. The works of the following authors are related to the individual components of such an accompaniment. For example, the peculiarities of the modeling of supportive systems are described by Hersh [5]; research on the development of personal e-learning recommendation systems was conducted, among others, by Starcic [6]; J. Mostow, J. Beck and others studied the problem of the selection of educational technology according to the individual characteristics of the student. Several other systems for collecting and analyzing students' achievement were implemented in IT and special software systems [7]. Teachers' methodological assistance is an open Internet resources for the development of a personal communication passport [8]; problem-oriented applications (for example, "Symbolic communication" for people with low reading skills) [9–11], etc. In Ukraine, the range of problem-oriented applications of IT support of IE includes the software for speech training and development [12]; fundamentally new computer technologies are developed to open the access to educational information resources for people with special needs [13]; modern computerized workplaces are being developed [14]; and a system for information and library services for users with special needs is being developed. To get the information and methodological assistance in solving the issues of inclusive and special education, one may use the educational web portals of regional and city education departments, on the website of the all-Ukrainian Foundation "Step by Step", on the website of the National Assembly of the Disabled of Ukraine, etc.

In general, IT support for the training of people with special needs varies, according to certain factors that determine the field of technology application. We will formally set the technology of supporting the training of people with special needs as:

$$Tech = Tech_1 \bigcup Tech_2 \bigcup Tech_3 \bigcup Tech_4$$

where $Tech_1$ is a set of general assisting information technologies; $Tech_2$ is a set of assistive information technology of special purpose; $Tech_3$ is a set of communication support technologies; $Tech_4$ is a set of access means [15].

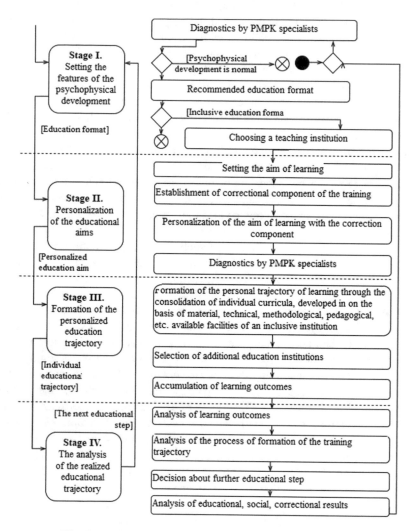

Fig. 1. Activity diagram for the inclusive education system

Based on the analysis of existing information technologies, developed to support IE, it was concluded that the IT support, available to domestic specialists, does not fully cover all phases of such education process, and to a certain extent, is partial and non-systematic. Consequently, the actual scientific problem is to develop information and technology support system for inclusive education in order to improve the maintenance of such process. In this paper, we propose certain approaches to solve basic aspects of the mentioned problem in ratio to the national policy in the field of inclusive education in Ukraine. These aspects are:

- Modeling the process of IT support of inclusive education from the viewpoint of the specialty of such education in Ukraine, taking into account both the order and the parallelism of considered educational tasks. Such model must cover all the stages of IE.
- Consider the big data amounts, that are accumulated in IE process, and design a data warehouse for the IT support of IE stage of featuring the person's psychophysical development.
- Using experts opinion, create a set of formal criteria, that will enable the evaluation of effects, caused by the implementation of IT support in inclusive educational. Using appropriate modeling technique, these criteria should reflect social, scientific, technological etc. impact.

2 Modeling of IT Support of IE

2.1 Model the Inclusive Education Process

To model the process of inclusive education, the partially ordered sets with non strict order were used, that allowed to determine the order of educational tasks that arise in inclusive education. Though, to reflect regular dependencies in IE process, the generative grammars were used. Its ability to take into account the context allows adequate reflection and implementation of specific dependencies that arise during the IE. A characteristic feature of inclusive education, in addition to a strict sequence of educational tasks, is the need for the parallel implementation of certain tasks at different stages of inclusive education. For the formal presentation of such requirements the Petri network was used.

The advantages of using this mathematical abstraction are the possibility of mapping it with the help of causation links in complex systems and the visual presentation of parallel phenomena in complex systems, such as the IE process.

Petri net is set as the $C = (P, T, I, O)$ and models the process of inclusive education, such as at Fig. 2, where the set of positions $P = \{p_0, p_2, ..., p_{22}\}$, the set of transitions $T = \{t_1, t_2, ..., t_{13}\}$; the initial marking μ_0 is one chip in the p_0 position.

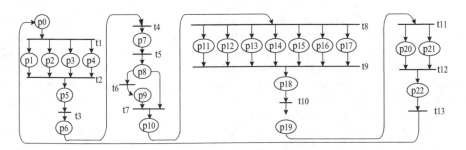

Fig. 2. Petri net as a model of the inclusive education process

$I(t_1)=\{p_0\}; I(t_2)=\{p_1, p_2, p_3, p_4\};$

$I(t_3)=\{p_5\}; I(t_4)=\{p_6\}; I(t_5)=\{p_7\}; I(t_6)=\{p_8\};$

$I(t_7)=\{p_8, p_9\}; I(t_8)=\{p_{10}\};$

$I(t_9)=\{p_{11}, p_{12}, p_{13}, p_{14}, p_{15}, p_{16}, p_{17}\};$

$I(t_{10})=\{p_{18}\}; I(t_{11})=\{p_{19}\};$

$I(t_{12})=\{p_{20}, p_{21}\}; I(t_{13})=\{p_{22}\};$

$O(t_1)=\{p_1, p_2, p_3, p_4\}; O(t_2)=\{p_5\};$

$O(t_3)=\{p_6\}; O(t_4)=\{p_7\}; O(t_5)=\{p_8\};$

$O(t_6)=\{p_9\}; O(t_7)=\{p_{10}\}; O(t_9)=\{p_{18}\};$

$O(t_8)=\{p_{11}, p_{12}, p_{13}, p_{14}, p_{15}, p_{16}, p_{17}\};$

$O(t_{10})=\{p_{19}\}; O(t_{11})=\{p_{20}, p_{21}\};$

$O(t_{12})=\{p_{22}\}; O(t_{13})=\{p_0\}.$

In the given Petri net, the positions are considered as the conditions for the events onset. Transitions, in their turn, are considered as events and are interpreted as processes.

Tha application of Petri nets formalism to modeling of educational processes allows mapping both graphically and analytically crucial features of educational processes, such as events order, interaction characteristics etc. [16]. The Petri net of displays not only the model's functionality but also structural properties, which greatly simplifies the process of modeling of parallelisms, inherent in educational processes. The advantage of the offered model is the unification into a single system of all stages of the inclusive educational process, which enables the comprehensive and holistic analysis of such a process. The model generally takes into account the current government policy in the education of people with special needs, and allows to use the experience of foreign scientists. Such model is the basis for developing a comprehensive system of IT support of educational processes of people with special needs.

2.2   Modeling the IT Support of IE

The IT system of IE support is to perform some functions, according to needs of the users, namely the PMPK specialists, the inclusive school faculty, not to mention the parents of a child with special needs. In respect to stages of IE, the set of information technologies (IT_{ij}), each to accomplish a specific task, were defined (index i is the number of IE stage). The IT_{11} is the information technology of psychophysical diag-nosing support, IT_{12} supports the formation of a comprehensive assessment of the person's psychophysical development, and the recommendations for the education format, IT_{13} supports the PMPK work analysis; IT_{21} supports the diagnostics of current basic skills and competencies, as well as accumulation of such data, IT_{22} supports the personalization of the educational plan; IT_{31} supports the formation of a personalized training trajectory, IT_{32} supports the accumulation of learning outcomes; IT_{41} supports the evaluation of learning outcomes, updating of the training trajectory, forming sug-gestions for the next step in education, IT_{42} supports the evaluation of the process of individual educational plan development, IT_{43} supports analysis of the work of spe-cialists in inclusive education. Using the proposed notation, let us set the finite state machine M^* to model the process of IT support of IE. $M^* = (S, S_0, I, \mu, F)$, $S = \{S_0, S_1, S_2, S_3, S_4, S_5, S_6, S_7, S_8, S_9, S_{10}\}$, S_0 – initial state, $I = \{IT_{11}, IT_{12}, IT_{13}, IT_{21}, IT_{22}, IT_{31}, IT_{32}; IT_{41}, IT_{42}, IT_{43}, continue\}$, $F = \{S_2, S_9, S_{10}\}$, $F \subset S$ [17]. The transition function μ is defined as in Table 1. At Fig. 3 is the state diagram of the appropriate machine.

Table 1. The transition function for a M* machine

State	Input	State	State	State	Input
S_0	IT_{11}	S_1	S_6	IT_{32}	S_7
S_1	IT_{12}	S_2	S_7	IT_{41}	S_8
S_1	IT_{13}	S_3	S_8	IT_{42}	S_9
S_2	Continue	S_0	S_8	IT_{43}	S_{10}
S_3	IT_{21}	S_4	S_9	Continue	S_0
S_4	IT_{22}	S_5	S_{10}	Continue	S_0
S_5	IT_{31}	S_6			

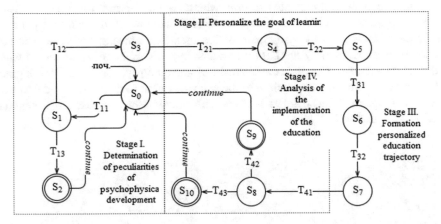

Fig. 3. Finite state machine as a model of the process of IT support of IE

As a result, we have a model of the process of IT support of IE, given in Fig. 3. This model takes into account the peculiarities of the inclusive education process in Ukraine, as well as covers all stages of the process. The developed model allows the software engineering community to navigate in IE process support, to choose at what stage their IT product will be helpful, which IE stage needs to be supported the most, which "gaps" their software might fill. When any separate information technology, that corresponds to IE in Ukraine, will be developed, its place (according to the proposed model) will be easily found in complex system of IT support of IE.

3   The Model of the Data Warehouse for Complex Assessment of Psychophysical Development

The special feature of the IT support of IE is the need of analysis of large data amounts. In order to develop it, it is necessary to take into account the main approaches to medical documentation conducting, the possibility of using the methods of structuring and accumulation of medical research data, the specifics of both monitoring and accumulation of psychophysical data and educational trajectory.

The model data warehouse for complex assessment of the person's psychophysical development is defined as

$$SD = <DB, RF, RM, rf, rm, func>$$

where *DB* is a set of relations, its schemes, and constraints, which contain information from incoming databases (the results of psychophysical diagnosing in PMPK, the results of diagnosing by other specialists etc.), *RF* is a scheme of the set of facts relations *rf*, *RM* is a scheme of a set of metadata relations *mf*, *func* is a set of decision-making procedures. Getting new solutions involves data mining from the warehouse by implementing appropriate data warehouse functions in facts relation, taking into account requirements *usr_prm*, which are put forward for such a decision: *Dc = func(rf, usr_prm)*. The set of requirements for the proposed solution depends on the needs of the user – the participant of the inclusive education process. The connection between relations *rf* and *DB* forms a data hypercube, the measurement of which is the set of relations of complex assessment of person's psychophysical development data warehouse. The implementation of the slice operation in the pre-formed data cube solves the problem of the next data analysis. The model of the complex assessment data warehouse is shown in Fig. 4.

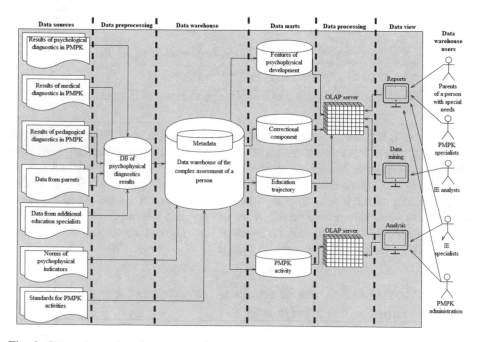

Fig. 4. The scheme of the data warehouse for complex assessment of psychophysical development

The proposed model underlines, that by now, all data sources of PMPK activities are stored in plain paper documents. It is crucial creating an information technology, that, according to the model, enables storage and processing of the data of person's psychophysical development. Methods of intellectual and multidimensional data analysis make it possible to identify the patterns in the complex assessment of the psychophysical development of a person and to professionally solve the problem of establishing the level of such development and the acceptability of inclusive education for a person with special needs. The research of such data by the analyst, who uses a range of data analysis methods [18, 19], will allow investigating the peculiarities of the development of the persons with special needs, not to mention the analysis of the work of PMPK specialists. The results of such data analysis will be used by parents of a child, who was diagnosed, by PMPK specialists themselves, the specialists of IE in schools, as well as might be used for maintenance and further improvement of the organization of PMPK work [20].

4 The Implementation of IT Support of IE: Evaluating Effects

The implementation of the complex information technologies to support the IN in Ukraine will cause a significant impact on different spheres, connected to the inclusive education process. The expected effects from such system implementation are scientific and technical, resource, social, cultural. To evaluate such difference, a system of criteria was developed, so both direct and indirect effects were taken into account.

The scientific and technical effect of such a system implementation lies in increasing the organizational level of production and labor – for specialists of inclusive education in schools, as well as PMPK specialists. The resource effect may be reflected by indicators of improved resource use, for example, by increasing productivity (or decreasing labor intensity). The social effect is to improve working conditions, as well as the skills of IE specialists. The most complex (from the methodological point of view) is the evaluation of the cultural effect.

In order to evaluate the effects of implementation of the IT system of IE support, a number of improving factors will be determined. It should be mentioned, that the IE support process is characterized by a lack of dynamics information, and at the same time, it has a significant number of interrelated factors. To determine the direction and force of impact of IT support implementation, the cognitive modeling will be used. It gives the opportunity to analyze the logic of interconnections and the consistency of events in case of a large number of interrelated factors [20].

4.1 The Cognitive Modeling of the Effects Impact

The process of IT support of IE implementation was analyzed using the PEST analysis (political, economic, socio-cultural and technological). A range of factors, that have a direct and indirect impact on the development of the IE system in case of IT support implementation, were defined. A group of effects, derived after the implementation of IT support of IE was distinguish, namely: scientific and technical effects (X), economic effects (W), resource effects (R), social effects (U), and cultural effects (C). To assess these groups of effects, experts from the field of information technology, personal-oriented education, inclusive education were involved, together with parents of children with special needs.

According to the cognitive modeling methodology [21, 22], to assess the relations between the concepts of the IT support of IE implementation, a linguistic scale has been chosen. Such scale corresponds to a number on the interval [0; 1]: 0 is for no effect, 0,1–0,2 is for very weak impact, 0,3–0,4 is for weak impact, 0,5–0,6 is for moderate impact, 0,7–0,8 is for strong impact and 0,9–1 is for very strong impact.

The weight coefficients, proposed by the experts, formed the appropriate matrix of the cognitive map (Fig. 5):

$$G = <V, E>$$

where $G = <V, E>$ is a directed graph, and V is a set of vertices (concepts), vertices $V_i \in V$, $i = 1,2,...,k$ are components of the system; E is a set of directed edges, $e_{ij} \in E$, i, $j = 1,2,...,n$ reflects the interconnections between vertices V_i and V_j. The adjacency matrix $\|a_{ij}\|$ is associated with the graph, and such matrix's elements characterize the impact the x_i concept over x_j concept.

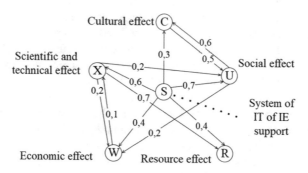

Fig. 5. Cognitive map of the effects of the IT support system implementation

The expert analysis of the effects of the IT support system implementation defined basic factors of such effects. The interrelations between the factors will be illustrated on the decomposed presentation of the impact *IT-support system of IE - Scientific and technical effect*.

The concepts (basic factors of scientific and technical effect) will be denoted as follows (Table 2).

Table 2. The concepts of the scientific and technical effect components

Denot.	Concept meaning
x_1	The informatization of PMPK and IE institutions activity
x_2	The accelerating the implementation of information and communication technologies of IE
x_3	The improvement of the PMPK administration and IE institutions activity
x_4	The improvement of the efficiency of the PMPK and IE specialists work
x_5	The increase of the analytical support level of IE institutions management

We will form a cognitive map of the decomposition of the *IT-support system of IE - Scientific and technical effect* impact on the basis of experts' assessments of the concepts' impact (Fig. 6). (In this study, the impacts of ITs in the considered system was not taken into account.) By its meanings, the considered concepts refer to information technologies, presented in the above mentioned model of IT support of IE (Fig. 3).

4.2   Impulse Modeling to Predict Scenario Development

The impulse on cognitive maps in the theoretical analysis is an ordered sequence of values $w_i(n)$, $w_i(n + 1)$ in vertex i without binding to the time, that is used to interpret the results of a computational experiment. The modeling of the impulse processes is conducted by the equation:

$$w_i(n+1) = w_i(n) + \sum_{j=1}^{k-1} f_{ij}P_j(n) + Q_i(n)$$

where $w_i(n)$ is an impulse in vertex i on the previous time of the modelling (n), $w_i(n + 1)$ is an impulse in vertex i at time $(n + 1)$; f_{ij} is a coefficient of impulse transformation, and $f_{ij} = 1$; $P(n)$ is an impulse in vertices, adjacent with vertex i; $Q_i(n)$ is a vector of vector of perturbations and control impacts that are included in the vertex i at time (n).

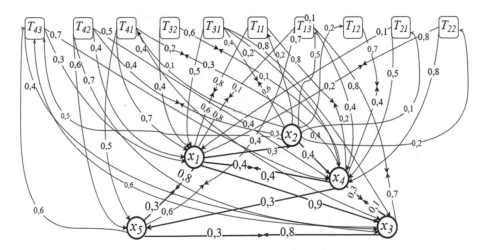

Fig. 6. Graph of the cognitive map of *IT-support system of IE - Scientific and technical effect* decomposition

As a control concepts, the T_{11}, ..., T_{43} are considered – those that are disturbed by improvement of the corresponding components of the system of IT support of IE. The target concepts include the effects of the scientific and technical impact of the implementation of the IT support system of IE. The control will take place through perturbation of the corresponding vertices of the cognitive map.

Scenario 1. Implementation (improvement) of IT for the formation of a comprehensive assessment of a person, the establishment of peculiarities of psychophysical development and recommendations for the format of education of a person. The impulse is delivered to the convex T_{13}. The results of the impulse modeling (Fig. 7) indicate that the implementation of the mentioned IT will be most effective in increasing the efficiency of the administration of the IE institutions and increasing the level of analytical support of the management of such institutions.

Scenario 2. Implementation (improvement) of IT for the analysis of the activities of specialists of PMPK and IE institutions. The impulse is delivered to the convex T_{13} and T_{43}. The results of the impulse modeling (Fig. 8) indicate that the implementation of the mentioned IT will have the greatest effect on increasing the efficiency of the improvement of the PMPK administration and IE institutions activity and increasing the level of the analytical support level of IE institutions management. This implies an increase in the effectiveness of the administration of the PMPK and IE institutions from 16.3 to 19.2, that is 17%. According to the modeling results, the level of the analytical support of IE institutions management will increase by 14% - from 15.5 to 17.8.

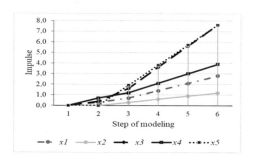

Fig. 7. Results of impulse modeling, Scenario 1

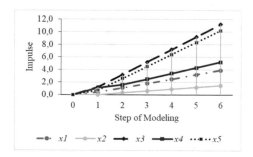

Fig. 8. Results of impulse modeling, Scenario 2

5 Conclusion

The model of IE, given in this paper as a Petri net, covers all stages of the inclusive education in Ukraine. Mathematical apparatus of Petri net allowed reflection of IE's special features, such as strict order of the educational tasks realization, as well as the parallelism of such tasks realization at appropriate stages of IE. That model became a basis to the model of IT support of inclusive education. The developed model conceptually shows ordered collection of IT, which are to be implemented at each stage of IE. The software developers, IE enthusiasts and IE specialists can use this model as a guide to development of IT, which will support an IE in Ukraine, at the same time giving a wider access to education for people with special needs.

In inclusive education, a big amount of data is produced at the first stage, when a person undertakes a psychophysical diagnosing by medicians, psychologists, educators etc. While developing a model of data warehouse, which should be used to process the results of psychophysical diagnosing, we took into account needs of all the participants of IE, engaged into this stage, as well as a wide range of data analysis and processing techniques, available for data analysts.

The experts asserted the relations between the concepts of the IT support system. Among the established components of scientific, technical, social, resource, economic and cultural effects, using the methods of expert analysis, we used the most relevant basic effects to form a cognitive map of impact. Using the impulse modeling technique, we analysed two scenarios of implementation (or improvement) of some IT's from the proposed system. The achieved results allow drawing a conclusion, that in order to improve the efficiency of the administration of the IE institutions, as well as increasing the level of analytical support of its management, it is necessary to implement the appropriate IT on the fist stage of IE, to make it possible to analyse the activities of inclusive education specialists.

References

1. Report of the 2012–2015 BFUG working group on the social dimension and lifelong learning to the BFUG. http://www.ehea.info/. Accessed 21 Nov 2017
2. Shestakevych, T.: Informational and technological support of educational processes for persons with special needs. Information systems and networks. Visnyk of the Lviv Polytechnic National University, Lviv No. 783, pp. 251–261 (2014). (in Ukrainian). http://ena.lp.edu.ua:8080/bitstream/ntb/26391/1/28-251-261.pdf. Accessed 10 Oct 2017
3. Rose, D., Meyer, A., Hitchcock, C.: The Universally Designed Classroom: Accessible Curriculum and Digital Technologies, 216 p. Harvard Education Press, Cambridge (2005)
4. Loreman, T. Deppeler, J., Harvey, D.: Inclusive Education: A Practical Guide to Supporting Diversity in the Classroom, 273 p. Psychology Press (2005)
5. Hersh, M.: Evaluation framework for ICT-based learning technologies for disabled people. Computers and Education, pp. 30–47 (2014). http://www.sciencedirect.com/science/article/pii/S0360131514001146. Accessed 08 Sept 2017
6. Starcic, A., Bagon, S.: ICT-supported learning for the inclusion of people with special needs: review of seven educational technology journals 1970–2011. Br. J. Educ. Technol. 45(2), 202–230 (2013)

7. Cen, H., Cuneo, A., Gouvea, E., Heiner, C.: An educational data mining tool to browse tutor-student interactions: time will tell! In Educational data mining. In: Proceedings of the Workshop, pp. 15–22 (2005)
8. IEP writer. http://www.iepwriter.co.uk/. Accessed 21 Nov 2017
9. Stem learning. http://www.nationalstemcentre.org.uk. Accessed 21 Nov 2017
10. Very special maths. http://veryspecialmaths.co.uk/. Accessed 21 Nov 2017
11. Widgit Software. http://www.widgit.com/. Accessed 21 Nov 2017
12. Lozynska, O., Demchuk, A.: Information technologies for blind and deaf people. In: Nauka i mir. International Scientific Journal, Volgograd, vol. 1(5), pp. 102–104 (2014). (in Ukrainian)
13. Kryvonos, Yu., Krak, Yu., Barmak, O.: Systems of gest communication: modeling of information processes, 228. Naukova Dumka, Kyiv (2014). (in Ukrainian)
14. Shakhovska, N., Vysotska, V., Chyrun, L.: Intelligent systems design of distance learning realization for modern youth promotion and involvement in independent scientific researches. In: Advances in Intelligent Systems and Computing, vol. 512. Springer, Cham (2017)
15. Pasichnyk, V., Shestakevych, T.: The model of data analysis of the psychophysiological survey results. In: Shakhovska, N. (ed.) Advances in Intelligent Systems and Computing, vol. 512. Springer, Cham (2017)
16. Larbi, S., Mohamed, S.: Modeling the scheduling problem of identical parallel machines with load balancing by Time Petri Nets. Int. J. Intell. Syst. Appl. (IJISA) 7(1), 42 (2015). https://doi.org/10.5815/ijisa.2015.01.04
17. Shestakevych, T.: The method of education format ascertaining in program system of inclusive education support. In: Computer Science and Information Technologies: Proceedings of the XIIth International Scientific and technical Conference CSIT, pp. 279–283. Lviv Polytechnic Publishing House, Lviv (2017)
18. Fatima, A., Nazir, N., Khan, M.G.: Data cleaning in data warehouse: a survey of data pre-processing techniques and tools. Int. J. Inf. Technol. Comput. Sci. (IJITCS) 9(3), 50–61 (2017). https://doi.org/10.5815/ijitcs.2017.03.06
19. Datta, D., Dey, K.N.: Application of materialized view in incremental data mining operation. Int. J. Inf. Technol. Comput. Sci. (IJITCS) 9(6), 43–49 (2017). https://doi.org/10.5815/ijitcs.2017.06.06
20. Pasichnyk, V., Shestakevych, T.: The application of multivariate data analysis technology to support inclusive education. In: Computer Science and Information Technologies: Proceedings of the Xth International Scientific and Technical Conference CSIT, pp. 88–90. Lviv Polytechnic Publishing House, Lviv (2015)
21. Kozlov, L.: Cognitive Modeling on the Early Stages of Project Activity. AltGTU, Barnaul (2008). (in Russian)
22. Narayanamoorthy, S., Kalaiselvan, S.: Adaptation of induced fuzzy cognitive maps to the problems faced by the power loom workers. Int. J. Intell. Syst. Appl. (IJISA) 4(9), 75–80 (2012)

Knowledge Representation and Formal Reasoning in Ontologies with Coq

Vasyl Lenko[1(✉)], Volodymyr Pasichnyk[1], Natalia Kunanets[1], and Yuriy Shcherbyna[2]

[1] Lviv Polytechnic National University, Lviv 79000, Ukraine
vs.lenko@gmail.com
[2] Ivan Franko National University of Lviv, Lviv 79000, Ukraine
yshcherbyna@yahoo.com

Abstract. The paper describes a modern type-theoretical approach to the knowledge representation and formal reasoning in ontologies. The current state and limitations of the adopted technology for reasoning in ontologies as well as the advantages of the proposed approach are highlighted. Curry-Howard correspondence and its role in the establishment of computational reasoning are emphasized. The main part is dedicated towards the representation of ontology elements in Coq proof assistant and the execution of a semi-automated reasoning over them.

Keywords: Ontology · Knowledge representation · Formal reasoning
Type theory · Coq

1 Introduction

An exponential growth of information and the increased structural complexity of knowledge raise the need for the development of auxiliary computer systems, which are supposed to ensure an efficient accumulation, logical consistency and convenient processing of the acquired knowledge. Recent years are characterized by a significant progress in Big Data and Data Analytics technologies that allow to store and process massive amounts of bits in a relatively short time. Despite the trend, the study of knowledge management, in particular, knowledge representation and reasoning (KR&R), wasn't affected by any crucial upgrade to its machinery. We envision semi-automation of knowledge management as a strategy for tackling the information overload problem.

According to the epistemology, knowledge is a justified true belief that resides within a human brain [1]. It could also be considered as a subjective interpretation of information. DIKW hierarchy defines knowledge as a third level of abstraction, with data and information on the lower levels and wisdom on the top [2]. Both definitions suggest its internal, subjective nature, while knowledge management requires some external entity. The study of KR&R provides several major classes of knowledge representation models, which serve as the tools for knowledge externalization. These models are typically evaluated from the point of their expressiveness as well as the properties of methods for reasoning.

© Springer International Publishing AG, part of Springer Nature 2019
Z. Hu et al. (Eds.): ICCSEEA 2018, AISC 754, pp. 759–770, 2019.
https://doi.org/10.1007/978-3-319-91008-6_74

Nowadays ontology de facto became a standard among the models of knowledge representation. In addition to its deep roots in philosophy, ontology also incorporates the strengths of the other KR models – hierarchy resembles a semantic network, the structure of classes and instances inherit a frame model, axioms and functions follow the logic model and production rules respectively [3, 4]. According to the refined definition, ontology is a formal, explicit specification of a shared conceptualization characterized by high semantic expressiveness required for increased complexity [5]. Here the conceptualization stands for an abstract simplified view of some selected part of the world that we wish to represent for some purpose [6]. Typically, it consists of concepts, objects, entities that are in the scope of interest and relationships between them. From the perspective of system analysis, the relationship between a conceptualization and ontology can be classified as "many-to-many".

An ontology that is used to specify the conceptualization of a real domain might contain hundreds of classes, instances, roles, axioms and tends to expand with a time. Keeping its logical consistency or testing the new hypothesis requires the presence of a reasoning mechanism that would help to avoid error-prone manual proof searching. Some reasoning mechanisms are tightly bound to the knowledge representation model, i.e. procedures in semantic networks, while the others exist as independent formal systems, like predicate calculus. Soundness, completeness, and decidability are the main characteristics of an inference engine. In ontologies, the most widely-used language for reasoning is OWL 2 [7] based on the description logic DL $SROIQ^D$. Description logics (DL) are a family of logics designed to be as expressive as possible while retaining decidability [8, 9]. Being a fragment of first-order logic the DL has limited expressiveness; in particular, the quantification of the classes is missing. In some cases, its expressiveness can be enhanced by the DL-safe SWRL rules [10], but in general, the operations on higher-order entities are still unavailable.

Despite the fact that first-order logic has good meta-properties, i.e. downward Löwenheim-Skolem property and countable compactness property, type theory as higher-order logic, provides more expressive and homogeneous framework for reasoning. In particular, typed λ-calculus provides a solid base for typed functional languages, which are expressive enough to serve both as syntax and as a deductive system. The main goal of the research is to present the theory and technology of an expressive knowledge representation and higher-order certified reasoning in Coq, applied to the widely adopted ontological KR model.

2   Related Works

The idea and foundations of type-theoretical reasoning in ontologies with Coq proof assistant have been presented in [11, 12]. Research on ontologically-based detection of underground networks by the means of a certified proof computation confirms an effectiveness of the proposed approach [13]. Calculus of inductive constructions, a formalism that is behind the Coq proof assistant, provides great insights into the functionality of technology [14]. The recent work on homotopy type theory (HoTT) aims to improve the properties of the defined equality in that formalism [15].

3 Curry-Howard Correspondence

The Curry-Howard correspondence is one of the most notable discoveries that established a relationship between two families of formalism, proof systems and models of computation, namely a natural deduction for intuitionistic propositional logic and a simply-typed λ-calculus. Later it was extended to the other systems of intuitionistic logic, where first-order logic corresponds to dependent types, second-order logic corresponds to polymorphic types, etc. The isomorphism has many aspects, even at the syntactic level: formulas correspond to types, proofs correspond to terms, provability corresponds to type inhabitation, proof normalization corresponds to term reduction [16]. It gave a birth to the new concepts "propositions as types", "proofs as programs", "simplification of proofs as evaluation of programs" and led to the development of a new kind of software called "proof assistant".

Typed λ-calculus is a formal language that provides a base for the most calculi in type theory and functional programming languages. Sometimes it is called "the smallest universal programming language of the world" since it consists only of a single transformation rule (variable substitution) and a single function definition scheme [17]. Yet, the untyped λ-calculus is Turing complete. Simply typed λ-calculus is the canonical example of a typed lambda calculus, with only one type constructor $\rightarrow$ that builds function types. Barendregt's λ-cube represents a classification of eight typed λ-calculi based on the forms of abstraction they possess: type operators, polymorphism and dependent types [18]. Calculus of inductive constructions, which underlies the Coq system, resides on top of the cube and has the richest calculus with all three abstractions. The framework of pure type systems (PTS) generalizes the λ-cube by allowing any number of sorts, and any type of relationships between them. Being a typed λ-calculus they provide a unified syntax for both terms and types.

Table 1. Correspondence between the elements of type theory and logic [15].

Type theory	Logic
A	Proposition
a : A	Proof
B(x)	Predicate
b(x) : B(x)	Conditional Proof
0, 1	$\bot$, $\top$
A + B	A $\vee$ B
A $\times$ B	A $\wedge$ B
A $\rightarrow$ B	A $\Rightarrow$ B
$\sum_{(x:A)} B(x)$	$\exists_{(x:A)} B(x)$
$\prod_{(x:A)} B(x)$	$\forall_{(x:A)} B(x)$
Id_A	equality $=$

4 The Coq Proof Assistant

The Coq is a formal proof management system. It provides an environment that facilitates the processes of context specification and hypothesis validation. Coq comes with the highly expressive specification language *Gallina*, the language of commands *The Vernacular*, the basic and standard libraries, a set of atomic tactics and tactic expressions for proof calculation, the tactic language L_{tac} and *SSReflect* proof language. All the developments are conducted in Coq IDE, which takes the burden of resource management and empowers the user with the graphical interface and utility tools, like forward and backward navigation in a vernacular file. Coq system is flexible: its default configuration can be easily enriched with the custom-tailored user modules, while the proof calculation hints can be stored and reused through *HintDb* databases. The current version of Coq 8.7.0 comes bundled with the CoqIDE in the installation files for Windows 32 bits (i686), Windows 64 bits (x86_64) and macOS. An installation of Coq and CoqIDE in Linux is performed with the command-line tool for package management "apt-get".

The specification language of Coq *Gallina* has a rich syntax for developing formalized theories that consist of logical objects like axioms, hypotheses, parameters, lemmas, theorems, definitions of constants, functions, predicates, and sets [19]. It provides a basic syntax of the terms of Calculus of Inductive Constructions (CIC), which can be extended with various primitives and customized according to the user's preferences. To ensure the logical correctness all the objects in Coq are typed, therefore enabling the development of proofs of specifications of programs with *Gallina*. To facilitate the higher-order reasoning, the language comes with three sorts *Prop*, *Set*, *Type* and requires every type to be well-formed. Terms construction and reduction are performed according to the inference rules of CIC. Integration with the execution environment is implemented through *Vernacular* commands. The presence of a typed module system allows managing and reusing the related specifications effectively.

The proof engine of Coq is represented by a set of tactics, the tactic language L_{tac} and *SSReflect* proof language. Tactics are deduction rules that implement the backward reasoning mechanism from a conclusion (goal) to the premises (subgoals). When applied to a conclusion, a tactic replaces the goal with the subgoals it generates [20]. Tactics are built from atomic tactics and tactic expressions that combine the former ones. Among the most used tactics are *intros*, *destruct*, *apply*, *assumption*, etc. The tactic language L_{tac} goes further and defines the syntax for error catching, repeating, branching, failing on tactics. It is mainly used in proof mode but can also be used in top-level definitions. A set of tactics *SSReflect* is a recent addition to the Coq system, which is designed to provide the support for the *small scale reflection* proof methodology. Despite its intersection with the default set of Coq tactics, it was developed independently and has a couple of notable difference, like hypothesis management approach, support of reflection steps, etc. In general, proofs written in *SSReflect* look quite different from the ones written using only tactics [19].

5 Ontology Representation in Coq

The development of Coq interactive proof assistant has been supported by the French Institute for Research in Computer Science and Automation (INRIA) since 1984. It provides an expressive functional language that implements a calculus of inductive constructions (CIC) within a pure type systems framework. CIC extents the Coquand's calculus of constructions with primitive inductive definitions, which help to prove some natural properties and improve an efficiency of computation of functions over inductive definitions [14]. The language of Coq allows an unlimited amount of higher-order types T_i, called sorts. Representation of logical propositions is performed according to the Curry-Howard isomorphism: every logical proposition corresponds to a type T_i that belongs to the sort *Prop*. Logical connectives are encoded according to the mapping presented in Table 1.

An ontology can be formally defined as a tuple $\langle C, R, F \rangle$, where C – denotes the finite set of concepts of a domain, R:C→C stands for the finite set of relations between concepts, F – denotes a finite set of interpretation functions (restrictions, axioms) [21]. The structure of ontology consists of the concepts organized in taxonomy, their properties or attributes, inference rules, related axioms and restrictions in the domain of discourse. In ontologies, it is also common to find a "part-whole" mereological relationship that is used for the description of concepts, which may serve as the aggregators of the lower-order concepts. Ontology together with a set of individual instances of classes constitutes a knowledge base [22]. An expressiveness of the Coq language is more than sufficient to represent the specified components of an ontology structure in a clear and concise manner (Table 2).

Table 2. Representation of an ontology structure in Coq.

Element of ontology	Coq representation
Concept	Class C : Type
Instance	Instance X : C
Properties, inference rules	Class C (att1:nat) (att2:bool): Type := {f1: attr1 –> bool;}
Binary relation	Parameter R : C –> C –> Prop.
Concept inheritance	Parameter D1 : SubClass_G – > C Coercion D1 : SubClass_G >–> C
"Part-whole" relation	Definition PartOf (x y : C): = R x y Axiom A1 : Reflexive PartOf Axiom A2 : Asymmetric PartOf Axiom A3 : Transitive PartOf
Quantifiers ∃,∀	exists X : C, forall X : C

The base building block, which represents a fact in the ontology, is called "triplet". It is a statement in a form $\langle subject, predicate, object \rangle$ that expresses a relationship denoted by the predicate between the subject and the object. In a triplet, a subject and a predicate are URIs pointing to Web resources, whereas an object may be either a URI or a literal representing a value [23]. The set of triples can be easily visualized as a directed graph, where each triple corresponds to an edge from the subject to the object. Below is presented a partial visualization of the triples set from air travel booking service ontology [24], captured with OntoGraf plugin in Protege 5.2.0 [25] (Fig. 1).

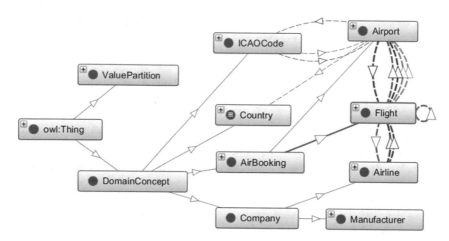

Fig. 1. Upper concepts in the air travel booking service ontology.

The whole ontology contains 69 classes, 62 individuals and lots of axioms. For the sake of convenience, let's select the fragment of ontology, which consists of the concepts "Airport", "Flight", "Airline" and the relations among them. To simplify it a bit more, the "Flight" concept is replaced with a transitive relation "Flight", and the set of relations is reduced to the "Flight", "AirlineHub" and "Part_of" relations. The resulting ontology fragment, captured with the VOWL plugin in Protege 5.2.0, is presented below (Fig. 2).

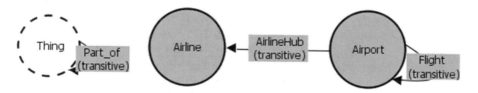

Fig. 2. Simplified fragment of the ontology.

Example 1. A simplified fragment of the "Air Travel Booking Service" ontology $O = \langle C, R, F \rangle$, where $C = \{Airline; Airport\}$; $R = \{Flight; AirlineHub; Part\_Of\}$; $F = \{Airline\_Airport; Flight\_Transitivity\}$ could be represented in Coq as follows:

1. Import used Coq modules:

Require Import Coq.Classes.RelationClasses.

2. Define a relation and a root concept in a hierarchy (or reuse an upper ontology, e.g. DOLCE [26]):

Definition Kind := Type.
Parameter Relation : Kind -> Kind -> Prop.

3. Define the taxonomy of concepts:

Class Airline : Type.
Parameter D1 : Airline -> Kind.
Coercion D1 : Airline >-> Kind.
Class Airport : Type.
Parameter D2 : Airport -> Kind.
Coercion D2 : Airport >-> Kind.

4. Define the relationships in ontology and their properties:

Definition Flight(a1 a2:Airport) := Relation a1 a2.
Axiom Flight_Transitivity:
forall (a1:Airport)(a2:Airport)(a3:Airport),
Flight(a1)(a2) ∧ Flight(a2)(a3) -> Flight(a1)(a3).
Definition Part_of (x y:Kind) := Relation x y.
Axiom r_of_part_of : Reflexive Part_of.
Axiom a_of_part_of : Asymmetric Part_of.
Axiom t_of_part_of : Transitive Part_of.
Definition AirlineHub(a1:Airport)(a2:Airline) := Part_of a1 a2.

5. Define the interpretation functions in ontology:

Axiom Airline_Airport:
forall(a1:Airport)(a2:Airport)(airline:Airline),
AirlineHub(a1)(airline) ∧ Flight(a1)(a2) -> Part_of(a2)(airline).

The example describes a relationship between an airport and airline with the *Air-lineHub* term, which states the fact that an airport is explicitly marked as a part of the airline network. To maintain the list of marked airports, *Flight_Transitivity* and *Air-line_Airport* relationships provide opportunity to do the inference whether an airport is connected to the airline network.

The other practical application comes from the domain of a software engineering and targets an access management issue [27]. Nowadays the development of a single software application is facilitated with a dozen of auxiliary tools intended for: version control, CI/CD, project management, security audit, quality assurance, logs monitoring, etc. When an enterprise employee joins a project, it usually takes from several hours to a couple of days to obtain access to the entire project's infrastructure. Having instead the access management ontology with Coq certified behavior not just saves man-hours, but also reduces the risk of human error (Fig. 3).

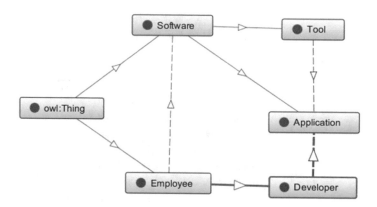

Fig. 3. Fragment of the enterprise software access management ontology.

Example 2. Fragment of the "Enterprise software access management" ontology $O = \langle C, R, F \rangle$, where $C = \{Software; Employee; Tool; Application; Developer\}$; $R = \{IsAppDeveloper; IsAppTool; AccessGranted\}$; $F = \{Employee\_App\_Access; App\_Developer\}$ has the following representation in Coq:

1. Define the taxonomy of concepts:

Class Software: Type.	*Class Employee: Type.*
Parameter D1: Software-> Kind.	*Parameter D4 : Employee-> Kind.*
Coercion D1: Software>-> Kind.	*Coercion D4 : Employee>-> Kind.*
Class Tool: Type.	*Class Developer: Type.*
Parameter D2 : Tool -> Software.	*Parameter D5 : Developer-> Employee.*
Coercion D2 : Tool >-> Software.	*Coercion D5 : Developer >->Employee.*

Class Application: Type.
Parameter D3 : Application-> Software.
Coercion D3 : Application>-> Software.

2. Define the relationships:

Definition isAppTool(t:Tool)(a:Application) := Relation t a.
Definition accessGranted(e:Employee)(s:Software) := Relation e s.
Definition isAppDeveloper(d:Developer)(a:Application) := Relation d a.

3. Define the interpretation functions:

Axiom App_Developer :
forall (d:Developer)(a:Application),
isAppDeveloper(d)(a) -> accessGranted(d)(a).

Axiom Employee_App_Access :
forall (e:Employee)(a:Application)(t:Tool),
accessGranted(e)(a) ∧ isAppTool(t)(a) -> accessGranted(e)(t).

The ontology fragment establishes relationships between enterprise *Employee*, namely *Developer*, and permissioned *Software* that includes *Application* and *Tool*. When an employee joins a project as an application developer, *isAppDeveloper*, it is granted access to the application, *accessGranted*, according to *App_Developer* axiom. Since the application development process is usually supported by multiple auxiliary tools, *isAppTool*, *Employee_App_Access* axiom grants access to these tools to everyone who is granted access to the application. If a developer or a tool is disconnected from an application, the corresponding access privileges are terminated as well. Next section highlights the details of the reasoning in Coq and extends the presented examples with the definitions and proofs of related lemmas.

6 Reasoning in Coq

Coq comes with the four modes of automated reasoning: type checking, Type Classes, pure reasoning with the domain rules, collection of the axioms and variables into HintDb databases [13]. Type checking verifies whether all terms are well-typed, otherwise, an error is thrown. Type Classes are dependent inductive types, which are especially convenient for the representation of a "part-whole" relation. Pure reasoning with the domain rules consists in the definition of axioms that are later used in inference as the transformation rules. Finally, the HintDb databases empower a user to make a request, which aims to be resolved by an automatic search mechanism. The language of tactics L_{tac} serves as a handy and reliable instrument for a proof calculation. It contains dozens of automated tactics, which save the time and avoid error-prone transformations. Since the underlying calculus of inductive constructions is proved to be strongly normalizing with respect to the computation rules, type checking in Coq is decidable.

Example 3. The definition and proof of the Air Travel Booking Service ontology lemma has the following representation in Coq:

Lemma Airline_Network:
forall (a1 a2 a3:Airport)(airline:Airline),
AirlineHub(a1)(airline) ∧ Flight(a1)(a2) ∧ Flight(a2)(a3) ->
Part_of(a3)(airline).
Proof.
intros Boryspil Chopin Dulles UIA H1.
destruct H1 as [H1 H2].
apply Flight_Transitivity in H2.
assert(H3: Part_of(Boryspil)(UIA) ∧ Flight(Boryspil)(Dulles)).
split; assumption.
apply Airline_Airport in H3.
assumption.
Qed.

The lemma states the "part-whole" relation between the higher-order concept *Airline* and lower-order concepts *Airport* and *Flight*. Going into details, every airport that is connected by flights to the marked airport is assumed to be a part of the airline network. The accompanying proof confirms the lemma assumption. It starts with *intros* tactic, which is used to provide the values of free variables into hypothesis in order to eliminate the universal quantification. Then the *destruct* tactic splits the hypothesis into two parts: *H1 : Part_of(Boryspil)(UIA)* and *H2 : Flight(Boryspil) (Chopin) ∧ Flight (Chopin)(Dulles)*. The appearance of H2 allows to *apply Flight_Transitivity*, thus reducing it to *Flight(Boryspil)(Dulles)*. By connecting the two parts of hypothesis with a conjunction, it is possible to apply *Airline_Airport* axiom, which finally reduces the construction to the desired goal.

Example 4. The definition and proof of the "Enterprise software access management" ontology lemma in Coq is as follows:

Lemma Tool_Dev_Access :
forall (d:Developer)(a:Application)(t:Tool),
isAppDeveloper(d)(a) ∧ isAppTool(t)(a) -> accessGranted(d)(t).
Proof.
intros d a t H1.
destruct H1 as [H1 H2].
apply App_Developer in H1.
assert(H3:accessGranted(d)(a) ∧ isAppTool(t)(a))..
split; assumption.
apply Employee_App_Access in H3.
assumption.
Qed.

The proof of lemma certifies the expected ontology behavior, where every application developer is automatically granted access to each tool used in the application's stack. The set of applied tactics is the same as in the previous example. As a result, both lemmas are proved with regards to the specified definitions of their components.

7   Conclusions

In this paper, we present a theory and technology that provide better expressiveness and higher-order reasoning capability into the field of KR&R. Typed λ-calculus is a fundamental underlying component of the improvement, since (1) it has a connection with a formal system of natural deduction in intuitionistic logic via Curry-Howard isomorphism; (2) it ensures the expressiveness of Coq language *Gallina*; (3) higher-order logic in Coq, Calculus of Inductive Constructions, is a typed λ-calculus. To foster an environment of certified reasoning, the proposed technology is applied to the ontological KR model. Additionally, we provide an explicit correspondence between the structural elements of ontology and their representations in Coq. The paper comes with four examples, which demonstrate an application of the technology in practice. The further research will concern a development of personal knowledge database systems based on the described technology.

References

1. Dutant, J.: The legend of the justified true belief analysis. Phil. Perspect. **29**(1), 95–145 (2016)
2. Baskarada, S., Koronios, A.: Data, Information, Knowledge, Wisdom (DIKW): a semiotic theoretical and empirical exploration of the hierarchy and its quality dimension. Australas. J. Inf. Syst. **18**(1), 5–24 (2013)
3. Mabel, V.H., Selwyn, J.: A review on the knowledge representation models and its implications. Int. J. Inf. Technol. Comput. Sci. **8**(10), 72–81 (2016). https://doi.org/10.5815/ijitcs.2016.10.09
4. Malhotra, M., Nair, T.R.G.: Evolution of knowledge representation and retrieval techniques. Int. J. Intell. Syst. Appl. (IJISA) **7**(7), 18–28 (2015). https://doi.org/10.5815/ijisa.2015.07.03
5. Feilmayr, C., Wöß, W.: An analysis of ontologies and their success factors for application to business. Data Knowl. Eng. **101**, 1–23 (2016)
6. Gruber, T.: A translation approach to portable ontology specifications. Knowl. Acquis. **5**(2), 199–220 (1993)
7. OWL 2 Web Ontology Language Document Overview (Second Edition). https://www.w3.org/TR/owl2-overview. Accessed 04 Dec 2017
8. Krötzsch, M., Simančik, F., Horrocks, I.: Description Logics. IEEE Intell. Syst. **29**(1), 12–19 (2014)
9. Yahiaoui, Y., Lehireche, A., Bouchiha, D.: Proposed representation approach based on description logics formalism. Int. J. Intell. Syst. Appl. **8**(5), 1–9 (2016). https://doi.org/10.5815/ijisa.2016.05.01
10. DL-Safe Rules. http://www.lesliesikos.com/dl-safe-rule. Accessed 04 Dec 2017
11. Dapoigny, R., Barlatier, P.: Modeling ontological structures with type classes in Coq. In: Pfeiffer, H.D., Ignatov, D.I., Poelmans, J., Gadiraju, N. (eds.) ICCS 2013. LNCS, vol. 7735, pp. 135–152. Springer, Heidelberg (2013)
12. Dapoigny, R., Barlatier, P.: Specifying well-formed part-whole relations in Coq. In: Hernandez, N., Jäschke, R., Croitoru, M. (eds.) ICCS 2014. LNCS, vol. 8577, pp. 159–173. Springer, Cham (2014)

13. Hafsi, M., Dapoigny, R., Bolon, P.: Toward a type-theoretical approach for an ontologically-based detection of underground networks. In: Zhang, S., Wirsing, M., Zhang, Z. (eds.) KSEM 2015. LNCS, vol. 9403, pp. 90–101. Springer, Cham (2015)
14. Paulin-Mohring, C.: Introduction to the calculus of inductive constructions. In: Delahaye, D., Paleo, B.W. (eds.) All about Proofs, Proofs for All. Mathematical Logic and Foundations. College Publications, London (2015)
15. The Univalent Foundations Program: Homotopy Type Theory: Univalent Foundations of Mathematics. Institute for Advanced Study (2013). https://homotopytypetheory.org/book
16. Sørensen, M., Urzyczyin, P.: Lectures on the Curry-Howard Isomorphism. Elsevier Science, Amsterdam (2006)
17. A Tutorial Introduction to the Lambda Calculus. https://arxiv.org/pdf/1503.09060.pdf. Accessed 18 Dec 2017
18. Barendregt, H.: Introduction to generalized type systems. J. Funct. Program. 1(2), 125–154 (1991)
19. The Coq Proof Assistant Reference Manual (v.8.7.0). https://coq.inria.fr/distrib/current/refman/. Accessed 18 Dec 2017
20. Chlipala, A.: Certified Programming with Dependent Types: A Pragmatic Introduction to the Coq Proof Assistant. The MIT Press, Cambridge (2013)
21. Lytvyn, V.: Pidkhid do pobudovy intelektualnykh system pidtrymky pryiniattia rishen na osnovi ontolohii. Problemy prohramuvannia 4, 43–52 (2013)
22. Ontology Development 101: A Guide to Creating Your First Ontology. https://protege.stanford.edu/publications/ontology_development/ontology101-noy-mcguinness.html. Accessed 18 Dec 2017
23. Abiteboul, S., Manolescu, I., Rigaux, P., Rousset, M., Senellart, P.: Web Data Management. Cambridge University Press, New York (2011)
24. Air travel booking service ontology. http://students.ecs.soton.ac.uk/cd8e10/airtravelbookingontology.owl. Accessed 18 Dec 2017
25. Protégé: a free, open-source ontology editor & framework for building intelligent systems. https://protege.stanford.edu/products.php. Accessed 18 Dec 2017
26. DOLCE: A Descriptive Ontology for Linguistic and Cognitive Engineering. http://www.loa.istc.cnr.it/old/DOLCE.html. Accessed 18 Dec 2017
27. Tayeb, S.H., Noureddine, M.: Measures for the ontological relations in enterprise. Int. J. Mod. Educ. Comput. Sci. 9(9), 13–23 (2017). https://doi.org/10.5815/ijmecs.2017.09.02

Author Index

Printed in the United States
By Bookmasters